MATHEMATICAL METHODS IN AERODYNAMICS

Mathematical Methods in Aerodynamics

by

LAZĂR DRAGOŞ,

Romanian Academy,
Bucharest, Romania

KLUWER ACADEMIC PUBLISHERS
DORDRECHT / BOSTON / LONDON

EDITURA ACADEMIEI ROMÂNE
BUCUREŞTI

A C.I.P. Catalogue record for this book is available from tha Library of Congress.

ISBN 1-4020-1663-8
ISBN 973-27-0986-3

Published by Kluwer Academic Publishers and Editura Academiei Române.

Kluwer Academic Publishers,
P.O. Box 17, 3300 AA Dordrecht, The Netherlands.
Editura Academiei Române,
P.O. Box 5–42, 050711 Bucureşti, România.

Sold and distributed in North, Central and South America
by Kluwer Academic Publishers,
101 Philip Drive, Norwell, MA 02061, U.S.A.

In all other countries, except for Romania and the Republic of Moldavia,
sold and distributed by Kluwer Academic Publishers,
P.O. Box 322, 3300 AH Dordrecht, The Netherlands.

In Romania and Republic of Moldavia
sold and distributed by Editura Academiei Române,
P.O. Box 5–42, 050711 Bucureşti, România.

Printed on acid-free paper

Printed in Romania

Table of Contents

Preface

The researchers in Aerodynamics know that there is not a unitary method of investigation in this field. The first mathematical model of the airplane wing, the model meaning the integral equation governing the phenomenon, was proposed by L. Prandtl in 1918. The integral equation deduced by Prandtl, on the basis of some assumptions which will be specified in the sequel, furnishes the circulation $C(y)$ (see Chapter 6). Using the circulation, one calculates the lift and moment coefficients, which are very important in Aerodynamics. The first hypothesis made by Prandtl consists in replacing the wing by a distribution of vortices on the plan-form D of the wing (i.e. the projection of the wing on the plane determined by the direction of the uniform stream at infinity and the direction of the span of the wing). Since such a distribution leads to a potential flow in the exterior of D and the experiences show that downstream the flow has not this character, Prandtl introduces as a supplementary hypothesis another vortices distribution on the trace of the domain D in the uniform stream. The first kind of vortices are called tied vortices and the second kind of vortices are called free vortices. On the basis of this model one developed later the main theories of Aerodynamics namely the lifting surface theory (after 1936, more precisely in 1950, when Multhopp gave the equation of this theory), the lifting surface theory for the supersonic flow (after 1946) and the lifting theory for oscillatory wings and surfaces for the subsonic, sonic and supersonic flow (after 1950). In the framework of the last theory the wing is replaced by doublets distributions. From a physical point of view, there is no reason for replacing the wing with vortices or doublets distributions. It is true that the vortices are detaching from the wing, but these are effects , not causes of the presence of the wing. The fact that these replacements lead to correct results shows how subtle was Prandtl's intuition. We specify that the distributions on D and its trace do not result from the equations of motion (they have been introduced outside the mathematical model). Taking into account this inconvenient, we have shown in [5.7] how it can be removed. We have to consider that the wing and the fluid constitute an interacting material system. If we want to study the fluid flow, then according to Cauchy's stress principle (the principle of the internal forces; see for example [1.11], p.35), we have to assume that there exists a forces distribution on the boundary, which has against the fluid the same action like the wing itself. We shall replace therefore the wing with a forces distribution instead of a vortices, sources or doublets distribution and we shall find out the density of this distribution such that it should have the same action against the fluid like the wing itself. We shall proceed by imposing to the fluid flow determined by the forces distribution to satisfy the slipping condition on the wing, condition which is also satisfied by the flow determined by the wing. In this way it follows an integral equation for determining the forces density. This equation constitutes the mathematical model for the wing we have in view. This method is an unitary one and it is based only on the classical principles of mechanics (in fact, Cauchy's stress principle). It may be applied to all configurations: see [5.7] for the wing in a subsonic stream, [8.4] for the wing in a supersonic stream, [10.15], [10.16], [10.17] for the oscillatory wings in subsonic, sonic or supersonic stream etc. All these results are given in this book (see chapters 5, 8, 10, 11). We called this method (in[5.7]): the fundamental solutions method. It may be utilized to all cases in which one can calculate the fundamental solutions of the equations of motion. We have to notice that in the framework of this method, the existence of the vortices downstream the wing follows from the model (i.e. from the equations of motion) and it must not be introduced artificially. In the sequel we shall present some of the models of aerodynamics. For two-dimensional configurations, in a subsonic stream, the models are one-dimensional singular integral equations considered in the sense of Cauchy's principal value. One may integrate analytically only the equation of thin profiles in a free stream. For other geometries one determines numerical solutions with the aid of Gauss-type quadrature formulas (see Chapter 3). For three-dimensional wings in a subsonic stream, the models are two-dimensional integral equations with strong singularities, which are defined in the sense of Finite Part (see Chapter 5). For other geometry (for example the wing in ground effects) the models are generalized equations. All these models are solved only numerically. For the wing in a free stream, Multhopp's method is available. In this book we introduce a more general method – the quadrature formulas method. In the last part of Chapter 5 one presents the theory of low aspect wings which was extended by the author to the general case of asymmetrical wings. The lifting line theory may be deduced from the lifting surface theory with the aid of Prandtl's assumptions (6). This theory is developed by presenting analytical and numerical methods for solving Prandtl's equation; one considers also extensions of this theory, all the methods representing one-dimensional integral-differential equations. The author shows how these equations may be reduced to integral equations with strong singularities and for this type of singularities he gives a Gauss-type quadrature formula, which allows the equation to be reduced to a linear algebraic system which is solved numerically. This method, which is very general, allows to obtain numerical solutions both in the case of the lifting line (Chapter 6) and the case of the lifting surface (Chapter 5). In the case of supersonic flow, the integral equations are solved analytically. For the three-dimensional wing (the lifting surface) we present in Chapter 8 a nice solution given by D. Homentcovschi in [8.16]. The integral equations describing the flow past oscillatory wings and profiles (chapter 10) have the same nature like the equations utilized in the case of steady flow but the kernels are more complicated. However for the sonic and supersonic flows these equations may be solved exactly by means of the Laplace transform, as it is shown in [10.17]. Chapter 9, devoted to the transonic motions, begins with a new asymptotic deduction of the equations of motion. The two and three-dimensional integral equations are obtained following the papers of the author and D. Homentcovschi. The theory of subsonic and supersonic flow past slender bodies (in Chapter 11) relies also on the fundamental solutions theory. In Chapter 2 one deduces the equations of the linear aerodynamics, on the basis of an asymptotic analysis, assuming that the small parameter depends on the thickness of the profile. In the classical aerodynamics this deduction is performed under the assumption that the unknowns and their derivatives have the same order of magnitude, but this fact cannot be *a priori* assumed. Then one calculates the fundamental solutions for the equation of the potential (paper [2.11]) and the fundamental solutions for the systems of equations of aerodynamics : the steady system[2.8], the oscillatory system [10.17], the unsteady system [2.6], [2.7]. On these solutions will rely the theories from the forthcoming chapters. The models we have already presented are the so called classical or linear models. They are suitable for the thin wings and thin profiles because they rely on the following assumptions: 1) one uses a linear boundary condition, 2) the boundary condition is imposed on the support of the wing (the segment $[-1, 1]$ for the profile, the plan-form D for the three-dimensional wing), 3) the equations of motion are linearized. The development of the scientific computing allows us to develop more exact methods. Indeed we can give up to the first two assumptions using the boundary integral equations method (BIEM), also called the boundary element method (BEM), which was employed for the first time by Hess and Smith [7.9], [7.10]. The integral equations on the boundary are obtained imposing the exact boundary condition on the boundary of the wing. The integral equation is discretized using, for example, the collocation method. One obtains an algebraic system which is solved numerically. The linearization of the equations of motion is necessary only in the case of compressible fluids. The theory that we have developed is thus valid for every body in an incompressible fluid and for a thin body in a compressible fluid. Two chapters from this book, Chapter 4 for the 2d airfoil and Chapter 7 for the 3d airfoil are based on our papers (L. Dragoş and A. Dinu). The comparison between the known analytical results and the numerical results shows a very good agreement. In the Appendices we give some results concerning The Distributions Theory, The Singular Integral Equations Theory, The Principal Value and The Finite Part, Gauss-type Quadrature Formulas, etc.

In every work one finds, in a certain measure, both the achievements of the predecessors and of the researchers contemporaneous with the author. Among the people which have directly collaborated with me, I have to mention at first my professors Victor Vâlcovici and Caius Iacob, who introduced me in the field of aerodynamics. I also mention my younger colleagues Nicolae Marcov, Liviu Dinu, Dorel Homentcovschi, Adrian Carabineanu, Victor Ţigoiu, Vladimir Cardoş, Gabriela Marinoschi, Stelian Ion and Adrian Dinu. They were my students at the University of Bucharest, but I learned a lot from their papers. Some of them were my fellow - workers in the aerodynamics research, many of them stimulated me with their youth and their way of thinking in our seminars from the Faculty of Mathematics of the University of Bucharest. I am very grateful to all of them.

My special gratitude goes to Adrian Carabineanu for his work in performing the English translation of the book, to Adrian Carabineanu and Stelian Ion for typesetting the monograph in Latex and to Victor Ţigoiu for his activity in finalizing the 195D Grant with the World Bank.

I acknowledge that the book was sponsored by MEC-CNCSIS Contract 49113/2000, Grant 195D with World Bank.

LAZĂR DRAGOŞ

Chapter 1

The Equations of Ideal Fluids

1.1 The Equations of Motion

1.1.1 Elements of Kinematics

In this Chapter we present the equations governing the flow of ideal fluids. On the basis of these equations we shall develop the theory in the forthcoming chapters of the book. It is well known (see, for example, [1.11]), that the fluid flow is defined by the *diffeomorphism*

$$\boldsymbol{x} = \chi(t, \boldsymbol{X}), \tag{1.1.1}$$

where $\boldsymbol{X}$ is the vector of position of a particle P in the reference configuration (for fluids this is the initial configuration), and $\boldsymbol{x}$ is the vector of position of the same particle at the moment t. For a fixed $\boldsymbol{X}$ and a variable t, the equation (1.1.1) furnishes the motion law for the particle having the vector of position $\boldsymbol{X}$. Hence the velocity and the acceleration of the particle will be given by the formulas

$$\boldsymbol{v}(t, \boldsymbol{X}) = \frac{\mathrm{d}}{\mathrm{d}t}\chi(t, \boldsymbol{X}), \quad \boldsymbol{a}(t, \boldsymbol{X}) = \frac{\mathrm{d}}{\mathrm{d}t}\boldsymbol{v}(t, \boldsymbol{X}). \tag{1.1.2}$$

The fluid is a *continuum medium.* It means that the support of the initial configuration is a domain $\mathcal{D}_0$. The image of this domain by the diffeomorphism (1.1.1) will be denoted by $\mathcal{D}$ and one demonstrates [1.11] that it is a domain. The functions $\chi_i(t, X_1, X_2, X_3)$ appearing in (1.1.1) belong to the class $C_2(\mathcal{D}_0)$ and the Jacobian is

$$J = \frac{\partial(\chi_1, \chi_2, \chi_3)}{\partial(X_1, X_2, X_3)} \neq 0. \tag{1.1.3}$$

The velocity field defined in (1.1.2) may be discontinuous in isolated points, on abstract curves or across abstract surfaces. This kind of surfaces will be named *shock waves.*

The fluid flow may be described by functions defined on $\mathcal{D}_0$, i.e. functions having the form $\phi(t, \boldsymbol{X})$, or by functions defined on $\mathcal{D}$, i.e. functions having the form $F(t, \boldsymbol{x})$. The first presentation is called *the material description*, because it utilizes quantities attached to the material particles, the second is called *the spatial description*, because it furnishes information about the particles which are located at the moment t in the points of a domain $\mathcal{D}$.

The *material derivative* of the quantity ϕ attached to the fixed particle $\boldsymbol{X}$, is denoted by $\dot{\phi}$ and it is given by the formula

$$\dot{\phi} = \frac{\partial}{\partial t}\phi(t, \boldsymbol{X}). \tag{1.1.4}$$

In order to obtain the derivative of F when $\boldsymbol{X}$ is fixed, we have to take into account that F depends on the coordinates X_α through the functions χ_i. Using the derivation rule for the composite functions, we obtain for the material derivative

$$\dot{F}(t, \boldsymbol{x}) = \frac{\partial F}{\partial t} + \frac{\partial F}{\partial x_i}\frac{\mathrm{d}\chi_i}{\mathrm{d}t} = \frac{\partial F}{\partial t} + v_i\frac{\partial F}{\partial x_i} = \frac{\partial F}{\partial t} + (\boldsymbol{v}\cdot\nabla)F. \tag{1.1.5}$$

For studying the motion of fluids we employ the spatial description. The main quantities are: the density (or specific mass) $\rho(t, \boldsymbol{x})$, the pressure $p(t, \boldsymbol{x})$, the velocity field $\boldsymbol{v}(t, \boldsymbol{x})$, the temperature $T(t, \boldsymbol{x})$, the entropy $s(t, \boldsymbol{x})$ etc. The acceleration which is the material derivative of the velocity is obtained by means of the formula:

$$\boldsymbol{a} = \frac{\partial \boldsymbol{v}}{\partial t} + (\boldsymbol{v}\cdot\nabla)\boldsymbol{v}. \tag{1.1.6}$$

Utilizing the derivation rule for the determinants, from (1.1.3) one obtains *Euler's theorem*

$$\dot{J} = J\,\mathrm{div}\,\boldsymbol{v} \tag{1.1.7}$$

and then (see for example [1.11]) Reynolds's formulas for *continuous motions* (i. e. motions characterized by fields belonging to the class) $C_1([t_0, t_1]\times \mathcal{D}_t)$

$$\begin{aligned}&\frac{\mathrm{d}}{\mathrm{d}t}\int_D F(t, \boldsymbol{x})\,\mathrm{d}v = \int_D (\dot{F} + F\,\mathrm{div}\,\boldsymbol{v})\,\mathrm{d}v = \\ &= \int_D \left[\frac{\partial F}{\partial t} + \mathrm{div}\,(F\boldsymbol{v})\right]\mathrm{d}v = \int_D \frac{\partial F}{\partial t}\,\mathrm{d}v + \int_{\partial D} F\boldsymbol{v}\cdot\boldsymbol{n}\,\mathrm{d}a\end{aligned} \tag{1.1.8}$$

and Reynolds's formula for motions with shock waves

$$\frac{\mathrm{d}}{\mathrm{d}t}\int_D F(t, \boldsymbol{x})\,\mathrm{d}v = \int_D \frac{\partial F}{\partial t}\,\mathrm{d}v + \int_{\Sigma_+ + \Sigma_-} F\boldsymbol{v}\cdot\boldsymbol{n}_e\,\mathrm{d}a - \int_S [\![F]\!]d\,\mathrm{d}a, \tag{1.1.9}$$

where $[\![F]\!] = F_+ - F_-$, and d is the displacement velocity of the surface of discontinuity S (see formula (2.5.1) and figure 2.5.1 from [1.11]).

1.1.2 The Equations of Motion

The principle of conservation of mass is

$$\frac{\mathrm{d}}{\mathrm{d}t}\int_D \rho \,\mathrm{d}v = 0 \quad (\forall)\, D \subset \mathcal{D}. \tag{1.1.10}$$

For continuous motions one utilizes the formula (1.1.8) and one obtains the equation of continuity

$$\dot{\rho} + \rho \operatorname{div} \boldsymbol{v} = 0. \tag{1.1.11}$$

The general expression for *the principle of variation of the momentum* is

$$\frac{\mathrm{d}}{\mathrm{d}t}\int_D \rho \boldsymbol{v} \,\mathrm{d}v = -\int_{\partial D} p \boldsymbol{n} \,\mathrm{d}a + \int_D \rho \boldsymbol{f} \,\mathrm{d}v \quad (\forall)\, D \subset \mathcal{D}, \tag{1.1.12}$$

$\boldsymbol{f}$ representing the force per unity of mass. For continuous motions, from (1.1.11) and (1.1.6), one obtains Euler's equation

$$\rho\left[\frac{\partial \boldsymbol{v}}{\partial t} + (\boldsymbol{v}\cdot\nabla)\boldsymbol{v}\right] = \rho \boldsymbol{f} - \operatorname{grad} p. \tag{1.1.13}$$

The balance equation of the energy (the first principle of the thermodynamics) is

$$\begin{aligned}\frac{\mathrm{d}}{\mathrm{d}t}\int_D \rho\left(e + \frac{1}{2}v^2\right)\mathrm{d}v = -\int_{\partial D} p\boldsymbol{v}\cdot\boldsymbol{n}\,\mathrm{d}a + \\ + \int_D \rho\boldsymbol{f}\cdot\boldsymbol{v}\,\mathrm{d}a - \int_{\partial D}\boldsymbol{q}\cdot\boldsymbol{n}\,\mathrm{d}a \quad (\forall)\, D\subset\mathcal{D},\end{aligned} \tag{1.1.14}$$

e representing the specific internal energy, and $\boldsymbol{q}$, the flux of heat vector. For continuous motions, taking into account (1.1.11) and (1.1.13), we deduce

$$\rho\dot{e} = -p \operatorname{div}\boldsymbol{v} - \operatorname{div}\boldsymbol{q}. \tag{1.1.15}$$

In ideal fluids the processes are *reversible*. Eliminating $\operatorname{div}\boldsymbol{v}$ by the aid of equation (2.1.11) and employing the second principle of thermodynamics [1.11], p.54, one obtains the fundamental equation of thermodynamics

$$\mathrm{d}e = T\mathrm{d}s - p\mathrm{d}v, \tag{1.1.16}$$

where s is the specific entropy and v is a notation for $1/\rho$.

Eliminating $\operatorname{div} \boldsymbol{v}$ from (1.1.15) and taking into account (1.1.16), we deduce the following remarkable form of the equation of energy

$$\rho T \dot{s} = -\operatorname{div} \boldsymbol{q} . \tag{1.1.17}$$

This shows that if it is possible to neglect the change of heat (it is the case of aerodynamics where the velocities are great), then

$$\dot{s} = 0 . \tag{1.1.18}$$

The equation (1.1.18) indicates that s is constant on trajectories, the constant varying from one trajectory to the other. One calls such a motion *isentropic motion.* If there exists a configuration where the entropy constant is the same everywhere, then in every configuration arising from the first one, the constant will be the same everywhere. Such a motion is called *homentropic* or *isentropic everywhere.*

The perfect gas is characterized by the following equations of state

$$p = \rho R T, \quad e = c_v T + C . \tag{1.1.19}$$

The first one (which is obtained from the laws of Boyle-Mariotte and Gay-Lussac) is *the thermic equation* (or Clapeyron's equation) and the second is *the caloric equation.* It is easy to prove (see for example [1.11] p. 57–58) that for this gas we have

$$s = c_v \ln\left(\frac{p}{\rho^\gamma}\right) + C_0 , \quad e = \frac{1}{\gamma - 1}\frac{p}{\rho} + C_1 . \tag{1.1.20}$$

c_p being the so called *specific heat at constant volume* and $\gamma = c_p/c_v$ where c_p is the specific heat at constant pressure. For the air $\gamma = 1.405$. From the expression of s it follows for the homentropic motion:

$$p = k\rho^\gamma , \tag{1.1.21}$$

k representing a constant. The quantity c, defined by the formula

$$\boldsymbol{c}^2 = \left(\frac{\mathrm{d}p}{\mathrm{d}\rho}\right)_s , \tag{1.1.22}$$

has the dimension of a velocity. One shows (see for example [1.10]) that it gives just the speed of propagation of the surfaces of discontinuity of the pressure (sound waves). For the ideal gas in homentropic motion it follows

$$c^2 = \gamma p/\rho . \tag{1.1.23}$$

1.2 The Potential Flow

1.2.1 Helmholtz's equation. Bernoulli's integral

If the massic forces possess a potential $\boldsymbol{f} = \operatorname{grad}\Pi$ (in aerodynamics these forces (representing the weight of the air) are neglected) and if the fluid is characterized by a thermodynamic law having the form $\rho = \hat{\rho}(p)$, where $\hat{\rho}$ is a derivable function, defined for $p \geq 0$ such that $\hat{\rho}'(p) \geq 0$, then we deduce

$$\boldsymbol{f} - \frac{1}{\rho}\operatorname{grad} p = \operatorname{grad}\left(\Pi - \int \frac{\mathrm{d}p}{\rho}\right).$$

Utilizing the identity $(\boldsymbol{v}\cdot\nabla)\boldsymbol{v} = \operatorname{curl}\boldsymbol{v}\times\boldsymbol{v} + \operatorname{grad}(\boldsymbol{v}^2/2)$, Euler's equation (1.1.13) becomes

$$\frac{\partial\boldsymbol{v}}{\partial t} + \operatorname{curl}\boldsymbol{v}\times\boldsymbol{v} = \operatorname{grad}\left(\Pi - \int\frac{\mathrm{d}p}{\rho} - \frac{1}{2}v^2\right). \tag{1.2.1}$$

This is *Helmholtz's equation.*

The flow of a fluid is *irrotational* in a domain $\mathcal{D}$ if everywhere in $\mathcal{D}$, we have $\operatorname{curl}\boldsymbol{v} = 0$. This equation constitutes the necessary and sufficient condition for the existence of a differentiable function $\varphi(x, \boldsymbol{x})$, such that

$$\boldsymbol{v} = \operatorname{grad}\phi. \tag{1.2.2}$$

Such a flow is called *potential.* Applying the operator *curl* in (1.2.1), we eliminate the term in the right hand side of the equation. The resulting equation is integrated [1.11] as follows

$$\frac{\boldsymbol{\omega}}{\rho} = \left(\frac{\boldsymbol{\omega}^0}{\rho_0}\cdot\nabla_{\boldsymbol{X}}\right)\boldsymbol{\chi}(t, \boldsymbol{X}), \tag{1.2.3}$$

$\boldsymbol{\omega}^0$ and ρ_0 representing the vortex $(2\boldsymbol{\omega} \equiv \operatorname{curl}\boldsymbol{v})$ and respectively the density in a reference configuration. The formula (1.2.3) shows that if the motion is potential in a configuration, it remains potential in every configuration arising from the first one (*Lagrange-Cauchy's theorem*).

We can therefore put $\operatorname{curl}\boldsymbol{v} = 0$ in (1.2.1). It results Bernoulli's integral

$$\phi_t + \frac{1}{2}v^2 + \int\frac{\mathrm{d}p}{\rho} - \Pi = C(t) \quad \phi_t = \partial\phi/\partial t. \tag{1.2.4}$$

One may give to the spatial constant $C(t)$ the value zero ([1.11], p.90). Neglecting the massic forces, Bernoulli's integral is

$$\phi_t + \frac{1}{2}v^2 + \int\frac{\mathrm{d}p}{\rho} = 0. \tag{1.2.5}$$

The motion of a fluid is *stationary* or *steady* if the velocity field does not depend explicitly on t. Now (1.2.4) becomes

$$\frac{1}{2}v^2 + \int \frac{\mathrm{d}p}{\rho} - \Pi = C\,. \tag{1.2.6}$$

In this case the constant cannot be zero.

For the perfect gas in homentropic motion we have the formula (1.1.21). With a convenient notation of the constant, from (1.2.6) we deduce

$$\frac{1}{2}v^2 + \frac{\gamma}{\gamma-1}\frac{p}{\rho} = \frac{\gamma}{\gamma-1}\frac{p_0}{\rho_0} = \frac{c_0^2}{\gamma-1}\,, \tag{1.2.7}$$

p_0, ρ_0 and c_0^2 representing the pressure, the density and the square of the sound velocity for the fluid at rest. From (1.2.7) one deduces that in the compressible fluid there is a superior limit $v_{\max}$ of the velocity which is obtained for $p = 0$ and a critical value v_{cr}, which is obtained for $v = c$. These values are

$$v_{\max} = c_0\sqrt{\frac{2}{\gamma-1}}\,, \qquad v_{\mathrm{cr}} = c_0\sqrt{\frac{2}{\gamma+1}}\,. \tag{1.2.8}$$

For $v < v_{\mathrm{cr}}$ the flow is subsonic and for $v_{\mathrm{cr}} < v < v_{\max}$, the flow is supersonic.

1.2.2 The Equation of the Potential

The equation of the potential is obtained from the equation of continuity (1.1.11), Bernoulli's integral (1.2.5) and the equation (1.1.21), assuming that $\boldsymbol{v}$ has the form (1.2.2). Calculating the material derivative of (1.2.5), we obtain:

$$\frac{\mathrm{d}}{\mathrm{d}t}\left(\phi_t + \frac{1}{2}v^2\right) + \frac{\dot{p}}{\rho} = 0\,. \tag{1.2.9}$$

Taking into account that ρ depends on t through the agency of p and utilizing the notation (1.1.22), we deduce

$$\dot{\rho} = \frac{\mathrm{d}\rho}{\mathrm{d}p}\dot{p} = -\frac{\rho}{c^2}\frac{\mathrm{d}}{\mathrm{d}t}\left(\phi_t + \frac{1}{2}v^2\right)\,. \tag{1.2.10}$$

Replacing in the equation of continuity, we obtain the equation of the potential

$$c^2\Delta\phi - (\boldsymbol{v}\cdot\nabla)(\boldsymbol{v}\cdot\nabla)\phi - 2(\boldsymbol{v}\cdot\nabla)\phi_t - \phi_{tt} = 0, \tag{1.2.11}$$

where c^2 depends on ϕ.

We find this dependence in the case of the perfect gas in homentropic flow. Taking into account (1.1.21), Bernoulli's integral (1.2.5) becomes

$$\phi_t + \frac{1}{2}v^2 + \frac{k\gamma}{\gamma - 1}\left(\rho^{\gamma-1} - \rho_0^{\gamma-1}\right) = 0\,, \tag{1.2.12}$$

ρ_0 representing the density in a reference state. From (1.1.21) and (1.1.22) it follows $c_0^2 = k\gamma\rho_0^{\gamma-1}$, and from (1.2.12) we deduce:

$$\rho = \rho_0\left[1 - \frac{\gamma-1}{c_0^2}\left(\phi_t + \frac{1}{2}v^2\right)\right]^{\frac{1}{\gamma-1}}. \tag{1.2.13}$$

From (1.1.21), we obtain:

$$p = p_0\left[1 - \frac{\gamma-1}{c_0^2}\left(\phi_t + \frac{1}{2}v^2\right)\right]^{\frac{\gamma}{\gamma-1}}, \tag{1.2.14}$$

$$c^2 = c_0^2 - (\gamma - 1)(\phi_t + \frac{1}{2}v^2)\,, \tag{1.2.15}$$

p_0 and c_0 representing the pressure and the sound velocity in the reference state of density ρ_0.

For the steady flow one obtains the equation

$$c^2\Delta\phi - (\boldsymbol{v}\cdot\nabla)(\boldsymbol{v}\cdot\nabla)\phi = 0 \tag{1.2.16}$$

and the formulas

$$\rho = \rho_0\left(1 - \frac{\gamma-1}{2}\frac{v^2}{c_0^2}\right)^{\frac{1}{\gamma-1}}, \quad p = p_0\left(1 - \frac{\gamma-1}{2}\frac{v^2}{c_0^2}\right)^{\frac{\gamma}{\gamma-1}}, \tag{1.2.17}$$

$$c^2 = c_0^2 - \frac{\gamma-1}{2}v^2 \tag{1.2.18}$$

where ρ_0, p_0 and c_0 are quantities corresponding to the fluid at rest $(v = 0)$.

One demonstrates [1.10], p. 207, that the equation (1.2.11) is hyperbolic and the equation (1.2.16) is elliptic for the *subsonic flow* $(v^2 < c^2)$ and hyperbolic for the *supersonic flow* $(c^2 < v^2)$. The equality $v^2 = c^2$ occurs only on curves in the two-dimensional flow or surfaces in the case of three-dimensional flow. These varieties are separating the domains where the flow is subsonic from the domains where the flow is supersonic. This kind of motion is called *transonic*. We denoted $v^2 = |\,\boldsymbol{v}\,|^2$.

1.2.3 The Linear Theory

When the equations (1.2.11) and (1.2.16) are utilized in order to determine the perturbation produced by a thin body in a fluid having a known flow, they may be *linearized.* Let us consider for example, that the uniform flow of a fluid having the velocity $U_\infty \boldsymbol{i}$, the pressure p_∞ and the density ρ_∞ is slightly perturbed by the presence of a body. We shall denote by

$$\boldsymbol{v} = U_\infty \boldsymbol{i} + \boldsymbol{v}' \tag{1.2.19}$$

the velocity field for the perturbed flow ($\boldsymbol{v}'$ is the perturbation of the velocity field). We assume that all the coordinates of $\boldsymbol{v}'$ have the same order of magnitude ε (ε representing a small parameter which characterizes the body). Hence we assume that

$$\boldsymbol{v}' = \varepsilon \overline{\boldsymbol{v}}\,, \tag{1.2.20}$$

where $|\overline{\boldsymbol{v}}|$ is bounded and $\varepsilon << 1$. According to Lagrange-Cauchy's theorem, the perturbed flow is potential, since it arises from a potential flow. Hence, $(\exists)\phi$, such that $\boldsymbol{v} = \operatorname{grad}\phi$ and from (1.2.19), it follows $\boldsymbol{v}' = \operatorname{grad}\varphi$ and

$$\phi = U_\infty x + \varphi. \tag{1.2.21}$$

The condition (1.2.20) is equivalent to

$$(\varphi_x, \varphi_y, \varphi_z) = \varepsilon(\overline{\varphi}_x, \overline{\varphi}_y, \overline{\varphi}_z)\,. \tag{1.2.22}$$

Replacing in the formulas above and neglecting the terms $\mathcal{O}(\varepsilon^2)$, we deduce

$$\boldsymbol{v} = (U_\infty + \varepsilon\overline{\varphi}_x, \varepsilon\overline{\varphi}_y, \varepsilon\overline{\varphi}_z),\ v^2 = U_\infty^2 + 2U_\infty\varphi_a + \mathcal{O}(\varepsilon^2) \tag{1.2.23}$$

and then

$$c_\infty^2\Delta\varphi - U_\infty^2\varphi_{xx} - 2U_\infty\varphi_{xt} - \varphi_{tt} = 0\,, \tag{1.2.24}$$

$$(1 - M^2)\varphi_{xx} + \varphi_{yy} + \varphi_{zz} = 0\,, \tag{1.2.25}$$

where $M(= U_\infty/c_\infty)$ is Mach's number in the perturbed flow. One easily find that the equation (1.2.24) is hyperbolic, and the equation (1.2.25) is elliptic if $M < 1$ and hyperbolic if $M > 1$. The motions characterized by $M < 1$ are named *subsonic*, and the motions with $M > 1$ are named *supersonic.*

The pressure is deduced from Bernoulli's integral, (1.2.5), which may be written as follows

$$\left(\phi_t + \frac{1}{2}v^2\right)_\infty^{\text{actual}} + \int_{p_\infty}^{p} \frac{dp}{\rho} = 0\,, \tag{1.2.26}$$

if we neglect the massic forces. It obviously results

$$\phi_t + \frac{1}{2}\left(v^2 - U^2\right) + \int_{p_\infty}^{p} \frac{dp}{\rho} = 0\,, \tag{1.2.27}$$

Let us assume that the relation

$$p = p_\infty + \mathcal{O}(\varepsilon)\,. \tag{1.2.28}$$

is also valid. Then we deduce

$$\rho(p) = \rho_\infty + \frac{d\rho}{dp}\Big|_\infty (p - p_\infty) + \mathcal{O}(\varepsilon^2) = \rho_\infty + \frac{p - p_\infty}{c_\infty^2} + \mathcal{O}(\varepsilon^2)$$

$$\int_{p_\infty}^{p} \frac{dp}{\rho} = \frac{1}{\rho_\infty}\int_{p_\infty}^{p}\left[1 - \frac{p - p_\infty}{c_\infty^2 \rho_\infty} + \mathcal{O}(\varepsilon^2)\right] dp = \frac{p - p_\infty}{\rho_\infty} + \mathcal{O}(\varepsilon^2)\,. \tag{1.2.29}$$

Replacing (1.2.21), (1.2.23) and (1.2.29) in (1.2.27), we find in the linear approximation the formula for the determination of the pressure

$$p - p_\infty = -\rho_\infty(\varphi_t + U_\infty \varphi_x)\,, \tag{1.2.30}$$

Since the previous formulas will be deduced in a different manner in Chapter 2, in dimensionless variables, we are going to introduce here these variables (x^*, y^*, z^*, t^*), by means of the formulas

$$\begin{aligned} (x, y, z) &= L_0(x^*, y^*, z^*), \quad t = (L_0/U_\infty)t^*, \\ \varphi &= U_\infty \varphi^*, \quad p - p_\infty = \rho_\infty U_\infty^2 p^*, \end{aligned} \tag{1.2.31}$$

L_0 representing a reference length which is not specified yet. So, the equation (1.2.22) becomes

$$\Delta_* \varphi^* - M^2 \varphi_{x^*x^*} - 2M\varphi_{x^*t^*} - M^2 \varphi^*_{t^*t^*} = 0, \tag{1.2.32}$$

and the formula (1.2.30),

$$p^* = -(\varphi^*_{t^*} + \varphi^*_{x^*})\,. \tag{1.2.33}$$

1.2.4 The Acceleration Potential

Another function, often utilized in aerodynamics, is the acceleration potential ψ, introduced by Prandtl. Its existence is ensured for the unlimited fluid characterized by the law $\rho = \widehat{\rho}(p)$, by the equation

$$\boldsymbol{a} = -\,\mathrm{grad} \int \frac{dp}{\rho} \tag{1.2.34}$$

resulting from (1.1.6), (1.1.13) and (1.2.1). Denoting

$$\boldsymbol{a} = \operatorname{grad} \psi \,, \tag{1.2.35}$$

we get

$$\psi = -\int \frac{\mathrm{d}p}{\rho} + C(t)\,, \tag{1.2.36}$$

where $C(t)$ is an arbitrary function depending on t, which can be determined by imposing the value of ψ for a certain state. For example, if the fluid is incompressible and one considers that ψ vanishes at infinity (where $p = p_\infty$), it follows $C = p_\infty/\rho$, whence

$$\psi = (p_\infty - p)/\rho. \tag{1.2.37}$$

Since ψ represents the perturbation of the pressure, it is usually called *the pressure function.*

The relation between ψ and ϕ may be deduced from the equations (1.2.5) and (1.2.26). We have

$$\operatorname{grad}\left[\phi_t + (1/2)v^2\right] = -\operatorname{grad}\int \frac{\mathrm{d}p}{\rho} = \operatorname{grad}\psi\,. \tag{1.2.38}$$

If the flow is uniform at infinity, with the velocity U_∞, it results

$$\psi = \phi_t + (1/2)(v^2 - U_\infty^2) \tag{1.2.39}$$

In the framework of the linearized theory this relation becomes

$$\psi = \varphi_t + U_\infty \varphi_x\,. \tag{1.2.40}$$

Taking into account the permutability of the operator $\partial/\partial t + U_\infty \partial/\partial x$ with the operators appearing in (1.2.24), it follows that ψ satisfies the same equations (1.2.24) or (1.2.25).

From (1.2.29) and (1.2.39), it results that ψ has the same expression (1.2.37) in the case of the linear approximation for compressible fluids.

In order to write the boundary conditions, we have to indicate the relation between ψ and the coordinates of the velocity. For steady motions, we obtain from (1.2.40), (taking into account that w must vanish at infinity) :

$$U_\infty \varphi = \int_{-\infty}^{x} \psi\,\mathrm{d}x, \quad U_\infty w = \int_{-\infty}^{x} \frac{\partial \psi}{\partial z}\,\mathrm{d}x\,. \tag{1.2.41}$$

For unsteady flow, deriving (1.2.40), we get

$$\partial\psi/\partial z = \partial w/\partial t + U_\infty \partial w/\partial x\,. \tag{1.2.42}$$

Performing the change of variables $t, x \longrightarrow \tau, \xi$:

$$\tau = t, \qquad \xi = x - U_\infty t \tag{1.2.43}$$

and denoting

$$\begin{aligned} w(t,x,y,z) &= w'(\tau,\xi,y,z), \\ \psi(t,x,y,z) &= \psi'(\tau,\xi,y,z)\,, \end{aligned} \tag{1.2.44}$$

we deduce $\partial\psi'/\partial z = \partial w'/\partial\tau$, whence, integrating

$$w'(\tau,\xi,y,z) = \int_{-\infty}^{\tau} \frac{\partial}{\partial t}\psi'(\tau',x',y,z)\mathrm{d}\tau'\,. \tag{1.2.45}$$

Returning to the variables t and x, we get

$$\begin{aligned} w(t,x,y,z) &= \int_{-\infty}^{\tau} \frac{\partial}{\partial z}\psi'(\tau',\xi+U_\infty\tau',y,z)\mathrm{d}\tau' = \\ &= \int_{-\infty}^{t} \frac{\partial}{\partial z}\psi(t',x-U_\infty(t-t'),y,z)\mathrm{d}t'\,. \end{aligned} \tag{1.2.46}$$

1.3 The Shock Waves Theory

1.3.1 The Jump Equations

The principles of motion for the continuous media are expressed in terms of *balance equations*. The general form of a balance equation is

$$\frac{\mathrm{d}}{\mathrm{d}t}\int_D \rho\boldsymbol{Q}\,\mathrm{d}v = \int_{\partial D} \boldsymbol{R}\,\boldsymbol{n}\,\mathrm{d}a + \int_D \rho\boldsymbol{S}\,\mathrm{d}v\,, \tag{1.3.1}$$

where D is a material domain, $\boldsymbol{Q}$ and $\boldsymbol{S}$ are tensors having the same order and $\boldsymbol{R}$ is a tensor having the order greater with a unity. For the ideal fluid, the principle of conservation of mass (1.1.10) has the form (1.3.1), with

$$\boldsymbol{Q} = \boldsymbol{I}\,,\ \ \boldsymbol{R} = 0,\ \boldsymbol{S} = 0\,. \tag{1.3.2}$$

The principle of variation of the momentum (1.1.12) is obtained for

$$\boldsymbol{Q} = \boldsymbol{v}\,,\ \ \boldsymbol{R} = -p\boldsymbol{I},\ \ \boldsymbol{S} = \boldsymbol{f}\,, \tag{1.3.3}$$

$\boldsymbol{I}$ representing the unity tensor, and the balance equation for the energy (1.1.14) for

$$Q = e + (1/2)v^2, \boldsymbol{R} = -p\boldsymbol{v} - \boldsymbol{q}, S = \boldsymbol{f}\cdot\boldsymbol{v}\,. \tag{1.3.4}$$

Let us see now the balance equation (1.3.1) in case that D is crossed by a surface of discontinuity with the displacement velocity d ([1.11], p.28). Applying the derivation formula (1.1.9), the equation (1.3.1) gives

$$\int_D \left[\frac{\partial}{\partial t}(\rho \boldsymbol{Q}) - \rho \boldsymbol{S}\right] \mathrm{d}v + \int_{\Sigma_+ + \Sigma_-} [\![\rho \boldsymbol{Q}\,(\boldsymbol{v}\cdot\boldsymbol{n}_\pm) - \boldsymbol{R}\boldsymbol{n}_\pm]\!] \mathrm{d}a -$$

$$- \int_S [\![\rho \boldsymbol{Q}]\!] d \;\mathrm{d}a = 0\,.$$

We pass to the limit superposing $\sum_+$ and $\sum_-$ on S (fig. 1.3.1). In this case, $\boldsymbol{n}_+$ becomes $\boldsymbol{n}$, and $\boldsymbol{n}_-$ becomes $-\boldsymbol{n}$.

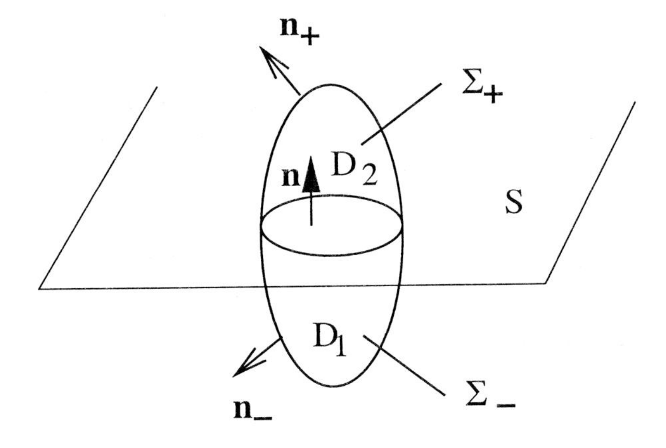

Fig. 1.3.1.

If we assume that the integrand from the first integral is bounded, the first term tends to zero, because vol $D_\pm \longrightarrow 0$. Hence one obtains:

$$\int_S [\![\rho \boldsymbol{Q}(\boldsymbol{v}\cdot\boldsymbol{n} - d) - \boldsymbol{R}\boldsymbol{n}]\!] \;\mathrm{d}a = 0\,, \tag{1.3.5}$$

where we denoted

$$[\![\rho \boldsymbol{Q}(\boldsymbol{v}\cdot\boldsymbol{n} - d) - \boldsymbol{R}\boldsymbol{n}]\!] = \rho_+ Q_+(\boldsymbol{v}_+\cdot\boldsymbol{n} - d) - \boldsymbol{R}_+\boldsymbol{n} - [\rho_- \boldsymbol{Q}_-(\boldsymbol{v}_-\cdot\boldsymbol{n} - d) - \boldsymbol{R}_-\boldsymbol{n}]\,. \tag{1.3.6}$$

The limit values encountered here are continuous on S (they are not continuous across S), and S is arbitrary, because the balance equations are formulated for every D. By virtue of the fundamental lemma ([1.11] p.31), from (1.3.5), we deduce

$$[\![\rho \boldsymbol{Q}(\boldsymbol{v}\cdot\boldsymbol{n} - d) - \boldsymbol{R}\boldsymbol{n}]\!] = 0\,. \tag{1.3.7}$$

This is the equation which leads to the balance equation. It establishes the connection between the limit values from the two parts of the shock waves. If across a surface the fields have discontinuities, this surface is a *shock wave.*

Utilizing (1.3.7) for (1.3.2), one obtains:

$$[\![\rho(\boldsymbol{v}\cdot\boldsymbol{n}-d)]\!]=0\,, \tag{1.3.8}$$

for (1.3.3)

$$[\![\rho\boldsymbol{v}(\boldsymbol{v}\cdot\boldsymbol{n}-d)+p\boldsymbol{n}]\!]=0\,, \tag{1.3.9}$$

and for (1.3.4),

$$[\![\rho(e+\frac{1}{2}v^2)(\boldsymbol{v}\cdot\boldsymbol{n}-d)+p\boldsymbol{v}\cdot\boldsymbol{n}+\boldsymbol{q}\cdot\boldsymbol{n}]\!]=0\,. \tag{1.3.10}$$

The conduction laws utilized in practice (particularly Fourier's law) make impossible the discontinuity for $\boldsymbol{q}$. Hence we shall utilize the equation (1.3.10) as follows

$$[\![\rho\left(e+\frac{1}{2}v^2\right)(\boldsymbol{v}\cdot\boldsymbol{n}-d)+p\boldsymbol{v}\cdot\boldsymbol{n}]\!]=0\,. \tag{1.3.11}$$

To the previous equations we shall add the inequation:

$$[\![\rho s(\boldsymbol{v}\cdot\boldsymbol{n}-d)]\!]\geq 0\,, \tag{1.3.12}$$

coming from the Second Principle of Thermodynamics ([1.11], p.50). For the ideal fluid we have an equality.

1.3.2 Hugoniot's Equation

The jump Equation (1.3.11) is quite complicated. It can be replaced by a simple equation (Hugoniot's equation), which establishes a connection only between the thermodynamic quantities from the two sides of the shock wave. In order to deduce these equations, for the sake of simplicity we introduce the propagation speed $P=d-\boldsymbol{v}\cdot\boldsymbol{n}$. In this way, the jump equations become

$$[\![\rho P]\!]=0\,, \tag{1.3.13}$$

$$[\![\rho P\boldsymbol{v}-p\boldsymbol{n}]\!]=0\,, \tag{1.3.14}$$

$$[\![\rho P\left(e+\frac{1}{2}v^2\right)P-p\boldsymbol{v}\cdot\boldsymbol{n}]\!]=0\,, \tag{1.3.15}$$

$$[\![\rho P s]\!] \leq 0, \tag{1.3.16}$$

From these equations we have to find the unknowns behind the shock wave and its propagation speed if we know the state of the fluid in front of the shock wave. For the sake of simplicity too, we mark by the index 1 the limit values (on S) in front of the shock wave and by the index 2 the limit values behind the shock wave. The normal to S is positively orientated from the region 2 toward region 1. We exclude the situation $P_1 = P_2 = 0$, because in this case the surface of discontinuity does not cross the fluid (it is a material surface moving together with the fluid).

For deducing Hogoniot's equation, we notice that from (1.3.13) it follows that the quantity ρP is continuous across the shock wave. We denote by

$$m = \rho_1 P_1 = \rho_2 P_2 \tag{1.3.17}$$

this quantity. Since in the jump equations we employ only the velocity on the shock wave in a current point of the wave, we write

$$\boldsymbol{v} = v_n \boldsymbol{n} + v_t \boldsymbol{t}, \quad v^2 = v_n^2 + v_t^2, \tag{1.3.18}$$

v_n representing the normal component and v_t the tangential component (situated in the plane determined by $\boldsymbol{n}$ and $\boldsymbol{v}$). We denoted by $\boldsymbol{t}$ the versor of the tangent.

Projecting (1.3.14) on the tangent and on the normal to the shock wave and taking into account that by hypothesis, $m \neq 0$, we deduce

$$[\![v_t]\!] = 0, \quad m[\![v_n]\!] - [\![p]\!] = 0. \tag{1.3.19}$$

The first equation shows that *the tangential component of the velocity remains constant across the shock wave.* From the definition of the propagation speed P and from the conservation relation (1.3.17), we deduce

$$v_n = d - P = d - m\tau, \quad \tau \equiv 1/\rho, \tag{1.3.20}$$

the second equation from (1.3.19) becoming

$$m^2[\![\tau]\!] + [\![p]\!] = 0. \tag{1.3.21}$$

From (1.3.15) taking into account the equations (1.3.19), (1.3.20) and (1.3.21), we notice that:

$$[\![\rho P v_t^2]\!] = m[\![v_t]\!](v_{t_1} + v_{t_2}) = 0,$$

$$[\![\rho P v_n^2]\!] = -2m^2 d[\![\tau]\!] + m^3[\![\tau^2]\!] = 2d[\![p]\!] - m[\![p]\!](\tau_1 + \tau_2).$$

From these equations and from (1.3.15), we get:

$$2[\![h]\!] = [\![p]\!](\tau_1 + \tau_2), \tag{1.3.22}$$

where $h = e + p\tau$ is *the enthalpy* ([1.11], p.55). This is the first form of *Hugoniot's equation.* Performing the calculations in (1.3.22), we get also the form

$$2[\![e]\!] + [\![\tau]\!](p_1 + p_2) = 0, \tag{1.3.23}$$

and, for the perfect gas, when e is given by (1.1.20), one finds the third expression:

$$\frac{p_2}{p_1} = \frac{(\gamma+1)\rho_2 - (\gamma-1)\rho_1}{(\gamma+1)\rho_1 - (\gamma-1)\rho_2} \tag{1.3.24}$$

which is useful in applications.

The inequality (1.3.2) shows that if $m > 0$ (the shock wave surpasses the fluid), then $s_1 \leq s_2$; if $m < 0$ (the fluid surpasses the shock wave), then $s_2 \leq s_1$. In all cases behind the shock wave *the entropy is not decreasing.* It increases or remains constant.

1.3.3 The Solution of the Jump Equations

In the sequel we are going to demonstrate, for the perfect gas, that giving the state of the medium in front of the shock wave and the propagation speed P_1, one may determine completely the state behind the shock wave. To this aim, we introduce the numbers:

$$\mathcal{M}_1 = P_1/c_1, \quad \mathcal{M}_2 = P_2/c_2, \tag{1.3.25}$$

where c_1 and c_2 represent the sound velocity in front of the shock wave respectively behind the shock wave ($c_1^2 = \gamma p_1/\rho_1$, $c_2^2 = \gamma p_2/\rho_2$). By hypothesis, $\mathcal{M}_1$ is known and $\mathcal{M}_2$ is unknown. From (1.3.21), we deduce:

$$p_2 - p_1 = m^2(\tau_1 - \tau_2) = \rho_1 p_1^2(1 - \tau_2/\tau_1) = \gamma p_1 \mathcal{M}_1^2(1 - \tau_2/\tau_1).$$

The function $(p_2/p_1) - 1$ may also be obtained from (1.3.24). Comparing the expressions we get:

$$\frac{\tau_2}{\tau_1} - 1 = \frac{2}{\gamma+1}\left(\frac{1}{\mathcal{M}_1^2} - 1\right), \quad \frac{p_2}{p_1} - 1 = \frac{2\gamma}{\gamma+1}(\mathcal{M}_1^2 - 1). \tag{1.3.26}$$

From the relation $P_1\tau_2 = P_2\tau_1$, it follows:

$$\left[1 + \frac{2\gamma}{\gamma+1}(\mathcal{M}_1^2 - 1)\right](1 - \mathcal{M}_2^2) = \mathcal{M}_1^2 - 1 \tag{1.3.27}$$

From the perfect gas law $p = \rho RT$ and from (1.1.20), we obtain:

$$\frac{T_2}{T_1} = \left(\frac{\tau_2}{\tau_1}\right)\left(\frac{p_2}{p_1}\right), \quad \frac{s_2 - s_1}{c_v} = \ln\left(\frac{p_2}{p_1}\right)\left(\frac{\tau_2}{\tau_1}\right)^{\gamma}, \tag{1.3.28}$$

τ_2/τ_1 and p_2/p_1 being replaced from (1.3.26).

1.3.4 Prandtl's Formula

For the stationary wave $(d = 0)$, we have $v_n = -m\tau$, such that the equation (1.3.11), for the perfect gas, becomes

$$[\![\frac{1}{2}v^2 + \frac{\gamma}{\gamma - 1}\frac{p}{\rho}]\!] = 0\,. \tag{1.3.29}$$

The expression written here intervenes in Bernoulli's integral (1.2.7). Taking into account the meaning given there to $v_{\max}$, we have the relation

$$\frac{1}{2}v_{1\,\max}^2 = \frac{1}{2}v_1^2 + \frac{\gamma}{\gamma - 1}\frac{p_1}{\rho_1} = \frac{1}{2}v_2^2 + \frac{\gamma}{\gamma - 1}\frac{p_2}{\rho_2} = \frac{1}{2}v_{2\,\max}^2\,, \tag{1.3.30}$$

which shows that $v_{\max}$ is invariant across the shock wave. Taking into account that $[\![v_t]\!] = 0$, from (1.3.30) we deduce:

$$\frac{1}{2}[\![v_n^2]\!] = \frac{\gamma}{\gamma - 1}\frac{p_1}{\rho_1}\left(\frac{p_2}{p_1}\frac{\rho_1}{\rho_2} - 1\right). \tag{1.3.31}$$

We eliminate p_2/p_1 by the aid of equation (1.3.24) and ρ_2/ρ_1, by the aid of relation $\rho_1 v_{1n} = \rho_2 v_{2n}$ which results from (1.3.17). Employing once again (1.3.30), we deduce:

$$(\gamma + 1)v_{1n}v_{2n} - (\gamma - 1)v_{1n}^2 = 2\gamma p_1/\rho_1 = (\gamma - 1)(v_{\max}^2 - v_1^2).$$

Writing $v_{1t} = v_{2t} = v_t$, we obtain Prandtl's formula

$$v_{1n}v_{2n} = \frac{\gamma - 1}{\gamma + 1}(v_{\max}^2 - v_t^2) = v_{cr}^2 - \frac{\gamma - 1}{\gamma + 1}v_t^2, \tag{1.3.32}$$

the significance of v_{cr} being that of (1.2.8). In the expressions $v_{\max}$ and v_{cr} utilized here, c_0 is the sound velocity for the fluid in front of the shock wave (at rest). The relation (1.3.32) gives v_{2n} when the state of the fluid in front of the shock wave is known. For the normal shock $(v_t = 0)$, the relation becomes $v_1 v_2 = v_{cr}^2$. According to this relation, if v_1 is supersonic, then v_2 is subsonic.

1.3.5 The Shock Polar

Busemann (*Vorträge aus dem Gebiete des Aerodynamik*, Aachen, 1929), gave a graphic method for constructing $\boldsymbol{v}_2$ if the angle with $\boldsymbol{v}_1$ is known. To this aim, we consider in the current point M from the shock wave (fig. 1.3.2) the MX axis, pointing in the direction and sense of $\boldsymbol{v}_1$ and the MY axis in the plane determined by $\boldsymbol{v}_1$ and the normal $\boldsymbol{n}$ to the shock wave.

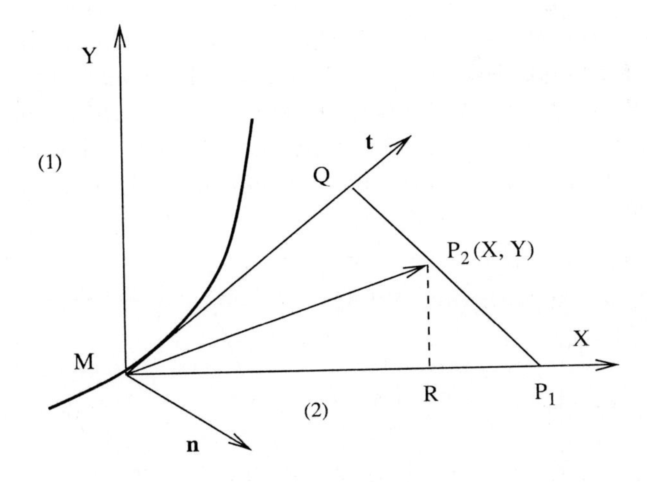

Fig. 1.3.2.

We use the normal pointing towards the state (2); the change of the sense of the normal does not affect the jump relations (excepting the relation (1.3.16), whose signification is known), because in these relations the normal intervenes linearly. Because of the continuity of the tangential component of the velocity, $\boldsymbol{v}_2$ lies also in the plane determined by $\boldsymbol{v}_1$ and $\boldsymbol{n}$ and it will have on the tangent versor $\boldsymbol{t}$ the same projection MQ like $\boldsymbol{v}_1$. We denoted by P_1 and P_2 the extremities of the vectors $\boldsymbol{v}_1$ and $\boldsymbol{v}_2$ and by θ, the angle between $\boldsymbol{v}_2$ and $\boldsymbol{v}_1$. Our aim is to find the geometric locus of the point P_2 when θ is varying. We shall use, obviously, Prandtl's formula (1.3.32). Denoting by (X, Y) the coordinates of the vector $\boldsymbol{v}_2$, from the figure, it results

$$\begin{gathered} X = v_2 \cos\theta, Y = v_2 \sin\theta, v_{1n} = v_1 \sin\sigma, v_t = v_1 \cos\theta\,, \\ v_{2n} = v_2 \sin(\sigma - \theta) = X \sin\sigma - Y \cos\sigma\,. \end{gathered} \tag{1.3.33}$$

So, Prandtl's formula becomes:

$$X \sin^2\sigma - Y \sin\sigma \cos\sigma = \frac{v_{cr}^2}{v_1} - \frac{\gamma - 1}{\gamma + 1} v_1 \cos^2\sigma\,. \tag{1.3.34}$$

From the triangle P_1P_2R, where the angle $P_1P_2R = \sigma$, it follows $\tan\sigma = (v_1 - X)/Y$, such that (1.3.34) becomes

$$Y^2(b - X) = (v_1 - X)^2(X - a)\,, \tag{1.3.35}$$

where we denoted

$$a = \frac{v_{cr}^2}{v_1}\,,\ b = \frac{v_{cr}^2}{v_1} + \frac{2}{\gamma+1}v_1\,. \tag{1.3.36}$$

From the assumption that v_1 is supersonic, it follows that $a < v_1$, and from $v_1 < v_{\max}$, we deduce $v_1 < b < v_{\max}$.

The curve representing eq. (1.3.35) is symmetric with respect to MX, it intersects the MX axis in the points $A(a)$ and $P_1(v_1)$ (which are inverse with respect to the sonic circle $x^2 + y^2 = v_{cr}^2$), the last point being double and has a vertical asymptote in $B(b)$. The curve is real for $a \leq X \leq b$ and it is represented in figure 1.3.3. It is called *folium of Descartes*. It is the *shock polar*.

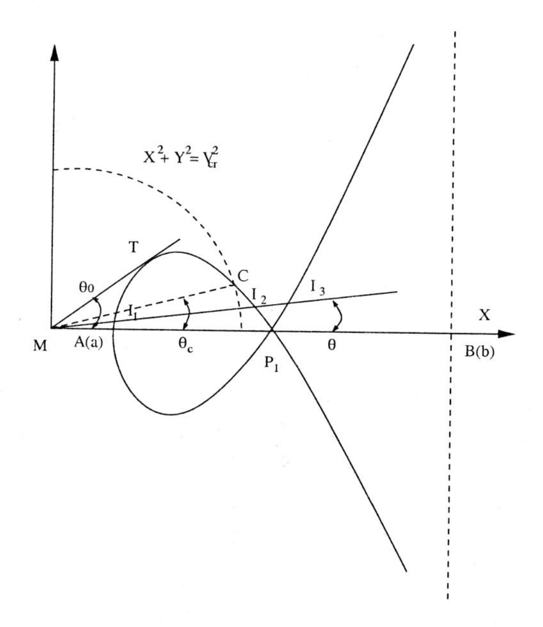

Fig. 1.3.3.

Let us show now how we construct $\boldsymbol{v}_2$ when we know the angle θ made with $\boldsymbol{v}_1$. First of all one constructs the shock polar. This is determined only by the state (1). The same state determines the angle θ_0 made by the tangent MT with the MX axis. If $\theta < \theta_0$, there exist three intersection points I_1, I_2, I_3, of the radius vector with the polar, but only one is P_2. The points on AT are eliminated because they correspond to instable shocks. One eliminates the points

I_3 because generally, the shocks are compressive. $(v_2 < v_1)$. P_2 will coincide therefore with I_2. If C is the intersection of the polar with the sonic circle and θ_c is the corresponding polar angle, then v_2 will be supersonic if $\theta < \theta_c$ and subsonic, if $\theta > \theta_c$.

Once P_2 determined, it results the direction of the versor $\boldsymbol{t}$ (the orthogonal from M on P_1P_2) and then the angle σ. The density, the pressure, the temperature and the entropy behind the shock wave is obtained from (1.3.26) and (1.3.28), setting $\mathcal{M}_1 = -v_{1n}/c_1 = -M_1 \sin\sigma$, where $M_1 = v_1/c_1$ is Mach's number in front of the shock wave.

1.3.6 The Compression Shock past a Concave Bend

We consider a supersonic flow having the velocity $v_1 \boldsymbol{i}$, the density ρ_1 and the pressure p_1 in the presence of a concave bend having the opening δ (fig. 1.3.4a)). The wall ME produces a compressive shock, the discontinuity line MM' being characterized by the unknown angle σ. Behind the shock, the velocity which has to be tangent to ME, will make the angle δ with MX . If $\delta < \theta_0$, the position of P_2 will be given by I_2 from the polar corresponding to this motion. It follows like above, σ and the flow behind the shock.

If $\delta > \theta_0$, one cannot satisfy for v_2 the condition to be parallel to ME. As the experience confirms, the assumption of a rectilinear shock wave MM' cannot be taken into consideration. In this case one admits the existence of a detached curvilinear shock wave which is formed in front of M (fig. 1.3.4b)). The experience confirms this assumption. Hence the detached shock waves are formed in front of the dihedron $\delta > \theta_0$ (fig. 1.3.5a)), or in front of the bodies with rounded leading edge (fig. 1.3.5b)). As an information we give the following values : $\theta_0 = 10°$ for $M_1 = 1.42$ and $\theta_0 = 22.55°$, for $M_1 = 2$. Depending on

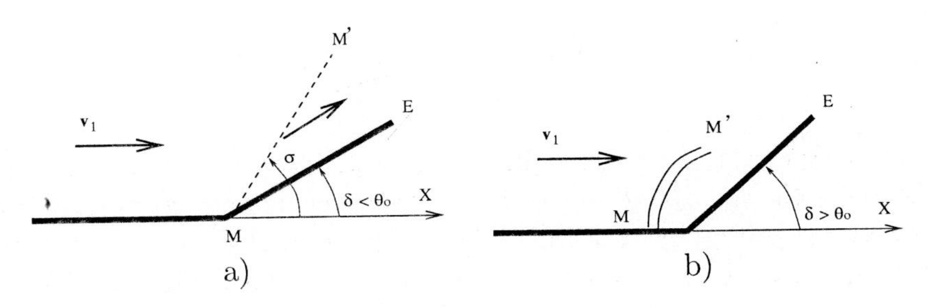

Fig. 1.3.4.

the shape of the body, i.e. on the values of θ, behind the shock wave,

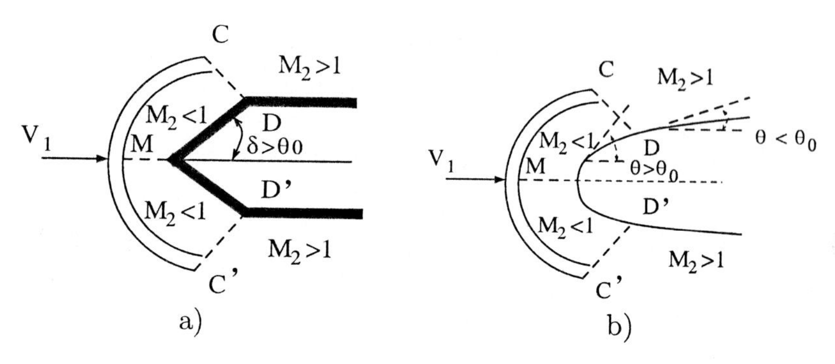

Fig. 1.3.5.

we meet regions where v_2 is subsonic ($M_2 < 1$), if $\theta > \theta_c$ and regions where v_2 is supersonic, if $\theta < \theta_0$. When θ is big, the compression is big. When θ is passing to small values, it appears a detente and the velocity becomes again supersonic. On the direction MV the shock is normal ($v_1 v_2 = v_{cr}^2$ and v_2 is subsonic. The subsonic regions are separated from the supersonic ones through sonic lines CD and $C'D'$. Behind the shock wave, the flow is transonic.

The shock waves theory will be present in the *transonic* flow and the *hypersonic* flow.

Chapter 2

The Equations of Linear Aerodynamics and its Fundamental Solutions

2.1 The Equations of Linear Aerodynamics

2.1.1 The Fundamental Problem of Aerodynamics

The fundamental problem of aerodynamics consists in determining the perturbation produced in a given state of a fluid by a certain motion of a body. The given state of the fluid is called *basic state* or *unperturbed state*. The unperturbed state of the fluid may be the rest state, the uniform flow state, or more generally, the state given by the flow with an imposed non-uniform velocity field. In this book the unperturbed state will be either the rest or the uniform flow.

In his turn, the body may be fixed, moving uniformly or may have a general imposed motion. Obviously, a fixed body in a rest state of a fluid does not produce any perturbation. The most common are the case when the unperturbed fluid moves uniformly and the body is fixed and the case when the fluid is at rest and the body has a given uniform motion. As we shall see in 2.1.6, these cases are equivalent from the mathematical point of view. In both cases, the resulting perturbation will be *stationary*.

If the perturbing body has a non-uniform motion, the perturbation will be non-stationary, whatever should be the state of the fluid. In the sequel we shall consider only the cases when the unperturbed fluid is at rest or has an uniform flow.

The problem of determining the perturbation may be practically solved only in the case of small perturbations, when we can neglect their products (we keep only the principal parts of the equations). In these cases, the linear systems of equations obtained for the perturbations may be investigated either with the methods of the classical analysis or with the methods of the theory of distributions. The systems will be linear with constant coefficients in case that the basic state is uniform and they will be linear with variable coefficients in case that the basic flow is not uniform. Since, as we have already mentioned, the last case will not be treated in this book we give here some references, [2.4], [2.14], [2.15], [2.20], [2.22], for the reader interested in this subject.

The order of magnitude of the perturbation is determined by the basic flow and by the shape of the body. For some basic flows, a slender body with a small incidence changes slightly the flow, i.e. produces small perturbations (governed by linear systems of equations). We cannot establish in advance the conditions of validity of the linear theory. This will be done after determining the solution of the linearized equations, imposing not to obtain results which cannot be accepted from a physical point of view. In this way, from Chapters 3 and 8, it will follow that in the case of steady flow, the linear theory is not valid when the basic flow has approximately the sonic velocity $(M \simeq 1)$, or has an hypersonic velocity $(M \geq 3)$ even if the body is slender with a small incidence. In these cases one has to employ the non-linear equations for determining the perturbation.

A special feature has the unsteady flow because the system of equations of motion is hyperbolic and it is well known (see, for example, [1.6]), that for this sort of problems, Cauchy's problem is correct, i.e. the solution depends continuously on the initial data. If these data are small, the perturbations will remain small at every instant.

2.1.2 The Equations of Motion

First of all we assume that the basic motion (the unperturbed motion) of a fluid is an uniform motion with the velocity $U_\infty \boldsymbol{i}$ (the Ox axis is taken to be on the direction and in the sense of the unperturbed stream and we denote by $\boldsymbol{i}$ the versor of this axis), the pressure p_∞ and the density ρ_∞. We denote by x_1, y_1, z_1 the generic spatial coordinates, by t_1 the time and we introduce the dimensionless coordinates x, y, z, t by means of the relations

$$(x_1, y_1, z_1) = L_0(x, y, z), \quad U_\infty t_1 = L_0 t, \tag{2.1.1}$$

where L_0 is a characteristic length which has to be specified in every problem we have in view.

We assume that the uniform motion defined above is perturbed by the presence of a body which has a prescribed motion. Let

$$F(t_1, x_1, y_1, z_1) = 0\,, \tag{2.1.2}$$

be the equation of the surface of the moving body.

Denoting by $\boldsymbol{V}_1, p_1$ and ρ_1 the velocity, the pressure and the density for the perturbed flow, we shall write:

$$\boldsymbol{V}_1 = U_\infty(\boldsymbol{i} + \boldsymbol{v}), \;\; p_1 = p_\infty + \rho_\infty U_\infty^2 p, \; \rho_1 = \rho_\infty(1+\rho), \qquad (2.1.3)$$

the first terms defining the basic motion, and the last ones, the perturbation. Obviously we have:

$$\lim_{x\to-\infty}(\boldsymbol{v}, p, \rho) = 0\,. \qquad (2.1.4)$$

We neglect the heat changes (in aerodynamics this assumption is plausible, because the variation of the phenomena are very rapid and there is not enough time for the heat change) and we assume that the fluid obeys to the perfect gas law. In these conditions, the perturbed flow will be determined by the following equations:

$$\dot\rho_1 + \rho_1 \mathrm{div}_1\, \boldsymbol{V}_1 = 0, \;\; \dot s_1 = 0, \;\; \rho_1 \dot{\boldsymbol{V}}_1 + \mathrm{grad}_1\, p_1 = 0\,, \qquad (2.1.5)$$

s_1 representing the specific entropy and “the point”, a notation for the material derivative:

$$s_1 = c_v \ln(p_1/\rho_1^\gamma) + C, \;\; \dot f = \partial f/\partial t_1 + (\boldsymbol{V}_1\cdot\nabla_1)f\,. \qquad (2.1.6)$$

The index 1 attached to the differential operators indicates that the derivatives are calculated with respect to x_1, y_1, z_1.

Taking into account the expression of the entropy, from the equation (2.1.6), we get:

$$\rho_1 \dot p_1 = \gamma p_1 \dot\rho_1, \qquad (2.1.7)$$

so that we can eliminate ρ_1 from the first equation (2.1.5). Hence, we have to take into consideration the system

$$\dot p_1 + \gamma p_1\, \mathrm{div}_1\, \boldsymbol{V}_1 = 0, \;\; \rho_1 \dot{\boldsymbol{V}}_1 + \mathrm{grad}_1\, p_1 = 0 \qquad (2.1.8)$$

which, taking into consideration (2.1.1) and (2.1.3) becomes:

$$M^2[p_t + (1+u)p_x + vp_y + wp_z] + (1+\gamma M^2 p)(u_x + v_y + w_z) = 0\,, \; (2.1.9)$$

$$(1+\rho)[u_t + (1+u)u_x + vu_y + wu_z] + p_x = 0\,, \qquad (2.1.10)$$

$$(1+\rho)[v_t + (1+u)v_x + vv_y + wv_z] + p_y = 0\,, \qquad (2.1.11)$$

$$(1+\rho)[w_t + (1+u)w_x + vw_y + ww_z] + p_z = 0\,, \qquad (2.1.12)$$

where $(u, v, w) = \boldsymbol{v}$, $p_t = \partial p/\partial t$, $p_x = \partial p/\partial x, \ldots$ and

$$M = U_\infty / c_\infty, \quad c_\infty = \sqrt{\gamma p_\infty / \rho_\infty}\,, \tag{2.1.13}$$

M representing Mach's number in the basic motion and γ, the ratio of the specific heats (c_p/c_v).

The condition for the perturbation surface (2.1.2) to be a material surface is $\dot{F} = 0$ (the Euler-Lagrange criterion)and it may be written as follows

$$F_t + (1 + u)F_x + vF_y + wF_z = 0\,. \tag{2.1.14}$$

This condition must be satisfied for $F(t, x, y, z) = 0$.

The system of equations (2.1.9) – (2.1.12), together with the boundary condition (2.1.14), will determine the perturbation $(p, \boldsymbol{v})$.

A. The Linearization around the Uniform Motion

2.1.3 The Equations of Linear Aerodynamics

We assume now that the equation of the perturbing surface is

$$z = h(t, x, y) = \varepsilon \overline{h}(t, x, y), \tag{2.1.15}$$

where ε is a small parameter and $h(t, x, y)$ is a known function with continuous first order derivatives (fig. 2.1.1a)). If the perturbation surface is cylindrical with generators parallel to Oz (fig. 2.1.1b)), then the equation of the profile determined in the xOy plane is assumed to have the form

$$y = h(t, x) = \varepsilon \overline{h}(t, x)\,. \tag{2.1.16}$$

In this case, the perturbation will be plane.

For the surface (2.1.15), we write $F = \varepsilon \overline{h}(t, x, y) - z$ in (2.1.14). One obtains

$$\varepsilon \overline{h}_t + \varepsilon(1 + u)\overline{h}_x + \varepsilon v \overline{h}_y = w \tag{2.1.17}$$

which has to be satisfied for $z = \varepsilon \overline{h}(t, x, y)$. The principal part from the left hand side of the equality (2.1.17) has the order of ε. We deduce that the right hand part must have the same order of magnitude. Hence it follows

$$w(t, x, y, \varepsilon \overline{h}) = \varepsilon \overline{w}(t, x, y, \varepsilon \overline{h})\,. \tag{2.1.18}$$

We consider that this relation is valid all over the fluid, whence

$$w(t, x, y, z) = \varepsilon \overline{w}(t, x, y, z)\,. \tag{2.1.19}$$

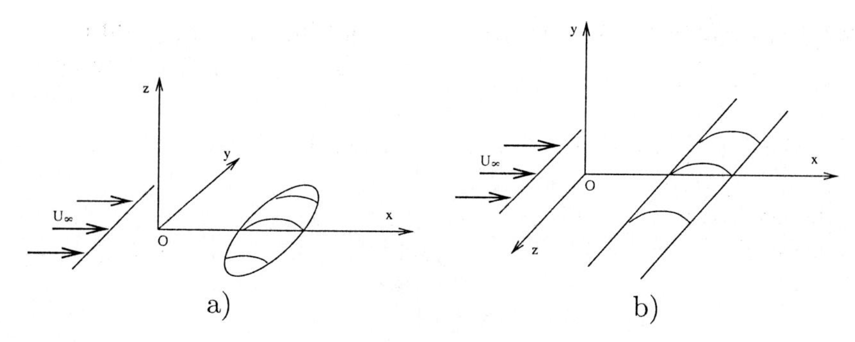

Fig. 2.1.1.

Taking into account (2.1.18), the principal part of the boundary condition (2.1.17) is

$$h_t(t,x,y) + h_x(t,x,y) = w(t,x,y,0)\,. \tag{2.1.20}$$

Taking into account (2.1.19), we deduce that the principal part of the product from (2.1.12) has the order ε. It follows therefore that p has the same order. Hence,

$$p(t,x,y,z) = \varepsilon\overline{p}(t,x,y,z)\,, \tag{2.1.21}$$

the residual equation from (2.1.12) being

$$Dw + p_z = 0, \tag{2.1.22}$$

where D is the material derivative operator for the unperturbed motion:

$$D = \partial/\partial t + \partial/\partial x\,. \tag{2.1.23}$$

Taking into account (2.1.21), from (2.1.10) and (2.1.11), we deduce

$$u(t,x,y,z) = \varepsilon\overline{u}(t,x,y,z)\,,\;\; v(t,x,y,z) = \varepsilon\overline{v}(t,x,y,z) \tag{2.1.24}$$

and the residual equations

$$Du + p_x = 0\,,\;\; Dv + p_y = 0\,. \tag{2.1.25}$$

Now, the equation (2.1.9) becomes

$$M^2 Dp + \operatorname{div} \boldsymbol{v} = 0\,. \tag{2.1.26}$$

This equation together with the equations (2.1.22) and (2.1.25) which have the vectorial form

$$D\boldsymbol{v} + \operatorname{grad} p = 0, \tag{2.1.27}$$

constitute *the fundamental system* of the linear aerodynamics. Obviously, this system has to be integrated with the boundary condition (2.1.20) and the condition at infinity upstream

$$\lim_{x \to -\infty} (p, \boldsymbol{v}) = 0 . \tag{2.1.28}$$

If the perturbation is plane, the condition (2.1.20) will be replaced by

$$h_t(t, x) + h_x(t, x) = v(t, x, 0) . \tag{2.1.29}$$

If the perturbation surface is fixed (i.e. h does not depend explicitly on t), the perturbation will be stationary, determined by the system

$$\begin{aligned} &M^2 p_x + \operatorname{div} \boldsymbol{v} = 0 , \quad \boldsymbol{v}_x + \operatorname{grad} p = 0 \\ &\lim_{x \to -\infty} (p, \boldsymbol{v}) = 0 \end{aligned} \tag{2.1.30}$$

and by the boundary conditions

$$w(x, y, 0) = h_x(x, y) , \tag{2.1.31}$$

$$v(x, 0) = h_x(x) . \tag{2.1.32}$$

Obviously, the equations (2.1.26), (2.1.27) and (2.1.29) could be easily obtained from (2.1.9)–(2.1.12) and (2.1.14), supposing that all the perturbations have the same order of magnitude. But this thing has to be demonstrated. In the case of transonic flow, for example, the perturbations have different orders of magnitude (see Chapter 9).

2.1.4 The Equation of the Potential

Applying the operator curl in (2.1.27), we obtain:

$$D \operatorname{curl} \boldsymbol{v} = 0 . \tag{2.1.33}$$

Taking into account the significance of the operator D, we deduce that $\operatorname{curl} \boldsymbol{v}$ is constant on the parallels to the Ox axis, the constant generally varying from one parallel to the other. Since at infinity upstream ($x = -\infty$) the constant vanishes on every parallel to Ox, it follows that $\operatorname{curl} \boldsymbol{v} = 0$ on every trajectory coming from $-\infty$. This property is not true for trajectories detaching downstream from the body. This fact will be better put into evidence by the fundamental solutions.

In the irrotational zone we have

$$\boldsymbol{v} = \operatorname{grad} \varphi(t, \boldsymbol{x}) . \tag{2.1.34}$$

Taking into account this representation of the field $\boldsymbol{v}$, from (2.1.27) we deduce

$$p = -D\varphi = -(\varphi_t + \varphi_x) . \tag{2.1.35}$$

The function $f(t)$ which should be added in the right hand side of the equation may be considered equal to zero, since φ is determined with the approximation of an arbitrary additive function of t. From (2.1.26) and (2.1.35) it follows $\Delta\varphi = M^2 D^2 \varphi$ or, explicitly

$$(1 - M^2)\varphi_{xx} + \varphi_{yy} + \varphi_{zz} - 2M^2\varphi_{tx} - M^2\varphi_{tt} = 0 . \tag{2.1.36}$$

This is the equation of the potential.

In case of the stationary flow, from (2.1.33) it results

$$\operatorname{curl} v = \boldsymbol{f}(y, z) ,$$

$\boldsymbol{f}$ representing an arbitrary field which must be considered zero because for $x \to -\infty$ we have $\operatorname{curl} \boldsymbol{v} = 0$. Again it is true the observation that $\operatorname{curl} \boldsymbol{v}$ is not zero on the parallels at the Ox axis detaching downstream from the body. In the irrotational zone we have therefore

$$\boldsymbol{v} = \operatorname{grad} \varphi(\boldsymbol{x}) . \tag{2.1.37}$$

From the first equation (2.1.25), taking into account (2.1.28), it follows

$$p(\boldsymbol{x}) = -u(\boldsymbol{x}) . \tag{2.1.38}$$

Replacing (2.1.37) and (2.1.38) in the first equation (2.1.30), it follows

$$(1 - M^2)\varphi_{xx} + \varphi_{yy} + \varphi_{zz} = 0 . \tag{2.1.39}$$

This equation may be obviously deduced from (2.1.36). For the incompressible fluid $(M = 0)$ these equations become

$$\Delta\varphi = 0 . \tag{2.1.40}$$

Applying the operator grad in (2.1.39), we get:

$$(1 - M^2)\boldsymbol{v}_{xx} + \boldsymbol{v}_{yy} + \boldsymbol{v}_{zz} = 0 . \tag{2.1.41}$$

Hence, the coordinates of the velocity (and the pressure) satisfy the same equation (2.1.39).

The equations (2.1.36) and (2.1.39) have been obtained in another way in Chapter 1. There we utilized the Lagrange-Cauchy theorem in order to prove that the perturbed flow is potential. Here, without utilizing this theorem, we demonstrated that in the first approximation the perturbation is potential. In addition, we see here that downstream the perturbing body, the perturbation does not possess this property any longer.

B. The Linearization around the Rest State

2.1.5 The Linear System

Let us assume now that the basic state of the fluid is the rest state, which is perturbed, as above, by a prescribed arbitrary motion of a body. We denote by p_0 and ρ_0 the pressure and the density of the fluid at rest. We denote also by $c^2 = \gamma p_0/\rho_0$ the square of the sound velocity for the same state. Since, in this case, there is no characteristic velocity, it is not recommended to use the variables (2.1.1) and (2.1.3). It is natural to use dimensional coordinates and to put

$$\boldsymbol{V}_1 = \boldsymbol{v},\ p_1 = p_0 + \rho_0 p,\ \rho_1 = \rho_0(1+\rho)\,, \tag{2.1.42}$$

$$\lim_{|\boldsymbol{x}|\to\infty} (p, \boldsymbol{v}, \rho) = 0\,. \tag{2.1.43}$$

The system of equations of motion has the shape

$$\begin{gathered}\dot{\rho} + (1+\rho)\,\mathrm{div}\,\boldsymbol{v} = 0,\ \ (1+\rho)\dot{p} = (c^2+\gamma p)\dot{\rho}\,,\\ (1+\rho)\dot{\boldsymbol{v}} + \mathrm{grad}\,p = 0\,.\end{gathered} \tag{2.1.44}$$

If the equations of the perturbing surface have the form (2.1.15), we shall have the boundary condition

$$\varepsilon\overline{h}_t + \varepsilon u\overline{h}_x + \varepsilon v\overline{h}_y = w \tag{2.1.45}$$

which has to be satisfied for $z = \varepsilon\overline{h}(t,x,y)$. We deduce

$$w(t,x,y,\varepsilon\overline{h}) = \varepsilon\overline{w}(t,x,y,\varepsilon\overline{h}) \tag{2.1.46}$$

and the boundary condition

$$w(t,x,y,0) = h_t(t,x,y)\,. \tag{2.1.47}$$

Assuming like in (2.1.3), that (2.1.46) is valid all over the fluid,

$$w(t,x,y,z) = \varepsilon\overline{w}(t,x,y,z),$$

from the projection of the second equation from (2.1.44) onto the Oz axis, we deduce

$$p(t,x,y,z) = \varepsilon \overline{p}(t,x,y,z)$$

and the residual equation

$$w_t + p_z = 0\,.$$

Acting in the sequel like in subsection (2.1.3), we obtain at last

$$\rho_t + \operatorname{div} \boldsymbol{v} = 0\,, \quad p_t = c^2 \rho_t\,, \quad \boldsymbol{v}_t + \operatorname{grad} p = 0 \tag{2.1.48}$$

We deduce the system

$$\begin{gathered} p_t + c^2 \operatorname{div} \boldsymbol{v} = 0\,, \quad \boldsymbol{v}_t + \operatorname{grad} p = 0 \\ \lim_{|\boldsymbol{x}| \to \infty} (p, \boldsymbol{v}) = 0 \end{gathered} \tag{2.1.49}$$

and the solution $p = c^2 \rho$. The system (2.1.49) has to be integrated with the boundary condition (2.1.47).

From (2.1.49) one obtains *the fundamental equation of acoustics*

$$p_{tt} - c^2 \Delta p = 0\,. \tag{2.1.50}$$

2.1.6 The Uniform Motion in the Fluid at Rest

Let us consider the particular case of the uniform motion of the perturbing body in the fluid at rest. We assume that the motion is performed with the velocity U_0 in the negative sense of the Ox axis.

Putting

$$\boldsymbol{V}_1 = U_0 \boldsymbol{v}, \quad \rho_1 = \rho_0 (1 + \rho)\,, \quad p_1 = p_0 + \rho_0 U_0^2 p \tag{2.1.51}$$

and utilizing the variables (2.1.1), we get the system:

$$\begin{gathered} M_0^2 \dot{p} + (1 + \gamma M_0^2 p) \operatorname{div} \boldsymbol{v} = 0\,, \quad (1+\rho) M_0^2 p = (1 + \gamma M_0^2 p) \dot{\rho}\,, \\ (1+\rho) \dot{\boldsymbol{v}} + \operatorname{grad} p = 0\,, \end{gathered} \tag{2.1.52}$$

where M_0 is Mach's number for the fluid at rest. As above, one deduces the boundary condition (2.1.47) and the system (2.1.52) reduces to the residual form

$$M^2 p_t + \operatorname{div} \boldsymbol{v} = 0\,, \quad \boldsymbol{v}_t + \operatorname{grad} p = 0 \tag{2.1.53}$$

and $M_0^2 p = \rho$.

In a frame of reference $R' = Cx'y'z'$ (fig. 2.1.2) solidary with the body, having the axes parallel and having the same sense with the axes of the frame $R = Oxyz$ (the frame R' is inertial), *the system* (2.1.53) *has the form of the steady system* (2.1.30). Indeed, we pass from the frame R to the frame R' by means of the Galilean transformation

$$t' = t, \quad x' = x + t, \quad y' = y, \quad z' = z\,. \tag{2.1.54}$$

With this transformation, the system (2.1.53) gains the form of the system (2.1.26), (2.1.27) and the condition (2.1.47), the form (2.1.20). But, since we deal with a translation of the body, h in (2.1.47) has the form $h(x + t, y)$ which is transformed in $h(x', y')$. The boundary condition in the variables x', y', z' will have therefore the form (2.1.31), so that it will determine a steady motion. The system of equations of motion in x', y', z' will have the form (2.1.30).

Hence we have demonstrated that the problem of determining the perturbation produced by a body moving uniformly with the velocity $-U_0\boldsymbol{i}$, in a fluid at rest, is equivalent to the problem of determining the perturbation produced by the same fixed body in an uniform stream with the velocity $U_0\boldsymbol{i}$.

2.2 The Fundamental Solutions of the Equation of the Potential

2.2.1 The Steady Solutions

This subsection is written on the basis of the paper [2.11]. The fundamental solutions of the equation (2.1.39) are the solutions of the equation

$$(1 - M^2)\mathcal{E}_{xx} + \mathcal{E}_{yy} + \mathcal{E}_{zz} = \delta(\boldsymbol{x})\,, \tag{2.2.1}$$

δ representing Dirac's distribution. These solutions are, obviously, distributions. Utilizing the Fourier transform method, we shall obtain temperate solutions. Taking into account (A.5.1), valid also for temperate solutions, we obtain using the notation $\alpha^2 = |\boldsymbol{\alpha}|^2$,

$$(\alpha^2 - M^2\alpha_1^2)\widehat{\mathcal{E}} = -1\,, \tag{2.2.2}$$

whence

$$\mathcal{E} = -\mathcal{F}^{-1}\left[\frac{1}{\alpha^2 - M^2\alpha_1^2}\right] = -\mathcal{F}^{-1}\left[\frac{1}{(1 - M^2)\alpha_1^2 + \alpha_2^2 + \alpha_3^2}\right]\,. \tag{2.2.3}$$

In the subsonic case $M < 1$, we shall denote

$$\beta = \sqrt{1 - M^2}\,, \tag{2.2.4}$$

and in the supersonic case $(M > 1)$,

$$k = \sqrt{M^2 - 1}\,. \tag{2.2.5}$$

In the *subsonic case* one utilizes the formulas (A.7.10) and (A.7.11). One obtains the following fundamental solutions:

$$\mathcal{E} = -\frac{1}{4\pi}\frac{1}{\sqrt{x^2 + \beta^2(y^2 + z^2)}}, \quad n = 3\,, \tag{2.2.6}$$

$$\mathcal{E} = \frac{1}{2\pi\beta}\ln\sqrt{x^2 + \beta^2 y^2}, \quad n = 2\,, \tag{2.2.7}$$

For the two-dimensional case we have not written the additive constant $C - \ln\beta$ appearing in (A.7.11), because the fundamental solution is determined with the approximation of a solution of the homogeneous equation.

In the *supersonic case*, one utilizes the formulas (A.7.14) and (A.7.15). One obtains the following fundamental solutions:

$$\mathcal{E} = -\frac{1}{2\pi}\frac{H(x - k\sqrt{y^2 + z^2})}{\sqrt{x^2 - k^2(y^2 + z^2)}}, \quad n = 3\,, \tag{2.2.8}$$

$$\mathcal{E} = -\frac{1}{2k}H(x - k|y|), \quad n = 2\,. \tag{2.2.9}$$

H representing, as we have considered in Appendix A, Heaviside's function. From the definition of this function (A.3.13), it follows that the three-dimensional solution is different from zero only for

$$x > k\sqrt{y^2 + z^2}\,. \tag{2.2.10}$$

This inequality implies

$$x > 0, \quad x^2 > k^2(y^2 + z^2)\,.$$

The set of points from the space verifying these inequalities constitutes the interior of the cone with the vertex in the origin of the system of coordinates (the perturbation point) and the symmetry axis along the Ox axis (fig. 2.2.1a)). This cone is called *Mach's cone.* In fact, it is the characteristic cone associated to the partial differential equation (2.1.39)

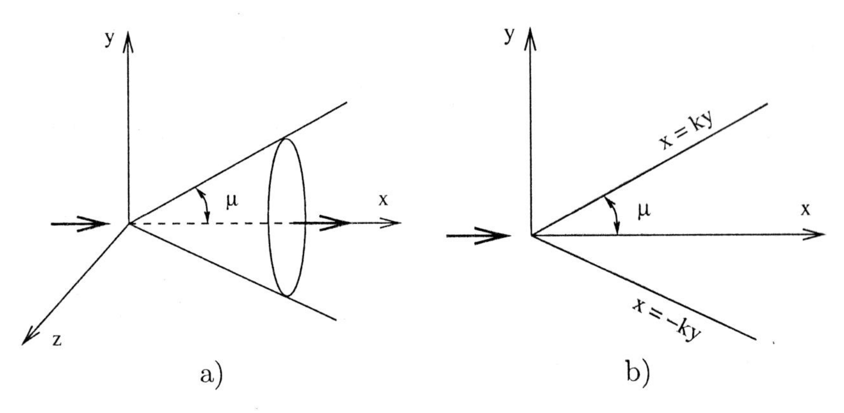

Fig. 2.2.1.

in the hyperbolic case. The radius of the cone is x/k, and the angle μ (the semi-opening) is determined by the formula

$$\tan\mu = \frac{x/k}{x} = \frac{1}{k} = \frac{1}{\sqrt{M^2-1}}, \quad \left(\sin\mu = \frac{1}{M}\right). \qquad (2.2.11)$$

In the two-dimensional case, the solution is different from zero only inside the dihedron made by the planes $x = \pm ky$ where $x > 0$ (fig. 2.2.1b)). This is *Mach's dihedron.*

2.2.2 Oscillatory Solutions

The fundamental oscillatory solutions are defined by the equation

$$(1-M^2)\mathcal{E}_{xx} + \mathcal{E}_{yy} + \mathcal{E}_{zz} - 2M^2\mathcal{E}_{tx} - M^2\mathcal{E}_{tt} = \delta(\boldsymbol{x})\exp(\mathrm{i}\omega t), \quad (2.2.12)$$

associated to the equation (2.1.38). They will have, obviously, the shape

$$\mathcal{E} = E(x)\exp(\mathrm{i}\omega t), \qquad (2.2.13)$$

where

$$(1-M^2)E_{xx} + E_{yy} + E_{zz} - 2\mathrm{i}\omega M^2 E_x + \omega^2 M^2 E = \delta(\boldsymbol{x}). \quad (2.2.14)$$

Performing the change of functions $E \to e$:

$$E = \exp(\lambda x)e \qquad (2.2.15)$$

and nullifying the coefficient of the derivative e_x, one obtains the equation:

$$(1-M^2)e_{xx} + e_{yy} + e_{zz} + ae = \exp(-\lambda x)\delta(\boldsymbol{x}), \qquad (2.2.16)$$

where

$$\lambda = \mathrm{i}kM\,, \qquad a = \frac{\omega^2 M^2}{1 - M^2}\,. \tag{2.2.17}$$

In the *subsonic case* $(M < 1)$ one performs the change of variable $(x, y, z) \to (X, Y, Z)$:

$$X = x, \;\; Y = \beta y, \;\; Z = \beta z \tag{2.2.18}$$

and one takes into account (A.3.11). One obtains the equations:

$$\begin{aligned} &e_{XX} + e_{YY} + e_{ZZ} + k^2 e = \exp(-\lambda X)\delta(\boldsymbol{X}), \quad n = 3\,, \\ &e_{XX} + e_{YY} + k^2 e = \beta^{-1}\exp(-\lambda X)\delta(\boldsymbol{X}), \quad n = 2\,, \end{aligned} \tag{2.2.19}$$

where $k = \omega M/\beta^2 = \varpi M$.

In (2.2.19) we have Helmholtz's non-homogeneous equations. It is well known (see, for example, [A.12], §9), that Helmholtz's equation has two fundamental solutions in the three-dimensional case and the same number of solutions in the two-dimensional case, the choice of the solution depending on the type of oscillation defined by the equation. Taking into account that we obtained Helmholtz's equation looking for oscillations having the form (2.2.13), it follows (see, for example, [1.41] Chapter 7, §2) that the following solutions have a physical meaning:

$$\overline{e}_3 = -\frac{\exp(-\mathrm{i}k|\boldsymbol{X}|)}{4\pi|\boldsymbol{X}|}\,, \qquad \overline{e}_2 = \frac{\mathrm{i}}{4}H_0^{(2)}(k|\boldsymbol{X}|)\,,$$

$\overline{e}_3$ representing the solution for the three-dimensional case and $\overline{e}_2$ the solution for the two-dimensional case. $H_0^{(2)}$ is Hankel's function.

Performing the convolutions of these solutions with the right hand member from (2.2.19) (A.4.6 formula), we get the following fundamental solutions

$$e_3 = -\frac{\exp(-\mathrm{i}k|\boldsymbol{X}|)}{4\pi|\boldsymbol{X}|}\,, \qquad e_2 = \frac{\mathrm{i}}{4}H_0^{(2)}(k|\boldsymbol{X}|)\,,$$

so that, taking into account the changes already made, we find:

$$\begin{aligned} E_3 &= -\frac{1}{4\pi R}\exp \mathrm{i}k(Mx - R)\,, \\ E_2 &= -\frac{1}{4\mathrm{i}\beta}H_0^2(k\overline{R})\exp\,(kMx)\,, \end{aligned} \tag{2.2.20}$$

with the notations

$$R = \sqrt{x^2 + \beta^2(y^2 + z^2)}\,, \;\; \overline{R} = \sqrt{x^2 + \beta^2 y^2}\,.$$

In the *supersonic case* ; $(M > 1)$ one performs the change of variable

$$X = x, \quad Y = ky, \quad Z = kz\,. \tag{2.2.21}$$

The equation (2.2.16) becomes

$$e_{XX} - e_{YY} - e_{ZZ} + \nu^2 e = -\exp(-\lambda X)\,\delta(\boldsymbol{X}), \tag{2.2.22}$$

where $\nu = \omega M/k^2$. In (2.2.22) we have the non-homogeneous Klein-Gordon-Fock equation. The fundamental solutions of this equation are also known ([A.8] [A.11]). Performing the convolutions of these equations with the right hand member of (2.2.22), we get:

$$e_3 = -\frac{1}{2\pi} H\left(X - \sqrt{Y^2 + Z^2}\right) \frac{\cos \nu\sqrt{X^2 - Y^2 - Z^2}}{\sqrt{X^2 - Y^2 - Z^2}}\,,$$

$$e_2 = -\frac{1}{2k} H(X - |Y|) J_0\left(\nu\sqrt{X^2 - Y^2}\right),$$

such that finally it follows

$$E_3 = -\frac{1}{2\pi} H\left(x - k\sqrt{y^2 + z^2}\right) \frac{\cos \nu\sqrt{x^2 - k^2(y^2 + z^2)}}{\sqrt{x^2 - k^2(y^2 + z^2)}} \exp\left(\frac{\mathrm{i}\omega M^2}{1 - M^2} x\right),$$

$$E_2 = -\frac{1}{2k} H(x - |y|) J_0\left(\nu\sqrt{x^2 - k^2 y^2}\right) \exp\left(\frac{\mathrm{i}\omega M^2}{1 - M^2} x\right). \tag{2.2.23}$$

Due to the presence of Heaviside's function in (2.2.23), it is obvious that these solutions will be different from zero only in the interior of Mach's cone with the vertex in the origin for $x > 0$, respectively in the interior of Mach's dihedron (with the edge on Oz) for $x > 0$.

2.2.3 Oscillatory Solutions for $M = 1$

If $M = 1$, the equation (2.2.14) becomes

$$E_{yy} + E_{zz} - 2\mathrm{i}\omega E_x + \omega^2 E = \delta(\boldsymbol{x})\,. \tag{2.2.24}$$

Introducing the Fourier transform $\widehat{E}$ with respect to the variables y and z one obtains

$$2\mathrm{i}\omega \widehat{E}_x - (\omega^2 - \alpha_2^2 - \alpha_3^2)\widehat{E} = -\delta(x)\,.$$

The solution of this equation (see [A.21]) has the shape $\widehat{E} = H(x)\widetilde{E}$, where

$$2\mathrm{i}\omega\widetilde{E}_x - (\omega^2 - \alpha_2^2 - \alpha_3^2)\widetilde{E} = 0\,, \quad \widetilde{E}(o) = -1/2\mathrm{i}\omega\,.$$

It follows

$$\widehat{E} = -\frac{H(x)}{2\mathrm{i}\omega}\exp\left[\frac{\omega^2 - \alpha_2^2 - \alpha_3^3}{2\mathrm{i}\omega}\right]. \tag{2.2.25}$$

For obtaining E we take into account that

$$\int_{-\infty}^{+\infty} \exp\left(-au^2\right)\mathrm{d}\,u = \sqrt{\pi/a} \tag{2.2.26}$$

and we notice that

$$\int_{-\infty}^{+\infty} \exp\left(-\mathrm{i}A\alpha - B\alpha^2\right)d\alpha = \int_{-\infty}^{+\infty} \exp\left[-B\left(\alpha + \frac{\mathrm{i}A}{2B}\right)^2 - \frac{A^2}{4B}\right]d\alpha =$$

$$= \sqrt{\frac{\pi}{B}}\exp\left(-\frac{A^2}{4B}\right).$$

One obtains:

$$E = \frac{H(x)}{4\pi x}\exp\left(-\frac{\mathrm{i}\omega}{2x}r^2\right), \tag{2.2.27}$$

with the notation $r^2 = x^2 + y^2 + z^2$.

In the two-dimensional case we have:

$$E = -a\frac{H(x)}{\sqrt{x}}\exp\left[-\frac{-\mathrm{i}\omega}{2x}(x^2 + y^2)\right], \tag{2.2.28}$$

where $2a\sqrt{2\pi\mathrm{i}\omega} = 1$.

The solutions (2.2.27) and (2.2.28) may also be obtained as *limits of the subsonic solutions* (2.2.20) for $M \to 1$. For obtaining (2.2.27), we notice that

$$\lim_{M\to 1}\frac{M}{\beta^2}(Mx - R) = -\lim\frac{Mr^2}{Mx + R} = -\frac{r^2}{2x}, \tag{2.2.29}$$

x being positive, as it follows from the supersonic solution (2.2.23). For obtaining (2.2.28) one takes into account the asymptotic behaviour of $H_0^{(2)}$ for great values of the argument ([1.42]).

$$H_0^{(2)}(k\overline{R}) \simeq \left(\frac{2}{\pi k\overline{R}}\right)^{1/2}\exp\left(-\mathrm{i}k\overline{R} + \mathrm{i}\frac{\pi}{4}\right) \tag{2.2.30}$$

and one performs a calculus similar to (2.2.29). In the two-dimensional case, the limit of the supersonic solution for $M \to 1$ is obtained in [10.17].

2.2.4 The Unsteady Solutions

In the sequel we are going to determine the fundamental solutions of the equation (2.2.14), i.e. the solutions of the equation

$$(1-M^2)\mathcal{E}_{xx}+\mathcal{E}_{yy}+\mathcal{E}_{zz}-2M^2\mathcal{E}_{tx}-M^2\mathcal{E}_{tt}=\delta(t,\boldsymbol{x})\,. \qquad (2.2.31)$$

This equation determines the perturbation produced in the uniform stream, defined in 2.1, by a source of potential acting at the moment $t=0$ in the origin. The problem is plane (two-dimensional) if the source is uniformly distributed on the Oz axis.

Applying the Fourier transform, we deduce

$$M^2\mathrm{d}^2\widehat{\mathcal{E}}/\mathrm{d}\,t^2-2M^2\mathrm{i}\alpha_1\,\mathrm{d}\,\widehat{\mathcal{E}}/\mathrm{d}\,t+(\alpha^2-M^2\alpha_1^2)\widehat{\mathcal{E}}=-\delta(t)\,. \qquad (2.2.32)$$

We know from Appendix A that the solution of this equation has the form $\widehat{\mathcal{E}}=H(t)\widehat{E}$, where $H(t)$ is Heaviside's function and $\widehat{E}$, is a solution of the problem

$$\begin{gathered}M^2\mathrm{d}^2\widehat{E}/\mathrm{d}t^2-2M^2\mathrm{i}\alpha_1\mathrm{d}\widehat{E}/\mathrm{d}\,t+(\alpha^2-M^2\alpha_1^2)\widehat{E}=0,\\ \widehat{E}(0)=0 \quad (\mathrm{d}\widehat{E}/\mathrm{d}\,t)(0)=-M^{-2}\,.\end{gathered} \qquad (2.2.33)$$

We get:

$$M\widehat{E}=-\frac{\sin|\boldsymbol{\alpha}|M^{-1}t}{|\boldsymbol{\alpha}|}\mathrm{e}^{i\alpha_1 t}\,. \qquad (2.2.34)$$

Utilizing the formulas (A.7.14) and (A.7.15), we deduce

$$\mathcal{E}(t,\boldsymbol{x})=-\frac{M}{4\pi t}\delta(t-MR), \quad n=3, \qquad (2.2.35)$$

$$\mathcal{E}(t,\boldsymbol{x})=-\frac{1}{2\pi}\frac{H(t-M\overline{R})}{\sqrt{t^2-M^2\overline{R}^2}}, \quad n=2, \qquad (2.2.36)$$

where

$$R=\sqrt{(x-t)^2+y^2+z^2}, \quad \overline{R}=\sqrt{(x-t)^2+y^2}\,.$$

We shall not write any longer the factor $H(t)$ from the right hand member, because obviously this member is different from zero only for $t>0$.

Further we shall perform a detailed investigation of the solutions (2.2.35), (2.2.36). We denote

$$h(t)=t-MR\,. \qquad (2.2.37)$$

We are interested to find the zeros t_i of this function, as we are going to utilize the formula:

$$\delta(h(t)) = \sum_i \frac{\delta(t - t_i)}{|h'(t_i)|}, \tag{2.2.38}$$

from [A.10], page 20. We have also to know the sign of the function h, because the two-dimensional solution differs from zero only for $h > 0$.

One notices in (2.2.37) that the zeros t_i are positive. Obviously, $h(0) < 0$ and $h(\infty) = (1 - M^2)\infty$. For the graphic representation of the function $h : [0, \infty) \to R$, we have to separate the cases $M < 1$ and $M > 1$. The zeros of the function $h(t)$ are

$$t_\pm = \frac{-M^2 x \pm M\sqrt{x^2 + (1 - M^2)(y^2 + z^2)}}{1 - M^2}. \tag{2.2.39}$$

In the *subsonic case* $(M < 1)$, $h(t)$ has a single positive root namely

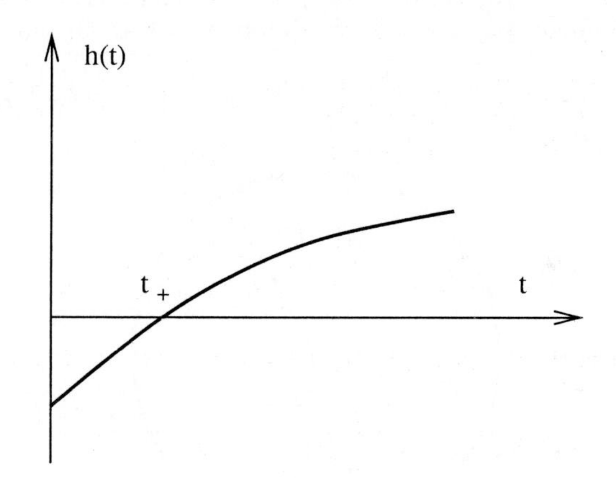

Fig. 2.2.2.

t_+. The graphic of the function h is represented in figure 2.2.2. Utilizing the formula (2.2.38) and taking into account that

$$t_+ = M\sqrt{(x - t_+)^2 + y^2 + z^2},$$

we deduce

$$\mathcal{E} = -\frac{1}{4\pi}\frac{t_+}{t}\frac{\delta(t - t_+)}{\sqrt{x^2 + \beta^2(y^2 + z^2)}}, \tag{2.2.40}$$

in the three-dimensional case and

$$\mathcal{E} = -\frac{1}{2\pi}\frac{H(t - t_+^0)}{\sqrt{t^2 - M^2(x - t)^2 - M^2 y^2}}, \tag{2.2.41}$$

in the two-dimensional case. Here we denoted $t_+^0 = t_+(z = 0)$.

The solution (2.2.40) was given for the first time in [2.11], and the solution (2.2.41) is given in [1.2]. A given point P (fig. 2.2.3) perceives in different manners the perturbation produced at the moment $t = 0$ in the origin of the system of coordinates in the three-dimensional case and the perturbation produced at the moment $t = 0$ uniformly on the Oz axis, in the two-dimensional case. In the three-dimensional case the perturbation is perceived at the moment t_+ and only at this moment (see the solution (2.2.40)). In the two-dimensional case, the perturbation is perceived by P continuously, beginning from the moment t_+^0 (see the solution (2.2.41)). The explanation of this difference is that in the case of the plane problem, one admits that, at the moment $t = 0$, the entire Oz axis emits perturbations, t^0 representing the moment when the perturbation emitted by the origin reaches the point $B(x, y, 0)$ and $t > t_+^0$, the period when the perturbations emitted at the moment $t = 0$ by the other points of the Oz axis reach P. In figure 2.2.3, Q is the position of the source (moving with the stream) at the moment t_+^0, the distance QP being given by the formula $\sqrt{(x - t_+^0)^2 + y^2} = M^{-1} t_+^0$.

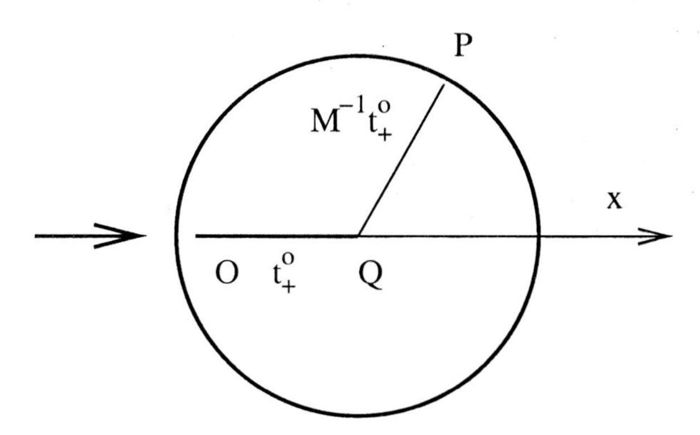

Fig. 2.2.3.

In the *supersonic case* , we have to determine the zeros of the function $h'(t)$. They are given by the equation

$$\sqrt{(x - t_0)^2 + y^2 + z^2} = M(t_0 - x) . \tag{2.2.42}$$

Noticing that $x < t_0$, it follows that there exists a single zero, namely

$$t_0 = x + \sqrt{y^2 + z^2}/k . \tag{2.2.43}$$

A simple calculation gives

$$h(t_0) = x - k\sqrt{y^2 + z^2} . \tag{2.2.44}$$

We shall distinguish three cases (fig. 2.2.4):

$$h(t_0) > 0, \quad h(t_0) = 0, \quad h(t_0) < 0\,. \tag{2.2.45}$$

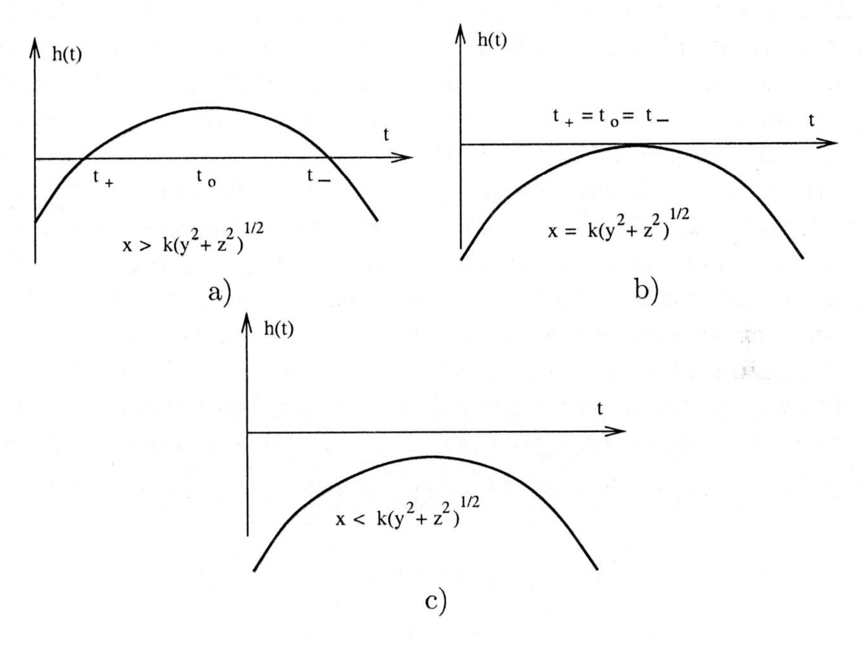

Fig. 2.2.4.

In the first case the fundamental solution is

$$\mathcal{E}_3 = -\frac{1}{4\pi}\frac{t_+\delta(t-t_+) + t_-\delta(t-t_-)}{t\sqrt{x^2 - k^2(y^2+z^2)}}\,, \tag{2.2.46}$$

$$\mathcal{E}_2 = \left\{\begin{array}{ll} 0, & t < t_+, t_- < t \\ -\dfrac{1}{2\pi}\dfrac{1}{\sqrt{t^2 - M^2(x-t)^2 - M^2y^2}}, & t_+ < t < t_-\,, \end{array}\right\}, \tag{2.2.47}$$

$\mathcal{E}_3$ was given for the first time in [2.11]. One interprets this solution as follows: given a point $P(x,y,z)$ in the interior of Mach's cone, there exist two and only two moments t_+ and t_- when the perturbation produced in the origin at the moment $t = 0$ affects this point (fig. 2.2.5). The moments t_+ and t_- are calculated as functions of the coordinates of P by means of the formula (2.2.39). In the two-dimensional case, $P(x,y,0)$ is affected during the interval $[t_+^0, t_-^0]$ the explanation being similar to that given in the subsonic case. In this interval, the expression under the square root is positive.

In the three-dimensional case, figure 2.2.5 presents the intersection of the $z = 0$ plane with Mach's cone and with the spheres having the centers Q_1 and Q_2. In the two-dimensional case the figure indicates the intersection of the same plane with Mach's dihedron and with the cylinders having the radii $M^{-}t^{0}_{+}$ and $M^{-1}t^{0}_{-1}$. Since from the relation $\sin\mu = Q_iP_i/OQ_i$ we deduce $\sin\mu = M^{-1}$ it follows that μ is Mach's angle, i.e. the two spheres (cylinders) are inscribed into Mach's cone (dihedron).

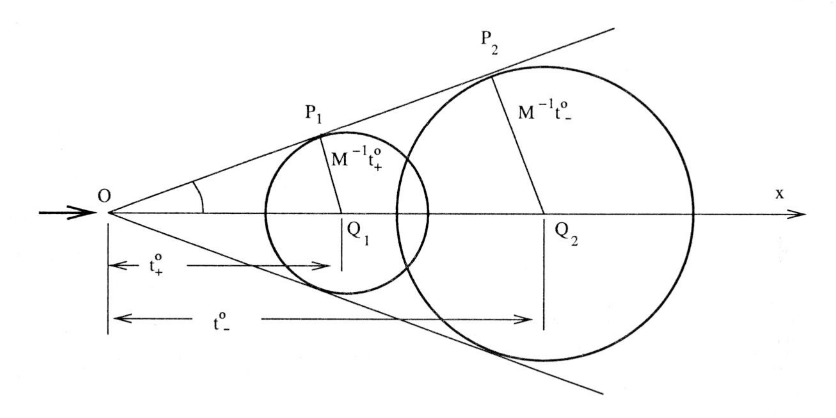

Fig. 2.2.5.

If P belongs to Mach's cone, there exists only one moment $t_+ = = t_- = t_0 = xM^2/k^2$, when the point P is affected by the perturbation. At that moment, the source is located in the intersection of the Ox axis with the normal at the cone in the point R .

If P is in the exterior of the cone, it is not affected by the perturbation.

Hence, for the unsteady flow, the perturbed zone is also in the interior of Mach's cone (respectively Mach's dihedron), but unlike the case of steady or oscillatory flow (when an arbitrary point P from the zone was affected all the time) in the unsteady motion it is affected only at two moments t_+ and t_- (in the interval between these moments in the two-dimensional flow) or in a single moment if P is on the cone (dihedron). This happens because in the first two cases the fundamental solutions correspond to sources acting all the time in the origin (on the Oz axis), while in the last case the solutions correspond to sources acting only in the moment $t = 0$.

2.2.5 The Unsteady Solutions for $M = 1$

If $M = 1$, the solutions of the equation (2.2.31) will be (2.2.35) and (2.2.36) with $M = 1$. The function $h(t) = t - R$ has an unique zero $(h' > 0)$ namely the moment $t_1 = r^2/2x$. It follows that the fundamental solutions

$$\mathcal{E}_3 = -\frac{1}{8\pi t}\frac{r^2}{x^2}\delta\left(t - \frac{r^2}{2x}\right) = -\frac{1}{4\pi x}\delta\left(t - \frac{r^2}{2x}\right),$$

$$\mathcal{E}_2 = -\frac{1}{2\pi}\frac{H(2xt - x^2 - y^2)}{\sqrt{2xt - x^2 - y^2}}. \tag{2.2.48}$$

vanish only for $x > 0$.

Every point P from the half-space (half-plane) $x > 0$ is affected by the perturbation. In the three-dimensional case, the perturbation emitted by the origin at the moment $t = 0$ is received in the point $P(x > 0, y, z)$ at the moment t_1, and only at this moment. If the perturbation is emitted by the Oz axis (the two-dimensional flow), an arbitrary point $P(x > 0, y, 0)$ receives this perturbation continuously, beginning with the moment $t_1 = (x^2 + y^2)/2x$.

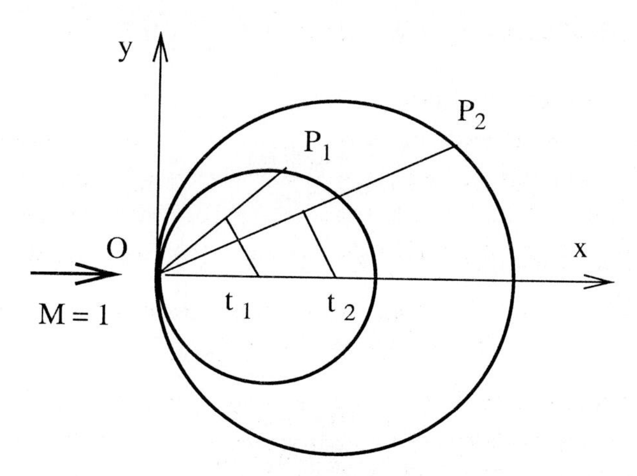

Fig. 2.2.6.

The moment of the first reception t_1 is given by the equation

$$(x - t_1)^2 + y^2 + z^2 = t_1^2. \tag{2.2.49}$$

The graphic of this equation is a sphere (a cylinder in the two-dimensional case) with the center on the Ox axis in the point t_1 and the radius

t_1 (figure 2.2.6). The point t_1 is obtained graphically, taking the intersection of the Ox axis with the mediator plane of the segment OP (the mediator line of OP in the two-dimensional case).

2.2.6 The Fundamental Solutions for the Fluid at Rest

As we have seen in subsection (2.1.5) , the propagation of a small perturbation in the fluid at rest, is described by the equation (2.1.50). The fundamental solutions of this equation will be defined by

$$\mathcal{E}_{tt} - c^2(\mathcal{E}_{xx} + \mathcal{E}_{yy} + \mathcal{E}_{zz}) = \delta(t, \boldsymbol{x}) . \tag{2.2.50}$$

They will give the pressure produced by an unitary perturbation in the origin of the system of coordinates at the moment $t = 0$. Applying the Fourier transform in (2.2.50), we get:

$$\mathrm{d}^2\widehat{\mathcal{E}}/\mathrm{d}\,t^2 + c^2\alpha^2\widehat{\mathcal{E}} = \delta(t) . \tag{2.2.51}$$

Utilizing (A.3.17), we deduce that the solution of the equation (2.2.51) is $\widehat{\mathcal{E}} = H(t)\widehat{E}$, where

$$\mathrm{d}^2\widehat{E}/\mathrm{d}\,t^2 + c^2\alpha^2\widehat{E} = 0, \quad \widehat{E}(0) = 0, \quad \widehat{E}'(0) = 1 . \tag{2.2.52}$$

Solving this problem, we find:

$$\widehat{E} = \frac{\sin c|\boldsymbol{\alpha}|t}{c|\boldsymbol{\alpha}|} .$$

By virtue of formulas (A.7.14) and (A.7.15), we obtain:

$$\begin{aligned} \mathcal{E}_3 &= \frac{\delta(ct - r)}{4\pi c^2 t} = \frac{\delta(ct - r)}{4\pi c r} \\ \mathcal{E}_2 &= \frac{1}{2\pi c} \frac{H(ct - r)}{\sqrt{c^2t^2 - r^2}} \end{aligned} \tag{2.2.53}$$

$\mathcal{E}_3$ representing the solution in the three-dimensional case and $\mathcal{E}_2$ the solution in the two-dimensional case. The solution $\mathcal{E}_3$ shows that the perturbation produced in the origin at the moment $t = 0$ is concentrated, for $t > 0$, on a spherical surface having the radius ct and the center in origin. The perturbation propagates as a spherical wave having the velocity c. On the sphere of radius r the perturbation is only at the moment $t = r/c$. We notice the validity of Huygens' principle. The solution $\mathcal{E}_2$ shows that the perturbation emitted on the Oz axis at the moment $t = 0$ lies at a moment $t > 0$ in the interior of the cylinder of radius ct. There is a foregoing front of the wave propagating with the velocity c, but there is no posterior front. Unlike the three-dimensional case, behind the foregoing front, the perturbation differs from zero at every moment t. In this case Huygens' principle is not valid any longer. We know from 2.2.4 the explanation of this fact.

2.2.7 On the Interpretation of the Fundamental Solution

The equation

$$\dot{\rho} + \rho \operatorname{div} \boldsymbol{v} = \rho q \delta(\boldsymbol{x}) \tag{2.2.54}$$

may be interpreted as the equation of continuity when there is a source with the intensity ρq in the origin. Indeed, integrating on every domain D containing the origin and taking into account (A.7.3), we get:

$$\int_D (\dot{\rho} + \rho \operatorname{div} \boldsymbol{v}) \mathrm{d}\, v = \rho q\,. \tag{2.2.55}$$

The integral gives the variation of the mass from D in the unity of time. This is given by the intensity of the source. In every domain D which does not contain the origin, the mass is preserved.

The presence of the term $\delta(\boldsymbol{x})$ in (2.2.54) represents the cause of the motion. Since the term has a spherical symmetry, it follows that the flow will have this property too. Hence $\boldsymbol{v} = F(r)\boldsymbol{x}/r$ and $\boldsymbol{v} = \operatorname{grad} \varphi$ with $\varphi = \int F(r) \mathrm{d}\, r$.

An uniform flow, with the velocity $U_\infty \boldsymbol{i}$ is also potential, hence the flow resulting by overlapping the uniform flow over the flow due to a source is also potential. According to the calculus from subsection 1.2.2, it follows

$$c^2 \Delta \phi - (\boldsymbol{V} \cdot \nabla)(\boldsymbol{V} \cdot \nabla)\phi = q\delta(\boldsymbol{x})\,.$$

Setting $q = \varepsilon \overline{q}$, the equation may be linearized and for $\overline{q} = 1$ one obtains (2.2.1). The solution of the equation (2.2.1) could therefore represent the perturbation produced into the uniform stream defined by M, by a mass source of intensity ρ, placed in the origin of the system of coordinates.

In the same way, the equation (2.2.31) could be obtained from the equation

$$\dot{\rho} + \rho \operatorname{div} \boldsymbol{v} = \rho q \delta(t, \boldsymbol{x})\,, \tag{2.2.56}$$

which would represent the equation of continuity in case that a source of intensity ρq is acting in the origin at the moment $t = 0$.

2.3 The Fundamental Solutions of the Steady System

2.3.1 The Significance of the Fundamental Solution

In order to put into evidence the physical significance of the fundamental solutions of the steady linear system, we consider the equations:

$$\begin{aligned} & M^2 p_x + \operatorname{div} \boldsymbol{v} = 0, \quad \boldsymbol{v}_x + \operatorname{grad} p = \boldsymbol{F}, \\ & \lim_{x \to -\infty} (p, \boldsymbol{v}) = 0. \end{aligned} \tag{2.3.1}$$

As it is already known, these equations determine, in the first approximation, the perturbation produced into the uniform stream defined in 2.1 by a force density $\boldsymbol{F}$. For being able to utilize these equations in the case of a force of intensity $\boldsymbol{f} = (f_1, f_2, f_3)$, applied in a point $\boldsymbol{\xi}$, we have to define a density $\boldsymbol{F}$ whose action against the fluid must have the same torsor (resultant and resultant moment) like the force $\boldsymbol{f}$. We state that this density is

$$\boldsymbol{F} = \boldsymbol{f}\delta(\boldsymbol{x} - \boldsymbol{\xi}), \tag{2.3.2}$$

where δ is Dirac's distribution. Indeed, taking into account (A.7.3), we deduce that the torsor of this density is

$$\int_{\mathbb{R}^3} \boldsymbol{f}\delta(\boldsymbol{x} - \boldsymbol{\xi})\mathrm{d}\,\boldsymbol{x} = \boldsymbol{f}, \quad \int_{\mathbb{R}^3} \boldsymbol{x} \times \boldsymbol{f}\delta(\boldsymbol{x} - \boldsymbol{\xi})\mathrm{d}\,\boldsymbol{x} = \boldsymbol{\xi} \times \boldsymbol{f}, \tag{2.3.3}$$

i.e. just the torsor of the force $\boldsymbol{f}$.

Hence the system

$$\begin{aligned} & M^2 p_x + \operatorname{div} \boldsymbol{v} = 0, \quad \boldsymbol{v}_x + \operatorname{grad} p = \boldsymbol{f}\delta(\boldsymbol{x}) \\ & \lim_{x \to -\infty} (p, \boldsymbol{v}) = 0, \end{aligned} \tag{2.3.4}$$

determines the perturbation produced in the uniform stream, defined in 2.1, by the force of intensity $\boldsymbol{f} = \varepsilon\overline{\boldsymbol{f}}$, applied in the origin of the system of coordinates. By *definition*, this system determines the fundamental solutions of the steady linearized system of aerodynamics.

One obtains the plane solutions if one considers that the force having the constant intensity $\boldsymbol{f}$ is uniformly distributed along the Oz axis, being parallel to the xOy plane, i.e. $\boldsymbol{f} = (f_1, f_2, 0)$. In this case we have the same conditions for every z; and the perturbation is plane.

2.3.2 The General Form of the Fundamental Solution

We are interested in those solutions of the system (2.3.4) which can be obtained by means of Fourier transform. They are, obviously, distributions. Utilizing the formulas (A.6.4), from (2.3.4) we deduce:

$$M^2\alpha_1\widehat{p} + \boldsymbol{\alpha}\cdot\widehat{\boldsymbol{v}} = 0\,, \quad \mathrm{i}\alpha_1\widehat{\boldsymbol{v}} + \mathrm{i}\boldsymbol{\alpha}\widehat{p} = -\boldsymbol{f}\,. \tag{2.3.5}$$

From (2.3.5) we deduce:

$$\widehat{p} = \frac{\mathrm{i}\boldsymbol{\alpha}\cdot\boldsymbol{f}}{\alpha^2 - M^2\alpha_1^2}\,. \tag{2.3.6}$$

Then, from the second equation (2.3.5) it follows

$$\widehat{\boldsymbol{v}} = -\frac{\boldsymbol{f}}{\mathrm{i}\alpha_1} - \frac{(-\mathrm{i}\boldsymbol{\alpha})(-\mathrm{i}\boldsymbol{\alpha}\cdot\boldsymbol{f})}{\mathrm{i}\alpha_1(\alpha^2 - M^2\alpha_1^2)}\,. \tag{2.3.7}$$

Utilizing (A.6.9) and (2.2.3) from (2.3.6) we deduce:

$$p(x,y,z) = -(\boldsymbol{f}\cdot\nabla)\mathcal{F}^{-1}\left[\frac{1}{\alpha^2 - M^2\alpha_1^2}\right] = (\boldsymbol{f}\cdot\nabla)\mathcal{E}\,. \tag{2.3.8}$$

Taking into account (A.7.7), and (2.3.7) we obtain:

$$\boldsymbol{v} = \boldsymbol{f}H(x)\delta(y,z) - \nabla(\boldsymbol{f}\cdot\nabla)\mathcal{F}^{-1}\left[\frac{1}{\mathrm{i}\alpha_1(\alpha^2 - M^2\alpha_1^2)}\right]\,. \tag{2.3.9}$$

But

$$\frac{\partial}{\partial x}\mathcal{F}^{-1}\left[\frac{1}{\mathrm{i}\alpha_1(\alpha^2 - M^2\alpha_1^2)}\right] = -\mathcal{F}^{-1}\left[\frac{1}{\alpha^2 - M^2\alpha_1^2}\right] = \mathcal{E}\,, \tag{2.3.10}$$

whence, integrating with respect to x, by virtue of Lebesgue's theorem [A.9], it follows

$$\mathcal{F}^{-1}\left[\frac{1}{\mathrm{i}\alpha_1(\alpha^2 - M^2\alpha_1^2)}\right] = \int_{-\infty}^{x}\mathcal{E}\,\mathrm{d}\,x\,. \tag{2.3.11}$$

The integration limits have been appropriately imposed in order to satisfy the last condition from (2.3.4). It results therefore

$$\boldsymbol{v}(x,y,z) = \boldsymbol{f}H(x)\delta(y,z) + \nabla\varphi\,, \tag{2.3.12}$$

where

$$\varphi = -(\boldsymbol{f}\cdot\nabla)\int_{-\infty}^{x}\mathcal{E}\,\mathrm{d}\,x\,. \tag{2.3.13}$$

The formula (2.3.12), which is valid both in the subsonic and supersonic cases, shows that the perturbation is potential, excepting the Ox axis for $x > 0$ (in the two-dimensional case, one excepts the xOz plane for $x > 0$), where the first term does not vanish. Hence the perturbation is not potential downstream the point (or the line) where the perturbing force is acting.

From (2.3.8) and (2.3.12) it also results:

$$u(x,y,z) = f_1 H(x)\delta(y,z) - p(x,y,z)\,. \tag{2.3.14}$$

With the exception mentioned above one obtains (2.1.38).

In the sequel we shall utilize also some expressions of the components of the velocity which do not result from (2.3.12). In the two-dimensional case, performing the change $\alpha_2^2 = \alpha^2 - M^2\alpha_1^2 - (1-M^2)\alpha_1^2$ in the component v resulting from (2.3.7), we deduce:

$$\widehat{v} = -\frac{\mathrm{i}\alpha_2 f_1}{\alpha^2 - M^2\alpha_1^2} + (1-M^2)\frac{\mathrm{i}\alpha_1 f_2}{\alpha^2 - M^2\alpha_1^2}\,, \tag{2.3.15}$$

whence

$$v(x,y) = -f_1\mathcal{E}_y + (1-M^2) f_2 \mathcal{E}_x\,. \tag{2.3.16}$$

Analogously, in the three-dimensional case, replacing $\alpha_3^2 = \alpha^2 - M^2\alpha_1^2 - (1-M^2)\alpha_1^2 - \alpha_2^2$, in the component w resulting from (2.3.7), we obtain:

$$\begin{aligned}\widehat{w} = &-\frac{\mathrm{i}\alpha_3 f_1}{\alpha^2 - M^2\alpha_1^2} + (1-M^2)\frac{\mathrm{i}\alpha_1 f_3}{\alpha^2 - M^2\alpha_1^2} - \\ &-\frac{\mathrm{i}\alpha_2\alpha_3 f_2}{\alpha_1(\alpha^2 - M^2\alpha_1^2)}\frac{\mathrm{i}\alpha_2^2 f_3}{\alpha_1(\alpha^2 - M^2\alpha_1^2)}\,,\end{aligned} \tag{2.3.17}$$

whence

$$w(x,y,z) = -f_1\mathcal{E}_z + (1-M^2) f_3\mathcal{E}_x - f_2\partial_{yz}^2\int_{-\infty}^{x}\mathcal{E}\,\mathrm{d}\,x + f_3\partial_{yy}^2\int_{-\infty}^{x}\mathcal{E}\,\mathrm{d}\,x\,, \tag{2.3.18}$$

with the notations $\partial_{yz}^2 = \partial^2/\partial y\partial z, \ldots$

2.3.3 The Subsonic Plane Solution

If the perturbation is plane and the free flow is subsonic, $\mathcal{E}$ has the expression (2.2.7). From (2.3.8) one obtains

$$p(x,y) = \frac{1}{2\pi\beta}\frac{xf_1 + \beta^2 y f_2}{x^2 + \beta^2 y^2}\,, \tag{2.3.19}$$

and from (2.3.16),

$$v(x,y) = \frac{\beta}{2\pi}\frac{xf_2 - yf_1}{x^2+\beta^2 y^2}\,. \tag{2.3.20}$$

If the force $\boldsymbol{f}$ is distributed on a straight line parallel to the Oz axis intersecting the xOy plane in the point $\boldsymbol{\xi} = (\xi, \eta)$, then the perturbation will be determined by the system

$$\begin{aligned} &M^2 p_x + \operatorname{div}\boldsymbol{v} = 0, \quad \boldsymbol{v}_x + \operatorname{grad} p = \boldsymbol{f}\delta(x-\xi, y-\eta), \\ &\lim_{x\to-\infty}(p,\boldsymbol{v}) = 0\,. \end{aligned} \tag{2.3.21}$$

After performing the change of variables $(x,y) \to (x_0, y_0)$:

$$x_0 = x - \xi, y_0 = y - \eta\,, \tag{2.3.22}$$

the system (2.3.21) is transformed into the system (2.3.4). Hence, for (2.3.21) one obtains

$$p(x,y) = \frac{1}{2\pi\beta}\frac{x_0 f_1 + \beta^2 y_0 f_2}{x_0^2+\beta^2 y_0^2}, \quad v(x,y) = \frac{\beta}{2\pi}\frac{x_0 f_2 - y_0 f_1}{x_0^2+\beta^2 y_0^2}\,. \tag{2.3.23}$$

These solutions have been obtained in [2.10].

2.3.4 The Three-Dimensional Subsonic Solution

In this case, $\mathcal{E}$ is determined by (2.2.6). From (2.3.8) one obtains

$$4\pi p(x,y,z) = -(f_1\partial_x + f_2\partial_y + f_3\partial_z)(1/R)\,, \tag{2.3.24}$$

where

$$R = \sqrt{x^2+\beta^2(y^2+z^2)}\,. \tag{2.3.25}$$

Taking into account that

$$\int \frac{\mathrm{d}\,x}{(x^2+a^2)^{3/2}} = \frac{x}{a^2(x^2+a^2)^{1/2}}\,, \tag{2.3.26}$$

we deduce

$$\begin{aligned} &\int_{-\infty}^{x} \partial_y\left(\frac{1}{R}\right)\mathrm{d}\,x = -\frac{y}{y^2+z^2}\left(1+\frac{x}{R}\right), \\ &\int_{-\infty}^{x} \partial_z\left(\frac{1}{R}\right)\mathrm{d}\,x = -\frac{z}{y^2+z^2}\left(1+\frac{x}{R}\right). \end{aligned} \tag{2.3.27}$$

Employing these results, we deduce from (2.3.13) the expression of the potential

$$\varphi(x,y,z)=\frac{1}{4\pi}\left[\frac{f_1}{R}-\frac{y}{y^2+z^2}\left(1+\frac{x}{R}\right)f_2-\frac{z}{y^2+z^2}\left(1+\frac{x}{R}\right)f_3\right] \tag{2.3.28}$$

which gives the possibility to calculate the components of the velocity. From (2.3.18) we also obtain:

$$\begin{aligned} w(x,y,z)= & \frac{f_1}{4\pi}\partial_z\left(\frac{1}{R}\right)-\frac{\beta^2}{4\pi}f_3\partial_x\left(\frac{1}{R}\right)- \\ & -\frac{1}{4\pi}(f_2\partial_z-f_3\partial_y)\frac{y}{y^2+z^2}\left(1+\frac{x}{R}\right). \end{aligned} \tag{2.3.29}$$

Considering $\beta=1$, we find the solutions for the incompressible fluid.
These results have been obtained in [2.8].

2.3.5 The Two-Dimensional Supersonic Solution

In this case, $\mathcal{E}$ is given by (2.2.9). Taking into account (A.7.17), from (2.3.8) we deduce

$$2kp(x,y)=(-f_1+kf_2\operatorname{sign}y)\delta(x-k|y|), \tag{2.3.30}$$

and from (2.3.16),

$$2v(x,y)=(-f_1\operatorname{sign}y+kf_2)\delta(x-k|y|). \tag{2.3.31}$$

Obviously, the perturbation differs from zero only in Mach's dihedron.
The solution was given and utilized in [2.10].

2.3.6 The Three-Dimensional Supersonic Solution

In this case, $\mathcal{E}$ is given by (2.2.8) and

$$p(x,y,z)=(f_1\partial_x+f_2\partial_y+f_3\partial_z)\mathcal{E}. \tag{2.3.32}$$

Denoting $s=k\sqrt{y^2+z^2}$ and taking into account (A.3.9) and (A.3.14), we deduce

$$\begin{aligned} \partial_y\int_{-\infty}^{x}\mathcal{E}\,\mathrm{d}\,x & =\partial_y\left[H(x-s)\int_s^x\frac{\mathrm{d}\,x}{\sqrt{x^2-s^2}}\right] \\ & =H(x-s)\partial_y\int_s^x\frac{\mathrm{d}\,x}{\sqrt{x^2-s^2}}=\frac{xy}{y^2+z^2}\mathcal{E}, \end{aligned} \tag{2.3.33}$$

such that,

$$\varphi(x,y,z) = \left(-f_1 + \frac{xy}{y^2+z^2} f_2 + \frac{xz}{y^2+z^2} f_3\right) \mathcal{E}\,. \tag{2.3.34}$$

By means of formula (2.3.12) we determine the velocity field. The similarity between this potential and the subsonic potential is striking (2.3.28).

Let us show now that both p given by (2.3.32) and the velocity field resulting from the potential (2.3.33), are different from zero only in the interior of Mach's cone. The assertion follows immediately if we use the formula

$$\frac{\mathrm{d}}{\mathrm{d}\,x}\left[\frac{H(x)}{x^\lambda}\right] = \frac{\mathrm{d}}{\mathrm{d}\,x}(x_+^{-\lambda}) = -\lambda \frac{H(x)}{x^{\lambda+1}}\,, \quad \lambda \neq 0,1,2,\dots\ , \tag{2.3.35}$$

demonstrated in the theory of distributions (see, for example, [A.5], §2.2). Indeed, we can write:

$$\mathcal{E} = -\frac{1}{2\pi}\frac{H(x-s)}{(x-s)^{1/2}}\frac{1}{(x+s)^{1/2}}\,, \tag{2.3.36}$$

such that the derivatives of $\mathcal{E}$ will have the factor $H(x-s)$.

From (2.3.18), it follows for w :

$$w(x,y,z) = -(f_1\partial_z + k^2 f_3 \partial_x)\mathcal{E} + (f_2\partial_z - f_3\partial_y)\frac{xy}{y^2+z^2}\mathcal{E}\,. \tag{2.3.37}$$

In the sequel, we shall utilize also another expression of the component w. This follows writing

$$\widehat{w} = -\frac{\mathrm{i}\alpha_3 f_1}{\alpha^2 - M^2\alpha_1^2} - \frac{\mathrm{i}\alpha_2\alpha_3 f_2}{\alpha_1(\alpha^2 - M^2\alpha_1^2)} - \frac{k^2\mathrm{i}\alpha_1^2 - \mathrm{i}\alpha_2^2}{\alpha_1(\alpha^2 - M^2\alpha_1^2)} f_3$$

instead of (2.3.17). Utilizing (2.3.11), one obtains:

$$w(x,y,z) = -f_1\partial_z\mathcal{E} - f_2\partial_{yz}^2 \int_{-\infty}^{x} \mathcal{E}\,\mathrm{d}\,x - f_3(k^2\partial_{xx}^2 - \partial_{yy}^2)\int_{-\infty}^{x}\mathcal{E}\,\mathrm{d}\,x\,. \tag{2.3.38}$$

2.4 The Fundamental Solutions of the Oscillatory System

2.4.1 The Determination of Pressure

The fundamental oscillatory solutions are defined by the system:

$$\begin{gathered} M^2(p_t + p_x) + \operatorname{div} \boldsymbol{v} = 0 \\ \boldsymbol{v}_t + \boldsymbol{v}_x + \operatorname{grad} p = \boldsymbol{f}\delta(\boldsymbol{x})\exp(\mathrm{i}\,\omega t) \\ \lim_{x\to-\infty} (p, \boldsymbol{v}) = 0 . \end{gathered} \tag{2.4.1}$$

They will be complex. Since the system is linear, the real part of the solutions will correspond to the case when $\exp(\mathrm{i}\,\omega t)$ will be replaced by $\cos\omega t$, and the imaginary part will correspond to the case when $\exp(\mathrm{i}\,\omega t)$ will be replaced by $\sin\omega t$. The solutions of the system (2.4.1) determine the perturbations produced in the uniform stream defined in 2.1, by a force having the periodic intensity $\boldsymbol{f}\exp(\mathrm{i}\,\omega t)$, applied into origin in the three-dimensional case and uniformly on the Oz axis in the two-dimensional case.

Obviously, the solution of the system (2.4.1) has the form

$$p = P(\boldsymbol{x})\exp(\mathrm{i}\omega t), \quad \boldsymbol{v} = \boldsymbol{V}(\boldsymbol{x})\exp(\mathrm{i}\omega t)\,, \tag{2.4.2}$$

P and $\boldsymbol{V}$ satisfying the system

$$\begin{gathered} M^2(\mathrm{i}\omega P + P_x) + \operatorname{div} \boldsymbol{V} = 0 \\ \mathrm{i}\omega \boldsymbol{V} + \boldsymbol{V}_x + \operatorname{grad} P = \boldsymbol{f}\delta(\boldsymbol{x}) \\ \lim_{x\to-\infty} (P, \boldsymbol{V}) = 0 . \end{gathered} \tag{2.4.3}$$

Applying the Fourier transform in (2.4.3), we get

$$M^2(\omega - \alpha_1)\widehat{P} = \boldsymbol{\alpha} \cdot \widehat{\boldsymbol{V}}\,, \quad (\omega - \alpha_1)\widehat{\boldsymbol{V}} - \boldsymbol{\alpha}\widehat{P} = -\mathrm{i}\boldsymbol{f}\,,$$

whence

$$P = -\frac{\mathrm{i}\boldsymbol{\alpha}\cdot\boldsymbol{f}}{M^2(\alpha_1-\omega)^2 - \alpha^2}, \boldsymbol{V} = \frac{\mathrm{i}\boldsymbol{f}}{\alpha_1 - \omega} + \frac{\boldsymbol{\alpha}}{\alpha_1-\omega}\,\frac{\mathrm{i}\boldsymbol{\alpha}\cdot\boldsymbol{f}}{M^2(\alpha_1-\omega)^2 - \alpha^2} \tag{2.4.4}$$

Applying the Fourier transform, from equation (2.2.14) we deduce:

$$E = -F^{-1}\left[\frac{1}{\alpha^2 - M^2(\alpha_1-\omega)^2}\right], \quad P = (\boldsymbol{f}\cdot\nabla)E\,. \tag{2.4.5}$$

The pressure P will be expressed by means of the solutions (2.2.9) and (2.2.23). In the *two-dimensional subsonic* case, we shall have:

$$P = -\frac{1}{4\mathrm{i}\beta}(\boldsymbol{f}\cdot\nabla)\left[H_0^{(2)}(k\overline{R})\exp(\mathrm{i}kMx)\right], \tag{2.4.6}$$

and in the *three-dimensional subsonic* case,

$$P = -\frac{1}{4\pi}(\boldsymbol{f}\cdot\nabla)\left[\frac{\exp \mathrm{i}k(Mx-R)}{R}\right], \tag{2.4.7}$$

k being given in (2.2.19) and R, $\overline{R}$ in (2.2.20).

In the *two-dimensional supersonic* case, we have:

$$P = -\frac{1}{2k}(\boldsymbol{f}\cdot\nabla)\left[H(x-k|y|)J_0\left(\nu\sqrt{x^2-k^2y^2}\right)\exp(-\mathrm{i}\nu Mx)\right], \tag{2.4.8}$$

and in the *three-dimensional supersonic* case,

$$P = -\frac{1}{2\pi}(\boldsymbol{f}\cdot\nabla)\left[H\left(x-k\sqrt{y^2+z^2}\right)\frac{\cos\nu\sqrt{x^2-k^2(y^2+z^2)}}{\sqrt{x^2-k^2(y^2+z^2)}}\exp(-\mathrm{i}\nu Mx)\right], \tag{2.4.9}$$

ν being given in (2.2.22).

Taking into account the formula (2.3), we deduce that the solution (2.4.9) differs from zero only in the interior of Mach's cone.

2.4.2 The Determination of the Velocity Field

Since the velocity field $\boldsymbol{V}$ vanishes at $-\infty$, we deduce like in (2.3.12):

$$\mathcal{F}^{-1}\left[\frac{1}{\mathrm{i}\alpha_1(\alpha^2-M^2\alpha_1^2+2\alpha_1\omega+\omega^2)}\right] = \int_{-\infty}^{x} G\,\mathrm{d}x,$$

where

$$G = -\mathcal{F}^{-1}\left[\frac{1}{\alpha^2-M^2\alpha_1^2+2\alpha_1\omega+\omega^2}\right]. \tag{2.4.10}$$

Using the change $\alpha_1-\omega\to\alpha_1$, we get:

$$\mathcal{F}^{-1}\left[\frac{1}{\alpha_1-\omega}\frac{1}{M^2(\alpha_1-\omega)^2-\alpha^2}\right] = -\mathrm{i}e^{\mathrm{i}\omega x}\int_{-\infty}^{x} G\,\mathrm{d}x. \tag{2.4.11}$$

If we replace α_1 by $\alpha_1-\omega$ in (A.7.9) we find

$$\mathcal{F}^{-1}\left[\frac{\mathrm{i}}{\alpha_1-\omega}\right] = \mathrm{e}^{-\mathrm{i}\omega x}H(x)\delta(y,z). \tag{2.4.12}$$

Utilizing (A.6.9), on the basis of formulas (2.4.11) and (2.4.12), from (2.4.4) we deduce:

$$\boldsymbol{V} = \boldsymbol{f}\mathrm{e}^{-\mathrm{i}\omega x} H(x)\delta(y,z) + \nabla\varphi, \tag{2.4.13}$$

where

$$\varphi = -(\boldsymbol{f}\cdot\nabla)\mathrm{e}^{-\mathrm{i}\omega x}\int_{-\infty}^{x} G\,\mathrm{d}x\,. \tag{2.4.14}$$

We notice that the perturbation is potential, excepting the Ox axis for $x > 0$, i.e. excepting the trace of the source in the uniform stream.

Let us determine now the distribution G. We notice, applying the Fourier transform, that it is the solution of the equation

$$(1-M^2)G_{xx} + G_{yy} + G_{zz} - 2\mathrm{i}\omega G_x - \omega^2 G = \delta(\boldsymbol{x})\,. \tag{2.4.15}$$

The solution of this equation may be obtained like in 2.2. So, in the two-dimensional subsonic case we find

$$G = -\frac{1}{4\mathrm{i}\beta}H_0^{(2)}(k\overline{R})\exp(\mathrm{i}\overline{\omega}x)\,, \tag{2.4.16}$$

and in the three-dimensional case

$$G = -\frac{1}{4\pi R}\exp\left[\mathrm{i}k\left(\frac{x}{M}-R\right)\right]. \tag{2.4.17}$$

In the two-dimensional supersonic case one obtains:

$$\begin{aligned} G &= -\frac{1}{2k}H(x-k|y|)J_0\left(\nu\sqrt{x^2-k^2y^2}\right)\exp(\mathrm{i}\overline{\omega}x) \overset{\text{not}}{=} \\ &\overset{\text{not}}{=} H(x-k|y|)g\,, \end{aligned} \tag{2.4.18}$$

and in the three-dimensional case

$$\begin{aligned} G &= -\frac{1}{2\pi}H\left(x-k\sqrt{y^2+z^2}\right)\frac{\cos\nu\sqrt{x^2-k^2(y^2+z^2)}}{\sqrt{x^2-k^2(y^2+z^2)}}\exp(\mathrm{i}\overline{\omega}x) \overset{\text{not}}{=} \\ &\overset{\text{not}}{=} H\left(x-k\sqrt{y^2+z^2}\right)g\,, \end{aligned} \tag{2.4.19}$$

$\overline{\omega}$ being defined in (2.2.19).

Taking into account the behaviour of $H_0^{(2)}$ for great values of the argument (2.2.30), we deduce that in all cases we have:

$$\lim_{x\to-\infty}(G,G_x) = 0\,. \tag{2.4.20}$$

2.4.3 Other Forms of the Components V and W

In the two-dimensional case from the component $\widehat{V}$ given by (2.4.4) one eliminates α_2^2 by means of the identity

$$\alpha_2^2 = \alpha^2 - M^2(\alpha_1 - \omega)^2 + M^2(\alpha_1 - \omega)^2 - \alpha_1^2 .$$

Thus one obtains

$$\widehat{V} = \frac{\mathrm{i}f_1}{\alpha_1 - \omega}\frac{\alpha_1\alpha_2}{M^2(\alpha_1 - \omega)^2 - \alpha^2} + \frac{\mathrm{i}f_2}{\alpha_1 - \omega}\frac{(M^2-1)\alpha_1^2 - 2M^2\omega\alpha_1 - M^2\omega^2}{M^2(\alpha_1 - \omega)^2 - \alpha^2},$$

whence, utilizing the inversion formulas (A.6.9) and (2.4.11),

$$V(x,y) = -\left\{f_1\partial_{xy}^2 + f_2\left[(M^2-1)\partial_{xx}^2 + 2M^2\mathrm{i}\omega\partial_x - M^2\omega^2\right]\right\}\cdot$$

$$\cdot\left(\mathrm{e}^{-\mathrm{i}\omega\mathrm{x}}\int_{-\infty}^{\mathrm{x}} \mathrm{G}\,\mathrm{d}\,\mathrm{x}\right) = -f_1\mathrm{e}^{-\mathrm{i}\omega\mathrm{x}}\left(\mathrm{G_y} - \mathrm{i}\omega\int_{-\infty}^{\mathrm{x}} \mathrm{G_y}\,\mathrm{d}\,\mathrm{x}\right) -$$

$$-f_2\mathrm{e}^{-\mathrm{i}\omega\mathrm{x}}\left[2\mathrm{i}\omega\mathrm{G} - (1-\mathrm{M}^2)\mathrm{G_x} + \omega^2\int_{-\infty}^{\mathrm{x}} \mathrm{G}\,\mathrm{d}\,\mathrm{x}\right]. \tag{2.4.21}$$

We also obtain this form using the expression of V given in (2.4.13) and (2.4.14),

$$V = f_2\mathrm{e}^{-\mathrm{i}\omega\mathrm{x}}\mathrm{H(x)}\delta(\mathrm{y,z}) - \mathrm{f_1}\partial_{\mathrm{xy}}^2\left(\mathrm{e}^{-\mathrm{i}\omega\mathrm{x}}\int_{-\infty}^{\mathrm{x}} \mathrm{G}\,\mathrm{d}\,\mathrm{x}\right) -$$

$$-f_2\mathrm{e}^{-\mathrm{i}\omega\mathrm{x}}\int_{-\infty}^{\mathrm{x}} \mathrm{G_{yy}}\,\mathrm{d}\,\mathrm{x},$$

and eliminating G_{yy} with the aid of the equation (2.4.15). Indeed, in the two-dimensional case, utilizing (2.4.20), we deduce

$$\int_{-\infty}^{x} G_{yy}\,\mathrm{d}\,x = H(x)\delta(y) - (1-M^2)G_x + 2\mathrm{i}\omega G + \omega^2\int_{-\infty}^{x} G\,\mathrm{d}\,x .$$

From the last two formulas we find again (2.4.21).

For the supersonic solution, taking into account (2.4.33), with the definition of g given in (2.4.18) we have

$$\int_{-\infty}^{x} G\,\mathrm{d}\,x = H(x - k|y|)\int_{k|y|}^{x} g\,\mathrm{d}\,x ,$$

$$\partial_y\int_{-\infty}^{x} G\,\mathrm{d}\,x = H(x - k|y|)\partial_y\int_{k|y|}^{x} g\,\mathrm{d}\,x , \tag{2.4.22}$$

such that V given by (2.4.21) is zero outside Mach's dihedron.

In a similar manner, in the three-dimensional case we deduce for the component W

$$W(x,y,z) = -f_1 \mathrm{e}^{-\mathrm{i}\omega x}\left(G_z - \mathrm{i}\omega \int_{-\infty}^{x} G_z \, \mathrm{d}x\right) - f_2 \mathrm{e}^{-\mathrm{i}\omega x} \int_{-\infty}^{x} G_{yz} \, \mathrm{d}x -$$

$$-f_3 \mathrm{e}^{-\mathrm{i}\omega x}\left[2\mathrm{i}\omega G - (1-M^2)G_x + (\omega^2 - \partial^2_{yy}) \int_{-\infty}^{x} G \, \mathrm{d}x\right]. \tag{2.4.23}$$

We notice that using the notation:

$$s = k\sqrt{y^2+z^2} \tag{2.4.24}$$

we may write in the supersonic case

$$W(x,y,z) = H(x-s)w(x,y,z), \tag{2.4.25}$$

where taking into account the definition of g from (2.4.19),

$$w = -f_1 \mathrm{e}^{-\mathrm{i}\omega x}\left(g_z - \mathrm{i}\omega \int_{s}^{x} g_z \, \mathrm{d}x\right) - f_2 \mathrm{e}^{-\mathrm{i}\omega x} \int_{s}^{x} g_{yz} \, \mathrm{d}x -$$

$$-f_3 \mathrm{e}^{-\mathrm{i}\omega x}\left[2\mathrm{i}\omega g + k^2 g_x + (\omega^2 - \partial^2_{yy}) \int_{s}^{x} g \, \mathrm{d}x\right],$$

whence we deduce that W is different from zero only in the interior of Mach's cone with the vertex in the origin. Indeed, utilizing the formulas (A.3.9), (2.3.35) and noticing that $s\partial/\partial y = k^2 y \partial/\partial s$, we may write :

$$\partial^2_{yy} \int_{-\infty}^{x} G \, \mathrm{d}x = \partial^2_{yy}\left[H(x-s)\int_{s}^{x} g \, \mathrm{d}x\right] =$$

$$= \partial_y\left[H(x-s)(k^2 y/s)\partial_s \int_{s}^{x} g \mathrm{d}x\right] = \partial_y\left[H(x-s)(k^2 y/s) \overset{*}{\int_{s}^{x}} \partial_s g \mathrm{d}x\right] =$$

$$= \left[H(x-s)\partial_y (k^2 y/s)\partial_s \int_{s}^{x} g \mathrm{d}x\right] = H(x-s)\partial^2_{yy} \int_{s}^{x} g \, \mathrm{d}x$$

and from

$$\partial_z\left(G - \mathrm{i}\omega \int_{-\infty}^{x} G \, \mathrm{d}x\right) = H(x-s)\left(g_z - \mathrm{i}\omega \int_{s}^{x} g_z \, \mathrm{d}x\right),$$

it follows the formula (2.4.25).

2.4.4 The Incompressible Fluid

The (oscillatory) solutions for the incompressible fluid may be obtained considering $M = 0$ in (2.4.4), or directly in the subsonic solution. So, in the two-dimensional case we deduce from (2.4.5):

$$P(x,y) = \frac{1}{2\pi}\frac{xf_1 + yf_2}{x^2+y^2}\,. \tag{2.4.26}$$

The general form of the equation (2.4.15) for $M = 0$ is

$$G = g\,\exp\left(\mathrm{i}\omega x\right), \tag{2.4.27}$$

where $\Delta g = 0$. One obtains therefore:

$$G(x,y) = \frac{\exp\left(\mathrm{i}\omega x\right)}{4\pi}\ln(x^2+y^2)\,. \tag{2.4.28}$$

For the three-dimensional problem we have:

$$P(x,y,z) = -\frac{1}{4\pi}(\boldsymbol{f}\cdot\nabla)\frac{1}{r}\,,\quad G(x,y,z) = -\frac{\exp\left(\mathrm{i}\omega x\right)}{4\pi r}\,, \tag{2.4.29}$$

with the notation $r = \sqrt{x^2+y^2+z^2}$.

2.4.5 The Fundamental Solutions in the Case $M = 1$

If $M = 1$, E is the solution of the equation (2.4.23), and G, the solution of the equation

$$G_{yy} + G_{zz} - 2\mathrm{i}\omega G_x - \omega^2 G = \partial(\boldsymbol{x})\,. \tag{2.4.30}$$

G is determined like E. They have the form:

$$E = H(x)e\,,\quad G = H(x)g\,. \tag{2.4.31}$$

In the two-dimensional case, e and g have the expressions:

$$e = -\frac{a}{\sqrt{x}}\exp\left[-\frac{\mathrm{i}\omega}{2}\left(x+\frac{y^2}{x}\right)\right.\,,\quad g = -\frac{a}{\sqrt{x}}\exp\left[\frac{\mathrm{i}\omega}{2}\left(x-\frac{y^2}{x}\right)\right]\,, \tag{2.4.32}$$

with the notation utilized in (2.4.26) for a. The solution given by (2.4.5) and (2.4.21) is

$$\begin{aligned} P(x,y) &= H(x)(f_1 e_x + f_2 e_y) \\ V(x,y) &= H(x)v(x,y)\,, \end{aligned} \tag{2.4.33}$$

where

$$v(x,y) = -f_1 \mathrm{e}^{-\mathrm{i}\omega \mathrm{x}} \left[\mathrm{g_y} - \mathrm{i}\omega \int_0^{\mathrm{x}} \mathrm{g_y}(\tau, \mathrm{y})\, \mathrm{d}\,\tau \right] - \\ -f_2 \mathrm{e}^{-\mathrm{i}\omega \mathrm{x}} \left[2\mathrm{i}\omega \mathrm{g} + \omega^2 \int_0^{\mathrm{x}} \mathrm{g}(\tau, \mathrm{y})\, \mathrm{d}\,\tau \right]. \tag{2.4.34}$$

In the three-dimensional case, utilizing (2.2.6) we deduce:

$$e = -\frac{1}{4\pi x} \exp\left(-\frac{\mathrm{i}\omega}{2x} r^2\right), \quad g = -\frac{1}{4\pi x} \exp\left[\frac{\mathrm{i}\omega}{2}\left(x - \frac{y^2+z^2}{x}\right)\right]. \tag{2.4.35}$$

Taking into account (2.4.5) and (2.4.23), it follows

$$P(x,y,z) = H(x)(f_1 e_x + f_2 e_y + f_3 e_z) \\ W(x,y,z) = H(x) w(x,y,z), \tag{2.4.36}$$

where

$$w(x,y,z) = -f_1 \mathrm{e}^{-\mathrm{i}\omega \mathrm{x}} \left(\mathrm{g_z} - \mathrm{i}\omega \int_0^{\mathrm{x}} \mathrm{g_z} \mathrm{d}\,\mathrm{x} \right) - \mathrm{f_2} \mathrm{e}^{-\mathrm{i}\omega \mathrm{x}} \int_0^{\mathrm{x}} \mathrm{g_{yz}} \mathrm{d}\,\mathrm{x} - \\ -f_3 \mathrm{e}^{-\mathrm{i}\omega \mathrm{x}} \left[2\mathrm{i}\omega \mathrm{g} + \left(\omega^2 - \partial^2_{\mathrm{yy}}\right) \int_0^{\mathrm{x}} \mathrm{g}\, \mathrm{d}\,\mathrm{x} \right]. \tag{2.4.37}$$

The integrals from (2.4.37) have a strong singularity in the origin, but they are convergent, as it results from the following calculus indicated by V.Iftimie:

$$\int_0^x g \mathrm{d}\,\tau = -\frac{1}{4\pi} \int_0^x \frac{1}{\tau} \exp\left[\mathrm{i}\left(b\tau - \frac{a^2}{\tau}\right)\right] \mathrm{d}\,\tau = \\ = -\frac{1}{4\pi} \lim_{\varepsilon \searrow 0} \int_\varepsilon^x \frac{1}{\tau} \exp\left[\mathrm{i}\left(b\tau - \frac{a^2}{\tau}\right)\right] \mathrm{d}\,\tau = \\ = \lim_{\varepsilon \searrow 0} \int_\varepsilon^x \frac{\exp(\mathrm{i}b\tau) - 1}{\tau} \exp\left(-\mathrm{i}\frac{c^2}{\tau}\right) \mathrm{d}\,\tau + \lim_{\varepsilon \searrow 0} \int_\varepsilon^x \frac{1}{\tau} \exp\left(-\frac{\mathrm{i}c^2}{\tau}\right) \mathrm{d}\,t.$$

The first limit exists because $\exp(-\mathrm{i}c^2/\tau)$ is a bounded function. In the second integral we perform the change of variable $\tau = 1/t$ and we integrate by parts. We obtain:

$$\lim_{\varepsilon \searrow 0} \int_\varepsilon^x \frac{1}{\tau} \exp\left(-\frac{\mathrm{i}c^2}{\tau}\right) \mathrm{d}\,\tau = \int_{1/x}^\infty \exp\left(-\mathrm{i}c^2 t\right) \frac{\mathrm{d}\,t}{t} =$$

$$= \frac{x}{c^2} \exp\left(\frac{\mathrm{i}c^2}{x}\right) - \frac{1}{\mathrm{i}c^2}\int_{1/x}^{\infty} \exp(-\mathrm{i}c^2 t)\frac{\mathrm{d}\,t}{t^2}\,.$$

The last integral is obviously convergent.

The solution (2.4.33) was given in [10.20] and the solution (2.4.36), in [10.17]. They can be also obtained as limits of the subsonic solutions. Indeed, for P given by (2.4.6) and (2.4.7) one utilizes (2.4.27) and (2.4.28). A similar calculus may be performed for G given by (2.4.16) and (2.4.17). Passing to the limit for $M \searrow 1$, we get g given by (2.4.32) respectively g given by (2.4.35). This section was written entirely on the basis of the papers [10.17]–[10.20].

2.5 Fundamental Solutions of the Unsteady System I

2.5.1 Fundamental Solutions

In this section we determine the fundamental solution of the system (2.1.51), i.e. the solution of the system

$$\begin{aligned} &p_t + c^2 \operatorname{div} \boldsymbol{v} = 0\,, \quad \boldsymbol{v}_t + \operatorname{grad} p = \boldsymbol{f}\delta(t, \boldsymbol{x}) \\ &\lim_{\infty}(p, \boldsymbol{v}) = 0\,. \end{aligned} \tag{2.5.1}$$

We already know that this system determines the perturbation produced in the fluid at rest, having the density ρ_0 and the pressure p_0, by a force having the intensity $\boldsymbol{f}$, applied instantaneously (at the moment $t = 0$) in the origin of the system of coordinates (on the Oz axis in the case of the two-dimensional problem).

Utilizing Duhamel's principle, we deduce that the solution of the system (2.5.1) has the form:

$$(p, \boldsymbol{v}) = H(t)(P, \boldsymbol{V})\,, \tag{2.5.2}$$

where $H(t)$ is Heaviside's function and $(P, \boldsymbol{V})$ is a solution of the system

$$\begin{aligned} &P_t + c^2 \operatorname{div} \boldsymbol{V} = 0, \qquad \boldsymbol{V}_t + \operatorname{grad} P = 0 \\ &P(0, \boldsymbol{x}) = 0, \qquad \boldsymbol{V}(0, \boldsymbol{x}) = \boldsymbol{f}\delta(\boldsymbol{x}) \\ &\lim_{\infty}(P, \boldsymbol{V}) = 0\,. \end{aligned} \tag{2.5.3}$$

Applying the Fourier transform, we deduce the system

$$\mathrm{d}\widehat{P}/\mathrm{d}t = c^2 \mathrm{i}\boldsymbol{\alpha}\cdot\widehat{\boldsymbol{V}}, \mathrm{d}\boldsymbol{V}/\mathrm{d}t = \mathrm{i}\boldsymbol{\alpha}\widehat{P} \tag{2.5.4}$$

which has to be integrated with the conditions

$$P(0) = 0\,, \quad \boldsymbol{V}(0) = \boldsymbol{f}\,. \tag{2.5.5}$$

The solution of this system is

$$\widehat{P} = c\mathrm{i}\boldsymbol{\alpha}\cdot\boldsymbol{f}\frac{\sin c|\boldsymbol{\alpha}|t}{|\boldsymbol{\alpha}|}\,, \quad \widehat{\boldsymbol{V}} = \boldsymbol{f} + \mathrm{i}\boldsymbol{\alpha}(\mathrm{i}\boldsymbol{\alpha}\cdot\boldsymbol{f})\left(\frac{1-\cos c|\boldsymbol{\alpha}|t}{\alpha^2}\right)\,,$$

such that, by inversion, we have:

$$\begin{aligned} P(t,\boldsymbol{x}) &= -c(\boldsymbol{f}\cdot\nabla)\mathcal{F}^{-1}\left(\frac{\sin c|\boldsymbol{\alpha}|t}{|\boldsymbol{\alpha}|}\right)\,, \\ V(t,\boldsymbol{x}) &= \boldsymbol{f}\delta(\boldsymbol{x}) + \nabla(\boldsymbol{f}\cdot\nabla)\mathcal{F}^{-1}\left(\frac{1-\cos c|\boldsymbol{\alpha}|t}{\alpha^2}\right)\,. \end{aligned} \tag{2.5.6}$$

Utilizing the formulas (A.7.14) and (A.7.22), with the notation $r = |\boldsymbol{x}|$, we deduce the following fundamental solution for the three-dimensional problem:

$$\begin{aligned} P(t,\boldsymbol{x}) &= -(\boldsymbol{f}\cdot\nabla)\frac{\delta(ct-r)}{4\pi t} \\ \boldsymbol{V}(t,\boldsymbol{x}) &= \boldsymbol{f}\delta(\boldsymbol{x}) + \nabla(\boldsymbol{f}\cdot\nabla)\frac{H(ct-r)}{4\pi r}\,, \end{aligned} \tag{2.5.7}$$

and for the two-dimensional problem,

$$\begin{aligned} P(t,\boldsymbol{x}) &= -\frac{c}{2\pi}(\boldsymbol{f}\cdot\nabla)\frac{H(ct-r)}{\sqrt{c^2t^2-r^2}} \\ \boldsymbol{V}(t,\boldsymbol{x}) &= \boldsymbol{f}\delta(\boldsymbol{x}) + \nabla(\boldsymbol{f}\cdot\nabla)\frac{H(ct-r)}{2\pi}\ln\frac{ct+\sqrt{c^2t^2-r^2}}{r}\,. \end{aligned} \tag{2.5.8}$$

These are the solutions of the system (2.5.3).

2.5.2 Fundamental Matrices

The solution of the system (2.5.1) has been determined for the first time by means of matrices in [2.6]. As we shall see, in some applications it is preferable to use fundamental matrices. We introduce therefore the matrices:

$$Q = \begin{bmatrix} 0 & c^2 & 0 & 0 \\ 1 & 0 & 0 & 0 \\ 0 & 0 & 0 & 0 \\ 0 & 0 & 0 & 0 \end{bmatrix}, \; R = \begin{bmatrix} 0 & 0 & c^2 & 0 \\ 0 & 0 & 0 & 0 \\ 1 & 0 & 0 & 0 \\ 0 & 0 & 0 & 0 \end{bmatrix}, \; S = \begin{bmatrix} 0 & 0 & 0 & c^2 \\ 0 & 0 & 0 & 0 \\ 0 & 0 & 0 & 0 \\ 1 & 0 & 0 & 0 \end{bmatrix}, \tag{2.5.9}$$

$$v^T = [p, u, v, w]\,, \quad F^T = [0, f_1, f_2, f_3]\,,$$

u, v, w representing the components of the vector $\boldsymbol{v}$ and f_1, f_2, f_3, the components of the vector $\boldsymbol{f}$. The system (2.5.1) is written as follows:

$$\begin{aligned} &V_t + QV_x + RV_y + SV_z = F\delta(t, \boldsymbol{x}) \\ &V(t, \infty) = 0\,. \end{aligned} \tag{2.5.10}$$

Introducing the matrix L, such that

$$V = LF\,, \tag{2.5.11}$$

we deduce:

$$\begin{aligned} &L_t + QL_x + RL_y + SL_z = E\delta(t, \boldsymbol{x}) \\ &L(t, \infty) = 0\,, \end{aligned} \tag{2.5.12}$$

E representing the unit matrix with 4×4 elements. Utilizing Duhamel's principle, we deduce that the solution of the equation (2.5.12) has the form

$$L = H(t)\, K(t, \boldsymbol{x})\,, \tag{2.5.13}$$

where

$$\begin{aligned} &K_t + QK_x + RK_y + SK_z = 0 \\ &K(0, \boldsymbol{x}) = E\delta(\boldsymbol{x})\,, \quad K(t, \infty) = 0\,. \end{aligned} \tag{2.5.14}$$

Applying the Fourier transform to the problem (2.5.14), we deduce:

$$\widehat{K}_t = \widehat{A}\widehat{K}\,, \quad \widehat{K}(0, \boldsymbol{\alpha}) = E\,, \tag{2.5.15}$$

where

$$\widehat{A} = \mathrm{i}\,\alpha_1 Q + \mathrm{i}\,\alpha_2 R + \mathrm{i}\,\alpha_3 S = \mathrm{i} \begin{bmatrix} 0 & c^2\alpha_1 & c^2\alpha_2 & c^2\alpha_3 \\ \alpha_1 & 0 & 0 & 0 \\ \alpha_2 & 0 & 0 & 0 \\ \alpha_3 & 0 & 0 & 0 \end{bmatrix}.$$

The solution of the problem (2.5.15) has the form:

$$\widehat{K} = E \exp{(\widehat{A}t)}\,. \tag{2.5.16}$$

It is well known that for determining the function $\exp{(\widehat{A}t)}$ one may employ two classical methods: the method of matrix functions and the method of the minimal polynomial. We shall utilize herein the method of the minimal polynomial described in 2.6.2. The eigenvalues of the matrix $\widehat{A}$ are $\lambda_1 = 0$, $\lambda_{2,3} = \pm \mathrm{i}\,|\boldsymbol{\alpha}|c$, the first one being multiple of order two. One deduces that the minimal polynomial has the form

$$\lambda(\lambda - \lambda_2)(\lambda - \lambda_3) = \lambda(\lambda^2 + \alpha^2 c^2)\,.$$

Comparing this form with (2.6.13), we obtain

$$a_0 = a_2 = 0\,, \quad a_1 = -\alpha^2 c^2\,.$$

Hence, the equations (2.6.18), (2.6.19) and (2.6.17) become

$$g_2''' - \alpha^2 c^2 g_2'\,, \; g_2(0) = g_2'(0) = 0\,, \; g_2''(0) = 1,$$

$$g_0' = 0, \quad g_0(0) = 1,$$

$$g_1' = -\alpha^2 c^2 g_2 + g_0, \quad g_1(0) = 0\,.$$

$\widehat{K}$ has the form $E g_0 + \widehat{A} g_1 + \widehat{A}^2 g_2$. We get:

$$\widehat{K} = E + A\frac{\sin c|\boldsymbol{\alpha}|t}{c|\boldsymbol{\alpha}|} + A^2\frac{1 - \cos c|\boldsymbol{\alpha}|t}{c^2\alpha^2}\,.$$

Using the inverse Fourier transform we obtain:

$$K(t, \boldsymbol{x}) = E\delta(\boldsymbol{x}) + A\mathcal{F}^{-1}\left[\frac{\sin c|\boldsymbol{\alpha}|t}{c|\boldsymbol{\alpha}|}\right] + A^2\mathcal{F}^{-1}\left[\frac{1 - \cos c|\boldsymbol{\alpha}|t}{c^2\alpha^2}\right],$$

where, for the three-dimensional case,

$$A = -\begin{bmatrix} 0 & c^2\partial_x & c^2\partial_y & c^2\partial_z \\ \partial_x & 0 & 0 & 0 \\ \partial_y & 0 & 0 & 0 \\ \partial_z & 0 & 0 & 0 \end{bmatrix}.$$

Utilizing the formulas (A.7.14) and (A.7.22), we deduce that the matrix K has the form:

$$K(t, \boldsymbol{x}) = E\delta(\boldsymbol{x}) + \frac{1}{4\pi c^2 t}A\delta(ct - r) + \frac{1}{4\pi c^2}A^2\frac{H(ct - r)}{r}\,. \qquad (2.5.17)$$

In the two-dimensional case we have:

$$K(t, \boldsymbol{x}) = E\delta(\boldsymbol{x}) + \frac{1}{2\pi c}A\frac{H(ct - r)}{\sqrt{c^2t^2 - r^2}} +$$

$$+\frac{1}{2\pi c^2}A^2\left[H(ct - r)\ln\frac{ct + \sqrt{c^2t^2 - r^2}}{r}\right] \qquad (2.5.18)$$

$$A = -\begin{bmatrix} 0 & c^2\partial_x & c^2\partial y \\ \partial_x & 0 & 0 \\ \partial_y & 0 & 0 \end{bmatrix}.$$

The formulas (2.5.13), (2.5.17) and (2.5.18) give the fundamental matrices.

2.5.3 Cauchy's Problem

We shall prove in the sequel that the solution of the problem

$$V_t + QV_x + RV_y + SV_z = F(t, \boldsymbol{x}) \tag{2.5.19}$$

$$V(0, \boldsymbol{x}) = 0, \quad V(t, \infty) = 0 \tag{2.5.20}$$

is

$$V(t, \boldsymbol{x}) = \int_0^t K * F \,\mathrm{d}t, \tag{2.5.21}$$

where

$$K * F = \int_{\mathbb{R}^3} K(t-\tau, \boldsymbol{x}-\boldsymbol{\xi}) F(\tau, \boldsymbol{\xi}) \,\mathrm{d}\boldsymbol{\xi} = \int_{\mathbb{R}^3} K(\tau, \boldsymbol{\xi}) F(t-\tau, \boldsymbol{x}-\boldsymbol{\xi}) \,\mathrm{d}\boldsymbol{\xi}$$

is a convolution.

Indeed, taking into account (2.5.14), we have:

$$V_t = K * F \Big|_{\tau=t} + \int_0^t K_t * F \,\mathrm{d}\tau,$$

$$QV_x + RV_y + SV_z = \int_0^t (QK_x + RK_y + SK_z) * F \,\mathrm{d}t,$$

$$V_t + QV_x + RV_y + SV_z = \int_{\mathbb{R}^3} \delta(\boldsymbol{x}-\boldsymbol{\xi}) F(t, \boldsymbol{\xi}) \,\mathrm{d}\boldsymbol{\xi} = F(t, \boldsymbol{x}).$$

The conditions (2.5.20) are obviously verified, because $K(t, \infty) = 0$.

Taking into account (2.5.17) and (2.5.18), with the notation $\rho = |\boldsymbol{\xi}|$, it follows for the solution of the problem (2.5.19), in the three-dimensional case,

$$\begin{aligned} V(t,\boldsymbol{x}) &= \int_0^t F(\tau,\boldsymbol{x})\,\mathrm{d}\tau + \frac{1}{4\pi c^2}A\int_0^t \frac{\mathrm{d}\tau}{\tau}\int_{\mathbb{R}^3} F(t- \\ &\quad -\tau,\boldsymbol{x}-\boldsymbol{\xi})\delta(c\tau-\rho)\,\mathrm{d}\boldsymbol{\xi} + \frac{1}{4\pi c^2}A^2\int_0^t \mathrm{d}\tau\int_{\mathbb{R}^3} F(t- \\ &\quad -\tau,\boldsymbol{x}-\boldsymbol{\xi})\frac{H(c\tau-\rho)}{\rho}\,\mathrm{d}\boldsymbol{\xi}, \end{aligned} \tag{2.5.22}$$

and in the two-dimensional case,

$$\begin{aligned} V(t,\boldsymbol{x}) &= \int_0^t F(\tau,\boldsymbol{x})\,\mathrm{d}\tau + \frac{1}{2\pi c}\;A\int_0^t \mathrm{d}\tau\int_{\mathbb{R}^2} F(t- \\ &\quad -\tau,\boldsymbol{x}-\boldsymbol{\xi})\frac{H(c\tau-\rho)}{\sqrt{c^2\tau^2-\rho^2}}\,\mathrm{d}\boldsymbol{\xi} + \frac{1}{2\pi c^2}A^2\int_0^t \mathrm{d}\tau\int_{\mathbb{R}^2} F(t- \\ &\quad -\tau,\boldsymbol{x}-\boldsymbol{\xi})H(c\tau-\rho)\ln\frac{c\tau+\sqrt{c^2\tau^2-\rho^2}}{\rho}\,\mathrm{d}\boldsymbol{\xi}\,. \end{aligned} \tag{2.5.23}$$

2.5.4 The Perturbation Produced by a Mobile Source

We shall apply the above formulas for determining the perturbation produced in the fluid at rest (having the density ρ_0 and the pressure p_0) by a force of intensity $\boldsymbol{f}$, whose application point moves uniformly, with the velocity v, in the direction of the Ox axis and in the negative sense. This problem, also considered in [2.6], is important in aerodynamics because it models the leading edge of an uniformly moving airplane in the air at rest.

The solution is obtained from (2.5.22) replacing

$$F(t,x) = F_0\delta(x+vt)\delta(y,x)\,, \tag{2.5.24}$$

where $F_0^T = (0, f_1, f_2, f_3)$. One finds

$$V(t,x) = F_0 \frac{H(x+vt)}{v}\delta(y,z) + \frac{1}{4\pi c^2} A F_0 \int_0^t \frac{\delta(c\tau - R_\tau)}{\tau}\,\mathrm{d}\,\tau +$$

$$+ \frac{1}{4\pi c^2} A^2 F_0 \int_0^t \frac{H(c\tau - R_\tau)}{R_\tau}\,\mathrm{d}\,\tau\,, \tag{2.5.25}$$

where

$$R_\tau = \sqrt{(x+vt-v\tau)^2 + y^2 + z^2}\,. \tag{2.5.26}$$

For calculating these integrals we have to study the behaviour of the function

$$h(\tau) = c\tau - R_\tau\,, \tag{2.5.27}$$

on the interval $(0,\infty)$. We shall proceed like in 2.2.4. Because $h(0) < 0$ and $h(\infty) = (c^2 - v^2)\infty$, we have to separate the subsonic flow case $(v < c)$ from the supersonic flow case $(v > c)$. The zeros of the function $h(\tau)$ are given by the formula

$$(c^2 - v^2)\tau_\pm = -v(x+vt) \pm R\,, \tag{2.5.28}$$

where

$$R = \left[(x+vt)^2 c^2 + (c^2 - v^2)(y^2 + z^2)\right]^{1/2}\,. \tag{2.5.29}$$

In the subsonic case, only τ_+ is positive. If the velocity of the source is supersonic $(v > c)$, the function $h(\tau)$ has two zeros $(\tau_+ < \tau_-)$ if $h_0 > 0$ and no zero if $h_0 < 0$. We denoted $h_0 = h(\tau_0)$, τ_0 representing the zero of the derivative of the function h,

$$v\tau_0 = x + vt + \frac{cr_0}{\sqrt{v^2 - c^2}}\,. \tag{2.5.30}$$

It results,

$$vh_0 = c(x+vt) - \sqrt{(v^2 - c^2)(y^2 + z^2)}\,. \tag{2.5.31}$$

Utilizing the relation $c\tau_\pm h'(\tau_\pm) = \pm R$ which may be verified directly, and the formula (2.2.38), we deduce

$$\int_0^t \frac{\delta(c\tau - R_\tau)}{\tau}\,\mathrm{d}\,\tau =$$

$$= \begin{cases} cR^{-1} H(t - \tau_+), & \text{if } v < c, \\ cR^{-1}\left[H(t-\tau_+) + H(t-\tau_-)\right], & \text{if } v > c \text{ and } h_0 > 0, \\ 0, & \text{if } v > c \text{ and } h_0 < 0. \end{cases} \tag{2.5.32}$$

In the same time, taking into account the definition of Heaviside's, function we have:

$$\int_0^t \frac{H(c\tau - R_\tau)}{R_\tau}\,\mathrm{d}\tau =$$

$$= \begin{cases} -\dfrac{H(t-\tau_+)}{v}\ln\dfrac{(c+v)(x+r)}{c(x+vt)+R}, & \text{if } (v<c)\cup(v>c, h_0>0, t<\tau_-), \\ -\dfrac{1}{v}\ln\dfrac{c(x+vt)-R}{c(x+vt)+R}, & \text{if } v>c, h_0>0, t>\tau_-, \\ 0, & \text{if } v>c, h_0<0\,. \end{cases} \tag{2.5.33}$$

Now the solution (2.5.25) is explicit. The solution of this problem may be also obtained directly (without using the formula (2.5.22)) [1.10]. The interpretation of the solution is similar to the interpretation from (2.2.4).

2.6 Fundamental Solutions of the Unsteady System II

2.6.1 The Fundamental Matrices

In this section we determine, using the papers [2.7] and [2.9], the fundamental solutions of the system (2.1.26), (2.5.27), i.e. the solutions of the system

$$\begin{cases} M^2(p_t + p_x) + \operatorname{div}\boldsymbol{v} = 0 \\ \boldsymbol{v}_t + \boldsymbol{v}_x + \operatorname{grad} p = \boldsymbol{f}\;\delta(t,\boldsymbol{x}) \\ \lim\limits_{\infty}(p,\boldsymbol{v}) = 0\,. \end{cases} \tag{2.6.1}$$

Introducing the matrices

$$Q = \begin{bmatrix} 1 & M^{-2} & 0 & 0 \\ 1 & 1 & 0 & 0 \\ 0 & 0 & 1 & 0 \\ 0 & 0 & 0 & 1 \end{bmatrix},\; R = \begin{bmatrix} 0 & 0 & M^{-2} & 0 \\ 0 & 0 & 0 & 0 \\ 1 & 0 & 0 & 0 \\ 0 & 0 & 0 & 0 \end{bmatrix},\; S = \begin{bmatrix} 0 & 0 & 0 & M^{-2} \\ 0 & 0 & 0 & 0 \\ 0 & 0 & 0 & 0 \\ 1 & 0 & 0 & 0 \end{bmatrix}$$

and the matrices V and F defined in (2.5.9), the system (2.6.1) may be written like in (2.5.10). We shall proceed like in the previous section. Considering

$$V = H(t)K(t,\boldsymbol{x})F, \tag{2.6.2}$$

one obtains the following equation for $\widehat{K}$:

$$\widehat{K}_t = \widehat{A}\widehat{K}, \quad \widehat{K}(0, \boldsymbol{\alpha}) = E, \tag{2.6.3}$$

where

$$\widehat{A} = \mathrm{i} \begin{bmatrix} \alpha_1 & \alpha_1 M^{-2} & \alpha_2 M^{-2} & \alpha_3 M^{-2} \\ \alpha_1 & 1 & 0 & 0 \\ \alpha_2 & 0 & 1 & 0 \\ \alpha_3 & 0 & 0 & 1 \end{bmatrix}$$

The roots of the characteristic polynomial are:

$$\lambda_1 = \mathrm{i}\,\alpha_1, \quad \lambda_{2,3} = \mathrm{i}\,(\alpha_1 \pm |\boldsymbol{\alpha}| M^{-1}), \tag{2.6.4}$$

the first root being double. One verifies that the minimal polynomial has the form

$$(\lambda - \lambda_1)(\lambda - \lambda_2)(\lambda - \lambda_3) = (\lambda - \mathrm{i}\,\alpha_1)^3 + \alpha^2 M^{-2}(\lambda - \mathrm{i}\,\alpha_1).$$

Comparing with (2.6.13) it results:

$$a_2 = 3\mathrm{i}\,\alpha_1, a_1 = 3\alpha_1^2 - \alpha^2 M^{-2}, \; a_0 = \mathrm{i}\,\alpha_1(\alpha^2 M^{-2} - \alpha_1^2). \tag{2.6.5}$$

$\widehat{K}$ has the form

$$\widehat{K} = E g_0(t) + \widehat{A} g_1(t) + \widehat{A}^2 g_2(t), \tag{2.6.6}$$

the functions g_0, g_1, g_2 being determined by the equations

$$\begin{aligned} &g g_2''' - a_2 g_2'' - a_1 g_2' - a_0 g_2 = 0, \; g_2(0) = g_2'(0) = 0, \; g_2''(0) = 1, \\ &g_0' = a_0 g_2, \; g_0(0) = 1, \\ &g_1' = a_1 g_2 + g_0, g_1(0) = 0. \end{aligned} \tag{2.6.7}$$

Determining these functions, we find:

$$\begin{aligned} g_0 &= \left[1 - \frac{\mathrm{i}\,\alpha_1 M}{|\boldsymbol{\alpha}|} \sin \frac{|\boldsymbol{\alpha}|}{M} t - \frac{\alpha_1^2 M^2}{\alpha^2}\left(1 - \cos \frac{|\boldsymbol{\alpha}|}{M} t\right)\right] \exp(\mathrm{i}\,\alpha_1 t), \\ g_1 &= \left[\frac{M}{|\boldsymbol{\alpha}|} \sin \frac{|\boldsymbol{\alpha}|}{M} t - \frac{2\mathrm{i}\,\alpha_1^2 M^2}{\alpha^2}\left(1 - \cos \frac{|\boldsymbol{\alpha}|}{M} t\right)\right] \exp(\mathrm{i}\,\alpha_1 t), \\ g_2 &= \frac{M^2}{\alpha^2}\left(1 - \cos \frac{|\boldsymbol{\alpha}|}{M} t\right) \exp(\mathrm{i}\,\alpha_1 t). \end{aligned} \tag{2.6.8}$$

In the three-dimensional case one utilizes the formulas (A.7.15) and (A.7.23). From (2.6.6) it follows the fundamental matrix:

$$
\begin{aligned}
K(t, \boldsymbol{x}) &= E\delta(x-t)\delta(y,z) + \frac{M^2}{4\pi t}(E\partial_x + A)\delta(tM^{-1} - R) + \\
&+ \frac{M^2}{4\pi}(E\partial^2_{xx} + 2A\partial_x + A^2)\frac{H(tM^{-1} - R)}{R},
\end{aligned}
\tag{2.6.9}
$$

where $R = \sqrt{(x-t)^2 + y^2 + z^2}$ and

$$
A = -\begin{bmatrix} \partial_x & M^{-2}\partial_x & M^{-2}\partial_y & M^{-2}\partial_z \\ \partial_x & \partial_x & 0 & 0 \\ \partial_y & 0 & \partial_x & 0 \\ \partial_z & 0 & 0 & \partial_x \end{bmatrix}.
$$

In the two-dimensional case one utilizes the formulas (A.7.15) and (A.7.23). From (2.6.6) one obtains the following fundamental matrix:

$$
\begin{aligned}
K(t, \boldsymbol{x}) &= E\delta(x-t, y) + \frac{M^2}{2\pi}(E\partial_x + A)\frac{H(t - M\overline{R})}{\sqrt{t^2 - M^2\overline{R}^2}} + \\
&+ \frac{M^2}{2\pi}(\partial^2_{xx} + 2A\partial_x + A^2)H(t - M\overline{R})\ln\frac{t + \sqrt{t^2 - M^2\overline{R}^2}}{M\overline{R}},
\end{aligned}
\tag{2.6.10}
$$

where $\overline{R} = \sqrt{(x-t)^2 + y^2}$ and

$$
A = -\begin{bmatrix} \partial_x & M^{-2}\partial_x & M^{-2}\partial_y \\ \partial_x & \partial_x & 0 \\ \partial_y & 0 & \partial_x \end{bmatrix}.
$$

In [1.10] and [2.7] one gives some applications which are important for aerodynamics. We mention the mobile source on the direction of the unperturbed stream and the mobile source on the direction perpendicular to the same stream.

2.6.2 The Method of the Minimal Polynomial

In the sequel we present the method utilized for determining the function $\exp(At)$. Let A be a matrix with $n \times n$ elements and let $\lambda_1, \ldots, \lambda_k (k \leq n)$ be its eigenvalues, i.e. the roots of the characteristic polynomial

$$
\Delta(\lambda) = \det(\lambda E - A).
$$

Obviously,

$$\Delta(\lambda) = (\lambda - \lambda_1)^{m_1} \dots (\lambda - \lambda_k)^{m_k}, \tag{2.6.11}$$

where $m_1 + \dots + m_k = n$.

The minimal polynomial associated to the matrix A is by definition the polynomial with the smallest degree which has the property

$$P(A) = 0. \tag{2.6.12}$$

Let

$$P(\lambda) = \lambda^m - a_{m-1}\lambda^{m-1} - \dots - a_1\lambda - a_0 \tag{2.6.13}$$

be this polynomial. Obviously, $m \leq n$, because $\Delta(A) = 0$.

From (2.6.12) and (2.6.13) it results

$$A^m = \sum_{j=0}^{m-1} a_j A^j$$

and step by step

$$A^{m+p} = \sum_{j=0}^{m-1} a_j^{(p)} A^j, \quad p = 0, 1, 2, \dots$$

Taking into account that the exponential is a power series, we deduce that there exist the functions

$$g_0(t), \dots, g_{m-1}(t),$$

such that

$$\exp(At) = g_0 E + g_1 A + \dots + g_{m-1} A^{m-1} \tag{2.6.14}$$

We have:

$$\begin{aligned} A\exp(At) = {} & g_0 A + \dots + g_{m-2}A^{m-1} + \\ & + g_{m-1}(a_0 E + a_1 A + \dots + a_{m-1}A^{m-1}), \end{aligned} \tag{2.6.15}$$

On the other hand,

$$A\exp(At) = (\mathrm{d}/\mathrm{d}t)\exp(At) = g_0' E + g_1' A + \dots + g_{m-1}' A^{m-1}. \tag{2.6.16}$$

Since $P(\lambda)$ is unique, from (2.6.15) and (2.6.16) we obtain:

$$g_0' = a_0 g_{m-1}, g_1' = a_1 g_{m-1} + g_0, \dots, g_{m-1}' = a_{m-1} g_{m-1} + g_{m-2}. \tag{2.6.17}$$

We deduce that g_{m-1} is the solution of the differential equation

$$g_{m-1}^{(m)} = a_{m-1} g_{m-1}^{(m-1)} + \ldots + a_1 g_{m-1}^{(1)} + a_0 g_{m-1}, \tag{2.6.18}$$

with the following initial conditions:

$$g_{m-1}^{(0)}(0) = g_{m-1}^{(1)}(0) = \ldots = g_{m-1}^{(m-2)}(0) = 0, g_{m-1}^{(m-1)}(0) = 1. \tag{2.6.19}$$

The functions $g_0, \ldots, g_{m-2}$ are determined from (2.6.17) with the cu conditions

$$g_0(0) = 1, g_1(0) = 0, \ldots, g_{m-2}(0) = 0, \tag{2.6.20}$$

imposed by (2.6.14).

In fact, for determining the minimal polynomial, one takes into account that it is a divisor of the characteristic polynomial, hence it has the form:

$$\begin{aligned} P(\lambda) &= (\lambda - \lambda_1)^{n_1}(\lambda - \lambda_2)^{n_2} \ldots (\lambda - \lambda_k)^{n_k} = \\ &= \lambda^m - a_{m-1}\lambda^{m-1} - \ldots - a_0, \end{aligned} \tag{2.6.21}$$

where $n_j \leq m_j$ and $n_1 + n_2 + \ldots + n_k = m$. To conclude, we try with divisors having the form (2.6.21), beginning with the simplest $(n_1 = n_2 = \ldots = n_k = 1)$ in order to satisfy the equation $P(A) = 0$.

We mention that dr. Şt. Mirică from the University of Bucharest has communicated us this method.

Chapter 3

The Infinite Span Airfoil in Subsonic Flow

3.1 The Airfoil in the Unlimited Fluid

3.1.1 The Statement of the Problem

In this subsection we determine the perturbation produced by a thin infinite cylindrical body in a subsonic uniform stream having the velocity $U_\infty \boldsymbol{i}$, the pressure p_∞ and the density ρ_∞. We assume that the generators of the cylinder are perpendicular on the velocity of the unperturbed stream which is parallel to the xOy reference plane. The cylindrical bodies are bodies with constant cross-section and the infinite cylinders are cylinders which are long enough, such that the effects of the end conditions over the flow in the xOy plane may be neglected. Under these assumptions (the conditions which determine the perturbation are not varying in time and they are the same in every plane parallel to xOy), the perturbed flow of the fluid will be stationary and plane.

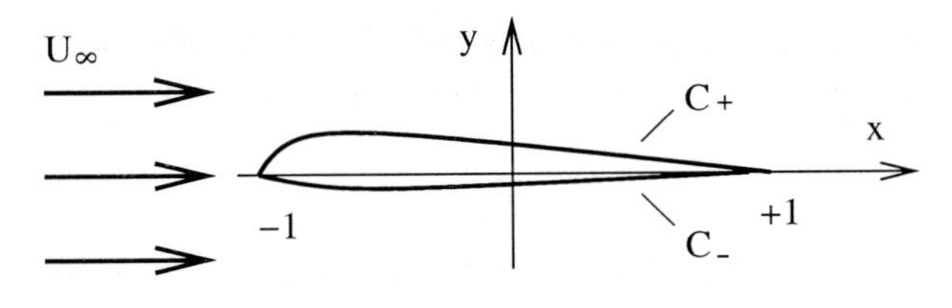

Fig. 3.1.1.

We utilize the variables x, y, z introduced in (2.1.1) and the perturbations p and $\boldsymbol{v}$ defined by (2.1.3). In figure 3.1.1 one presents the profile determined by the intersection of the cylindric body with the xOy plane. We denote by

$$y = h(x) \pm h_1(x) \tag{3.1.1}$$

the equations of the curves C_+ (the upper surface) and C_- (the lower surface). The curve having the equation $y = h(x)$ gives the so called *skeleton of the profile.* For imposing the boundary conditions we use the projection of the profile onto the Ox axis (the direction of the unperturbed stream). Defining L_0 from (2.1.1) as the half length of this projection and taking the origin of the axes of coordinates in the middle of the projection, it follows that the functions $h(x)$ and $h_1(x)$ have to be defined on the segment $[-1,+1]$. We assume that these functions have continuous first order derivatives and that the derivative $h'(x)$ satisfies Hölder's condition (see Appendix B). The cylinder is thin if

$$h(x) = \varepsilon\overline{h}(x), h_1(x) = \varepsilon\overline{h}_1(x)\,,$$

ε representing a small parameter.

We notice that whatever would be the equations of the curves C_+ and C_-, they may be written like in (3.1.1). Indeed, if the equations would be $y = h_\pm(x) = \varepsilon\overline{h}_\pm(x)$, they might take the form (3.1.1) putting

$$h_+(x) + h_-(x) = 2h(x), h_+(x) - h_-(x) = 2h_1(x)\,.$$

Taking into account the equations (3.1.1) the following conditions have to be satisfied

$$v(x,\pm 0) = h(x) \pm h_1'(x), |x| < 1 \tag{3.1.2}$$

as it follows from (2.1.32).

3.1.2 A Classical Method

The classical methods for solving this problem rely on the assumption that the perturbation is potential. We have therefore

$$v = \operatorname{grad}\varphi(x,y). \tag{3.1.3}$$

The potential φ which is defined in the exterior of the segment $[-1,+1]$ from the xOy plane, is the solution of the equation

$$\beta^2\varphi_{xx} + \varphi_{yy} = 0 \tag{3.1.4}$$

with the boundary conditions

$$\varphi_y(x,\pm 0) = h'(x) \pm h_1'(x)\,, \quad |x| < 1 \tag{3.1.5}$$

and the condition at infinity

$$\lim_{x\to-\infty}(\varphi_x,\varphi_y) = 0 \tag{3.1.6}$$

which follows from (2.1.28).

Using Glauert's transformation $x, y \to X, Y$:

$$X = x, \qquad Y = \beta y, \tag{3.1.7}$$

one obtains from (3.1.3) and (3.1.4)

$$u = \varphi_X \equiv U, \qquad v = \beta \varphi_Y \equiv \beta V \tag{3.1.8}$$

$$\varphi_{XX} + \varphi_{YY} = 0 \tag{3.1.9}$$

in the exterior of the segment $[-1, +1]$ from the XOY plane. The boundary conditions (3.1.5) and the conditions at infinity (3.1.6) become

$$\begin{gathered} V(X, \pm 0) = h'(X) \pm h_1'(X), \quad |X| \leq 1 \\ \lim_{X \to -\infty} (U, V) = 0 \,. \end{gathered} \tag{3.1.10}$$

The problem was solved by many authors (see for example Birnbaum [3.5], [3.6], Söhngen [A.34], Sedov [1.38], p.47, Iacob [1.21], p.661), by means of one of the following two methods: reduction of the problem to an ntegral equation which is then solved or the reduction of the problem to a boundary value problem in the complex plane. We also mention the methods relying on the expansion of the solution in a Fourier series. (Glauert [3.21] and Weissinger [3.49]).

In the sequel we give one of the most natural methods for reducing the problem to an integral equation. We represent the holomorphic function $W = U - \mathrm{i}V$ in the $Z = X + \mathrm{i}Y$ plane with the cut $[-1, +1]$ on the real axis, by means of the Cauchy integral

$$W(Z) = \frac{1}{2\pi \mathrm{i}} \int_{-1}^{+1} \frac{-f(t) + \mathrm{i}\, f_1(t)}{t - Z} \mathrm{d}t \,, \tag{3.1.11}$$

where $f(t)$ and $f_1(t)$ are unknown real Hölder functions. It is easy to understand the significance of this representation. $W(Z)$ is the sum of the complex velocity determined by a continuous superposition of sources having the intensity f_1 on the segment $[-1, +1]$ and the complex velocity determined by a continuous superposition of vortices having the circulation f, on the same segment. In fact, the integral containing f may also be regarded as a continuous superposition of doublets on the segment $[-1, +1]$ from the (Z) plane.

Obviously, $W(Z)$ is a holomorphic function in the exterior of the segment $[-1, +1]$ and it vanishes for $Z \to \infty$. Hence, U and V

satisfy the equation (3.1.9) and the condition (3.1.10). In order to prove this we pass to the limit considering $Z \to X \pm \mathrm{i}\,0$, where $X \in (-1,+1)$. Using Plemelj's formulas (B.3.1) and separating the real part from the imaginary one we obtain:

$$U(X,\pm 0) = \mp \frac{1}{2} f(X) + \frac{1}{2\pi} {\int_{-1}^{+1}}' \frac{f_1(t)}{t-X} \mathrm{d}\,t \tag{3.1.12}$$

$$V(X,\pm 0) = \mp \frac{1}{2} f_1(X) - \frac{1}{2\pi} {\int_{-1}^{+1}}' \frac{f(t)}{t-X} \mathrm{d}\,t\,. \tag{3.1.13}$$

The mark " $'$ " indicates the principal value in Cauchy's sense (Appendix B). From (2.1.38), (3.1.8) and (3.1.12) it follows

$$p(x,+0) - p(x,-0) = U(X,-0) - U(X,+0) = f(x). \tag{3.1.14}$$

This relation puts into evidence the significance of the function $f(x)$. It will be fundamental for the determination of the lift and moment coefficients.

Imposing the conditions (3.1.10) in (3.1.13), adding and subtracting the relations just obtained, we find

$$\frac{\beta}{\pi} {\int_{-1}^{+1}}' \frac{f(t)}{t-X} \mathrm{d}\,t = -2h'(X) \quad |X| < 1, \tag{3.1.15}$$

$$\beta f_1(X) = -2h_1'(X) \quad |X| < 1. \tag{3.1.16}$$

The relation (3.1.16) determines the function βf_1 and (3.1.15) represents an integral singular equation for the determination of the function βf. It is the well known equation of thin profiles which was put into evidence by Birnbaum in 1923 and solved by Söhngen in 1939. Its solution, given in Appendix C, will be written in (3.1.25).

We notice that the thickness of the profile is ensured by the distribution of sources $(h_1 = 0 \iff f_1 = 0)$ and the discontinuity of the pressure (hence the lift and moment coefficients) is ensured by the distribution of vortices.

3.1.3 The Fundamental Solutions Method

There is no physical justification for replacing the wing by a continuous distribution of sources and vortices. It is more natural to replace the wing by a continuous distribution of forces and to determine the intensity of these forces, in order to obtain the same action against the fluid like in the case of the wing. Indeed, the fluid and the wing are two interacting systems. If we want to determine the motion of the first subsystem, we have to replace the action of the second subsystem by a distribution of forces. This is the basic idea of the fundamental solutions method.

We shall replace the wing by a continuous distribution of forces $\boldsymbol{f} = (f_1, f)$ on the segment $[+1, -1]$, having the intensity $\boldsymbol{f}$ a priori unknown (the intensity will depend on the point). In fact, for obtaining a plane perturbation, the forces have to be uniformly applied on parallels to the Oz axis, intersecting the xOy plane in the points of the segment $[-1, +1]$. Taking into account the formulas (2.3.23), it will result the following perturbation in the fluid:

$$
\begin{aligned}
p(x,y) &= \frac{1}{2\pi\beta}\int_{-1}^{+1}\frac{x_0 f_1(\xi) + \beta^2 y f(\xi)}{x_0^2+\beta^2y^2}\,\mathrm{d}\,\xi \\
v(x,y) &= \frac{\beta}{2\pi}\int_{-1}^{+1}\frac{x_0 f(\xi) - y f_1(\xi)}{x_0^2+\beta^2y^2}\,\mathrm{d}\,\xi\,.
\end{aligned}
\tag{3.1.17}
$$

In view of imposing the boundary conditions (3.1.2) we have to calculate $\lim\limits_{y\to\pm 0} v(x,y)$ for $|x| < 1$. Denoting by $v(x,\pm 0)$ these limits, we have

$$
v(x,\pm 0) = \frac{\beta}{2\pi}\lim_{y\to\pm 0}\int_{-1}^{+1}\frac{x_0 f(\xi) - y f_1(\xi)}{x_0^2+\beta^2y^2}\,\mathrm{d}\,\xi \tag{3.1.18}
$$

The first integral represents the tangential derivative of a simple layer potential and the second the normal derivative of the same potential.

Although the formulas which give the limits of these derivatives are known, we prefer to calculate them directly.

If we pass to the limit in the first integral making $y = 0$, we cannot simplify with x_0 because for $\xi = x$ it is zero. We shall divide this integral in three parts. Since for a ε small enough, we may approximate $f(\xi)$ on the segment $(x-\varepsilon, x+\varepsilon)$ with $f(x)$, we deduce performing the substitution $\xi - x = u$,

$$
\int_{x-\varepsilon}^{x+\varepsilon}\frac{x_0 f(\xi)}{x_0^2+\beta^2y^2}\mathrm{d}\,\varepsilon \simeq -f(x)\int_{-\varepsilon}^{+\varepsilon}\frac{u\,\mathrm{d}\,u}{u^2+\beta^2y^2} = 0
$$

the integrand being an odd function. It remains

$$\lim_{y\to\pm 0}\int_{-1}^{+1}\frac{x_0 f(\xi)}{x_0^2+\beta^2y^2}\mathrm{d}\,\xi=\lim_{\varepsilon\to 0}\left[\int_{-1}^{x-\varepsilon}\frac{f(\xi)}{x_0}\mathrm{d}\,\xi+\int_{x+\varepsilon}^{1}\frac{f(\xi)}{x_0}\mathrm{d}\,\xi\right]=$$

$$=\int_{-1}^{+1}{}' \frac{f(\xi)}{x-\xi}\mathrm{d}\,\xi\,. \tag{3.1.19}$$

The second limit from (3.1.18) is zero, excepting the interval $(x-\varepsilon, x+\varepsilon)$ where the integrand is infinite for $y=0$ and $\xi=x$. Approximating again $f_1(\xi)$ by $f_1(x)$, or applying the mean formula, since f_1 is integrable and multiplied by an integrable factor having a constant sign, we get:

$$\lim_{y\to\pm 0}\int_{-1}^{+1}\frac{f_1(\xi)}{x_0^2+\beta^2y^2}\mathrm{d}\,\xi=f_1(x)\lim_{y\to\pm 0}y\int_{x-\varepsilon}^{x+\varepsilon}\frac{\mathrm{d}\,\xi}{x_0^2+\beta^2y^2}, \tag{3.1.20}$$

$$f_1(x)\lim_{y\to\pm 0}y\int_{-\varepsilon}^{+\varepsilon}\frac{\mathrm{d}\,u}{u^2+\beta^2y^2}=\pm(\pi/\beta)f_1(x).$$

Hence,

$$v(x,\mp 0)=\pm\frac{1}{2}f_1(x)-\frac{\beta}{2\pi}\int_{-1}^{+1}{}'\frac{f(\xi)}{\xi-x}\mathrm{d}\,\xi.$$

Imposing the boundary conditions (3.1.2), adding and subtracting, we get:

$$\frac{\beta}{\pi}\int_{-1}^{+1}{}'\frac{f(\xi)}{\xi-x}\mathrm{d}\,\xi=-2h'(x), \tag{3.1.21}$$

$$f_1(x)=-2h_1'(x). \tag{3.1.22}$$

We also have

$$p(x,\pm 0)=\pm\frac{1}{2}f(x)+\frac{1}{2\pi\beta}\int_{-1}^{+1}{}'\frac{f_1(\xi)}{x-\xi}\mathrm{d}\,\xi\,, \tag{3.1.23}$$

such that $f(x)$ has the same significance like in (3.1.14). The sntegral equations (3.1.15) and (3.1.21) coincide.

In fact, the representation (3.1.11) may also be obtained from the fundamental solution. This representation is valid for the points which do not belong to the semi-axis Ox $(x>1)$. Indeed in these points we have $u=-p$, such that

$$u-\frac{\mathrm{i}}{\beta}v=\frac{1}{2\pi\mathrm{i}}\int_{-1}^{+1}\frac{-f(\xi)+\mathrm{i}\,\beta^{-1}f_1(\xi)}{\xi-Z}\mathrm{d}\,\xi\,, \tag{3.1.24}$$

where $Z = x + \mathrm{i}\beta y$. We see from (3.1.16) and (3.1.22) that instead of βf_1 from (3.1.11) we put here f_1. The equivalence just established shows that the formula (3.1.11) is not valid on Ox $(x > 1)$. On this half-line the flow is not irrotational.

We mention finally that the proof of Plemelj's formulas (which are necessary for obtaining the integral equation (3.1.15)) is anyway more difficult than the proof of the formulas (3.1.19) and (3.1.20).

3.1.4 The Function $f(x)$. The Complex Velocity in the Fluid

The solution (bounded in the trailing edge) of the integral equations (3.1.15) and (3.1.21) are obtained from the formula (C.1.9). We have therefore:

$$\beta f(x) = \frac{2}{\pi}\sqrt{\frac{1-x}{1+x}}\int_{-1}^{+1}{}' \sqrt{\frac{1+t}{1-t}}\frac{h'(t)}{t-x}\mathrm{d}t. \tag{3.1.25}$$

Obviously, for the symmetric profile $(h = 0)$ it results $f = 0$.

One obtains the complex velocity substituting (3.1.16) and (3.1.25) in (3.1.11). Since, taking into account the formulas (B.5.5) and (B.5.7), we deduce

$$\int_{-1}^{+1}\sqrt{\frac{1-t}{1+t}}\frac{\mathrm{d}t}{(t-Z)(s-t)} =$$

$$= \frac{1}{s-Z}\int_{-1}^{+1}\sqrt{\frac{1-t}{1+t}}\left(\frac{1}{t-Z}+\frac{1}{s-t}\right)\mathrm{d}t = \frac{\pi}{s-Z}\sqrt{\frac{Z-1}{Z+1}},$$

it follows

$$\begin{aligned} W(Z) = &-\frac{1}{\pi\beta}\int_{-1}^{+1}\frac{h_1'(t)}{t-Z}\mathrm{d}t + \\ &+\frac{\mathrm{i}}{\pi\beta}\sqrt{\frac{Z-1}{Z+1}}\int_{-1}^{+1}\sqrt{\frac{1+t}{1-t}}\frac{h'(t)}{t-Z}\mathrm{d}t \end{aligned} \tag{3.1.26}$$

For the incompressible fluid $(\beta = 1)$ one obtains the formula

$$\begin{aligned} u - \mathrm{i}v = &-\frac{1}{\pi}\int_{-1}^{+1}\frac{h_1'(t)}{t-z}\mathrm{d}t + \\ &+\frac{\mathrm{i}}{\pi}\sqrt{\frac{z-1}{z+1}}\int_{-1}^{+1}\sqrt{\frac{1+t}{1-t}}\frac{h'(t)}{t-z}\mathrm{d}t, \end{aligned} \tag{3.1.27}$$

where $z = x + \mathrm{i}y$. The formula (3.1.27) was given for the first time by Sedov [1.38], p.51 and deduced in a different manner by Iacob [1.21], p.664, which solved the problem (3.1.9), (3.1.10) reducing it to a boundary value problem in the complex plane. It is not simpler to solve the boundary value problem than solving the singular integral equation (C.1.1). Moreover, the complex velocity field is not of much interest in aerodynamics. It is utilized only for determining the jump of the pressure on the profile (which is calculated directly in the framework of the method of the integral equations).

3.1.5 The Calculation of the Aerodynamic Action

In some papers the aerodynamic action is calculated by means of a curvilinear integral on the contour of the profile. This calculation is wrong for contours with angular points. A correct calculation is performed using a control contour (surface in the three-dimensional case) surrounding the profile (wing). We shall perform this calculation where it will be absolutely necessary. Here we shall give a simple calculation, observing that in the first approximation, the action of the fluid comes from the jump of the pressure $p_1(x_1,-0) - p_1(x_1,+0) = [\![p_1]\!]$, which gives the lifting force, parallel to the Oy axis. Taking into account the formulas (2.1.1) and (2.1.3) on the unity of length of the cylinder, it follows the lifting force

$$L = \int_{-L_0}^{L_0} [\![p_1]\!]\,\mathrm{d}x_1 = \rho_\infty U_\infty^2 L_0 \int_{-1}^{+1} [\![p]\!]\,\mathrm{d}\,x \tag{3.1.28}$$

and the following momentcalculated with respect to the point $\boldsymbol{x}_1^0$:

$$\boldsymbol{M}_1 = \int_{-L_0}^{+L_0} (\boldsymbol{x}_1 - \boldsymbol{x}_1^0) \times [\![p_1]\!]\boldsymbol{j}\mathrm{d}\,x_1\,,$$

where $\boldsymbol{j}$ represents the versor of the Oy axis. We obtain $\boldsymbol{M}_1 = M\boldsymbol{k}$, where

$$M = \rho_\infty U_\infty^2 L_0^2 \int_{-1}^{+1} (x - x^0)\,[\![p]\!]\,\mathrm{d}\,x \tag{3.1.29}$$

We denoted

$$[\![p]\!] = p(x,-0) - p(x,+0) \tag{3.1.30}$$

Instead of the dimensional quantities L and M, it is preferable to use the dimensionless quantities c_L and c_M, named *the lift coefficient* respectively *the moment coefficient*. These aerodynamic coefficients are defined by the formulas:

$$c_L = \frac{L}{(1/2)\rho_\infty U_\infty^2 (2L_0)}\,, \quad c_M = \frac{M}{(1/2)\rho_\infty U_\infty^2 (2L_0)^2}\,. \tag{3.1.31}$$

It is preferable to use the dimensionless aerodynamic coefficients because the numerical calculations are performed for dimensionless quantities. From (3.1.14), (3.1.28)–(3.1.30), and (3.1.31) it results

$$c_L = -\int_{-1}^{+1} f(x)\mathrm{d}\,x, \quad c_M = -\frac{1}{2}\int_{-1}^{+1} (x - x^0)f(x)\mathrm{d}\,x\,. \tag{3.1.32}$$

Finally, utilizing the solution (3.1.25), (B.5.4) and (B.6.9) we get

$$c_L = -\frac{2}{\beta}\int_{-1}^{+1} \sqrt{\frac{1+t}{1-t}}h'(t)\mathrm{d}\,t\,, \tag{3.1.33}$$

$$c_M = -\frac{1}{\beta}\int_{-1}^{+1} \sqrt{\frac{1+t}{1-t}}th'(t)\mathrm{d}\,t - \frac{1+x^0}{2}c_L\,. \tag{3.1.34}$$

In the case of the profiles which are symmetric with respect to the Ox axis, $(h = 0), c_L$ and c_M vanish.

Obviously we cannot use the method for calculating the drag because it has the order ε^2.

3.1.6 Examples

The flat plate. For the flat plate having the angle of attack ε (fig. 3.1.2a)), the equation (3.1.1) is $y = -xtg\varepsilon \simeq -\varepsilon x$. It follows $h(x) = -\varepsilon x$ whence

$$c_L = \frac{2\pi\varepsilon}{\beta}\,, \quad c_M = -(x^0 + \frac{1}{2})\frac{\pi\varepsilon}{\beta} \tag{3.1.35}$$

These formulas were given for the first time by Glauert (1928) and Prandtl (1930). For the incompressible fluid $(M = 0)$ one obtains

$$c_L = 2\pi\varepsilon\,, \quad c_M = -\left(x^0 + \frac{1}{2}\right)\pi\varepsilon \tag{3.1.36}$$

We notice that c_L is increasing (c_M is decreasing) for $M \nearrow 1$ (fig. 3.1.3) (the lift is increasing because of the compressibility). In the vicinity of $M = 1$ (starting approximately with $M = 0.8$) the lift has very great values, in contradiction with the reality. We deduce that in the vicinity of $M = 1$ the linear theory is not valid any longer. Therefore, for the transonic flow (Chapter 9) we shall utilize other equations. From (3.1.25) one obtains the jump of the pressure

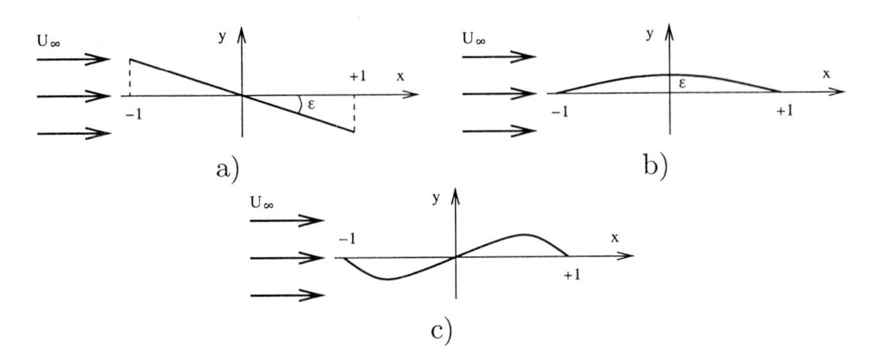

Fig. 3.1.2.

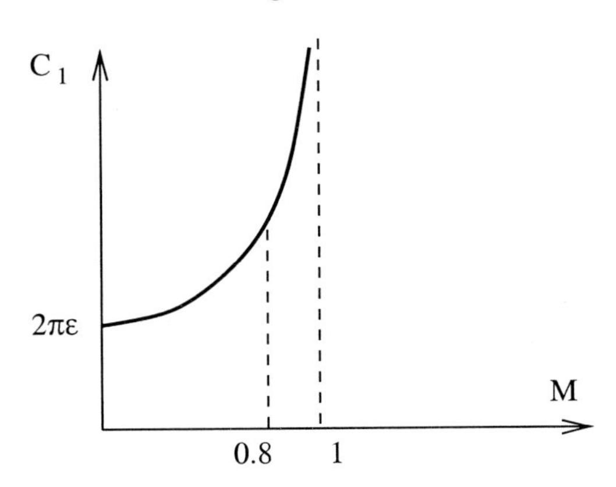

Fig. 3.1.3.

$$\beta f = -2\varepsilon\sqrt{\frac{1-x}{1+x}} \tag{3.1.37}$$

The parabolic profile (fig. 3.1.2b)). This profile is obtained for $h(x) = \varepsilon(1-x^2), h_1 = 0$. Utilizing (B.6.9) we find

$$c_L = \frac{2\pi\varepsilon}{\beta}, \; c_M = -\frac{\pi}{\beta}\varepsilon x^0, \; f = -\frac{4\varepsilon}{\beta}\sqrt{1-x^2} \tag{3.1.38}$$

with the same interpretation for c_L.

The S-profile (fig. 3.1.2c)). For $h(x) = \varepsilon(x - x^3)$ and $h_1(x) = 0$, we obtain

$$c_L = \frac{\pi\varepsilon}{\beta}, \; c_M = \frac{\pi\varepsilon}{4\beta}\left(\frac{1}{4} - x^0\right), \; f = \frac{\varepsilon}{\beta}\sqrt{\frac{1-x}{1+x}} + \frac{6\varepsilon x}{\beta}\sqrt{1-x^2} \tag{3.1.39}$$

The profile with thickness, having the shape of an elliptic sector For a profile having the shape of a sector bounded by two arcs of ellipse having the small semi-axes $\varepsilon_2 < \varepsilon_1$ (fig. 3.1.4), we have

$$\begin{aligned} &h_+(x) = -\varepsilon_2\sqrt{1-x^2}, \quad h_-(x) = -\varepsilon_1\sqrt{1-x^2}, \\ &2h(x) = -(\varepsilon_1+\varepsilon_2)\sqrt{1-x^2}, \quad 2h_1(x) = (\varepsilon_1-\varepsilon_2)\sqrt{1-x^2}. \end{aligned} \tag{3.1.40}$$

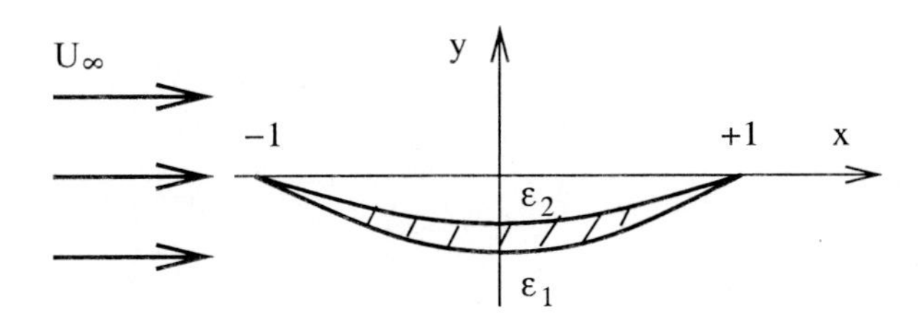

Fig. 3.1.4.

For f_1 and f, taking into account the representation (3.1.16) and (3.1.25) we deduce

$$\beta f_1 = (\varepsilon_1-\varepsilon_2)\frac{x}{\sqrt{1-x^2}}, \quad \beta f = \frac{\varepsilon_1+\varepsilon_2}{\pi}\sqrt{\frac{1-x}{1+x}}\,I \tag{3.1.41}$$

where

$$I = {}'\!\!\int_{-1}^{+1}\sqrt{\frac{1+t}{1-t}}\frac{t}{\sqrt{1-t^2}}\frac{\mathrm{d}t}{t-x} = \frac{x}{1-x}\,{}'\!\!\int_{-1}^{+1}\frac{\mathrm{d}t}{t-x} + \frac{1}{1-x}\,{}^*\!\!\int_{-1}^{+1}\frac{\mathrm{d}t}{1-t}$$

Performing the change of variable $1-t=u$ and taking into account (D.2.3), we get

$${}^*\!\!\int_{-1}^{+1}\frac{\mathrm{d}t}{1-t} = {}^*\!\!\int_{0}^{2}\frac{\mathrm{d}u}{u} = \ln 2. \tag{3.1.42}$$

Utilizing (B.5.8) we get

$$\beta f = \frac{\varepsilon_1+\varepsilon_2}{\pi\sqrt{1-x^2}}\left(\ln 2 + x\ln\frac{1-x}{1+x}\right). \tag{3.1.43}$$

We also have

$$\begin{aligned} c_L &= \frac{\varepsilon_1+\varepsilon_2}{\beta}\left(2 - {}^*\!\!\int_{-1}^{+1}\frac{\mathrm{d}t}{1-t}\right) = \frac{\varepsilon_1+\varepsilon_2}{\beta}(2-\ln 2), \\ c_M &= -\frac{\varepsilon_1+\varepsilon_2}{2\beta}x^0(2-\ln 2). \end{aligned} \tag{3.1.44}$$

The action of the fluid is equivalent to a lifting force passing through the origin.

3.1.7 The General Case

It is well known that the Chebyshev polynomialsof first kind

$$T_n(t) = \cos(n \arccos t) \tag{3.1.45}$$

constitute a basis on the interval $(-1,+1)$. We shall consider the series expansion of $h'(t)$ in this basis:

$$h'(t) = \sum a_n \cos(n \arccos t). \tag{3.1.46}$$

Putting

$$t = \cos\sigma \tag{3.1.47}$$

it follows

$$h'(\cos\sigma) = \sum_{n=0}^{\infty} a_n \cos n\sigma$$

whence

$$a_0 = \frac{1}{\pi}\int_0^{\pi} h'(\cos\theta)\mathrm{d}\,\theta\,,\ a_n = \frac{2}{\pi}\int_0^{\pi} h'(\cos\theta)\cos n\theta \mathrm{d}\,\theta\,, \tag{3.1.48}$$

$$n = 1, 2, \ldots$$

Substituting (3.1.46) in (3.1.33) and (3.1.34) and performing the change of variable (3.1.47), one obtains

$$c_L = -\frac{2\pi}{\beta}\left(a_0 + \frac{a_1}{2}\right),\ c_M = -\frac{\pi}{\beta}\left(a_0 + a_1 + \frac{a_2}{2}\right). \tag{3.1.49}$$

Theses formulas were given by Homentcovschi in [A.20].

It is important to calculate the distribution of the jump of the pressure on the profile. Using the change of variable (3.1.47) and Glauert's formula (B.6.6) we deduce

$$\int_{-1}^{+1}\sqrt{\frac{1+t}{1-t}}\frac{h'(t)}{t-x}\mathrm{d}\,t = \sum_{n=0}^{\infty} a_n \int_0^{\pi}\frac{(1+\cos\sigma)\cos n\sigma}{\cos\sigma - \cos\theta}\mathrm{d}\sigma = \tag{3.1.50}$$

$$= \pi \cot\frac{\theta}{2}\sum_{n=0}^{\infty} a_n \sin n\theta, \quad x = \cos\theta.$$

Replacing in (3.1.25) it follows:

$$\beta f(x) = 2\sum_{n=0}^{\infty} a_n \sin n\theta \tag{3.1.51}$$

It is *Glauert* who proposed this type of solution.

3.1.8 Numerical Integrations

In the case of an arbitrary profile, for calculating the integrals (3.1.33) and (3.1.34) one may employ quadrature formulas (F.2.24). Using the notations

$$t_\alpha = \cos\frac{2\alpha - 1}{2n+1}, \quad \alpha = 1, \ldots, n, \tag{3.1.52}$$

one obtains

$$\begin{aligned} c_L &= -\frac{4\pi}{\beta(2n+1)}\sum_{\alpha=1}^{n}(1+t_\alpha)h'(t_\alpha), \\ c_M &= -\frac{2\pi}{\beta(2n+1)}\sum_{\alpha=1}^{n}(1+t_\alpha)t_\alpha h'(t_\alpha) - \frac{1+x^0}{2}c_L. \end{aligned} \tag{3.1.53}$$

For calculating the integral from (3.1.25) one utilizes the formula (F.3.1). In the collocation points

$$x_i = \cos\frac{2i\pi}{2n+1}, \quad i = 1, \ldots, n, \tag{3.1.54}$$

it results

$$f(x_i) = \frac{4}{\beta(2n+1)}\sqrt{\frac{1-x_i}{1+x_i}}\sum_{\alpha=1}^{n}\frac{1+t_\alpha}{t_\alpha - x_j}h'(t_\alpha). \tag{3.1.55}$$

The results obtained with these formulas are much more accurate than the results obtained with other methods.

3.1.9 The Integration of the Thin Airfoil Equation with the Aid of Gauss-type Quadrature Formulas

We may use the quadrature formulas from Appendix F to create and extremely efficient method for determining the solution of the integral equation (3.1.15). There were made many such attempts in various papers but nowhere one can find the good solution because there is not prescribed the behaviour of the solution in the points ± 1. We know from Appendix C that the solution of the above mentioned equation depends on the behaviour imposed in ± 1. The solution satisfying the Kutta-Joukovsky condition in the trailing edge is

$$\beta f(t) = -\varepsilon\sqrt{\frac{1-t}{1+t}}F(t) \tag{3.1.56}$$

Using (F.2.19) and posing $h = -\varepsilon\overline{h}$, we reduce the equation (3.1.15) to the algebraic system

$$\sum_{\alpha=1}^{n} A_{j\alpha}F_\alpha = \overline{h}'_j, \quad j = 1,\ldots,n, \tag{3.1.57}$$

where $F_\alpha = F(t_\alpha)$, $\overline{h}'_j = \overline{h}'(x_j), A_{j\alpha} = \dfrac{1}{2n+1}\dfrac{t_\alpha - 1}{t_\alpha - x_j}$,

$$t_\alpha = \cos\frac{2\alpha\pi}{2n+1}, \quad x_j = \cos\frac{2j-1}{2n+1}\pi. \tag{3.1.58}$$

In (3.1.57) we have a linear algebraic system with n equations for n unknowns $F_1,\ldots,F_n$. We can create a computer code for solving the system. The coefficients $A_{j\alpha}$, the weight points t_α and the collocation points x_j are the same for every profile. For a given profiles we have to change only the column with the elements $\overline{h}'_1,\ldots,\overline{h}'_n$. After determining the unknowns F_α, the lift and moment coefficients result from the formulas (3.1.32) and (F.2.18).

$$\begin{aligned} c_L &= -\int_{-1}^{+1} f(t)\mathrm{d}t = \frac{\varepsilon}{\beta}\int_{-1}^{+1}\sqrt{\frac{1-t}{1+t}}F(t)\mathrm{d}t = \frac{2\pi\varepsilon}{\beta}N_L, \\ c_M &= -\frac{1}{2}\int_{-1}^{+1} tf(t)\mathrm{d}t = -\frac{\pi\varepsilon}{2\beta}N_M, \end{aligned} \tag{3.1.59}$$

where

$$N_L = \frac{1}{2n+1}\sum_{\alpha=1}^{n}(1-t_\alpha)F_\alpha, \quad N_M = \frac{2}{2n+1}\sum_{\alpha=1}^{n} t_\alpha(t_\alpha - 1)F_\alpha. \tag{3.1.60}$$

For verifying the method we utilized the analytic solution for the flat plate (3.1.37). From (3.1.56), it follows that for $\overline{h}'_j = 1$ we must have $F = 2$. The results obtained with the numerical method described above, with $n = 20$ gave for F_α values situated between 0.999 and 2.001 and for N_L and N_M the value 0.999. Hence the method is extremely efficient [3.12].

3.2 The Airfoil in Ground Effects

3.2.1 The Integral Equation

When an airplane is landing or taking off we have to take into account the ground effects. Some of the first papers devoted to these effects belong to Tomotika and his fellow-workers [3.44], [3.45]. We have also to mention the papers of Pancenkov [1.33], [1.34], [3.36], and especially the papers of Plotkin and his fellow-workers [3.37]–[3.39], where one gives the integral equation of thin profiles in ground effects and one proposes approximate solutions. The is considered a small parameter. Widnall and Barows [5.37] and Tuck [3.46] used asymptotic methods for investigating the problem. In fact one encountered two small parameters: the arrow of the airfoil and the distance from the airfoil to the ground. The fluid was considered incompressible.

In the sequel, following [3.13], we shall utilize the method of fundamental solutions, in order to obtain the integral equation for the compressible fluid and for the airfoil with thickness. The small parameter is the arrow of the airfoil. In case that the distance from the airfoil to the ground is also small, we have to elaborate a new theory.

We use the notations from the previous subsection. We denote by $a/2$ the distance from the airfoil to the ground (fig. 3.2.1). The perturbation has to satisfy the following boundary conditions:

$$v(x, \pm 0) = h(x) \pm h_1(x), \quad |x| < 1\,, \tag{3.2.1}$$

$$v(x, -a/2) = 0 \qquad -\infty < x < \infty\,. \tag{3.2.2}$$

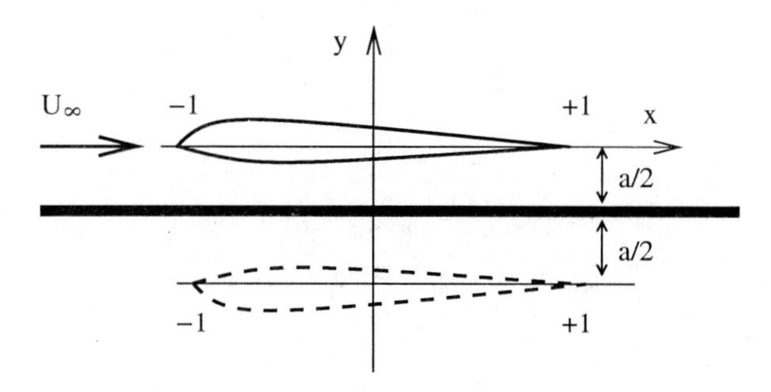

Fig. 3.2.1.

According to the method of fundamental solutions, we have to replace the airfoil by a continuous distribution of forces $(f_1, f)(\xi)$ on the segment $[-1, +1]$ from the $y = 0$ axis. For satisfying the boundary condition (3.2.2) we shall consider a symmetric distribution of forces $(f_1, -f)(\xi)$ on the symmetric segment $[-1, +1]$ of the $y = -a$ axis and we shall determine the intensity of the distributions from the condition (3.2.1). Taking into account that the perturbation produced within the uniform stream having the velocity $U_\infty \boldsymbol{i}$ by the force (f_1, f_2) is (2.3.23), it follows that the two distributions will determine in the fluid the perturbation

$$
\begin{aligned}
p(x,y) &= \frac{1}{2\pi\beta}\int_{-1}^{+1}\frac{x_0 f_1(\xi)+\beta^2 y f(\xi)}{x_0^2+\beta^2 y^2}\,\mathrm{d}\,\xi + \\
&+\frac{1}{2\pi\beta}\int_{-1}^{+1}\frac{x_0 f_1(\xi)-\beta^2 (y+a) f(\xi)}{x_0^2+\beta^2 (y+a)^2}\,\mathrm{d}\,\xi \\
v(x,y) &= \frac{\beta}{2\pi}\int_{-1}^{+1}\frac{x_0 f(\xi)-y f_1(\xi)}{x_0^2+\beta^2 y^2}\,\mathrm{d}\,\xi - \\
&-\frac{\beta}{2\pi}\int_{-1}^{+1}\frac{x_0 f(\xi)+(y+a) f_1(\xi)}{x_0^2+\beta^2 (y+a)^2}\,\mathrm{d}\,\xi\,,
\end{aligned}
\tag{3.2.3}
$$

f and f_1 having to be determined. One easily verifies that this representation satisfies the condition (3.2.2). We denoted $x_0 = x - \xi$.

Only the first integrals become singular when passing to limit for $y \to \pm 0$, Using the formulas (3.1.19) and (3.1.20), we deduce

$$
p(x,+0) - p(x,-0) = f(x)\,, \tag{3.2.4}
$$

$$
\begin{aligned}
v(x,\pm 0) &= \mp\frac{1}{2} f_1(x) + \frac{\beta}{2\pi}\int_{-1}^{+1}{}' \frac{f(\xi)}{x-\xi}\,\mathrm{d}\,\xi - \\
&-\frac{\beta}{2\pi}\int_{-1}^{+1}\frac{x_0 f(\xi)+a f_1(\xi)}{x_0^2+m^2}\,\mathrm{d}\,\xi\,,
\end{aligned}
\tag{3.2.5}
$$

where $m = \beta a$. Imposing the conditions (3.2.1), subtracting and adding, we get:

$$
f_1(x) = -2h_1'(x) \qquad |x|<1 \tag{3.2.6}
$$

$$
\frac{\beta}{2\pi}\int_{-1}^{+1}{}' \frac{f(\xi)}{x_0} - \frac{\beta}{2\pi}\int_{-1}^{+1}\frac{x_0 f(\xi)}{x_0^2+m^2}\,\mathrm{d}\,\xi = H(x)\,, \tag{3.2.7}
$$

where

$$
H(x) = h'(x) - \frac{m}{\pi}\int_{-1}^{+1}\frac{h_1'(\xi)}{x_0^2+m^2}\,\mathrm{d}\,\xi \tag{3.2.8}
$$

The equation (3.2.7) is the integral equation In Appendix C it was called the generalized equation of the thin airfoil. For the incompressible fluid ($\beta = 1$) and for the airfoil without thickness ($h_1 = 0$) it coincides with the equation given by Pancenkov and Plotkin.

3.2.2 A Numerical Method

The equation (3.2.7) is obviously a singular integral equation. As we have shown in Appendix C, it may be reduced to a Fredholm equation, but the problem remains still unsolved because there are not available general methods for solving this type of equations (excepting the method of successive approximations). As we have already shown in [3.13], the equation (3.2.7) may be solved numerically utilizing the quadrature formulas from Appendix F. Looking for the solutions of (3.2.7) having the form (3.1.56), putting $H = -\varepsilon\overline{H}$ and using (F.2.18) and (F.2.19), one obtains the linear algebraic system

$$\sum_{\alpha=1}^{n}(A_{j\alpha} + B_{j\alpha})F_\alpha = \overline{H}_j\,, \qquad j = 1,\ldots,n\,, \tag{3.2.9}$$

where $F_\alpha = F(t_\alpha)$, $\overline{H}_j = \overline{H}(x_j)$,

$$B_{j\alpha} = \frac{1}{2n+1}\frac{(t_\alpha - 1)(x_j - t_\alpha)}{(x_j - t_\alpha)^2 + m^2}\,, \tag{3.2.10}$$

$A_{j\alpha}, t_\alpha$ and x_j being given in (3.1.58).

From the system (3.2.9) we determine the unknowns $F_1,\ldots,F_n$. Now the lift and moment coefficients (3.1.32) (for $x^0 = 0$) will be obtained from the formulas

$$\begin{aligned} c_L &= \frac{\varepsilon}{\beta}\int_{-1}^{+1}\sqrt{\frac{1-t}{1+t}}F(t)\mathrm{d}\,t = \frac{2\pi\varepsilon}{\beta}N_L\,, \\ c_M &= \frac{\varepsilon}{2\beta}\int_{-1}^{+1}\sqrt{\frac{1-t}{1+t}}tF(t)\mathrm{d}\,t = -\frac{\pi\varepsilon}{2\beta}N_M\,, \end{aligned} \tag{3.2.11}$$

where we have utilized the notations (3.1.60).

One may write computer programs for solving the system (3.2.9). The coefficients $A_{j\alpha}$ and $B_{j\alpha}$ do not depend on the shape of the airfoil, such that the program may be utilized for every airfoil. One changes only the matrix with one column $\overline{H}_j$.

3.2.3 The Flat Plate

In the case of the flat plate with the angle of attack ε, we have $\overline{H}_j = 1$. We solve the system (3.2.9) and then we determine N_L and

N_M with the formulas (3.1.60). We write the aerodynamic coefficients (3.2.11) as follows

$$c_L = c_L^{\infty} N_L, \quad c_M = c_M^{\infty} N_M \tag{3.2.12}$$

where c_L^{∞} and c_M^{∞} are the coefficients for the free stream $(a = \infty)$.

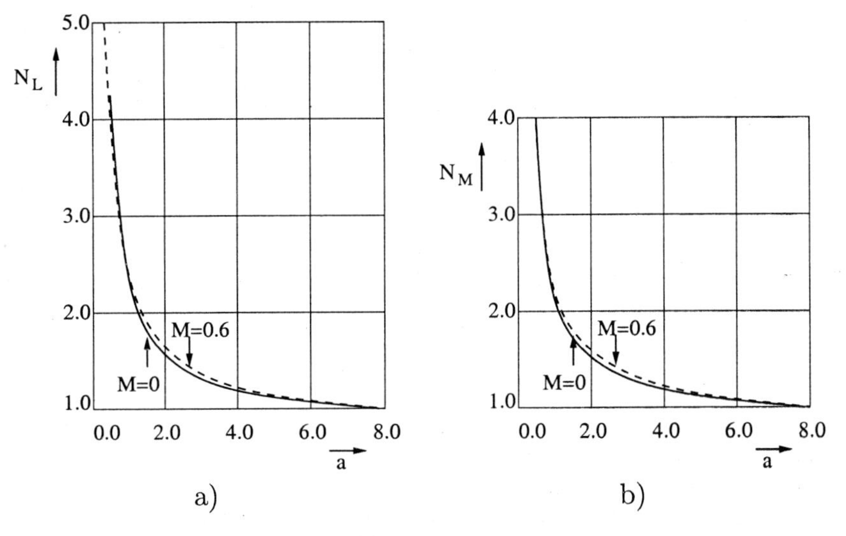

Fig. 3.2.2.

The variation of N_L and N_M versus a is shown in figures 3.2.2a) and 3.2.2b). The compressibility effects on the lift and moment are greater than it is shown in the above mentioned figures, because β also intervenes in c_L^{∞} and c_M^{∞}. Obviously, the ground effect is great when a is small and it quickly decreases when a increases.

3.2.4 The Symmetric Airfoil

As we have seen in (3.1.5), for the symmetric airfoil in a free stream with zero angle of attack, the lift and the moment vanish. The situation is different for the wing in ground effects. As we may observe from (3.2.7) and (3.2.8), in the case of symmetric airfoil $(h = 0, h_1 \neq 0)$, H does not vanish, hence the solution of the equation (3.2.7) is different from zero. For the Joukovsky symmetric profile considered in [1.28], [3.37], [3.13]

$$y = \pm\varepsilon(1 - x)\sqrt{1 - x^2} \tag{3.2.13}$$

we obtain using the notation $\overline{h}_1 = -(1-x)\sqrt{1-x^2}$ and the formulas (F.2.12) and (F.2.18):

$$\overline{H}_j = -\frac{m}{\pi}\int_{-1}^{+1} \frac{h_1'(\xi)}{(x_j-\xi)^2+m^2}\mathrm{d}\,\xi =$$

$$= -\frac{m}{p+1}\sum_{\gamma=1}^{p}\frac{1-\zeta_\gamma^2}{(x_j-\zeta_\gamma)^2+m^2} - \frac{2m}{2n+1}\sum_{\alpha=1}^{n}\frac{t_\alpha(1-t_\alpha)}{(x_j-t_\alpha)^2)+m^2}, \tag{3.2.14}$$

where $\zeta_\gamma = \cos\dfrac{\gamma\pi}{p+1}$. In this case one presents N_L and N_M versus a in figures 3.2.3a) and 3.2.3b) for $n = p = 20$. The lift coefficient is negative i.e. the resultant is a force pointing towards the ground. Hence when the airplane is landing or taking off it becomes heavier. The pilot has to take into account this fact.

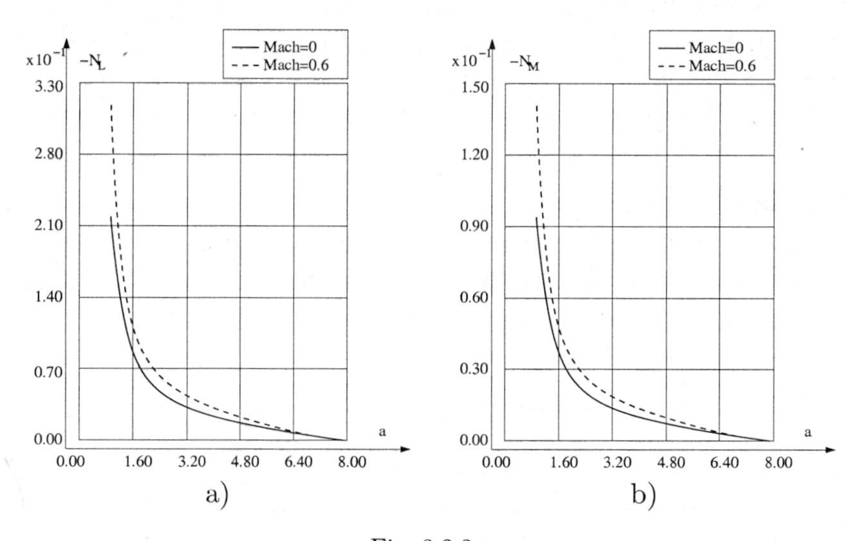

Fig. 3.2.3.

The fact that both in the case of the flat plate and the case of the symmetric profile, the lift and the moment become very great when a is very small, is not true in reality. Hence, for small values of the parameter a we have to elaborate a new theory based on two small parameters (see [5.37]).

3.3 The Airfoil in Tunnel Effects

3.3.1 The Integral Equation

The experiments for determining the aerodynamic parameters are performed in wind tunnels. We have therefore to take into account the influence of the walls of the tunnel on the aerodynamic characteristics of the airfoil. Since many papers dedicated to this subject are secret, we cannot give the history of the research in this field. We have cited in the bibliography some authors, without consulting their papers. We shall present therefore, only the model that we gave in [3.14]. This model can be easily obtained with the method of fundamental solutions and it is in the spirit of the theory previously presented in this book.

We formulate the problem as follows: an uniform stream, having the velocity $U_\infty \boldsymbol{i}$, the pressure p_∞ and the density ρ_∞, flowing between two infinite flat plates parallel to the Ox axis, encounters a thin airfoil of infinite span with the generatrices parallel to the Oz axis. The fluid is compressible, and the velocity of the uniform stream is subsonic. One requires to determine the perturbation and the influence of the stream on the airfoil. We utilize the variables (2.1.1) and (2.1.3). Let (3.1.1) represent the equations and a the distance between the plates (walls) (fig. 3.2.1). For determining the perturbation, we have to impose the following boundary condition:

$$v(x, \pm 0) = h'(x) \pm h_1'(x), |x| < 1 \tag{3.3.1}$$

$$v(x, \pm a/2) = 0, -\infty < x < \infty \tag{3.3.2}$$

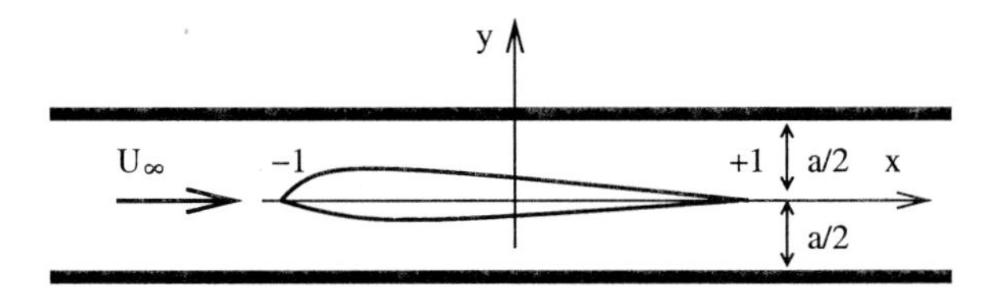

Fig. 3.3.1.

In order to utilize the method of fundamental solutions, we shall replace the airfoil by a continuous distribution of forces $(f_1, f)(\xi)$ defined on the segment $[-1, +1]$. For satisfying the boundary conditions (3.3.2), we have also to take into account symmetric distributions on the images of the strip $[-1, +1]$ in the planes $y = \pm a/2$ and symmetric distributions on the images of the images in the planes $y = \pm 3a/2$ etc.

(the method of images). In this way one obtains the following general representation of the perturbation:

$$p(x,y) = \frac{1}{2\pi\beta}\sum_{-\infty}^{+\infty}\int_{-1}^{+1}\frac{x_0 f_1(\xi) + (-1)^n\beta^2(y-na)f(\xi)}{x_0^2+\beta^2(y-na)^2}\mathrm{d}\,\xi\,,$$

$$v(x,y) = \frac{\beta}{2\pi}\sum_{-\infty}^{+\infty}\int_{-1}^{+1}\frac{(-1)^n x_0 f(\xi) - (y-na)f_1(\xi)}{x_0^2+\beta^2(y-na)^2}\mathrm{d}\,\xi\,, \tag{3.3.3}$$

with $x_0 = x-\xi$. We can easily verify that v given above satisfies (3.3.2). For imposing the conditions (3.3.1) we have to pass to the limit considering $y \to \pm 0$, $-1 < x < 1$. The only singular integrals correspond to $n = 0$ and they are calculated using the formulas (3.1.19) and (3.1.20). Taking into account the equalities [1.16]:

$$\sum_{n=1}^{\infty}\frac{1}{k^2+n^2} = \frac{\pi}{2k}\frac{\cosh k\pi}{\sinh k\pi} - \frac{1}{2k^2}\,,$$

$$\sum_{n=1}^{\infty}\frac{(-1)^n}{k^2+n^2} = \frac{\pi}{2k}\frac{1}{\sinh k\pi} - \frac{1}{2k^2}\,,$$

we deduce:

$$p(x,\pm 0) = \pm\frac{1}{2}f(x) + \frac{1}{2\pi\beta}{\int_{-1}^{+1}}' \frac{f_1(\xi)}{x_0}\mathrm{d}\,\xi + \frac{1}{2\pi\beta}\int_{-1}^{+1}K_1(x_0)f_1(\xi)\mathrm{d}\,\xi\,, \tag{3.3.4}$$

$$v(x,\pm 0) = \mp\frac{1}{2}f_1(x) + \frac{\beta}{2\pi}{\int_{-1}^{+1}}' \frac{f(\xi)}{x_0}\mathrm{d}\,\xi + \frac{\beta}{2\pi}\int_{-1}^{+1}K(x_0)f(\xi)\mathrm{d}\,\xi\,, \tag{3.3.5}$$

where $m = a\beta$,

$$K_1(x_0) = \frac{\pi}{m}\coth\left(\frac{\pi}{m}x_0\right) - \frac{1}{x_0}\,,\quad K(x_0) = \frac{\pi}{m}\sinh^{-1}\left(\frac{\pi}{m}x_0\right) - \frac{1}{x_0}\,. \tag{3.3.6}$$

From (3.3.4) we deduce again the significance of the function f:

$$f(x) = p(x,+0) - p(x,-0)\,, \tag{3.3.7}$$

and from (3.3.5) and (3.3.1),

$$f_1(x) = -2h_1'(x)\,, \tag{3.3.8}$$

$$\frac{\beta}{2\pi}{\int_{-1}^{+1}}' \frac{f(\xi)}{x_0}\mathrm{d}\,\xi + \frac{\beta}{2\pi}\int_{-1}^{+1} K(x_0)f(\xi)\mathrm{d}\,\xi = h'(x)\,. \tag{3.3.9}$$

The formula (3.3.8) determines f_1 and the equation (3.3.9), the function f. The integral equation of the problem is in fact the generalized equation of the thin profiles. The kernel K has no singularity for $\xi = x$. We notice that for $a \to \infty$ $(m \to \infty)$ the equations (3.3.4) and (3.3.9) are reduced to the equations corresponding to the airfoil in a free (unlimited) airfoil.

3.3.2 The Integration of the Equation (3.3.9)

We utilize the quadrature formulas from Appendix F for integrating the equation (3.3.9). Putting $h = -\varepsilon\overline{h}$, we shall look for solutions f having the form (3.1.56). Utilizing (F.2.18) and (F.2.19), we reduce the equation (3.3.9) to the system

$$\sum_{\alpha=1}^{n} C_{j\alpha}F_\alpha = \overline{h}_j'\,, \qquad j = 1,\ldots,n\,, \tag{3.3.10}$$

where $F_\alpha = F(t_\alpha)$, $\overline{h}_j' = \overline{h}(x_j)$,

$$C_{j\alpha} = \frac{\pi}{m}\frac{1-t_\alpha}{2n+1}\sinh^{-1}\left[\frac{\pi}{m}(x_j - t_\alpha)\right], \tag{3.3.11}$$

t_α şi x_j being given in (3.1.58). After determining the unknowns $F_1,\ldots,F_n$ from (3.3.10), the lift and moment coefficients may be obtained by means of the formulas:

$$c_L = \frac{2\pi\varepsilon}{\beta}N_L\,, \quad c_M = -\frac{\pi\varepsilon}{2\beta}N_M\,, \tag{3.3.12}$$

N_L and N_M having the expressions given in (3.1.60). The distribution of the pressure on the airfoil may be obtained from (3.3.4), taking into account (3.3.8). We get

$$p(t_\alpha, \mp 0) = \pm\frac{\varepsilon}{2\beta}F_\alpha\sqrt{\frac{1-t_\alpha}{1+t_\alpha}} - \frac{1}{\pi\beta}{\int_{-1}^{+1}}' \frac{h_1'(x)}{t_\alpha - x}\mathrm{d}\,x -$$

$$-\frac{1}{\pi\beta}\int_{-1}^{+1} K_1(t_\alpha - x)h_1'(x)\mathrm{d}\,x\,.$$

The values of the pressure in the points t_α may be determined in the wind tunnel by means of pressure plugs. We have thus the possibility to verify the theory presented herein.

We may use a computer for solving the system (3.3.10). Since the coefficients $C_{j\alpha}$ do not depend on the shape of the airfoil, we have to change in the program only the column containing the elements $\overline{h}_j'$.

3.3.3 Numerical Results

For the flat plate having the angle of attack ε we have $\overline{h}_j' = 1$. In this case, the formulas (3.3.12) become $c_L = c_L^\infty N_L$, $c_M = c_M^\infty N_M$, c_L^∞ and c_M^∞ representing the coefficients corresponding to the unlimited fluid.

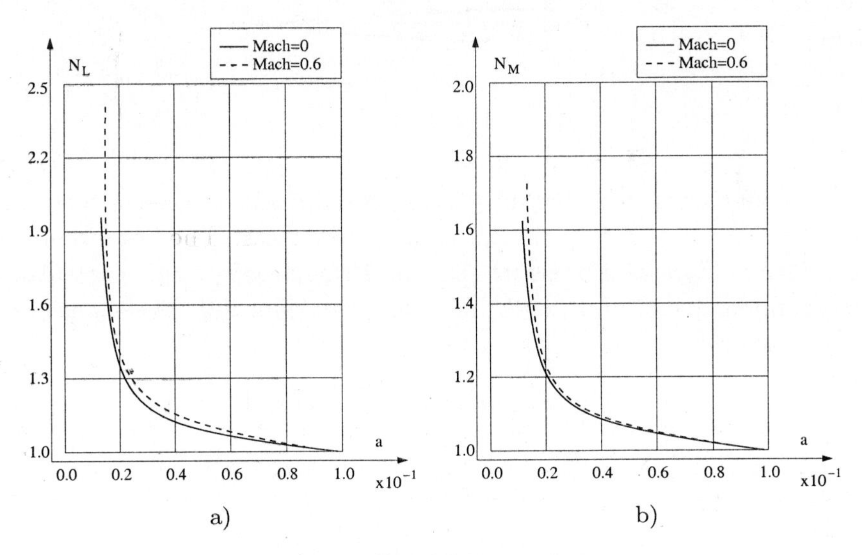

Fig. 3.3.2.

The figures 3.3.2 present the variations of N_L and N_M versus the width of the tunnel for $M = 0$ (incompressible fluid) and $M = 0.6$. We notice that the lift in the wind tunnel is greater then the lift in the unlimited fluid and it decreases when the width of the tunnel increases. For the same width, the lift is an increasing function of M. The theory presented herein allows us to determine these variations.

3.4 Airfoils Parallel to the Undisturbed Stream

3.4.1 The Integral Equations

In the literature, the problem of the uniform flow past a configuration of airfoils has been solved for the incompressible fluid and for airfoils whose exterior may be mapped conformally on the exterior of a circle. There were given especially solutions for the biplane [1.36], [3.38]. We shall consider in the three forthcoming subsections, following the paper [3.15], the general problem of the compressible fluid in subsonic flow and airfoils with thickness of arbitrary shape.

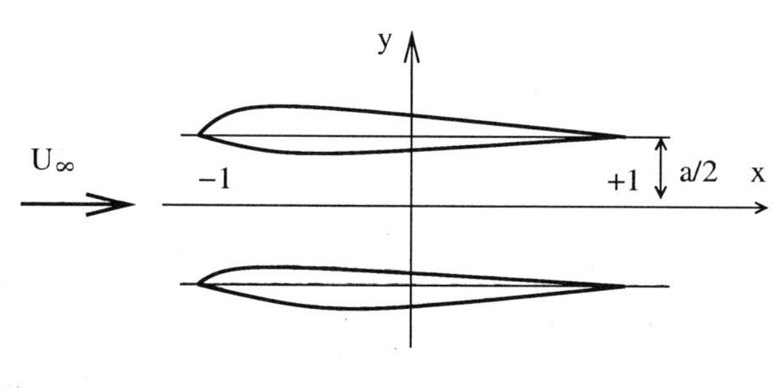

Fig. 3.4.1.

We shall consider in this subsection, two airfoils parallel to the undisturbed stream (fig. 3.4.1), not necessarily identical. The free stream has the velocity $U_\infty \boldsymbol{i}$, the pressure p_∞ and the density ρ_∞. One utilizes the variables (2.1.1) and (2.1.3) and one considers that the equations of the airfoils are:

$$y = h(x) \pm h_1(x), \qquad |x| < 1 \tag{3.4.1}$$

$$y = l(x) \pm l_1(x), \qquad |x| < 1 \tag{3.4.2}$$

the first corresponding to the upper airfoil and the second to the lower airfoil. The dimensionless distance between the projections of the airfoils is denoted by a. Utilizing the method of fundamental solutions, we shall replace the upper airfoil by a distribution of forces of intensity $(f_1, f)(\xi)$ and the lower airfoil by a distribution $(g_1, g)(\xi)$, both of them defined on the segment $[-1, +1]$. Starting from the above mentioned perturbation produced by a force in the free stream (the fundamental solution (2.3.23)), we deduce that the perturbation produced by the two distributions is:

$$
\begin{aligned}
p(x,y) &= \frac{1}{2\pi\beta}\int_{-1}^{+1}\frac{x_0 f_1(\xi)+\beta^2(y-a/2)f(\xi)}{x_0^2+\beta^2(y-a/2)^2}\,\mathrm{d}\,\xi+ \\
&+\frac{1}{2\pi\beta}\int_{-1}^{+1}\frac{x_0 g_1(\xi)+\beta^2(y+a/2)g(\xi)}{x_0^2+\beta^2(y+a/2)^2}\,\mathrm{d}\,\xi\,, \\
v(x,y) &= \frac{\beta}{2\pi}\int_{-1}^{+1}\frac{x_0 f(\xi)-(y-a/2)f_1(\xi)}{x_0^2+\beta^2(y-a/2)^2}\,\mathrm{d}\,\xi+ \\
&+\frac{\beta}{2\pi}\int_{-1}^{+1}\frac{x_0 g(\xi)-(y+a/2)g_1(\xi)}{x_0^2+\beta^2(y+a/2)^2}\,\mathrm{d}\,\xi\,,
\end{aligned}
\tag{3.4.3}
$$

where $x_0 = x - \xi$. The distributions of forces will be determined from the boundary conditions

$$v(x,(a/2)\pm 0) = h'(x) \pm h_1'(x), \quad |x|<1 \tag{3.4.4}$$

$$v(x,-(a/2)\pm 0) = l'(x) \pm l_1'(x), \quad |x|<1 \tag{3.4.5}$$

When $y \to (a/2)\pm 0)$, the first integrals from the representation (3.4.3) become singular and they have to be calculated with the formulas (3.1.19) and (3.1.20). One obtains

$$
\begin{aligned}
p(x,(a/2)\pm 0) &= \pm\frac{1}{2}f(x)+\frac{1}{2\pi\beta}{\int_{-1}^{+1}}'\frac{f_1(\xi)}{x_0}\,\mathrm{d}\,\xi+ \\
&+\frac{1}{2\pi\beta}\int_{-1}^{+1}\frac{x_0 g_1+m\beta g}{x_0^2+m^2}\,\mathrm{d}\,\xi\,,
\end{aligned}
\tag{3.4.6}
$$

$$
\begin{aligned}
v(x,(a/2)\pm 0) &= \mp\frac{1}{2}f_1(x)+\frac{\beta}{2\pi}{\int_{-1}^{+1}}'\frac{f(\xi)}{x_0}\,\mathrm{d}\,\xi+ \\
&+\frac{1}{2\pi}\int_{-1}^{+1}\frac{\beta x_0 g-m g_1}{x_0^2+m^2}\,\mathrm{d}\,\xi\,,
\end{aligned}
\tag{3.4.7}
$$

where $m = a\beta$. From (3.4.6) it follows

$$f(x) = p(x,(a/2)+0) - p(x,(a/2)-0)\,, \tag{3.4.8}$$

and from (3.4.4) and (3.4.7)

$$f_1(x) = -2h_1'(x) \tag{3.4.9}$$

$$\frac{\beta}{2\pi}\int_{-1}^{+1}{}' \frac{f(\xi)}{x_0}\mathrm{d}\,\xi + \frac{\beta}{2\pi}\int_{-1}^{+1} \frac{x_0 g(\xi)}{x_0^2+m^2}\mathrm{d}\,\xi = H(x)\,, \tag{3.4.10}$$

where

$$H(x) = h'(x) + \frac{m}{2\pi}\int_{-1}^{+1} \frac{g_1(\xi)}{x_0^2+m^2}\mathrm{d}\,\xi\,. \tag{3.4.11}$$

In the same way, passing to the limit in (3.4.3) when $y \to -(a/2)\pm 0$ (in this case the second integrals become singular), from the boundary condition (3.4.5), we obtain

$$g(x) = p(x, -(a/2)+0) - p(x, -(a/2)-0)\,, \tag{3.4.12}$$

$$g_1(x) = -2l_1'(x)\,, \tag{3.4.13}$$

$$\frac{\beta}{2\pi}\int_{-1}^{+1}{}' \frac{g(\xi)}{x_0}\mathrm{d}\,\xi + \frac{\beta}{2\pi}\int_{-1}^{+1} \frac{x_0 f(\xi)}{x_0^2+m^2}\mathrm{d}\,\xi = L(x)\,, \tag{3.4.14}$$

where

$$L(x) = l'(x) - \frac{m}{2\pi}\int_{-1}^{+1} \frac{f_1(\xi)}{x_0^2+m^2}\mathrm{d}\,\xi\,. \tag{3.4.15}$$

The distributions of forces on the two chords are determined by solving the system of integral singular equations (3.4.10) and (3.4.14). From (3.4.9) and (3.4.13) the functions $H(x)$ and $L(x)$ are known. The field of pressure on the profiles are given by the formula (3.4.6) and the corresponding formula for $p(x,-(a/2)\pm 0)$.

Symmetric Airfoils. If the two airfoils are symmetric with respect to the Ox axis, then

$$l(x) = -h(x), \quad l_1(x) = h_1(x)\,. \tag{3.4.16}$$

It follows $L(x) = -H(x)$ whence

$$\int_{-1}^{+1} \frac{f+g}{x_0}\mathrm{d}\,\xi + \int_{-1}^{+1} \frac{x_0(f+g)}{x_0^2+m^2}\mathrm{d}\,\xi = 0\,. \tag{3.4.17}$$

Since the solution of the generalized equation of thin profiles is unique, it results $f = -g$, and the system of equations (3.4.10) and (3.4.14) reduces to a single equation

$$\frac{\beta}{2\pi}\int_{-1}^{+1}{}' \frac{f(\xi)}{x_0}\mathrm{d}\,\xi - \frac{\beta}{2\pi}\int_{-1}^{+1} \frac{x_0 f(\xi)}{x_0^2+m^2}\mathrm{d}\,\xi = H(x)\,, \tag{3.4.18}$$

which is just the equation (3.4.7) of the airfoil in ground effects. The result is natural. The lift and moment coefficients for the upper airfoil are given by the formulas (3.2.11). For the lower airfoil they have the opposite sign. For the entire configuration we have $c_L = c_M = 0$.

Identical Airfoils. If the airfoils are identical, then

$$l(x) = h(x) - 2a\,, \quad l_1(x) = h_1(x)\,. \tag{3.4.19}$$

Moreover, if the airfoils have no thickness, then $L = H$. Subtracting the equations (3.4.10) and (3.4.14), it results $g = f$ whence

$$\frac{\beta}{2\pi}\,{'}\!\!\int_{-1}^{+1} \frac{f(\xi)}{x_0}\mathrm{d}\,\xi + \frac{\beta}{2\pi}\int_{-1}^{+1} \frac{x_0 f(\xi)}{x_0^2+m^2}\mathrm{d}\,\xi = h'(x)\,. \tag{3.4.20}$$

This equation has the form of the generalized equation of thin profiles and it can be integrated numerically like in (3.2.2) and (3.3.2).

3.4.2 The Numerical Integration

In order to solve numerically the system (3.4.10) and (3.4.14), we shall use the quadrature formulas from Appendix F. For thin airfoils, the functions $H(x)$ and $L(x)$ have the form $H(x) = -\varepsilon\overline{H}(x), L(x) =$ $= -\varepsilon\overline{L}(x)$, hence we shall look for the following type of solutions

$$\beta f(t) = -\varepsilon\sqrt{\frac{1-t}{1+t}}F(t)\,,\ \beta g(t) = -\varepsilon\sqrt{\frac{1-t}{1+t}}G(t) \tag{3.4.21}$$

which satisfy the Kutta-Joukowsky condition on the trailing edge. Utilizing (F.2.18) and (F.2.19), the system (3.4.10) and (3.4.14) is reduced to

$$\begin{aligned} &\sum_{\alpha=1}^{n}(A_{j\alpha}F_\alpha - B_{j\alpha}G_\alpha) = \overline{H}_j\,,\ \ j = 1,\ldots,n\,,\\ &\sum_{\alpha=1}^{n}(A_{j\alpha}G_\alpha - B_{j\alpha}F_\alpha) = \overline{L}_j\,,\ \ j = 1,\ldots,n\,. \end{aligned} \tag{3.4.22}$$

$A_{j\alpha}, t_\alpha$ and x_j are given in (3.1.58) and $B_{j\alpha}$ in (3.2.10). Like always, $F_\alpha = F(t_\alpha)$, $G_\alpha = G(t_\alpha)$, $\overline{H}_j = \overline{H}(x_j)$, $\overline{L}_j = L(x_j)$. We have to find out the unknowns $F_1,\ldots,F_n, G_1,\ldots,G_n$ from the system (3.4.22). The coefficients $A_{j\alpha}$ and $B_{j\alpha}$ are the same for all the airfoils; only $\overline{H}$ and $\overline{L}_j$ are varying.

The lift and moment coefficients for the entire configuration are given by the formulas

$$c_L = -\int_{-1}^{+1}(f+g)\mathrm{d}\,t = \frac{\varepsilon}{\beta}\int_{-1}^{+1}\sqrt{\frac{1-t}{1+t}}(F+G)\mathrm{d}\,t = \frac{2\pi\varepsilon}{\beta}N_1$$

$$c_M = -\frac{1}{2}\int_{-1}^{+1}t(f+g)\mathrm{d}\,t = \frac{\varepsilon}{2\beta}\int_{-1}^{+1}t\sqrt{\frac{1-t}{1+t}}(F+G)\mathrm{d}\,t = -\frac{\pi\varepsilon}{2\beta}N_2\,, \tag{3.4.23}$$

where

$$N_1 = \frac{1}{2n+1}\sum_{\alpha=1}^{n}(1-t_\alpha)(F_\alpha+G_\alpha),$$

$$N_2 = \frac{2}{2n+1}\sum_{\alpha=1}^{n}t_\alpha(t_\alpha-1)(F_\alpha+G_\alpha)\,. \tag{3.4.24}$$

We notice that N_1 and N_2 contain only the unknowns $F_\alpha + G_\alpha$ which may be determined from the system

$$\sum_{\alpha=1}^{n}A_{j\alpha}(F_\alpha+G_\alpha) - B_{j\alpha}(F_\alpha+G_\alpha) = \overline{H}_j + \overline{L}_j \tag{3.4.25}$$

of n equations $(j = 1,\ldots,n)$ with n unknowns.

The unknowns F_α and G_α are separating only when we calculate the pressure on the two airfoils. For example, the field of the pressure on the upper airfoil is obtained from (3.4.6) with the formula

$$p(t_\alpha,(a/2)\pm 0) = \pm\frac{\varepsilon}{2\beta}\sqrt{\frac{1-t_\alpha}{1-t_\alpha}}F_\alpha - \frac{m\varepsilon}{\beta(2n+1)}\sum_{i=1}^{n}\frac{(1-t_i)G(t_i)}{(t_\alpha-t_i)^2+m^2} -$$

$$-\frac{1}{\pi\beta}{\int_{-1}^{+1}}'\frac{h_1'(x)}{t_\alpha-x}\mathrm{d}\,x - \frac{1}{\pi\beta}\int_{-1}^{+1}\frac{(t_\alpha-x)l_1'(x)}{(t_\alpha-x)^2+m^2}\mathrm{d}\,x\,,$$

$$t_i = \cos\frac{2i\,\pi}{2n+1}\,, \tag{3.4.26}$$

t_α being given in (3.1.58). In the same way we may determine the field of the pressure on the lower airfoil.

The numerical determinations have been performed for the biplane having the angle of attack $\varepsilon(h' = l' = -\varepsilon)$ (fig. 3.4.2a)) and for the symmetric biplane $(h' = -l' = -\varepsilon)$ (fig. 3.4.2b)) taking $n = 10$. In

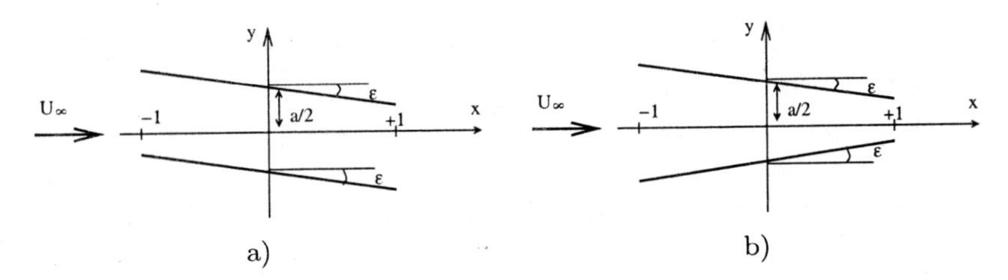

Fig. 3.4.2.

the first case $\overline{H}_j = \overline{L}_j = 1$, and in the second $\overline{H}_j = -\overline{L}_j = 1$. The coefficients $A_{j\alpha}$, $B_{j\alpha}$ depend only on the parameter $m = a\beta$.

In the second case we obtain $N_1 = N_2 = 0$ for all the values given to m (it is natural). In the first case, the values of $N_1\beta^{-1}$ and $N_2\beta^{-1}$ depend on a and M. They are given in tables 1 and 2. In the first line we may find the values of $\beta^{-1}N_1$ and $\beta^{-1}N_2$ for the monoplane. We notice that the lift coefficient for the biplane is much greater then the lift coefficient for the monoplane and it increases with a. The lift increases also with M. The same conclusions are true for the moment coefficient.

Table 1

The values of $N_1\beta^{-1}$

	M=0	M=0.6
	1.00	1.25
a=0.5	1.461886	1.741425
a=1	1.70842	2.034937
a=5	1.981013	2.46087

Table 2

The values of $N_2\beta^{-1}$

	M=0	M=0.6
	1.00	1.25
a=0.5	1.624288	1.941250
a=1	1.827670	2.209987
a=5	1.991025	2.477736

3.5 Grids of Profiles

3.5.1 The Integral Equation

The classical problem may be also solved in a simple manner by means of the method of fundamental solutions. One obtains again the generalized equation of thin profiles. Let us consider a grid of identical airfoils having the equations

$$y = na + h(x) \pm h_1(x), \quad n = 0, \pm 1, \pm 2, \dots, \tag{3.5.1}$$

which perturb the uniform subsonic stream defined in section 2.1.1. Since every airfoil produces the same perturbation, we shall replace every profile by the same distribution of forces $(f_1, f)(\xi)$. Since the perturbation produced by the force (f_1, f_2) in the uniform stream is (2.3.23), we deduce that the perturbation produced by the grid is

$$
\begin{aligned}
p(x,y) &= \frac{1}{2\pi\beta}\sum_{-\infty}^{+\infty}\int_{-1}^{+1}\frac{x_0 f_1(\xi)+\beta^2(y-na)f(\xi)}{x_0^2+\beta^2(y-na)^2}\mathrm{d}\,\xi\,,\\
v(x,y) &= \frac{\beta}{2\pi}\sum_{-\infty}^{+\infty}\int_{-1}^{+1}\frac{x_0 f(\xi)-(y-na)f_1(\xi)}{x_0^2+\beta^2(y-na)^2}\mathrm{d}\,\xi\,,
\end{aligned}
\tag{3.5.2}
$$

where $x_0 = x - \xi$.

The profile corresponding to $n = 0$ will be called *the reference profile.* Passing to limit, when $y \to \pm 0$ the integrals corresponding to $n = 0$ become singular and they are calculated with the formulas (3.1.19) and (3.1.20). Employing the notation $k = x_0/m$. we get

$$
\begin{aligned}
&\sum_{-\infty}^{+\infty}{}^{n\neq 0}\frac{x_0}{x_0^2+m^2n^2} = \frac{2x_0}{m^2}\sum_{n=1}^{\infty}\frac{1}{k^2+n^2} = \frac{2x_0}{m^2}\left(\frac{\pi}{2k}\coth k\pi - \frac{1}{2k^2}\right) =\\
&\qquad = \frac{\pi}{m}\coth\frac{\pi}{m}x_0 - \frac{1}{x_0} \equiv K(x_0)\,,\\
&\sum_{-\infty}^{+\infty}{}^{n\neq 0}\frac{n}{x_0^2+m^2n^2} = 0\,,
\end{aligned}
\tag{3.5.3}
$$

whence:

$$
\begin{aligned}
p(x,\pm 0) &= \pm\frac{1}{2}f(x) + \frac{1}{2\pi\beta}{\int_{-1}^{+1}}'\frac{f_1(\xi)}{x_0}\mathrm{d}\,\xi +\\
&+\frac{1}{2\pi\beta}\int_{-1}^{+1}K(x_0)f_1(\xi)\mathrm{d}\,\xi\,,
\end{aligned}
\tag{3.5.4}
$$

$$
p(x,+0) - p(x,-0) = f(x)\,,
\tag{3.5.5}
$$

$$
\begin{aligned}
v(x,\pm 0) &= \mp\frac{1}{2}f_1(x) + \frac{\beta}{2\pi}{\int_{-1}^{+1}}'\frac{f(\xi)}{x_0}\mathrm{d}\,\xi +\\
&+\frac{\beta}{2\pi}\int_{-1}^{+1}K(x_0)f_1(\xi)\mathrm{d}\,\xi\,.
\end{aligned}
\tag{3.5.6}
$$

Imposing the boundary condition

$$v(x,\pm 0) = h'(x) \pm h_1'(x) \tag{3.5.7}$$

we deduce

$$f_1(x) = -2h_1'(x) \tag{3.5.8}$$

$$\frac{\beta}{2\pi}{\int_{-1}^{+1}}' \frac{f(\xi)}{x_0}\mathrm{d}\,\xi + \frac{\beta}{2\pi}\int_{-1}^{+1} K(x_0)f(\xi)\mathrm{d}\,\xi = h'(x) \tag{3.5.9}$$

This is the integral equation of the problem. It has the same form like the equations (3.2.7), (3.3.9) and (3.4.10), (3.4.14), excepting the non - singular kernel. Obviously, from (3.5.3) we deduce that the kernel $K(x_0)$ is not singular. After determining f by means of (3.5.9), the field of pressure on the reference profile will be obtained from (3.5.4), and (3.5.5) will be utilized for the calculation of lift and moment coefficients.

In order to obtain the pressure on the profile corresponding to $n = 1$ and to impose the boundary condition on this profile, we must pass to the limit in (3.5.2) $y \to a \pm 0$. In this case, the integrals corresponding to $n = 1$ become singular. With the change of variable $y - a = Y$ we have for example

$$\begin{aligned} p(x, a \pm 0) = & \frac{1}{2\pi\beta}\lim_{Y\to\pm 0}\int_{-1}^{+1}\frac{x_0 f_1 + \beta^2 Y f}{x_0^2 + \beta^2 Y^2}\mathrm{d}\,\xi + \\ & + \frac{1}{2\pi\beta}\sum_{-\infty}^{+\infty}{}^{n\neq 1}\int_{-1}^{+1}\frac{x_0 f_1 + \beta^2(1-n)af}{x_0^2 + m^2(1-n)^2}\mathrm{d}\,\xi\,. \end{aligned}$$

Putting $1 - n = n_1$ in the last integral we deduce

$$\sum_{-\infty}^{+\infty}{}^{n\neq 1}\frac{1}{x_0^2 + m^2(1-n)^2} = \sum_{-\infty}^{+\infty}{}^{n_1\neq 0}\frac{1}{x_0^2 + m^2 n_1^2},$$

$$\sum_{-\infty}^{+\infty}{}^{n\neq 1}\frac{1-n}{x_0^2 + m^2(1-n)^2} = 0$$

such that, taking into account (3.5.3), we obtain for $p(x, a \pm 0)$ the same expression like for $p(x, \pm 0)$. In the same way, for $v(x, a \pm 0)$ one obtains the same formula like for $v(x, \pm 0)$. In this way, imposing the boundary condition

$$v(x, a \pm 0) = h'(x) \pm h_1'(x)\,, \tag{3.5.10}$$

one obtains the equations (3.5.8) and (3.5.9). We draw the conclusion that if f_1 and f are determined by (3.5.8) and (3.5.9), the boundary conditions are satisfied on each profile and the field of pressure coincides with the field on the reference airfoil.

The lift and moment coefficients are calculated on every profile by means of the formulas

$$c_L = -\int_{-1}^{+1} f(t)\mathrm{d}t, \quad c_M = -\frac{1}{2}\int_{-1}^{+1} tf(t)\mathrm{d}t. \tag{3.5.11}$$

3.5.2 The Numerical Integration

If f has the form (3.1.56) and h has the form $-\varepsilon\overline{h}$, using the quadrature formulas (F.2.18) and (F.2.19), we obtain from the integral equation (3.5.9) the system

$$\sum_{\alpha=1}^{n} D_{j\alpha}F_\alpha = \overline{h}_j, \quad j = 1, \ldots, n, \tag{3.5.12}$$

where

$$D_{j\alpha} = \frac{\pi}{m}\frac{1-t_\alpha}{2n+1}\coth\frac{\pi}{m}(x_j - t_\alpha), \tag{3.5.13}$$

t_α and x_j being given in (3.1.58). Considering for c_L and c_M the form (3.1.59), we obtain for N_L and N_M the expressions (3.1.60). For example, for a grid of flat plates having the angle of attack $\varepsilon(\overline{h}_j = 1)$, the variations of the coefficients N_L and N_M versus a are given in figures 3.5.1.

We notice that the coefficients c_L and c_M are increasing functions of a and M. They tend asymptotically to 1 when $a \to \infty$, hence the lift and moment coefficients tend asymptotically to the values taken in the case of a single profile.

The pressure in the control points (nodes) is obtained from the formula

$$p(t_\alpha, \pm 0) = \mp \frac{\varepsilon}{2\beta}\sqrt{\frac{1-t_\alpha}{1+t_\alpha}}F_\alpha + \frac{1}{\pi\beta}{}'\!\!\int_{-1}^{+1}\frac{h_1'(x)}{x-t_\alpha}\mathrm{d}x - \frac{1}{\pi\beta}\int_{-1}^{+1}K(t_\alpha - x)h_1'(x)\mathrm{d}x. \tag{3.5.14}$$

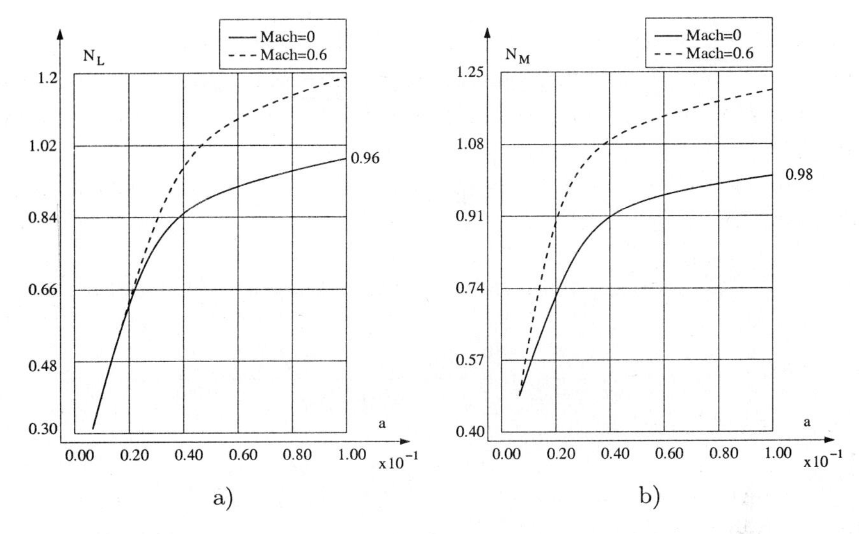

a) b)

Fig. 3.5.1.

3.6 Airfoils in Tandem

3.6.1 The Integral Equations

In the sequel we shall determine the perturbation produced in the uniform stream defined in 2.1 (under the assumption that it is subsonic ($M < 1$)), by a configuration of airfoils in tandem (fig. 3.6.1), having the equations

$$y = l_1(x) \pm h_1(x), \quad a_1 < x < b_1\,, \tag{3.6.1}$$

$$y = l_2(x) \pm h_2(x), \quad a_2 < x < b_2\,. \tag{3.6.2}$$

For the incompressible fluid this problem was studied by Chaplygin and Sedov [1.38] who determined the complex velocity. If we reduce the problem to integral equations and we solve them like in [3.15], we can determine directly the quantities of interest in aerodynamics (the field and the jump of the pressure on the airfoils, the lift and moment coefficients). Moreover, the fluid may be compressible and we don't need to use elliptic integrals. In the sequel, we shall present the results from the last paper cited above.

Replacing the airfoils by distributions of forces having the intensities (g_1, f_1) respectively (g_2, f_2) and taking into account the fundamental

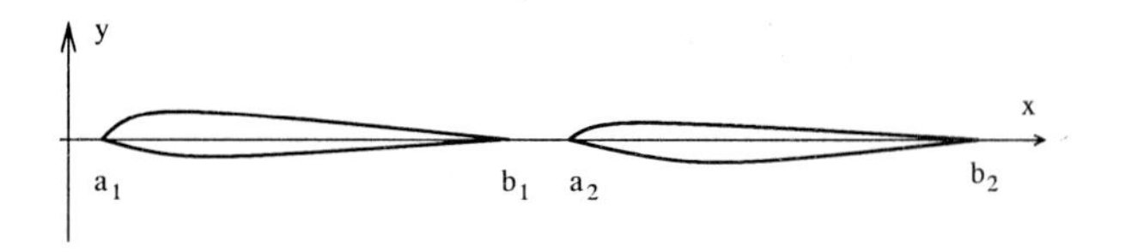

Fig. 3.6.1.

solution (2.3.23), we get the following representation of the perturbation:

$$\begin{aligned} p(x,y) &= \frac{1}{2\pi\beta}\int_{a_1}^{b_1} \frac{x_0 g_1(\xi)+\beta^2 y f_1(\xi)}{x_0^2+\beta^2 y^2}\mathrm{d}\,\xi + \\ &\quad + \frac{1}{2\pi\beta}\int_{a_2}^{b_2} \frac{x_0 g_2(\xi)+\beta^2 y f_2(\xi)}{x_0^2+\beta^2 y^2}\mathrm{d}\,\xi\,, \\ v(x,y) &= \frac{\beta}{2\pi}\int_{a_1}^{b_1} \frac{x_0 f_1(\xi)-y g_1(\xi)}{x_0^2+\beta^2 y^2}\mathrm{d}\,\xi + \\ &\quad + \frac{\beta}{2\pi}\int_{a_2}^{b_2} \frac{x_0 f_2(\xi)-y g_2(\xi)}{x_0^2+\beta^2 y^2}\mathrm{d}\,\xi\,, \end{aligned} \tag{3.6.3}$$

We kept the same notations like above. Passing to limit for $a_1 < x < b_1$ and taking into account (3.1.19) and (3.1.20), one obtains:

$$p(x,\pm 0) = \pm\frac{1}{2}f_1(x) + \frac{1}{2\pi\beta}{\int_{a_1}^{b_1}}' \frac{g_1(\xi)}{x_0}\mathrm{d}\,\xi + \frac{1}{2\pi\beta}\int_{a_2}^{b_2} \frac{g_2(\xi)}{x_0}\mathrm{d}\,\xi\,, \tag{3.6.4}$$

$$p(x,+0) - p(x,-0) = f_1(x) \tag{3.6.5}$$

$$v(x,\pm 0) = \mp\frac{1}{2}g_1(x) + \frac{\beta}{2\pi}{\int_{a_1}^{b_1}}' \frac{f_1(\xi)}{x_0}\mathrm{d}\,\xi + \frac{\beta}{2\pi}\int_{a_2}^{b_2} \frac{f_2(\xi)}{x_0}\mathrm{d}\,\xi\,. \tag{3.6.6}$$

Imposing the boundary condition

$$v(x,\pm 0) = l_1'(x) \pm h_1'(x)$$

it follows

$$g_1(x) = -2h_1'(x) \tag{3.6.7}$$

$$\frac{\beta}{2\pi}{\int_{a_1}^{b_1}}' \frac{f_1(\xi)}{x_0}\mathrm{d}\,\xi + \frac{\beta}{2\pi}\int_{a_2}^{b_2} \frac{f_2(\xi)}{x_0}\mathrm{d}\,\xi + \frac{\beta}{2\pi}\int_{a_2}^{b_2} \frac{f_2(\xi)}{x_0}\mathrm{d}\,\xi = l_1'(x) \tag{3.6.8}$$

In the same way, for $a_2 < x < b_2$ we deduce:

$$p(x,\pm 0) = \pm\frac{1}{2}f_2(x) + \frac{1}{2\pi\beta}\int_{a_1}^{b_1}\frac{g_1(\xi)}{x_0}\mathrm{d}\,\xi + \frac{1}{2\pi\beta}\int'^{b_2}_{a_2}\frac{g_2(\xi)}{x_0}\mathrm{d}\,\xi \quad (3.6.9)$$

$$v(x,\pm 0) = \mp\frac{1}{2}g_2(x) + \frac{\beta}{2\pi}\int_{a_1}^{b_1}\frac{f_1(\xi)}{x_0}\mathrm{d}\,\xi + \frac{\beta}{2\pi}\int'^{b_2}_{a_2}\frac{f_2(\xi)}{x_0}\mathrm{d}\,\xi\,,$$

whence, utilizing the boundary condition

$$v(x,\pm 0) = l_2'(x) \pm h_2'(x)$$

we obtain

$$p(x,+0) - p(x,-0) = f_2(x)\,, \quad (3.6.10)$$

$$g_2(x) = -2h_2'(x)\,, \quad (3.6.11)$$

$$\frac{\beta}{2\pi}\int_{a_1}^{b_1}\frac{f_1(\xi)}{x_0}\mathrm{d}\,\xi + \frac{\beta}{2\pi}\int'^{b_2}_{a_2}\frac{f_2(\xi)}{x_0}\mathrm{d}\,\xi = l_2'(x)\,. \quad (3.6.12)$$

Hence the functions $f_1(x)$ and $f_2(x)$ will be defined by the system (3.6.8) and (3.6.12). This is an interesting mathematical problem, since the first equation is defined for $a_1 < x < b_1$ and the second is defined for $a_2 < x < b_2$.

3.6.2 The Determination of the Functions f_1 and f_2

For f_1 the equation (3.6.8) has the structure (C.1.5) which has the solution (C.1.9). Utilizing (B.6.11) and the last formula from (B.5.4) we deduce:

$$\beta f_1(x) = L_1(x) + \frac{\beta}{\pi}\sqrt{\frac{b_1 - x}{x - a_1}}\int_{a_2}^{b_2}\sqrt{\frac{\xi - a_1}{\xi - b_1}}\,\frac{f_2(\xi)}{\xi - x}\mathrm{d}\,\xi\,, \quad (3.6.13)$$

where

$$L_1(x) = \frac{2}{\pi}\sqrt{\frac{b_1 - x}{x - a_1}}\int'^{b_1}_{a_1}\sqrt{\frac{t - a_1}{b_1 - t}}\,\frac{l_1'(t)}{t - x}\mathrm{d}\,t\,. \quad (3.6.14)$$

Substituting f_1 given by (3.6.13) in (3.6.12) and taking (B.6.11) and the last formula from (B.5.4) into account, one obtains the following integral equation for f_2:

$$\frac{\beta}{2\pi}\int'^{b_2}_{a_2}\sqrt{\frac{t - a_1}{t - b_1}}\,\frac{f_2(t)}{t - x}\mathrm{d}\,t = L_2(x),\ a_2 < x < b_2\,, \quad (3.6.15)$$

$$L_2(x) = -l_2'(x)\sqrt{\frac{x-a_1}{x-b_1}} - \frac{1}{\pi}\int_{a_1}^{b_1}\sqrt{\frac{\xi-a_1}{b_1-\xi}}\,\frac{l_1'(\xi)}{\xi-x}\mathrm{d}\,\xi\,. \qquad (3.6.16)$$

The type of the equation (3.6.15) is (C.1.1) and its solution has the form (C.1.9). One obtains:

$$\beta f_2(x) = -\frac{2}{\pi}\left[\frac{(x-b_1)(b_2-x)}{(x-a_1)(x-a_2)}\right]^{1/2}{}'\!\!\int_{a_2}^{b_2}\sqrt{\frac{t-a_2}{b_2-t}}\,\frac{L_2(t)}{t-x}\mathrm{d}\,t$$

or, taking (3.6.16) and the first formula from (B.6.11) into account,

$$\beta f_2(x) = \frac{2}{\pi}\sqrt{\frac{(x-b_1)(b_2-x)}{(x-a_1)(x-a_2)}}\left\{{}'\!\!\int_{a_1}^{b_1}\sqrt{\frac{(\xi-a_1)(a_2-\xi)}{(b_1-\xi)(b_2-\xi)}}\,\frac{l_1'(\xi)}{\xi-x}\mathrm{d}\xi + \right.$$

$$\left. + {}'\!\!\int_{a_2}^{b_2}\sqrt{\frac{(t-a_1)(t-a_2)}{(t-b_1)(b_2-t)}}\,\frac{l_2'(\xi)}{t-x}\mathrm{d}\,t\right\}, \qquad a_2 < x < b_2\,. \qquad (3.6.17)$$

Once βf_2 determined, from (3.6.13) it results βf_1:

$$\beta f_1(x) = \frac{2}{\pi}\sqrt{\frac{(b_1-x)(x-b_2)}{(x-a_1)(x-a_2)}}\left\{{}'\!\!\int_{a_1}^{b_1}\sqrt{\frac{(\xi-a_1)(a_2-\xi)}{(b_1-\xi)(b_2-\xi)}}\,\frac{l_1'(\xi)}{\xi-x}\mathrm{d}\,\xi + \right.$$

$$\left. + \int_{a_2}^{b_2}\sqrt{\frac{(t-a_1)(t-a_2)}{(t-b_1)(b_2-t)}}\,\frac{l_2'(\xi)}{t-x}\mathrm{d}\,t\right\}, \qquad a_1 < x < b_1\,. \qquad (3.6.18)$$

In this way we determined the solution of the system (3.6.8) and (3.6.12).

3.6.3 The Lift and Moment Coefficients

From the formulas (3.6.5) and (3.6.10), we deduce for the lift and moment coefficients:

$$c_L = -\sum_{i=1,2}\int_{a_i}^{b_i} f_i(x)\mathrm{d}\,x,\; c_M = -\frac{1}{2}\sum_i\int_{a_i}^{b_i} x f_i(x)\mathrm{d}\,x \qquad (3.6.19)$$

Utilizing the solutions (3.6.17) and (3.6.18), changing the order of integration and taking (B.6.5) into account, we find

$$c_L = -\frac{2}{\beta}\left\{\int_{a_1}^{b_1}\sqrt{\frac{(\xi-a_1)(a_2-\xi)}{(b_1-\xi)(b_2-\xi)}}\,l_1'(\xi)\mathrm{d}\,\xi + \right.$$

$$\left. + \int_{a_2}^{b_2}\sqrt{\frac{(t-a_1)(t-a_2)}{(t-b_1)(b_2-t)}}\,l_2'(t)\mathrm{d}\,t\right\}. \qquad (3.6.20)$$

Utilizing the same solutions, changing the order of integration and taking (B.6.6) into account, we get for the moment coefficient:

$$c_M = -\frac{1}{\beta}\left\{ \int_{a_1}^{b_1} \sqrt{\frac{(\xi - a_1)(a_2 - \xi)}{(b_1 - \xi)(b_2 - \xi)}} \xi l_1'(\xi) \mathrm{d}\,\xi + \right.$$
$$\left. + \int_{a_2}^{b_2} \sqrt{\frac{(t - a_1)(t - a_2)}{(t - b_1)(b_2 - t)}} t l_2'(t) \mathrm{d}\,t \right\} + \frac{a_1 + a_2 - b_1 - b_2}{4} c_L \,. \quad (3.6.21)$$

Obviously, the formulas (3.6.20) and (3.6.21) are generalizations of the formulas (3.1.33) and (3.1.34).

Utilizing (B.5.14) we may prove by induction that in the case of n profiles in tandem we have:

$$c_L = -\frac{2}{\mathrm{i}\beta} \sum_{j=1}^{n} \int_{a_j}^{b_j} \prod_{k=1}^{n} \sqrt{\frac{t - a_k}{t - b_k}} l_j'(t) \mathrm{d}\,t \,,$$
$$c_M = -\frac{1}{\mathrm{i}\beta} \sum_{j=1}^{n} \int_{a_j}^{b_j} \prod_{k=1}^{n} \sqrt{\frac{t - a_k}{t - b_k}} t l_j'(t) \mathrm{d}\,t + \frac{1}{4} \sum_{k=1}^{n} (a_k - b_k) \,. \quad (3.6.22)$$

3.6.4 Numerical Values

In the case of two flat plates having the angle of attack $\pm\varepsilon$ (fig. 3.6.2a) or 3.6.2b)), the integrals (3.6.20) may be expressed with the aid of elliptic functions. The first one has the form of the integral 252.21 from [1.4], page 105, and the second has the form of the integral 256.21 from the same book, page 122. For more complex configurations, these kinds of expressions are not available. However, in all cases, the integrals intervening in the expressions of c_L and c_M (for every n), may be calculated numerically by means of the quadrature formulas (F.2.24).

For example, if $n = 2$, performing the change of variable $\xi \to t$:

$$\xi = \frac{a_1 + b_1}{2} + \frac{b_1 - a_1}{2} t, \quad -1 < t < +1 \quad (3.6.23)$$

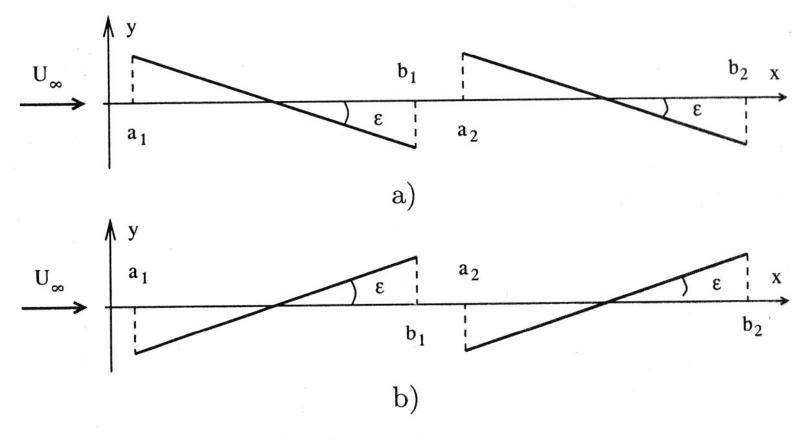

Fig. 3.6.2.

and utilizing (F.2.24) we deduce:

$$
\begin{aligned}
I_1 &= \frac{1}{\pi}\int_{a_1}^{b_1}\sqrt{\frac{(\xi-a_1)(a_2-\xi)}{(b_1-\xi)(b_2-\xi)}}\,\mathrm{d}\,\xi = \frac{b_1-a_1}{2n+1}\sum_{\alpha=1}^{n}(1+t_\alpha)\sqrt{\frac{a-t_\alpha}{b-t_\alpha}}, \\
J_1 &= \frac{1}{\pi}\int_{a_1}^{b_1}\sqrt{\frac{(\xi-a_1)(a_2-\xi)}{(b_1-\xi)(b_2-\xi)}}\,\xi\mathrm{d}\,\xi = \\
&= \frac{a_1+b_1}{2}I_1 + \frac{(b_1-a_1)^2}{2(2n+1)}\sum_{\alpha=1}^{n}t_\alpha(1+t_\alpha)\sqrt{\frac{a-t_\alpha}{b-t_\alpha}}, \\
a &= \frac{2a_2-(a_1+b_1)}{b_1-a_1}, \quad b = \frac{2b_2-(a_1+b_1)}{b_1-a_1},
\end{aligned}
\tag{3.6.24}
$$

where t_α are given by (3.1.52).

Similarly we get:

$$
\begin{aligned}
I_2 &= \frac{1}{\pi}\int_{a_2}^{b_2}\sqrt{\frac{(t-a_1)(t-a_2)}{(t-b_1)(b_2-t)}} = \frac{b_2-a_2}{2n+1}\sum_{\alpha=1}^{n}(1+t_\alpha)\sqrt{\frac{c-t_\alpha}{d-t_\alpha}}, \\
J_2 &= \frac{1}{\pi}\int_{a_2}^{b_2}\sqrt{\frac{(t-a_1)(t-a_2)}{(t-b_1)(b_2-t)}}\,t\mathrm{d}\,t = \\
&= \frac{a_2+b_2}{2}I_2 + \frac{(b_2-a_2)^2}{2(2n+1)}\sum_{\alpha=1}^{n}t_\alpha(1+t_\alpha)\sqrt{\frac{c-t_\alpha}{d-t_\alpha}}, \\
c &= \frac{2a_1-(a_2+b_2)}{b_2-a_2}, \quad d = \frac{2b_1-(a_2+b_2)}{b_2-a_2}.
\end{aligned}
\tag{3.6.25}
$$

Hence for two flat plates in tandem having the angle of attack ε we have

$$c_L = \frac{2\pi\varepsilon}{\beta}(I_1 + I_2), \quad c_M = \frac{\pi\varepsilon}{\beta}(J_1 + J_2) \tag{3.6.26}$$

where I_1, I_2, J_1, J_2 may be calculated numerically using a computer. For example, for $a_1 = 1,\ b_1 = 2,\ a_2 = 3$ and $b_2 = 4$ one obtains: $I_1 = 0.370671$ and $I_2 = 0.629328$. For $a_1 = 1,\ b_1 = 2,\ a_2 = 4$ and $b_2 = 5$ one obtains: $I_1 = 0.415458$ and $I_2 = 0.584541$. For $a_1 = 1,\ b_1 = 2,\ a_2 = 6$ and $b_2 = 7$ one obtains: $I_1 = 0.449746$ and $I_2 = 0.550253$. In all the cases we have $I_1 + I_2 = 0.999999$, hence the lift has practically the value of the lift for a single plate and this is true even if the distance between plates increases. Since $-I_1 + I_2$ takes the values 0.258657, 0.169083, 0.100507 it results that if the first plate has the angle of attack $-\varepsilon$ and the second ε, the lift decreases when the distance between plates increases. If the first plate has the angle of attack $-\varepsilon$ and the first has the angle of attack ε, the lift becomes negative.

The previous formulas enable us to obtain numerical values for c_L and c_M in the case of n airfoils with different shapes.

Chapter 4

The Application of the Boundary Element Method to the Theory of the Infinite Span Airfoil in Subsonic Flow

4.1 The Equations of Motion

4.1.1 Introduction

The theory exposed in the previous chapter relies on the following three assumptions:

1^0 The linearization of the boundary condition.

2^0 The linearized condition is imposed on the chord of the profile, not on the boundary (as it is natural to do).

3^0 The linearization of the equations of motion.

The assumptions are plausible for thin airfoils, otherwise they may be the cause of great errors.

The application of the boundary integral equations method (BIEM), which is also called the boundary elements method (BEM), enable us to give up the first two assumptions. Hence we shall utilize the non-linear boundary condition which will be imposed on the contour (boundary) of the airfoil.

For the incompressible fluid, we shall employ the exact equations (4.1.4) i.e. the equation of continuity and the equation of irrotationality, such that, in this case, the mathematical model (the equations of motion and the boundary conditions) will be valid for every airfoil (not only for the thin ones). In the case of the compressible fluid, we shall utilize the linearized equations, hence the results will be valid only for thin profiles, but even in this case the results are better then the results obtained by adopting the first two assumptions.

The first step in solving a problem with the boundary integral equations method (or the boundary elements method) consists in reducing the boundary value problem to a boundary integral equation (whence it follows the first name of the method); the second step consists in solving approximately the integral equation, replacing the boundary with a polygonal line (whence it follows the second name). In addition we have to mention that there exists a *direct* method and an *indirect* method. In the direct method we utilize the equations of motion for deducing a representation of the solution in the domain occupied by the fluid with the aid of the solution on the boundary. Then we pass to the limit on the boundary. In the indirect method, we replace the wing with a distribution of fundamental solutions (sources, vortices, doublets) on the boundary and and we determine their intensity imposing the boundary conditions to be verified; in this way one obtains an integral equation for the intensity of the fundamental solutions; the method is therefore indirect. As we shall see in 4.2, it is easier to use the indirect method, but the results are less accurate. It is more difficult to utilize the direct method (4.3) but the results are more accurate. Therefore, in the forthcoming subsections we shall use mainly this method.

4.1.2 The Statement of the Problem

The problem was already formulated in the previous chapter. A subsonic stream having the velocity $U_\infty \boldsymbol{i}$ (the Ox_1 axis has the direction of the unperturbed stream), the pressure p_∞ and the density ρ_∞ is perturbed by the presence of a cylindric body with the generatrices perpendicular on the x_1Oy_1 plane, having a given cross section (see fig. 4.1.1). One requires to determine the perturbation and the action of the fluid against the body. Denoting by X, Y the dimensionless coordinates introduced by means of the relation

$$(x_1, y_1) = L_0(X, Y)\,, \tag{4.1.1}$$

L_0 being an unspecified reference length, and by $U_\infty \boldsymbol{V}$ and $\rho_\infty U_\infty^2 P$, the perturbation of the velocity and the perturbation of the pressure, we have:

$$\boldsymbol{V}_1 = U_\infty(\boldsymbol{i} + \boldsymbol{V}), \quad P_1 = p_\infty + \rho_\infty U_\infty^2 P\,. \tag{4.1.2}$$

The system which determines the perturbation is (2.1.30), written with capital letters. Projecting the second equation on OX and taking the damping condition at infinity into account, we deduce

$$P = -U\,. \tag{4.1.3}$$

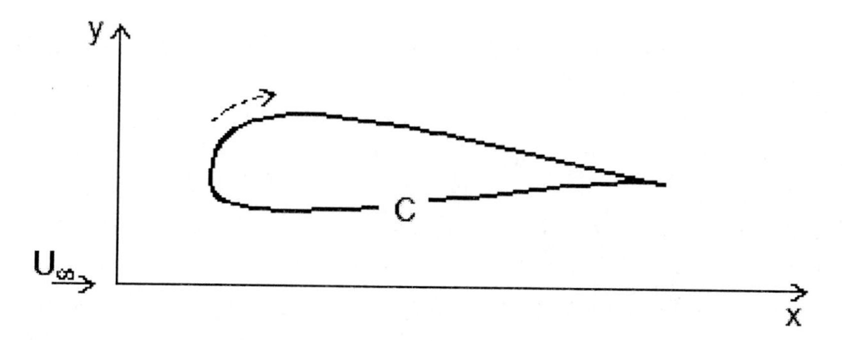

Fig. 4.1.1.

In this way we deduce the equations which determine the perturbation:

$$\beta^2 \partial U/\partial X + \partial V/\partial Y = 0, \quad \partial V/\partial X - \partial U/\partial Y = 0\,, \tag{4.1.4}$$

U and V representing the coordinates of the vector $\boldsymbol{V}$, and β having the usual significance ($\beta = \sqrt{1-M^2}$).

On the boundary C_1 of the airfoil we shall impose the condition

$$(1+U)N_X + VN_Y = 0\,, \tag{4.1.5}$$

N_X and N_Y representing the coordinates of the inward normal to C_1.

Performing the change of variables

$$\begin{aligned} x = X, \quad y = \beta Y \\ u = \beta U, \quad v = V, \end{aligned} \tag{4.1.6}$$

the system (4.1.4) becomes

$$\partial u/\partial x + \partial v/\partial y = 0, \quad \partial v/\partial x - \partial u/\partial y\,. \tag{4.1.7}$$

In order to transform the condition (4.1.5), we notice that if the boundary C_1 has the parametric equations $X = X(S)$, $Y = Y(S)$ with S increasing when the curve C_1 is traversed like in figure 4.1.1, then

$$\boldsymbol{N} = (\mathrm{d}Y/\mathrm{d}S, \ -\mathrm{d}X/\mathrm{d}S)\,. \tag{4.1.8}$$

Performing the change of variable (4.1.6), we have $x = x(X(S))$, $y = y(Y(S))$ and

$$\boldsymbol{n} = \left(\frac{\mathrm{d}y}{\mathrm{d}s}, -\frac{\mathrm{d}x}{\mathrm{d}s}\right) = \left(\frac{\mathrm{d}y}{\mathrm{d}Y}N_X, \frac{\mathrm{d}x}{\mathrm{d}X}N_Y\right)\frac{\mathrm{d}S}{\mathrm{d}s}\,,$$

C_1 and the transformed curve C being traversed in the same sense (s increases with S). It follows

$$N_X = \frac{n_x}{\beta}\frac{\mathrm{d}s}{\mathrm{d}S}, \quad N_Y = n_y\frac{\mathrm{d}s}{\mathrm{d}S}, \tag{4.1.9}$$

such that (4.1.5) becomes

$$(\beta + u)n_x + \beta^2 v n_y = 0 \quad \text{on } C. \tag{4.1.10}$$

One also imposes the damping condition at infinity

$$\lim_{\infty}(u, v) = 0. \tag{4.1.11}$$

4.1.3 The Fundamental Solutions

The first step in applying the BIEM consists in determining the fundamental solution of the system of equations (in our case, the system (4.1.7)). We call *source-type solution* the solution of the system

$$\begin{aligned} &\partial u^*/\partial x + \partial v^*/\partial y = \delta(x - \xi, y - \eta), \\ &\partial v^*/\partial x - \partial u^*/\partial y = 0. \end{aligned} \tag{4.1.12}$$

The fact that the perturbation represented by δ intervenes in the equation of continuity justifies the name of the solution. We apply the Fourier transform and like in Chapter 2 we obtain:

$$u^* = \frac{1}{2\pi}\frac{x - \xi}{(x - \xi)^2 + (y - \eta)^2}, \quad v^* = \frac{1}{2\pi}\frac{y - \eta}{(x - \xi)^2 + (y - \eta)^2}. \tag{4.1.13}$$

We call *vortex-type solution* the solution of the system

$$\begin{aligned} &\partial \overline{u}^*/\partial x + \partial \overline{v}^*/\partial y = 0, \\ &\partial \overline{v}^*/\partial x - \partial \overline{u}^*/\partial y = \delta(x - \xi, y - \eta). \end{aligned} \tag{4.1.14}$$

Obviously one obtains:

$$\overline{u}^* = -v^*, \quad \overline{v}^* = u^*. \tag{4.1.15}$$

4.2 Indirect Methods for the Unlimited Fluid Case

4.2.1 The integral equation for the Distribution of Sources

Taking (4.2.13) into account, we deduce that if we replace the airfoil having the contour C (fig 4.1.1) with a distribution of sources of intensity $f(\boldsymbol{x})$ apriori unknown, the perturbation of the velocity in the point $M(\boldsymbol{\xi})$ from the fluid will be given by the formulas:

$$\boldsymbol{v}(\boldsymbol{\xi}) = -\frac{1}{2\pi}\oint_C f(\boldsymbol{x})\frac{\boldsymbol{x}-\boldsymbol{\xi}}{|\boldsymbol{x}-\boldsymbol{\xi}|^2}\,\mathrm{d}\,s\,. \tag{4.2.1}$$

In order to determine the intensity $f(\boldsymbol{x})$ we impose the boundary condition (4.1.10). To this aim, we have to calculate the boundary values in (4.2.1) when $\boldsymbol{\xi} \to \boldsymbol{x}_0$ a current point Q_0 of the boundary C.

We are going to prove that if $f(\boldsymbol{x})$ satisfies the Hölder condition on C, i.e. there exist two positive constants A and μ $(\mu \leq 1)$ such that, for every two points $Q(\boldsymbol{x}_1)$ and $Q(\boldsymbol{x}_2)$ belonging to C, we have:

$$|f(\boldsymbol{x}_1) - f(\boldsymbol{x}_2)| < A|\boldsymbol{x}_1 - \boldsymbol{x}_2|^{\mu}, \tag{4.2.2}$$

then

$$\begin{aligned}\boldsymbol{v}(\boldsymbol{x}_0) &= \lim_{\boldsymbol{\xi}\to\boldsymbol{x}_0}\left(-\frac{1}{2\pi}\oint_C f(\boldsymbol{x})\frac{\boldsymbol{x}-\boldsymbol{\xi}}{|\boldsymbol{x}-\boldsymbol{\xi}|^2}\mathrm{d}\,s\right) = \\ &= -\frac{1}{2}f(\boldsymbol{x}_0)\boldsymbol{n}^0 - \frac{1}{2\pi}{\oint_C}' f(\boldsymbol{x})\frac{\boldsymbol{x}-\boldsymbol{x}_0}{|\boldsymbol{x}-\boldsymbol{x}_0|^2}\mathrm{d}\,s\,,\end{aligned} \tag{4.2.3}$$

in every regular point $\boldsymbol{x}_0$. We denoted

$${\oint_C}' = \lim_{\varepsilon\to 0}\oint_{C-c}, \tag{4.2.4}$$

where c is the arc cut out from C by the circle of radius ε and center Q_0 (the arc Q_1Q_2 from fig. 4.2.1).

We denoted by $\boldsymbol{n}^0$ the inward normal at C in Q_0.

Indeed, taking the definition (4.2.4) into account, we notice that it suffices to evaluate the integral on c. We have

$$\begin{aligned}&\lim_{\boldsymbol{\xi}\to\boldsymbol{x}_0}\left(-\frac{1}{2\pi}\int_c f(\boldsymbol{x})\frac{\boldsymbol{x}-\boldsymbol{\xi}}{|\boldsymbol{x}-\boldsymbol{\xi}|^2}\mathrm{d}\,s\right) = \\ &= \lim_{\boldsymbol{\xi}\to\boldsymbol{x}_0}\left(-\frac{1}{2\pi}\oint_c [f(\boldsymbol{x}) - f(\boldsymbol{x}_0)]\frac{\boldsymbol{x}-\boldsymbol{\xi}}{|\boldsymbol{x}-\boldsymbol{\xi}|^2}\mathrm{d}\,s\right) - \frac{f(\boldsymbol{x}_0)}{2\pi}L, \\ &L = \lim_{\boldsymbol{\xi}\to\boldsymbol{x}_0}\oint_c \frac{\boldsymbol{x}-\boldsymbol{\xi}}{|\boldsymbol{x}-\boldsymbol{\xi}|^2}\mathrm{d}\,s\,.\end{aligned} \tag{4.2.5}$$

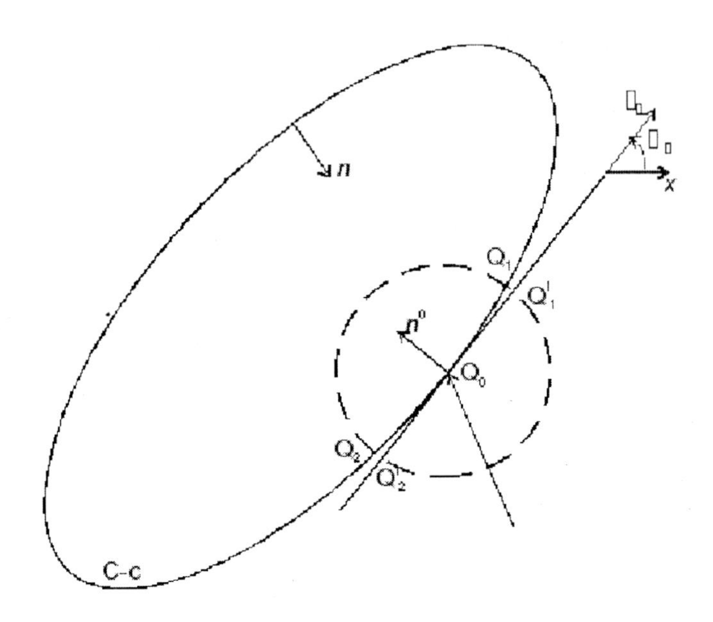

Fig. 4.2.1.

Taking into account that f satisfies the condition (4.2.2), we deduce that the first integral from the right hand side of the equality (4.2.5) vanishes when $\varepsilon \to 0$. For calculating L we shall replace the arc Q_1Q_2 with the segment on the tangent $Q_1'Q_2'$ (fig. 4.2.1) and we shall put on this segment:

$$\boldsymbol{x} = \boldsymbol{x}_0 + s\boldsymbol{\tau}_0, \quad s \in [-\varepsilon, +\varepsilon]. \tag{4.2.6}$$

We assume, for the sake of simplicity, that $M(\boldsymbol{\xi})$ tends to $Q_0(\boldsymbol{x}_0)$ on the direction of the normal $\boldsymbol{n}^0$. We have therefore

$$\boldsymbol{\xi} = \boldsymbol{x}_0 - \eta\boldsymbol{n}^0 \quad \text{cu } \eta > 0. \tag{4.2.7}$$

It follows

$$L = \lim_{\eta \to 0} \int_{-\varepsilon}^{+\varepsilon} \frac{s\boldsymbol{\tau}_0 + \eta\boldsymbol{n}^0}{s^2 + \eta^2} \mathrm{d}\, s = \pi\boldsymbol{n}^0,$$

q.e.d.

Imposing the condition (4.1.10), we obtain from (4.2.1) and (4.2.3) the following integral equation:

$$\left(n_x^{0\,2} + \beta^2 n_y^{0\,2}\right) f(\boldsymbol{x}_0) + \\ + \frac{1}{\pi} \oint_C{}' f(\boldsymbol{x}) \frac{(x - x_0)n_x^0 + \beta^2 (y - y_0) n_y^0}{|\boldsymbol{x} - \boldsymbol{x}_0|^2} \mathrm{d}\, s = 2\beta n_x^0 . \tag{4.2.8}$$

From this integral equation we are going to determine the intensity of the sources. For the incompressible fluid $(M = 0)$ it may be solved exactly for the circular obstacle [4.12].

4.2.2 The Integral Equation for the Distribution of Vortices

If we replace the airfoil with a distribution of vortices having the intensity (circulation) $g(\boldsymbol{x})$, defined on C, then, according to the formulas (4.1.15), we obtain the following expressions for the components of the perturbation u and v in a point $M(\xi)$ from the fluid:

$$u(\boldsymbol{\xi}) = \frac{1}{2\pi} \oint_C g(\boldsymbol{x}) \frac{y-\eta}{|\boldsymbol{x}-\boldsymbol{\xi}|^2} \mathrm{d}\, s\,, \qquad v(\boldsymbol{\xi}) = -\frac{1}{2\pi} \oint_C g(\boldsymbol{x}) \frac{x-\xi}{|\boldsymbol{x}-\boldsymbol{\xi}|^2} \mathrm{d}\, s\,, \tag{4.2.9}$$

where $\boldsymbol{\xi} = (\xi, \eta), \boldsymbol{x} = (x, y)$.

Taking the formula (4.2.6) into account, we obtain:

$$u(\boldsymbol{x}_0) = \frac{1}{2} g(\boldsymbol{x}_0) n_y^0 + \frac{1}{2\pi} \oint_C{}' g(\boldsymbol{x}) \frac{y-y_0}{|\boldsymbol{x}-\boldsymbol{x}_0|^2} \mathrm{d}\, s\,, \qquad v(\boldsymbol{x}_0) = -\frac{1}{2} g(\boldsymbol{x}_0) n_x^0 - \frac{1}{2\pi} \oint_C{}' g(\boldsymbol{x}) \frac{x-x_0}{|\boldsymbol{x}-\boldsymbol{x}_0|^2} \mathrm{d}\, s\,, \tag{4.2.10}$$

and from the boundary condition (4.1.10):

$$-M^2 g(\boldsymbol{x}_0) n_x^0 n_y^0 + \frac{1}{\pi} \oint_C{}' g(\boldsymbol{x}) \frac{\beta^2 (x-x_0) n_y^0 - (y-y_0) n_x^0}{|\boldsymbol{x}-\boldsymbol{x}_0|^2} \mathrm{d}\, s = 2\beta n_x^0\,, \tag{4.2.11}$$

where M is Mach's number in the unperturbed flow. For the incompressible fluid $(M = 0)$ we have a first kind integral equation.

4.2.3 The Boundary Elements Method

One utilizes the following collocation method in order to solve the equations (4.2.8) and (4.2.11): one approximates the boundary C by a polygonal line $\{L_j\}(j = \overline{1, N})$, with the end points on C (fig.

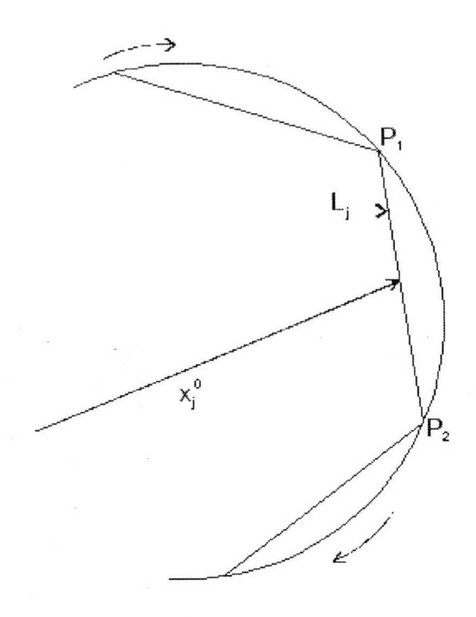

Fig. 4.2.2.

4.2.2) and one approximates on every segment L_j, the unknown f (respectively g) by the value f_j (resp. g_j) from the midpoint of the segment. Denoting by x_j^0 the vector of position of the midpoint, we have $f_j = f(\boldsymbol{x}_j^0)$, $(j = \overline{1,N})$. We may also consider a linear variation of f on L_j. In this case we have to determine the two constants from the values of f in the end points of the segment L_j, i.e. from the values of f on C.

Now the equation (4.2.8) may be written as follows:

$$\left(n_x^{0\,2} + \beta^2 n_y^{0\,2}\right) f(\boldsymbol{x}_0) + \\ + \frac{1}{\pi}\sum_{j=1}^{N} f_j \int_{L_j} \frac{(x-x_0)n_x^0 + \beta^2(y-y_0)n_y^0}{|\boldsymbol{x}-\boldsymbol{x}_0|^2}\,\mathrm{d}\,s = 2\beta n_x^0\,.$$

Imposing the equation to be satisfied in every midpoint, i.e. putting successively $\boldsymbol{x}_0 = \boldsymbol{x}_i^0$, $i = \overline{1N}$, we get the system:

$$a_i f_i + \sum_{j=1}^{N} A_{ij} f_j = A_i, \quad i = \overline{1,N}, \tag{4.2.12}$$

where

$$\begin{aligned} 2a_i &= n_x^2(\boldsymbol{x}_i^0) + \beta^2 n_y^2(\boldsymbol{x}_i^0)\,, \\ A_{ij} &= n_x(\boldsymbol{x}_i^0)U_{ij} + \beta^2 n_y(\boldsymbol{x}_i^0)V_{ij}\,, \quad A_i = \beta n_x(\boldsymbol{x}_i^0)\,, \end{aligned} \tag{4.2.13}$$

with the notations

$$U_{ij} = \int_{L_j} u^*(\boldsymbol{x}, \boldsymbol{x}_i^0)\,\mathrm{d}\,s$$
$$V_{ij} = \int_{L_j} v^*(\boldsymbol{x}, \boldsymbol{x}_i^0)\,\mathrm{d}\,s\,, \tag{4.2.14}$$

u^* and v^* being given in (4.1.13). The linear algebraic system (4.2.12) consisting in N equations will determine the N unknowns f_i.

Now the perturbation of the velocity in the points $P_0(\boldsymbol{x}_i^0)$ follows from (4.2.3)

$$\boldsymbol{v}_i = -\frac{1}{2} f_i\, \boldsymbol{n}(\boldsymbol{x}_i^0) - \sum_{j=1}^{N} \boldsymbol{V}_{ij} f_j \quad i = \overline{1, N}\,,$$
$$\boldsymbol{V}_{ij} = (U_{ij}, V_{ij})\,. \tag{4.2.15}$$

In the same way, from the equation (4.2.11) we get the system

$$b_i\, g_i + \sum_{j=1}^{N} B_{ij} g_j = A_i\,, \quad i = \overline{1, N}\,, \tag{4.2.16}$$

where

$$2b_i = -M^2 n_x(\boldsymbol{x}_i^0) n_y(\boldsymbol{x}_i^0)\,,$$
$$B_{ij} = \beta^2 n_y(\boldsymbol{x}_i^0) U_{ij} - n_x(\boldsymbol{x}_i^0) V_{ij}\,. \tag{4.2.17}$$

From (4.2.10) we obtain the following velocity field

$$u_i = \frac{1}{2} g_i\, n_y(\boldsymbol{x}_i^0) + \sum_{j=1}^{N} V_{ij} g_j$$
$$v_i = -\frac{1}{2} g_i\, n_x(\boldsymbol{x}_i^0) - \sum_{j=1}^{N} U_{ij} g_j\,. \tag{4.2.18}$$

4.2.4 The Determination of the Unknowns

The integrals (4.2.14) may be computed numerically using quadrature formulas. They may be also calculated exactly for every shape of the contour C. Since the integrals U_{jj}, V_{jj} are singular we shall calculate their Finite Parts. For avoiding the singular integrals, we may utilize the regularization method 4.6.

For obtaining a parametrization of the segment L_j, we shall denote by (x_1, y_1) and (x_2, y_2) the coordinates of the end points P_1, respectively P_2, considered in the sense of traversing C (fig. 4.2.2). Then, the coordinates of the generic point on L_j will be

$$x = x_j^0 + \frac{x_2 - x_1}{2} t, \quad y = y_j^0 + \frac{y_2 - y_1}{2} t, \quad -1 \leq t \leq 1, \tag{4.2.19}$$

where, obviously,

$$x_j^0 = \frac{x_2 - x_1}{2}, \quad y_j^0 = \frac{y_2 - y_1}{2}. \tag{4.2.20}$$

Utilizing the formula

$$\mathrm{d}s = \sqrt{(\mathrm{d}x)^2 + (\mathrm{d}y)^2},$$

we deduce

$$2\mathrm{d}s = l_j \mathrm{d}t, \quad l_j \equiv \sqrt{(x_2 - x_1)^2 + (y_2 - y_1)^2}. \tag{4.2.21}$$

Denoting

$$\begin{aligned} &4a = l_j^2\,, b = (x_j^0 - x_i^0)(x_2 - x_1) + (y_j^0 - y_i^0)(y_2 - y_1)\,, \\ &c = (x_j^0 - x_i^0)^2 + (y_j^0 - y_i^0)^2\,, \\ &I_k = \int_{-1}^{+1} \frac{t^k \mathrm{d}t}{at^2 + bt + c}\,, \quad k = 0, 1\,, \end{aligned} \tag{4.2.22}$$

we get

$$(x - x_i^0)^2 + (y - y_i^0)^2 = at^2 + bt + c \tag{4.2.23}$$

whence,

$$\begin{aligned} U_{ij} &= \frac{l_j}{4\pi}\left[(x_j^0 - x_i^0) I_0 + \frac{x_2 - x_1}{2} I_1\right], \\ V_{ij} &= \frac{l_j}{4\pi}\left[(y_j^0 - y_i^0) I_0 + \frac{y_2 - y_1}{2} I_1\right]. \end{aligned} \tag{4.2.24}$$

Taking the running sense on C into account, we shall have in (4.2.13), (4.2.15), (4.2.17) and (4.2.18):

$$n_x(x_j^0) = -\frac{y_1 - y_2}{l_j}, \quad n_y(x_j^0) = \frac{x_1 - x_2}{l_j},$$
$$n_x(x_i^0) = -\frac{y_1^{(i)} - y_2^{(i)}}{l_i}, \quad n_y(x_i^0) = \frac{x_1^{(i)} - x_2^{(i)}}{l_i}, \tag{4.2.25}$$

where $(x_1^{(i)}, y_1^{(i)})$ and $(x_2^{(i)}, y_2^{(i)})$ are the coordinates of the end points of the segment L_i and l_i is the length of this segment, i.e.

$$l_i = [(x_2^{(i)} - x_1^{(i)})^2 + (y_2^{(i)} - y_1^{(i)})^2]^{1/2}. \tag{4.2.26}$$

We have to notice that in (4.2.24) the integrals I_k may be calculated exactly. Indeed, since

$$\Delta \equiv b^2 - 4ac = -[(y_j^0 - y_i^0)(x_2 - x_1) - (x_j^0 - x_i^0)(y_2 - y_1)]^2 < 0, \tag{4.2.27}$$

we have

$$I_0 = \frac{2}{\sqrt{-\Delta}} \arctan \frac{\sqrt{-\Delta}}{c - a},$$
$$I_1 = \frac{1}{2a} \ln \frac{a + b + c}{a - b + c} - \frac{b}{a\sqrt{-\Delta}} \arctan \frac{\sqrt{-\Delta}}{c - a}. \tag{4.2.28}$$

Taking (4.2.23) into account, we deduce

$$a + b + c = (x_2 - x_i^0) + (y_2 - y_i^0)^2,$$
$$a - b + c = (x_1 - x_i^0)^2 + (y_1 - y_i^0)^2. \tag{4.2.29}$$

For $i = j$ we have $b = c = 0$, such that the integrals I_k become singular. Utilizing the formulas (D.2.3) we obtain

$$I_0 = -2/a, \quad I_1 = 0. \tag{4.2.30}$$

So, the coefficients a_i, A_{ij}, A_i, b_i and B_{ij} depend on the coordinates of the end points of the segments L_j on C. Hence, we may solve the systems (4.2.12) and (4.2.16) using a computer.

4.2.5 The Circular Obstacle

For testing the two methods presented above and the direct method from the following subsection, we shall use the exact solution for the incompressible flow past a circular obstacle. This solution is already known. If O is the origin (center) of the obstacle and R its radius, then the complex velocity in dimensionless variables is (see (5.2.4) from [1.11])

$$U_1 - \mathrm{i}V_1 = U_\infty\left(1 - \frac{R^2}{Z_1^2}\right). \tag{4.2.31}$$

Taking R as reference length and putting $z = \exp(\mathrm{i}\theta)$ it results

$$u = -\cos 2\theta\,, \quad y = -\sin 2\theta \tag{4.2.32}$$

In figures 4.2.3 and 4.2.4 we compare the numerical values obtained for u by mens of (4.2.15) and (4.2.18) with the values obtained by means of (4.2.32). We observe that the solution obtained by means of the distribution of vortices is better. It is also useful to compare the numerical values obtained on the three ways for a quantity of aerodynamic interest. This is the local pressure coefficient defined by the formula

$$C_p = \frac{P_1 - p_\infty}{(1/2)\rho_\infty U_\infty^2}\,, \tag{4.2.33}$$

with the notations from (4.1.2). Taking into account that for the incompressible fluid, the system (4.1.7), for which we have determined the approximate solutions, is the exact system of the equations of motion, we shall calculate also C_p exactly. This may be obtained from Bernoulli's integral

$$\frac{1}{2}U_\infty^2 + \frac{p_\infty}{\rho_\infty} = \frac{1}{2}V_1^2 + \frac{P_1}{\rho_\infty}. \tag{4.2.34}$$

It follows

$$C_p = 1 - V^2 = 1 - (1+u)^2 - v^2\,. \tag{4.2.35}$$

Using this formula, from (4.2.15) we get C_p in the case of the distribution of sources, from (4.2.18) we get C_p in the case of the distribution of vortices and from (4.2.32) we get C_p exactly. We also calculate C_p with the direct method from the following subsection. The graphic representations show that the direct method gives better results than the indirect methods and among these ones, the method based on the distributions of vortices is better.

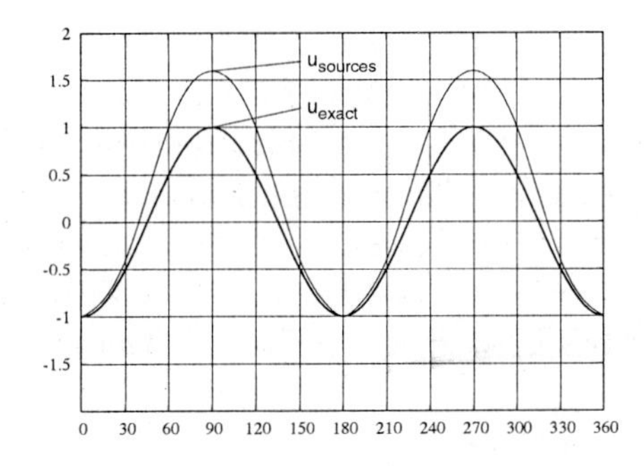

Fig. 4.2.3.

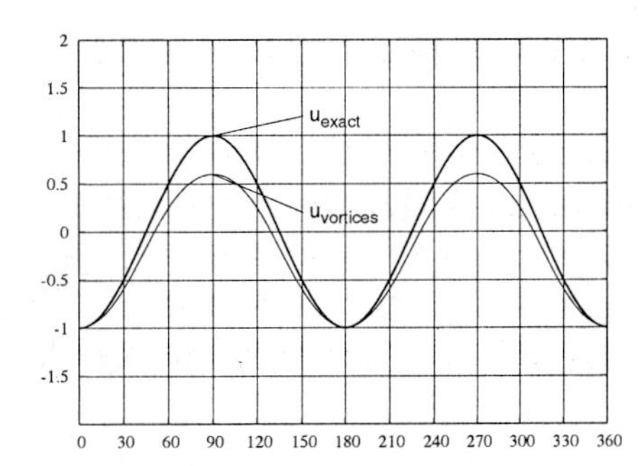

Fig. 4.2.4.

4.2.6 The Elliptical Obstacle

The profiles with smooth boundaries (for example the circle or the ellipse) are *non-lifting* profiles. The profiles with angular points determine a circulatory motion of the fluid and therefore it appears the lift. The circulation (and the lift) are determining from the Kutta-Joukovsky condition (see for example, [1.10] p. 179), hence these profiles are *lifting profiles*. In the first case we encounter d'Alembert's paradox. As we shall see in the following subsection, the indirect methods are utilized mainly for non-lifting bodies, while the direct formulation is suited for writing the $K-J$ condition.

It is well known the exact solution for the elliptic obstacle in the incompressible fluid. The complex potential from the (Z_1) plane is given by the formula (see for example, [1.10] p.178).

$$F(z_1) = \frac{U_\infty}{2}\left[\left(z_1 + \sqrt{z_1^2 - c_1^2}\right)\mathrm{e}^{-\mathrm{i}\alpha} + \right.$$
$$\left. + \left(z_1 - \sqrt{z_1^2 - c_1^2}\right)\frac{a_1 + b_1}{a_1 - b_1}\mathrm{e}^{\mathrm{i}\alpha}\right], \tag{4.2.36}$$

where a_1 and b_1 are the semiaxes, $c_1^2 = a_1^2 - b_1^2$, and α is the angle of attack. Deriving (4.2.36) we deduce the complex velocity which has the same form in dimensionless variables $z = Z, a, b, c$. Hence it follows

$$1 + u - \mathrm{i}v = \frac{1}{2}\left[\left(1 + \frac{z}{\sqrt{z^2 - c^2}}\right)\mathrm{e}^{-\mathrm{i}\alpha} + \right.$$
$$\left. + \left(1 - \frac{z}{\sqrt{z^2 - c^2}}\right)\frac{a + b}{a - b}\mathrm{e}^{\mathrm{i}\alpha}\right]. \tag{4.2.37}$$

We determine the components u and v separating the real part from the imaginary one. The graphic representations for the exact solution and for the solution obtained with the first indirect method, are given in [4.7]. We notice that the results are almost identical.

4.3 The Direct Method for the Unlimited Fluid Case

4.3.1 The representation of the solution

As we have specified in 4.1, in the framework of the direct method we represent the velocity field in the fluid with the aid of its values on the boundary of the body and then we pass to the limit. In order to obtain this representation, we shall consider a domain D exterior to the boundary C and limited by a circle C_R having the radius big enough, such that C should be situated in the interior of C_R. For f and g continuously differentiable function C^1, we have the identity

$$\int_D [f(\partial u/\partial x + \partial v/\partial y) + g(\partial v/\partial x - \partial u/\partial y)]\mathrm{d}a = 0 \tag{4.3.1}$$

which follows from (4.1.7). Noticing that

$$f\partial u/\partial x = \partial(fu)/\partial x - u\partial f/\partial x, \ldots$$

and applying Gauss's formula, we obtain the identity

$$\begin{aligned}\int_D [u(\partial f/\partial x - \partial g/\partial y) + v(\partial f/\partial y + \partial g/\partial x)]\mathrm{d}a = \\ = \int_C [u(fn_x - gn_y) + v(gn_x + fn_y)]\mathrm{d}s\,.\end{aligned} \tag{4.3.2}$$

the integral on C_R vanishing when $R \to \infty$, because of the behaviour at infinity of f and g which will be identified with u^* and v^* and because of the condition (4.1.11).

For $f = u^*$ and $g = -v^*$ (4.1.13), taking (4.1.12) into account, we deduce:

$$\begin{aligned}u(\boldsymbol{\xi}) = \int_C \{u(\boldsymbol{x})\big[u^*(\boldsymbol{x},\boldsymbol{\xi})n_x(\boldsymbol{x}) + v^*(\boldsymbol{x},\boldsymbol{\xi})n_y(\boldsymbol{x})\big] + \\ +v(\boldsymbol{x})\big[u^*(\boldsymbol{x},\boldsymbol{\xi})n_y(\boldsymbol{x}) - v^*(\boldsymbol{x},\boldsymbol{\xi})n_x(\boldsymbol{x})\big]\}\mathrm{d}s\,,\end{aligned} \tag{4.3.3}$$

and for $f = v^*$ and $g = u^*$

$$v(\boldsymbol{\xi}) = \int_C \{u(\boldsymbol{x})\big[v^*(\boldsymbol{x},\boldsymbol{\xi})n_x(\boldsymbol{x}) - u^*(\boldsymbol{x},\boldsymbol{\xi})n_y(\boldsymbol{x})\big] + \\ + v(\boldsymbol{x})\big[u^*(\boldsymbol{x},\boldsymbol{\xi})n_x(\boldsymbol{x}) + v^*(\boldsymbol{x},\boldsymbol{\xi})n_y(\boldsymbol{x})\big]\}\mathrm{d}s\,. \tag{4.3.4}$$

This is the integral representation of the solution, valid for every point $M(\boldsymbol{\xi})$ from the exterior of C.

4.3.2 The Integral Equation

For obtaining the boundary integral equation, we have to pass to the limit in (4.3.3) and (4.3.4), letting the point $M(\boldsymbol{\xi})$ to tend to the generic point $Q_0(\boldsymbol{x}_0)$ belonging to C. To the limit, the integrals containing the fundamental solution (u^*, v^*) become singular. Hence we have to adopt the definition (4.2.4). We calculate the integrals on c, as follows:

$$\lim_{\boldsymbol{\xi}\to\boldsymbol{x}_0} \int_c u(\boldsymbol{x})\big[u^*(\boldsymbol{x},\boldsymbol{\xi})n_x(\boldsymbol{x}) + v^*(\boldsymbol{x},\boldsymbol{\xi})n_y(\boldsymbol{x})\big]\mathrm{d}s =$$

$$= \lim_{\boldsymbol{\xi}\to\boldsymbol{x}_0} \int_c \big[u(\boldsymbol{x}) - u(\boldsymbol{x}_0)\big]\big[u^*(\boldsymbol{x},\boldsymbol{\xi})n_x(\boldsymbol{x}) + \\ + v^*(\boldsymbol{x},\boldsymbol{\xi})n_y(\boldsymbol{x})\big]\mathrm{d}s + u(\boldsymbol{x}_0)L_1\,, \quad \text{where} \tag{4.3.5}$$

$$L_1 = \lim_{\boldsymbol{\xi}\to\boldsymbol{x}_0} \int_c \big[u^*(\boldsymbol{x},\boldsymbol{\xi})n_x(\boldsymbol{x}) + v^*(\boldsymbol{x},\boldsymbol{\xi})n_y(\boldsymbol{x})\big]\mathrm{d}s\,.$$

If $c = \overline{Q_1Q_2}$ is replaced by the segment $Q_1'Q_2'$ (fig. 4.2.1), then we have the parametrizations (4.2.6) and (4.2.7) and $\boldsymbol{n} = \boldsymbol{n}^0$. Hence,

$$L_1 = \frac{1}{2\pi}\lim_{\eta\to 0}\int_{-\varepsilon}^{+\varepsilon} \frac{\eta}{s^2+\eta^2}\mathrm{d}s = \frac{1}{2}\,.$$

Passing to the limit, the first integral from the right hand side of the equality (4.3.5) vanishes when $\varepsilon \to 0$.

Similarly we have:

$$\lim_{\boldsymbol{\xi}\to\boldsymbol{x}_0}\int_c v(\boldsymbol{x})\big[u^*(\boldsymbol{x},\boldsymbol{\xi})n_y(\boldsymbol{x})-v^*(\boldsymbol{x},\boldsymbol{\xi})n_x(\boldsymbol{x})\big]\mathrm{d}s$$

$$=\lim_{\boldsymbol{\xi}\to\boldsymbol{x}_0}\int_c\big[v(\boldsymbol{x})-v(\boldsymbol{x}_0)\big]\big[u^*(\boldsymbol{x},\boldsymbol{\xi})n_y(\boldsymbol{x})-$$

$$-v^*(\boldsymbol{x},\boldsymbol{\xi})n_x(\boldsymbol{x})\big]\mathrm{d}s+v(\boldsymbol{x}_0)L_2\,,\quad\text{where}\tag{4.3.6}$$

$$L_2=\lim_{\boldsymbol{\xi}\to\boldsymbol{x}_0}\int_c\big[u^*(\boldsymbol{x},\boldsymbol{\xi})n_y(\boldsymbol{x})+v^*(\boldsymbol{x},\boldsymbol{\xi})n_x(\boldsymbol{x})\big]\mathrm{d}s=$$

$$=\frac{1}{2\pi}(\tau_x^0n_y^0-\tau_y^0n_x^0)\int_{-\varepsilon}^{+\varepsilon}\frac{s}{s^2+\eta^2}\mathrm{d}s=0\,.$$

Therefore, from (4.3.3) we deduce:

$$\frac{1}{2}u(\boldsymbol{x}_0)=\oint_C'\big\{u(\boldsymbol{x})\big[u^*(\boldsymbol{x},\boldsymbol{x}_0)n_x(\boldsymbol{x})+v^*(\boldsymbol{x},\boldsymbol{x}_0)n_y(\boldsymbol{x})\big]+$$
$$+v(\boldsymbol{x})\big[u^*(\boldsymbol{x},\boldsymbol{x}_0)n_y(\boldsymbol{x})-v^*(\boldsymbol{x},\boldsymbol{x}_0)n_x(\boldsymbol{x})\big]\big\}\mathrm{d}s\,.\tag{4.3.7}$$

In the same way, from (4.3.4) we obtain:

$$\frac{1}{2}v(\boldsymbol{x}_0)=\oint_C'\big\{u(\boldsymbol{x})\big[u^*(\boldsymbol{x},\boldsymbol{x}_0)n_x(\boldsymbol{x})-u^*(\boldsymbol{x},\boldsymbol{x}_0)n_y(\boldsymbol{x})\big]+$$
$$+v(\boldsymbol{x})\big[u^*(\boldsymbol{x},\boldsymbol{x}_0)n_x(\boldsymbol{x})+v^*(\boldsymbol{x},\boldsymbol{x}_0)n_y(\boldsymbol{x})\big]\big\}\mathrm{d}s\,.\tag{4.3.8}$$

The formulas (4.3.7) and (4.3.8) constitute a system of two singular integral equations on C for the unknowns $u(\boldsymbol{x}_0)$ and $v(\boldsymbol{x}_0)$.

Now we introduce the function

$$G=(\beta+u)n_y-vn_x\,.\tag{4.3.9}$$

We may utilize on C the condition (4.1.10). It follows

$$\beta+u=\frac{\beta^2n_y}{n_x^2+\beta^2n_y^2}G\,,\qquad v=-\frac{n_x}{n_x^2+\beta^2n_y^2}G\,.\tag{4.3.10}$$

Replacing these relations in (4.3.7) and (4.3.8) and taking into account the relations

$$\oint_C'(u^*n_x+v^*n_y)\mathrm{d}s=-1/2;\quad\oint_C'(u^*n_y-v^*n_x)\mathrm{d}s=0\,,\tag{4.3.11}$$

which will be demonstrated in (4.3.7), we get:

$$u(\boldsymbol{x}_0) = \beta + 2\oint_C (v^* - M^2 \frac{n_x n_y}{n_x^2 + \beta^2 n_y^2} u^*) G \, \mathrm{d}s$$

$$v(\boldsymbol{x}_0) = 2\oint_C \left(-u^* - M^2 \frac{n_x n_y}{n_x^2 + \beta^2 n_y^2} v^* \right) G \, \mathrm{d}s \tag{4.3.12}$$

and then

$$G(\boldsymbol{x}_0) - 2\oint_C [u^* n_x^0 + v^* n_y^0 + \\ + M^2 \frac{n_x n_y}{n_x^2 + \beta^2 n_y^2} (v^* n_x^0 - u^* n_y^0)] G \, \mathrm{d}s = 2\beta n_y^0 \, , \tag{4.3.13}$$

with the notation $\boldsymbol{n}^0 = \boldsymbol{n}(\boldsymbol{x}_0)$. The equation (4.3.13) is the singular integral equation of the problem. Putting $M = 0$, one obtains the equation for the incompressible fluid.

4.3.3 The Circulation

In the case of the profiles with angular trailing edge, we have to utilize the circulation. Taking the sense defined on C_1 into account and denoting by $\boldsymbol{s}^{(1)}$ and $\boldsymbol{n}^{(1)}$ the (dimensional) versors of the tangent and inward normal, we have

$$s_x^{(1)} = -n_y^{(1)} \, , \; s_y^{(1)} = n_x^{(1)} \, , \tag{4.3.14}$$

such that

$$\Gamma_1 = \oint_{C_1} \boldsymbol{V}_1 \cdot \boldsymbol{s}^{(1)} \mathrm{d}s^{(1)} = -\oint_{C_1} (U_1 n_y^{(1)} - V_1 n_x^{(1)}) \mathrm{d}s^{(1)} = \\ = -U_\infty L_0 \oint_C [(1+U)N_Y - V N_X] \mathrm{d}s \, . \tag{4.3.15}$$

With he changes of variables (4.1.6) and (4.1.9), we deduce

$$\Gamma_1 = -\frac{U_\infty L_0}{\beta} \oint_C [(\beta + u) n_y - v n_x] \mathrm{d}s \, . \tag{4.3.16}$$

Denoting $\Gamma_1 = U_\infty L_0 \Gamma$, Γ representing the dimensionless circulation and utilizing (4.3.9), it results the simple formula:

$$\Gamma = -\frac{1}{\beta} \oint_C G \, \mathrm{d}s \, . \tag{4.3.17}$$

For the incompressible fluid one obtains:

$$\Gamma = -\oint_C G \, \mathrm{d}s \, . \tag{4.3.18}$$

These formulas also emphasize the significance of the function G.

4.3.4 The Discretization of the Equations

Acting like in 4.2.3, we reduce the equation (4.3.13) to the algebraic system

$$(1/2)G_i + \sum_{j=1}^{N} A_{ij}G_j = B_i, \quad i = 1, 2, \ldots, N, \tag{4.3.19}$$

where

$$\begin{aligned} G_i &= G(\boldsymbol{x}_i^0), \; B_i = \beta n_y(\boldsymbol{x}_i^0), \\ A_{ij} &= -n_x(\boldsymbol{x}_i^0)U_{ij} - n_y(\boldsymbol{x}_i^0)V_{ij} - \\ & -M^2 \frac{n_x(\boldsymbol{x}_j^0)n_y(\boldsymbol{x}_j^0)}{n_x^2(\boldsymbol{x}_j^0) + \beta^2 n^2(\boldsymbol{x}_j^0)} \left[n_x(\boldsymbol{x}_i^0)V_{ij} - n_y(\boldsymbol{x}_i^0)U_{ij} \right], \end{aligned} \tag{4.3.20}$$

U_{ij} and V_{ij} being defined (4.2.14).

After determining the unknowns G_1 from (4.3.19), we may determine the circulation from the formula:

$$-\beta\Gamma = \sum_{j=1}^{N} l_j G_j, \tag{4.3.21}$$

l_j representing the length of the segment L_j. We deduce the components of the velocity

$$\beta + u_i = \frac{\beta^2 n_y(\boldsymbol{x}_i^0)G_i}{n_x^2(\boldsymbol{x}_i^0) + \beta^2 n_y^2(\boldsymbol{x}_i^0)}, \quad v_i = -\frac{n_x(\boldsymbol{x}_i^0)G_i}{n_x^2(\boldsymbol{x}_i^0) + \beta^2 n_y^2(\boldsymbol{x}_i^0)}, \tag{4.3.22}$$

from (4.3.10). Obviously, for $\boldsymbol{n}(\boldsymbol{x}_i^0)$ and $\boldsymbol{n}(\boldsymbol{x}_j^0)$ we shall utilize the expressions (4.2.25) and for U_{ij} and V_{ij}, (4.2.24).

4.3.5 The Lifting Profile

If the profile has an angular point (fig. 4.3.1), or a cusp, like the Joukovsky-type profiles , then we have to determine the circulation using the Kutta-Joukovsky rule.

The circulation determines the lift. For determining the circulation, we impose the equality of the pressure in the points P_s (on the upper side of the angle) and P_i (on the lower side of the angle) when

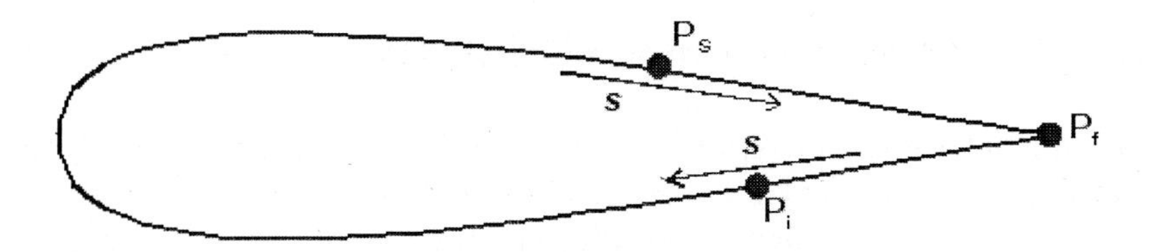

Fig. 4.3.1.

these points (P_s and P_i) tend to the angular trailing edge P_f. From Bernoulli's integral it follows that the equality of the pressures implies the equality of the tangential components of the velocity, (according to the slipping condition the normal components are null). Taking into account the orientation of the versor of the tangent $\boldsymbol{s}^{(1)}$ on the upper and lower sides of the angle (fig. 4.3.1), we deduce the condition:

$$V_1(P_s) \cdot \boldsymbol{s}^{(1)} + \boldsymbol{V}_1(P_i) \cdot \boldsymbol{s}^{(1)} = 0 \,. \tag{4.3.23}$$

With the notation (4.3.9), we have:

$$\begin{aligned}\boldsymbol{V}_1 \cdot \boldsymbol{s}^{(1)} &= U_\infty[(1+U)s_X^{(1)} + Vs_Y^{(1)}] = \\ &= U_\infty\left[-\left(1+\frac{u}{\beta}\right)n_y + v\frac{n_x}{\beta}\right]\frac{\mathrm{d}s}{\mathrm{d}S} = -\frac{U_\infty}{\beta}G\frac{\mathrm{d}s}{\mathrm{d}S}\,,\end{aligned}$$

such that (4.3.23) implies

$$G(P_s) + G(P_i) = 0 \,. \tag{4.3.24}$$

Consequently, for such a profile, we have to add the condition (4.3.24) to the system (4.3.19). In order to solve this problem we introduce [4.18] the regularization variable λ and we obtain the system:

$$\begin{aligned}&\lambda + (1/2)G_i + \sum_j A_{ij}G_j = \beta_i \,, \quad i = 1,\ldots,N \,, \\ &G(P_s) + G(P_i) = 0 \,,\end{aligned} \tag{4.3.25}$$

for the unknowns $G_1, \ldots, G_N$, λ. In this way the number of equations equals the number of unknowns. For big values of N we obtain a small λ i.e. $(\lambda \to 0)$.

4.3.6 The Local Pressure Coefficient

As we have seen in 4.2.5, the local pressure coefficient for the incompressible fluid (4.2.33) is (4.2.35). Let us deduce the expression of the local pressure coefficient for the compressible fluid. For the potential flow we have the formulas [1.10] page 206 which become, using the notations from this chapter:

$$\rho_1 = \rho_0 \left(1 - \frac{\gamma - 1}{2}\frac{V_1^2}{c_0^2}\right)^{\frac{1}{\gamma-1}},$$

$$P_1 = p_0 \left(1 - \frac{\gamma - 1}{2}\frac{V_1^2}{c_0^2}\right)^{\frac{\gamma}{\gamma-1}}, \tag{4.3.26}$$

$$c_1^2 = c_0^2 \left(1 - \frac{\gamma - 1}{2}\frac{V_1^2}{c_0^2}\right).$$

For the uniform flow that we have in view, these formulas give

$$\rho_\infty = \rho_0 \left(1 - \frac{\gamma - 1}{2}\frac{U_\infty^2}{c_0^2}\right)^{\frac{1}{\gamma-1}},$$

$$p_\infty = p_0 \left(1 - \frac{\gamma - 1}{2}\frac{U_\infty^2}{c_0^2}\right)^{\frac{\gamma}{\gamma-1}}, \tag{4.3.27}$$

$$c_\infty^2 = c_0^2 \left(1 - \frac{\gamma - 1}{2}\frac{U_\infty^2}{c_0^2}\right).$$

From the last relation we deduce

$$k \equiv \frac{U_\infty^2}{c_0^2} = \left(\frac{1}{M^2} + \frac{\gamma - 1}{2}\right)^{-1}, \tag{4.3.28}$$

such that, denoting

$$\tau = \frac{\gamma - 1}{2} M^2, \tag{4.3.29}$$

it follows

$$\frac{1}{k(\gamma - 1)} = 1 + \frac{1}{\tau}. \tag{4.3.30}$$

Introducing the parameter k in the formulas (4.3.26) and (4.3.27), we get:

$$P_1 = p_0 \left(1 - \frac{\gamma - 1}{2} \frac{V_1^2}{U_\infty^2} k\right)^{\frac{1}{\gamma-1}},$$

$$p_\infty = p_0 \left(1 - \frac{\gamma - 1}{2} k\right)^{\frac{\gamma}{\gamma-1}}, \tag{4.3.31}$$

$$\rho_\infty = \rho_0 \left(1 - \frac{\gamma - 1}{2} k\right)^{\frac{1}{\gamma-1}}.$$

These formulas will be replaced in (4.2.27). Taking out the factor $(\gamma - 1)k/2$ and utilizing (4.3.30), it results

$$C_p = \frac{2}{\gamma M^2} \left\{ -1 + \frac{\gamma - 1}{2} M^2 \left(1 - \frac{V_1^2}{U_\infty^2}\right) \right]^{\frac{\gamma}{\gamma-1}} - 1 \right\}. \tag{4.3.32}$$

For the incompressible fluid $(M = 0)$ one obtains the formula (4.2.29). For the compressible fluid, V_1^2/U_∞^2 from (4.3.32) will be replaced by $(1 + u/\beta)^2 + v^2$.

We shall employ the local pressure coefficient in order to define *the lift coefficient* which gives a global measure of the action of the fluid against the profile

$$c_L = \frac{1}{L_1} \int_{C_1} C_p n_y^{(1)} \mathrm{d}s_1, \tag{4.3.33}$$

where L_1 is the length of the contour C_1.

It is also important to get the expression of the critical velocity v_{cr} because the subsonic character of the flow is maintained only if $V_1 < v_{cr}$. This velocity is obtained from (4.3.26) putting $v_1 = c_1$ and utilizing the definition of k. We obtain:

$$v_{cr}^2 = U_\infty^2 \left[1 + \frac{2}{\gamma - 1} \left(\frac{1}{M^2} - 1\right)\right]. \tag{4.3.34}$$

4.3.7 Appendix

In this subsection we shall prove the formulas (4.3.11). To this aim, we denote by C_ε the arc (semicircle) of the circle of radius ε with the center in Q_0, interior to the profile (fig. 4.3.2) and we notice that we have the parametrization

$$x - x_0 = \varepsilon \cos\theta, \quad y - y_0 = \varepsilon \sin\theta$$
$$\theta_0 \le \theta < \theta_0 + \pi \tag{4.3.35}$$

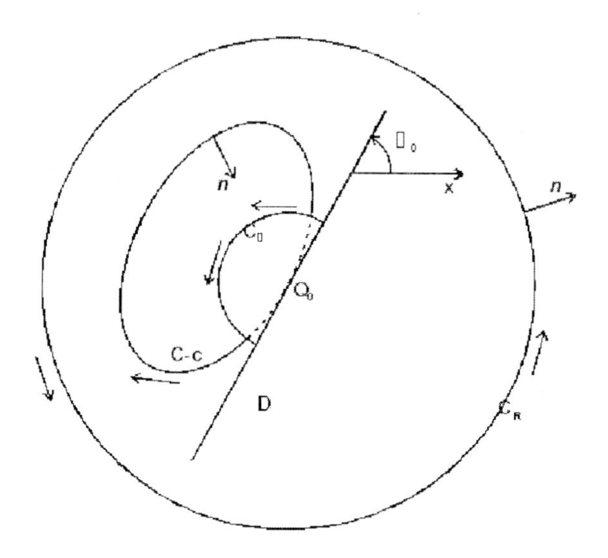

Fig. 4.3.2.

which, for u^*, v^* (4.1.13) implies:

$$u^*(\boldsymbol{x}, \boldsymbol{x}_0) = \frac{1}{2\pi} \frac{\cos\theta}{\varepsilon}, v^*(\boldsymbol{x}, \boldsymbol{x}) = \frac{1}{2\pi} \frac{\sin\theta}{\varepsilon}. \tag{4.3.36}$$

We also consider the circle C_R with the center in Q_0 and the radius R big enough, such that the profile C is in the interior of the circle. On this circle we have the parametrization

$$x - x_0 = R\cos\varphi, \quad y - y_0 = R\sin\varphi, \tag{4.3.37}$$

such that

$$u^* = \frac{1}{2\pi} \frac{\cos\varphi}{R}, \quad v^* = \frac{1}{2\pi} \frac{\sin\varphi}{R}.$$

In the domain D, exterior to the contour $C - c + C_\varepsilon$ and interior to the circle C_R, we may integrate the equation:

$$\partial u^*/\partial x + \partial v^*/\partial y = \delta(x - x_0, y - y_0). \tag{4.3.38}$$

Applying Green's formula we get:

$$\begin{aligned} &\oint_C (u^* n_x + v^* n_y) \mathrm{d}s + \oint_{C_\varepsilon} (u^* n_x + v^* n_y) \mathrm{d}s + \\ &+ \oint_{C_R} (u^* n_x + v^* n_y) \mathrm{d}s = 1, \end{aligned} \tag{4.3.39}$$

or,

$$\oint_C [u^*(\boldsymbol{x}, \boldsymbol{x}_0) n_x(\boldsymbol{x}) + v^*(\boldsymbol{x}, \boldsymbol{x}_0) n_y(\boldsymbol{x})] \mathrm{d}s + \\ + \frac{1}{2\pi} \int_{\theta_0}^{\theta_0+\pi} \mathrm{d}\theta + \frac{1}{2\pi} \int_0^{2\pi} \mathrm{d}\varphi = 1 .$$

One obtains the first of the relations (4.3.11). Integrating on D the equation

$$\partial u^*/\partial y - \partial v^*/\partial x = 0 ,$$

one obtains the second of the relations (4.3.11).

4.3.8 Numerical Determinations

In figure 4.3.3 we compare u calculated by means of the direct method with u calculated exactly. We observe that the values are almost the same. In figure 4.3.4 we give 4 graphic representations: u exact, u direct, u sources, u vortices. We notice that u obtained by means of distributions of sources (4.2.15) does not give very good results and u obtained by means of distributions of vortices (4.2.18) gives better results. The values of u obtained with the direct method are constantly in the vicinity of exact values of u .

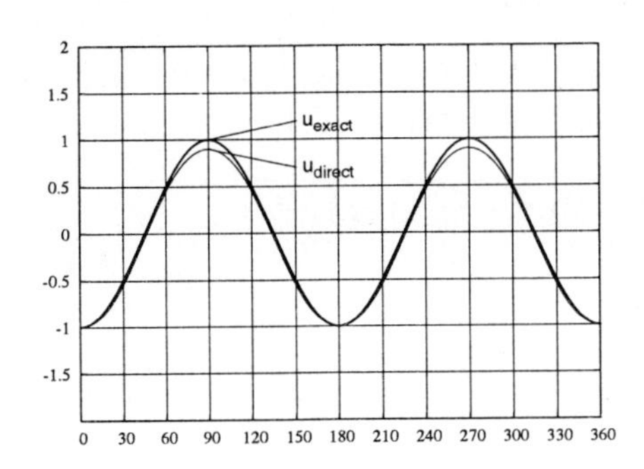

Fig. 4.3.3.

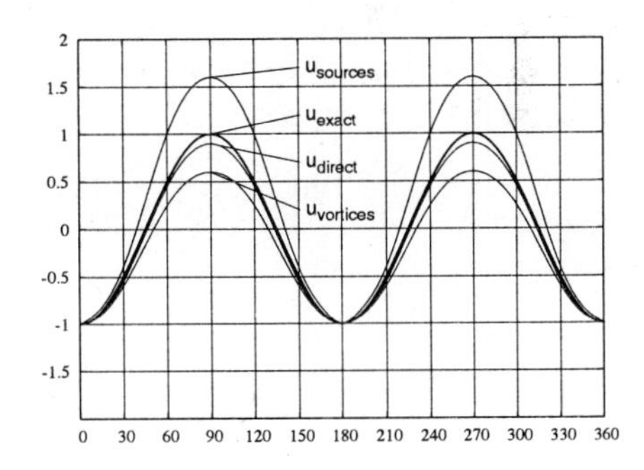

Fig. 4.3.4.

4.4 The Airfoil in Ground Effects

4.4.1 The Representation of the Solution

When the airplane is landing or taking off we have to take into account the influence of the ground. We shall consider this problem herein

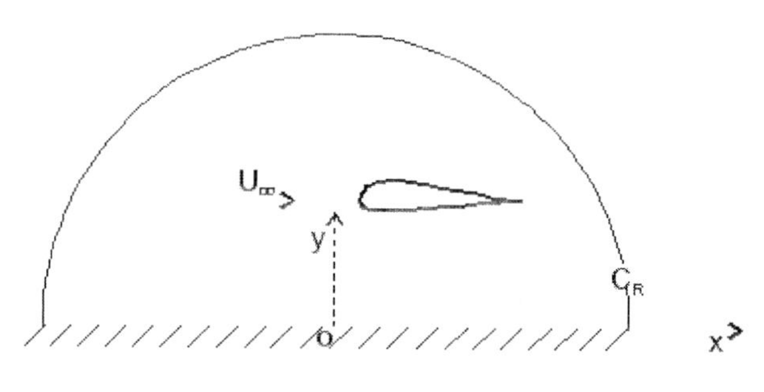

Fig. 4.4.1.

replacing the ground with the x_1Oz_1 plane (fig. 4.4.1). We have the same problem like in the previous sections. An uniform subsonic stream, having the velocity $U_\infty \boldsymbol{i}$, the pressure p_∞ and the density ρ_∞, flowing in the $y_1 > 0$ half-space, is perturbed by the presence of an infinite cylindrical body, having the generatrices parallel to x_1Oz_1, the Ox_1 axis having the direction of the stream. The x_1Oy_1 plane determines a cross section whose boundary C_1 is assumed to be a thin profile. Using the same notations like in 4.1 we have to integrate the system (4.1.7) in the half-plane $y > 0$, with the boundary condition (4.1.10) on C and the supplementary condition

$$v(x,0) = 0, \quad (\forall)x. \tag{4.4.1}$$

In this case, instead of the fundamental solution (u^*, v^*) from 4.1.3 we shall utilize Green's distribution (U^*, V^*) determined by the system

$$\begin{aligned} &\partial U^*_\pm/\partial x + \partial V^*_\pm/\partial y = \delta(x-\xi, y-\eta) \\ &\partial V_\pm/\partial x - \partial U_\pm/\partial y = 0\,, \end{aligned} \tag{4.4.2}$$

$\boldsymbol{\xi} = (\xi, \eta)$ representing a point from the $y > 0$ half-plane, and the boundary condition

$$U^*_-(x,0) = 0\,, \quad V^*_+(x,0) = 0\,, \quad (\forall)x. \tag{4.4.3}$$

This distribution is obtained by means on the fundamental solution (4.1.13) using the method of images. Denoting

$$\begin{aligned} u^*(x,y,\xi,\eta) &= \frac{1}{2\pi}\,\frac{x-\xi}{(x-\xi)^2+(y-\eta)^2}\,, \\ v^*(x,y,\xi,\eta) &= \frac{1}{2\pi}\,\frac{y-\eta}{(x-\xi)^2+(y-\eta)^2}\,, \end{aligned} \tag{4.4.4}$$

we have

$$U_{\pm}^{*} = u^{*}(x,y,\xi,\eta) \pm u^{*}(x,y,\xi,-\eta)$$
$$V_{\pm}^{*} = v^{*}(x,y,\xi,\eta) \pm v^{*}(x,y,\xi,-\eta)\,. \tag{4.4.5}$$

Indeed, $u^{*}(x,y,\xi,\eta)$ and $v^{*}(x,y,\xi,\eta)$ verify the system (4.1.12) for every $\boldsymbol{\xi}$ in the xOy plane. Obviously, $u^{*}(x,y,\xi,-\eta)$ and $v^{*}(x,y,\xi,-\eta)$ verify the system

$$\partial u^{*}/\partial x + \partial v^{*}/\partial y = \delta(x-\xi, y+\eta)$$
$$\partial v^{*}/\partial x - \partial u^{*}/\partial y = 0\,. \tag{4.4.6}$$

When $\boldsymbol{\xi}$ is in the superior half-plane $(y > 0)$ this system becomes

$$\partial u^{*}/\partial x + \partial v^{*}/\partial y = 0$$
$$\partial v^{*}/\partial x + \partial u^{*}/\partial y = 0\,. \tag{4.4.7}$$

With these specifications one easily finds that (4.4.5) verifies (4.4.2). It is also very easy to verify the conditions (4.4.3).

Let us return now to the representation of the solution. Denoting by D the domain from the superior half-plane $y > 0$ which is exterior to C and delimited by a half-circle C_R with the center in the origin, the diameter on Ox and the radius big enough in order to contain C in the interior (fig. 4.4.1), we deduce like in 4.3, for $R \to \infty$, the identity

$$\int_D \big[u(\partial f/\partial x - \partial g/\partial y) + v(\partial f/\partial y + \partial g/\partial x)\big]\mathrm{d}s = \tag{4.4.8}$$
$$= \oint_C \big[(uf+vg)n_x + (vf-ug)n_y\big]\mathrm{d}s + \int_{-\infty}^{+\infty} u(x,0)g(x,0)\mathrm{d}x\,.$$

We took into account the fact that on the diameter of the half-circle we have $\boldsymbol{n} = (0,-1)$ and that the integral on C_R vanishes for $R \to \infty$ when f and g are distributions having the form $U_{\pm}^{*}, V_{\pm}^{*}$.

Putting now $f = U_{+}^{*}$ and $g = -V_{+}^{*}$, from (4.4.2) and (4.4.3), we deduce

$$u(\boldsymbol{\xi}) = \int_C \big\{u(\boldsymbol{x})\big[U_{+}^{*}(\boldsymbol{x},\boldsymbol{\xi})n_x(\boldsymbol{x}) + V_{+}^{*}(\boldsymbol{x},\boldsymbol{\xi})n_y(\boldsymbol{x})\big] +$$
$$+ v(\boldsymbol{x})\big[U_{+}^{*}(\boldsymbol{x},\boldsymbol{\xi})n_y(\boldsymbol{x}) - V_{+}^{*}(\boldsymbol{x},\boldsymbol{\xi})n_x(\boldsymbol{x})\big]\big\}\mathrm{d}s\,. \tag{4.4.9}$$

In the same way, for $f = V_-^*$ and $g = U_-^*$, from (4.4.9) we obtain

$$v(\boldsymbol{\xi}) = \oint_C \{u(\boldsymbol{x})[V_-^*(\boldsymbol{x},\boldsymbol{\xi})n_x(\boldsymbol{x}) - V_-^*(\boldsymbol{x},\boldsymbol{\xi})n_y(\boldsymbol{x})] + \\ + v(\boldsymbol{x})[U_-^*(\boldsymbol{x},\boldsymbol{\xi})n_x(\boldsymbol{x}) + V_-^*(\boldsymbol{x},\boldsymbol{\xi})n_y(\boldsymbol{x})]\}\mathrm{d}s\,. \tag{4.4.10}$$

Formally, the solution is identical to (4.3.3) and (4.3.4). It differs only by $U_\pm^*, V_\pm^*$.

4.4.2 The Integral Equation

Acting like in 4.3, we deduce

$$\begin{aligned}(1/2)u(\boldsymbol{x}_0) &= \oint_C{}' \{u(\boldsymbol{x})[U_+(\boldsymbol{x},\boldsymbol{x}_0)n_z(\boldsymbol{x}) + V_+(\boldsymbol{x},\boldsymbol{x}_0)n_y(\boldsymbol{x})] + \\ &\quad + v(\boldsymbol{x})[U_+^*(\boldsymbol{x},\boldsymbol{x}_0)n_y(\boldsymbol{x}) - V_+^*(\boldsymbol{x},\boldsymbol{x}_0)n_y(\boldsymbol{x})]\}\mathrm{d}s\,,\\ (1/2)v(\boldsymbol{x}_0) &= \oint_C{}' \{u(\boldsymbol{x})[V_-^*(\boldsymbol{x},\boldsymbol{x}_0)n_x(\boldsymbol{x}) - U_-^*(\boldsymbol{x},\boldsymbol{x}_0)n_y(\boldsymbol{x})] + \\ &\quad + v(\boldsymbol{x})[U_-^*(\boldsymbol{x},\boldsymbol{x}_0)n_x(\boldsymbol{x}) + V_-^*(\boldsymbol{x},\boldsymbol{x}_0)n_y(\boldsymbol{x})]\}\mathrm{d}s\,.\end{aligned} \tag{4.4.11}$$

Utilizing the identities

$$\oint_C (U_+^* n_x + V_+^* n_y)\mathrm{d}s = -\frac{1}{2}, \quad \oint_C{}' (U_-^* n_y - V_-^* n_x)\mathrm{d}s = 0 \tag{4.4.12}$$

which will be demonstrated at the end of this section, we obtain for G defined by (4.3.9), the following integral equation:

$$\begin{aligned}(1/2)G(\boldsymbol{x}_0) - \oint_C{}' &\{U_-^*(\boldsymbol{x},\boldsymbol{x}_0)n_x(\boldsymbol{x}_0) + V_+^*(\boldsymbol{x},\boldsymbol{x}_0)n_y(\boldsymbol{x}_0) + \\ &+ M^2 \frac{n_x(\boldsymbol{x})n_y(\boldsymbol{x})}{n_x^2(\boldsymbol{x}) + \beta^2 n_y^2(\boldsymbol{x})}[V_-^*(\boldsymbol{x},\boldsymbol{x}_0)n_x(\boldsymbol{x}) - \\ &- U_+^*(\boldsymbol{x},\boldsymbol{x}_0)n_y(\boldsymbol{x})]\}G\mathrm{d}s = \beta n_y(\boldsymbol{x}_0)\,,\end{aligned} \tag{4.4.13}$$

where the mark "prime" has the same significance like in (4.2.4).

The formulas (4.3.10) which determine u and v in function of G, remain valid.

4.4.3 The Computer Implementation

The equation (4.4.13) is discretized like in 4.3.4. One obtains the system

$$(1/2)G_i + \sum_{j=1}^{N} A_{ij}G_j = B_i, \quad i = 1, \ldots, N, \tag{4.4.14}$$

where

$$\begin{aligned} A_{ij} &= -n_x(\boldsymbol{x}_i^0)U_{ij}^- - n_y(\boldsymbol{x}_i^0)V_{ij}^+ - \\ &-M^2 \frac{n_x(\boldsymbol{x}_j^0)n_y(\boldsymbol{x}_j^0)}{n_x^2(\boldsymbol{x}_j^0) + \beta^2 n_y^2(\boldsymbol{x}_j^0)} \left[n_x(\boldsymbol{x}_i^0)V_{ij}^- - n_y(\boldsymbol{x}_i^0)U_{ij}^+\right], \end{aligned} \tag{4.4.15}$$

with the notations

$$U_{ij}^{\pm} = \int_{L_j} U_{\pm}^*(\boldsymbol{x}, \boldsymbol{x}_i^0)\mathrm{d}s\,, \quad V_{ij}^{\pm} = \int_{L_j} V_{\pm}(\boldsymbol{x}, \boldsymbol{x}_i^0)\mathrm{d}s\,, \tag{4.4.16}$$

B_i remaining unchanged. The system (4.4.14) will determine the unknowns $G_1, \ldots, G_N$. From (4.3.22) it follows $u_i, v_i, (i = 1, \ldots, N)$. The circulation is obtained with the formula (4.3.21).

For lifting airfoils, we also consider the Kutta-Joukovsky condition (4.3.24). We solve the problem introducing the regularization variable λ and acting like in 4.3.5.

In order to obtain the expressions of the unknowns $U_{ij}^{\pm}$ and $V_{ij}^{\pm}$, we use the notations (4.2.22) and the relations

$$\begin{aligned} b' &= (x_j^0 - x_i^0)(x_2 - x_1) + (y_j^0 + y_i^0)(y_2 - y_1)\,, \\ c' &= (x_j^0 - x_i^0)^2 + (y_j^0 + y_i^0)^2 \quad I_k' = \int_{-1}^{+1} \frac{t^k \mathrm{d}t}{at^2 + b't + c'}\,. \end{aligned} \tag{4.4.17}$$

In this way one obtains

$$(x - x_i^0)^2 + (y + y_i^0)^2 = at^2 + b't + c'\,,$$

$$U_{ij}^{\pm} = \frac{l_j}{4\pi}\left[(x_j^0 - x_i^0)(I_0 \pm I_0') + \frac{x_2 - x_1}{2}(I_1 \pm I_1')\right], \tag{4.4.18}$$

$$V_{ij}^{\pm} = \frac{l_j}{4\pi}\left[(y_j^0 - y_i^0)I_0 \pm (y_j^0 + y_i^0)I_0' + \frac{y_2 - y_1}{2}(I_1 \pm I_1')\right],$$

$\boldsymbol{n}(x_i^0)$ and $\boldsymbol{n}(x_j^0)$ being defined in (4.2.25).

For I_0 and I_1 one obtains the expressions (4.2.28). Since

$$\Delta \equiv b'^2 - 4ac' = -\left[(y_j^0 + y_i^0)(x_2 - x_1) - (x_j^0 - x_i^0)(y_2 - y_1)\right]^2 < 0\,,$$

it follows

$$\begin{aligned} I_0' &= \frac{2}{\sqrt{-\Delta'}} \arctan \frac{\sqrt{-\Delta'}}{c' - a}\,, \\ I_1' &= \frac{1}{2a} \ln \frac{a + b' + c'}{a - b' + c'} - \frac{b'}{a\sqrt{-\Delta'}} \arctan \frac{\sqrt{-\Delta'}}{c' - a}\,, \end{aligned} \tag{4.4.19}$$

where

$$\begin{aligned} a + b' + c' &= (x_2 - x_i^0)^2 + (y_2 + y_i^0)^2 \\ a - b' + c' &= (x_1 - x_i^0)^2 + (y_1 + y_i^0)^2\,. \end{aligned} \tag{4.4.20}$$

For $i = j$ the integrals I_0 and I_1 become singular and they have the values (4.2.30).

4.4.4 The Treatment of the Method

The problem presented in this subsection was studied in [4.10] and we present with small modifications the solution given in that paper. For testing the method we use the exact solution which is available for the circular obstacle in ground effects (fig. 4.4.2), [4.12].

If the center of the circle is in the point $(0, b)$ and the radius is a $(a < b)$, then, employing the usual notations, we have the exact solution:

$$u - \mathrm{i}v = -2 \sum_{n \geq 1} \frac{c_n}{z^2 + b_n^2} + 4 \sum_{n \geq 1} \frac{c_n b_n^2}{z^2 + b_n^2}\,, \tag{4.4.21}$$

where the sequences $\{b_n\}$ and $\{c_n\}$ are defined by the following formulas:

$$\begin{aligned} b_{n+1} &= b - \frac{a^2}{b + b_n}\,, \quad c_{n+1} = \frac{a^2 c_n}{(b + b_n)^2} n \geq 0\,, \\ b_1 &= b\,, \quad c_1 = a^2\,. \end{aligned} \tag{4.4.22}$$

Decomposing (4.4.21) in simple fractions, we get

$$u - \mathrm{i}v = -\sum_{n \geq 1} c_n \left[\frac{1}{(z - \mathrm{i}b_n)^2} + \frac{1}{(z + \mathrm{i}b_n)^2}\right] \tag{4.4.23}$$

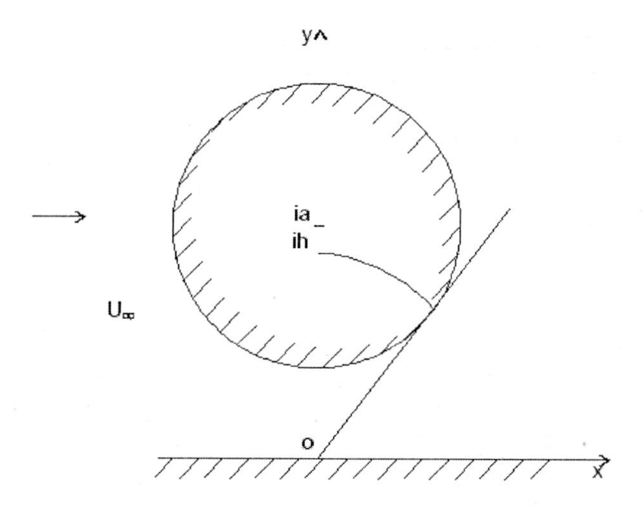

Fig. 4.4.2.

whence, separating the real part from the imaginary one, we deduce

$$
\begin{aligned}
u &= -\sum_{n\geq 1} c_n \left\{ \frac{x^2-(y-b_n)^2}{[x^2+(y-b_n)^2]^2} + \frac{x^2-(y+b_n)^2}{[x^2+(y+b_n)^2]^2} \right. , \\
v &= -2x \sum_{n\geq 1} c_n \left\{ \frac{y-b_n}{[x^2+(y-b_n)^2]^2} + \frac{y+b_n}{[x^2+(y+b_n)^2]^2} \right. .
\end{aligned}
\tag{4.4.24}
$$

Putting

$$x = a\cos\theta\,, \quad y = b + a\sin\theta\,,$$

we find $u|_c$ and $v|_c$.

Using the formula (4.2.35), we represent in figure 4.4.3 the exact solution by a continuous curve and the approximate solution (in the case of the incompressible fluid, $M = 0$) by small squares. We considered $\Gamma = 0$, $a = 1$ and $b = 1.1$. Since the two solutions practically coincide, we conclude that the method presented herein is very good. It may be applied for every shape of the airfoil.

4.4.5 The Circular Obstacle in a Compressible Fluid

In this case there is not available any exact solution, such that we have only the information provided by the BIEM. We have to take care to use

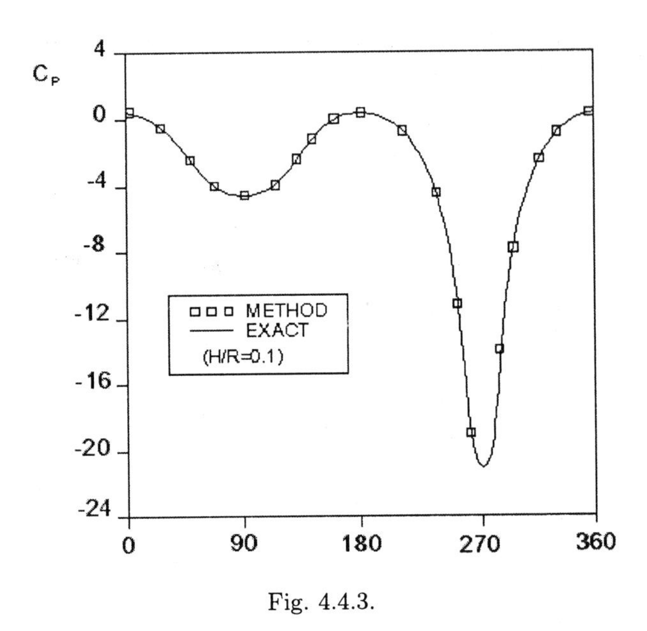

Fig. 4.4.3.

for the radius of the circle values that are small with respect to the other characteristic lengths (for example, the distance to the ground) because only in these circumstances we can employ the linearized equations. In figure 4.4.4, we represent the coefficient C_L defined by (4.3.33) for the circular obstacle with the radius $a = 1$, considering 60 nodes on the boundary, . Obviously, the compressibility determines the increase of the lift (as we have already seen in the case of the free fluid). We considered $\gamma = 1.405$. Giving the distance to the ground, we may calculate (in the absence of the circulation) the values of M for which the flow remains subsonic (one utilizes $v_{\rm cr}$ given by the formula (4.3.34).

4.4.6 Appendix

In order to prove the formulas (4.4.12) we take into account that in the entire plane, the solutions $(U_\pm^*, V_\pm^*)$ satisfy the systems

$$\begin{aligned} &\partial U_\pm^*/\partial x + \partial V_\pm^*/\partial y = \delta(x-\xi, y-\eta) \pm \delta(x-\xi, y+\eta) \\ &\partial V_\pm^*/\partial x - \partial U_\pm^*/\partial y = 0\,, \end{aligned} \tag{4.4.25}$$

where $\eta > 0$.

Like in 4.3.7, we denote by C_ε the arc of circle interior to the profile C, having the center in $Q_0(\boldsymbol{x}_0)$ and the radius ε and by C_R the

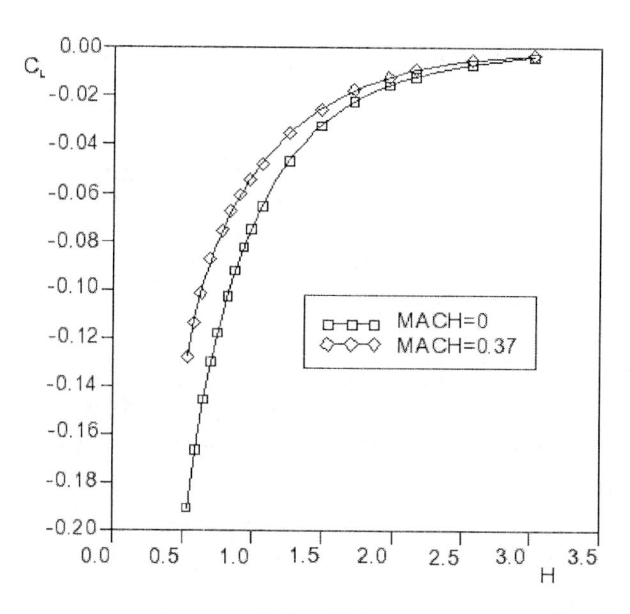

Fig. 4.4.4.

circle having the center in Q_0 and the radius R big enough, such that the profile C belong to the interior of the circle. The domain exterior to the curve $C - c + C_\varepsilon$ and interior to the circle C_R will be denoted by D. We shall integrate on this domain the first equation from (4.4.25) and and we shall apply Green's formula. One obtains:

$$2 = \oint_C [U_+^*(\boldsymbol{x}, \boldsymbol{x}_0) n_x(\boldsymbol{x}) + V_+^*(\boldsymbol{x}, \boldsymbol{x}_0) n_y(\boldsymbol{x})] \mathrm{d}s + \\ + \lim_{\varepsilon \to 0} \oint (U_+^* n_x + V_+^* n_y) \mathrm{d}s + \lim_{R \to \infty} \oint_{C_R} (U_+^* n_x + V_+^* n_y) \mathrm{d}s \,. \tag{4.4.26}$$

With the parametrization (4.3.37) we deduce on C_R:

$$\frac{x - x_0}{(x - x_0)^2 + (y - y_0)^2} = \frac{\cos\varphi}{R},$$

$$\frac{x - x_0}{(x - x_0)^2 + (y + y_0)^2} = \frac{x - x_0}{R^2 + 4y_0(y - y_0) + 4y_0^2} = \\ = \frac{\cos\varphi}{R}\left(1 + 4\frac{y_0}{R}\sin\varphi + 4\frac{y_0^2}{R^2}\right)^{-1} = \frac{\cos\varphi}{R} + O(R^{-2})$$

whence,

$$U_+^* = \frac{1}{\pi}\frac{\cos\varphi}{R} + O(R^{-2}),\ V_+^* = \frac{1}{\pi}\frac{\sin\varphi}{R} + O(R^{-2})$$

$$U_-^* = O(R^{-2}), \quad V_-^* = O(R^{-2}).$$

Utilizing the parametrization (4.3.35) on C_ε we have for example

$$\frac{y+y_0}{(x-x_0)^2(y+y_0)^2} = \frac{y-y_0+2y_0}{(x-x_0)^2+(y-y_0)^2+4y_0(y-y_0)+4y_0^2} =$$

$$= \frac{1}{2y_0}\left(1+\varepsilon\frac{\sin\theta}{2y_0}\right)\left(1+\varepsilon\frac{\sin\theta}{y_0}+\frac{\varepsilon^2}{4y_0^2}\right)^{-1} = \frac{1}{2y_0}+O(\varepsilon).$$

Hence,

$$U_\pm^* = \frac{1}{2\pi}\frac{\cos\theta}{\varepsilon}+O(\varepsilon), \; V_\pm^* = \frac{1}{2\pi}\frac{\sin\theta}{\varepsilon} \pm \frac{1}{4\pi y_0}+O(\varepsilon).$$

Taking into account that on C_ε we have $\mathrm{d}\,s = \varepsilon\mathrm{d}\,\theta$ and on C_R, $\mathrm{d}\,s = R\mathrm{d}\varphi$, we get:

$$\lim_{\varepsilon\to 0}\oint_{C_\varepsilon}(U_+^* n_x + V_+^* n_y)\mathrm{d}\,s = \frac{1}{2\pi}\int_{\theta_0}^{\theta_0+\pi}\mathrm{d}\theta = \frac{1}{2},$$

$$\lim_{R\to 0}\oint_{C_R}(U_+^* n_x + V_+^* n_y)\mathrm{d}\,s = \frac{1}{\pi}\int_0^{2\pi}\mathrm{d}\varepsilon = 2.$$

Replacing all these in (4.4.26) it follows the first relation (4.4.12).

4.5 The Airfoil in Tunnel Effects

4.5.1 The Representation of the Solution

Assuming that the airfoil is situated between two planes parallel to the direction of the unperturbed stream (fig. 4.5.1), we have to integrate the system (4.1.7) in the domain exterior to the contour C and bounded by the two planes, with the boundary conditions (4.1.10) and the supplementary condition

$$v(x,0) = 0, v(x,a) = 0 \quad (\forall)x, \tag{4.5.1}$$

Denoting by D the domain of motion delimited by the sides $x = \pm d$ (fig. 4.5.2), we deduce for $d \to \infty$ the identity

$$\int_D [u(\partial f/\partial x - \partial g/\partial y) + v(\partial f/\partial y + \partial g/\partial x)]\mathrm{d}a =$$

$$= \int_C [(fu+gv)n_x + (fv-gu)n_y]\mathrm{d}s + \tag{4.5.2}$$

$$+ \int_{-\infty}^{+\infty} g(x,0)u(x,0)\mathrm{d}\,x - \int_{+\infty}^{-\infty} g(x,a)u(x,a)\mathrm{d}\,x.$$

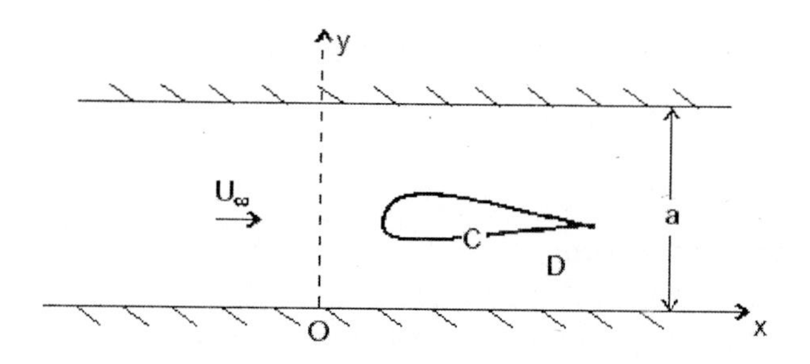

Fig. 4.5.1.

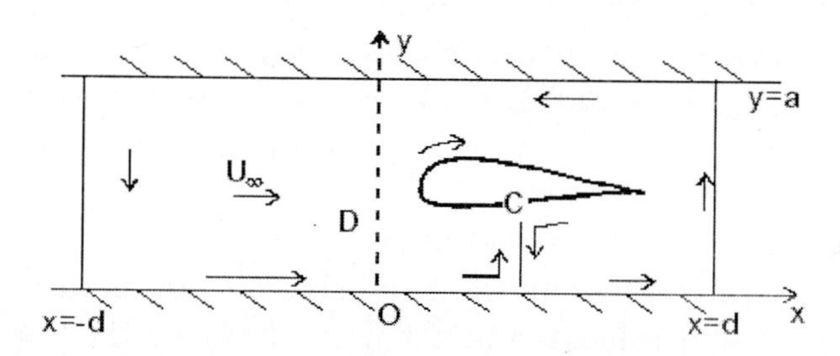

Fig. 4.5.2.

We took into account (4.5.1) and the fact that for $x = \pm d$ the integrals vanish when $d \to \infty$ if f and g behave like the solution of (4.1.13) and u and v have the property (4.1.11).

For $f = \overline{U}_+$ and $g = -\overline{V}_+$, where $(\overline{U}_+, \overline{V}_+)$ verifies in the domain occupied by the fluid, the system (4.1.12) and the relations

$$\overline{V}_+(x, 0) = 0, \quad \overline{V}_+(x, a) = 0\,, \tag{4.5.3}$$

we get from (4.5.2):

$$\begin{aligned} u(\boldsymbol{\xi}) = \oint_C \{&u(\boldsymbol{x})\left[\overline{U}_+(\boldsymbol{x}, \boldsymbol{\xi})n_x(\boldsymbol{x}) + \overline{V}_+(\boldsymbol{x}, \boldsymbol{\xi})n_y(\boldsymbol{x})\right] + \\ &+ v(\boldsymbol{x})\left[\overline{U}_+(\boldsymbol{x}, \boldsymbol{\xi})n_y(\boldsymbol{x}) - \overline{V}_+(\boldsymbol{x}, \boldsymbol{\xi})n_x(\boldsymbol{x})\right]\}\mathrm{d}\,s\,. \end{aligned} \tag{4.5.4}$$

Analogously, for $f = \overline{V}_-$ and $g = \overline{U}_-$, where $(\overline{U}_-, \overline{V}_-)$ verify the system (4.1.12) and have the supplementary property

$$\overline{U}_-(x, 0) = 0, \quad \overline{U}_-(x, a) = 0\,, \tag{4.5.5}$$

we get from (4.5.2):

$$\begin{aligned} v(\boldsymbol{\xi}) = \oint_C \{&u(\boldsymbol{x})\left[\overline{V}_-(\boldsymbol{x}, \boldsymbol{\xi})n_x(\boldsymbol{x}) + \overline{U}_-(\boldsymbol{x}, \boldsymbol{\xi})n_y(\boldsymbol{x})\right] + \\ &+ v(\boldsymbol{x})\left[\overline{U}_-(\boldsymbol{x}, \boldsymbol{\xi})n_x(\boldsymbol{x}) - \overline{V}_-(\boldsymbol{x}, \boldsymbol{\xi})n_y(\boldsymbol{x})\right]\}\mathrm{d}\,s\,. \end{aligned} \tag{4.5.6}$$

Hence we obtain for $u(\boldsymbol{x})$ and $v(\boldsymbol{\xi})$ the same representation like in (4.4.1), with the difference that $(U_{\pm}^{*}, V_{\pm}^{*})$ become $(\overline{U}_{\pm}, \overline{V}_{\pm})$.

4.5.2 Green Functions

The system (4.1.12) determines the fundamental solutions of the system (4.1.7). These are distributions. The fundamental solutions which are defined only in a portion of the plane and satisfy the boundary conditions are named Green functions, or, more properly, Green distributions. The solutions $(\overline{U}_{+}, \overline{V}_{+})$ and $(\overline{U}_{-}, \overline{V}_{-})$ are therefore Green functions. For obtaining these solutions we employ the method of images. If $M(\boldsymbol{\xi})$ is a point of the strip $-\infty < x < +\infty \quad 0 < y < a$ (fig. 4.5.1), then the solution (without restrictions) of the system (4.1.12) is (4.1.13). As we already know, it represents the perturbation produced in the entire plane xOy by a source located in M. For satisfying the first condition of (4.5.3), we have to place a source having the same intensity (taken to be equal to the unity) in the symmetric point $\overline{M}(\xi, -\eta)$. For satisfying the second condition of (4.5.3), we have to place sources having the same intensity in the points which are the symmetric points of M and $\overline{M}$ with respect to the $y = a$ axis, i.e. in the points $M_1^-(\xi, 2a - \eta)$, $M_1^+(\boldsymbol{\xi}, 2a + \eta)$. These sources disturb the condition on $y = 0$, such that we have to place sources in the symmetrical points of M_1^- and M_1^+ with respect to the $y = 0$ axis, i.e. in the points $M_{-1}^+(\xi, -2a + \eta)$ and $M_{-1}^-(\xi, -2a - \eta)$ etc. The perturbations produced by these sources are

$$
\begin{aligned}
U_{\pm}^{(n)} &= \frac{1}{2\pi} \frac{x - \xi}{(x - \xi)^2 + [(y - (2na \pm \eta)]^2} \\
V_{\pm}^{(n)} &= \frac{1}{2\pi} \frac{y - (2na \pm \eta)}{(x - \xi)^2 + [y - (2na \pm \eta)]^2}
\end{aligned} \tag{4.5.7}
$$

and they satisfy the system

$$
\begin{aligned}
&\partial U_{\pm}^{(n)}/\partial x + \partial V_{\pm}^{(n)}/\partial y = \delta(x - \xi, y - (2na \pm \eta)] \\
&\partial V_{\pm}^{(n)}/\partial x + \partial U_{\pm}^{(n)}/\partial y = 0\,.
\end{aligned} \tag{4.5.8}
$$

In the strip $-\infty < x < +\infty, \quad 0 < y < a$, all the perturbations satisfy the homogeneous system, excepting the perturbation (U_0^+, V_0^+)

which satisfies the corresponding non-homogeneous system.

$$\overline{U}_{+}(\boldsymbol{x},\boldsymbol{\xi}) = \frac{1}{2\pi}\sum_{n=-\infty}^{+\infty}\left\{\frac{x-\xi}{(x-\xi)^2+[y-(2na+\eta)]^2}+\right.$$

$$\left.+\frac{x-\xi}{(x-\xi)^2+[y-(2na-\eta)]^2}\right\}$$

$$\overline{V}_{+}(\boldsymbol{x},\boldsymbol{\xi}) = \frac{1}{2\pi}\sum_{n=-\infty}^{+\infty}\left\{\frac{y-(2na+\eta)}{(x-\xi)^2+[y-(2na+\eta)]^2}+\right. \tag{4.5.9}$$

$$\left.+\frac{y-(2na-\eta)}{(x-\xi)^2+[y-(2na-\eta)]^2}\right\}$$

satisfies the system (4.1.12) and the boundary conditions (4.5.3). It is just the solution utilized in (4.5.4).

In the same way we deduce that the solution

$$\overline{U}_{-}(\boldsymbol{x},\boldsymbol{\xi}) = \frac{1}{2\pi}\sum_{n=-\infty}^{+\infty}\left\{\frac{x-\xi}{(x-\xi)^2+[y-(2na+\eta)]^2}-\right.$$

$$\left.-\frac{x-\xi}{(x-\xi)^2+[y-(2na-\eta)]^2}\right\}$$

$$\overline{V}_{-}(\boldsymbol{x},\boldsymbol{\xi}) = \frac{1}{2\pi}\sum_{n=-\infty}^{+\infty}\left\{\frac{y-(2na+\eta)}{(x-\xi)^2+[y-(2na+\eta)]^2}+\right. \tag{4.5.10}$$

$$\left.+\frac{y-(2na-\eta)}{(x-\xi)^2+[y-(2na-\eta)]^2}\right\}$$

satisfies the system (4.1.12) and the boundary condition (4.5.5). This is just the solution utilized in (4.5.6). The series (4.5.9) and (4.5.10) are the Green functions of the problem.

The sums intervening in (4.5.9) and (4.5.10) may be calculated using the formula

$$\sum_{-\infty}^{+\infty}\frac{1}{z-n} = \pi\cot\pi z\,, \tag{4.5.11}$$

where $z = x + iy$. Separating the real part from the imaginary one, we get from (4.5.11):

$$\sum_{n=-\infty}^{+\infty} \frac{y}{y^2+(x-n)^2} = \pi \frac{\sinh 2\pi y}{\cosh 2\pi y - \cos 2\pi x}$$

$$\sum_{n=-\infty}^{+\infty} \frac{x-n}{y^2+(x-n)^2} = \pi \frac{\sin 2\pi x}{\cosh 2\pi y - \cos 2\pi x}\,. \tag{4.5.12}$$

Using these relations we obtain the expressions of the Green functions:

$$\overline{U}_{\pm}(\boldsymbol{x},\boldsymbol{\xi}) = \frac{1}{4a}\,\frac{\sinh\frac{\pi}{a}(x-\xi)}{\cosh\frac{\pi}{a}(x-\xi)-\cos\frac{\pi}{a}(y-\eta)} \pm$$

$$\pm\frac{1}{4a}\,\frac{\sinh\frac{\pi}{a}(x-\xi)}{\cosh\frac{\pi}{a}(x-\xi)-\cos\frac{\pi}{a}(y+\eta)}\,,$$

$$\overline{V}_{\pm}(\boldsymbol{x},\boldsymbol{\xi}) = \frac{1}{4a}\,\frac{\sin\frac{\pi}{a}(y-\eta)}{\cosh\frac{\pi}{a}(x-\xi)-\cos\frac{\pi}{a}(y-\eta)} \pm \tag{4.5.13}$$

$$\pm\frac{1}{4a}\,\frac{\sin\frac{\pi}{a}(y+\eta)}{\cosh\frac{\pi}{a}(x-\xi)-\cos\frac{\pi}{a}(y+\eta)}\,.$$

4.5.3 The Integral Equation

In the sequel we shall proceed like in 4.3.2, i.e. we pass to the limit $\boldsymbol{\xi} \to \boldsymbol{x}_0 \in C$ in (4.5.4) and (4.5.6), we introduce the function G by means of the formula (4.3.9) and we utilize the relations

$$\oint_C [\overline{U}_+(\boldsymbol{x},\boldsymbol{x}_0)n_x(\boldsymbol{x}) + \overline{V}_+(\boldsymbol{x},\boldsymbol{x}_0)n_y(\boldsymbol{x})]\,\mathrm{d}\,s = -\frac{1}{2}$$

$$\oint_C [\overline{U}_-(\boldsymbol{x},\boldsymbol{x}_0)n_y(\boldsymbol{x}) - \overline{V}_-(\boldsymbol{x},\boldsymbol{x}_0)n_x(\boldsymbol{x})]\,\mathrm{d}\,s = 0\,, \tag{4.5.14}$$

which will be demonstrated in Appendix 4.5.5. One obtains the following integral equation:

$$(1/2)G(\boldsymbol{x}_0) - \oint_C \Big\{ \overline{U}_-(\boldsymbol{x}, \boldsymbol{x}_0) n_x(\boldsymbol{x}) + \overline{V}_+(\boldsymbol{x}, \boldsymbol{x}_0) n_y(\boldsymbol{x}) +$$

$$+ M^2 \frac{n_x(\boldsymbol{x}) n_y(\boldsymbol{x})}{n_x^2(\boldsymbol{x}) + \beta^2 n_y^2(\boldsymbol{x})} \big[\overline{V}_-(\boldsymbol{x}, \boldsymbol{x}_0) n_x(\boldsymbol{x}_0) - \tag{4.5.15}$$

$$- \overline{U}_+(\boldsymbol{x}, \boldsymbol{x}_0) n_y(\boldsymbol{x}_0) \big] \Big\} G \mathrm{d}\, s = \beta n_y(\boldsymbol{x}_0) ,$$

and discretizing, we deduce the system

$$(1/2)G_i + \sum_{i=1}^{N} A_{ij} G_j = B_i, j = 1, \ldots, N , \tag{4.5.16}$$

where A_{ij} is A_{ij}^* from (4.5.15) with the notations

$$U_{ij}^{\pm} = \int_{L_j} \overline{U}_{\pm}(\boldsymbol{x}, \boldsymbol{x}_i^0) \mathrm{d}\, s, \quad V_{ij}^{\pm} = \int_{L_j} \overline{V}_{\pm}(\boldsymbol{x}, \boldsymbol{x}_i^0) \mathrm{d}\, s . \tag{4.5.17}$$

In order to determine these coefficients we consider on L_j the parametrization (4.2.19). Further we have two possibilities: the exact calculus like in 4.2.4, or the approximate calculus, with Gauss-type quadrature formulas [A.55], for the resulting integrals

$$U_{ij}^{\pm} = \frac{l_j}{2} \int_{-1}^{+1} \overline{U}_{\pm}(t) \mathrm{d}\, t, \quad V_{ij}^{\pm} = \frac{l_j}{2} \int_{-1}^{+1} \overline{V}_{\pm}(t) \mathrm{d}\, t . \tag{4.5.18}$$

In case that $i = j$, these integrals become singular for $t = 0$, because of the terms

$$\int_{-1}^{+1} \frac{\sinh \frac{\pi}{a} [x(t) - x_i^0] \mathrm{d}\, t}{\cosh \frac{\pi}{a} [x(t) - x_i^0] - \cos \frac{\pi}{a} [y(t) - y_i^0]}$$

$$\int_{-1}^{+1} \frac{\sin \frac{\pi}{a} [y(t) - y_i^0] \mathrm{d}\, t}{\cosh \frac{\pi}{a} [x(t) - x_i^0] - \cos \frac{\pi}{a} [y(t) - y_i^0]}$$

but, taking into account that the integrands are odd functions, we deduce that they vanish. We have therefore

$$U_{ii}^{\pm} = \pm\frac{l_i}{8a}\int_{-1}^{+1}\frac{\sinh\frac{\pi}{a}[x(t)-x_i^0]\mathrm{d}\,t}{\cosh\frac{\pi}{a}[x(t)-x_i^0]-\cos\frac{\pi}{a}[y(t)-y_i^0]},$$

$$V_{ii}^{\pm} = \int_{-1}^{+1}\frac{\sin\frac{\pi}{a}[y(t)-y_i^0]\mathrm{d}\,t}{\cosh\frac{\pi}{a}[x(t)-x_i^0]-\cos\frac{\pi}{a}[y(t)-y_i^0]}. \tag{4.5.19}$$

These integrals may be also calculated numerically, using the Gauss-type quadrature formulas.

4.5.4 The Verification of the Method

In order to perform this verification we shall use the solution given by the complex potential

$$f(z) = z + C\tanh\frac{\pi}{a}z\,, \tag{4.5.20}$$

where C and a are real constants. Let us study the flow determined by this potential. Employing Euler's formulas we deduce the identities

$$\sinh \mathrm{i}y = \mathrm{i}\,\sin y, \quad \cosh \mathrm{i}y = \cos y$$

$$\cosh^2 Ax\cos^2 Ay + \sinh^2 Ax\sin^2 Ay = \frac{1}{2}(\cosh 2Ax + \cos 2Ay)\,,$$

$$\cosh^2 Ax\cos^2 Ay - \sinh^2 Ax\sin Ay = \frac{1}{2}(1 + \cosh 2Ax\cos 2Ay)\,, \tag{4.5.21}$$

such that, separating the imaginary parts, we get from (4.5.20):

$$\Psi = y + \frac{C\sin 2Ay}{\cosh 2Ax + \cos 2Ay}\,, \tag{4.5.22}$$

We denoted $A = \pi/a$. Obviously the straight lines $y = 0$ and $y = a$ are streamlines ($\Psi = 0$ and $\Psi = a$). Another streamline is the line $y = a/2 (\Psi = a/2)$ but not entirely because for $x = 0$ the fraction from (4.5.22) is not determined.

Further we shall study the velocity field. From (4.5.20) it results

$$w(z) = u - \mathrm{i}\,v = 1 + \frac{AC}{\cosh^2 Az}\,, \tag{4.5.23}$$

whence, utilizing (4.5.21),

$$u = 1 + 2AC\frac{1 + \mathrm{ch}^2\, Ax \cos 2Ay}{(\cosh 2Ax + \cos 2Ay)^2}$$

$$v = 2AC\frac{\sinh^2 Ax \sin 2Ay}{(\cosh 2Ax + \cos 2Ay)^2}\,. \tag{4.5.24}$$

Now we return to the streamline $y = a/2$. This line may be continued $(\Psi = a/2)$ on the set of points (x, y) where we have:

$$\frac{a}{z} = y + \frac{C \sin 2Ay}{\cosh^2 Ax + \cos 2Ay}\,. \tag{4.5.25}$$

This curve is symmetric with respect to the Oy axis, because the equation is even in the x variable. It is also symmetric with respect to the $y = a/2$ axis, because the points having the ordinates $y = a/2 + y_0$ and $y = a/2 - y_0$, where

$$y_0 = \frac{C \sin 2Ay_0}{\cosh 2Ax - \cos^2 Ay_0} \tag{4.5.26}$$

simultaneously belong to the curve. Hence the curve having the equations (4.5.25) is an oval which orthogonally intersects the axes $y = a/2$ and $x = 0$. In the point $P(-\alpha, a/2)$ we have therefore $u = 0$. From (4.5.24) it follows the equation

$$\sinh^2 A\alpha = AC\,, \tag{4.5.27}$$

for the half-diameter α as a function of C. The semi-diameter β on the Oy axis may be determined from (4.5.26) as follows

$$\beta = \frac{C \sin 2A\beta}{1 - \cos 2A\beta}\,,$$

whence it results a simpler equation

$$\beta \tan A\beta = C\,. \tag{4.5.28}$$

We draw the conclusion that the complex potential (4.5.20) characterizes the uniform flow with the velocity $(1, 0)$, in the channel having the sides $y = 0$ and $y = a$, in the presence of the oval having the diameters 2α and 2β with the center on the Oy axis (fig. 4.5.3).

It is not difficult to see how this potential was obtained. One knows that the uniform flow at infinity, in the presence of a doublet, determines

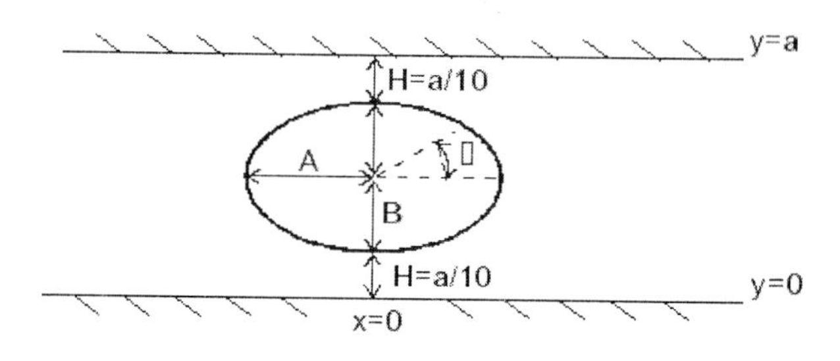

Fig. 4.5.3.

the flow past a circular obstacle. Let us study now the circular flow in the presence of a doublet in a channel. For the sake of simplicity, we consider the doublet situated in the origin and $y = a/2$, $y = -a/2$, the equations of the lines which are the walls of the channel.

These right lines become streamlines for the flow determined by the doublet, if we add, according to the method of images, symmetric doublets. The potential $f_0(z)$ describing the flow produced by these doublets is

$$f_0(z) = \ldots + \frac{1}{z + \mathrm{i}\, a} + \frac{1}{z} + \frac{1}{z - \mathrm{i}\, a} + \ldots = \sum_{-\infty}^{+\infty} \frac{1}{z - n\mathrm{i}\, a} =$$

$$= \frac{\pi}{\mathrm{i}\, a} \cot \frac{\pi z}{\mathrm{i}\, a} = \frac{\pi}{a} \coth \frac{\pi z}{a}\,.$$

We take

$$f_0(z) = C \coth Az\,, \quad A = \pi/a\,,$$

C representing a real o constant. If the Ox axis coincides with the lower wall, as we have previously considered, then the potential becomes

$$f_0(z) = C \coth A \left(z - \frac{\mathrm{i}\, a}{2} \right) = C \tanh Az\,.$$

Adding the potential of the uniform flow having the velocity $(1, 0)$ we obtain (4.5.20).

In the paper [4.9] that we utilized for writing this subsection, we made tests for the local pressure coefficient C_p, defined by (4.2.35). In figure 4.5.4 we present the values of C_p calculated exactly (by means of (4.5.24) and the values calculated with the numerical method presented above with $H = a/10$. The maximum relative error is 0.73% in the case of a discretization with 80 nodes. For the circular obstacle, symmetrically placed in a tunnel, we present the compressibility effect in figure 4.5.5 where one indicates the variation of the maximum velocity

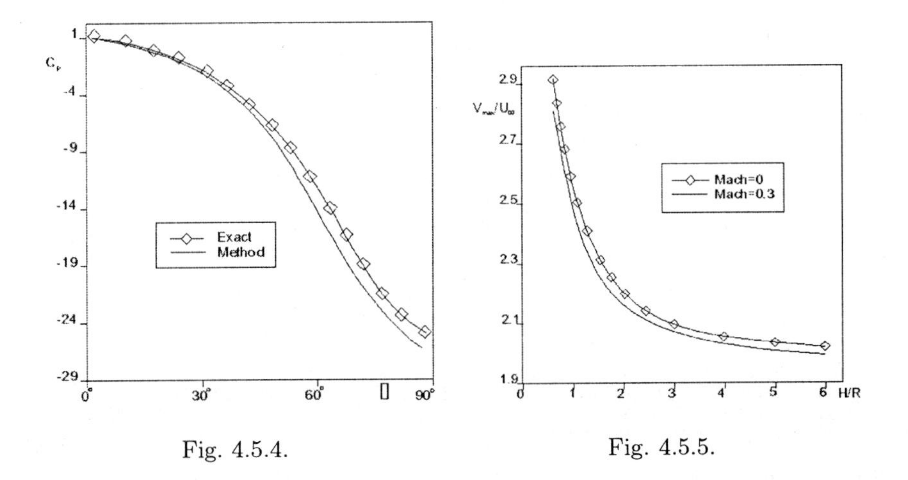

Fig. 4.5.4.

Fig. 4.5.5.

against the distance to the walls for $\gamma = 1.405$ in the cases $M = 0$ and $M = 0.3$ and in the absence of the circulation ($\Gamma = 0$). We considered 60 equidistant nodes on the circumference.

In [4.9] one presents the graphic representations for C_p in the case of the NACA-4412 profile at 0 angle of attack and Mach=0.5 in a free stream and in a tunnel. The example is instructive, because, in this case the trailing edge is angular, and we have to impose the equality of the pressures on the two sides of the edge in order to determine the circulation.

4.5.5 Appendix

In order to prove the first formula (4.5.14) we shall notice that $(\overline{U}_+, \overline{V}_+)$ (x, y, x_0, y_0) satisfy in the domain D, exterior to the contour $C - c + C_\varepsilon$, and bounded by the walls $y = 0$ and $y = a$ and by the segments $x = \pm d$, the equation:

$$\partial \overline{U}_+ / \partial x + \partial \overline{V}_+ / \partial y = \delta(x - x_0, y - y_0) .$$

Integrating this equation over D and applying Gauss's formula one obtains:

$$\int_C (\overline{U}_+ n_x + \overline{V}_+ n_y) \mathrm{d}\, s + \lim_{\varepsilon \to 0} \int_{C_\varepsilon} (\overline{U}_+ n_x + \overline{V}_+ n_y) \mathrm{d}\, s +$$

$$+ \lim_{d \to \infty} \int_0^a \overline{U}_+(d, y, x_0, y_0) \mathrm{d}\, y + \lim_{d \to \infty} \int_a^0 \overline{U}_+(-d, y, x_0, y_0) \mathrm{d}\, y = 1 , \tag{4.5.29}$$

the integrals on $y = 0$ and $y = a$ vanishing by virtue of the relations (4.5.1). The integrals on C_ε are calculated utilizing the parametrizations (4.3.35) which imply

$$\overline{U}_\pm = \frac{1}{2\pi}\,\frac{\cos\theta}{\varepsilon} + 0(\varepsilon),\ \overline{V}_\pm = \frac{1}{2\pi}\,\frac{\sin\theta}{\varepsilon} \pm C\,,$$

where

$$C = \frac{1}{2\pi y_0}\left[1 + \frac{y_0^2}{a^2}\sum_{n=0}^{\infty}\frac{1}{(y_0/a)^2 - n^2}\right].$$

To the limit, when $\varepsilon \to 0$, this is $1/2$. One calculate the last two integrals taking into account that

$$\int \frac{\mathrm{d}t}{a + b\cos t} = \frac{2}{\sqrt{a^2 - b^2}}\arctan\left[\frac{(a-b)\tan(t/2)}{\sqrt{a^2 - b^2}}\right].$$

To the limit, when $d \to 0$, every integral is $1/2$. One obtains the first formula (4.5.14). Analogously, we obtain the second formula if we integrate the equation

$$\partial\overline{U}_-/\partial y - \partial\overline{V}_-/\partial x = 0$$

on the same domain D.

4.6 Other Methods. The Intrinsic Integral Equation

4.6.1 The Method of Regularization

The methods that we have already utilized have been given by L. Dragoş [4.6] and by L. Dragoş and A. Dinu [4.7], [4.8], [4.9], [4.10]. Specific to these methods is the fact that one utilizes physical variables (the velocity and pressure fields), such that from the solution one obtains directly the elements of interest in aerodynamics (the velocity and the pressure on the profile). Other methods (Bindolino a.o. [4.2], Morino and Luo [4.20]) utilize the real potential and others (Griello a.o. [4.15], Carabineanu [4.4]), the stream function. All these methods may be utilized only for incompressible fluid. The theory that we are going to present in the sequel will be also applied to the incompressible fluid, but it has the advantage to utilize the physical fields. Moreover, the singular integrals are avoided and one utilizes the intrinsic elements of the flow (the tangential component of the velocity). Therefore it is suitable to call this method *the method of regularization* [4.11].

Denoting

$$\boldsymbol{V} = U_\infty(\boldsymbol{i} + \boldsymbol{v}), \tag{4.6.1}$$

we shall determine the perturbation from the equations

$$\operatorname{div} \boldsymbol{v} = 0, \ \operatorname{rot} \boldsymbol{v} = 0 \tag{4.6.2}$$

with the boundary condition

$$\boldsymbol{v} \cdot \boldsymbol{n} = -n_x \tag{4.6.3}$$

and the condition at infinity

$$\lim_{x \to \infty} \boldsymbol{v}(\boldsymbol{x}) = 0. \tag{4.6.4}$$

We may write the equations (4.6.2) as follows

$$\operatorname{div} (\boldsymbol{v} - \boldsymbol{c}) = 0, \ \operatorname{rot} (\boldsymbol{v} - \boldsymbol{c}) = 0, \tag{4.6.5}$$

$\boldsymbol{c}$ representing a constant vector. We put into evidence the normal and tangential components of the velocity:

$$\boldsymbol{v} = (\boldsymbol{v} \cdot \boldsymbol{n})\boldsymbol{n} + (\boldsymbol{v} \cdot \boldsymbol{s})\boldsymbol{s} = v_n \boldsymbol{n} + v_s \boldsymbol{s}. \tag{4.6.6}$$

By virtue of the condition (4.6.3) we may write

$$\boldsymbol{v} = -n_x \boldsymbol{n} + v_s \boldsymbol{s}. \tag{4.6.7}$$

Rewriting the scalar formulas from 4.3 as a vectorial formula we deduce that for every two continuously differentiable functions or distributions f and g, by virtue of the equations (4.6.5) ($\boldsymbol{k}$ – the versor of the Oz axis), we have:

$$\int_D [f \operatorname{div} (\boldsymbol{v} - \boldsymbol{c}) + (g\boldsymbol{k}) \cdot \operatorname{rot}(\boldsymbol{v} - \boldsymbol{c})] \mathrm{d}\, a = 0, \tag{4.6.8}$$

D representing, like in 4.3.1, the domain exterior to C and interior to the circle C_R having the radius R big enough. Utilizing the formulas

$$\begin{aligned} &\operatorname{div} f(\boldsymbol{v} - \boldsymbol{c}) = (\boldsymbol{v} - \boldsymbol{c}) \cdot \operatorname{grad} f + f \operatorname{div} (\boldsymbol{v} - \boldsymbol{c}) \\ &\operatorname{div} [g\boldsymbol{k} \times (\boldsymbol{v} - \boldsymbol{c})] = (\boldsymbol{v} - \boldsymbol{c}) \cdot \operatorname{rot} (g\boldsymbol{k}) - g\boldsymbol{k} \cdot \operatorname{rot} (\boldsymbol{v} - \boldsymbol{c}) \end{aligned} \tag{4.6.9}$$

and applying Gauss's theorem, we obtain the identity:

$$\int_D (\boldsymbol{v} - \boldsymbol{c}) \cdot [\operatorname{grad} f - \operatorname{rot} (g\boldsymbol{k})] \mathrm{d}\, a = \int_{C + C_R} (\boldsymbol{v} - \boldsymbol{c}) \cdot (f\boldsymbol{n} - \boldsymbol{n} \times g\boldsymbol{k}) \mathrm{d}\, s, \tag{4.6.10}$$

$\boldsymbol{n}$ being the normal (pointing outwards the domain D (i.e. the inward pointing normal with respect to C and the outward pointing normal with respect to C_R).

We shall write the distribution (4.1.13) as follows

$$\boldsymbol{v}^* = \frac{1}{2\pi} \frac{\boldsymbol{x} - \boldsymbol{x}_0}{|\boldsymbol{x} - \boldsymbol{x}_0|^2} \tag{4.6.11}$$

and the system (4.1.12) that it satisfies:

$$\operatorname{div} \boldsymbol{v}^* = \delta(\boldsymbol{x} - \boldsymbol{x}_0), \quad \operatorname{rot} \boldsymbol{v}^* = 0, \tag{4.6.12}$$

the equations being valid for every point $\boldsymbol{x}_0$ from $D+C$. For $(f, g) \to (u^*, -v^*)$, one obtains from (4.6.10) the projection of the identity

$$\int_D (\boldsymbol{v} - \boldsymbol{c}) \operatorname{div} \boldsymbol{v}^* \mathrm{d}\, a = \int_{C+C_R} \{[\boldsymbol{n} \cdot (\boldsymbol{v} - \boldsymbol{c})]\boldsymbol{v} + [\boldsymbol{n} \times (\boldsymbol{v} - \boldsymbol{c})] \times \boldsymbol{v}\} \mathrm{d}\, s\,, \tag{4.6.13}$$

on the Ox axis. For $(f, g) \to (v^*, u^*)$ we obtain the projection of the same identity on the Oy axis. Hence the relation (4.6.13) is valid. By virtue of (4.6.12) we obtain from (4.6.13)

$$\boldsymbol{v}(\boldsymbol{x}_0) - \boldsymbol{c} = \int_{C+C_R} \{[\boldsymbol{n} \cdot (\boldsymbol{v} - \boldsymbol{c})]\boldsymbol{v}^* + [\boldsymbol{n} \times (\boldsymbol{v} - \boldsymbol{c})] \times \boldsymbol{v}\} \mathrm{d}\, s\,, \tag{4.6.14}$$

Here we shall put $\boldsymbol{c} \equiv \boldsymbol{v}(\boldsymbol{x}_0) \equiv \boldsymbol{v}_0$ and we shall evaluate the integral on C_R. Utilizing the parametrization (4.3.37), we shall deduce for the projection of this integral on the Ox axis

$$\int_{C_R} \{\boldsymbol{n} \cdot (\boldsymbol{v} - \boldsymbol{v}_0) u^* + (u - u_0)(u^* n_x + v^* n_y) -$$

$$-n_x[(u - u_0)u^* + (v - v_0)v^*]\} \mathrm{d}\, s = \frac{1}{2\pi} \int_0^{2\pi} (u - u_0) \mathrm{d}\, \varphi = -u_0\,,$$

because for $R \to \infty, u$ vanishes according to the condition (4.6.4). One deduces a similar formula for the projection on the Oy axis.

Utilizing these results one obtains from (4.6.14) the following representation

$$\boldsymbol{v}(\boldsymbol{x}_0) = \oint_C \{[\boldsymbol{n} \cdot (\boldsymbol{v} - \boldsymbol{v}_0)]\boldsymbol{v}^* + [\boldsymbol{n} \times (\boldsymbol{v} - \boldsymbol{v}_0)] \times \boldsymbol{v}^*\} \mathrm{d}\, s\,. \tag{4.6.15}$$

The integral is not singular because the factor $\boldsymbol{v} - \boldsymbol{v}_0$ tends to zero when $\boldsymbol{x} \to \boldsymbol{x}_0 \in C$. This is a regularized integral. The formula (4.6.15) is valid for both $\boldsymbol{x}_0$ in the fluid and on the boundary C.

We notice utilizing (4.6.7) that on C we have

$$\boldsymbol{n} \times \boldsymbol{v} = v_s \boldsymbol{k}$$
$$[\boldsymbol{n} \cdot (\boldsymbol{v} - \boldsymbol{v}_0)]\, v^* = [-n_x + n_x^0(\boldsymbol{n} \cdot \boldsymbol{n}^0) - v_s^0(\boldsymbol{n} \cdot \boldsymbol{s}_0)]\boldsymbol{v}^* \,, \tag{4.6.16}$$

such that, after elementary calculations, from (4.6.15) one obtains

$$\begin{aligned}\boldsymbol{v}(\boldsymbol{x}_0) = \oint_C \{ & v_s \boldsymbol{k} \times \boldsymbol{v}^* + v_s^0[(\boldsymbol{v}^* \cdot \boldsymbol{s}^0)\boldsymbol{n} - \\ & -(\boldsymbol{n} \cdot \boldsymbol{v}^*)\boldsymbol{s}^0 - (\boldsymbol{n} \cdot \boldsymbol{s}^0)\boldsymbol{v}^*] - n_x \boldsymbol{v}^* - \\ & -n_x^0[(\boldsymbol{n}^0 \cdot \boldsymbol{v}^*)\boldsymbol{n} - (\boldsymbol{n} \cdot \boldsymbol{v}^*)\boldsymbol{n}^0 - (\boldsymbol{n} \cdot \boldsymbol{n}^0)\boldsymbol{v}^*]\} \mathrm{d}\, s \,.\end{aligned} \tag{4.6.17}$$

For obtaining the unknown v_s outside the integral, we shall perform the operation $\boldsymbol{n}^0 \times (4.6.17)$ and we shall take into account that we have

$$n_x^0 = s_y^0 \,, \quad n_y^0 = -s_x^0 \,, \tag{4.6.18}$$

We deduce

$$\begin{aligned}& \boldsymbol{n}^0 \times \boldsymbol{v} = (\boldsymbol{v} \cdot \boldsymbol{s}^0)\boldsymbol{k} \,, \quad \boldsymbol{n}^0 \times \boldsymbol{n} = (\boldsymbol{n} \cdot \boldsymbol{s}^0)\boldsymbol{k} \,, \\ & \boldsymbol{n}^0 \times \boldsymbol{s}^0 = \boldsymbol{k}, \quad \boldsymbol{n}^0 \times \boldsymbol{v}^* = (\boldsymbol{s}^0 \cdot \boldsymbol{v}^*)\boldsymbol{k} \\ & (\boldsymbol{n} \cdot \boldsymbol{n}^0)(\boldsymbol{s}^0 \cdot \boldsymbol{v}^*)(\boldsymbol{n} \cdot \boldsymbol{v}^*)(\boldsymbol{n} \cdot \boldsymbol{s}^0) = \boldsymbol{s} \cdot \boldsymbol{v}\end{aligned} \tag{4.6.19}$$

whence

$$\begin{aligned}v_s^0 & = \oint_C [(\boldsymbol{n}^0 \cdot \boldsymbol{v}^*)v_s - (\boldsymbol{n}^0 \cdot \boldsymbol{v}^*)v_s^0] \mathrm{d}\, s = \\ & = \oint_C [(\boldsymbol{s}^0 \cdot \boldsymbol{v}^*)n_x^0 - (\boldsymbol{s}^0 \cdot \boldsymbol{v}^*)n_x] \mathrm{d}\, s \,.\end{aligned} \tag{4.6.20}$$

This is the integral equation of the problem. It is a regularized (non-singular) integral equation because when $\boldsymbol{x} \to \boldsymbol{x}_0$, the denominators and the numerators ($(\boldsymbol{n} \cdot \boldsymbol{v}^*)v_s \to (\boldsymbol{n}^0 \cdot \boldsymbol{v}^*)v_s^0$) of the integrand simultaneously vanish (see (4.6.11)). The equation (4.6.20) was obtained in a different manner by V.Cardoş [4.5]. It is, obviously an intrinsic equation. This equation may be also solved by means of the boundary elements method. One obtains, like above, a linear algebraic system of N equations with N unknowns.

In the case of lifting airfoils the following boundary condition is added (see also (4.3.23)).

$$\boldsymbol{V}(P_s) \cdot \boldsymbol{s} + \boldsymbol{V}(P_i) \cdot \boldsymbol{s} \to 0 \,, \quad P_s, P_i \to P_f \,. \tag{4.6.21}$$

Taking (4.6.1) and (4.6.18) into account, we apply this condition as follows:

$$v_s(P_s) + v_s(P_i) = n_y(P_s) + n_y(P_i)\,, \tag{4.6.22}$$

P_s and P_i being sufficiently close of P_f. From the equation (4.6.20) and the condition (4.6.22) we shall determine v_s on C. Since we have more equations than unknowns, we shall treat the problem like in 4.3.5, introducing the auxiliary variable λ.

The reader may find examples in [4.11].

Chapter 5

The Theory of Finite Span Airfoil in Subsonic Flow. The Lifting Surface Theory

5.1 The Lifting Surface Equation

5.1.1 The Statement of the Problem

We assume that an uniform flow, having the velocity $U_\infty \boldsymbol{i}$ (the Ox - axis has the direction and the sense of the uniform flow), the pressure p_∞ , the density ρ_∞ and $M(= U_\infty/c_\infty) < 1$, is perturbed by the presence of a finite span wing, perpendicular on the flow direction. Any body which has a characteristic dimension much larger then the two other dimensions is considered to be a wing. We call the *span* of the wing, the length of the wing taken along the direction of the large characteristic dimension (figure 5.1.1).

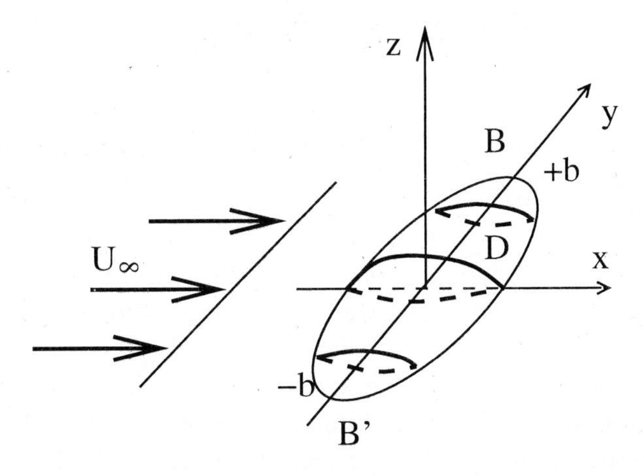

Fig. 5.1.1.

One requires to determine the action of the fluid against the wing. In fact, as we have already seen when we studied the two-dimensional case, for determining the action, one has to calculate the perturbation fields $p(\boldsymbol{x})$ and $\boldsymbol{v}(\boldsymbol{x})$.

We employ the Cartesian variables x, y, z introduced in (2.1.1) and the fields p and $\boldsymbol{v}$ introduced by (2.1.3). The Oy-axis has the span direction and the origin O is situated in the middle of the wing. The Oz-axis is taken perpendicular to the Ox and Oy-axes in order to determine a Cartesian, positive oriented frame.

We assume that the projection of the wing onto the xOy-plane is a simple connected domain D, with a piecewise smooth boundary ∂D such that every straight line parallel to the Ox-axis (which has the direction of the unperturbed flow) should intersect the boundary in almost two points Q_a and Q_f. As we may see in figures 5.1.2 and 5.1.3, *the lateral edges* of the wing may be exceptions from this assumption if they are parallel to the Ox -axis . The intersection of ∂D with a lateral edge parallel to the Ox -axis consists of a point, two confounded points or a straight segment. Assuming that the two lateral edges consist of a point (denoted respectively by B and B') we notice that they divide the boundary into two arcs

The front arc BB' which is attacked by the stream is called *the leading edge* and the rear arc is called *the trailing edge*. The equation of the leading edge (consisting of the points Q_a) is $x = x_-(y)$, and the equation of the trailing edge is $x = x_+(y)$. Hence Q_a has the Cartesian coordinates $(x_-(y), y)$ and $Q_f(x_+(y), y)$.

For wings with lateral edges consisting of simple or confounded points we obviously have:

$$x_+(\pm b) = x_-(\pm b)\,, \tag{5.1.1}$$

where $2b$ is the span (in dimensionless variables). We shall call these ones I wings, and the wings whose lateral edges consist of straight segments will be called II wings. For the II wings the condition (5.1.1) will be replaced by the condition of the continuity of the pressure along the edge (the Kutta-Joukowski condition (see (5.1.30)). In figures 5.1.2 we indicated the domain D for some I wings having lateral edges consisting of simple points (the *delta* wing, figure a), the *gothic* wing, figure b), the *trapezoidal* wing, figure c), the *rhombic* wing, figure d), the *swallow tail* wing, figure e) or lateral edges consisting of double confounded points (the *elliptical* wing, figure f)).

In figure 5.1.3 we present the domain D for the *arrow shaped* wing (with lateral edges consisting of straight segments).

The domain D is named the *plane form of the wing* or the *planar form of the wing.*

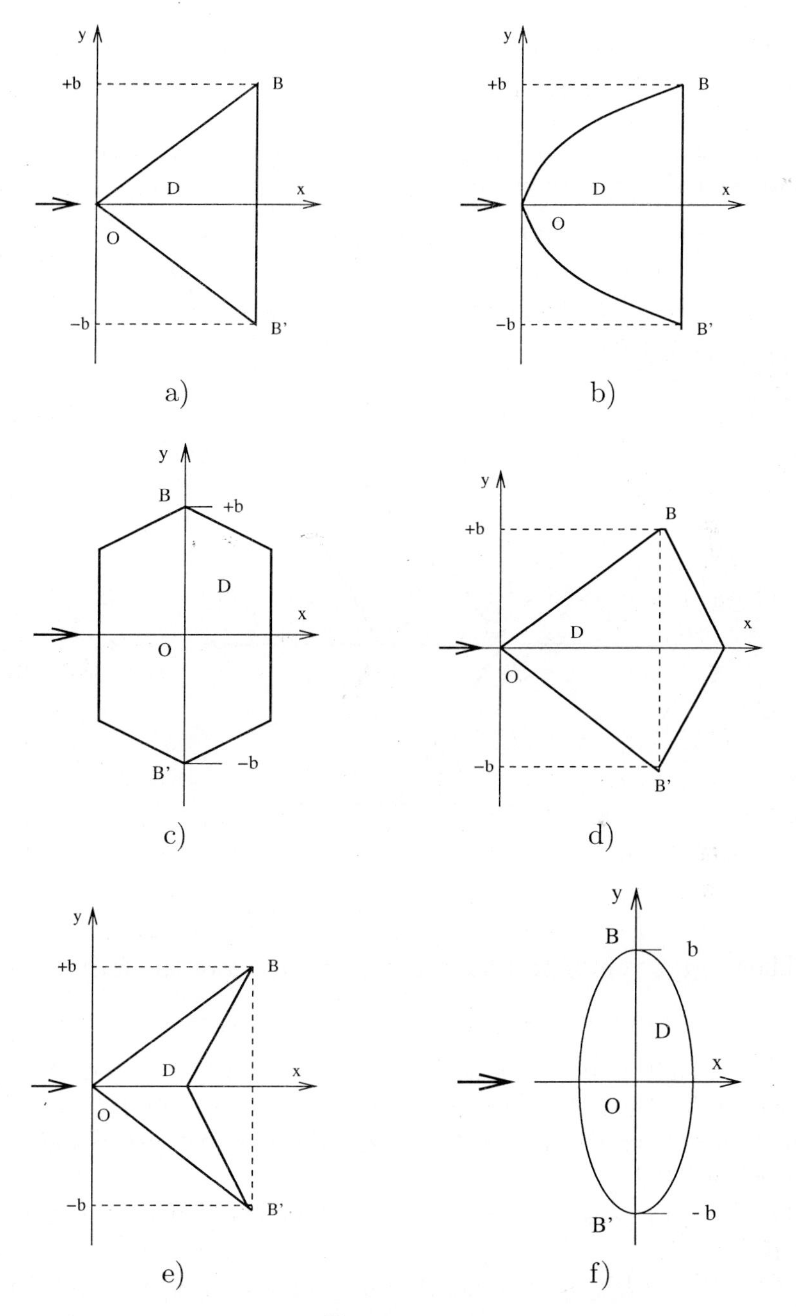

Fig. 5.1.2.

We notice that the wing has two surfaces: the *upper surface*, denoted by S_+ and the *lower surface* denoted by S_-. The equations of these surfaces may be written (for the sake of simplicity) in the following form

$$z = h(x,y) \pm h_1(x,y)\,, \quad (x,y) \in D\,. \tag{5.1.2}$$

Indeed, considering $z = f(x,y)$ the equation for S_+ and $z = g(x,y)$ the equation for S_-, then, setting

$$\begin{aligned} 2h(x,y) &= f(x,y) + g(x,y) \\ 2h_1(x,y) &= f(x,y) - g(x,y)\,, \end{aligned} \tag{5.1.3}$$

we obtain (5.1.2).

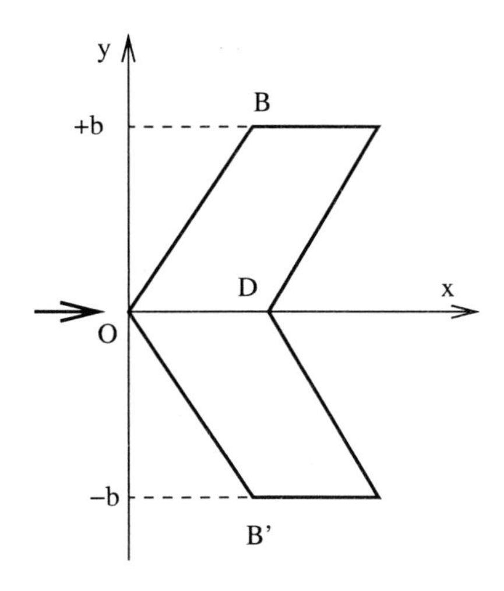

Fig. 5.1.3.

The wing is thin if and only if h and h_1 have the form

$$h(x,y) = \varepsilon\overline{h}(x,y), \quad h_1(x,y) = \varepsilon\overline{h}_1(x,y)\,, \tag{5.1.4}$$

$\varepsilon(<< 1)$ being a real parameter (max h, h_1 în D), and $\overline{h}$ şi $\overline{h}_1$ bounded functions. We also assume that $h(x,y)$ and $h_1(x,y)$ are defined on D and have the first order derivatives with respect to x, denoted by h'_x, h'_{1x}.

The two surfaces S_+ and S_- do not intersect each other if and only if

$$h(x,y) + h_1(x,y) \geq h(x,y) - h_1(x,y)\,. \tag{5.1.5}$$

5.1.2 Bibliographical Comments

The bibliography concerning this subject is extremely rich and it is not possible to mention it entirely. In fact, this is not necessary because the fundamental solutions method, utilized herein is different from the previously utilized methods. As we have already mentioned in the introduction, the methods utilized so far rely on the equation of the potential (we assume that the perturbation is irrotational, the wing being assimilated to a distributions of bound vortices or doublets. Downstream, behind the wing, where the experience shows that the flow is not irrotational, one introduces in an artificial manner a free vortices distribution. The perturbation of the velocity results from the velocity field induced by the two distributions. This is Prandtl's model [6.21] for the *lifting line theory*, which assumes that the wing may be replaced by the segment BB'. *The lifting surface theory*, which does not replace the domain D with the segment BB', was developed after 1936 by Prandtl [5.26], [5.27], Weissinger [5.35], Reissner [5.29], Multhopp [5.24], Flax, Lawrence [5.12], Truckenbrodt [5.32], Mangler and Spencer [5.22], etc. We also mention the unpublished paper [5.6] of P.Cocârlan. Some synthesis of the research in this domain may be found for example in [1.1], [1.2], [1.38] and [1.41]. D. Homentcovschi [5.16] presents, for the compressible fluid, a theory relying on the equations of motion written in distributions, without assuming that the flow is potential. Another theory is given in [5.7]. Here one introduces the fundamental solutions method. It is not necessary to assume that the flow is potential and a vortices sheet is formed behind the the wing. These properties result from equations.

5.1.3 The General Solution

As we have already mentioned in 3.1.3, there is no physical reason to replace the wing by a distribution of vortices. It is reasonable to replace the wing by a distribution of forces which should have against the fluid the same action as the wing. Indeed, the fluid and the wing must be considered as an interacting material system. According to Cauchy's tension principle (see [1.11], p.35), there exists a forces distribution $\boldsymbol{f}$ on the wing (or on D) which has the same action against the fluid like the wing itself. The perturbation of the fluid flow will be determined by this distribution and by the slipping condition on the surface of the wing. We try to employ a forces distribution on D, having the form

$$\boldsymbol{f} = (f_1, 0, f)(\xi, \eta)\,.$$

If we manage to satisfy the boundary conditions, on the basis of the uniqueness theorem, we deduce that we found the solution of the problem. Obviously, there exist various distributions which enable us to reach the solution. In chapter 7 we shall employ distributions on the boundary of the body.

From (2.3.24) we deduce that the perturbation of the pressure due to the force $\boldsymbol{f}$ applied in the point (ξ, η) is

$$p(x,y,z) = -\frac{1}{4\pi}\left[f_1(\xi,\eta)\frac{\partial}{\partial x} + f(\xi,\eta)\frac{\partial}{\partial z}\right]\left(\frac{1}{R_1}\right), \tag{5.1.6}$$

where

$$x_0 = x - \xi, y_0 = y - \eta, \quad R_1 = \sqrt{x_0^2 + \beta^2(y_0^2 + z^2)}, \tag{5.1.7}$$

A continuous superposition of such forces on D, will give

$$p(x,y,z) = -\frac{1}{4\pi}\iint_D \left[f_1(\xi,\eta)\frac{\partial}{\partial x} + f(\xi,\eta)\frac{\partial}{\partial z}\right]\left(\frac{1}{R_1}\right) \mathrm{d}\xi\, \mathrm{d}\eta, \tag{5.1.8}$$

since the problem is linear.

For the potential φ, from (2.3.28) we deduce

$$\varphi(x,y,z) = \frac{1}{4\pi}\iint_D \left[\frac{f_1(\xi,\eta)}{R_1} - \frac{z}{y_0^2+z^2}\left(1+\frac{x_0}{R_1}\right)f(\xi,\eta)\right]\mathrm{d}\xi\, \mathrm{d}\eta, \tag{5.1.9}$$

and for the velocity field from (2.3.12)

$$\boldsymbol{v}(x,y,z) = \delta(z)\iint_D \boldsymbol{f}(\xi,\eta)H(x_0)\delta(y_0)\mathrm{d}\xi\, \mathrm{d}\eta + \nabla\varphi \tag{5.1.10}$$

which can be written explicitly:

$$\begin{aligned}
u(x,y,z) &= \delta(z)\iint_D f_1(\xi,\eta)H(x_0)\delta(y_0)\mathrm{d}\xi \mathrm{d}\eta - p(x,y,z),\\
v(x,y,z) &= \frac{1}{4\pi}\iint_D \left\{f_1(\xi,\eta)\frac{\partial}{\partial y}\left(\frac{1}{R_1}\right) - \right.\\
&\quad \left. -f(\xi,\eta)\frac{\partial}{\partial y}\left[\frac{z}{y_0^2+z^2}\left(1+\frac{x_0}{R_1}\right)\right]\right\}\mathrm{d}\xi\, \mathrm{d}\eta,\\
w(x,y,z) &= \delta(z)\iint_D f(\xi,\eta)H(x_0)\delta(y_0)\mathrm{d}\xi\, \mathrm{d}\eta +\\
&\quad +\frac{1}{4\pi}\iint_D f_1(\xi,\eta)\frac{\partial}{\partial z}\left(\frac{1}{R_1}\right)\mathrm{d}\xi\, \mathrm{d}\eta -\\
&\quad -\frac{1}{4\pi}\iint_D f(\xi,\eta)\frac{\partial}{\partial z}\left[\frac{z}{y_0^2+z^2}\left(1+\frac{x_0}{R_1}\right)\right]\mathrm{d}\xi\, \mathrm{d}\eta.
\end{aligned} \tag{5.1.11}$$

For w, from (2.3.29) we get:

$$
\begin{aligned}
w(x,y,z) = \frac{1}{4\pi}\iint_D \Big[& f_1(\xi,\eta)\frac{\partial}{\partial z}\left(\frac{1}{R_1}\right) - \\
& -\beta^2 f(\xi,\eta)\frac{\partial}{\partial x}\left(\frac{1}{R_1}\right)\Big] \mathrm{d}\xi\,\mathrm{d}\eta + \\
& + \frac{1}{4\pi}\iint_D f(\xi,\eta)\frac{\partial}{\partial y}\left[\frac{y_0}{y_0^2+z^2}\left(1+\frac{x_0}{R_1}\right)\right]\mathrm{d}\xi\,\mathrm{d}\eta .
\end{aligned}
\tag{5.1.12}
$$

From formula (5.1.10) we deduce that the perturbation is potential, excepting the strip from the xOy - plane, $-b < y < b$, $x > x_-(y)$, where the first term does not vanish. We notice that the existence of the vortices sheet behind the wing, which was a hypothesis in the classical theories (see Prandtl [6.21]), represents herein a consequence following from the equations of motion. As the experience confirms the existence of such a sheet, it means that the experience confirms the mathematical model based on the fundamental solutions method.

5.1.4 The Boundary Values of the Pressure

In view of calculating the boundary values of the pressure when $z \to \pm 0$ and $(x,y) \in D$, we notice that the first term from (5.1.8) represents the tangential derivative of a simple layer potential and the second, the normal derivative of the same potential. From the potential theory one knows (see for example, [1.6], [1.39]) how to obtain the boundary values of these derivatives but we prefer to calculate them directly. After deriving we have to calculate

$$
\begin{aligned}
I_1 &\equiv \lim_{\substack{z \to \pm 0 \\ (x,y)\in D}} \iint_D f(\xi,\eta)\frac{x_0}{R_1^3}\mathrm{d}\xi\,\mathrm{d}\eta , \\
I_2 &\equiv \lim_{\substack{z \to \pm 0 \\ (x,y)\in D}} \iint_D f(\xi,\eta)\frac{z}{R_1^3}\mathrm{d}\xi\,\mathrm{d}\eta .
\end{aligned}
\tag{5.1.13}
$$

We notice that setting $z = \pm 0$ and $(x,y) \in D, R_1$ vanishes when the generic integration point $Q(\xi,\eta)$ coincides with $M(x,y) \in D$. The integral I_1 becomes singular. Denoting by D_ε the disk with the radius

ε and the center M, we shall write and we shall consider by definition

$$\iint_D = \iint_{D-D_\varepsilon} + \iint_{D_\varepsilon}, \quad \lim_{\varepsilon \to 0} \iint_{D-D_\varepsilon} = \overset{\odot}{\iint_D}, \tag{5.1.14}$$

if the limit exists. For calculating the integral on D_ε, one makes the change of variable $\xi, \eta \to r, \theta$:

$$\begin{gathered} \xi - x = \beta r \cos\theta, \eta - y = r \sin\theta \\ 0 < r \leq \varepsilon, \qquad 0 \leq \theta < 2\pi \end{gathered} \tag{5.1.15}$$

and one notices that if f is a *continuous* function, for ε small enough, we may approximate f on the disk with $f(x, y)$. We deduce:

$$\iint_{D_\varepsilon} f(\xi, \eta) \frac{x_0}{R_1^3} \mathrm{d}\,\xi\, \mathrm{d}\,\eta = \frac{f(x,y)}{\beta} \int_0^{2\pi} \int_0^{\varepsilon} \frac{r^2 \cos\theta}{(r^2 + z^2)^{3/2}} \mathrm{d}\,\theta\, \mathrm{d}\,r = 0$$

and then, the formula

$$I_1 = \overset{\odot}{\iint_D} f(\xi, \eta) \frac{x_0}{R^3} \mathrm{d}\,\xi\, \mathrm{d}\,\eta, \tag{5.1.16}$$

where

$$R = \sqrt{x_0^2 + \beta^2 y_0^2} \tag{5.1.17}$$

As it is shown in (E.0.12), the last integral exists.

As regarding I_2, it vanishes outside a vicinity D_ε of M. We have therefore

$$\begin{aligned} I_2 = & \lim_{\substack{z \to \pm 0 \\ (x,y) \in D}} \iint_{D_\varepsilon} f(\xi, \eta) \frac{z}{R_1^3} \mathrm{d}\,\xi\, \mathrm{d}\,\eta \\ & = f(\xi, \eta) \lim_{\substack{z \to \pm 0 \\ (x,y) \in D}} z \int_0^{2\pi} \int_0^{\varepsilon} \frac{r \,\mathrm{d}\,\theta \mathrm{d}\,r}{(r^2 + z^2)^{3/2}} = \pm 2\pi f(x, y)\,. \end{aligned} \tag{5.1.18}$$

Taking into account these results, from (5.1.8) we deduce:

$$p(x, y, \pm 0) = \frac{1}{4\pi} \overset{\odot}{\iint_D} f_1(\xi, \eta) \frac{x_0}{R^3} \mathrm{d}\,\xi\, \mathrm{d}\,\eta \pm \frac{1}{2} f(x, y), \quad (x, y) \in D\,. \tag{5.1.19}$$

This is the formula for the pressure on the upper and lower surfaces of the wing. We deduce the formula

$$p(x, y, +0) - p(x, y, -0) = f(x, y)\,, \tag{5.1.20}$$

which gives us the significance of the function $f(x, y)$. Like in the two-dimensional case, f is the jump of the pressure on the wing. This is a fundamental result for the calculus of the aerodynamic action.

5.1.5 The Boundary Values of the Component w

If one utilizes (5.1.11), then taking the limit, when $z \to \pm 0$, the first term disappears because $\delta(\pm 0) = 0$. It is more difficult to calculate the limit of the last term because it contains the derivative with respect to z. The calculation is performed in [5.22] and in [1.11], p.208, 209. We utilize here the expression (5.1.12) for w. In this expression, the limits of the first two terms are given by the formulas (5.1.17) and (5.1.18) and the derivative with respect to y from the last term commutes with the limit. We have therefore,

$$I_3 \equiv \lim_{\substack{z \to \pm 0 \\ (x,y) \in D}} \iint_D f(\xi,\eta) \frac{\partial}{\partial y} \left[\frac{y_0}{y_0^2 + z^2} \left(1 + \frac{x_0}{R_1}\right)\right] \mathrm{d}\xi \, \mathrm{d}\eta =$$

$$= \frac{\partial}{\partial y} \lim_{\substack{z \to \pm 0 \\ (x,y) \in D}} \int_{-b}^{+b} \frac{y_0}{y_0^2 + z^2} K(x,y,z,\eta) \mathrm{d}\eta \,, \tag{5.1.21}$$

with the notation

$$K(x,y,z,\eta) = \int_{x_-(\eta)}^{x_+(\eta)} f(\xi,\eta) \left(1 + \frac{x_0}{R}\right) \mathrm{d}\xi \,. \tag{5.1.22}$$

After setting $z = 0$ in the last expression of I_3, we notice that we may not simplify with y_0, because it vanishes for $\eta = y$. We shall divide the integral into three parts:

$$\int_{-b}^{+b} = \int_{-b}^{y-\varepsilon} + \int_{y+\varepsilon}^{+b} + \int_{y-\varepsilon}^{y+\varepsilon} .$$

The last integral vanishes because for ε small enough we may consider $K(x,y,z,\eta) = K(x,y,z,y)$ whence

$$\int_{y-\varepsilon}^{y+\varepsilon} \frac{y_0}{y_0^2 + z^2} K(x,y,z,\eta) \mathrm{d}\eta = -K(x,y,z,y) \int_{-\varepsilon}^{+\varepsilon} \frac{u \, \mathrm{d}u}{u^2 + z^2} = 0 \,.$$

For the integrals on $(-b, y - \varepsilon)$ and $(y + \varepsilon, +b)$ we may pass to the limit, setting $z \pm 0$ and we may then simplify by y_0, because it does not vanish. It follows

$$I_3 = \frac{\partial}{\partial y} \lim_{\varepsilon \to 0} \left(\int_{-b}^{y-\varepsilon} + \int_{y+\varepsilon}^{+b} \right) \frac{K(x,y,0,\eta)}{y_0} \mathrm{d}\,\eta =$$

$$= \lim_{\varepsilon \to 0} \left[\frac{K(x,y,0,y-\varepsilon)}{\varepsilon} + \frac{K(x,y,0,y+\varepsilon)}{\varepsilon} + \right. \tag{5.1.23}$$

$$\left. + \left(\int_{-b}^{y-\varepsilon} + \int_{y+\varepsilon}^{+b} \right) \frac{\partial}{\partial y} \left(\frac{K}{y_0} \right) \mathrm{d}\,\eta \right].$$

Expanding into a Taylor series the first two terms and taking into account the definition of the "Principal Value" (D.3.1) and the definition of the "Finite Part" (D.3.6), we deduce

$$I_3 = -\,{}^{*}\!\!\int_{-b}^{+b} \frac{K(x,y,0,\eta)}{y_0^2} \mathrm{d}\,\eta + {}'\!\!\int_{-b}^{+b} \frac{1}{y_0} \frac{\partial}{\partial y} K(x,y,0,\eta) \mathrm{d}\,\eta =$$

$$= -\,{}^{*}\!\!\iint_D \frac{f(\xi,\eta)}{y_0^2} \left(1 + \frac{x_0}{R}\right) \mathrm{d}\,\xi\, \mathrm{d}\,\eta + \beta^2 \iint_D f(\xi,\eta) \frac{\partial}{\partial x} \left(\frac{1}{R} \right) \mathrm{d}\,\xi\, \mathrm{d}\,\eta\,. \tag{5.1.24}$$

Taking into account the expressions of I_1, I_2 and I_3 from (5.1.12) we obtain:

$$w(x,y,\pm 0) = \pm \frac{1}{2} f_1(x,y) - \frac{1}{4\pi}\, {}^{*}\!\!\iint_D \frac{f(\xi,\eta)}{y_0^2} \left(1 + \frac{x_0}{R}\right) \mathrm{d}\,\xi\, \mathrm{d}\,\eta\,. \tag{5.1.25}$$

5.1.6 The Integral Equation

From the boundary conditions (2.1.33) and the equations (5.1.2) we deduce:

$$w(x,y,\pm 0) = h'_x(x,y) \pm h'_{1x}(x,y)\,, \quad (x,y) \in D\,. \tag{5.1.26}$$

Replacing the last relation in (5.1.25), subtracting and adding the two relations, the first corresponding to the upper sign (the condition on the upper surface) and the second corresponding to the lower sign, we get

$$f_1(x,y) = -2h'_{1x}(x,y) \tag{5.1.27}$$

$$-\frac{1}{4\pi}\, {}^{*}\!\!\iint_D \frac{f(\xi,\eta)}{y_0^2} \left(1 + \frac{x_0}{R}\right) \mathrm{d}\,\xi\, \mathrm{d}\,\eta = h'_x(x,y)\,, \quad (x,y) \in D\,. \tag{5.1.28}$$

The relation (5.1.27) determines directly $f_1(x,y)$ and the relation (5.1.28) constitutes an integral equation for determining the function f. This is the well known equation of *the lifting surface.* It is obviously a singular integral equation with a strong singularity. The mark $*$ is for the "Finite Part". For the incompressible fluid, The equation has been written for the first time by Multhopp in 1950 [5.24]. A rigorous demonstration is given by Mangler in the Appendix of the above mentioned article. The demonstration is quite complicated because the author utilizes the representation

$$w(x,y,z) = -\frac{1}{4\pi}\iint_D f(\xi,\eta)\frac{\partial}{\partial z}\left[\frac{z}{y_0^2+z^2}\left(1+\frac{x_0}{R_1}\right)\right]\mathrm{d}\xi\,\mathrm{d}\eta \quad (5.1.29)$$

which follows from (5.1.11). Short deductions are presented in [1.1], [1.38].

For wings with lateral edges consisting from straight segments (figure 5.1.3) the equation (5.1.28) is integrated with the conditions

$$f(x,\pm b) = 0 \quad (5.1.30)$$

which, taking into account (5.1.20), means the continuity of the pressure (the Kutta-Joukowski condition). On the trailing edge

$$x = x_+(y)\,,\quad -b < y < +b \quad (5.1.31)$$

one imposes the boundedness condition for the function $f(x,y)$. An equation equivalent to (5.1.28) was given by Lawrence and Flax in 1951 [6.11]. *Formally,* it can be obtained from (5.1.28) using the identity

$$\frac{1}{y_0^2}\left(1+\frac{x_0}{R}\right) = -\frac{\partial}{\partial y}\left(\frac{x_0+R}{x_0y_0}\right) = \frac{\partial}{\partial\eta}\left(\frac{x_0+R_0}{x_0y_0}\right). \quad (5.1.32)$$

It follows:

$$\frac{1}{4\pi}\frac{\partial}{\partial y}\,{}'\!\!\iint_D \frac{f(\xi,\eta)}{y_0}\left(1+\frac{R}{x_0}\right)\mathrm{d}\xi\,\mathrm{d}\eta = h_x'(x,y)\,,\ (x,y)\in D\,. \quad (5.1.33)$$

Sometimes it is preferable to use this equation instead of (5.1.28), because it has weaker singularities (Cauchy-type singularities). A rigorous deduction of equation (5.1.33) from (5.1.28) must start from the definitions of the Finite Part and Principal Value . Using the function

$$C(y) = -\int_{x_-(\eta)}^{x_+(\eta)} f(\xi,\eta)\mathrm{d}\xi\,, \quad (5.1.34)$$

whose significance will be given in 6.1, one demonstrates the identities

$$\frac{\partial}{\partial y} {\iint_D}' \frac{f(\xi,\eta)}{y_0} \mathrm{d}\,\xi\,\mathrm{d}\,\eta = - \iint_D \frac{f(\xi,\eta)}{y_0^2} \mathrm{d}\,\xi\,\mathrm{d}\,\eta, \tag{5.1.35}$$

$$\frac{\partial}{\partial y} {\iint_D}'' f(\xi,\eta) \frac{R}{x_0 y_0} \mathrm{d}\,\xi\,\mathrm{d}\,\eta = - {\iint_D}^{*} \frac{f(\xi,\eta)}{y_0^2} \frac{x_0}{R} \mathrm{d}\,\xi\,\mathrm{d}\,\eta, \tag{5.1.36}$$

which lead to the equivalence in view (see also 6.1).

5.1.7 Other Forms of the Integral Equation

1. We are going to give another form for the expression (5.1.12) of the component w. In this expression the integrals are not singular, so we may employ Green-type formulas. We have

$$\iint_D f(\xi,\eta) \frac{\partial}{\partial y} \left[\frac{y_0}{y_0^2+z^2} \left(1+\frac{x_0}{R_1}\right) \right] \mathrm{d}\,\xi\,\mathrm{d}\,\eta =$$

$$= - \iint_D f(\xi,\eta) \frac{\partial}{\partial \eta} \left[\frac{y_0}{y_0^2+z^2} \left(1+\frac{x_0}{R_1}\right) \right] \mathrm{d}\,\xi \mathrm{d}\,\eta =$$

$$= - \iint_D \frac{\partial}{\partial \eta} \left[f \frac{y_0}{y_0^2+z^2} \left(1+\frac{x_0}{R_1}\right) \right] \mathrm{d}\,\xi\,\mathrm{d}\,\eta +$$

$$+ \iint_D \frac{\partial f}{\partial \eta} \frac{y_0}{y_0^2+z^2} \left(1+\frac{x_0}{R_1}\right) \mathrm{d}\,\xi\,\mathrm{d}\,\eta =$$

$$= \oint_{\partial D} f(\xi,\eta) \frac{y_0}{y_0^2+z^2} \left(1+\frac{x_0}{R_1}\right) \mathrm{d}\,\xi + \iint_D \frac{\partial f}{\partial \eta} \frac{y_0}{y_0^2+z^2} \left(1+\frac{x_0}{R_1}\right) \mathrm{d}\,\xi\,\mathrm{d}\,\eta \tag{5.1.37}$$

Using the identity:

$$f \frac{\partial}{\partial x} \left(\frac{1}{R_1}\right) = \frac{\partial}{\partial \eta} \left(\frac{f}{R_1} \frac{y_0}{x_0} \right) + \beta^2 \frac{f z^2}{x_0 R_1^3} - \frac{\partial f}{\partial \eta} \frac{y_0}{x_0 R} \tag{5.1.38}$$

and Green's formula, we deduce

$$\begin{aligned} \iint_D f(\xi,\eta) \frac{\partial}{\partial x} \left(\frac{1}{R_1}\right) \mathrm{d}\,\xi\,\mathrm{d}\,\eta = & - \oint_{\partial D} \frac{f(\xi,\eta)}{R} \frac{y_0}{x_0} \mathrm{d}\,\xi + \\ & + \beta^2 z^2 \iint_D \frac{f(\xi,\eta)}{x_0 R_1^3} \mathrm{d}\,\xi\,\mathrm{d}\,\eta - \iint_D \frac{\partial f}{\partial \eta} \frac{y_0}{x_0 R_1} \mathrm{d}\,\xi\,\mathrm{d}\,\eta. \end{aligned} \tag{5.1.39}$$

Replacing (5.1.37) and (5.1.39) in (5.1.12) and passing to the limit, we obtain

$$w(x,y,\pm 0) = \pm\frac{1}{2}f_1(x,y) + \frac{1}{4\pi}\oint_{\partial D}' \frac{f(\xi,\eta)}{y_0}\left(1+\frac{R}{x_0}\right)\mathrm{d}\,\xi + \\ + \frac{1}{4\pi}\iint_D'' \frac{\partial f}{\partial \eta}\frac{1}{y_0}\left(1+\frac{R}{x_0}\right)\mathrm{d}\,\xi\,\mathrm{d}\,\eta\,, \tag{5.1.40}$$

whence it follows the integral equation:

$$\frac{1}{4\pi}\oint_{\partial D}' \frac{f(\xi,\eta)}{y_0}\left(1+\frac{R}{x_0}\right)\mathrm{d}\,\xi + \\ + \frac{1}{4\pi}\iint_D'' \frac{\partial f}{\partial \eta}\frac{1}{y_0}\left(1+\frac{R}{x_0}\right)\mathrm{d}\,\xi\,\mathrm{d}\,\eta = h_x'(x,y). \tag{5.1.41}$$

This is the equation given by Homentcovschi equation herein, utilizing Green's formulas before passing to the limit is preferable to that of Homentcovschi which utilizes Green's formulas for singular integrals. *Formally*, the equation (5.1.41) is deduced from (5.1.28) by means of the identity (5.1.32) as follows:

$$\iint_D^* \frac{f(\xi,\eta)}{y_0^2}\left(1+\frac{x_0}{R}\right)\mathrm{d}\,\xi\,\mathrm{d}\,\eta = \iint_D^* f(\xi,\eta)\frac{\partial}{\partial\eta}\left[\frac{1}{y_0^2}\left(1+\frac{R}{x_0}\right)\right]\mathrm{d}\,\xi\,\mathrm{d}\,\eta = \\ = \iint_D^* \left\{\frac{\partial}{\partial\eta}\left[\frac{f}{y_0}\left(1+\frac{R}{x_0}\right)\right] - \frac{\partial}{\partial\eta}\frac{1}{y_0^2}\left(1+\frac{R}{x_0}\right)\right\}\mathrm{d}\,\xi\,\mathrm{d}\,\eta = \\ = \oint_{\partial D}' \frac{f(\xi,\eta)}{y_0}\left(1+\frac{R}{x_0}\right)\mathrm{d}\,\xi - \iint_D'' \frac{\partial f}{\partial\eta}\frac{1}{y_0}\left(1+\frac{R}{x_0}\right)\mathrm{d}\,\xi\,\mathrm{d}\,\eta\,. \tag{5.1.42}$$

2. In other papers (see for example, [1.2]) we find the equation:

$$\frac{1}{4\pi}\iint_D'' \frac{\partial f}{\partial\eta}\frac{1}{y_0}\left(1+\frac{R}{x_0}\right)\mathrm{d}\,\xi\,\mathrm{d}\,\eta = h_x'(x,y)\,. \tag{5.1.43}$$

It can be obtained from Homentcovschi's equation, imposing the condition

$$f(x,y) = 0 \quad \text{on } \partial D\,. \tag{5.1.44}$$

Usually this condition may be imposed either on the trailing edge or on the leading edge and on the lateral edges. Indeed, imposing the condition (5.1.44) both on the leading and trailing edges means that for the plane profile obtained through a section of the wing with a plane parallel to the xOz-plane, one has to determine the jump of the pressure which vanishes at the both edges. But it is well known that the jump of the pressure for the plane profile satisfies the equation (C.1.1) which has a bounded solution at both edges only if the free term satisfies the condition (C.1.13). Hence the condition (5.1.44) is valid only after imposing some restrictions for $h_x(x, y)$.

However, for wings whose trailing (leading) edge is a straight line perpendicular on the Ox-axis, like, for example, the delta wing with the base representing the trailing (leading) edge perpendicular on Ox, or the trapezoidal wing with the big base representing the trailing (leading) edge perpendicular to Ox, the entire integral equation (5.1.41) reduces to the integral equation (5.1.43), because on the perpendicular trailing (leading) edge we have $\mathrm{d}\,\xi = 0$, and on the lateral edges we may impose the condition (5.1.41). For the rectangular wing the equation (5.1.43) is exact.

5.1.8 The Plane Problem

In order to obtain the solution of the plane problem from the three-dimensional solution we assume that the intersection of the wing with each plan parallel to the xOz -plane is a profile having the same shape (figure 5.1.1). This means that the equations (5.1.2) have the form

$$z = h(x) \pm h_1(x)\,, \quad (\forall)\, y\,, \tag{5.1.45}$$

and the domain D is the rectangle $(-1 < x < 1, -b < y < b)$ with $b \to \infty$ (half of the chord of the wing was taken as the reference length L_0). From these hypotheses it follows the general representations of the solutions (5.1.8) and (5.1.12),

$$f_1 = f_1(\xi), \quad f = f(\xi)\,.$$

Taking into account that

$$\int \frac{\mathrm{d}\,y}{(a^2 + \beta^2 y^2)^{3/2}} = \frac{y}{a^2\sqrt{a^2 + \beta^2 y^2}}$$

we deduce

$$\lim_{b\to\infty} \int_{-b}^{+b} \frac{\partial}{\partial x}\left(\frac{1}{R}\right) \mathrm{d}\,\eta = -\frac{2x_0}{\beta(x_0^2 + \beta^2 z^2)} \tag{5.1.46}$$

$$\lim_{b\to\infty}\int_{-b}^{+b}\frac{\partial}{\partial y}\left[\frac{y_0}{y_0^2+z^2}\left(1+\frac{x_0}{R}\right)\right]\mathrm{d}\,\eta=$$

$$=-\lim_{b\to\infty}\int_{-b}^{+b}\frac{\partial}{\partial y}\left[\frac{y_0}{y_0^2+z}\left(1+\frac{x_0}{R}\right)\right]\mathrm{d}\,\eta=0\,. \tag{5.1.47}$$

So, the above mentioned representation becomes

$$p(x,z)=\frac{1}{2\pi\beta}\int_{-1}^{+1}\frac{x_0f_1(\xi)+\beta^2zf(\xi)}{x_0^2+\beta^2z^2}\mathrm{d}\,\xi$$

$$w(x,z)=\frac{\beta}{2\pi}\int_{-1}^{+1}\frac{x_0f(\xi)-zf_1(\xi)}{x_0^2+\beta^2z^2}\mathrm{d}\,\xi \tag{5.1.48}$$

and $v(x,z)=0$. Taking into account that here Oz has the position of the Oy -axis from the problem considered in 3.1,we find again the formulas (3.1.17).

The integral equation is obtained from (5.1.28). For calculating the integral

$$\overset{*}{\int}_{-b}^{+b}\frac{1}{y_0^2}\left(1+\frac{x_0}{R}\right)\mathrm{d}\,\eta\,,$$

we use the definition of the Finite Part and the identity (5.1.32) valid on each of the intervals $(-b,y-\varepsilon)$, $(y+\varepsilon,b)$. Thus we obtain

$$\int_{-b}^{+b}\frac{1}{y_0^2}\left(1+\frac{x_0}{R}\right)\mathrm{d}\eta=\lim_{\varepsilon\to0}\left\{\frac{x_0+R}{x_0y_0}\Bigg|_{-b}^{y-\varepsilon}+\right.$$

$$\left.+\frac{x_0+R}{x_0y_0}\Bigg|_{y+\varepsilon}^{+b}-\frac{2}{\varepsilon}\left(1+\frac{x_0}{|x_0|}\right)\right\},$$

whence

$$\lim_{b\to\infty}\int_{-b}^{+b}\frac{1}{y_0^2}\left(1+\frac{x_0}{R}\right)\mathrm{d}\eta=-\frac{2\beta}{x_0}\,. \tag{5.1.49}$$

The integral equation reduces to (3.1.21).

5.1.9 The Aerodynamic Action in the First Approximation

We indicated in [1.11], p. 79 the way to calculate the action of the fluids against the bodies by applying the transport equations of the momentum and the moment of momentum to the fluid filling the domain bounded by the body surface Σ and the control surface Σ_0 surrounding Σ. This calculus is absolutely necessary when the body has corners, like in the case of thin bodies. However, in a first approximation the calculations may be done taking into account that the action of the fluid is given by the jump of the total pressure

$$[\![p_1]\!] = p_1(x_1, y_1, -0) - p_1(x, y_1, +0)$$

which determines the lift L. On the unit of area the force is $[\![p_1]\!]\boldsymbol{k}$ (fig. 5.1.4), $\boldsymbol{k}$ representing the versor of the Oz_1-axis, and on the entire area of the domain $D_1, \boldsymbol{L} = L\boldsymbol{k}$, where

$$L = \iint_{D_1} [\![p_1]\!] \mathrm{d}x_1 \mathrm{d}y_1 = \rho_\infty U_\infty^2 L_0^2 \iint_D [\![p]\!] \mathrm{d}x \mathrm{d}y \,. \qquad (5.1.50)$$

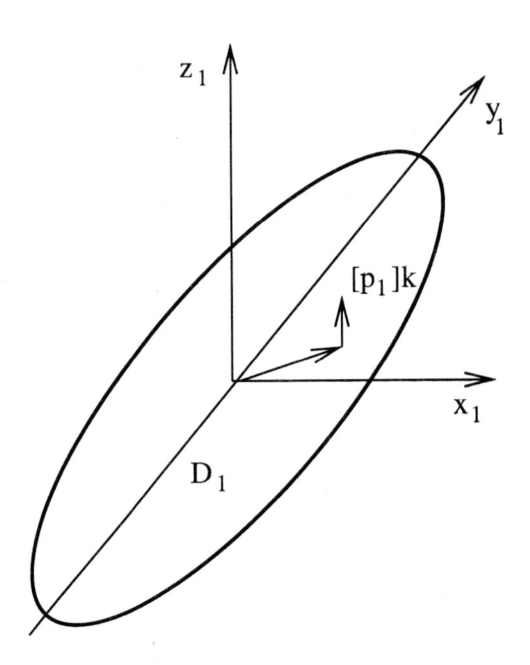

Fig. 5.1.4.

For the resulting moment one obtains the formula

$$\boldsymbol{M} = \iint_{D_1} \boldsymbol{x}_1 \times [\![p_1]\!] \boldsymbol{k} \mathrm{d}x_1 \mathrm{d}y_1 = \rho_\infty U_\infty^2 L_0^3 \iint_D \boldsymbol{x} \times [\![p]\!] \boldsymbol{k} \mathrm{d}x \mathrm{d}y \,, \quad (5.1.51)$$

where according to (5.1.20),

$$[\![p]\!] = p(x, y, -0) - p(x, y, +0) = -f(x, y) \,. \qquad (5.1.52)$$

Denoting by A_1 the area of the domain D_1 (which is the domain D in dimensional coordinates) we have

$$A_1 = \iint_{D_1} \mathrm{d}x_1 \mathrm{d}y_1 = L_0^2 \iint_D \mathrm{d}x \mathrm{d}y = L_0^2 A \,. \tag{5.1.53}$$

For the lifting coefficient c_L one obtains the formula

$$c_L \stackrel{\text{def}}{=} \frac{L}{(1/2)\rho_\infty U_\infty^2 A_1} = -\frac{2}{A} \iint_D f(x,y) \mathrm{d}x \mathrm{d}y \,, \tag{5.1.54}$$

and for the moment coefficients

$$c_x \stackrel{\text{def}}{=} \frac{M_x}{(1/2)\rho_\infty U_\infty^2 A_1 a_1} = -\frac{2}{A a_0} \iint_D y f(x,y) \mathrm{d}x \mathrm{d}y \,, \tag{5.1.55}$$

$$c_y \stackrel{\text{def}}{=} \frac{M_y}{(1/2)\rho_\infty U_\infty^2 A_1 a_1} = \frac{2}{A a_0} \iint x f(x,y) \mathrm{d}x \mathrm{d}y \,, \tag{5.1.56}$$

where $a_1 = a_0 L_0$ is the medium chord of the domain D_1 on the direction of the unperturbed stream. As we shall see in the sequel, *the drag* (the projection of the resultant on the direction of the unperturbed stream) and *the gyration moment* (the projection of the moment Oz_1-axis) have the order of magnitude ε^2, for this reason they are not present in this calculation. M_x is *the rolling moment*, and M_y, *the pitching moment.*

5.1.10 A More Accurate Calculation

We obtain a more accurate calculation if we have in view that the resultant and the resultant moment are determined by the jump of the tension vector i.e. by the formulas

$$\boldsymbol{R} = -\iint_{D_1} [\![p_1 \boldsymbol{n}]\!] \mathrm{d}x_1 \mathrm{d}y_1 \,, \quad \boldsymbol{M} = -\iint_{D_1} \boldsymbol{x}_1 \times [\![p_1 \boldsymbol{n}]\!] \mathrm{d}x_1 \mathrm{d}y_1 \,, \tag{5.1.57}$$

where $\boldsymbol{n}$ is the outward normal on Σ. The presence of the normal enables us to take into account the shape of the wing. If the equation of the wing is $h(x,y) - z = 0 (h = \varepsilon \overline{h})$, the normal $\boldsymbol{n}$ is

$$\boldsymbol{n} = \frac{h_x \boldsymbol{i} + h_y \boldsymbol{j} - \boldsymbol{k}}{\sqrt{1 + h_x^2 + h_y^2}} = h_x \boldsymbol{i} + h_y \boldsymbol{j} - \boldsymbol{k} + \frac{h_x^2 + h_y^2}{2} \boldsymbol{k} + O(\varepsilon^3) \,.$$

Neglecting the terms of order $O(\varepsilon^3)$ and taking into account that h is $h(x,y) \pm h_1(x,y)$, it follows

$$[\![-p_1 \boldsymbol{n}]\!] = 2p_\infty \left[h_{1x}\boldsymbol{i} + h_{1y}\boldsymbol{j} + (h_x h_{1x} + h_y h_{1y})\boldsymbol{k} \right] -$$

$$-\rho_\infty U_\infty^2 [\![p]\!] (h_x \boldsymbol{i} + h_y \boldsymbol{j} - \boldsymbol{k}) + \rho_\infty U_\infty^2 < p > (h_{1x}\boldsymbol{i} + h_{1y}\boldsymbol{j}),$$

where we denoted

$$< p >= p(x,y,+0) + p(x,y,-0). \tag{5.1.58}$$

Hence *the lift* is:

$$R_z = -\rho_\infty^2 U_\infty^2 L_0^2 \iint_D f(y)\mathrm{d}x\mathrm{d}y + 2p_\infty L_0^2 \iint_D (h_x h_{1x} + h_y h_{1y})\mathrm{d}x\mathrm{d}y, \tag{5.1.59}$$

and *the drag*

$$R_x = 2p_\infty L_0^2 \iint_D h_{1x}\mathrm{d}x\mathrm{d}y + \rho_\infty U_\infty^2 L_0^2 \iint_D f(x,y) h_x \mathrm{d}x\mathrm{d}y + \tag{5.1.60}$$

$$+\rho_\infty U_\infty^2 L_0^2 \iint_D < p > h_{1x}\mathrm{d}x\mathrm{d}y.$$

For *the rolling moment* one obtains the formula:

$$M_x = -\rho_\infty U_\infty^2 L_0^3 \iint_D y f(x,y)\mathrm{d}x\mathrm{d}y + 2p_\infty L_0^3 \iint_D y(h_x h_{1x} + h_y h_{1y})\mathrm{d}x\mathrm{d}y, \tag{5.1.61}$$

for *the pitching moment* one obtains

$$M_y = \rho_\infty U_\infty^2 L_0^3 \iint_D x f(x,y)\mathrm{d}x\mathrm{d}y - 2p_\infty L_0^3 \iint_D x(h_x h_{1x} + h_y h_{1y})\mathrm{d}x\mathrm{d}y, \tag{5.1.62}$$

and for *the gyration moment,*

$$M_z = 2p_\infty L_0^3 \iint_D (x h_{1y} - y h_{1x})\mathrm{d}x\mathrm{d}y +$$

$$+\rho_\infty U_\infty^2 L_0^3 \iint_D f(x h_y - y h_x)\mathrm{d}x\mathrm{d}y + \tag{5.1.63}$$

$$+\rho_\infty U_\infty^2 L_0^3 \iint_D < p > (x h_{1y} - y h_{1x})\mathrm{d}x\mathrm{d}y,$$

where

$$< p >= -\frac{1}{\pi} \overset{\odot}{\iint_D} h_{1x} \frac{x_0}{R^3} \mathrm{d}\xi \mathrm{d}\eta, \tag{5.1.64}$$

as it results from (5.1.19) and (5.1.27).

If we keep only the terms of order $O(\varepsilon)$, it follows:

$$
\begin{aligned}
R_z &= -\rho_\infty U_\infty^2 L_0^2 \iint_D f(x,y)\mathrm{d}x\mathrm{d}y\,, \\
R_x &= 2p_\infty L_0^2 \iint_D h_{1x}(x,y)\mathrm{d}x\mathrm{d}y, \\
(M_x, M_y) &= \rho_\infty U_\infty^3 L_0^3 \iint_D (-y,x) f(x,y)\mathrm{d}x\mathrm{d}y, \\
M_z &= 2p_\infty L_0^3 \iint_D (xh_{1y} - yh_{1x})\mathrm{d}x\mathrm{d}y\,.
\end{aligned}
\tag{5.1.65}
$$

For the lifting surface $(h_1 = 0)$ the formulas (5.1.55)–(5.1.59) become

$$
\begin{aligned}
R_z &= -\rho_\infty U_\infty^2 L_0^2 \iint_D f(x,y)\mathrm{d}x\mathrm{d}y\,, \\
R_x &= -\rho_\infty U_\infty^2 L_0^2 \iint_D f(x,y) h_x(x,y)\mathrm{d}x\mathrm{d}y, \\
(M_x, M_y) &= \rho_\infty U_\infty^3 L_0^3 \iint_D (-y,x) f(x,y)\mathrm{d}x\mathrm{d}y, \\
M_z &= 2p_\infty L_0^3 \iint_D (xh_y - yh_x)\mathrm{d}x\mathrm{d}y\,.
\end{aligned}
\tag{5.1.66}
$$

R_z, M_x şi M_y being $O(\varepsilon)$, and R_x and $M_z, O(\varepsilon^2)$.

5.1.11 Another Deduction of the Representation of the General Solution

In the sequel we shall deduce again the representation of the solution (5.1.8), (5.1.12). We start from D. Homentcovschi's idea (exposed in [A.8]) to utilize the Fourier transform for bounded domains [A.6]. The method synthesizes the problem of determination of the fundamental solution and the problem of replacing the wing with a forces distribution. In addition it justifies the assimilation of the wing with a distribution of forces having the form $(f_1, 0, f)$. Indeed, employing the formulas

(A.8.2) to the system of linearized aerodynamics (2.1.30) and taking into account that D is a surface of discontinuity, we get

$$\begin{aligned} 0 &= -\mathrm{i}\alpha_1 M^2 \widehat{p} - \mathrm{i}\boldsymbol{\alpha}\cdot\widehat{\boldsymbol{v}} - W \\ 0 &= -\mathrm{i}\alpha_1 \widehat{\boldsymbol{v}} - \mathrm{i}\boldsymbol{\alpha}\widehat{p} - P\boldsymbol{k}\,, \end{aligned} \tag{5.1.67}$$

with the notations

$$(W, P) = \iint_D ([\![w]\!], [\![p]\!]) \mathrm{e}^{\mathrm{i}(\alpha_1 \mathrm{x} + \alpha_2 \mathrm{y})} \mathrm{d}\,\mathrm{x}\,\mathrm{d}\,\mathrm{y}\,,$$

where, taking into account (5.1.20), (5.1.25) and (5.1.27),

$$\begin{aligned} [\![w]\!] &= w(x, y, +0) - w(x, y, -0) = 2h_{1x}(x, y) \\ [\![p]\!] &= p(x, y, +0) - p(x, y, -0) = f(x, y)\,. \end{aligned} \tag{5.1.68}$$

From (5.1.67) one determines first $\widehat{p}$ and then $\widehat{w}$. One obtains

$$\begin{aligned} \widehat{p} &= \frac{\mathrm{i}\alpha_3 P - \mathrm{i}\alpha_1 W}{\alpha^2 - M^2\alpha_1^2}\,, \\ \widehat{w} &= \frac{\mathrm{i}\alpha_3 W}{\alpha^2 - M^2\alpha_1^2} + \beta^2 \frac{\mathrm{i}\alpha_1 P}{\alpha^2 - M^2\alpha_1^2} + \frac{\mathrm{i}\alpha_2^2 P}{\alpha_1(\alpha^2 - M^2\alpha_1^2)}\,. \end{aligned} \tag{5.1.69}$$

Utilizing the expressions of P and W, we find:

$$\begin{aligned} \widehat{p} = & \iint_D f(\xi, \eta) \frac{\mathrm{i}\alpha_3}{\alpha^2 - M^2\alpha_1^2} \mathrm{e}^{\mathrm{i}(\alpha_1\xi + \alpha_2\eta)} \mathrm{d}\xi\,\mathrm{d}\eta - \\ & -2 \iint_D h_{1\xi}(\xi, \eta) \frac{\mathrm{i}\alpha_1}{\alpha^2 - M^2\alpha^2} \mathrm{e}^{\mathrm{i}(\alpha_1\xi + \alpha_2\eta)} \mathrm{d}\xi\,\mathrm{d}\eta\,, \end{aligned}$$

whence, taking into account (A.6.9),

$$\begin{aligned} p(x, y, z) = & -\iint_D f(\xi, \eta) \frac{\partial}{\partial z} \mathcal{F}^{-1} \left[\frac{\mathrm{e}^{\mathrm{i}(\alpha_1\xi + \alpha_2\eta)}}{\alpha^2 - M^2\alpha_1^2} \right] \mathrm{d}\xi\,\mathrm{d}\eta + \\ & +2 \iint_D h_{1\xi}(\xi, \eta) \frac{\partial}{\partial z} \mathcal{F}^{-1} \left[\frac{\mathrm{e}^{\mathrm{i}(\alpha_1\xi + \alpha_2\eta)}}{\alpha^2 - M^2\alpha_1^2} \right] \mathrm{d}\xi\,\mathrm{d}\eta\,. \end{aligned} \tag{5.1.70}$$

Employing (A.7.2) we obtain

$$\begin{aligned} p(x, y, z) = & -\frac{1}{4\pi} \iint_D f(\xi, \eta) \frac{\partial}{\partial z} \left(\frac{1}{R_1} \right) \mathrm{d}\xi\,\mathrm{d}\eta + \\ & +\frac{1}{2\pi} \iint_D h_{1\xi}(\xi, \eta) \frac{\partial}{\partial x} \left(\frac{1}{R_1} \right) \mathrm{d}\xi\,\mathrm{d}\eta\,, \end{aligned} \tag{5.1.71}$$

where R_1 is that from (5.1.7). We obtained in this way the representation (5.1.8).

Similarly, utilizing (2.3.11) and (2.3.27), we have:

$$\frac{\partial}{\partial y}\mathcal{F}^{-1}\left[\frac{\mathrm{e}^{\mathrm{i}(\alpha_1\xi+\alpha_2\eta)}}{\mathrm{i}\alpha_1(\alpha^2-M^2\alpha_1^2)}\right]=$$
$$=-\frac{1}{4\pi}\int_{-\infty}^{x}\frac{\partial}{\partial y}\left(\frac{1}{R_1}\right)\mathrm{d}x=\frac{1}{4\pi}\frac{y_0}{y_0^2+z^2}\left(1+\frac{x_0}{R_1}\right), \tag{5.1.72}$$

such that R, having the same expression, it follows:

$$\begin{aligned}w(x,y,z)=&-\frac{1}{2\pi}\iint_D h_{1\xi}(\xi,\eta)\frac{\partial}{\partial z}\left(\frac{1}{R}\right)\mathrm{d}\xi\,\mathrm{d}\eta-\\&-\frac{\beta^2}{4\pi}\iint_D f(\xi,\eta)\frac{\partial}{\partial x}\left(\frac{1}{R_1}\right)\mathrm{d}\xi\,\mathrm{d}\eta+\\&+\frac{1}{4\pi}\iint_D f(\xi,\eta)\frac{\partial}{\partial y}\left[\frac{y_0}{y_0^2+z^2}\left(1+\frac{x_0}{R_1}\right)\right]\mathrm{d}\xi\,\mathrm{d}\eta,\end{aligned} \tag{5.1.73}$$

which is just the representation (5.1.12). We have to notice that from (5.1.70) one deduces the inversion formulas

$$\mathcal{F}^{-1}\left[\frac{\alpha_2}{\alpha_1(\alpha^2-M^2\alpha_1^2)}\right]=-\frac{1}{4\pi}\frac{y}{y^2+z^2}\left[1+\frac{x}{\sqrt{x^2+\beta^2(x^2+z^2)}}\right] \tag{5.1.74}$$

which could be utilized in 2.3.

5.2 Methods for the Numerical Integration of the Lifting Surface Equation

5.2.1 The General Theory

There are not yet known exact solutions of the equation (5.1.28). The first numerical solution was given by Multhopp in 1950 [5.24]. Previously the same author had given in 1937 the approximate solution (to which one assigned his name), for the lifting line equation (Prandtl's equation (6.1.16)). Multhopp's method relies on the Gauss-type quadrature formulas for non-singular integrals. At that time there were not available quadrature formulas for singular equations. For the singularity appearing in the lifting surface equation, Multhopp utilizes a series expansion with Chebyshev polynomials of first kind, which is truncated in order to obtain an algebraic system. The method is analogous to Glauert's method for Prandtl's equations, except that the sin functions are replaced by Chebyshev polynomials. In 1958, at a Meeting in Fort Worth, Hsu gave a quadrature formula for integrals with a strong singularity and employed this formula for the singularity from the equation (5.1.28). However, in Hsu's formula the unknowns are present even in the collocation points and this is a drawback. Starting from a formula given by Monegato [A.52], Dragoş gives [6.5] the formula (F.3.5) where there are present supplementary unknowns in the collocation points. One utilizes successfully this formula for solving Prandtl's equation in 6.5 and for solving the lifting surface equation in [5.10] and [5.11]. These solution will be presented in 5.2 and 5.3. In 5.2 we shall sketch the solution of the equation S via the collocation method.

We have to solve the equation:

$$\frac{1}{4\pi} \, {}^{*}\!\!\iint_D \frac{f(\xi,\eta)}{y_0^2} N(x_0,y_0) \mathrm{d}\xi \, \mathrm{d}\eta = -h_x'(x,y) \,, \tag{5.2.1}$$

where

$$N(x_0,y_0) = 1 + \frac{x_0}{\sqrt{x_0^2 + k^2 y_0^2}} \,, \tag{5.2.2}$$

with the following conditions:

$$f(\xi, \pm b) = 0 \,, \quad x_-(\pm b) < \xi < x_+(\pm b) \tag{5.2.3}$$

$$f(x_+(\eta),\eta) = 0 \,, \quad -b < \eta < +b \,. \tag{5.2.4}$$

The first two conditions mean the continuity of the pressure on the lateral edges in case that they are straight segments (if the lateral edges are represented by points the conditions disappear by virtue of (5.1.1)). The condition (5.2.4) ensures the boundedness of the pressure along the trailing edge. One performs the following reasoning: each intersection of the wing with a plane parallel to xOz determines a thin profile; as it is known from 3.1 for such a profile one imposes the boundedness condition on the trailing edge.

In order to utilize the quadrature formulas we shall perform the change of variables

$$(x,y) \longrightarrow (u,v), (\xi,\eta) \longrightarrow (\alpha,\beta) \,,$$

defined by the equations

$$\begin{aligned} x &= a(y)u + c(y) \quad & \xi &= a(\eta)\alpha + c(\eta) \\ y &= bv & \eta &= b\beta \,, \end{aligned} \tag{5.2.5}$$

with

$$a(y) = \frac{x_+(y) - x_-(y)}{2}, \quad c(y) = \frac{x_+(y) + x_-(y)}{2}. \tag{5.2.6}$$

Writing the equation (5.2.1) as follows:

$$\int_{-1}^{*+1} \frac{1}{y_0^2} \left[\int_{x_-(\eta)}^{x_+(\eta)} f(\xi,\eta) N(x_0, y_0) \mathrm{d}\xi \right] \mathrm{d}\eta = -4\pi h'_x(x,y)$$

and taking into account that

$$\frac{\partial(\xi,\eta)}{\partial(\alpha,\beta)} = a(\beta) b \,,$$

we deduce

$$\int_{-1}^{*+1} \frac{a(\beta)}{(v-\beta)^2} \left[\int_{-1}^{+1} f(\alpha,\beta) N(u,v,\alpha,\beta) \mathrm{d}\alpha \right] \mathrm{d}\beta = g(u,v)\,, \tag{5.2.7}$$

where denoting by $a(v), c(v), f(\alpha,\beta)$ etc., the functions $a(y), c(y)$, $f(\xi,\eta)$ in the new variables, we have

$$N(u,v,\alpha,\beta) = 1 + \frac{a(v)u + c(v) - a(\beta)\alpha - c(\beta)}{\left\{ [a(v)u + c(v) - a(\beta)\alpha - c(\beta)]^2 + k^2 b^2 (v-\beta)^2 \right\}^{1/2}} \tag{5.2.8}$$

$$g(u,v) = -4\pi b h'_x(u,v)\,, \tag{5.2.9}$$

with the notation $k = \sqrt{1-M^2}$. The Kutta-Joukowski conditions (5.2.3) become

$$f(\alpha, \pm 1) = 0\,, \quad -1 < \alpha < +1)\,, \tag{5.2.10}$$

$$f(1,\beta) = 0\,, \quad -1 < \beta < +1)\,. \tag{5.2.11}$$

The solution of the equation (5.2.7) depends on the behaviour that we impose at the extremities of the intervals (± 1). For a fixed β , one obtains within the wing a thin profile. For such a profile we have imposed in 3.1 the behavior given by $\sqrt{\dfrac{1-\alpha}{1+\alpha}}$. Along the span one imposes in the behaviour $\sqrt{1-\beta^2}$. Such a behaviour ensures that the conditions (5.2.10) are satisfied. We shall seek for solutions $f(\alpha\beta)$ having the shape

$$f(\alpha,\beta) = \sqrt{1-\beta^2}\sqrt{\frac{1-\alpha}{1+\alpha}} F(\alpha,\beta)\,, \tag{5.2.12}$$

with F bounded. For F one obtain the equation

$$\int_{-1}^{*+1} \frac{\sqrt{1-\beta^2}}{(\beta-v)^2} \dot{a}(\beta)\left[\int_{-1}^{+1} \sqrt{\frac{1-\alpha}{1+\alpha}} F(\alpha,\beta) N(u,v,\alpha,\beta)\mathrm{d}\alpha\right]\mathrm{d}\beta =$$

$$= g(u,v)\,, \qquad -1 < u,v < +1\,. \tag{5.2.13}$$

Utilizing the quadrature formula (F.2.18), this equation becomes

$$2\pi \sum_{i=1}^{m} (1-\alpha_i) \int_{-1}^{*+1} \frac{\sqrt{1-\beta^2}}{(\beta-v)^2} a(\beta) N(u,v,\alpha_i,\beta) F(\alpha_i,\beta)\mathrm{d}\,\beta = \\ = (2m+1) g(u,v)\,, \tag{5.2.14}$$

where

$$\alpha_i = \cos\frac{2i\pi}{2m+1}\,, \qquad i = \overline{1m}\,. \tag{5.2.15}$$

From now on Multhopp's method and the quadrature formulas method are separating.

5.2.2 Multhopp's Method

Using the formula (F.4.2) for $(\beta-v)^{-2}$, equation (5.2.14) becomes

$$-4\pi \sum_{i=1}^{m}\sum_{k=1}^{p} (1+k)(1- \\ -\alpha_1) U_k(v) \int_{-1}^{+1} \sqrt{1-\beta^2} a(\beta) N(u,v,\alpha_i,\beta) U_k(\beta) F(\alpha_i,\beta)\mathrm{d}\beta = \\ = (2m+1) g(u,v)\,. \tag{5.2.16}$$

One calculates the integral from this equation utilizing (F.2.12). One obtains

$$-4\pi \sum_{i=1}^{m}\sum_{k=1}^{p}\sum_{j=1}^{n} (1+k)(1-\alpha_i) U_k(v)(1- \\ -\beta_j^2) N(u,v,\alpha_i,\beta_j) U_k(\beta_j) F(\alpha_i,\beta_j) = \tag{5.2.17} \\ = (2m+1)(n+1) g(u,v)\,, \quad -1 < u,v < +1\,,$$

with

$$\beta_j = \cos\frac{j\pi}{n+1}, \quad j=\overline{1,n}. \tag{5.2.18}$$

Introducing the notation

$$H(\alpha_i,\beta_j) = (1-\alpha_i)(1-\beta_j^2)a(\beta_j)F(\alpha_i,\beta_j), \tag{5.2.19}$$

the system (5.2.17) becomes

$$\begin{aligned}-4\pi\sum_{i=1}^{m}\sum_{j=1}^{n}\sum_{k=1}^{p}(1+k)U_k(v)U_k(\beta_j)N(u,v,\alpha_i,\beta_j)H(\alpha_i,\beta_j) = \\ = (2m+1)(n+1)g(u,v), \quad -1<u,v<+1.\end{aligned} \tag{5.2.20}$$

There are $m\times n$ unknowns, $H(\alpha_i,\beta_j)$. In order to obtain the same number of equations in (5.2.20) we shall assign m values to u and n values to v, obviously, all of them in the interval $(-1,+1)$.

As we shall see later, the aerodynamic coefficients are functions of $H(\alpha_i,\beta_j)$. It is sufficient to find out these unknowns. We may write computer programs for solving the system (5.2.20).

5.2.3 The Quadrature Formulas Method

Utilizing in equation (5.2.14) the formula (F.3.5), we obtain the equation

$$\begin{aligned}&2\pi\sum_{i=1}^{m}\sum_{j=1}^{n}{}' \, [1- \\ &-(-1)^{j+k}]\frac{(1-\alpha_i)(1-\beta_j^2)}{(\beta_j-\beta_k)^2}a(\beta_j)N(u,\beta_k,\alpha_i,\beta_j)F(\alpha_i,\beta_j)- \\ &-\pi(2m+1)(n+1)^2\sum_{i=1}^{m}(1-\alpha_i)a(\beta_k)N(u,\beta_k,\alpha_i,\beta_k)F(\alpha_i,\beta_k) = \\ &= (2m+1)(n+1)g(u,\beta_k),\end{aligned} \tag{5.2.21}$$

where

$$\beta_j = \cos\frac{j\pi}{n+1}, \quad j=\overline{1,n}; \tag{5.2.22}$$

for $k=\overline{1,n}$.

In (5.2.21) we have a system of n linear algebraic equations with $m \times n$ unknowns $F(\alpha_i, \beta_j)$. Imposing the system to be verified in m points

$$u_\ell = \cos\frac{2\ell\pi}{2m+1} = \alpha_\ell\,, \quad \ell = \overline{1,m}\,, \tag{5.2.23}$$

the number of equations equals the number of unknowns. The system will be written as follows

$$\sum_{i=1}^{m}\sum_{j=1}^{n}{}' A_{\ell kij} H(\alpha_i, \beta_j) = B_{\ell k}, \quad \ell = \overline{1,m}\,,\ k = \overline{1n}\,, \tag{5.2.24}$$

where we denoted

$$\begin{aligned} A_{\ell kij} &= \pi \left\{ 2\frac{\left[1-(-1)^{j+k}\right]}{(\beta_j - \beta_k)^2} - \right. \\ &\left. -(2m+1)(n+1)^2 \frac{\delta_{jk}}{1-\beta_j^2} \right\} N(\alpha_\ell, \beta_k, \alpha_i, \beta_j) \\ B_{\ell k} &= (2m+1)(n+1) g(\alpha_\ell, \beta_k)\,. \end{aligned} \tag{5.2.25}$$

5.2.4 The Aerodynamic Action

The lift and moment coefficients are calculated by the aid of formulas (5.1.54)–(5.1.56). Using (5.2.12), (F.2.18) and (F.2.12) we obtain:

$$\begin{aligned} c_L &= \frac{2}{A}\iint_D f(\xi,\eta)\mathrm{d}\xi\mathrm{d}\eta = \frac{2b}{A}\int_{-1}^{+1}\int_{-1}^{+1} f(\alpha,\beta)a(\beta)\mathrm{d}\alpha\,\mathrm{d}\beta = \\ &= \frac{2b}{A}\int_{-1}^{+1}\sqrt{1-\beta^2}a(\beta)\left[\int_{-1}^{+1}\sqrt{\frac{1-\alpha}{1+\alpha}}F(\alpha,\beta)\mathrm{d}\alpha\right]\mathrm{d}\beta = \\ &= \frac{4\pi^2 b}{A(2m+1)(n+1)}\sum_{i=1}^{m}\sum_{j=1}^{n} H(\alpha_i,\beta_j)\,, \end{aligned} \tag{5.2.26}$$

where $H(\alpha_i, \beta_j)$ are (5.2.19). Formula (5.2.26) is valid for both Multhopp's method and the quadrature formulas method.

In a similar way we obtain:

$$c_x = \frac{2}{Aa_0}\iint_D \eta f(\xi,\eta)\mathrm{d}\xi\mathrm{d}\eta = \frac{2b^2}{Aa_0}\int_{-1}^{+1}\int_{-1}^{+1}\beta a(\beta)f(\alpha,\beta)\mathrm{d}\alpha\,\mathrm{d}\beta =$$
$$= \frac{4\pi^2 b^2}{a_0 A(2m+1)(n+1)}\sum_{i=1}^{m}\sum_{j=1}^{n}\beta_j H(\alpha_i,\beta_j)\,, \tag{5.2.27}$$

$$c_y = -\frac{2}{Aa_0}\iint_D \xi f(\xi,\eta)\mathrm{d}\xi\mathrm{d}\eta =$$
$$= -\frac{2b}{Aa_0}\int_{-1}^{+1}\int_{-1}^{+1}[a(\beta)\alpha + c(\beta)]a(\beta)f(\alpha,\beta)\mathrm{d}\alpha\,\mathrm{d}\beta = \tag{5.2.28}$$
$$= -\frac{4\pi^2 b}{a_0 A(2m+1)(n+1)}\sum_{i=1}^{m}\sum_{j=1}^{n}[a(\beta_j)\alpha_i + c(\beta_j)]H(\alpha_i,\beta_j)\,,$$

a_0 representing the length of the medium chord and

$$c_D = -\frac{2}{A}\iint_D f(\xi,\eta)h(\xi,\eta)\mathrm{d}\xi\,\mathrm{d}\eta =$$
$$= -\frac{4\pi^2 b}{A(2n+1)(n+1)}\sum_{i=1}^{n}\sum_{j=1}^{n}h(\alpha_i,\beta_j)H(\alpha_i,\beta_j)\,. \tag{5.2.29}$$

For the flat plate of incidence ε, we have $h(x,y) = -\varepsilon x$ then $g(u,v) = 4\pi b$. For the rectangular flat plate we have $x_- = -1$, $x_+ = 1$, $a = 1$, $c = 0$,

$$N(u,v,\alpha,\beta) = 1 + \frac{u-\alpha}{\sqrt{(u-\alpha)^2 + k^2b^2(v-\beta)^2}}\,. \tag{5.2.30}$$

5.2.5 The Third Method

Some numerical experiments show that it is not always indicated to impose the behaviour along the direction α under the form $\sqrt{\frac{1-\alpha}{1+\alpha}}$. We shall use in (5.2.7) the following quadrature formula

$$\int_{-1}^{+1} f(\alpha,\beta)N(u,v,\alpha,\beta)\mathrm{d}\alpha = \sum_{i=1}^{m} f(\alpha_i,\beta)K_i(u,v,\beta)\,, \tag{5.2.31}$$

where $\alpha_i = \dfrac{2i-m}{m}$, $i = 0, 1, \ldots, m$ are equidistant nodes on the interval $[-1, +1]$, and

$$K_i(u,v,\beta) = \int_{\alpha_i - 1}^{\alpha_i} N(u,v,\alpha,\beta)\mathrm{d}\alpha = \frac{2}{m} -$$

$$-\frac{1}{a(\beta)}\sqrt{[a(v)u + c(v) - a(\beta)\alpha_i - c(\beta)]^2 + k^2b^2(v-\beta)^2} +$$

$$+\frac{1}{a(\beta)}\sqrt{[a(v)u + c(v) - a(\beta)\alpha_{i-1} - c(\beta)]^2 + k^2b^2(v-\beta)^2}\,. \tag{5.2.32}$$

The behaviour in the span direction remains the same (like in the previous methods), hence:

$$f(\alpha,\beta) = \sqrt{1-\beta^2}F(\alpha,\beta)\,, \tag{5.2.33}$$

where $F(\alpha,\beta)$ is finite for $\beta = \pm 1$. We shall use the same formula (F.3.5) with respect to β. So, the equation (5.2.7) furnishes the following discretized form of the equation (5.2.7)

$$\sum_{i=1}^{m}\sum_{j=1}^{n} H_{\ell kij}F(\alpha_i,\beta_i) = bh'_x(\alpha_\ell\beta_k)\,, \tag{5.2.34}$$

where

$$H_{\ell kij} = \frac{1}{4(n+1)}\left[1-(-1)^{j+k}\right]\frac{1-\beta_j^2}{(\beta_j-\beta_k)^2}a(\beta_j)H_i(\alpha_\ell,\beta_k,\beta_j)\,, \quad j \neq k \tag{5.2.35}$$

and

$$H_{\ell kik} = -\frac{n+1}{8}a(\beta_k)H_i(\alpha_\ell,\beta_k,\beta_k)\,. \tag{5.2.36}$$

The algebraic system (5.2.34) has $m \times n$ linear equations with $m \times n$ unknowns $F(\alpha_i,\beta_j)$, where $\beta_j = \cos\dfrac{j\pi}{n+1}$.

The lift and moment coefficients are calculated by means of formulas

$$c_L = \frac{2b}{A}\int_{-1}^{+1}\int_{-1}^{+1}\sqrt{1-\beta^2}a(\beta)F(\alpha,\beta)\mathrm{d}\alpha\,\mathrm{d}\beta =$$

$$= \frac{2b\pi}{(n+1)A}\sum_{j=1}^{n}(1-\beta_j^2)a(\beta_j)\int_{-1}^{+1}F(\alpha,\beta_j)\mathrm{d}\alpha = \qquad (5.2.37)$$

$$= \frac{4b\pi}{(n+1)mA}\sum_{j=1}^{n}\sum_{i=1}^{m}(1-\beta_j^2)a(\beta_j)F(\alpha_i,\beta_j)$$

$$c_x = \frac{2b^2}{Aa_0}\int_{-1}^{+1}\int_{-1}^{+1}\beta a(\beta)\sqrt{1-\beta^2}F(\alpha,\beta)\mathrm{d}\alpha\,\mathrm{d}\beta =$$

$$= \frac{4b^2\pi}{(n+1)mAa_0}\sum_{j=1}^{n}\sum_{i=1}^{m}\beta_j a(\beta_j)(1-\beta_j^2)F(\alpha_i,\beta_j) \qquad (5.2.38)$$

$$c_y = -\frac{2b}{Aa_0}\int_{-1}^{+1}\int_{-1}^{+1}a(\beta)[a(\beta)a+c(\beta)]\sqrt{1-\beta^2}F(\alpha,\beta)\mathrm{d}\alpha\,\mathrm{d}\beta =$$

$$= \frac{-4b\pi}{(n+1)mAa_0}\sum_{j=1}^{n}\sum_{i=1}^{m}a(\beta_j)(1-$$

$$-\beta_j^2)\left[\frac{2i-m-1}{m}a(\beta_j)+c(\beta_j)\right]F(\alpha_i,\beta_j)\,. \qquad (5.2.39)$$

In the paper [5.10], from which we presented this method, there are presented computer programs for the elliptical flat plate

$$h(x,y) = -\varepsilon x\,, \quad x^2+\frac{y^2}{b^2}\le 1, b=2$$

and for the wing whose projection on the Oxy - plane is an rectangle

$$-b\le y\le b\,,\ 0\le x\le 1\,,\ x(y)=0\,,\ x_+(y)=1\,,$$

and whose normal section is an arc of parabola having the equation

$$h(x,y)=\varepsilon(1-x^2)\,, \quad \varepsilon << 1\,, \qquad (5.2.40)$$

The fluid in considered incompressible. For the rectangular wing one obtains an analytic solutions in the framework of the theory dealing with the wings of low aspect ratio. In this way we have a test solution for the third method. One deduces that this method furnishes very good results.

5.3 Ground Effects in the Lifting Surface Theory

5.3.1 The General Solution

The ground effects in the lifting surface theory were taken into consideration in [5.31] and [5.37] where one utilizes asymptotic methods. An approach of this subject in the framework of curvilinear lifting line theory may be found in [1.32], [1.33]. The present subsection is elaborated following [5.8], where one gives the general theory. One utilizes the fundamental solutions method.

The geometry of the problem is presented in figure 5.3.1. The origin of the reference frame is located in the middle of the span, the Ox-axis has the direction of the unperturbed stream and the Oy-axis has the span direction. The ground is considered to be the plane Π having the equation $z = -d/2$. The unperturbed flow is characterized by the velocity $U\boldsymbol{i}$, the pressure p_∞ and the density ρ_0 and it is considered to be subsonic like in 5.1. The field of velocity $\boldsymbol{v}_1$, the pressure p_1 and

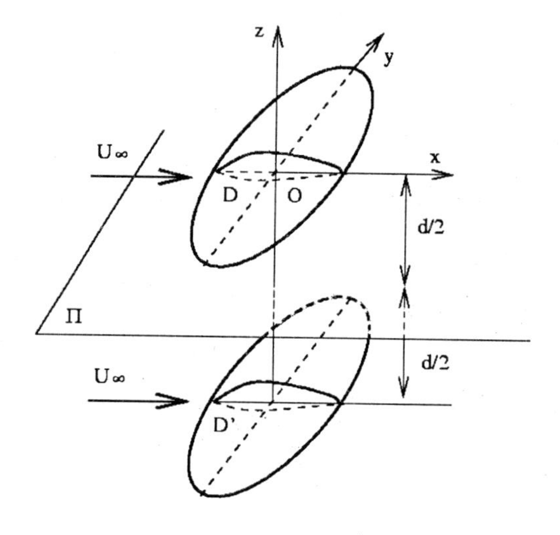

Fig. 5.3.1.

the density ρ_1 for the perturbed flow have the form (2.1.3). One utilizes dimensionless variables. The equation of the upper and lower surfaces are given by (5.1.2). It results the following boundary conditions

$$w(x,y,\pm 0) = h'_x(x,y,\pm h'_{1x}(x,y) \quad (x,y)\in D \tag{5.3.1}$$

$$w(x,y,-d/2) = 0, (\forall) x,y\,. \tag{5.3.2}$$

For satisfying the last condition we shall utilize the images method. This means to replace the wing with a forces distribution $\boldsymbol{f}^+ = (f_1,0,f)$ defined on D and with a symmetric distribution $\boldsymbol{f}^- = (f_1,0,-f)$ defined on the domain D', which is the symmetric of D with respect to the plane Π. In order to write the general form of the perturbation fields p and w we have at first to write the form determined by the concentrated forces $\boldsymbol{f}^+$ in $P_+(\boldsymbol{\xi}^+)$ and $\boldsymbol{f}^-$ in $P_-(\boldsymbol{\xi}^-)$. We have therefore to determine the solution of the equations

$$\begin{aligned} &M^2\partial p/\partial x + \mathrm{div}\boldsymbol{v} = 0 \\ &\partial \boldsymbol{v}/\partial x + \mathrm{grad}\, p = \boldsymbol{f}^+\delta(\boldsymbol{x}-\boldsymbol{\xi}^+) + \boldsymbol{f}^-\delta(\boldsymbol{x}-\boldsymbol{\xi}^-) \end{aligned} \tag{5.3.3}$$

corresponding to the system (2.3.4). The system (5.3.3) is linear and its solution is the sum of the corresponding solutions from (2.3.4). Taking into account (2.3.24) and (2.3.29), and the form of the fields $\boldsymbol{f}^+$ and $\boldsymbol{f}^-$, we deduce:

$$p(x,y,z) = -\frac{1}{4\pi}\left(f_1\frac{\partial}{\partial x} + f\frac{\partial}{\partial z}\right)\left(\frac{1}{R_+}\right) - \frac{1}{4\pi}\left(f_1\frac{\partial}{\partial x} - f\frac{\partial}{\partial z}\right)\left(\frac{1}{R_-}\right),$$

$$\begin{aligned} w(x,y,z) = &\frac{f_1}{4\pi}\frac{\partial}{\partial z}\left(\frac{1}{R_+}\right) - \frac{\beta^2}{4\pi}f\frac{\partial}{\partial x}\left(\frac{1}{R_+}\right) + \\ &+\frac{f}{4\pi}\frac{\partial}{\partial y}\left[\frac{y-\eta^+}{(y-\eta^+)^2+(z-z^+)^2}\left(1+\frac{x-\xi^+}{R_+}\right)\right] + \\ &+\frac{f_1}{4\pi}\frac{\partial}{\partial z}\left(\frac{1}{R_-}\right) + \frac{\beta^2}{4\pi}f\frac{\partial}{\partial x}\left(\frac{1}{R_-}\right) - \\ &-\frac{f}{4\pi}\frac{\partial}{\partial y}\left[\frac{y-\eta^-}{(y-\eta^-)^2+(z-\zeta^-)^2}\left(1+\frac{x-\xi^-}{R_-}\right)\right], \end{aligned} \tag{5.3.4}$$

where

$$R_\pm = \sqrt{(x-\xi^\pm)^2 + \beta^2[(y-\eta^\pm)^2+(z-\zeta^\pm)^2]}\,. \tag{5.3.5}$$

The points P_+ and P_- are symmetric if $\xi^+ = \xi^- = \xi$, $\eta^+ = \eta^- = \eta$, $\xi^+ = \zeta$, $\zeta^- = \zeta - d$. When P_+ are in $D(\zeta = 0)$, the symmetric points $P_-(\zeta^- = -d)$ are in D'.

Assimilating the wing with a continuous superposition of forces defined on on D, it results the following general solution:

$$p(x,y,z) = -\frac{1}{4\pi}\iint_D \left[f_1(\xi,\eta)\frac{\partial}{\partial x} + f(\xi,\eta)\frac{\partial}{\partial z}\right]\left(\frac{1}{R_1}\right) d\xi\, d\eta - \\ -\frac{1}{4\pi}\iint_D \left[f_1(\xi,\eta)\frac{\partial}{\partial x} - f(\xi,\eta)\frac{\partial}{\partial z}\right]\left(\frac{1}{R_{1d}}\right) d\xi\, d\eta \tag{5.3.6}$$

$$w(x,y,z) = \frac{1}{4\pi}\iint_D \left[f_1(\xi,\eta)\frac{\partial}{\partial z} - \beta^2 f(\xi,\eta)\frac{\partial}{\partial x}\right]\left(\frac{1}{R_1}\right) d\xi\, d\eta + \\ +\frac{1}{4\pi}\iint_D \left[f_1(\xi,\eta)\frac{\partial}{\partial z} + \beta^2 f(\xi,\eta)\frac{\partial}{\partial x}\right]\left(\frac{1}{R_{1d}}\right) d\xi\, d\eta + \\ +\frac{1}{4\pi}\iint_D f(\xi,\eta)\frac{\partial}{\partial y}\left[\frac{y_0}{y_0^2 + z^2}\left(1 + \frac{x_0}{R_1}\right)\right] d\xi\, d\eta - \\ -\frac{1}{4\pi}\iint_D f(\xi,\eta)\frac{\partial}{\partial y}\left[\frac{y_0}{y_0^2 + (z+d)^2}\left(1 + \frac{x_0}{R_{1d}}\right)\right] d\xi\, d\eta\,, \tag{5.3.7}$$

where $x_0 = x - \xi$, $y_0 = y - \eta$ and

$$R_1 = \sqrt{x_0^2 + \beta^2(y_0^2 + z^2)}\,,\quad R_{1d} = \sqrt{x_0^2 + \beta^2[y_0^2 + (z+d)^2]}\,. \tag{5.3.8}$$

This is the general representation of the perturbation, the functions $f_1(\xi,\eta)$ and $f(\xi,\eta)$ being determined from the boundary conditions (5.3.1). The condition (5.3.2) is obviously satisfied.

5.3.2 The Integral Equation

Acting like in 5.1, we shall pass to limit considering $z \to \pm 0$, $(x,y) \in$ $\in D$. To the limit, only the integrals related to R are singular. Using the formulas (5.1.16), (5.1.18) and (5.1.24), from (5.3.6) and (5.3.7) we get

$$p(x,y,\pm 0) = \frac{1}{4\pi} \overset{\odot}{\iint_D} f_1(\xi,\eta)\frac{x_0}{R^3} d\xi\, d\eta \pm \frac{1}{2} f(x,y) +$$

$$+ \frac{1}{4\pi}\iint_D f_1(\xi,\eta)\frac{x_0}{R^3} d\xi\, d\eta - \frac{\beta^2}{4\pi}\iint_D f(\xi,\eta)\frac{z}{R^3} d\xi\, d\eta,$$

$$p(x,y,+0) - p(x,y,-0) = f(x,y), \tag{5.3.9}$$

and

$$w(x,y,\pm 0) = \mp \frac{1}{2} f_1(x,y) - \frac{1}{4\pi} \overset{*}{\iint_D} \frac{f(\xi,\eta)}{y_0^2}\left(1 + \frac{x_0}{R}\right) d\xi\, d\eta -$$

$$-\frac{d\beta^2}{4\pi}\iint_D \frac{f_1(x,y)}{R_d^3} d\xi\, d\eta - \frac{1}{4\pi}\iint_D f(\xi,\eta) N(x_0,y_0) d\xi\, d\eta\,, \tag{5.3.10}$$

where

$$N(x_0,y_0) = \frac{d^2\beta^2 x_0}{(d^2+y_0^2)R_d^3} + \frac{d^2 - y_0^2}{(d^2+y_0^2)^2}\left(1 + \frac{x_0}{R_d}\right), \tag{5.3.11}$$

with the notations

$$R = \sqrt{x_0^2 + \beta^2 y_0^2}\,,\quad R_d = \sqrt{x_0^2 + \beta^2(y_0^2 + d^2)}\,. \tag{5.3.12}$$

Imposing the boundary conditions (5.3.1) we deduce

$$f_1(x,y) = -2h'_{1x}(x,y) \tag{5.3.13}$$

and then

$$\frac{1}{4\pi} \overset{*}{\iint_D} \frac{f(\xi,\eta)}{y_0^2}\left(1 + \frac{x_0}{R}\right) d\xi\, d\eta + \frac{1}{4\pi}\iint_D f(\xi,\eta) N(x_0,y_0) d\xi\, dy =$$

$$= -h'_x(x,y) + \frac{\beta^2 d}{2\pi}\iint_D \frac{h'_{1\xi}(\xi,\eta)}{R^3} d\xi\, d\eta\,. \tag{5.3.14}$$

The equation (5.3.14) may be called *the generalized lifting surface equation.* It was given in [5.8]. The sign "$*$" is for the Finite Part. Using the identity (5.1.32) we may write the equation as follows:

$$\frac{1}{4\pi}\frac{\partial}{\partial y}\iint_D \frac{f(\xi,\eta)}{y_0}\left(1+\frac{R}{x_0}\right)\mathrm{d}\xi\,\mathrm{d}\eta - \frac{1}{4\pi}\iint_D f(\xi,\eta)N(x_0,y_0)\mathrm{d}\xi\,\mathrm{d}\eta =$$

$$= h'_x(x,y) - \frac{d\beta^2}{2\pi}\iint_D \frac{h'_{1\xi}(\xi,1)}{R^3}\mathrm{d}\xi\,\mathrm{d}\eta\,. \tag{5.3.15}$$

5.3.3 The Two-Dimensional Problem

Like in 5.1.7, in order to obtain the representation of the solution and the integral equation for the wing of infinite span we assume that the normal sections in the wing determine profiles having the same shape. It means that the equations (5.1.2) have the form

$$z = h(x) \pm h_1(x) \quad (\forall)y$$

and the domain D is a rectangle $(-1 < x < 1,\ -b < y < b)$ with $b \to \infty$. We notice that in the representation of the general solution (5.3.6) and (5.3.7) f_1 and f have the form $f_1(\xi)$ and $f(\xi)$.

Relying on formulas like (5.1.46), from (5.3.6) we deduce

$$p(x,z) = \frac{1}{2\pi\beta}\int_{-1}^{+1}\frac{x_0 f_1(\xi)+\beta^2 z f(\xi)}{x_0^2+\beta^2 z^2}\mathrm{d}\xi +$$

$$+\frac{1}{2\pi\beta}\int_{-1}^{+1}\frac{x_0 f_1(\xi)+\beta^2(z+d)f(\xi)}{x_0^2+\beta^2(z^2+d)^2}\mathrm{d}\,\xi\,,$$

which are just $(3.2.3)_1$. Using formulas like (5.1.47), from (5.3.7) it results:

$$w(x,z) = \frac{\beta}{2\pi}\int_{-1}^{+1}\frac{x_0 f(\xi) - z f_1(\xi)}{x_0^2+\beta^2 z^2}\mathrm{d}\xi - \frac{\beta}{2\pi}\int_{-1}^{+1}\frac{x_0 f(\xi)+(z+d)f_1(\xi)}{x_0^2+\beta^2(z+d)^2}\mathrm{d}\xi,$$

i.e. just $(3.2.3)_2$.

For obtaining the integral equation we shall use (5.1.45). We also have

$$\int_{-\infty}^{+\infty}\frac{\mathrm{d}\eta}{R_d^3} = \frac{2}{\beta(x_0^2+d^2\beta^2)}\,,\ \int_{-\infty}^{+\infty}\frac{d^2-y_0^2}{(y_0^2+d^2)^2}\mathrm{d}\eta = 0\,. \tag{5.3.16}$$

The integral

$$I \equiv \int_{-\infty}^{+\infty}\frac{d^2-y_0^2}{(y_0^2+d^2)^2}\,\frac{x_0}{R_1}\mathrm{d}\eta$$

reduces to integrals having the shape

$$\int_0^\infty \frac{du}{(u^2+d^2)\sqrt{u^2+d^2}}\,, \qquad \int_\partial^\infty \frac{du}{(u^2+d^2)^2\sqrt{u^2+d^2}}$$

which may be calculated using the change of variable $u \to v$ where

$$v = \frac{u}{\sqrt{u^2+d^2}}\,. \tag{5.3.17}$$

One obtains

$$I = \frac{2\beta}{x_0}\left(1 - \frac{d}{x_0}\arctan\frac{x_0}{d\beta}\right). \tag{5.3.18}$$

Using the same change of variable (5.3.17) one obtains also

$$\int_{-\infty}^{+\infty} \frac{x_0 \mathrm{d}\eta}{(y_0^2+d^2)R_d^3} = -\frac{2\beta}{x_0}\frac{1}{x_0^2+d^2\beta^2} + \frac{2}{dx_0^2}\arctan\frac{x_0}{d\beta}\,, \tag{5.3.19}$$

such that we finally obtain

$$\int_{-\infty}^{+\infty} N(x_0,y_0)\mathrm{d}\eta = \frac{2\beta x_0}{x_0^2+d^2\beta^2}\,. \tag{5.3.20}$$

Taking into account the previous results, the integral equation (5.4.12) becomes

$$\begin{aligned} &\frac{\beta}{2\pi}{\int_{-1}^{+1}}' \frac{f(\xi)}{x_0}\mathrm{d}\xi - \frac{\beta}{2\pi}\int_{-1}^{+1} \frac{x_0 f(\xi)}{x_0^2+d^2\beta^2}\mathrm{d}\xi = \\ &= -h'(x) + \frac{d\beta}{\pi}\int_{-1}^{+1} \frac{h_1'(\xi)}{x_0^2+d^2\beta^2}\mathrm{d}\xi\,, \end{aligned} \tag{5.3.21}$$

which is just (3.2.7).

In [5.11] one extends the method from 5.2.3 also to the equation (5.3.14).

5.4 The Wing of Low Aspect Ratio

5.4.1 The Integral Equation

In the sequel we shall pay attention to the ratio

$$\lambda = (2b)^2/A\,, \tag{5.4.1}$$

called *the aspect ratio.* We denoted by $2b$ the span and by A the area of the domain D. For wings characterized by a small λ (wings of low aspect ratio) one develops herein *a theory* which leads to the integration of the lifting surface equation (5.1.33). For wings characterized by a big λ we shall develop in the following chapter *the lifting line theory.* These are the two asymptotic theories of the lifting surface equation. If the square of the span is small with respect to the area A, we deduce that on the greatest part of the wing we have $y_0^2 << x_0^2$ (see fig. 5.4.1) and we may write

$$\frac{R}{x_0} = \frac{|x_0|}{x_0}\left[1 + O\left(\frac{y_0^2}{x_0^2}\right)\right], \tag{5.4.2}$$

where R is (5.1.17). On the contrary, when the square of the span is big with respect to A (see figure 6.1.1), on the greatest part of the wing we have $x_0^2 << y_0^2$, so that we can make the hypothesis (see 6.1).

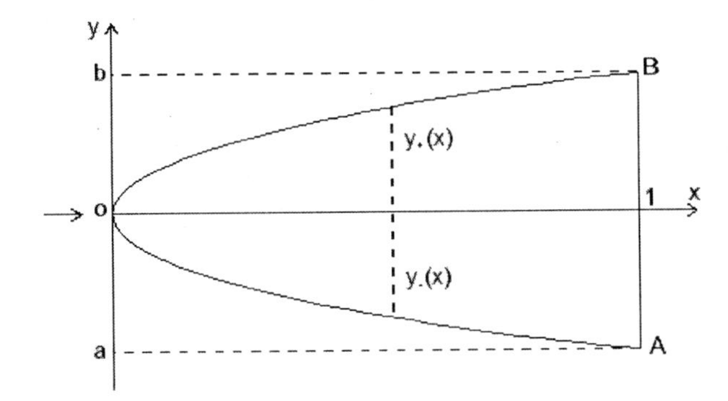

Fig. 5.4.1.

$$R = k|y_0| \tag{5.4.3}$$

Returning to the wings of low aspect ratio, we notice that under the hypothesis (5.4.2) the equation (5.1.33) may be approximated by

$$\frac{1}{4\pi}\frac{\partial}{\partial y}\iint_D' \frac{f(\xi,\eta)}{y_0}\left(1 + \frac{|x_0|}{x_0}\right) \mathrm{d}\xi\, \mathrm{d}\eta = h_x(x,y)\,. \tag{5.4.4}$$

Noticing that the integrand of this equation is different from zero only on $\xi < x$, we shall consider only wings whose trailing edge is a straight line perpendicular on Ox like in 5.4.1. Introducing the unknown function

$$F(x,\eta) = \int_{x_-(\eta)}^{x} f(\xi,\eta)\mathrm{d}\xi\,, \tag{5.4.5}$$

utilized for the first time by Jones [5.17], [5.18], the equation (5.4.4) becomes

$$\frac{1}{2\pi}\frac{\partial}{\partial y}\,{}'\!\!\int_{y_-(x)}^{y_+(x)}\frac{F(x,\eta)}{y_0}\mathrm{d}\eta = h_x(x,y)\,, \tag{5.4.6}$$

where $y = y_+(x)$ represents the equation of the leading edge OB, and $y = y_-(x)$, the equation of the leading edge OA.

From (5.4.5) and from the figure one notices that

$$F(x, y_-(x)) = F(x, y_+(x)) = 0\,. \tag{5.4.7}$$

In (5.4.6), utilizing the definition of the principal value, we derive taking into account (5.4.7) and then we integrate by parts. It follows

$$\frac{\partial}{\partial y}\,{}'\!\!\int_{y_-(x)}^{y_+(x)}\frac{F(x,\eta)}{y_0}\mathrm{d}\eta = {}'\!\!\int_{y_-}^{y_+}F\frac{\partial}{\partial y}\left(\frac{1}{y_0}\right)\mathrm{d}\eta = {}'\!\!\int_{y_-}^{y_+}\frac{\partial F}{\partial \eta}\frac{\mathrm{d}\eta}{y_0}\,, \tag{5.4.8}$$

where, from now on,

$$y_- = y_-(x)\,, \quad y_+ = y_+(x)\,.$$

In this way equation (5.4.6) becomes

$$\frac{1}{2\pi}\int_{y_-}^{y_+}\frac{\partial F}{\partial \eta}\,\frac{\mathrm{d}\eta}{y_0} = h_x(x,y)\,. \tag{5.4.9}$$

This equation can be also obtained starting from Homentcovsch's equation (5.1.41) which, in case that one imposes the condition (5.1.44) on the leading edge, becomes equation (5.1.43). Then we can see that from (5.4.5) it follows

$$\frac{\partial F}{\partial \eta} = -\int_{x_-(\eta)}^{x}\frac{\partial f}{\partial \eta}\mathrm{d}\xi\,. \tag{5.4.10}$$

The equation (5.4.8) is a thin profiles-type equation (C.1.1) and it has one of the solutions (C.1.9), (C.1.10), (C.1.11) or (C.1.14). The solution of the equation (5.4.8) must satisfy the condition

$$\int_{y_-}^{y_+}\frac{\partial F}{\partial \eta}\mathrm{d}\eta = 0 \tag{5.4.11}$$

arising from (5.4.7). The only solution satisfying this solution is (C.1.11) with $C = 0$. Hence,

$$\frac{\partial F}{\partial \eta}(x,\eta) = \frac{2}{\pi}\,\frac{1}{\sqrt{(\eta - y_-)(y_+ - \eta)}}\,{}'\!\!\int_{y_-}^{y_+}\frac{\sqrt{(t-y_-)(y_+ - t)}}{t-\eta}h_x(x,t)\mathrm{d}t\,. \tag{5.4.12}$$

Integrating this solution on the interval (y_-, y) and taking into account (5.4.7) and (B.6.11), we get

$$F(x,y) = \frac{2}{\pi}\int_{y_-}^{y} \frac{\mathrm{d}\eta}{\sqrt{(\eta - y_-)(y_+ - \eta)}} {\int_{y_-}^{y_+}}' \frac{\sqrt{(t-y_-)(y_+-t)}}{t-\eta} h_x(x,t)\mathrm{d}t\,. \tag{5.4.13}$$

As from (5.4.5) it results $f(x,y) = F_x(x,y)$, we deduce

$$f(x,y) =$$

$$= \frac{2}{\pi}\frac{\partial}{\partial x}\int_{y_-}^{y} \frac{\mathrm{d}\eta}{\sqrt{(\eta - y_-)(y_+ - \eta)}} \int_{y_-}^{y_+} \frac{\sqrt{(t-y_-)(y_+-t)}}{t-\eta} h_x(x,t)\mathrm{d}t\,. \tag{5.4.14}$$

Employing (B.6.11) ones easily check that this solution satisfies the vanishing condition on the leading edge

$$f(x,y_-) = f(x,y_+) = 0\,. \tag{5.4.15}$$

5.4.2 The Case $h = h(x)$

In case that h does not depend on the variable y, taking into account (B.5.6) and performing the change of variable

$$\eta = y_- \cos^2\theta + y_+ \sin^2\theta\,, \quad 0 < \theta < \pi/2\,, \tag{5.4.16}$$

we get

$$\int_{y_-}^{y} \frac{\mathrm{d}\eta}{\sqrt{(\eta - y_-)(y_+ - \eta)}} {\int_{y_-}^{y_+}}' \frac{\sqrt{(t-y_-)(y_+-t)}}{t-\eta}\mathrm{d}t =$$

$$= \pi\frac{y_+ - y_-}{2}\int_{y_-}^{y} \frac{\mathrm{d}\eta}{\sqrt{(\eta - y_-)(y_+ - \eta)}} -$$

$$-\pi\int_{y_-}^{y}\sqrt{\frac{\eta - y_-}{y_+ - \eta}}\mathrm{d}\eta = \pi\sqrt{(y-y_-)(y_+-y)}\,.$$

Hence,

$$f(x,y) = 2\frac{\partial}{\partial x}\left[h_x'\sqrt{(y-y_-)(y_+-y)}\right]\,. \tag{5.4.17}$$

Utilizing (5.1.54)–(5.1.56) and Green's formula we deduce:

$$c_L = -\frac{4}{A}\iint_D \frac{\partial}{\partial x}\left[h'(x)\sqrt{(y-y_-)(y_+-y)}\right] \mathrm{d}x\mathrm{d}y =$$

$$= -\frac{4}{A}h'(1)\int_a^b \sqrt{(y-a)(b-y)}\mathrm{d}y = -\frac{\pi}{2A}h'(1)(b-a)^2,$$

$$c_x = -\frac{4}{A}\iint_D y\frac{\partial}{\partial x}\left[h'(x)\sqrt{(y-y_-)(y_+-y)}\right] \mathrm{d}x\mathrm{d}y =$$

$$= -\frac{4}{A}h'(1)\int_a^b y\sqrt{(y-a)(b-y)}\mathrm{d}y = -\frac{a+b}{4A}\pi h'(1)(b-a)^2,$$

$$c_y = \frac{4}{A}\iint_D \frac{\partial}{\partial x}\left[xh'\sqrt{(y-y_-)(y_+-y)}\right] \mathrm{d}x\, \mathrm{d}y -$$

$$-\frac{4}{A}\iint_D h'(x)\sqrt{(y-y_-)(y_+-y)}\mathrm{d}x\mathrm{d}y =$$

$$= -c_L - \frac{\pi}{2A}\int_0^1 \left[y_+(x) - y_-(x)\right]^2 h'(x)\mathrm{d}x\,, \tag{5.4.18}$$

the reference chord a_0 being 1.

For example, for the rectangular flat plate of incidence $-\varepsilon$ with the span $2b$ one obtains

$$c_L = \pi b\varepsilon\,, \quad c_x = 0\,, \quad c_y = -\frac{\pi b\varepsilon}{2}\,, \tag{5.4.19}$$

and for the triangular flat plate of incidence $-\varepsilon$, we have,

$$c_L = \pi\varepsilon(b-a)\,, \quad c_x = (b^2-a^2)\frac{\pi\varepsilon}{2}\,, \quad c_y = -\frac{2}{3}\pi\varepsilon(b-a)\,, \tag{5.4.20}$$

because $y_+ = bx\,, \quad y_- = ax\,.$

5.4.3 The General Case

We shall write (5.4.14) as follows

$$f(x,y) = \frac{2}{\pi}\frac{\partial}{\partial x}\int_{y_-}^{y_+} \sqrt{(t-y_-)(y_+-t)}h_x(x,t)I(y,t)\mathrm{d}t\,, \tag{5.4.21}$$

where

$$I(y,t) = \int_{y_-}^{y} \frac{1}{\sqrt{(\eta - y_-)(y_+ - \eta)}} \frac{d\eta}{t-\eta}, \tag{5.4.22}$$

with $y_- < y < y_+$, $y_- < t < y_+$. This integral may be calculated explicitly. For $y < t$ the integral is not singular. For $t < y$ the integral is singular but noticing that we have $I(y_+, t) = 0$ (it results from (B.5.8)) we get

$$I(y,t) = -\int_{y}^{y_+} \frac{1}{\sqrt{(\eta - y_-)(y_+ - \eta)}} \frac{d\eta}{t-\eta} \tag{5.4.23}$$

which also is not a singular integral.

Since

$$\int_a^b \frac{1}{\sqrt{(\eta-a)(b-\eta)}} \frac{d\eta}{t-\eta} =$$

$$= \frac{1}{\sqrt{(t-a)(b-t)}} \ln \left| \frac{\sqrt{(t-a)(b-\eta)} + \sqrt{(b-t)(\eta-a)}}{\sqrt{(t-a)(b-\eta)} - \sqrt{(b-t)(\eta-a)}} \right|, \tag{5.4.24}$$

we deduce the fundamental formula

$$f(x,y) =$$

$$= \frac{2}{\pi} \frac{\partial}{\partial x} \int_{y_-}^{y_+} h_x(x,t) \ln \left| \frac{\sqrt{(t-y_-)(y_+-y)} + \sqrt{(y_+-t)(y-y_-)}}{\sqrt{(t-y_-)(y_+-y)} - \sqrt{(y_+-t)(y-y_-)}} \right| dt\,. \tag{5.4.25}$$

For determining the lift and moment coefficients we have to calculate the integral

$$G(x,t) = \int_{y_-}^{y_+} \ln \left| \frac{\sqrt{(t-y_-)(y_+-y)} + \sqrt{(y_+-t)(y-y_-)}}{\sqrt{(t-y_-)(y_+-y)} - \sqrt{(y_+-t)(y-y_-)}} \right| dy\,. \tag{5.4.26}$$

Performing the changes of variables

$$t = \frac{y_+ + y_-}{2} + \frac{y_+ - y_-}{2}\cos\sigma\,, \quad y = \frac{y_+ + y_-}{2} + \frac{y_+ - y_-}{2}\cos\theta\,, \tag{5.4.27}$$

we deduce

$$G(x,t) = -\frac{y_+ - y_-}{2} \int_0^{\pi} \ln \left| \frac{\sin\dfrac{\theta-\sigma}{2}}{\sin\dfrac{\theta+\sigma}{2}} \right| \sin\theta\, d\theta\,, \tag{5.4.28}$$

the integral having an integrable singularity for $\theta = \sigma$. The integrand from (5.4.28) is the kernel from the equation (6.2.15). Using for $\theta \neq \sigma$ the expansion (6.2.17), we get:

$$G(x,t) = \pi \frac{y_+ - y_-}{2} \sin\sigma = \pi\sqrt{(y_+ - t)(t - y_-)}\,. \tag{5.4.29}$$

It follows

$$G(1,t) = \int_a^b \ln\left|\frac{\sqrt{(t-a)(b-y)} + \sqrt{(b-t)(y-a)}}{\sqrt{(t-a)(b-y)} - \sqrt{(b-t)(y-a)}}\right| \mathrm{d}\,y$$
$$= \pi\sqrt{(b-t)(t-a)}\,. \tag{5.4.30}$$

Similarly one deduces

$$\int_a^b y \ln\left|\frac{\sqrt{(t-a)(b-y)} + \sqrt{(b-t)(y-a)}}{\sqrt{(t-a)(b-y)} - \sqrt{(b-t)(y-a)}}\right| \mathrm{d}\,y =$$
$$= \pi\left(t + \frac{b+a}{2}\right)\sqrt{(b-t)(t-a)}\,. \tag{5.4.31}$$

Utilizing Green's formula and the expression (5.4.25) for $f(x,y)$, taking into account (5.4.30), (5.4.31), we get:

$$c_L = -\frac{4}{\pi A}\int_a^b h'(1,t)G(1,t)\mathrm{d}\,t =$$
$$= -\frac{4}{A}\int_a^b \sqrt{(b-t)(t-a)}h_x(1,t)\mathrm{d}\,t,$$
$$c_x = -\frac{2}{A}\int_a^b t\sqrt{(b-t)(t-a)}h_x(1,t)\mathrm{d}\,t + \frac{a+b}{4}c_L, \tag{5.4.32}$$
$$c_y = -c_L - \frac{4}{\pi A}\int_0^1 \mathrm{d}\,x\int_{y_-}^{y_+} h_x(x,t)G(x,t)\mathrm{d}\,t =$$
$$= -c_L - \frac{4}{A}\int_0^1 \mathrm{d}\,x\int_{y_-}^{y_+} h_2(x,t)\sqrt{(y_+ - t)(t - y_-)}\mathrm{d}\,t\,.$$

When $h(x,t)$ does not depend on t we find again the formulas (5.4.18).

This problem, for wings symmetric with respect to the Ox-axis, was studied in [5.17]. It was also included in [1.1], [1.2]. The general problem for the arbitrary wing as it was presented herein, was solved in [5.9].

Chapter 6

The Lifting Line Theory

6.1 Prandtl's Theory

6.1.1 The Lifting Line Hypotheses. The Velocity Field

Prandtl's theory is the first mathematical model for the three - dimensional wing (the finite span airfoil). It was elaborated in 1918 [6.21] and it remained until the years '40 the only theory for this wing. The german scientist, gifted with an extraordinary engineering intuition, guessed very well the simplifications which may be performed. Prandtl's method consists in replacing the wing with a distribution of vortices on its plane - form (the domain D from 5.1). Since the experience indicates that downstream the wing the flow is not potential, Prandtl introduced a vortical distribution defined on S, the trace of the plane-form D in the uniform stream (figure 6.1.1), the velocity field in the fluid being determined by the two distributions. The vortices on D are called *tied vortices* and the vortices on S are called *free vortices.*

This idea continued to dominate the aerodynamics, the models concerning the subsonic, supersonic and transonic, steady or oscillatory flows, elaborated in the years '50, '60, being conceived on the basis of this method. In 1975 D. Homentcovschi, utilizing the theory of distributions proved that the hypothesis of the existence of free vortices is not necessary because it follows from equations. L. Dragoş [5.7] obtained the same result utilizing the method of the fundamental solutions. In this subsection, we deduce the lifting line theory from the lifting surface theory, as we proceeded in [5.7], utilizing the following three hypotheses (the hypotheses of Prandtl's theory):

1^0 One neglects the thickness of the wing, therefore in (5.1.2) one considers $h_1 = 0$. From (5.1.27) it results $f_1 = 0$, such that the representations (5.1.8), (5.1.11) and (5.1.12) give

$$p(x,y,z) = -\frac{1}{4\pi}\int\int_D f(\xi,\eta)\frac{\partial}{\partial z}\left(\frac{1}{R_1}\right)\mathrm{d}\xi\mathrm{d}\eta \qquad (6.1.1)$$

$$u(x,y,z) = -p(x,y,z) \tag{6.1.2}$$

$$v(x,y,z) = -\frac{1}{4\pi}\int\int_D f(\xi,\eta)\frac{\partial}{\partial y}\left[\frac{z}{y_0^2+z^2}\left(1+\frac{x_0}{R_1}\right)\right]\mathrm{d}\,\xi\,\mathrm{d}\,\eta \tag{6.1.3}$$

$$\begin{aligned} w(x,y,z) = & -\frac{\beta^2}{4\pi}\int\int_D f(\xi,\eta)\frac{\partial}{\partial x}\left(\frac{1}{R_1}\right)\mathrm{d}\,\xi\,\mathrm{d}\,\eta + \\ & +\frac{1}{4\pi}\int\int_D f(\xi,\eta)\frac{\partial}{\partial y}\left[\frac{y_0}{y_0^2+z^2}\left(1+\frac{x_0}{R_1}\right)\right]\mathrm{d}\,\xi\,\mathrm{d}\,\eta\,; \end{aligned} \tag{6.1.4}$$

2^0 One considers that the unknown is the circulation C_y along the contour c, resulting from the intersection of the wing with an arbitrary plane Π parallel to the xOz plane at the distance $y(-b < y < b)$ (fig. 6.1.1). This will be, obviously, a function of y

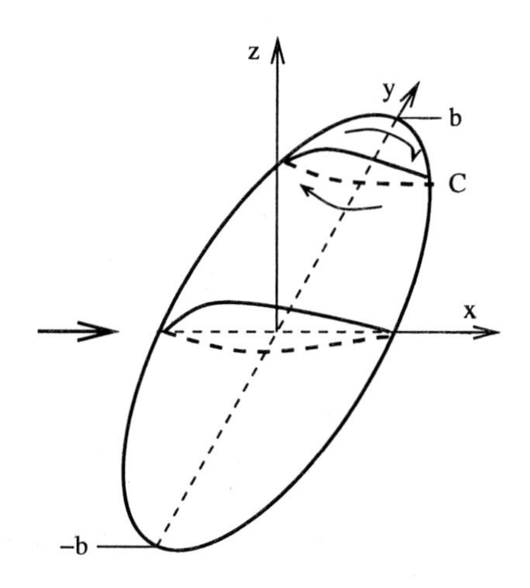

Fig. 6.1.1.

$$C(y) = \oint_C \boldsymbol{v}\cdot\mathrm{d}\,\boldsymbol{x} = \oint_C u\mathrm{d}\,x\,,$$

because on this contour $\mathrm{d}\,y = 0$ and $w\mathrm{d}z = O(\varepsilon^2)$ (it results from (2.1.17) and (2.1.21)). Taking into account (6.1.2) and (5.1.20), we obtain

$$C(y) = \int_{x_+}^{x_-} u(x,y,-0)\mathrm{d}\,x + \int_{x_-}^{x_+} u(x,y,+0)\mathrm{d}\,x = -\int_{x_-(y)}^{x_+(y)} f(x,y)\mathrm{d}\,x\,. \tag{6.1.5}$$

We have also

$$C(\eta) = -\int_{x_-(y)}^{x_+(y)} f(\xi,\eta)\mathrm{d}\,\xi\,. \tag{6.1.6}$$

From (5.1.1) or (5.1.30) it results

$$C(\pm b) = 0\,; \tag{6.1.7}$$

3^0 The domain D, i.e. the projection of the wing on the xOy plane, is replaced by the segment $[-b, +b]$ (fig. 6.7.1); for this reason we call this theory the *lifting line theory*.

For studying the behaviour of the integrals (6.1.1)–(6.1.4) when $x_-(\eta) \to 0 \leftarrow x_+(\eta)$, we notice that in the vicinity of this segment, for a given η , the function $f(\xi,\eta)$ keeps a constant sign, such that we may apply the mean formula. For a function $h(x,y,z,\xi,\eta)$, continuous in ξ, when $x_-(\eta) \to 0 \leftarrow x_+(\eta)$, we have therefore

$$\begin{aligned}
&\lim \iint_D f(\xi,\eta)k(x,y,z,\xi,\eta)\mathrm{d}\,\xi\;\mathrm{d}\,\eta = \\
&= \lim \int_{-b}^{+b}\left[\int_{x_-(\eta)}^{x_+(\eta)} f(\xi,\eta)k(x,y,z,\xi,\eta)\mathrm{d}\,\xi\right]\mathrm{d}\,\eta = \\
&= -\lim \int_{-b}^{+b} k(x,y,z,\xi^*,\eta)C(\eta)\mathrm{d}\,\eta = -\int_{-b}^{+b} k(x,y,z,0,\eta)C(\eta)\mathrm{d}\,\eta\,,
\end{aligned} \tag{6.1.8}$$

where $\xi^* \in (x_-(\eta), x_+(\eta))$. Applying this formula in (6.1.1)–(6.1.4) and taking into account (6.1.7), we obtain

$$p(x,y,z) = -\frac{\beta^2}{4\pi}\int_{-b}^{+b} C(\eta)\frac{z}{R_0^3}\mathrm{d}\,\eta = -u(x,y,z)\,,$$

$$\begin{aligned}
v(x,y,z) &= \frac{1}{4\pi}\int_{-b}^{+b} C(\eta)\frac{\partial}{\partial y}\left[\frac{z}{y_0^2+z^2}\left(1+\frac{x}{R_0}\right)\right]\mathrm{d}\,\eta = \\
&= -\frac{1}{4\pi}\int_{-b}^{+b} C(\eta)\frac{\partial}{\partial \eta}\left[\frac{z}{y_0^2+z^2}\left(1+\frac{x}{R_0}\right)\right]\mathrm{d}\,\eta = \\
&= \frac{1}{4\pi}\int_{-b}^{+b} C'(\eta)\frac{z}{y_0^2+z^2}\left(1+\frac{x}{R_0}\right)\mathrm{d}\,\eta\,,
\end{aligned} \tag{6.1.9}$$

$$\begin{aligned}
w(x,y,z) = &-\frac{\beta^2}{4\pi}\int_{-b}^{+b} C(\eta)\frac{x}{R_0^3}\mathrm{d}\,\eta - \\
&-\frac{1}{4\pi}\int_{-b}^{+b} C'(\eta)\frac{y_0}{y_0^2+z^2}\left(1+\frac{x}{R_0}\right)\mathrm{d}\,\eta\,,
\end{aligned}$$

where we denoted

$$R_0 = \sqrt{x^2 + \beta^2(y_0^2 + z^2)} \tag{6.1.10}$$

For $\beta = 1$ we obtain Prandtl's representation ([1.21], p. 708).

6.1.2 Prandtl's Equation

In the sequel we shall deduce the equation for $C(y)$. This will obviously result from the lifting surface equation by virtue of the above hypotheses. We shall employ the equation (5.1.33). Taking (6.1.6) and (D.3.6) into account it results

$$\begin{aligned} \frac{\partial}{\partial y} {\int\!\!\int_D}' \frac{f(\xi,\eta)}{y-\eta} \mathrm{d}\,\xi \mathrm{d}\,\eta &= \frac{\mathrm{d}}{\mathrm{d}y} {\int_{-b}^{+b}}' \frac{C(\eta)}{\eta - y} \mathrm{d} = \eta \\ &= {\int_{-b}^{+b}}^{*} \frac{C(\eta)}{(\eta-y)^2} \mathrm{d}\,\eta. \end{aligned} \tag{6.1.11}$$

Utilizing (D.3.7) and (6.1.7), we also deduce:

$$ {\int_{-b}^{+b}}^{*} \frac{C(\eta)}{(\eta-y)^2} \mathrm{d}\,\eta = {\int_{-b}^{+b}}' \frac{C'(\eta)}{\eta - y} \mathrm{d}\,\eta\,. \tag{6.1.12}$$

In the second term from (5.1.33) we take into account that when $x_-(\eta) \to 0 \leftarrow x_+(\eta)$, on the greatest part of the domain D, we have $x_0^2 << y_0^2$, before ϵ which intervenes in the definition of the principal value of the integral with respect to η becomes zero, i.e. before y_0 becomes zero. In the principal value we shall perform therefore the approximation $R = \beta |y_0|$. An exact evaluation of this approximation is made by D. Homentcovschi in [6.9] as follows

$$ {\int\!\!\int}' f(\xi,\eta) \frac{\sqrt{\varepsilon^2 x_0^2 + \beta^2 y_0^2}}{x_0 y_0} \mathrm{d}\,\xi \mathrm{d}\,\eta = {\int\!\!\int}' f(\xi,\eta) \frac{\beta |y_0|}{x_0 y_0} \mathrm{d}\,\xi \mathrm{d}\,\eta + O(\varepsilon^2 \ln \varepsilon)\,.$$

Taking into account that

$$\operatorname{sign} y_0 = 2H(y_0) - 1\,, \quad H'(y_0) = \delta(y - \eta)\,,$$

H representing Heaviside's function and δ Dirac's distribution we have:

$$\frac{\partial}{\partial \eta} {\int\!\!\int}' f(\xi,\eta)\frac{R}{x_0 y_0}\mathrm{d}\,\xi\mathrm{d}\,\eta \simeq \beta\frac{\partial}{\partial y}{\int\!\!\int}' \frac{f(\xi,\eta)}{x_0}\operatorname{sign} y_0 \mathrm{d}\,\xi\mathrm{d}\,\eta =$$

$$= 2\beta\int_{-b}^{+b}\left[{\int'}_{x_-(\eta)}^{x_+(\eta)}\frac{f(\xi,\eta)}{x_0}\mathrm{d}\,\xi\right]\delta(y-\eta)\mathrm{d}\,\eta = 2\beta{\int'}_{x_-(y)}^{x_+(y)}\frac{f(\xi,\eta)}{x_0}\mathrm{d}\,\xi\,. \tag{6.1.13}$$

Hence the equation (5.1.33) becomes:

$$\frac{1}{2\pi}{\int'}_{-b}^{+b}\frac{C'(\eta)}{\eta-y}\mathrm{d}\,\eta + \frac{\beta}{\pi}\int_{x_-(y)}^{x_+(y)}\frac{f(\xi,y)}{x_0}\mathrm{d}\,\xi = 2h'_x(x,y)\,. \tag{6.1.14}$$

Multiplying this equation by

$$[x-x_-(y)]^{1/2}[x_+(y)-x]^{-1/2} \tag{6.1.15}$$

and integrating it with respect to x on the interval $(x_-(y),x_+(y))$, we obtain from the first formula (B.5.4):

$$\beta C(y) = \frac{a(y)}{2}{\int'}_{-b}^{+b}\frac{C'(\eta)}{\eta-y}\mathrm{d}\,\eta + j(y)\,, \tag{6.1.16}$$

where we denoted

$$a(y) = \frac{x_+(y)-x_-(y)}{2}\,, \tag{6.1.17}$$

$$j(y) = -2\int_{x_-(y)}^{x_+(y)}\sqrt{\frac{x-x_-(y)}{x_+(y)-x}}h'_x(x,y)\mathrm{d}\,x\,, \tag{6.1.18}$$

$a(y)$ representing a half of the chord of the profile at the distance y. The equation (6.1.16) is the well known *Prandtl's equation* (*the lifting line equation*). This equation, together with the conditions (6.1.7), has to determine the unknown $C(y)$. It is an *integro-differential singular equation*, with a Cauchy-type singularity. Utilizing (6.1.12), the equation for $C(y)$ may be written

$$\beta C(y) = \frac{a(y)}{2}{\int^*}_{-b}^{+b}\frac{C(\eta)}{(\eta-y)^2}\mathrm{d}\,\eta + j(y)\,. \tag{6.1.19}$$

In this form, the equation is not integro-differential any longer, it is only integral, but with a stronger singularity. The mark "*" is for the Finite Part. As we have already shown in [6.5] [6.6], this form is more adequate for the numerical integrations (see also 6.4.2, 6.4.3).

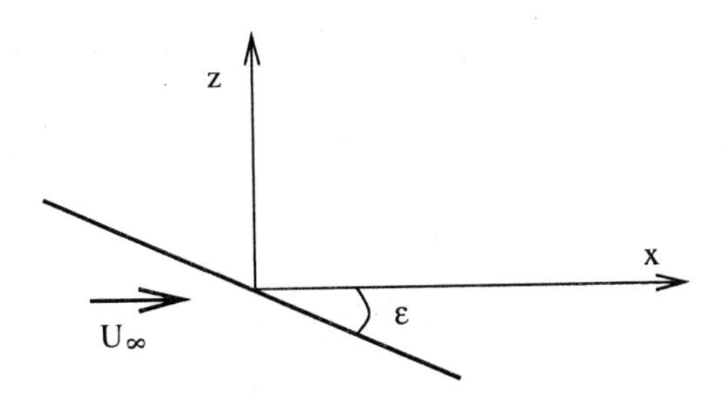

Fig. 6.1.2.

The significance of the function j (6.1.18) is given in [1.2]. In the case of the flat plates having the angle of attack ε (fig. 6.1.2) we have $h_x = -\varepsilon$ whence

$$j(y) = 2\pi\varepsilon a(y)\,. \tag{6.1.20}$$

6.1.3 The Aerodynamic Action

Taking into account (6.1.6), from (5.1.54) and (5.1.56) we deduce

$$c_L = \frac{2}{A}\int_{-b}^{+b} C(y)\mathrm{d}\,y,\quad c_x = \frac{2}{Aa_0}\int_{-b}^{+b} yC(y)\mathrm{d}\,y,\quad c_y = 0\,. \tag{6.1.21}$$

The expression of c_y is natural because for the wing reduced to the segment $[-b,+b]$ the fluid cannot create a moment which should rotate this axis. By virtue of the first hypothesis $(h_1 = 0)$, in the framework of the lifting surface theory, the quantities R_x and M_z which have the order of magnitude $O(\varepsilon^2)$ are reduced to those given in (5.1.56) and (5.1.59). Utilizing (6.1.6) and (6.1.7), we obtain:

$$\begin{aligned} R_x &= -\rho_\infty U_\infty^2 L_0^2 \int_{-b}^{+b} C(y)h_x'(0,y)\mathrm{d}\,y\,,\\ M_z &= \rho_\infty U_\infty^2 L_0^3 \int_{-b}^{+b} yC(y)h_x'(0,y)\mathrm{d}\,y \end{aligned} \tag{6.1.22}$$

But, from the boundary condition

$$w(x,y,0) = h_x'(x,y) \tag{6.1.23}$$

and from (6.1.9) it results

$$h_x'(0,y) = w(0,y,0) = -\frac{1}{4\pi}\lim_{\varepsilon\to 0}\left(\int_{-b}^{y-\varepsilon} + \int_{y+\varepsilon}^{b}\right)\frac{C'(\eta)}{y_0}\mathrm{d}\,\eta\,,$$

because, with the change of variable $\eta - y = u$, we have

$$\lim_{\varepsilon\to 0}\int_{y-\varepsilon}^{y+\varepsilon} C'(\eta)\frac{y_0}{y_0^2+z^2}\mathrm{d}\,\eta = -C'(y)\lim_{\varepsilon\to 0}\int_{-\varepsilon}^{+\varepsilon}\frac{\mathrm{d}\,u}{u^2+z^2} = 0\,.$$

Denoting now

$$w(y) = w(0,y,0) = \frac{1}{4\pi}{\int_{-b}^{+b}}' \frac{C'(\eta)}{\eta - y}\mathrm{d}\,\eta\,, \tag{6.1.24}$$

we deduce

$$\begin{aligned} c_D &= \frac{R_x}{(1/2)\rho_\infty U_\infty^2 A_1} = -\frac{2}{A}\int_{-b}^{+b} C(y)w(y)\mathrm{d}\,y\,,\\ c_z &= \frac{M_z}{(1/2)\rho_\infty U_\infty^2 A_1 a_1} = \frac{2}{Aa_0}\int_{-b}^{+b} yC(y)w(y)\mathrm{d}\,y\,, \end{aligned} \tag{6.1.25}$$

A representing the area of the domain D and a_0, the length of the dimensionless medium chord along the direction of the unperturbed stream. In fact, for w we may also utilize the expression

$$w(y) = \frac{\beta C(y) - j(y)}{2\pi a(y)} \tag{6.1.26}$$

which results from (6.1.18) and (6.1.26).

6.1.4 The Elliptical Flat Plate

Until the appearance of the papers of Magnaradze [6.16] and Vekua [6.28] in the years 1942, 1945 it was available only one exact solution of Prandtl's equation corresponding to the elliptical flat plate wing. Usually this solution is obtained as an answer to the following minimum problem: "To determine among the wings having the same lift, the wing corresponding to the minimum drag" (see, for example, [1.20]). Sometimes one utilizes Glauert's approximate method (see 6.2.4). In the sequel we shall present a simple method which does not need any special considerations on Prandtl's equation. Let us consider a flat plate with the angle of attack ε, whose projection on the x_1Oy_1 plane is an ellipse having the semi-axes L_0 and bL_0 (figure 6.1.3).

In dimensionless coordinates, the equation of the ellipse is x^2+ $+y^2/b^2 = 1$. For the edges and the chord it results

$$x_\pm = \pm\sqrt{1-\frac{y^2}{b^2}}\,,\quad a(y) = \sqrt{1-\frac{y^2}{b^2}}\,. \tag{6.1.27}$$

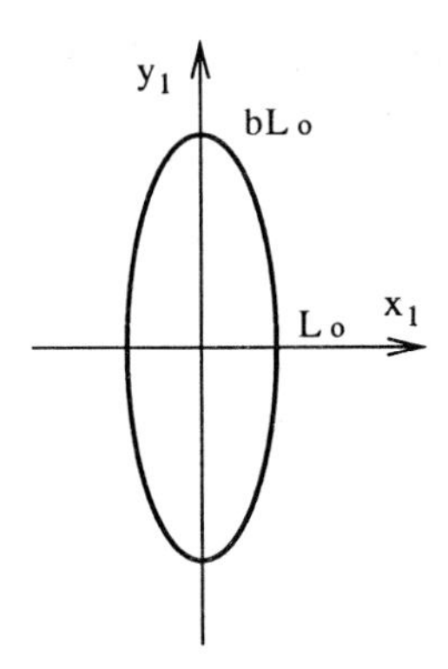

Fig. 6.1.3.

For a flat plate $j(y)$ has the form (6.1.20) hence it is proportional to $a(y)$. From Prandtl's equation it results that $C(y)$ is also proportional to $a(y)$. We shall look therefore for solutions having the form:

$$C(y) = k\sqrt{1 - \frac{y^2}{b^2}}\,, \tag{6.1.28}$$

k being a constant which will be determined by imposing (6.1.23) to verify (6.1.27). Using the change of variables $\eta = b\cos\theta$, $y = b\cos\sigma$, and Glauert's formulas (B.6.6), it results:

$$\int_{-b}^{+b}{}' \frac{\eta}{\sqrt{b^2-\eta^2}}\frac{d\eta}{\eta - y} = \pi\,, \tag{6.1.29}$$

such that

$$k = \frac{2\pi\varepsilon}{\beta + (\pi/2b)}\,. \tag{6.1.30}$$

Since the area of the ellipse is πb, from the formulas (6.1.23) and (6.1.27) we obtain

$$c_L = k\,, \quad c_D = \frac{k^2}{4b}\,, \quad c_x = c_z = 0\,. \tag{6.1.31}$$

Obviously, $c_D = O(\varepsilon^2)$. For $b \to \infty$ one obtains the infinite span flat plate. From (6.1.23) and (6.1.27) it results:

$$c_L = \frac{2\pi\varepsilon}{\beta} \tag{6.1.32}$$

6.2 The Theory of Integration of Prandtl's Equation. The Reduction to Fredholm-Type Integral Equations

6.2.1 The Equation of Trefftz and Schmidt

The general method of solving the integro-differential equations consists in reducing them to Fredholm-type integral equations. As it is well known, for the last ones a general theory is available (existence and uniqueness theorems, exact and approximate methods for solving the equation). The first investigation of Prandtl's equation was performed by Trefftz in 1921. He reduced the problem of solving Prandtl's equation to the problem of determination of a harmonic function in the superior half-plane with mixed conditions on the boundary. We shall prove in the sequel that such a problem may be reduced to a Fredholm-type equation. To this aim we shall consider the harmonic function $U(y,z)$ in the superior half-plane $z>0$ from the yOz plane, with mixed boundary conditions in the $Z=y+\mathrm{i}z$ complex plane (fig. 6.2.1). We shall also consider the complex function

$$F(Z)=U(y,z)+\mathrm{i}V(y,z)=\frac{1}{2\pi\mathrm{i}}\int_{-b}^{+b}\frac{C(\eta)}{\eta-Z}\mathrm{d}\,\eta\,. \qquad (6.2.1)$$

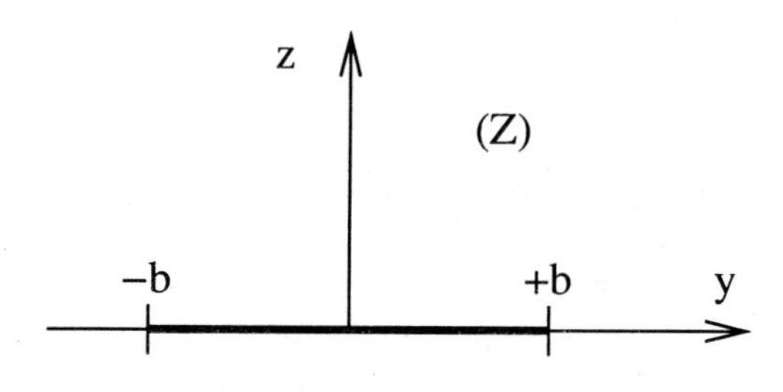

Fig. 6.2.1.

Obviously this is an holomorphic function in the (Z) plane with the cut $[-b,+b]$ and it vanishes to infinity. $U(y,z)$ is therefore a harmonic function in the half-plane $z>0$. It results obviously

$$U(y,+0)=0\,,\quad y\in(-\infty,-b)\cup(b,\infty)\,. \qquad (6.2.2)$$

The limit values on the segment $(-b,+b)$ are obtained by means of Plemelj's formulas. We deduce

$$2U(y,+0)=C(y)\,,\quad y\in(-b,+b)\,. \qquad (6.2.3)$$

Taking the conditions (6.1.7) into account, we get

$$\frac{\mathrm{d}}{\mathrm{d}\,Z}\int_{-b}^{+b}\frac{C(\eta)}{\eta-Z}\mathrm{d}\,\eta=\int_{-b}^{+b}C(\eta)\frac{\partial}{\partial Z}\left(\frac{1}{\eta-Z}\right)\mathrm{d}\,\eta=$$

$$=-\int_{-b}^{+b}C(\eta)\frac{\partial}{\partial\eta}\left(\frac{1}{\eta-Z}\right)\mathrm{d}\,\eta=\int_{-b}^{+b}\frac{C'(\eta)}{\eta-Z}\mathrm{d}\,\eta$$

Deriving we deduce from (6.2.1):

$$F'(Z)=\frac{\partial U}{\partial y}-\mathrm{i}\frac{\partial U}{\partial z}=\frac{1}{2\pi\mathrm{i}}\int_{-b}^{+b}\frac{C'(\eta)}{\eta-Z}\mathrm{d}\,\eta \tag{6.2.4}$$

and with Plemelj's formulas

$$\frac{\partial U}{\partial z}(y,+0)=\frac{\mathrm{d}\,U}{\mathrm{d}\,n}(y,+0)=\frac{1}{2\pi}{}'\!\!\int_{-b}^{+b}\frac{C'(\eta)}{\eta-y}\mathrm{d}\,\eta\,, \tag{6.2.5}$$

n representing the inward pointing normal of the superior half-plane on the segment $(-b,+b)$ of the real axis.

Taking (6.2.3) and (6.2.5) into account, Prandtl's equation is transformed into the following condition on the segment $(-b,+b)$:

$$\frac{\mathrm{d}\,U}{\mathrm{d}\,n}(y,+0)=\frac{2\beta U(y,+0)}{A(y)}-J(y)\,, \tag{6.2.6}$$

where, from now on, we denote

$$A(y)=\pi a(y)\,,\quad A(y)J(y)=\jmath(y)\,, \tag{6.2.7}$$

So, Prandtl's equation was replaced by the following *mixed boundary value problem*: "*to determine the harmonic function* $U(y,z)$ *in the* $z>0$ *half-plane, vanishing at infinity and satisfying the boundary conditions* (6.2.2) *and* (6.2.6)". Then, the function $C(y)$ will result from (6.2.3).

The mixed problem is reduced to a Fredholm-type integral equation. Proceeding like in [1.21], p. 713, we notice that because of the condition (6.2.2), the function $F(Z)$ may be extended by symmetry in the lower half-plane $z<0$. In this way, on the lower margin of the cut $[-b,+b]$ we shall have the condition

$$\frac{\mathrm{d}\,U}{\mathrm{d}\,n}(y,-0)=\frac{2\beta U(y,-0)}{A(y)}+J(y)\,, \tag{6.2.8}$$

So, we reduced the above mixed problem to the problem of *determination of the harmonic function* $U(y,z)$, *in the* yOz *plane, with the cut*

$[-b,+b]$ *and the conditions* (6.2.6) *and* (6.2.8) on the two margins of the cut.

It is known (see for example, [1.11] [1.20] [1.31]) that the Joukovsky-type conformal mapping $Z \to W$:

$$Z = \frac{1}{2}\left(W + \frac{b^2}{W}\right), \tag{6.2.9}$$

maps the exterior of the cut $[-b,+b]$ from the (Z) plane onto the exterior of the circle of radius b and the center in origin from the (W) plane (fig. 6.2.2), the superior margin of the cut being mapped on the superior half-circle from the superior half-plane. The points $\pm b$ are double and singular. One obtains the correspondence of the boundaries putting $W = b\mathrm{e}^{\mathrm{i}\sigma}$. We have

$$y = b\cos\sigma\,, \quad z = 0\,. \tag{6.2.10}$$

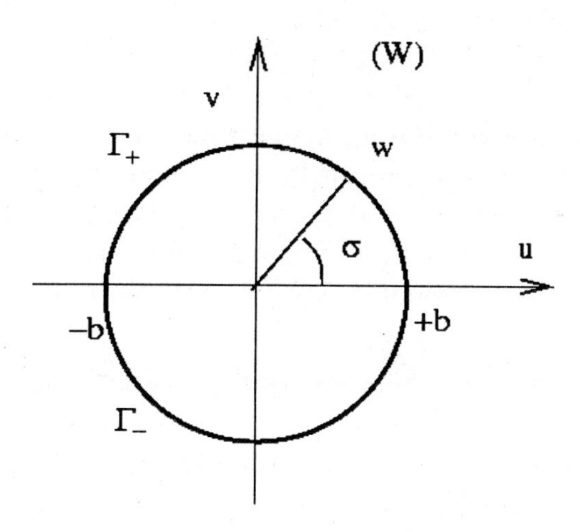

Fig. 6.2.2.

For $0 \le \sigma \le \pi$, (6.2.10) gives the correspondence between the half-circle Γ_+ and the superior margin of the cut, and for $-\pi \le \sigma \le 0$, the correspondence between Γ_- and the inferior margin (fig. 6.2.2). Since from the extension by symmetry it results $U(y,+0) = -U(y,-0)$, we deduce that $U(\sigma)$ is an odd function on Γ. With the same application (6.2.10), the functions $A(y)$ and $j(y)$ become even functions. We denote them by $A(\sigma)$, respectively $j(\sigma)$. Now we shall see how the boundary conditions (6.2.6) and (6.2.8) are transformed. To this aim we remind to the reader that, after performing a conformal mapping, the ratio of the lengths is given by the modulus of the derivative. More precisely, let $W = f(Z)$ be a conformal mapping and let $M(Z)$ and $N(Z+\Delta Z)$ be two neighboring points and $M_1(W)$ and $N_1(W+\Delta W)$ their images.

Obviously we have

$$\lim \frac{|M_1 N_1|}{MN} = \lim \frac{|\Delta W|}{|\Delta Z|} = \lim \left| \frac{\Delta W}{\Delta Z} \right| = |f'(Z)|$$

Hence, returning to our problem and denoting by $\boldsymbol{N}$ the outward normal to Γ, we shall have

$$\frac{\mathrm{d}U}{\mathrm{d}n} = \frac{\mathrm{d}U}{\mathrm{d}N}(\sigma)\frac{\mathrm{d}N}{\mathrm{d}n} = \frac{\mathrm{d}U}{\mathrm{d}N}\left|\frac{\mathrm{d}W}{\mathrm{d}Z}\right| \tag{6.2.11}$$

Taking (6.2.10) into account, the boundary conditions (6.2.6) and (6.2.8) on the two half - circles Γ_+ and Γ_- give

$$\frac{\mathrm{d}U}{\mathrm{d}N}(\sigma) = \frac{2\beta|\sin\sigma|}{A(\sigma)}U(\sigma) - J(\sigma)\sin\sigma \tag{6.2.12}$$

Denoting $W = u + \mathrm{i}\,v$, the harmonic function $U(y,z)$ in the yOz plane becomes the harmonic function $U(u,v)$ in the exterior of the circle Γ. This one vanishes at infinity and has the normal derivative (6.2.12) known on Γ. It is a Neumann problem. Its solution is given by Dini's formula (see for example [1.20] p. 31). We obtain

$$U(u,v) = \frac{b}{\pi}\int_{-\pi}^{+\pi} \frac{\mathrm{d}U}{\mathrm{d}N}(\sigma) \ln |b\mathrm{e}^{\mathrm{i}\,\sigma} - W| \mathrm{d}\sigma + k_0 \tag{6.2.13}$$

k_0 being an unknown constant. Considering that W tends to a point $b\mathrm{e}^{\mathrm{i}\,s}$ from C, one obtains

$$U(s) = \frac{b}{\pi}\int_{-\pi}^{+\pi} \frac{\mathrm{d}U}{\mathrm{d}N}(\sigma) \ln \left| 2\sin\frac{s-\sigma}{2} \right| \mathrm{d}\sigma + k_0\,. \tag{6.2.14}$$

Since $U(\sigma)$ is an odd function it results

$$\frac{\mathrm{d}U}{\mathrm{d}N}(\sigma) = -\frac{\mathrm{d}U}{\mathrm{d}N}(-\sigma)\,,$$

and integrating only on the interval $(0,\pi)$, we deduce

$$U(s) = \frac{b}{\pi}\int_0^{\pi} \frac{\mathrm{d}U}{\mathrm{d}N}(\sigma) S(s,\sigma)\mathrm{d}\,\sigma\,, \tag{6.2.15}$$

where we denoted

$$S(s,\sigma) = \ln \left| \frac{\sin \dfrac{s-\sigma}{2}}{\sin \dfrac{s+\sigma}{2}} \right| . \tag{6.2.16}$$

We have (see, for example, [1.16]):

$$S(s,\sigma) = -2 \sum_{k=1}^{\infty} \frac{\sin ks \sin k\sigma}{k} , \tag{6.2.17}$$

the series being absolutely and uniformly convergent $(s \neq \sigma)$.

In (6.2.15) we did not encounter k_0, because we have $U(s) = -U(-s)$ whence $U(0) = 0$.

Taking into account the relation (6.2.12) in which $2U(\sigma)$ is replaced by $C(\sigma)$ according to the condition (6.2.3), one obtains the following Fredholm-type integral equation

$$C(s) = \lambda \int_0^{\pi} C(\sigma) \frac{\sin \sigma}{A(\sigma)} S(s,\sigma) \mathrm{d}\,\sigma + J_0(s) , \tag{6.2.18}$$

where

$$\lambda = \frac{2b\beta}{\pi} , \quad J_0(s) = -\frac{2b}{\pi} \int_0^{\pi} J(\sigma) S(s,\sigma) \sin \sigma \mathrm{d}\,\sigma . \tag{6.2.19}$$

Hence we reduced Prandtl's integro-differential equation to the Fredholm-type equation (6.2.18).

The kernel of this equation has an integrable singularity. Equivalent integral equations have been given by Betz and Gebelein in 1936 and Trefftz in 1938.

6.2.2 Existence and Uniqueness Theorems

For proving the existence and uniqueness of the solution of Prandtl's equation , we shall use the first theorem of Fredholm. This may be enunciated as follows: *the equation*

$$\varphi(s) = \lambda \int_a^b K(s,\sigma) \varphi(\sigma) \mathrm{d}\,s + f(s) , \tag{6.2.20}$$

has an unique solution for a given value of λ *and for every free term* f *if and only if the corresponding homogeneous equation admits only the trivial solution* $\varphi(s) \equiv 0$. Hence, we must show that the equation

(6.2.18) which is homogeneous ($J_0 = 0$) has only the trivial solution. But the homogeneous equation corresponds to the boundary problem (6.2.12) which is homogeneous ($J = 0$). Applying Green's formula

$$\int\int_D (\operatorname{grad} U)^2 \mathrm{d}\, u\, \mathrm{d}\, v = -\int_{\partial D} U \frac{\mathrm{d}U}{\mathrm{d}N} \mathrm{d}\, S\,, \tag{6.2.21}$$

where D is an annulus, exterior to the circle Γ, bounded by an concentric circle of radius $R > b$ and observing that for $R \to \infty$ the last term vanishes (see, for example, §6.1 from [1.11]), we deduce

$$\int_{-\pi}^{+\pi} U \frac{\mathrm{d}U}{\mathrm{d}N} \mathrm{d}\,\sigma \leq 0\,.$$

Utilizing the homogeneous condition (6.2.12), we obtain:

$$\int_{-\pi}^{+\pi} \frac{|\sin\sigma|}{A(\sigma)} U^2(\sigma) \mathrm{d}\,\sigma \leq 0\,. \tag{6.2.22}$$

This inequality implies $U = 0$, because, as it results from the definition (6.1.17), we have $A > 0$.

Hence, Prandtl's equation has an unique solution. This result is very important, because, as we have already seen in 6.1.4 and as we shall see in the sequel, we manage, on various ways, to determine a solution of this equations. The above result ensures that if we find a solution, this is the unique solution of the equation.

6.2.3 Foundation of Glauert's Method

The integral equation (6.2.20) is a Fredholm-type equation of the second kind. This equation has a symmetric kernel if

$$K(s, \sigma) = K(\sigma, s)\,. \tag{6.2.23}$$

The equation (6.2.18) has not a symmetric kernel, but it can be symmetrized. Indeed, multiplying the equation with $\sqrt{\dfrac{\sin s}{A(s)}}$ (on the integration interval the quantity under the radical is positive) and, taking the function as an unknown

$$c(s) = \sqrt{\frac{\sin s}{A(s)}} C(s)\,, \tag{6.2.24}$$

one obtains the following equation:

$$c(s) = -2\lambda \int_0^{\pi} c(\sigma) \sqrt{\frac{\sin s}{A(s)} \frac{\sin \sigma}{A(\sigma)}} \sum_{k=1}^{\infty} \frac{\sin ks \sin k\sigma}{k} \mathrm{d}\sigma + \sqrt{\frac{\sin s}{A(s)}} J_0(s), \tag{6.2.25}$$

The kernel

$$K \equiv -2 \sum_{k=1}^{\infty} \sqrt{\frac{\sin s}{A(s)} \frac{\sin \sigma}{A(\sigma)}} \frac{\sin ks \sin k\sigma}{k}$$

is obviously *symmetric*. The kernel is even *degenerate*, but not of finite rank. As it is known (see, for example, [1.22], vol.3, p.193), the integral equations with degenerate kernel may be reduced to infinite algebraic systems and the equations with degenerate kernel of finite rank may be reduced to linear algebraic systems with a finite number of equations. We are not in this situation, and according to (6.2.25)we shall take only the property of symmetry of the kernel into account. According to the theory of Hilbert and Schmidt (see, for example, [1.22] v. 3, p. 243) the solution of the integral equation may be expanded, with respect to the eigenfunctions of the kernel, into absolutely and uniformly convergent series. Hence the solution of the equation (6.2.25) has the form:

$$c(\sigma) = \sqrt{\frac{\sin \sigma}{A(\sigma)}} \sum_{k=1}^{\infty} A_k \sin k\sigma .$$

Taking (6.2.24) into account, it results:

$$C(\sigma) = \sum_{k=1}^{\infty} A_k \sin k\sigma . \tag{6.2.26}$$

The (constant) coefficients A_k will be determined replacing (6.2.26) in the equation (6.2.18), or easier, performing this replacement in Prandtl' equation (6.1.18), which, with the change of variables

$$y = b \cos s, \quad \eta = b \cos \sigma \tag{6.2.27}$$

and with the notation $C(s)$ for the function $C(y)$ composed with $(6.2.27)_1$ etc., becomes

$$2b\beta C(s) + a(s) \int_0^{\pi} \frac{C'(\sigma)\mathrm{d}\sigma}{\cos \sigma - \cos s} = 2\pi b a(s) j(s) . \tag{6.2.28}$$

Before discussing about how to determine the coefficients A_i from (6.2.26) and (6.2.27), we must notice that the form (6.2.26) of the solution $C(\sigma)$ may result directly from (6.2.18), without utilizing the theory of Hilbert and Schmidt. Indeed, taking (6.2.17) into account, the equation (6.2.18) becomes:

$$C(s) = -2\lambda \sum_{k=1}^{\infty} \frac{\sin ks}{k} \int_0^{\pi} \frac{C(\sigma)}{kA(\sigma)} \sin k \sin \sigma \, d\sigma +$$

$$+\frac{4b}{\pi} \sin ks \int_0^{\pi} \frac{j(\sigma)}{kA(\sigma)} \sin k\sigma \sin \sigma d\sigma \, .$$

The integrals are constants and the solution will have the form (6.2.26).

6.2.4 Glauert's Approximation

We shall return now to the problem of determination of the coefficients A_i from (6.2.26). Replacing C from (6.2.26) in (6.2.28) and using Glauert's formula (B.6.6) (herein is the origin of this formula), we deduce:

$$\sum_{k=1}^{\infty} A_k [2b\beta \sin s + k\pi a(s)] \sin ks = 2\pi b a(s) j(s) \sin s \, . \qquad (6.2.29)$$

Glauert's approximation consists in keeping the first n terms from the expansion (6.2.26) and then imposing (6.2.29) to be satisfied for n distinct values of the variable s. The coefficients A_k are the solution of an linear algebraic system, but we cannot evaluate the error of the approximation. Many other approximations have been given in the literature (see Lotz in [1.24], Carafoli in [1.5]).

6.2.5 The Minimal Drag Airfoil

The foundation of Glauert's method,which consists in establishing the formula (6.2.26) gives the possibility to give an answer to the following problem of practical interest: *to determine among the wings with the same lift, that one which has the minimum drag.* In view of this determination we shall calculate, utilizing the formulas (6.1.21), (6.1.25) and (6.2.26), the lift and drag coefficients c_L and c_D. For determining the lift and the drag, we multiply these coefficients by the same factor $\frac{1}{2}\rho_\infty U_\infty^2 A_1$. Since

$$2 \int_0^{\pi} \sin k\sigma \sin l\sigma d\sigma = \pi \delta_{kl} \, , \quad k, l = 1, 2, \ldots$$

we deduce

$$c_L \equiv \frac{2}{A}\int_{-b}^{+b} C(y)\mathrm{d}\,y = \frac{2b}{A}\int_0^{\pi} C(\sigma)\sin\sigma\mathrm{d}\,\sigma = \frac{b}{A}A_1\pi. \tag{6.2.30}$$

Utilizing Glauert's formula (B.6.6), we obtain:

$$w(\sigma) = -\frac{1}{4\pi b}\int_0^{\pi}\frac{C'(s)\mathrm{s}\,s}{\cos s - \cos\sigma} = -\frac{1}{4b}\sum_{k=1}^{\infty} kA_k\frac{\sin k\sigma}{\sin\sigma},$$

such that:

$$c_D \equiv -\frac{2}{A}\int_{-b}^{+b} C(y)w(y)\mathrm{d}\,y = -\frac{2b}{A}\int_0^{\pi} w(\sigma)C(\sigma)\sin\sigma\mathrm{d}\,\sigma =$$

$$= \frac{1}{2A}\int_0^{\pi}\left(\sum_{k=1}^{\infty} kA_k\sin k\sigma\right)\left(\sum_{l=1}^{\infty} A_l\sin l\sigma\right)\mathrm{d}\,\sigma = \tag{6.2.31}$$

$$= \frac{\pi}{4A}\sum_{k=1}^{\infty} kA_k^2\,.$$

The formulas (6.2.30) and (6.2.31) indicate that among all the wings with the same lift (with the same A_1), the minimum drag corresponds to the wings for which $A_2 = A_3 = \ldots = 0$.

The solution of Prandtl's equation for these wings is

$$C(\sigma) = A_1\sin\sigma \Longrightarrow C(y) = A_1\sqrt{1-\frac{y^2}{b^2}}, \tag{6.2.32}$$

where $A_1 = C(0)$ may be determined obviously from the equation (6.1.26). We have

$$\frac{\beta C(y)}{A(y)} = 2w(y) + J(y) = -\frac{1}{2b}A_1 + J(y)\,. \tag{6.2.33}$$

In the case of the flat plate, j has the form (6.1.20). It results $J = 2\varepsilon$ whence we deduce that the member from the left hand side of (6.2.33) is constant. Hence,

$$a(y) = a_0\sqrt{1-\frac{y^2}{b^2}}, \tag{6.2.34}$$

the constant a_0 being determined by the relation

$$\left(\frac{\beta}{\pi a_0} + \frac{1}{2b}\right)A_1 = 2\varepsilon\,, \tag{6.2.35}$$

if one gives A_1. The same relation determines A_1 if one gives a_0. For example, when $A_1 = k$ like in (6.1.28), it results $a_0 = 1$, like in (6.1.27) and vice versa.

The expression (6.2.33) shows that the wings which have the above property are the elliptical flat plates.

6.3 The Symmetrical Wing. Vekua's Equation. A Larger Class of Exact Solutions

6.3.1 Symmetry Properties

Very often in aerodynamics we encounter the case when the wing is symmetric with respect to the xOz plane. In this situation we have

$$x_\pm(y) = x_\pm(-y)\,, \quad h(x,y) = h(x,-y)\,, \quad -b \leq y \leq +b \tag{6.3.1}$$

From (6.1.19) and (6.1.20) it results

$$a(y) = a(-y)\,, \quad j(y) = j(-y)\,. \tag{6.3.2}$$

Let us prove that we also have

$$C(y) = C(-y)\,. \tag{6.3.3}$$

Indeed, changing in Prandtl's equation (6.1.18) y by $-y$ and taking (6.3.2) into account, it results

$$\beta C(-y) = \frac{a(y)}{2} {\int_{-b}^{+b}}' \frac{C(\eta)}{\eta + y} \mathrm{d}\,\eta + j(y)\,. \tag{6.3.4}$$

Putting in the integral $\eta = -u$ and observing that

$$C'(\eta)\mathrm{d}\,\eta = \mathrm{d}\,C = C'(u)\mathrm{d}\,u\,,$$

we deduce

$${\int_{-b}^{+b}}' \frac{C'(\eta)}{\eta + y} \mathrm{d}\,\eta = {\int_{+b}^{-b}}' \frac{C'(u)\mathrm{d}\,u}{-u + y} = {\int_{-b}^{+b}}' \frac{C'(\eta)\mathrm{d}\,\eta}{\eta - y}\,. \tag{6.3.5}$$

Introducing this relation in (6.3.4) and comparing with (6.1.18), we get (6.3.3).

6.3.2 The Integral Equation

We shall present in the sequel the simplest method for obtaining the equation (6.2.18). The demonstration is inspired from [A.27], where, on his turn, it was taken from Magnaradze [6.16] and Vekua [6.28]. With the notations (6.2.7) Prandtl's equation is

$$\frac{1}{2\pi}\int_{-b}^{+b}{}' \frac{C'(\eta)}{\eta - y}\mathrm{d}\,\eta = \beta\frac{C(y)}{A(y)} - J(y)\,. \tag{6.3.6}$$

For the existence of the principal value we have to assume that $C'(y)$ satisfies Hölder's condition on the segment $(-b,+b)$. We shall invert this equation assuming that the right hand member is known. As it is known from (C.1.1) the solution $C'(y)$ depends on the behaviour imposed in the points $\pm b$. We know that we cannot obtain a bounded solution in the two points without imposing a restriction to the right hand member. In the same time, because of the symmetry of the wing, we cannot consider C' bounded only in an extremity. Hence C' is unbounded in the two extremities, i.e. the solution has the form (C.1.11). Further, for inverting the equation (6.3.6), $A(y)$ and $J(y)$ have to satisfy Hölder's condition on $[-b,+b]$. If $A(y)$ and $h'_x(x,y)$ (with respect to the y variable) have this property we deduce the same thing for $J(y)$. Moreover, $a(y)$ must not vanish on $(-b,+b)$. If all these conditions are satisfied, then, using the formula (C.1.11), we obtain

$$C'(y) = -\frac{2}{\pi}\frac{1}{\sqrt{b^2-y^2}}\int_{-b}^{+b}{}' \frac{\sqrt{b^2-\eta^2}}{\eta - y}\left[\beta\frac{C(\eta)}{A(\eta)} - J(\eta)\right]\mathrm{d}\,\eta + \frac{B}{\sqrt{b^2-y^2}}, \tag{6.3.7}$$

B representing a constant which has to be determined. It is zero because from (6.3.3) we have $C'(y) = -C'(-y)$, whence $C'(0) = 0$. Imposing this in (6.3.7) and observing that the integrand is an odd function, it results the assertion.

Utilizing now the identity

$$-\frac{\sqrt{b^2-\eta^2}}{\sqrt{b^2-y^2}(\eta - y)} = \frac{\mathrm{d}}{\mathrm{d}y}\ln\left|\frac{\mathrm{i}\,(y-\eta)+\sqrt{b^2-y^2}-\sqrt{b^2-\eta^2}}{\mathrm{i}\,(y-\eta)+\sqrt{b^2-y^2}+\sqrt{b^2-\eta^2}}\right| \tag{6.3.8}$$

and integrating (6.3.7) on the interval $(-b, y)$, from (6.3.3) one obtains:

$$C(y) = \frac{2}{\pi} \int_{-b}^{+b} \left[\beta \frac{C(\eta)}{A(\eta)} - \right.$$
$$\left. - J(\eta) \right] \ln \left| \frac{\mathrm{i}\,(y-\eta) + \sqrt{b^2 - y^2} - \sqrt{b^2 - \eta^2}}{\mathrm{i}\,(y-\eta) + \sqrt{b^2 - y^2} + \sqrt{b^2 - \eta^2}} \right| \mathrm{d}\,\eta\,, \tag{6.3.9}$$

because the modulus is equal to the unity for $y = -b$.

Performing the change of variable

$$y = b \cos s\,, \qquad \eta = b \cos \sigma \tag{6.3.10}$$

and taking into account that

$$\ln \left| \frac{\mathrm{i}\,(y-\eta) + \sqrt{b^2 - y^2} - \sqrt{b^2 - \eta^2}}{\mathrm{i}\,(y-\eta) + \sqrt{b^2 - y^2} + \sqrt{b^2 - \eta^2}} \right| = \ln \left| \frac{\mathrm{e}^{-\mathrm{i}\,s} - \mathrm{e}^{-\mathrm{i}\,\sigma}}{\mathrm{e}^{-\mathrm{i}\,s} - \mathrm{e}^{\mathrm{i}\,\sigma}} \right| = S(s, \sigma)\,,$$

we obtain obviously the equation (6.2.18).

Using the notation $A(\pi - \sigma)$ we have $A(b \cos(\pi - \sigma)) = A(-b \cos \sigma) = A(-y)$. Hence, taking (6.3.2) and (6.3.3) into account, it results

$$A(\pi - \sigma) = A(\sigma), \quad j(\pi - \sigma) = j(\sigma)\,,$$
$$C(\pi - \sigma) = C(\sigma) \tag{6.3.11}$$

whence:

$$\int_{\pi/2}^{\pi} \left[\beta \frac{C(\sigma)}{A(\sigma)} - J(\sigma) \right] \ln \left| \frac{\sin \dfrac{s - \sigma}{2}}{\sin \dfrac{s + \sigma}{2}} \right| \sin \sigma \mathrm{d}\,\sigma =$$
$$= \int_{0}^{\pi/2} \left[\beta \frac{C(\sigma)}{A(\sigma)} - J(\sigma) \right] \ln \left| \frac{\cos \dfrac{s + \sigma}{2}}{\cos \dfrac{s - \sigma}{2}} \right| \sin \sigma \mathrm{d}\,\sigma\,.$$

But,

$$\ln \left| \frac{\sin \dfrac{s - \sigma}{2}}{\sin \dfrac{s + \sigma}{2}} \cdot \frac{\cos \dfrac{s + \sigma}{2}}{\cos \dfrac{s - \sigma}{2}} \right| = \ln \left| \frac{\sin s - \sin \sigma}{\sin s + \sin \sigma} \right|\,.$$

In this way, the equation (6.2.18) for the symmetric wing becomes

$$C(s) = \frac{2b}{\pi} \int_0^{\pi/2} \left[\beta \frac{C(\sigma)}{A(\sigma)} - \right.$$
$$\left. - J(\sigma) \right] \ln \left| \frac{\sin s - \sin \sigma}{\sin s + \sin \sigma} \right| \sin \sigma \mathrm{d} \sigma \,, \tag{6.3.12}$$

with s in the interval $(0, \pi/2)$.

6.3.3 Vekua's Equation

In 1945, I.N.Vekua [6.28] gave for the symmetric profile whose chord has the form

$$a(y) = \frac{\sqrt{b^2 - y^2}}{p(y)} \,, \quad \text{with } p(y) = p(-y) > 0 \,, \tag{6.3.13}$$

(where $p(y)$ is an analytic function on $[-b, +b]$), a Fredholm-type integral equation which has the great advantage that it may be integrated exactly for a large class of profiles.

Vekua's method was extended immediately by Magnaradze [6.16] to wings for which the function $p(y)$ is not necessarily analytic on the interval $[-b, +b]$. Since we had not the occasion to read this papers, we present herein a a synthesis due to Muschelisvili [A.27]. To this aim, we write the equation (6.3.7), where we considered $B = 0$, as follows

$$A(y)\, C'(y) + \frac{2\beta}{\pi} {\int_{-b}^{+b}}' \frac{C(\eta)}{\eta - y} \mathrm{d}\, \eta =$$
$$= A(y) J_1(y) - \frac{a(y)}{\sqrt{b^2 - y^2}} \int_{-b}^{+b} R(y, \eta) C(\eta) \mathrm{d}\, \eta \,, \tag{6.3.14}$$

where

$$R(y, \eta) = \frac{2\beta}{\pi} \frac{1}{\eta - y} \left[\frac{\sqrt{\beta^2 - \eta^2}}{a(\eta)} - \frac{\sqrt{b^2 - y^2}}{a(y)} \right] , \tag{6.3.15}$$

$$J_1(y) = \frac{2}{\pi \sqrt{b^2 - y^2}} {\int_{-b}^{+b}}' \frac{\sqrt{b^2 - y^2}}{\eta - y} J(\eta) \mathrm{d}\, \eta \,. \tag{6.3.16}$$

Obviously we have:

$$R(y, \eta) = -R(-y, -\eta) \,, \quad J_1(y) = -J_1(-y) \,. \tag{6.3.17}$$

Further we shall assume the continuity of the first order derivative of the function

$$p(y)=\frac{\sqrt{b^2-y^2}}{a(y)}. \tag{6.3.18}$$

In this case, $R(y,\eta)$ will be a continuous function.

Since according to (6.1.7) we have:

$$\frac{\mathrm{d}}{\mathrm{d}y}{\int_{-b}^{+b}}' \frac{C(\eta)}{\eta-y}\mathrm{d}\,\eta=\lim_{\varepsilon\to 0}\frac{\mathrm{d}}{\mathrm{d}y}\left(\int_{-b}^{y-\varepsilon}+\int_{y+\varepsilon}^{+b}\right)\frac{C(\eta)}{\eta-y}\mathrm{d}\,\eta={\int_{-b}^{+b}}'\frac{C'(\eta)}{\eta-y}\mathrm{d}\,\eta\,,$$

from (6.3.14) it results

$$\frac{\mathrm{d}}{\mathrm{d}y}\left[A(y)C'(y)\right]+\frac{2\beta}{\pi}{\int_{-b}^{+b}}'\frac{C'(\eta)}{\eta-y}\mathrm{d}\,\eta=B(y)\,, \tag{6.3.19}$$

where

$$B(y)=\frac{\mathrm{d}}{\mathrm{d}y}\left[A(y)J_1(y)-\frac{a(y)}{\sqrt{b^2-y^2}}\int_{-b}^{+b}R(y,\eta)C(\eta)\mathrm{d}\,\eta\right]. \tag{6.3.20}$$

Obviously,

$$B(y)=B(-y) \tag{6.3.21}$$

Eliminating the integral from (6.3.19) by means of Prandtl's equation, we obtain the following differential equation:

$$A(y)\frac{\mathrm{d}}{\mathrm{d}y}\left[A(y)C'(y)\right]+4\beta^2C(y)=A(y)\left[B(y)+4\beta J(y)\right]. \tag{6.3.22}$$

Assuming that the right hand member is known, we have in (6.3.22) a differential linear equation for $C(y)$.

The homogeneous equation has the linear independent solutions $\cos s(y), \sin s(y)$, where

$$s(y)=\frac{2\beta}{\pi}\int_0^y\frac{\mathrm{d}\,\eta}{a(\eta)}. \tag{6.3.23}$$

Utilizing Lagrange's method of variation of constants, we deduce that the equation (6.3.22) has the following solution:

$$C(y)=C_0\cos s(y)+C_1\sin s(y)+ \\ +\frac{1}{2\beta}\int_0^y\left[B(\eta)+4\beta J(\eta)\right]\sin\left[s(y)-s(\eta)\right]\mathrm{d}\,\eta\,, \tag{6.3.24}$$

C_0 and C_1 being constants. Obviously, $C_0 = C(0)$.

Calculating $C(-y)$, taking into account that $s(y)$ is an odd function (its derivative is an even function) and $B(\eta)$ and $J(\eta)$ are odd functions, and imposing (6.3.3), it results $C_1 = 0$.

Introducing B given by (6.3.20) in (6.3.24), performing an integration by parts and observing that the integrated term is zero because

$$J_1(0) = 0\,, \quad \int_{-b}^{+b} R(0,\eta_1)C(\eta_1)\mathrm{d}\,\eta_1 = 0\,,$$

it results the following integral equation:

$$C(y) + \frac{1}{\pi}\int_{-b}^{+b} K(y,\eta)C(\eta)\mathrm{d}\,\eta = g(y)\,, \tag{6.3.25}$$

$$K(y,\eta) = \int_0^y \frac{R(\eta_1,\eta)}{\sqrt{b^2-\eta_1^2}}\cos\left[s(y)-s(\eta_1)\right]\mathrm{d}\,\eta_1\,, \tag{6.3.26}$$

$$\begin{aligned} g(y) = C_0\cos s(y) &+ 2\int_0^y J(\eta)\sin[s(y)-s(\eta)]\mathrm{d}\,\eta + \\ &+ \int_0^y J_1(\eta)\cos[s(y)-s(\eta)]\mathrm{d}\,\eta\,. \end{aligned} \tag{6.3.27}$$

The equation (6.3.25) for $\beta = 1$ is the equation given by Vekua and Magnaradze. Unlike the equation of Trefftz (6.2.18), this is regular (the kernel has no singularity). Moreover, in case that the function $p(y)$ given by (6.3.18) is a rational function, more precisely in case that $a(y)$ has the form:

$$a(y) = a\cdot\sqrt{b^2-y^2}\,\frac{1+p_1y^2+\ldots+p_ny^{2n}}{1+q_1y^2+\ldots+q_ny^{2n}}\,, \tag{6.3.28}$$

as we shall see in an example, the equation of Vekua and Magnaradze reduces to an algebraic finite system. This form for $a(y)$ is suitable for approximating every wing of practical interest. We have to mention that, for the wings having the form (6.3.28), the case when $q_1 = q_2 = \ldots = q_n = 0$, has been solved by H.Schmidt in 1937, [6.24], and the case when $p_1 = p_2 = \ldots = p_n = 0$ belongs to a larger class, considered by the author of the present book in 1958, [6.4]. For this class one obtains the exact solution.

Before passing to applications we notice that if $J(\eta) = k$ is a constant, then, taking (B.5.6) into account, we deduce

$$g(y) = C_0\cos s(y) + 2kI, \tag{6.3.29}$$

where

$$I(y) = \int_0^y \left\{ \sin[s(y) - s(\eta)] - \frac{\eta}{\sqrt{b^2 - \eta^2}} \cos[s(y) - s(\eta)] \right\} \mathrm{d}\eta \,. \tag{6.3.30}$$

6.3.4 The Elliptical Wing

Denoting by a and b the semi-axes of the ellipse from the xOy plane, we deduce

$$x_{\pm} = \pm a\sqrt{1 - \frac{y^2}{b^2}}\,, \quad a(y) = a_0\sqrt{b^2 - y^2}\,, \quad a_0 \equiv a/b\,.$$

Obviously, $R = 0$ whence $K(y, \eta) = 0$. The equation (6.3.25) gives directly the solution $C(y) = g(y)$, where g is calculated with the formula (6.3.29). In I one performs an integration by parts. Since from (6.3.23) it results

$$s'(y) = \frac{2\beta}{\pi a(y)}\,,$$

we deduce

$$\left(1 + \frac{2\beta}{\pi a_0}\right) I = \sqrt{b^2 - y^2} - b\cos s(y)\,.$$

Since from (6.1.20) and (6.2.7) it results $J = 2\varepsilon$, using the notation

$$k = \frac{4\pi\varepsilon a_0}{\pi a_0 + 2\beta}\,,$$

we deduce

$$C(y) = C_0 \cos s(y) + k\sqrt{b^2 - y^2} - kb\cos s(y)\,. \tag{6.3.31}$$

For determining the constant C_0 we shall employ the condition $C(b) = 0$. Since from (6.3.23) it results

$$s(y) = \frac{2\beta}{\pi a_0} \arcsin\frac{y}{b}\,,$$

we deduce $C_0 = kb$ whence

$$C(y) = k\sqrt{b^2 - y^2}\,. \tag{6.3.32}$$

For $a_0 = 1/b$ one obtains exactly the solution (6.1.30).

6.3.5 The Rectangular Wing

We shall consider now that $a(y)$ has the form

$$a(y) = a_0\sqrt{b^2 - y^2}\frac{1 + py^2}{1 + qy^2}, \tag{6.3.33}$$

the real numbers p and q being chosen in order to ensure only positive values of the fraction one $[-b, +b]$. In the sequel we shall see that one imposes $pb^2 > -1$ whence $qb^2 > -1$. From (6.3.15) and (6.3.33) we deduce

$$R(\eta_1, \eta) = \frac{c(\eta + \eta_1)}{(1 + p\eta^2)(1 + p\eta_1^2)}, \quad C \equiv \frac{2\beta(q - p)}{\pi a_0}, \tag{6.3.34}$$

and from (6.3.26)

$$K(y, \eta) = \frac{\eta\varphi_0(y) + \varphi_1(y)}{1 + p\eta^2}, \tag{6.3.35}$$

where

$$\varphi_i(y) \equiv c\int_0^y \frac{\cos[s(y) - s(\eta_1)]}{\sqrt{b^2 - \eta_1^2}} \frac{\eta_1^i}{1 + p\eta_1^2} \mathrm{d}\,\eta_1 \tag{6.3.36}$$

Taking into account that for $pb^2 > -1$ we have

$$\int \frac{\mathrm{d}\,\eta}{(1 + p\eta^2)\sqrt{b^2 - \eta^2}} = \frac{1}{\sqrt{1 + pb^2}} \arctan \frac{\eta\sqrt{1 + pb^2}}{\sqrt{b^2 - \eta^2}},$$

from (6.3.23), for $p \neq 0$, we deduce

$$s(y) = \frac{2\beta}{\pi a_0}\left[\frac{q}{p}\arcsin\frac{y}{b} + \frac{p - q}{p\sqrt{1 + pb^2}} \arctan \frac{y\sqrt{1 + pb^2}}{\sqrt{b^2 - y^2}}\right] \tag{6.3.37}$$

and for $p = 0$,

$$s(y) = \frac{2\beta}{\pi a_0}\left[\left(1 + \frac{qb^2}{2}\right) \arcsin\frac{y}{b} - \frac{q}{2}y\sqrt{b^2 - y^2}\right]. \tag{6.3.38}$$

Replacing $K(y, \eta)$ given by (6.3.35) in the integral equation (6.3.25) and observing that the first term vanishes because the integrand is an odd function, we deduce:

$$C(y) + \frac{\varphi_1(y)}{\pi}\int_{-b}^{+b} \frac{C(\eta)}{1 + p\eta^2} \mathrm{d}\,\eta = C_0 \cos s(y) + 2kI(y). \tag{6.3.39}$$

The integral is a constant C_1 which may be determined by multiplying (6.3.39) with $(1+py^2)^{-1}$ and integrating with respect to y on the interval $(-b,+b)$. We obtain

$$C_1\left[1+\frac{1}{\pi}\int_{-b}^{+b}\frac{\varphi_1(y)}{1+py^2}\mathrm{d}\,y\right]- \\ -C_0\int_{-b}^{+b}\frac{\cos s(y)}{1+py^2}\mathrm{d}\,y = 2k\int_{-b}^{+b}\frac{I(y)}{1+py^2}\mathrm{d}\,y\,. \tag{6.3.40}$$

Imposing (6.3.39) and the condition $C(b)=0$, we deduce the relation

$$C_0\cos s(b)-\frac{\varphi_1(b)}{\pi}C_1=-2kI(b)\,. \tag{6.3.41}$$

Determining the constants C_0 and C_1 from the system (6.3.40) and (6.3.41), we find the exact solution of Prandtl's equation

$$C(y)=C_0\cos s(y)-\frac{1}{\pi}C_1\varphi_1(y)+2kI(y)\,. \tag{6.3.42}$$

Using the inverse method, i.e. considering various values for the constants p and q and calculating the form of the chord, we may find important wings for which the exact solution (6.3.42) is valid. So, in [A.27], considering $q=0$ and $pb^2=0,9$, one obtains an almost rectangular wing (the variation of the chord versus the span is very small). Indeed, we have

y/b	0.1	0.2	0.3	0.4	0.5	0.6	0.7	0.8	0.9
a/ba_0	1.00	1.02	1.03	1.05	1.06	1.06	1.03	0.95	0.75

Important results of this method are given in [1.33].

6.3.6 Extensions

Modifying Vekua's method, we managed in [6.4] to give the exact solution of Prandtl's equation for wings whose chord satisfies the relation

$$a(y)=\frac{\sqrt{b^2-y^2}}{p(y)}\,, \tag{6.3.43}$$

where $p(Z)$ is a holomorphic function in the $Z=y+\mathrm{i}\,z$ complex plane, excepting the vicinity of the point at infinity where one admits the following series expansion:

$$p(Z) = \sum_{n=-\infty}^{k\geq 0} p_n Z^n . \tag{6.3.44}$$

The determination of the solution of Prandtl's equation is reduced to solving a Hilbert-type problem, whose exact solution is given.

The polynomials belong to the class (6.3.44), whence the great importance of this solution. According to Weierstrass's theorem (see for example, [6.13], p.61), every continuous function on the interval $[-b, +b]$ (i.e., every possible form of the wing) may be approximated by polynomials. Practically, for this one may employ an interpolation method (for example, Newton or Lagrange's method). Even if $p(y)$ is a polynomial, in this method it is not necessary to be symmetric, like in the theory of Vekua and Magnaradze.

6.4 Numerical Methods

6.4.1 Multhopp's Method

The idea of Multhopp's method consists in approximating the function $C(\sigma)$ by the trigonometric polynomial $P_n(\sigma)$ obtained by the Lagrange interpolation in the basis $\{\sin k\sigma\}_{k=1,\ldots,n}$. For determining P_n, we notice that after introducing the matrices

$$s^T = (\sin\sigma, \ldots, \sin n\sigma), \quad a^T = (a_1, \ldots, a_n),$$

it may be written as follows

$$P_n(\sigma) = \sum_{k=1}^{n} a_k \sin k\sigma = s^T a . \tag{6.4.1}$$

The points where one imposes for $P_n(\sigma)$ to coincide with $C(\sigma)$ (the nodes) are given by the uniform grid

$$\sigma_1 = \frac{\pi}{n+1} \equiv \alpha, \sigma_2 = \frac{2\pi}{n+1} = 2\alpha, \ldots, \sigma_n = \frac{n\pi}{n+1} = n\alpha , \tag{6.4.2}$$

which is usual in the theory of interpolation [6.13] p.20. These are equidistant on the half-circle with the diameter on the span (fig. 6.4.1). The points

$$x_j = \cos\sigma_j \tag{6.4.3}$$

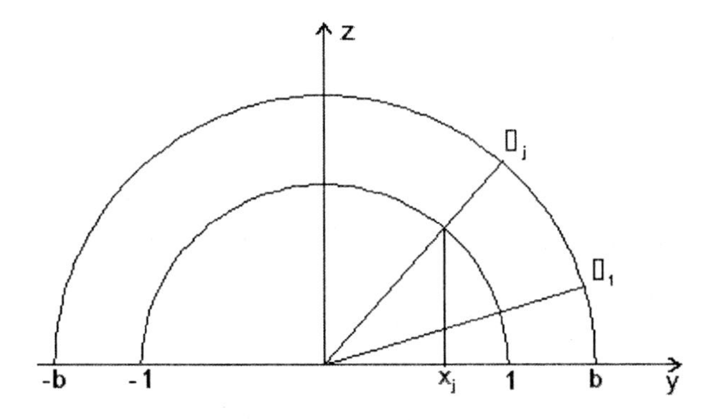

Fig. 6.4.1.

are the zeros of the Chebyshev polynomial of order two (F.2.8) on the interval $(0, \pi)$.

Denoting $c_j = C(\sigma_j)$ we have to determine the matrix a from the system

$$Sa = c\,, \tag{6.4.4}$$

where

$$S = \begin{bmatrix} \sin\sigma_1 \sin 2\sigma_1 \ldots \sin n\sigma_1 \\ \sin\sigma_2 \sin 2\sigma_2 \ldots \sin n\sigma_2 \\ \cdots\cdots \\ \sin\sigma_n \sin 2\sigma_n \ldots \sin n\sigma_n \end{bmatrix}, \quad c = \begin{bmatrix} c_1 \\ c_2 \\ \vdots \\ c_n \end{bmatrix}. \tag{6.4.5}$$

One obtains [6.13], p.21

$$S^{-1} = \frac{2}{n+1} S\,. \tag{6.4.6}$$

We shall present at the end of the section this calculus. Hence,

$$P_n(\sigma) = \frac{2}{n+1} s^T S c =$$

$$= \frac{2}{n+1}(\sin\sigma \ldots \sin n\sigma) \begin{bmatrix} \sin\sigma_1 \ldots \sin n\sigma_1 \\ \vdots \quad . \quad \vdots \\ \sin\sigma_n \ldots \sin n\sigma_n \end{bmatrix} \begin{bmatrix} c_1 \\ \vdots \\ c_n \end{bmatrix}$$

$$= \frac{2}{n+1}(\sin\sigma \ldots \sin n\sigma) \begin{bmatrix} \sum_{k=1}^{n} c_k \sin k\sigma_1 \\ \vdots \\ \sum_{k=1}^{n} c_k \sin k\sigma_n \end{bmatrix} \tag{6.4.7}$$

$$= \frac{2}{n+1} \sum_{k=1}^{n} c_k \sum_{j=1}^{n} \sin k\sigma_j \sin j\sigma \, .$$

For $C(\sigma)$ we have the following expansion:

$$C(\sigma) = \frac{2}{n+1} \sum_{k=1}^{n} c_k \sum_{j=1}^{n} \sin k\sigma_j \; \sin j\sigma \, . \tag{6.4.8}$$

We must notice that this expression could be also obtained from (6.2.26) approximating the Fourier coefficients according to the definition of the integral for an equidistant division α. Indeed, from (6.2.26) we have:

$$A_j = \frac{2}{\pi} \int_0^{\pi} C(\sigma) \sin j\sigma \mathrm{d}\,\sigma = \frac{2}{n+1} \sum_{k=1}^{n} C(\sigma_k) \sin j\sigma_k \, .$$

Since $j\sigma_k = jk\alpha = k\sigma_j$, employing Glauert's approximation

$$C(\sigma) = \sum_{j=1}^{n} A_j \sin j\sigma \, ,$$

one obtains (6.4.8).

Utilizing Multhopp's expansion (6.4.8), from Prandtl's equation (6.2.28) and from Glauert's formula, we deduce:

$$\sin s \frac{4b\beta}{n+1}\sum_{k=1}^{n} c_k \sum_{j=1}^{n} \sin jk\alpha \sin js+$$
$$+a(s)\frac{2\pi}{n+1}\sum_{k=1}^{n} c_k \sum_{j=1}^{n} j\sin jk\alpha \sin js = 2\pi b a(s) j(s)\sin s\,. \tag{6.4.9}$$

Giving to s the successive values $\ell\alpha(\ell = 1,\dots,n)$ and taking into account the formulas (6.4.17) and (6.4.19) from below, we deduce the system $(\ell = 1,\dots,n)$,

$$\beta c_\ell + \sum_{k=1}^{n} B_{\ell k} c_k = B_\ell\,, \tag{6.4.10}$$

where

$$2B_\ell = (n+1)\pi a(\ell k) j(\ell\alpha)\,. \tag{6.4.11}$$

For $k \neq \ell$,

$$B_{\ell k} = \frac{\pi a(\ell\alpha)}{2b\sin\ell\alpha}\,\frac{1-(-1)^{n-\ell}}{8}\left[\frac{1}{\sin^2\dfrac{(k+\ell)\alpha}{2}} - \frac{1}{\sin^2\dfrac{(k-\ell)\alpha}{2}}\right], \tag{6.4.12}$$

and for $k = \ell$,

$$A_{\ell\ell} = \frac{\pi a(\ell\alpha)}{2b\sin\ell\alpha}\cdot\frac{n(n+1)}{4}\,. \tag{6.4.13}$$

Determining the unknowns $c_1,\dots,c_n$ from the system (6.4.10), one obtains the solution of Prandtl's equation from (6.4.8).

Since $A_{\ell k}$ vanishes when $k-\ell$ is an even number, the system (6.4.10) may be separated; more precisely, the unknowns with odd indices may be expressed by means of the unknowns with odd indices and vice versa. This fact was proved in [1.2]. As it is already mentioned in [1.2] and in [6.17] an iterative procedure for determining the unknowns is also established. The author had not at his disposition this paper. The procedure simplifies in the case of the symmetric wing $(C(y) = C(-y))$, i.e. when $c_k = c_{n+1-k}$ $(k = 1,2,\dots,[(n+1)/2])$ or in the case of the antisymmetric wing $(c_k = -c_{n+1-k})$. In the first case the system reduces to $[(n+1)/2]$ equations, and in the second to $[(n-1)/2]$. The square brackets indicate the integer part of the number from the interior.

In the sequel we shall calculate the sums that intervened in the above formulas. Let it be for the beginning

$$I_1 \equiv \sum_{j=1}^{n} \cos rj\alpha\,, \quad I_2 \equiv \sum_{j=1}^{n} \sin rj\alpha\,. \tag{6.4.14}$$

Denoting $z = \mathrm{e}^{\mathrm{i}r\alpha}$, for $r \neq 0$, we obtain

$$I_1 + \mathrm{i}I_2 = \sum_{j=1}^{n} z^j = \frac{z - z^{n+1}}{1 - z} = \frac{z - (-1)^r}{1 - z}\,,$$

whence, separating the real part from the imaginary one,

$$I_1 = -1 + \frac{1 - (-1)^r}{2}\,, \quad I_2 = \frac{1 - (-1)^r}{2} \cot\frac{r\alpha}{2}\,. \tag{6.4.15}$$

One obtains the sum

$$\sum_{j=1}^{n} j \cos rj\alpha = -\frac{1 - (-1)^r}{4} \frac{1}{\sin^2 \dfrac{r\alpha}{2}} \tag{6.4.16}$$

deriving I_2 with respect to α. We deduce therefore, noticing that $k - \ell$ and $k + \ell$ are odd or even simultaneously,

$$2\sum_{j=1}^{n} \sin kj\alpha \sin \ell j\alpha = \sum_{j=1}^{n} \cos(k - \ell)j\alpha - \sum_{j=1}^{n} \cos(k + \ell)j\alpha = (n + 1)\delta_{k\ell}\,. \tag{6.4.17}$$

With these formulas, the relation (6.4.6) written as follows

$$S^{-1}S = \frac{2}{n + 1}S^2 = I \tag{6.4.18}$$

may be immediately proved, since we have:

$$S^2 = \begin{bmatrix} \sum_{j=1}^{n} \sin^2 j\alpha \sum_{j=1}^{n} \sin^2 j\alpha \sin 2j\alpha \ldots \sum_{j=1}^{n} \sin^2 j\alpha \sin nj\alpha \\ \ldots\ldots\ldots \ldots\ldots\ldots\ldots\ldots\ldots \ldots \ldots\ldots\ldots\ldots\ldots\ldots \end{bmatrix}$$

Utilizing now (6.4.16), for $k \neq \ell$ we deduce

$$2\sum_{j=1}^{n} j \sin kj\alpha \sin \ell j\alpha =$$

$$= \frac{1-(-1)^{k-l}}{4}\left[\frac{1}{\sin^2\dfrac{(k+\ell)\alpha}{2}} - \frac{1}{\sin^2\dfrac{(k-\ell)\alpha}{2}}\right] \tag{6.4.19}$$

and for $k = \ell$:

$$4\sum_{j=1}^{n} j \sin^2 kj\alpha = n(n+1)\,. \tag{6.4.20}$$

Analogously it results:

$$2\sum_{j=1}^{n} \sin jk\alpha \sin j\sigma = \sum_{j=1}^{n} \cos j(k\alpha-\sigma) - \sum_{j} \cos j(k\alpha+\sigma) =$$

$$= \mathrm{Re}\, \frac{z-z^{n+1}}{1-z}\bigg|_{z=\mathrm{e}^{\mathrm{i}(k\alpha-\sigma)}} - \mathrm{Re}\, \frac{z-z^{n+1}}{1-z}\bigg|_{z=\mathrm{e}^{\mathrm{i}(k\alpha+\sigma)}}$$

$$= \frac{(-1)^{k+1}\sin(n+1)\sigma \sin \sigma_k}{\cos\sigma - \cos\sigma_k} \tag{6.4.21}$$

If we utilize this identity, for the formula (6.4.8) which gives the solution of Prandtl's equation , we obtain the final form

$$C(\sigma) = \frac{1}{n+1}\sum_{k=1}^{n}(-1)^{k+1}c_k \frac{\sin(n+1)\sigma \sin k\alpha}{\cos\sigma - \cos k\alpha}\,. \tag{6.4.22}$$

In [A.23], p. 98–111 one gives a mathematical justification of this method. More precisely, one demonstrates that under certain circumstances, the iterative procedure that one utilizes for solving the system (6.4.10) is convergent and the solution (6.4.22) converges uniformly to the solution of Prandtl's equation.

6.4.2 The Quadrature Formulas Method

In [6.5] we gave a numerical method for solving Prandtl's equation by means of Gauss-type quadrature formulas. It is well known that these formulas give the best approximation. The key of this method consists in writing Prandtl's equation in the form (6.1.21). With the change of variables

$$y = bs, \quad \eta = bx \tag{6.4.23}$$

and, keeping the notations $C(s)$, $a(s)$ and $j(s)$ for $C(bs)$, $a(bs)$ and respectively $j(bs)$, the equation (6.1.21) becomes

$$\beta C(s) = \frac{a(s)}{2b} \, {}^{*}\!\!\int_{-1}^{+1} \frac{C(x)}{(x-s)^2} \mathrm{d}\,x + j(s), \tag{6.4.24}$$

and the conditions (6.1.7),

$$C(\pm 1) = 0. \tag{6.4.25}$$

The solution of the equation (6.4.24) has the form

$$C(s) = \sqrt{1-s^2}\, c(s). \tag{6.4.26}$$

Employing the quadrature formulas (F.3.5), the equation (6.4.24) reduces to the algebraic system

$$\sum_{i=1}^{n} A_{ki} c_i = j_k, \quad k = \overline{1,n}, \tag{6.4.27}$$

where we denoted

$$a_k = a(x_k), \quad c_k = c(x_k), \quad j_k = j(x_k),$$

$$A_{ki} = \frac{\pi a_k}{2b(n+1)} \left[1 - (-1)^{i+k}\right] \frac{x_i^2 - 1}{(x_i - x_k)^2}, \quad i \neq k, \tag{6.4.28}$$

$$A_{kk} = \beta\sqrt{1 - x_k^2} + \frac{\pi}{b}\frac{n+1}{4}; \quad x_i = \cos\frac{i\pi}{n+1}, i = \overline{1,n},$$

the unknowns being $c_1, \ldots, c_n$. The system has to be studied theoretically and solved numerically using a computer. After determining the unknowns, the lift, drag and moment coefficients, defined in (6.1.21) and (6.1.25), written in the form

$$c_L = \frac{2b}{A}\int_{-1}^{+1} \sqrt{1-s^2}\, c(s)\mathrm{d}\,s, \; c_D = -\frac{2b}{A}\int_{-1}^{+1} \sqrt{1-s^2}\, w(s)c(s)\,\mathrm{d}s,$$

$$c_x = \frac{2b^2}{mA}\int_{-1}^{+1} s\sqrt{1-s^2}\, c(s)\mathrm{d}\,s, \; c_z = -\frac{2b^2}{mA}\int_{-1}^{+1} s\sqrt{1-s^2}\, w(s)c(s)\,\mathrm{d}s, \tag{6.4.29}$$

give with the formula (F.2.12)

$$c_L = \frac{2\pi b}{(n+1)A}\sum_{i=1}^{n}(1-x_i^2)c_i, \quad c_D = -\frac{2\pi b}{(n+1)A}\sum_{i=1}^{n}(1-x_i^2)w_i c_i$$

$$c_x = \frac{2\pi b^2}{(n+1)mA}\sum_{i=1}^{n}(1-x_i^2)x_i c_i, \quad c_z \frac{2\pi b^2}{(n+1)mA}\sum_{i=1}^{n}(1-x_i^2)x_i w_i c_i\,, \tag{6.4.30}$$

where

$$w_i = \frac{\beta\sqrt{1-x_i^2}\;c_i - j_i}{2\pi a_i}\,. \tag{6.4.31}$$

In (6.4.30) we used, like in (6.1.21), the notations A for the area of the domain D and m for dimensionless length of the mean chord (in the direction of the unperturbed stream).

For verifying the method, we applied it in [6.5] to the elliptical flat wing, for which the exact solution is known (6.1.28), (6.1.30). With the notation (6.4.26) it results $c(s) = k$. Putting in (6.4.27)

$$A_{ki} = \sqrt{1-x_k^2}A'_{ki}, \quad c_i = kc'_i\,,$$

it results that the system

$$\sum_{i=1}^{n} A'_{ki}c'_i = \beta + \pi/(2b) \tag{6.4.32}$$

must have the solution $c'_1 = c'_2 = \ldots = c'_n = 1$. For $b = 10$ one obtains numerically $c'_1 = c'_2 = \ldots = c'_n = 1000$. We deduce therefore that the proposed method is very good.

In the sequel we shall give the numeric solution for the rectangular wing (in this case there exists no exact solution). Taking the reference length L_0 in the definition of the dimensionless variables (2.1.1), to coincide with the half of the chord, we deduce $x_\pm = \pm 1$. It results $a(y) = 1$ and $j(y) = 2\pi\varepsilon$. Putting $c_i = 2\pi\varepsilon c''$, the system (6.4.27) becomes

$$\sum_{i=1}^{n} A_{ki}c''_i = 1\,, \quad k = \overline{1n}\,, \tag{6.4.33}$$

where A_{ki} are given by (6.4.28) where we put $a_k = 1$. The quantities of interest in (6.4.30) are

$$k_1 = \frac{1}{n+1}\sum_{i=1}^{n}(1-x_i^2)c_i'', \quad k_2 = \frac{\beta}{n+1}\sum_{i=1}^{n}(1-x_i^2)^{3/2}(c_i'')^2,$$

$$k_3 = \frac{1}{n+1}\sum_{i=1}^{n}(1-x_i^2)x_i c_i'', \ k_4 = \frac{\beta}{n+1}\sum_{i=1}^{n}(1-x_i^2)^{3/2}x_i(c_i'')^2. \qquad (6.4.34)$$

One obtains

$$c_L = \pi^2\varepsilon k_1, \quad c_D = \pi^2\varepsilon^2(k_1 - k_2),$$
$$2c_x = \pi^2 b\varepsilon k_3, 2c_z = \pi^2 b\varepsilon^2(k_4 - k_3). \qquad (6.4.35)$$

For $b = 10$ we obtain the following values:

β	k_1	k_2	k_3	k_4
1	0.2347	0.0907	0	0
0.8	0.2544	0.0856	0	0

The result $c_x = c_z = 0$ is natural because of the symmetry of the wing. The lift and the drag increase because of the compressibility. This result is also natural. The value $2.307\,\varepsilon$ obtained here for c_L in the case of the incompressible fluid is smaller than the values $7.29\,\varepsilon, 5.28\,\varepsilon, \ldots,$ obtained with Glauert's method [1.12], but the values obtained with Glauert's method come closer to the values given here if $\lambda(= 2b/m)$ increases, i.e. the span is great with respect to the chord. Just in this situation the lifting line theory is valid.

We may think therefore that the method we have just exposed is at the same time very simple and very efficient.

6.4.3 The Collocation Method

The simplest numerical integration method is certainly *the collocation method* [6.6]. In case that Prandtl's equation has the form (6.1.19) and satisfies the conditions (6.1.7), the solution has the form

$$C(y) = \sqrt{b^2 - y^2}\, c(y). \qquad (6.4.36)$$

According to the collocation method, the segment $[-b, +b]$ is divided into N elements L_i and the function $c(y)$ is approximated on each element with its value c_i from the mid - point y_i^0 of the segment. So, the equation (6.1.19) gives

$$2\beta\sqrt{b^2-y^2}c(y) = a(y)\sum_{i=1}^{N} c_i \,{}^*\!\!\int_{L_i} \frac{\sqrt{b^2-\eta^2}}{(\eta-y)^2}\mathrm{d}\,\eta + 2j(y)\,, \qquad (6.4.37)$$

where, as we have already stated, $c_i = c(y_i^0)$. Imposing this equality to be satisfied in every point y_k^0, $k=1,\ldots,N$, one obtains

$$2\beta\sqrt{b^2-y_k^{0\,2}}\,c_k = a_k\sum_{i=1}^{N} A_{ki}c_i + 2j_k\,, \quad k=\overline{1N}\,, \qquad (6.4.38)$$

where

$$c_k = c(y_k^0)\,,\ a_k = a(y_k^0)\,,\ j_k = j(y_k^0)\,,$$

$$A_{ki} = \int_{y_i}^{y_{i+1}} \frac{\sqrt{b^2-\eta^2}}{(y-y_k^0)^2}\mathrm{d}\,\eta\,, \qquad (6.4.39)$$

y_i, y_{i+1} representing the extremities of the segment $L_i(y_1=-b, y_{N+1}= = b)$. So, (6.4.35) represents an algebraic linear system consisting of N equations with N unknowns c_i.

For calculating A_{ki}, we notice that for $e\notin[\alpha,\gamma]$ we have

$$I = \int_\alpha^\gamma \frac{\sqrt{b^2-\eta^2}}{(\eta-e)^2}\mathrm{d}\,\eta = \frac{\sqrt{b^2-\alpha^2}}{\alpha-e} - \frac{\sqrt{b^2-\gamma^2}}{\gamma-e} +$$

$$+\frac{e}{\sqrt{b^2-e^2}}\ln\left[\frac{(\gamma-e)^2+(\sqrt{b^2-e^2}+\sqrt{b^2-\gamma^2})^2}{(\alpha-e)^2+(\sqrt{b^2-e^2}+\sqrt{b^2-\alpha^2})^2}\,\frac{(\alpha-e)}{(\gamma-e)}\right]+$$

$$+\arcsin\frac{\alpha\sqrt{b^2-\gamma^2}-\gamma\sqrt{b^2-\alpha^2}}{b^2}\,. \qquad (6.4.40)$$

For $e\in(\alpha,\gamma)$ we shall write

$$I_1 \equiv \int_\alpha^\gamma \frac{\sqrt{b^2-\eta^2}}{(\eta-e)^2}\mathrm{d}\,\eta = {}^*\!\!\int_{-b}^{+b} \frac{\sqrt{b^2-\eta^2}}{(\eta-e)^2}\mathrm{d}\,\eta -$$

$$-\int_{-b}^{\alpha} \frac{\sqrt{b^2-\eta^2}}{(\eta-e)^2}\mathrm{d}\,\eta - \int_{\gamma}^{+b} \frac{\sqrt{b^2-\eta^2}}{(\eta-e)^2}\mathrm{d}\,\eta\,. \qquad (6.4.41)$$

We use the formula (6.4.37) for calculating the last two integrals. Since (D.3.9)

$${}^*\!\!\int_{-b}^{+b} \frac{\sqrt{b^2-\eta^2}}{(\eta-e)^2}\mathrm{d}\,\eta = -\pi\,, \qquad (6.4.42)$$

it results that I_1 is also determined. Utilizing these formulas, for $k \neq i$ we obtain:

$$A_{ki} = \frac{\sqrt{b^2 - y_i^2}}{y_i - y_k^0} - \frac{\sqrt{b^2 - y_{i+1}^2}}{y_{i+1} - y_k^0} + \frac{y_k^0}{\sqrt{b^2 - y_k^{02}}}\cdot$$

$$\cdot \ln \frac{(y_{i+1} - y_k^0)^2 + \left(\sqrt{b^2 - y_k^{02}} + \sqrt{b^2 - y_{i+1}^2}\right)^2}{(y_i - y_k^0)^2 + \left(\sqrt{b^2 - y_k^{02}} + \sqrt{b^2 - y_i^2}\right)^2} \frac{y_i - y_k^0}{y_{i+1} - y_k^0} +$$

$$+\arcsin\left(\frac{y_i\sqrt{b^2 - y_{i+1}^2} - y_{i+1}\sqrt{b^2 - y_i^2}}{b^2}\right), \tag{6.4.43}$$

and for $k = i$:

$$A_{ii} = -\pi - \frac{\sqrt{b^2 - y_i^2}}{\ell_i} - \frac{\sqrt{b^2 - y_{i+1}^2}}{\ell_i} +$$

$$+\arcsin\left(\frac{|y_i|\sqrt{b^2 - y_{i+1}^2} + |y_{i+1}|\sqrt{b^2 - y_i^2}}{b^2}\right) - \tag{6.4.44}$$

$$-\frac{y_i^0}{\sqrt{b^2 - y_i^{02}}} \ln \frac{\ell_i^2 + \left(\sqrt{b^2 - y_i^{02}} + \sqrt{b^2 - y_i^2}\right)^2}{\ell_i^2 + \left(\sqrt{b^2 - y_i^{02}} + \sqrt{b^2 - y_{i+1}^2}\right)^2},$$

where we denoted $2\ell_i = y_{i+1} - y_i$.

For testing the method one utilizes also the elliptic flat wing. For this wing the exact solution is (6.1.28) with k determined in(6.1.30). Comparing (6.4.33) with (6.1.28), it results $c = k/b$. For $\varepsilon = 0.1, \beta = 1$ (the incompressible fluid) $b = 10$ and $N = 50$, the numerical values obtained for c_i are situated between 0.050 and 0.053, and the value of k/b is 0.052. For $\varepsilon = 0.0872$, $\beta = 1$, $b = 20$ and $N = 50$ the values obtained for c_i are situated between 0.0243 and 0.0252 and the value of k/b is 0.0252. The results given by this method are more accurate when the angle of attack is small. The accuracy also increases when N increases.

Since the method is simple and can be easily applied one can accept the error of order 10^{-3} that it contains.

6.5 Various Extensions of the Lifting Line Theory

6.5.1 The Equation of Weissinger and Reissner

In the sequel we shall study new integral equations in order to improve Prandtl's model. In the past the researchers did not pay too much attention to these equations because the integrands contain both the unknown C and the derivative C'. Now we may use successfully these equations because some fundamental results concerning the Finite Part are known and, as we have mentioned in [6.5], after replacing the Principal Value of an integral which contains C' and has a Cauchy - type singular kernel by the Finite Part of an integral which contains C and has an hypersingular kernel, we may employ the numerical methods.

First we shall study the equation of Weissinger [6.30] and Reissner [5.29]. For obtaining it, we return to the lifting surface equation (5.1.28).

Taking (6.1.6), (6.1.11) and (6.1.12) into account, we deduce

$$\frac{\partial}{\partial y}\int\!\!\int_D \frac{f}{y_0}\mathrm{d}\,\xi\mathrm{d}\,\eta = \int\!\!\int_D \frac{f}{y_0^2}\mathrm{d}\,\xi\mathrm{d}\,\eta + 2\int_{-b}^{+b}\frac{C'(\eta)}{\eta-y}\mathrm{d}\,\eta\,. \qquad (6.5.1)$$

Utilizing (5.1.35), from (5.1.28) it results the first form of the integral equation

$$\frac{1}{2\pi}\int_{-b}^{+b}\frac{C'(\eta)}{\eta-y}\mathrm{d}\,\eta + \frac{1}{4\pi}\,{}^{*}\!\!\int\!\!\int_D \frac{f}{y_0^2}\left(1-\frac{x_0}{R}\right)\mathrm{d}\,\xi\mathrm{d}\,\eta = h_x(x,y)\,. \qquad (6.5.2)$$

Multiplying this equation by (6.1.15) and integrating it on the interval $[x_-(y), x_+(y)]$ one obtains

$$a\,(y)\int_{-b}^{+b}\frac{C'(\eta)\mathrm{d}\,\eta}{\eta-y}+$$

$$+\frac{1}{2\pi}\int_{-b}^{+b}\int_{x_-(y)}^{x_+(y)}\int_{x_-(\eta)}^{x_+(\eta)}\sqrt{\frac{x-x_-(y)}{x_+(y)-x}}\,\frac{f(\xi,\eta)}{y_0^2}\left(1-\frac{x_0}{R}\right)\mathrm{d}\,x\mathrm{d}\,\xi\mathrm{d}\,\eta = j(y)\,, \qquad (6.5.3)$$

$j(y)$ being (6.1.18).

Further we shall perform an approximation which consists in replacing $f(\xi, \eta)$ by

$$f(\xi, \eta) = -\frac{C(\eta)}{A(\eta)} \sqrt{\frac{x_+(\eta) - \xi}{\xi - x_-(\eta)}} \tag{6.5.4}$$

This is just the solution of the two-dimensional problem when h' is constant. Indeed, according to the formula (C.1.9), for a given y, the equation of the two-dimensional problem

$$\frac{1}{\pi} \int_{x_-(y)}^{x_+(y)} \frac{f(\xi, y)}{\xi - x} d\xi = h'_x(x, y), \tag{6.5.5}$$

becomes, in case that h' does not depend on x:

$$f(x, y) = -h'(y) \sqrt{\frac{x_+(y) - x}{x - x_-(y)}}. \tag{6.5.6}$$

Multiplying (6.5.5) by (6.1.15), integrating the result over $[x_-(y), x_+(y)]$ and utilizing the formulas (B.5.4) one obtains:

$$C(y) \equiv - \int_{x_-(y)}^{x_+(y)} f(\xi, y) \mathrm{d}\xi = \pi a(y) h'(y). \tag{6.5.7}$$

Eliminating h' from (6.5.6) and (6.5.7) one obtains (6.5.4). With this approximation, the equation (6.5.3) becomes:

$$a(y) {}' \!\!\int_{-b}^{+b} \frac{C'(\eta)}{y_0} \mathrm{d}\eta + \frac{1}{2\pi} {}^* \!\!\int_{-b}^{+b} \frac{C(\eta)}{y_0^2} \frac{N(y, \eta)}{A(\eta)} \mathrm{d}\eta = j(y), \tag{6.5.8}$$

where

$$N(y, \eta) = \int_{x_-(y)}^{x_+(y)} \int_{x_-(\eta)}^{x_+(\eta)} \left(1 - \frac{x_0}{R}\right) \sqrt{\frac{x_+(\eta) - \xi}{\xi - x_-(\eta)}} \sqrt{\frac{x - x_-(y)}{x_+(y) - x}} \mathrm{d}x \, \mathrm{d}\xi. \tag{6.5.9}$$

Using the notation

$$N_0(y, \eta) = \frac{N(y, \eta)}{A(y) A(\eta)} = 1 -$$

$$- \frac{1}{A(y) A(\eta)} \int_{x_-(y)}^{x_+(y)} \int_{x_-(\eta)}^{x_+(\eta)} \sqrt{\frac{x_+(\eta) - \xi}{\xi - x_-(\eta)}} \sqrt{\frac{x - x_-(y)}{x_+(y) - x}} \mathrm{d}x \, \mathrm{d}\xi, \tag{6.5.10}$$

we may write the equation (6.5.8) as follows:

$$\frac{1}{\pi}{}'\!\!\int_{-b}^{+b}\frac{C'(\eta)}{y_0}\mathrm{d}\,\eta+\frac{1}{2\pi}{}^*\!\!\int_{-b}^{+b}\frac{C(\eta)}{y_0^2}N_0(y,\eta)\mathrm{d}\,\eta=J(y)\,. \tag{6.5.11}$$

$A(y)$ and $J(y)$ being defined in (6.2.7). This equation has been deduced by Reissner [5.29]. Other simplifications are performed in the papers [5.24] [6.30] in order to obtain an approximate solution. The difficulty consists in the presence of both $C(\eta)$ and $C'(\eta)$ in the integrand (the equation is integro-differential). Employing, as we have already shown, the Finite Part, i.e. taking (6.1.12) into account, we may write the equation (6.5.11) as follows

$$\frac{1}{2\pi}{}^*\!\!\int_{-1}^{+b}\frac{C(\eta)}{y_0^2}N_1(y,\eta)\mathrm{d}\,\eta=J(y)\,, \tag{6.5.12}$$

where we denoted $N_1 = N_0 - 2$. This is the equation we are going to use in applications. This is an integral equation.

6.5.2 Weissinger's Equation. The Rectangular Wing

In the case of the wings for which the lifting surface equation cannot have the form (5.1.41) (this class is established in 5.1.7, and it contains the rectangular wing) one obtains an equation which may be easily utilized in applications. This equation was given by Weissinger [6.31], and received his name. Starting from the definition (6.1.6), we have:

$$\begin{aligned}C'(\eta)&=-x'_+(\eta)f(x_+(\eta),\eta)+x'_-(\eta)f(x_-(\eta),\eta)-\\&-\int_{x_-(\eta)}^{x_+(\eta)}\frac{\partial f}{\partial\eta}\mathrm{d}\,\xi\simeq-\int_{x_-(\eta)}^{x_+(\eta)}\frac{\partial f}{\partial\eta}\mathrm{d}\,\xi\,.\end{aligned} \tag{6.5.13}$$

We are going to show why the sum

$$-x'_+(\eta)f(x_+(\eta),\eta)+x'_-(\eta)f(x_-(\eta),\eta)$$

can be neglected. The first term vanishes by virtue of Kutta-Joukowski condition. The last term vanishes for a straight leading edge, perpendicular on the Ox axis (for example this is the case of the rectangular wing). It also can be neglected when the leading edge is slightly curved ($x'_-(\eta)\simeq 0$). In general we can neglect this term considering that we have prolonged the theoretic leading edge in front of the real leading edge and in this zone $f = 0$. With this approximation, the equation (5.1.41) may be written as follows:

$$\frac{1}{2\pi}{\int_{-b}^{+b}}' \frac{C'(\eta)}{y_0}\mathrm{d}\,\eta - \frac{1}{2\pi}{\iint_D}' \frac{\partial f}{\partial \eta}\frac{R}{x_0 y_0}\mathrm{d}\,\xi\mathrm{d}\,\eta = -2h'_x(x,y)\,. \qquad (6.5.14)$$

We make a second hypothesis, having in view to be satisfied in the case of the rectangular wing of width, let's say, $2a$, namely we substitute a^2 instead of x_0^2 from R. This approximation is justified by the fact that $|x-\xi|$ varies from 0 to $2a$, hence on the greatest part of the domain D we have $|x_0| = a$. So, the equation (6.5.14) becomes:

$$\frac{1}{2\pi}{\int_{-b}^{+b}}' \frac{C'(\eta)}{y_0}\mathrm{d}\,\eta - $$

$$-\frac{1}{2\pi}\int_{-b}^{+b}\int_{-a}^{+a} \frac{\partial f}{\partial \eta}\frac{\sqrt{a^2+\beta^2 y_0^2}}{x_0 y_0}\mathrm{d}\,\xi\mathrm{d}\,\eta = -2h'_x(x,y)\,. \qquad (6.5.15)$$

Multiplying by $\sqrt{(a+x)(a-x)}$ and integrating with respect to x on the interval $(-a,+a)$ one obtains

$$\frac{a}{2}{\int_{-b}^{+b}}' \frac{C'(\eta)}{y_0}\mathrm{d}\,\eta + \frac{1}{2}\int_{-b}^{+b} \frac{\sqrt{a^2+\beta^2 y_0^2}}{y_0}C'(\eta)\mathrm{d}\,\eta = \pi a j_0(y)\,, \qquad (6.5.16)$$

$$j_0(y) = -\frac{2}{a}\int_{-a}^{+a}\sqrt{\frac{a+x}{a-x}}\,h'_x(x,y)\mathrm{d}\,x\,. \qquad (6.5.17)$$

Introducing the non-singular kernel

$$K(y_0) = \frac{\sqrt{a^2+\beta^2 y_0^2}}{y_0} - \frac{a}{y_0}\,, \qquad (6.5.18)$$

the equation (6.5.4) may be written as follows

$$\frac{1}{\pi}{\int_{-b}^{+b}}' \frac{C'(\eta)}{y_0}\mathrm{d}\,\eta + \frac{1}{2a\pi}\int_{-b}^{+b} C'(\eta)K(y_0)\mathrm{d}\,\eta = j_0(y)\,. \qquad (6.5.19)$$

This is the definitive form of Weissinger's equation. It was created mainly for the rectangular wing. Before presenting the method of integration of this equation, we have to notice that it is not necessary to determine $C(\eta)$. It suffices to find $C'(\eta)$ because from (6.1.9), integrating by parts, it results

$$c_L = -\frac{2}{A}\int_{-b}^{+b} yC'(y)\mathrm{d}\,y\,,\quad c_x = -\frac{1}{Aa_0}\int_{-b}^{+b} y^2 C'(y)\mathrm{d}\,y\,. \qquad (6.5.20)$$

The solution of the equation (6.5.19) has the form

$$C'(y) = \frac{\varphi(y)}{\sqrt{b^2 - y^2}} \tag{6.5.21}$$

with φ satisfying the equation

$$\frac{1}{\pi} {\int_{-b}^{+b}}' \frac{\varphi(\eta)}{\sqrt{b^2-\eta^2}} \frac{\mathrm{d}\,\eta}{y_0} + \frac{1}{2\pi a} \int_{-b}^{+b} \frac{\varphi(\eta)}{\sqrt{b^2-\eta^2}} K(y_0)\mathrm{d}\,\eta = j_0(y) \,. \tag{6.5.22}$$

Performing the change of variables $\eta = b\eta'$, $y = by'$ one obtains

$$\frac{1}{\pi b} {\int_{-1}^{+1}}' \frac{\varphi(\eta)}{\sqrt{1-\eta^2}} \frac{\mathrm{d}\,\eta}{\eta - y} + \frac{1}{2\pi a} \int_{-1}^{+1} \frac{\varphi(\eta)}{\sqrt{1-\eta^2}} K_0(by_0)\mathrm{d}\,\eta = j_0(by) \,. \tag{6.5.23}$$

This is Weissinger's equation. It looks like the generalized equation of thin profiles (C.2.1). The numeric solution is determined by means of the formulas (F.2.5) and (F.2.6).

6.6 The Lifting Line Theory in Ground Effects

6.6.1 The Integral Equation

The lifting line theory in ground effects is obtained from 5.3, as well as Prandtl's theory is obtained from the lifting surface theory. We are not interested in finding the velocity and pressure fields. The integral equation is obtained from (5.3.13) utilizing the formulas (6.1.11)-(6.1.13) and the relation (6.1.8). One finds

$$\begin{aligned} &\frac{1}{2\pi} {\int_{-b}^{+b}}' \frac{C'(\eta)}{\eta - y} \mathrm{d}\,\eta + \frac{\beta}{\pi} \int_{x_-(y)}^{x_+(y)} \frac{f(\xi, y)}{x_0} \mathrm{d}\,\xi + \\ &+ \frac{1}{2\pi} \int_{-b}^{+b} C(\eta) N(x, y_0) \mathrm{d}\,\eta = 2h'_x(x, y) \,, \end{aligned} \tag{6.6.1}$$

where

$$\begin{aligned} N(x, y_0) &= \frac{d^2 \beta^2 x}{(d^2 + y_0^2) R_2^3} + \frac{d^2 - y_0^2}{(d^2 + y_0^2)^2} \left(1 + \frac{x}{R_2}\right) , \\ R_2 &= \sqrt{x^2 + e^2}\,, \quad e^2 = \beta^2 (y_0^2 + d^2) \,. \end{aligned} \tag{6.6.2}$$

Multiplying (6.6.1) by (6.1.15) and integrating the relation just obtained with respect to x on the interval $(x_-(\eta),\ x_+(\eta))$, we obtain, taking (B.5.4) into account:

$$2\beta C(y) + a(y) \int_{-b}^{+b} \frac{C'(\eta)}{y-\eta} \mathrm{d}\,\eta + \int_{-b}^{+b} C(\eta) N_0(y, y_0) \mathrm{d}\,\eta = 2j(y)\,, \quad (6.6.3)$$

where we denoted

$$N_0(y, y_0) = -\frac{1}{\pi} \int_{x_-(y)}^{x_+(y)} \sqrt{\frac{x - x_-(y)}{x_+(y) - x}} N(x, y_0) \mathrm{d}\,x\,, \quad (6.6.4)$$

j being (6.1.18).

The equation (6.6.3) is the lifting line equation in ground effects. It was given in [5.8]. It is obviously an integro-differential equation which generalizes Prandtl's equation (6.1.16) $(\lim N = 0)$.

Further we shall make some considerations concerning the kernel $N_0(y, y_0)$. Taking (5.3.11) into account and introducing the functions

$$I_\nu(y, y_0) = -\frac{1}{\pi} \int_{x_-(y)}^{x_+(y)} \sqrt{\frac{x - x_-(y)}{x_+(y) - x}} \frac{x}{(x^2 + e^2)^{\nu/2}} \mathrm{d}\,x,\ \nu = \overline{1,3}\,, \quad (6.6.5)$$

we obtain:

$$N_0(y, y_0) = \frac{d^2 - y_0^2}{(a^2 + y^2)^2} [I_1(y, y_0) - a(y)] + \frac{d^2 \beta^2}{d^2 + y_0^2} I_3(y, y_0)\,. \quad (6.6.6)$$

Performing the usual substitution

$$x = s(y) + a(y)t\,, \quad (6.6.7)$$

where

$$s(y) = \frac{x_+(y) + x_-(y)}{2}\,,\ a(y) = \frac{x_+(y) - x_-(y)}{2} \quad (6.6.8)$$

the integrals I_ν become

$$I_\nu = -\frac{1}{\pi} a(y) s(y) \int_{-1}^{+1} \sqrt{\frac{1+t}{1-t}} \frac{\mathrm{d}\,t}{\sqrt{P(t)^\nu}} - \frac{a^2(y)}{\pi} \int_{-1}^{+1} \sqrt{\frac{1+t}{1-t}} \frac{t \mathrm{d}\,t}{\sqrt{P(t)^\nu}}\,, \quad (6.6.9)$$

where

$$P(t) = a^2 t^2 + 2ast + e^2 + s^2\,. \quad (6.6.10)$$

It is well known that the best approximation of these integrals is given by the Gauss-type quadrature formulas (F.2.24). With this form (6.6.9) one may study the asymptotic behaviour of the kernel $N_0(y, y_0)$ given by (6.6.6) depending on the parameter $\lambda^2 = d^2/b^2$.

6.6.2 The Elliptical Flat Plate

Without considering the ground effects, the solution of this problem is (6.1.28), (6.1.30). In the sequel we shall determine the influence of the ground. We assume that the wing has the equation $x^2/\ell^2 + y^2/b^2 = 1$. We deduce

$$x_{\pm}(y) = \pm\frac{\ell}{b}\sqrt{b^2 - y^2}\,,\quad a(y) = \frac{\ell}{b}\sqrt{b^2 - y^2}\,,\quad s(y) = 0\,. \qquad (6.6.11)$$

Assuming that the span is much larger than the chord $(\ell << b)$, we shall neglect the terms of order $(\ell/b)^2$. From (6.6.9) we deduce $I_\nu = 0$, whence

$$N_0(y, y_0) = -a(y)\frac{d^2 - y^2}{(d^2 + y_0^2)^2} = a(y)\frac{\partial}{\partial\eta}\left(\frac{y_0}{d^2 + y_0^2}\right).$$

Integrating by parts, we deduce:

$$\int_{-b}^{+b} C(\eta)N_0(y, y_0)\mathrm{d}\,\eta = -a(y)\int_{-b}^{+b} C'(\eta)\frac{y_0}{d^2 + y_0^2}\mathrm{d}\,\eta\,,$$

and the integral equation (6.6.3) becomes:

$$2\beta C(y) + a(y)\int_{-b}^{+b}{}' \frac{C'(\eta)}{y_0}\mathrm{d}\,\eta - 2a(y)\int_{-b}^{+b} C'(\eta)\frac{y_0}{d^2 + y_0^2}\mathrm{d}\,\eta = 4\varepsilon\pi a(y)\,. \qquad (6.6.12)$$

For $d \to \infty$ one obtains Prandtl's equation. (6.1.16).

Taking into account the shape of $a(y)$ from (6.6.11), we shall look for solutions of (6.6.12) having the form

$$C(y) = k\sqrt{b^2 - y^2}\,. \qquad (6.6.13)$$

The first integral was calculated in (6.1.29). The second may be calculated with the substitution $y = b\overline{y}, \eta = b\overline{\eta}$. If $b << d$ we neglect the terms $O(b/d)^4$ and we obtain

$$\int_{-b}^{+b} C'(\eta)\frac{y_0}{d^2 + y_0^2}\mathrm{d}\,y = k\left(\frac{b}{d}\right)^2\frac{\pi}{2}\,.$$

We assumed that the span is much smaller than the distance to the ground. Replacing (6.6.13) in (6.6.12), we obtain the relation which determines k:

$$k\left[2\beta + \frac{\ell}{b}\pi - \frac{\ell}{b}\left(\frac{b}{a}\right)^2\pi\right] = 4\varepsilon\pi\frac{\ell}{b}\,. \qquad (6.6.14)$$

Denoting by k the value of the constant when there are no ground effects, i.e. (6.1.30), we deduce

$$k = k_0 k\,, \tag{6.6.15}$$

where

$$k_0 = \frac{2\beta + (\ell/b)\pi}{2\beta + (\ell/b)\pi - (\ell/b)(b/d)^2\pi}\,. \tag{6.6.16}$$

Obviously, $k_0 > 1$.

The lift, drag and moment coefficients are given by the formulas (6.1.21) and (6.1.25), where for $w(y) = w(0, y, 0)$ we have [5.8]:

$$w(y) = -\frac{1}{4\pi}\,{}'\!\!\int_{-b}^{+b} \frac{C'(\eta)}{y_0}\mathrm{d}\,\eta - \frac{1}{4\pi}\int_{-b}^{+b} C'(\eta)\frac{y_0}{d^2 + y_0^2}\mathrm{d}\,\eta\,. \tag{6.6.17}$$

These formulas (6.1.21), (6.1.25) and (6.6.17) are valid for every shape of the wing. For the elliptical flat plate wing we obtain:

$$w(y) = -\frac{k}{4}\left(1 + \frac{b^2}{2d^2}\right)\,, \tag{6.6.18}$$

such that

$$c_L = k_0 c_L^\infty\,,\quad c_D = k_1 c_D^\infty\,,\quad c_x = c_z = c\,, \tag{6.6.19}$$

with the notation

$$k_1 = k_0^2\left(1 + \frac{b^2}{2d^2}\right) > 1\,, \tag{6.6.20}$$

c_L^∞ and c_D^∞ representing the lift respectively drag coefficients in the absence of ground. Obviously, both the lift and the drag are increasing in the presence of the ground.

The coefficients k_0 and k_1 depend on $M, \ell/b, b/d$. The numerical calculations from [5.8] show that for the lift the increase is not significant but for the drag it is considerable. The ground effect is a decreasing function of d. For the same values of the ratios ℓ/b and b/d the influence coefficient k_0 is an increasing function of Mach's number M.

6.6.3 Numerical Solutions in the General Case

Utilizing the formula (6.1.12), one may write the equation (6.6.3) as follows:

$$2\beta C(y) - a(y)\,{}'\!\!\int_{-b}^{+b} \frac{C(\eta)}{(\eta - y)^2}\mathrm{d}\,\eta + \int_{-b}^{+b} C(\eta)N(y, y_0)\mathrm{d}\,\eta = 2j(y)\,. \tag{6.6.21}$$

As it is known, this is an integral equation, not an integro-differential one, but the singularity is stronger than in (6.6.3). In (6.6.3) we have a Cauchy - type singularity and in (6.6.21) we have to consider the Finite Part of a hypersingularity. But for this kind of equations there are available quadrature formulas. In order to apply this method, we have to perform the change of variables $y = by'$, $\eta = b\eta'$ for calculating the integrals on the interval $(-1, +1)$. We obtain

$$2b\beta C(y) - a(y) {}' \int_{-1}^{+1} \frac{C(\eta)}{(\eta - y)^2} \mathrm{d}\,\eta + b^2 \int_{-1}^{+1} C(\eta) N(y, y_0) \mathrm{d}\,\eta = 2bj(y)\,. \tag{6.6.22}$$

Since the solution of this equation has the form

$$C(y) = \sqrt{1 - y^2}\, c(y)\,, \tag{6.6.23}$$

we obtain

$$2b\beta\sqrt{1 - y_j^2}\, c_j - a_j {}^* \int_{-1}^{+1} \frac{\sqrt{1 - \eta^2} C(\eta)}{(\eta - y_j)^2} \mathrm{d}\,\eta +$$

$$+ b^2 \int_{-1}^{+1} \sqrt{1 - \eta^2} c(\eta) N(y_j, y_j - \eta) \mathrm{d}\,\eta = 2bj_j\,,\ y_j = \cos\frac{j\pi}{n+1}\,,\ j = \overline{1n}\,. \tag{6.6.24}$$

Using (F.3.5) and (F.2.12) one obtains the system:

$$A_j c_j + \sum_{k=1}^{n} A_{jk} c_k = 2bj_j\,,\ \ j = \overline{1n}\,, \tag{6.6.25}$$

where

$$A_j = 2b\beta\sqrt{1 - y_j^2} + a_j \pi \frac{n+1}{2}\,,$$

$$A_{jk} = \left[-a_j \frac{\pi}{n+1} \frac{1 - (-1)^{j+k}}{(y_k - y_j)^2} + \frac{b^2 k}{n+1} N(y_j, y_j - y_k) \right] (1 - y_k^2)\,. \tag{6.6.26}$$

In the first term from A_{jk} one excepts $k = j$. The system (6.6.25) is solved numerically.

6.7 The Curved Lifting Line

6.7.1 The Pressure and Velocity Fields

In this subsection, we shall pay a special attention to the aspect ratio $\lambda = (2b)^2/A$ introduced in (5.4.1). Usually, if λ is small, one applies the theory from 5.4 concerning the wings of low aspect ratio. If λ is large one applies the lifting line theory. These are the two asymptotic theories of the lifting surface theory.

As it is known, one of Prandtl's hypotheses consists in replacing the domain D by the segment $[-b, +b]$ taken along the span (the Oy axis). This hypothesis is plausible for the wings having the shape of an ellipse, triangle, trapezium or rhombus (see fig. 6.7.1) but it can be the source of great errors in the case of the wings having the shape of a swallow tail or the shape of an arrow. In the first case it is natural to replace the wing by the curvilinear median (see fig. 6.7.2), and in the second case one approximates the wing by *the median broken line* (fig. 6.7.3). For birds, the nature preferred the curvilinear median. These are enough reasons for studying in this subsection the curved lifting line. In

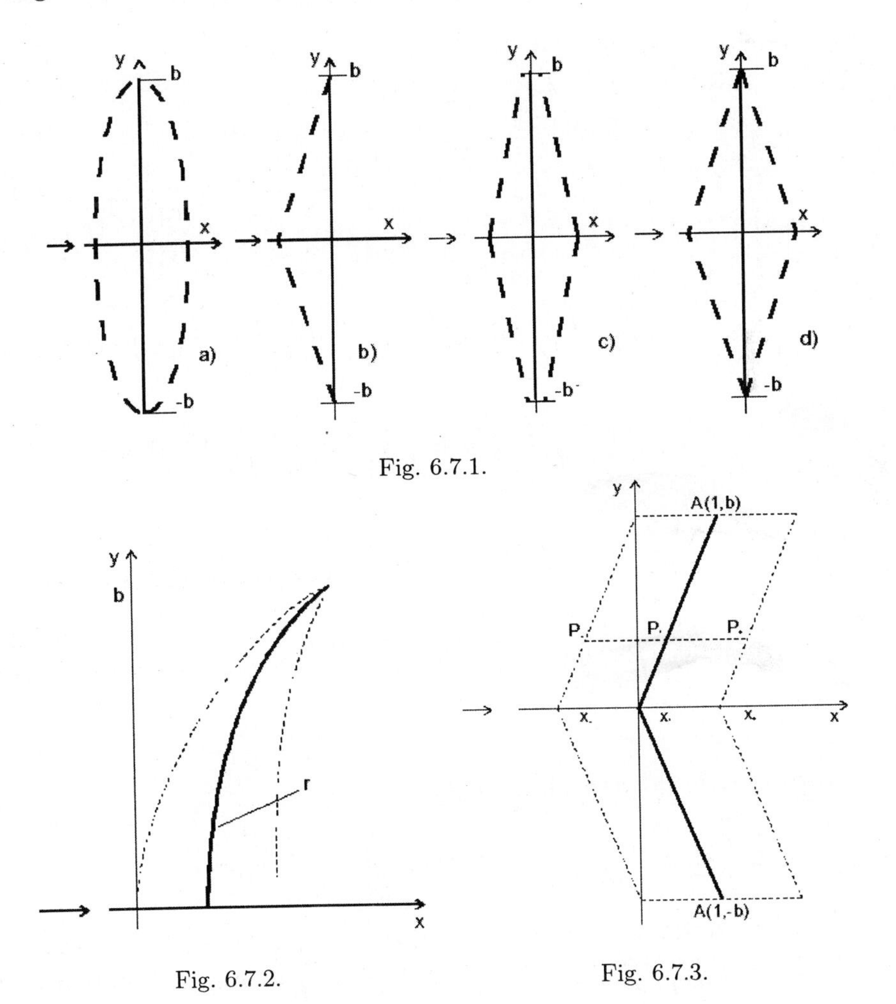

Fig. 6.7.1.

Fig. 6.7.2.

Fig. 6.7.3.

this case, one starts too from the general representation (5.1.8)–(5.1.12) and Prandtl's hypotheses. We assume therefore that the wing is without thickness ($h_1 = 0 \Longleftrightarrow f_1 = 0$) and that the unknown is $C(y)$ defined by (6.1.5), with the conditions (6.1.7). When the domain D reduces to the curved line Γ (fig. 6.7.2) having the equation $x = x_*(y)$, the formula (6.1.8) is replaced by

$$\lim \iint_D f(\xi,\eta)k(x,y,z,\xi,\eta)\mathrm{d}\,\xi\mathrm{d}\,y = -\int_{-b}^{+b} k(x,y,z,x_*(\eta),\eta)C(\eta)\mathrm{d}\,\eta\,, \tag{6.7.1}$$

for $x_-(\eta) \to x_*(\eta) \leftarrow x_+(\eta)$.

So, with the notation

$$R_* = \sqrt{[x - x_*(\eta)]^2 + \beta^2(y_0^2 + z^2)}\,, \tag{6.7.2}$$

the formulas (6.1.9) become

$$p(x,y,z) = -\frac{\beta^2}{4\pi}\int_{-b}^{+b} C(\eta)\frac{z}{R_*^3}\mathrm{d}\,\eta\,. \tag{6.7.3}$$

$$\begin{aligned}
v(x,y,z) &= -\frac{1}{4\pi}\lim_{x_\pm \to x_*}\iint_D f(\xi,\eta)\frac{\partial}{\partial y}\left[\frac{z}{y_0^2+z^2}\left(1+\frac{x_0}{R_1}\right)\right]\mathrm{d}\,\xi\mathrm{d}\,\eta = \\
&= \frac{1}{4\pi}\lim_{x_\pm \to x_*}\iint_D f(\xi,\eta)\left(1+\frac{x_0}{R_1}\right)\frac{\partial}{\partial\eta}\left(\frac{z}{y_0^2+z^2}\right)\mathrm{d}\,\xi\mathrm{d}\,\eta + \\
&+ \frac{1}{4\pi}\lim_{x_\pm \to x_*}\iint_D f(\xi,\eta)\frac{x_0 z}{y^2+z^2}\frac{\partial}{\partial\eta}\left(\frac{1}{R_1}\right)\mathrm{d}\,\xi\mathrm{d}\,\eta = \\
&= -\frac{1}{4\pi}\int_{-b}^{+b} C(\eta)\left[1+\frac{x-x_*(\eta)}{R_*}\right]\frac{\partial}{\partial\eta}\left(\frac{z}{y_0^2+z^2}\right)\mathrm{d}\,\eta - \\
&- \frac{1}{4\pi}\int_{-b}^{+b} C(\eta)\frac{[x-x_*(\eta)]z}{y_0^2+z^2}\cdot\frac{\beta^2 y_0}{R_*^3}\mathrm{d}\,\eta\,.
\end{aligned} \tag{6.7.4}$$

Utilizing now the identity

$$\begin{aligned}
&\left(1+\frac{x-x^*}{R_*}\right)\frac{\partial}{\partial\eta}\left(\frac{z}{y_0^2+z^2}\right) + \frac{z(x-x^*)}{y_0^2+z^2}\,\frac{\beta^2 y_0}{R_*^3} = \\
&= \frac{\partial}{\partial\eta}\left[\frac{z}{y_0^2+z^2}\left(1+\frac{x-x_*}{R_*}\right)\right] + \frac{\beta^2 z}{R_*^3}x_*'\,,
\end{aligned} \tag{6.7.5}$$

we obtain, after performing an integration by parts,

$$v(x,y,z) = \frac{1}{4\pi}\int_{-b}^{+b} C'(\eta)\frac{z}{y_0^2+z^2}\left[1+\frac{x-x_*(\eta)}{R_*}\right]\mathrm{d}\,\eta - \\ -\frac{\beta^2}{4\pi}\int_{-b}^{+b} C(\eta)\frac{zx'_*}{R_*^3}\mathrm{d}\,\eta\,. \tag{6.7.6}$$

Analogously,

$$w(x,y,z) = w_1 + w_2\,, \tag{6.7.7}$$

where

$$w_1 = \frac{\beta^2}{4\pi}\lim_{x_\pm\to x_*}\iint_D f(\xi,\eta)\frac{\partial}{\partial x}\left(\frac{1}{R_1}\right)\mathrm{d}\,\xi\mathrm{d}\,\eta = \\ = -\frac{\beta^2}{4\pi}\int_{-b}^{+b} C(\eta)\frac{x-x_*(\eta)}{R_*^3}\mathrm{d}\,\eta$$

$$w_2 = \frac{1}{4\pi}\lim_{x_\pm\to x_*}\iint_D f(\xi,\eta)\frac{\partial}{\partial y}\left[\frac{y_0}{y_0^2+z^2}\left(1+\frac{x_0}{R_1}\right)\right]\mathrm{d}\,\xi\mathrm{d}\,\eta = \\ = -\frac{1}{4\pi}\lim_{x_\pm\to x_*}\iint_D f(\xi,\eta)\left(1+\frac{x_0}{R_1}\right)\frac{\partial}{\partial\eta}\left(\frac{y_0}{y_0^2+z^2}\right)\mathrm{d}\,\xi\mathrm{d}\,\eta - \\ -\frac{1}{4\pi}\lim_{x_\pm\to x_*}\iint_D f(\xi,\eta)\frac{y_0x_0}{y'_0+z^2}\frac{\partial}{\partial\eta}\left(\frac{1}{R_1}\right)\mathrm{d}\,\xi\mathrm{d}\,\eta = \\ = \frac{1}{4\pi}\int_{-b}^{+b} C(\eta)\left[1+\frac{x-x_*(\eta)}{R_*}\right]\frac{\partial}{\partial\eta}\left(\frac{y_0}{y_0^2+z^2}\right)\mathrm{d}\,\eta + \\ +\frac{\beta^2}{4\pi}\int_{-b}^{+b} C(\eta)\frac{x-x_*(\eta)}{y_0^2+z^2}\,\frac{y_0}{R_*^3}\mathrm{d}\,\eta\,. \tag{6.7.8}$$

Introducing the identity

$$\left[1+\frac{x-x_*(\eta)}{R_*}\right]\frac{\partial}{\partial\eta}\left(\frac{y_0}{y_0^2+z^2}\right) + \frac{\beta^2y_0^2[x-x_*(\eta)]}{y_0^2+z^2}\,\frac{1}{R_*^3} = \\ = \frac{\partial}{\partial\eta}\left\{\frac{y_0}{y_0^2+z^2}\left[1+\frac{x-x_*(\eta)}{R_1}\right]\right\} + \frac{\beta^2y_0^2x'_*(\eta)}{R_*^3} \tag{6.7.9}$$

and, integrating by parts, we obtain w_2. In fact, it results:

$$w(x,y,z) = -\frac{\beta^2}{4\pi}\int_{-b}^{+b} C(\eta)\frac{x-x_*(\eta)}{R_*^3}\,\mathrm{d}\,\eta + \frac{\beta^2}{4\pi}\int_{-b}^{+b} C(\eta)\frac{y_0 x_*'(\eta)}{R_*^3}\,\mathrm{d}\,\eta - $$
$$-\frac{1}{4\pi}\int_{-b}^{+b} C'(\eta)\frac{y_0}{y_0^2+z^2}\left[1+\frac{x-x_*(\eta)}{R_*}\right]\mathrm{d}\,\eta\,. \tag{6.7.10}$$

6.7.2 The Integral Equation

We start from the lifting surface equation having the form (5.1.28). Utilizing (6.1.6) and (6.1.12), we deduce:

$$I_1 \equiv -\frac{1}{4\pi}\,{}^*\!\!\int_{-b}^{+b}\int_{x_-(\eta)}^{x_+(\eta)} \frac{f(\xi,\eta)}{y_0^2}\,\mathrm{d}\,\xi\mathrm{d}\,\eta =$$
$$= \frac{1}{4\pi}\,{}^*\!\!\int_{-b}^{+b}\frac{C(\eta)}{y_0^2}\,\mathrm{d}\,\eta = \frac{1}{4\pi}\,{}'\!\!\int_{-b}^{+b}\frac{C'(\eta)}{\eta-y}\,\mathrm{d}\,\eta\,. \tag{6.7.11}$$

Taking (6.2.1) into account, we obtain:

$$I_2 \equiv -\frac{1}{4\pi}\lim_{x_-\to x_*\leftarrow x_+}\,{}^*\!\!\iint_D \frac{f(\xi,\eta)}{y_0^2}\,\frac{x_0}{R}\,\mathrm{d}\,\xi\mathrm{d}\,\eta =$$
$$= \frac{1}{4\pi}\,{}^*\!\!\int_{-b}^{+b}\frac{C(\eta)}{y_0^2}\,\frac{x-x_*(\eta)}{R_*^0}\,\mathrm{d}\,\eta\,, \tag{6.7.12}$$

where

$$R_*^0 = \sqrt{[x-x_*(\eta)]^2+\beta^2 y_0^2}\,. \tag{6.7.13}$$

Utilizing the formula (D.3.7) and taking into account that $C(\pm b) = = 0$, we deduce

$$I_2 = \frac{1}{4\pi}\,{}'\!\!\int_{-b}^{+b}\frac{1}{\eta-y}\,\frac{\partial}{\partial\eta}\left[C(\eta)\frac{x-x_*(\eta)}{R_*^0}\right]\mathrm{d}\,\eta =$$
$$= \frac{1}{4\pi}\,{}'\!\!\int_{-b}^{+b}\frac{C'(\eta)}{\eta-y}\,\frac{x-x_*(\eta)}{R_*^0}\,\mathrm{d}\,\eta + I_3\,, \tag{6.7.14}$$

where

$$I_3 = \frac{1}{4\pi}\,{}' \!\!\int_{-b}^{+b} \frac{C(\eta)}{\eta - y}\frac{\partial}{\partial \eta}\left[\frac{x - x_*(\eta)}{R_*^0}\right] \mathrm{d}\,\eta$$
$$= -\frac{\beta^2}{4\pi}\int_{-b}^{+b} C(\eta)\frac{x - x_*(\eta) - y_0 x_*'(\eta)}{(R_*^0)^3}\mathrm{d}\,\eta\,. \tag{6.7.15}$$

From (5.1.28), (6.7.11), (6.7.12) and (6.7.14) we obtain the equation:

$$\frac{1}{4\pi}\,{}' \!\!\int_{-b}^{+b} \frac{C'(\eta)}{\eta - y}K(x, y, \eta)\mathrm{d}\,\eta + \frac{1}{4\pi}\int_{-b}^{+b} C(\eta)L(x, y, \eta)\mathrm{d}\,\eta = h_x'(x, y)\,, \tag{6.7.16}$$

where

$$K(x, y, \eta) = 1 + \frac{x - x_*(\eta)}{R_*^0}\,,$$
$$L(x, y, \eta) = -\beta^2\frac{x - x_*(\eta) - y_0 x_*'(\eta)}{(R_*^0)^3} \tag{6.7.17}$$

are non-singular kernels. The equation (6.7.10) was obtained in another way by Prössdorf and Tordella [6.22]. It is a singular integro-differential equation.

For the straight line $(x_*(\eta) \equiv 0)$ one deduces

$$\frac{1}{4\pi}\,{}' \!\!\int_{-b}^{+b} \frac{C'(\eta)}{\eta - y}\mathrm{d}\,\eta + \frac{1}{4\pi}\,{}' \!\!\int_{-b}^{+b} \frac{C'(\eta)}{\eta - y}\frac{x}{R}\mathrm{d}\,\eta - \frac{\beta^2}{4\pi}\int_{-b}^{+b} C(\eta)\frac{x}{R^3}\mathrm{d}\,\eta = h_x'\,, \tag{6.7.18}$$

where $R = \sqrt{x^2 + \beta^2 y_0^2}$. This equation is a first approximation of the lifting line equation. For deducing the equation (6.1.16) we had $x_0^2 << y_0^2$ on the greatest part of the domain D, such that we might consider $R_0 \simeq \beta|y_0|$. Here we cannot perform this approximation.

6.7.3 The Numerical Method

Using (D.3.7), one demonstrates the identity

$$^*\!\!\int_{-b}^{+b} \frac{C(\eta)}{(\eta - y)^2}K(x, y, \eta)\mathrm{d}\,\eta = {}' \!\!\int_{-b}^{+b} \frac{C'(\eta)}{\eta - y}K(x, y, \eta)\mathrm{d}\,\eta +$$
$$+ {}' \!\!\int_{-b}^{+b} \frac{C(\eta)}{\eta - y}\frac{\partial}{\partial \eta}K(x, y, y)\mathrm{d}\,\eta\,, \tag{6.7.19}$$

such that (6.7.16) becomes:

$$\frac{1}{\pi}{}^{*}\!\!\int_{-b}^{+b}\frac{C(\eta)}{(\eta-y)^2}K(x,y,\eta)\mathrm{d}\,\eta-\frac{1}{\pi}{}'\!\!\int_{-b}^{+b}\frac{C(\eta)}{\eta-y}\frac{\partial}{\partial\eta}K(x,y,y)\mathrm{d}\,\eta+$$
$$+\frac{1}{\pi}\int_{-b}^{+b}C(\eta)L(x,y,\eta)\mathrm{d}\,\eta=4h'_x(x,y)\,. \tag{6.7.20}$$

This is an integral equation (not an integro-differential one) but with a strong singularity, for which the Finite Part is considered.

We denote

$$M(x,y,y)=\frac{\partial}{\partial\eta}K(x,y,y) \tag{6.7.21}$$

and we perform the substitution $y=by', \eta=b\eta'$. The equation (6.2.14) becomes:

$$\frac{1}{\pi}{}^{*}\!\!\int_{-1}^{+1}\frac{C(eta)}{(\eta-y)^2}K(x,y,\eta)\mathrm{d}\,\eta-\frac{b}{\pi}{}'\!\!\int_{-1}^{+1}\frac{C(\eta)}{\eta-y}M(x,y,y)\mathrm{d}\,\eta+$$
$$+\frac{b^2}{\pi}\int_{-1}^{+1}C(\eta)L(x,y,\eta)\mathrm{d}\,\eta=4bh'_x(x,y)\,, \tag{6.7.22}$$

where

$$K(x,y,\eta)=1+\frac{x-x_*(\eta)}{[(x-x_*)^2+b^2\beta^2y_0^2]^{1/2}}\,,$$

$$L(x,y,\eta)=-\beta^2\frac{x-x_*(\eta)-by_0x'_*}{[(x-x_*)^2+b^2\beta^2y_0^2]^{3/2}} \tag{6.7.23}$$

$$M(x,y,y)=\frac{1}{b}\frac{\partial K}{\partial\eta}(x,y,y)\,.$$

Utilizing the *quadrature formulas* method, we shall take into account that the solution of the equation (6.7.22) has the form:

$$C(\eta)=\sqrt{1-\eta^2}\,c(\eta) \tag{6.7.24}$$

and we shall utilize the formulas (F.2.12), (F.3.4) and (F.3.5). Denoting

$$\eta_n=\cos\frac{k\pi}{n+1}\,,\quad k=\overline{1n}\,, \tag{6.7.25}$$

one obtains from (6.2.16) the algebraic system

$$A_jc_j+\sum_{k=1}^{n}{}'A_{jk}c_k=4bh_x(x,\eta_j)\,,\quad j=\overline{1n}\,, \tag{6.7.26}$$

where

$$A_j = -\frac{n+1}{2} K(x, \eta_j, \eta_i),$$

$$A_{jk} = \frac{(1-\eta_k^2)}{n+1} \left\{ \left[1-(-1)^{j+k}\right] \frac{K(x,\eta_j,\eta_k)}{(\eta_k-\eta_j)^2} + \right. \tag{6.7.27}$$

$$\left. +b\left[1-(-1)^{j+k}\right] \frac{M(x,\eta_j,\eta_k)}{\eta_k-\eta_j} + b^2 \delta_{jk} L(x,\eta_j,\eta_k) \right\}.$$

For writing explicitly this system we have to know the shape of the wing. For example, for the flat plate having the shape of an arrow with the angle of attack ε we adopt the broken line model (fig. 6.7.3). We have $h_x = -\varepsilon$ and utilizing the substitution $y = by'$,

$$x = x^*(y) = \begin{cases} y, & 0 < y < 1 \\ -y, & -1 < y < 0, \end{cases}$$

$$K(x,y,\eta) = \begin{cases} K_1 = 1 + \dfrac{x-\eta}{R}, & 0 < \eta < 1 \\ K_2 = 1 + \dfrac{x+\eta}{R}, & -1 < \eta < 0, \end{cases} \tag{6.7.28}$$

$$L(x,y,\eta) = \begin{cases} L_1 = -\beta^2 \dfrac{x-\eta-y_0}{R^3}, & 0 < \eta < 1 \\ L_2 = -\beta^2 \dfrac{x+\eta+y_0}{R^3}, & -1 < \eta < 0, \end{cases}$$

where

$$R = \left[(x-y)^2 + b^2\beta^2(y-\eta)^2\right]^{1/2}$$

The system (6.7.26) may be solved numerically using a computer.

Chapter 7

The Application of the Boundary Integral Equations Method to the Theory of the Three-Dimensional Airfoil in Subsonic Flow

7.1 The First Indirect Method (Sources Distributions)

7.1.1 The General Equations

The superiority of this method in comparison with the classical methods has been exposed in Chapter 4. Using this method we succeed to impose the non-linear boundary condition just on the boundary of the wing. Moreover, it allows to solve numerically the integral equation of the problem, approximating the boundary by a polygonal line in the two-dimensional case, or by a polyhedral surface (consisting of panels) in the three-dimensional case.

We deal with the problem considered everywhere in this book. A subsonic stream, having the velocity $U_\infty \boldsymbol{i}$, the pressure p_∞ and the density ρ_∞, is perturbed by the presence of a fixed body, having a known surface Σ. One requires to determine the perturbed flow and the action of the fluid against the body. Introducing the dimensionless variables X, Y, Z related by the dimensional variables x_1, y_1, z_1 as follows

$$(x_1, y_1, z_1) = L_0(X, Y, Z)$$

and putting

$$\boldsymbol{V}_1 = U_\infty(\boldsymbol{i} + \boldsymbol{V}), \quad P_1 = p_\infty + \rho_\infty U_\infty^2 P \tag{7.1.1}$$

we obtain the system for the perturbed fields:

$$M^2 \partial P/\partial X + \operatorname{Div} \boldsymbol{V} = 0, \quad \partial \boldsymbol{V}/\partial X + \operatorname{Grad} P = 0\,. \tag{7.1.2}$$

Projecting the last equation on the OX axis, we deduce:

$$P = -U \tag{7.1.3}$$

Taking into account this result, the first equation from (7.1.2) and the projections of the second on OY and OZ, give

$$\begin{aligned} &\beta^2 \partial U/\partial X + \partial V/\partial Y + \partial W/\partial Z = 0 \\ &\partial V/\partial X - \partial U/\partial Y = 0\,, \quad \partial W/\partial X - \partial U/\partial Z = 0\,, \end{aligned} \tag{7.1.4}$$

U, V, W representing the coordinates of the vector $\boldsymbol{V}$. Denoting by $F(X, Y, Z) = 0$ the equation of the boundary Σ, we have to impose the condition

$$(1+U)N_X + VN_Y + WN_Z = 0\,, \quad F = 0\,, \tag{7.1.5}$$

where

$$\boldsymbol{N} = \frac{\operatorname{Grad} F}{|\operatorname{Grad} F|}\,. \tag{7.1.6}$$

With the change of variables

$$\begin{aligned} &x = X, \;\; y = \beta Y, z = \beta Z \\ &u = \beta U, v = V, \;\; w = W\,, \end{aligned} \tag{7.1.7}$$

the system (7.1.4) becomes

$$\partial u/\partial x + \partial v/\partial y + \partial w/\partial z = 0 \tag{7.1.8}$$

$$\partial v/\partial x + \partial u/\partial y = 0, \quad \partial w/\partial x - \partial u/\partial z = 0\,. \tag{7.1.9}$$

Performing also a change of variables in F we have

$$\partial F/\partial X = \partial F/\partial x\,, \quad \partial F/\partial Y = \beta \partial F/\partial y \;\; \partial F/\partial Z = \beta \partial F/\partial z\,,$$

such that the boundary equation (7.1.5) becomes

$$un_x + \beta^2(vn_y + wn_z) = -\beta b_x, F = 0\,, \tag{7.1.10}$$

where

$$\boldsymbol{n} = \frac{\operatorname{grad} F}{|\operatorname{grad} F|}\,. \tag{7.1.11}$$

We agree to utilize the inward pointing normal to the body. We also impose the damping conditions at infinity

$$\lim_{\infty}(u, v, w) = 0\,. \tag{7.1.12}$$

The first equation from (7.1.9) represents a necessary and sufficient condition for the existence of a function $\varphi(x, y, z)$ such that

$$u = \partial\varphi/\partial x\,, \quad v = \partial\varphi/\partial y\,.$$

From the second equation and from the damping condition it results

$$w = \partial\varphi/\partial z\,.$$

With this representation, the equation (7.1.8) gives

$$\Delta\varphi = 0\,.$$

It is well known that the fundamental solution of this equation,

$$\varphi(x) = -\frac{f}{4\pi}\frac{1}{r}\,, \quad r = |\boldsymbol{x} - \boldsymbol{\xi}|\,, \tag{7.1.13}$$

represents the potential of the flow determined by a source of intensity f, having the position vector $\boldsymbol{\xi}$. The velocity field is

$$\boldsymbol{v} = \operatorname{grad}\varphi = \frac{f}{4\pi}\frac{\boldsymbol{x} - \boldsymbol{\xi}}{r}\,. \tag{7.1.14}$$

7.1.2 The Integral Equation

Replacing the body with a continuous distribution of sources on Σ, having the unknown intensity $f(\boldsymbol{x})$, the velocity field in fluid will be

$$v(\boldsymbol{\xi}) = -\frac{1}{4\pi}\int\int_{\Sigma} f(\boldsymbol{x})\frac{\boldsymbol{x} - \boldsymbol{\xi}}{|\boldsymbol{x} - \boldsymbol{\xi}|^3}\mathrm{d}\,a\,, \tag{7.1.15}$$

$\boldsymbol{\xi}$ representing the position vector of the generic point M in the fluid.

7.1.3 The Integral Equation

In order to impose the boundary condition (7.1.10) we have to pass to the limit in (7.1.15) considering that $M(\boldsymbol{\xi})$ tends to the generic point $Q_0(\boldsymbol{x}_0) \in \Sigma$. To the limit, the integral from (7.1.15) becomes singular. Following the procedure from the two - dimensional case (see 4.2.1), we shall prove that if $f(\boldsymbol{x})$ satisfies Hölder's condition on Σ, then

$$\begin{aligned}\boldsymbol{v}(\boldsymbol{x}_0) &= \lim_{\boldsymbol{\xi}\to\boldsymbol{x}_0}\left(-\frac{1}{4\pi}\int\int_{\Sigma} f(\boldsymbol{x})\frac{\boldsymbol{x} - \boldsymbol{\xi}}{|\boldsymbol{x} - \boldsymbol{\xi}|^3}\mathrm{d}\,a\right) = \\ &= -\frac{1}{2}f(\boldsymbol{x}_0)\boldsymbol{n}_0 - \frac{1}{4\pi}{\int\int_{\Sigma}}' f(\boldsymbol{x})\frac{\boldsymbol{x} - \boldsymbol{x}_0}{|\boldsymbol{x} - \boldsymbol{x}_0|}\mathrm{d}\,a\,,\end{aligned} \tag{7.1.16}$$

where

$$\iint_{\Sigma}' = \lim_{\varepsilon \to 0} \iint_{\Sigma - \sigma}, \tag{7.1.17}$$

σ representing the surface cut from Σ by a sphere Σ_ε having the center in Q_0 and the radius ε.

Indeed, writing

$$\iint_{\Sigma} = \iint_{\Sigma - \sigma + \sigma}, \tag{7.1.18}$$

we have to calculate

$$L = \lim_{\boldsymbol{\xi} \to \boldsymbol{x}_0} \iint_{\sigma} f(\boldsymbol{x}) \frac{\boldsymbol{x} - \boldsymbol{\xi}}{|\boldsymbol{x} - \boldsymbol{\xi}|^3} \mathrm{d}\, a\,,$$

i.e. the last term from (7.1.18). Writing this term as follows

$$L = \lim_{\boldsymbol{\xi} \to \boldsymbol{x}_0} \iint_{\sigma} [f(\boldsymbol{x}) - f(\boldsymbol{x}_0)] \frac{\boldsymbol{x} - \boldsymbol{\xi}}{|\boldsymbol{x} - \boldsymbol{\xi}|^3} \mathrm{d}\, a + f(\boldsymbol{x}_0) L_0\,,$$

where

$$L_0 = \lim_{\boldsymbol{\xi} \to \boldsymbol{x}_0} \iint_{\sigma} f(\boldsymbol{x}) \frac{\boldsymbol{x} - \boldsymbol{\xi}}{|\boldsymbol{x} - \boldsymbol{\xi}|^3} \mathrm{d}\, a\,, \tag{7.1.19}$$

we notice that the first integral from the expression of L tends to zero when $\varepsilon \to 0$ because f satisfies Hölder's condition. For calculating L_0 we shall replace σ by A, the projection of the surface σ on the plane Π which is tangent to Σ in Q_0 (fig. 7.1.1).

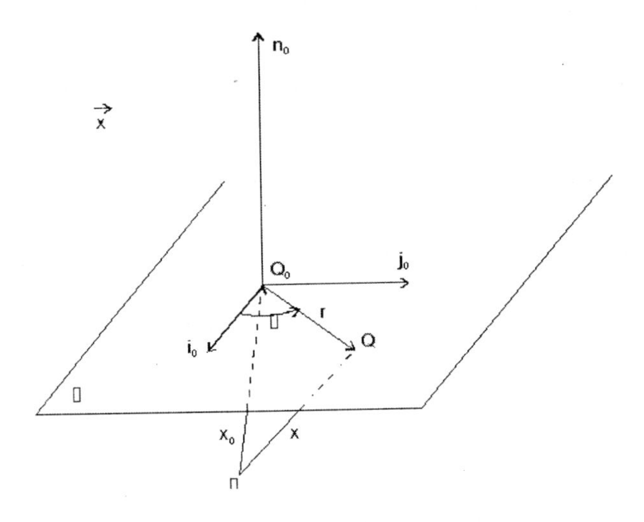

Fig. 7.1.1.

On this projection we shall use the parametrization

$$\boldsymbol{x} = \boldsymbol{x}_0 + r(\cos\theta \boldsymbol{i}_0 + \sin\theta \boldsymbol{j}_0), \quad 0 \le r \le \varepsilon, \ 0 \le \theta < 2\pi\,, \tag{7.1.20}$$

$\boldsymbol{i}_0$ and $\boldsymbol{j}_0$ being versors orthogonal to the plane π.

Also, taking into account that the limit is the same on every path on which $\boldsymbol{\xi} \to \boldsymbol{x}_0$, we consider the limit on the direction of the inward normal $\boldsymbol{n}_0 \equiv \boldsymbol{n}(\boldsymbol{x}_0)$ to Σ in Q_0. We have therefore

$$\boldsymbol{\xi} = \boldsymbol{x}_0 - \eta \boldsymbol{n}_0\,, \quad \eta > 0\,. \tag{7.1.21}$$

Hence

$$L_0 = \lim_{\eta \to 0} \int_0^{\varepsilon} \int_0^{2\pi} \frac{r(\cos\theta \boldsymbol{i}_0 + \sin\theta \boldsymbol{j}_0) + \eta \boldsymbol{n}_0}{(r^2 + \eta^2)^{3/2}} r \,\mathrm{d}\, r \,\mathrm{d}\,\theta = 2\pi n\,. \tag{7.1.22}$$

Now the formula (7.1.16) is demonstrated.

Imposing the condition (7.1.10), we deduce the following integral equation

$$\{n_x^2(\boldsymbol{x}_0) + \beta^2[n_y^2(\boldsymbol{x}_0) + n_z^2(\boldsymbol{x}_0)]\} f(\boldsymbol{x}_0) +$$
$$+ \frac{1}{2\pi} \iint_{\Sigma}' z f(\boldsymbol{x}) \frac{(x - x_0)n_x^0 + \beta^2[(y - y_0)n_y^0 + (z - z_0)n_z^0]}{|\boldsymbol{x} - \boldsymbol{x}_0|^3} \,\mathrm{d}\, a = 2\beta n_x^0\,. \tag{7.1.23}$$

For the incompressible fluid it becomes

$$Lf \equiv f(\boldsymbol{x}_0) + \frac{1}{2\pi} \iint_{\Sigma}' f(\boldsymbol{x}) \frac{(\boldsymbol{x} - \boldsymbol{x}_0) \cdot \boldsymbol{n}^0}{|\boldsymbol{x} - \boldsymbol{x}_0|^3} \,\mathrm{d}\, a = 2n_x^0\,. \tag{7.1.24}$$

The integrals (7.1.23) and (7.1.24) are singular.

7.1.4 The Discretization of the Integral Equation

We shall approximate the surface Σ of the body by a set of triangular panels $T_j\,(j = 1, \ldots, N)$ and we shall approximate on every panel T_j, the function f by the value f_j of the function in the center of mass $G_j(x_j^0)$ of the triangle. The equation (7.1.23) reduces to

$$\{(n_x^0)^2 + \beta^2[(\boldsymbol{n}_y^0)^2 + (\boldsymbol{n}_z^0)^2]\} f(\boldsymbol{x}_0) +$$
$$+ \frac{1}{2\pi} \iint_{\Sigma}' z f(\boldsymbol{x}) \frac{(x - x_0)n_x^0 + \beta^2[(y - y_0)n_y^0 + (z - z_0)n_z^0]}{|\boldsymbol{x} - \boldsymbol{x}_0|^3} \,\mathrm{d}\, a = 2\beta n_x^0\,.$$

Imposing this equation to be satisfied in the centers of mass G_i, i.e. putting $\boldsymbol{x}_0 = \boldsymbol{x}_i^0 (i = 1, \ldots, N)$, we obtain the linear algebraic system

$$a_i f_i + \sum_{j=1}^{N} A_{ij} f_j = b_i \,, \quad i = 1, \ldots, N \,, \tag{7.1.25}$$

where we have no summation with respect to i. We denoted

$$a_i = n_x^2(\boldsymbol{x}_i^0) + \beta^2 n_y^2(\boldsymbol{x}_i^0) + \beta^2 n_z^2(\boldsymbol{x}_i^0), \;\; b_i = 2\beta n_x(\boldsymbol{x}_i^0) \tag{7.1.26}$$

$$A_{ij} = X_{ij} n_x(\boldsymbol{x}_i^0) + \beta^2 Y_{ij} n_y(\boldsymbol{x}_i^0) + \beta^2 Z_{ij} n_z(\boldsymbol{x}_i^0) \tag{7.1.27}$$

$$\boldsymbol{X}_{ij} = \frac{1}{2\pi} \int\int_{T_j} \frac{\boldsymbol{x} - \boldsymbol{x}_i^0}{|\boldsymbol{x} - \boldsymbol{x}_i^0|^3} \mathrm{d}\, a \,. \tag{7.1.28}$$

For determining the quantities $n(\boldsymbol{x}_i^0)$ and X_{ij} we shall denote by $\boldsymbol{x}_{i1}$, $\boldsymbol{x}_{i2}$ and $\boldsymbol{x}_{i3}$ the vectors of position of the vertices of the triangle T_i ; we choose the sense on the sides of the triangle such that the normal $\boldsymbol{n}(\boldsymbol{x}_i^0)$ is positively oriented towards the interior of the body. Taking into account the definition of the vector product, we obviously have

$$\boldsymbol{n}(\boldsymbol{x}_i^0) = \frac{(\boldsymbol{x}_{i2} - \boldsymbol{x}_{i1}) \times (\boldsymbol{x}_{i3} - \boldsymbol{x}_{i1})}{2S_i} \,, \tag{7.1.29}$$

S_i being the area of the triangle T_i expressed by means of the coordinates of the vectors $\boldsymbol{x}_{i1}$, $\boldsymbol{x}_{i2}$, $\boldsymbol{x}_{i3}$.

The integrals X_{ij} are singular when $i = j$. We consider at first the case $i \neq j$. Denoting by $\boldsymbol{x}_{j1}$, $\boldsymbol{x}_{j2}$, $\boldsymbol{x}_{j3}$ the vectors of position of the vertices of the triangle T_{ij}, we shall consider the parametrization of the triangle

$$\boldsymbol{x} = \boldsymbol{x}_{j1} + (\boldsymbol{x}_{j2} - \boldsymbol{x}_{j1})\lambda_1 + (\boldsymbol{x}_{j3} - \boldsymbol{x}_{j1})\lambda_2 \,. \tag{7.1.30}$$

Further we shall proceed like in 5.3. Introducing the polar coordinates by means of the formulas

$$\lambda_1 = r\cos\theta \,, \quad \lambda_2 = r\sin\theta \quad 0 \leq \theta < \pi/2 \,, \; 0 < r \leq \rho \,, \tag{7.1.31}$$

where ρ is defined (fig. 7.1.2) by the relation

$$(\cos\theta + \sin\theta)\rho = 1 \tag{7.1.32}$$

and denoting

$$\boldsymbol{e}(\theta) = (\boldsymbol{x}_{j2} - \boldsymbol{x}_{j1})\cos\theta + (\boldsymbol{x}_{j3} - \boldsymbol{x}_{j1})\sin\theta \,, \tag{7.1.33}$$

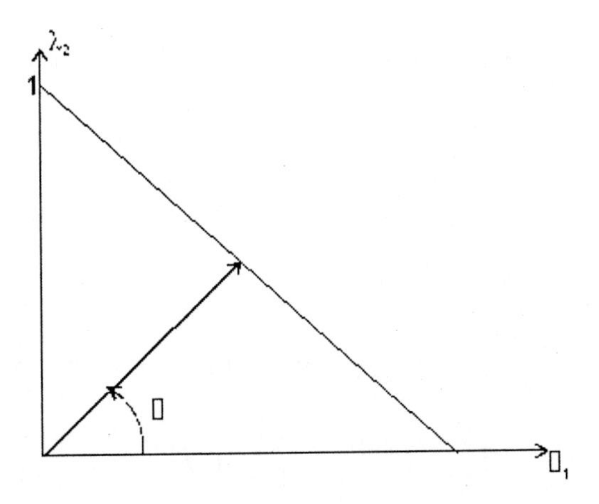

Fig. 7.1.2.

we deduce

$$\begin{aligned} \boldsymbol{x} - \boldsymbol{x}_i^0 &= \boldsymbol{x}_{j1} - \boldsymbol{x}_i^0 + r\boldsymbol{e}(\theta) \\ |\boldsymbol{x} - \boldsymbol{x}_i^0|^2 &= ar^2 + br + c\,, \end{aligned} \tag{7.1.34}$$

where

$$\begin{aligned} &a = |\boldsymbol{e}|^2,\ b = 2(\boldsymbol{x}_{j1} - \boldsymbol{x}_i^0)\cdot \boldsymbol{e}\,,\ c = |\boldsymbol{x}_{j1} - \boldsymbol{x}_i^0|^2 \\ &\delta \equiv b^2 - 4ac = -4[\boldsymbol{x}_{ji} - \boldsymbol{x}_i^0)e_y - (\boldsymbol{y}_{j1} - \boldsymbol{y}_i^0)e_x]^2 < 0\,. \end{aligned} \tag{7.1.35}$$

For the element of area of the triangle T_j one obtains

$$\mathrm{d}\,a = 2S_j \mathrm{d}\,\lambda_1\,\mathrm{d}\,\lambda_2 = 2rS_j \mathrm{d}\,r\mathrm{d}\,\theta\,, \tag{7.1.36}$$

S_j being the area of the triangle. Hence,

$$\boldsymbol{X}_{ij} = \frac{S_j}{\pi}\int_0^{\pi/2}[(\boldsymbol{x}_{j1} - \boldsymbol{x}_i^0)I_1(\theta) + \boldsymbol{e}(\theta)I_2(\theta)]\mathrm{d}\,\theta\,, \tag{7.1.37}$$

where

$$\begin{aligned} I_1(\theta) &= \int_0^{\rho}\frac{r\mathrm{d}r}{(ar^2 + br + c)^{3/2}} = \frac{2b\rho + 4c}{\delta(a\rho^2 + b\rho + c)^{1/2}} - \frac{4\sqrt{c}}{\delta} \\ aI_2(\theta) &= \int_0^{\rho}\frac{ar^2\mathrm{d}r}{(ar^2 + br + c)^{3/2}} = J_1 - bI_1 - ck\,, \end{aligned} \tag{7.1.38}$$

with the notations

$$J = \int_0^{\rho}\frac{\mathrm{d}r}{ar^2 + br + c} = \frac{2}{\sqrt{-\delta}}\left(\arctan\frac{2a\rho + b}{\sqrt{-\delta}} - \arctan\frac{b}{\sqrt{-\delta}}\right)$$

$$K = \int_0^\rho \frac{\mathrm{d}r}{(ar^2+br+c)^{3/2}} = -\frac{4a\rho+2b}{\delta(a\rho^2+b\rho+c)^{1/2}} + \frac{2b}{\delta\sqrt{c}}\,.$$

Ones employ the formulas (7.1.29) and (7.1.37) for calculating the coefficients A_{ij}.

7.1.5 The Singular Integrals

The integrals (7.1.28) are singular when $i=j$. Following the model from 5.3 we shall write:

$$T_j = T_j^{(12)} + T_j^{(23)} + T_j^{(31)}\,, \tag{7.1.39}$$

where $T_j^{(kl)}$ is the triangle $G_j P_k P_l$ and we shall consider the parametrization of the triangle $T_j^{(12)}$ putting

$$\boldsymbol{x} - \boldsymbol{x}_j^0 = (\boldsymbol{x}_{jl} - \boldsymbol{x}_j^0)\lambda_1 + (\boldsymbol{x}_{j2} - \boldsymbol{x}_j^0)\lambda_2\,. \tag{7.1.40}$$

Passing to polar coordinates and denoting

$$\boldsymbol{E}_{12} = (\boldsymbol{x}_{j1} - \boldsymbol{x}_j^0)\cos\theta + (\boldsymbol{x}_{j2} - \boldsymbol{x}_j^0)\sin\theta\,, \tag{7.1.41}$$

we deduce

$$\boldsymbol{x} - \boldsymbol{x}_j^0 = r\boldsymbol{E}_{12}, \quad |\boldsymbol{x} - \boldsymbol{x}_j^0| = r|\boldsymbol{E}_{12}|\,. \tag{7.1.42}$$

Utilizing also (D.2.3) it results

$$\begin{aligned} \boldsymbol{X}_{jj}^{(12)} &= \frac{1}{2\pi}\int\!\!\int_{T_j^{(12)}} \frac{\boldsymbol{x} - \boldsymbol{x}_j^0}{|\boldsymbol{x} - \boldsymbol{x}_j^0|^3}\,\mathrm{d}\,a = \\ &= \frac{1}{3\pi}S_j \int\!\!\int_0^{\pi/2} \frac{\boldsymbol{E}_{12}(\theta)\ln\rho(\theta)}{|\boldsymbol{E}_{12}|^3}\,\mathrm{d}\,\theta \end{aligned} \tag{7.1.43}$$

$$\boldsymbol{X}_{jj} = \boldsymbol{X}_{jj}^{(12)} + \boldsymbol{X}_{jj}^{(23)} + \boldsymbol{X}_{jj}^{(31)}\,. \tag{7.1.44}$$

These expressions are utilized for determining the coefficients A_{jj}.

7.1.6 The Velocity Field. The Validation of the Method

The numerical values of the velocity field are obtained from (7.1.16) with the formula

$$2\boldsymbol{v}(\boldsymbol{x}_i^0) = -f(\boldsymbol{x}_i^0)\boldsymbol{n}(\boldsymbol{x}_i^0) - \sum_{j=1}^{N} \boldsymbol{X}_{ij} f_j\,. \tag{7.1.45}$$

For testing the method we shall use the exact solution in the case of the sphere placed in an uniform incompressible stream. We know (see, for example, [1.11], p.163) that if the sphere has the radius a and the center in the origin of the coordinate axes, and the uniform stream has the velocity $U_\infty \boldsymbol{k}$, then the potential of the perturbed flow is

$$\phi = U_\infty z \left(1 + \frac{1}{2}\frac{a^3}{r^3}\right) . \tag{7.1.46}$$

Calculating $\boldsymbol{V}_1 = \text{grad}\,\phi$ and $\boldsymbol{V}_1 = U_\infty(\boldsymbol{i}+ \boldsymbol{v})$ which results from (7.1.1) under the hypothesis that the fluid is incompressible $(\beta = 1)$ and using for v the values (7.1.45), it results the comparison from figure 7.1.3 for $|\boldsymbol{V}_1|/U_\infty$. We notice that the approximate method and

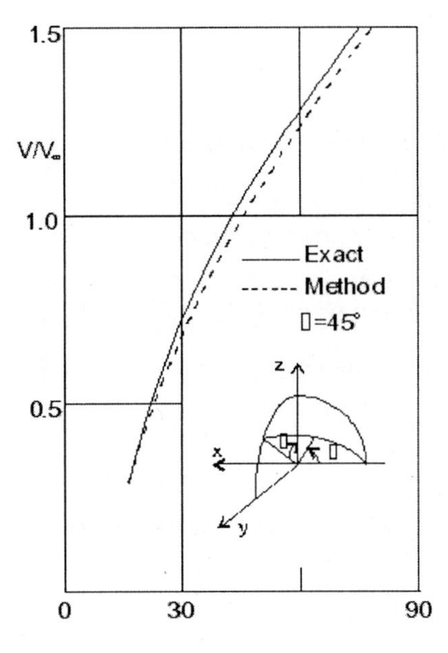

Fig. 7.1.3.

the exact one give very closed results.

In the paper [7.2], that we have utilized for writing this subsection, we may find the approximate results for the ellipsoid for various angles of attack and for various Mach numbers. We may also find approximate results for the wings whose cross section is a NACA - 64 - A - 008 profile.

7.1.7 The Incompressible Fluid. An Exact Solution

We have seen that, for the incompressible fluid the integral equation is (7.1.24). In this subsection we shall determine the exact solution of the equation in case that the perturbing body is a sphere [7.4].

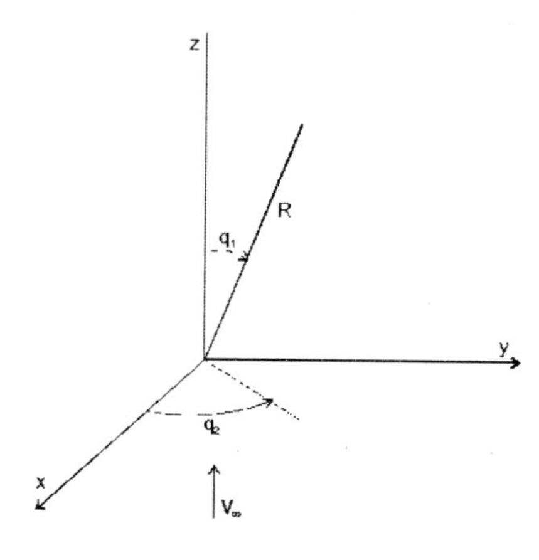

Fig. 7.1.4.

Considering that the points on the sphere ($\boldsymbol{x}$ and $\boldsymbol{Q}_0(\boldsymbol{x}_0)$ have the coordinates (fig. 7.1.4):

$$\begin{aligned} \boldsymbol{x} &= R(\sin q, \cos q_2, \sin q_1 \sin q_2, \cos q_1) \\ 0 &< q_1 < \pi, \quad 0 \leq q_2 < 2\pi, \\ \boldsymbol{x}_0 &= R(\sin q_1^0 \cos q_2^0, \sin q_1^0 \sin q_2^0, \cos q_1^0)\,, \end{aligned} \tag{7.1.47}$$

we deduce for $\boldsymbol{n}^0$ (the inward pointing normal to the sphere in Q_0)

$$\boldsymbol{n}^0 = -(\sin q_1^0 \cos q_2^0, \sin q_1^0 \sin q_2^0, \cos q_1^0)\,. \tag{7.1.48}$$

We have therefore:

$$\begin{aligned} & f(\boldsymbol{x}) \to f(q_1, q_2), \quad f(\boldsymbol{x}_0) \to f(q_1^0, q_2^0) \\ & \mathrm{d}a = R^2 \sin q_1 \mathrm{d}q_1 \mathrm{d}q_2, \quad n_x^0 \to n_z^0 = -\cos q_1^0 \end{aligned} \tag{7.1.49}$$

and

$$|\boldsymbol{x} - \boldsymbol{x}_0|^2 = |\boldsymbol{x}|^2 + |\boldsymbol{x}_0|^2 - 2\boldsymbol{x}_0 \cdot \boldsymbol{x}_0 = 2R^2(1 - \cos\theta)\,, \tag{7.1.50}$$

where

$$\cos\theta = \sin q_1 \sin q_1^0 - \cos(q_2 - q_2^0) + \cos q_1 \cos q_1^0\,. \tag{7.1.51}$$

Hence it results

$$(\boldsymbol{x} - \boldsymbol{x}_0) \cdot (\boldsymbol{n}^0) - (\boldsymbol{x} - \boldsymbol{x}_0) \cdot \frac{\boldsymbol{x}_0}{R} = R(1 - \cos\theta)\,,$$

$$\frac{(\boldsymbol{x} - \boldsymbol{x}_0) \cdot \boldsymbol{n}^0}{|\boldsymbol{x} - \boldsymbol{x}_0|^3} \mathrm{d}\,a = \frac{(1 - \cos\theta) \sin q_1 \mathrm{d}q_1 \mathrm{d}q_2}{2(1 - \cos\theta)\sqrt{2(1 - \cos\theta)}} = \frac{\sin q_1 \mathrm{d}q_1 \mathrm{d}q_2}{2\sqrt{1 - \cos\theta)}}\,. \tag{7.1.52}$$

With these formulas the equation (7.1.24) becomes

$$f(q_1^0, q_2^0) + \frac{1}{4\pi} {\int_0^\pi}' \int_0^{2\pi} f(q_1, q_2) \frac{\sin q_1 \mathrm{d}\, q_1 \mathrm{d}\, q_2}{\sqrt{2(1 \cos\theta)}} = -2\cos q_1^0 \,. \qquad (7.1.53)$$

the sign "$'$" indicating that one eliminates the vicinity of the point Q_0, i.e. $\theta = 0$ according to the definition (7.1.17).

Denoting

$$K(q_1^0, q_2^0, q_1, q_2) = \frac{1}{4\pi\sqrt{2(1-\cos\theta)}}\,,$$
$$D = (0, \pi) \times (0, 2\pi)\,, \qquad (7.1.54)$$

we deduce that the integral operator of the equation which determines the density $f(q_1, q_2)$, is a bounded operator with respect to the uniform convergence norm for real functions, continuous on the closed rectangle $\overline{D}$. This boundedness follows from the property

$$\int\int_D K \sin q_1 \mathrm{d}\, q_1 \mathrm{d}\, q_2 = 1\,. \qquad (7.1.55)$$

We shall also prove that the integral operator has the invariance property

$$\int\int (\cos q_1) K \sin q_1 \mathrm{d}\, q_1 \mathrm{d}\, q_2 = \frac{1}{3}\cos q_1^0\,. \qquad (7.1.56)$$

For proving these properties we shall perform the change of variables $(q_1, q_2) \to (\theta, \lambda)$ defined by (7.1.51) and by the following relation, which is a direct consequence of the sine rule from the spheric trigonometry

$$\sin\lambda \sin\theta = \sin q_1 \sin(q_2^0 - q_2), \quad (\theta, \lambda) \in D\,. \qquad (7.1.57)$$

Hence we choose the spherical coordinates relative to the point $\boldsymbol{x}$ for which, instead of the Oz axis we take the direction $-\boldsymbol{n}_0$, and instead of the xOz plane we take the plane of the versors $\boldsymbol{k}$ and $\boldsymbol{n}^0$ (fig. 7.1.5). In this way, for the element of area in the generic point Q one obtains

$$\mathrm{d}\, a = R^2 \sin q_1 \mathrm{d}q_1 \mathrm{d}q_2 = R^2 \sin\theta \mathrm{d}\theta \mathrm{d}\lambda\,. \qquad (7.1.58)$$

Using the cosine theorem from the spherical trigonometry, we obtain

$$\cos q_1 = \cos q_1^0 \cos\theta + \sin q_1^0 \sin\theta \cos\lambda\,. \qquad (7.1.59)$$

Hence,

$$\int\int_D K \sin q_1 \mathrm{d}\, q_1 \mathrm{d}\, q_2 = \int\int_D K \sin\theta \mathrm{d}\,\theta \mathrm{d}\,\lambda =$$

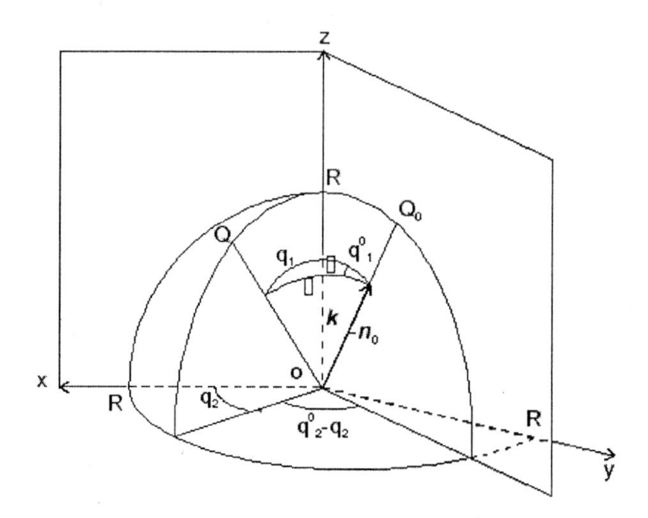

Fig. 7.1.5.

$$= \frac{1}{4\pi}\int_0^{2\pi} \mathrm{d}\,\lambda \int_0^{\pi} \frac{\sin\theta \mathrm{d}\,\theta}{\sqrt{2(1-\cos\theta)}} = 1\,,$$

$$\int\int_D (\cos q_1) K \sin q_1 \mathrm{d}q_1 \mathrm{d}q_2 = \int\int_D (\cos q_1) K \sin\theta \mathrm{d}\,\theta \mathrm{d}\,\lambda =$$

$$= (\cos q_1^0) \int\int_D \cos\theta K \sin\theta \mathrm{d}\theta \mathrm{d}\lambda +$$

$$+ (\sin q_1^0)\left(\int_0^{2\pi} \cos\lambda \mathrm{d}\,\lambda\right)\int_0^{\pi} (\sin\theta) K \sin\theta \mathrm{d}\,\theta =$$

$$= \frac{1}{2}(\cos q_1^0)\int_0^{\pi} \frac{\sin\theta\cos\theta}{\sqrt{2(1-\cos\theta)}} \mathrm{d}\,\theta = \frac{1}{3}\cos q_1^0$$

Utilizing the identity (7.1.56), we may solve the equation (7.1.53) by means of the successive approximations method. Indeed, putting

$$f_1 = -2\cos q_1^0\,,$$

we deduce

$$f_2 = -\int\int_D f_1(q_1, q_2) K(q_1^0, q_2^0, q_1, q_2) \sin q_1 \mathrm{d}q_1 \mathrm{d}q_2 =$$

$$= -\frac{1}{3}(-2\cos q_1^0)$$

$$f_3 = -\int\int_D f_2(q_1, q_2) K \sin q_1 \mathrm{d}\,q_1 \mathrm{d}\,q_2 = \left(\frac{1}{3}\right)^2 (-2\cos q_1^0)$$

. . . .

$$f_{k+1} = -\int\int_D f_k(q_1, q_2) K \sin q_1 \mathrm{d}\, q_1 \mathrm{d}\, q_2 = \left(\frac{1}{3}\right)^k (-2\cos q_1^0)$$

whence

$$\begin{aligned} f(q_1, q_2) &= \sum_{k=0}^{\infty} f_{k+1}(q_1, q_2) = \\ &= (-2\cos q_1) \sum_{k=0}^{\infty} \left(-\frac{1}{3}\right)^k = -\frac{3}{2}\cos q_1 \,. \end{aligned} \tag{7.1.60}$$

This is the exact solution for the spherical obstacle.

7.1.8 The Expression of the Potential

For testing the integral equation, we must prove that the potential calculated with the density (7.1.60) coincides with the exact potential given by (7.1.46). The potential in the generic point $M(\boldsymbol{\xi})$ determined by a source having the intensity f placed in $Q(\boldsymbol{x})$ is according to (7.1.13),

$$\varphi(\boldsymbol{\xi}) = -\frac{f}{4\pi} \frac{1}{|\boldsymbol{x} - \boldsymbol{\xi}|} \,,$$

whence it results that the potential determined by a continuous distribution, of intensity $f(\boldsymbol{x})$, on the surface Σ, will have the expression

$$\varphi(\boldsymbol{\xi}) = -\frac{1}{4\pi} \int\int_\Sigma \frac{f(\boldsymbol{x})}{|\boldsymbol{x} - \boldsymbol{\xi}|} \mathrm{d}\, a \,. \tag{7.1.61}$$

In the case of the sphere $\Sigma = S(O, R)$, with the density (7.1.60), the potential is

$$\varphi(\boldsymbol{\xi}) = \frac{3}{8\pi} \int\int_\Sigma \frac{\cos q_1}{|\boldsymbol{x} - \boldsymbol{\xi}|} \mathrm{d}\, a \,. \tag{7.1.62}$$

Considering the point $M(\boldsymbol{\xi})$ exterior to the sphere, having the coordinates

$$\boldsymbol{\xi} = r(\sin q_1^0 \cos q_2^0, \sin q_1^0 \sin q_2^0 \cos q_1^0), r > R \,, \tag{7.1.63}$$

we get

$$|\boldsymbol{x} - \boldsymbol{\xi}|^2 = R^2 - 2rR\cos\theta + r^2$$

where $\cos\theta$ is defined by (7.1.51). Taking (7.1.58) into account, it results

$$\varphi(\boldsymbol{\xi}) = \frac{3}{8\pi}R^2 \int\int_D \frac{\cos q_1 \sin q_1 \mathrm{d}\, q_1 \mathrm{d}\, q_2}{\sqrt{R^2 - 2rR\cos\theta + r^2}} =$$

$$= \frac{3}{8\pi}R^2 \int\int_D \frac{\cos q_1 \sin\theta \mathrm{d}\,\theta \mathrm{d}\,\lambda}{\sqrt{R^2 - 2rR\cos\theta + r^2}}\,.$$

On the basis of the formula (7.1.59) and the periodicity of $\cos\lambda$, we obtain

$$\varphi(\boldsymbol{\xi}) = \frac{3}{8\pi}R^2(\cos q_1^0)\int_0^{2\pi}\mathrm{d}\,\lambda\int_0^{\pi}\frac{\cos\theta\sin\theta\mathrm{d}\,\theta}{\sqrt{R^2 - 2rR\cos\theta + r^2}} =$$

$$= \frac{3}{8}R\frac{\cos q_1^0}{r}\int_0^{\pi}\Big[(R^2 + r^2)(R^2 - 2rR\cos\theta + r^2)^{1/2} -$$

$$-(R^2 - 2rR\cos\theta + r^2)^{1/2}\Big]\sin\theta\mathrm{d}\,\theta =$$

$$= \frac{3}{16}\frac{\xi_3}{r^3}\Big[2(R^2 + r^2)(r + R - |r - R|) - \frac{2}{3}(r + R)^3 - |r - R|^3\Big]$$

and finally, since $r > R$,

$$\varphi(\xi) = \frac{1}{2}\frac{R^3}{r^3}\xi_3\,. \tag{7.1.64}$$

This is also the perturbation potential given by (7.1.46).

7.2 The Second Indirect Method (Doublet Distributions). The Incompressible Fluid

7.2.1 The Integral Equation

The theory in this subsection follows the paper [1.11]. We denote by $\varphi_0(\boldsymbol{x})$ the potential of a known flow in the entire space. We assume that this flow is perturbed by the presence of a body whose support is the simple connected and bounded domain D. We assume that the boundary of this domain, Σ is smooth, such that we may apply Poisson's formula. We denote by $\boldsymbol{n}$ the outward pointing normal to Σ. The potential φ_0 is a harmonic function, such that we may write [1.11] (D.3.2)

$$-\frac{1}{2\pi}\int\int_{\Sigma}\left[\varphi_0(\boldsymbol{x})\frac{\partial}{\partial n}\frac{1}{|\boldsymbol{x}-\boldsymbol{\xi}|}-\right.$$
$$\left.-\frac{1}{|\boldsymbol{x}-\boldsymbol{\xi}|}\frac{\partial\varphi_0}{\partial n}(\boldsymbol{x})\right]\mathrm{d}\,a=\begin{cases}2\varphi_0(\boldsymbol{\xi}), & \boldsymbol{\xi}\in D\\ \varphi_0(\boldsymbol{\xi}), & \boldsymbol{\xi}\in\Sigma\\ 0 & \boldsymbol{\xi}\in E\,,\end{cases} \tag{7.2.1}$$

E representing the exterior of $\overline{D}$. When we write $\boldsymbol{\xi}\in D$ we understand that the point M whose vector of position is $\boldsymbol{\xi}$ belongs to D.

Now we assume that the perturbation is produced by the body having the support D and let $\varphi(\boldsymbol{x})$ be the potential of the perturbation in E. According to the equation of continuity (7.1.8), φ will be a harmonic function which has to satisfy to infinity the damping condition

$$\lim_{|\boldsymbol{\xi}|\to\infty}\varphi(\boldsymbol{\xi})=0\,.$$

Under these conditions, φ will have the representation (D.3.10) from [1.11]

$$\frac{1}{2\pi}\int\int_{\Sigma}\left[\varphi(\boldsymbol{x})\frac{\partial}{\partial n}\frac{1}{|\boldsymbol{x}-\boldsymbol{\xi}|}-\right.$$
$$\left.-\frac{1}{|\boldsymbol{x}-\boldsymbol{\xi}|}\frac{\partial}{\partial n}\varphi(\boldsymbol{x})\right]\mathrm{d}\,a=\begin{cases}2\varphi(\boldsymbol{\xi}), & \boldsymbol{\xi}\in E\\ \varphi(\boldsymbol{\xi}), & \boldsymbol{\xi}=\boldsymbol{x}_0\in\Sigma\\ 0 & \boldsymbol{\xi}\in D\,,\end{cases} \tag{7.2.2}$$

the normal being the same like in (7.2.1).

Adding $(7.2.1)_3$ to $(7.2.2)_1$ we obtain

$$\varphi(\boldsymbol{\xi}) = \frac{1}{4\pi}\int\int_{\Sigma}\{[\varphi_0(\boldsymbol{x}) + \varphi(\boldsymbol{x})]\frac{\partial}{\partial n}\frac{1}{|\boldsymbol{x}-\boldsymbol{\xi}|} - \\ -\frac{1}{|\boldsymbol{x}-\boldsymbol{\xi}|}\frac{\partial}{\partial n}(\varphi_0+\varphi)(\boldsymbol{x})\}\mathrm{d}\,a\,, \quad \xi \in E\,. \tag{7.2.3}$$

Determining φ in order to have

$$\frac{\partial}{\partial n}\varphi(\boldsymbol{x}) = -\frac{\partial}{\partial n}\varphi_0(\boldsymbol{x}), \quad \boldsymbol{x}\in\Sigma\,, \tag{7.2.4}$$

the representation (7.2.3) becomes

$$\varphi(\boldsymbol{\xi}) = \frac{1}{4\pi}\int\int_{\Sigma}[\varphi_0(\boldsymbol{x}) + \varphi(\boldsymbol{x})]\frac{\partial}{\partial n}\left(\frac{1}{|\boldsymbol{x}-\boldsymbol{\xi}|}\right)\mathrm{d}\,a\,. \tag{7.2.5}$$

Subtracting $(7.2.1)_2$ from $(7.2.2)_2$ one obtains

$$\varphi(\boldsymbol{x}_0) - \varphi_0(\boldsymbol{x}_0) = \frac{1}{2\pi}\int\int_{\Sigma}[\varphi_0(\boldsymbol{x}) + \varphi(\boldsymbol{x})]\frac{\partial}{\partial n}\left(\frac{1}{|\boldsymbol{x}-\boldsymbol{x}_0|}\right)\mathrm{d}\,a, \boldsymbol{x}_0\in\Sigma\,. \tag{7.2.6}$$

Introducing the function $\mu:\Sigma\to R$ by means of the formula

$$\mu(\boldsymbol{x}) = \varphi_0(\boldsymbol{x}) + \varphi(\boldsymbol{x}), \quad (\forall)\boldsymbol{x}\in\Sigma\,, \tag{7.2.7}$$

the equation (7.2.6) reduces to

$$\mu(\boldsymbol{x}_0) - \frac{1}{2\pi}\int\int_{\Sigma}\mu(\boldsymbol{x})\frac{\partial}{\partial n}\left(\frac{1}{|\boldsymbol{x}-\boldsymbol{x}_0|}\right)\mathrm{d}\,a = \varphi_0(\boldsymbol{x}_0), (\forall)\boldsymbol{x}_0\in\Sigma\,, \tag{7.2.8}$$

and (7.2.5) to

$$\varphi(\boldsymbol{\xi}) = \frac{1}{4\pi}\int\int_{\Sigma}\mu(\boldsymbol{\xi})\frac{\partial}{\partial n}\left(\frac{1}{|\boldsymbol{x}-\boldsymbol{\xi}|}\right)\mathrm{d}\,a\,. \tag{7.2.9}$$

The equation (7.2.8) is the integral equation which determines the function μ, and the equation (7.2.9) shows that this function is just the doublets density Σ which replaces the body. We may write the equation (7.2.8) as follows

$$\mu(\boldsymbol{x}_0) + \frac{1}{2\pi}\int\int_{\Sigma}\mu(\boldsymbol{x})\boldsymbol{n}(\boldsymbol{x})\cdot\frac{\boldsymbol{x}-\boldsymbol{x}_0}{|\boldsymbol{x}-\boldsymbol{x}_0|^3}\mathrm{d}\,a = 2\varphi(\boldsymbol{x}_0), (\forall)\boldsymbol{x}_0\in\Sigma\,, \tag{7.2.10}$$

$\boldsymbol{n}$ representing the normal pointing towards the fluid (the outward normal to Σ). This equation which determines the doublets density, is similar to the equation (7.2.21) which determines the sources density. Only the free terms and the normals from the kernels are different. In (7.1.24) it appears $\boldsymbol{n}(\boldsymbol{x}_0)$ while in (7.2.10) this is replaced by $\boldsymbol{n}(\boldsymbol{x})$.

7.2.2 The Flow past the Sphere. The Exact Solution

Considering the problem from (7.1.7), i.e. the integral equation in the case of the uniform flow with the velocity $U_\infty \boldsymbol{k}$ past the sphere $S(O, R)$ and utilizing the formulas (7.1.47) we have:

$$\varphi_0 = z = R\cos q_1^0 \tag{7.2.11}$$

such that the equation (7.2.10) becomes

$$\mu(q_1^0, q_2^0) + \frac{1}{4\pi}{}' \int_0^\pi \int_0^{2\pi} \mu(q_1, q_2) \frac{\sin q_1 \mathrm{d}\, q_1 \mathrm{d}\, q_2}{\sqrt{2(1-\cos\theta)}} = 2R\cos q_1^0 \,. \tag{7.2.12}$$

This equation differs from (7.1.53) only by the right hand side term. The equation (7.2.12) has therefore the exact solution

$$\mu = \frac{3}{2} R\cos q_1 \,. \tag{7.2.13}$$

For testing the integral equation (7.2.10), we shall prove that the potential φ, obtained from (7.2.9) with the density (7.2.13), coincides with the exact perturbation potential (7.1.64). Indeed, (7.2.9) becomes

$$\varphi(\boldsymbol{\xi}) = -\frac{1}{4\pi} \int \int_\Sigma \mu(\boldsymbol{x}) \frac{\boldsymbol{x} - \boldsymbol{\xi}}{|\boldsymbol{x} - \boldsymbol{\xi}|^3} \mathrm{d}\, a \,. \tag{7.2.14}$$

Using the notations (7.1.47) and (7.1.63) we have

$$\boldsymbol{x} \cdot \boldsymbol{\xi} = Rr\cos\theta, \quad (\boldsymbol{x} - \boldsymbol{\xi}) \cdot \boldsymbol{n} = R - r\cos\theta$$

$$da = R^2 \sin q_1 \mathrm{d}q_1 \mathrm{d}q_2 \,,$$

such that, utilizing (7.2.13), we deduce

$$\varphi = -\frac{1}{4\pi}\frac{3}{2}R^3 \int_0^\pi \int_0^{2\pi} \frac{(R - r\cos\theta)\cos q_1 \sin q_1}{(R^2 - 2Rr\cos\theta + r^2)^{3/2}} \mathrm{d}\, q_1 \mathrm{d}\, q_2 \,.$$

Passing to the variables θ and λ and taking the formulas (7.1.58), (7.1.59) and the periodicity of the function $\cos\lambda$ into account, it results

$$\varphi = -\frac{3}{4}R^3 \cos q_1^0 \int_0^\pi \frac{\sin\theta\cos\theta(R - r\cos\theta)}{(R^2 - 2Rr\cos\theta + r^2)^{3/2}} \mathrm{d}\, \theta \,.$$

Performing the change of variable $\theta \to u$:

$$\mu = (R^2 - 2Rr\cos\theta + r^2)^{1/2} \,,$$

we deduce

$$\int_0^\pi \frac{\sin\theta\cos\theta(R - r\cos\theta)}{(R^2 - 2Rr\cos\theta + r^2)^{3/2}} \mathrm{d}\, \theta = -\frac{2}{3r^2}$$

and finally, (7.1.64).

7.2.3 The Velocity Field

Since the potential has the expression (7.2.14) we deduce:

$$\boldsymbol{v}(\boldsymbol{\xi}) = \operatorname{grad}_{\boldsymbol{\xi}}\varphi = -\frac{1}{4\pi}\int\int_{\Sigma}\mu(\boldsymbol{x})\operatorname{grad}_{\boldsymbol{\xi}}\left[\frac{\boldsymbol{x}-\boldsymbol{\xi}}{|\boldsymbol{x}-\boldsymbol{\xi}|^3}\cdot\boldsymbol{n}(\boldsymbol{x})\right]\mathrm{d}\,a =$$

$$= -\frac{1}{4\pi}\int\int_{\Sigma}\mu(\boldsymbol{x})(\boldsymbol{n}\cdot\operatorname{grad}_{\boldsymbol{\xi}})\left(\frac{\boldsymbol{x}-\boldsymbol{\xi}}{|\boldsymbol{x}-\boldsymbol{\xi}|^3}\right)\mathrm{d}\,a$$

and finally

$$\boldsymbol{v}(\boldsymbol{\xi}) = \frac{1}{4\pi}\int\int_{\Sigma}\mu(\boldsymbol{x})\left[\boldsymbol{n}(\boldsymbol{x}) - 3(\boldsymbol{x}-\boldsymbol{\xi})\frac{(\boldsymbol{x}-\boldsymbol{\xi})\cdot\boldsymbol{n}}{|\boldsymbol{x}-\boldsymbol{\xi}|^2}\right]\frac{\mathrm{d}\,a}{|\boldsymbol{x}-\boldsymbol{\xi}|^3}\,. \quad (7.2.15)$$

This formula gives the velocity field in the fluid. We deduce that far away the kernel has the order $|\boldsymbol{\xi}|^{-3}$, as it is natural to be (see for example (6.1.16) from [1.11]). On the boundary of the body, the integrals from (7.2.15) have strong singularities. We have therefore to transform these integrals.

7.2.4 The Velocity Field on the Body. N. Marcov's Formula

In the beginning we have to notice that the double layer potential of constant density equal to 1 defines a piecewise constant function whose gradient is obviously zero in the continuity points. Indeed, from (7.2.1) it results:

$$\frac{1}{4\pi}\int\int_{\Sigma}\frac{\partial}{\partial n}\left(\frac{1}{|\boldsymbol{x}-\boldsymbol{\xi}|}\right)\mathrm{d}\,a = \begin{cases} 1, & \boldsymbol{\xi}\in D \\ 1/2, & \boldsymbol{\xi}\in\Sigma \\ 0 & \boldsymbol{\xi}\in E\,, \end{cases} \quad (7.2.16)$$

and for the gradient we have

$$\frac{1}{4\pi}\int\int_{\Sigma}\left[\boldsymbol{n}(\boldsymbol{x}) - 3(\boldsymbol{x}-\boldsymbol{\xi})\frac{(\boldsymbol{x}-\boldsymbol{\xi})\cdot\boldsymbol{n}}{|\boldsymbol{x}-\boldsymbol{\xi}|^2}\right]\frac{\mathrm{d}\,a}{|\boldsymbol{x}-\boldsymbol{\xi}|^3}\,, \quad (7.2.17)$$

in E (and D). Multiplying this identity by the constant $\mu(\boldsymbol{x}_0)$ and subtracting it from (7.2.15), it results

$$\boldsymbol{v}(\boldsymbol{\xi}) = \frac{1}{4\pi}\int\int_{\Sigma}\frac{\mu(\boldsymbol{x})-\mu(\boldsymbol{x}_0)}{|\boldsymbol{x}-\boldsymbol{\xi}|}\left[\boldsymbol{n} - 3(\boldsymbol{x}-\boldsymbol{\xi})\frac{(\boldsymbol{x}-\boldsymbol{\xi})\cdot\boldsymbol{n}}{|\boldsymbol{x}-\boldsymbol{\xi}|^2}\right]\frac{\mathrm{d}\,a}{|\boldsymbol{x}-\boldsymbol{\xi}|^2}\,. \quad (7.2.18)$$

for $M(\boldsymbol{\xi})$ in E.

For calculating the velocity when $M(\boldsymbol{\xi}) \to Q_0(\boldsymbol{x}_0)$, a point on the boundary Σ, we shall denote again by σ the portion cut from Σ by the sphere with the center in Q_0 and the radius ε and we shall use the definition of Cauchy's principal value

$$\lim_{\varepsilon \to 0} \int\int_{\Sigma - \sigma} = \int\int_{\Sigma}' \tag{7.2.19}$$

if it exists. With $\boldsymbol{n}_0$ the outward normal in Q_0, we put $\boldsymbol{\xi} = \boldsymbol{x}_0 + \eta \boldsymbol{n}_0$ $(\eta > 0)$. For the limit value of the velocity in this point, we have:

$$\boldsymbol{v}_\varepsilon(\boldsymbol{x}_0) = \lim_{\eta \to 0} \boldsymbol{v}(\boldsymbol{x}_0 + \eta \boldsymbol{n}_0) = I_1 + I_2 \,, \tag{7.2.20}$$

for every $\varepsilon > 0$. We denoted

$$\begin{aligned} I_1 = \lim_{\eta \to 0} \frac{1}{4\pi} \int\int_{\Sigma - \sigma} \frac{\mu(\boldsymbol{x}) - \mu(\boldsymbol{x}_0)}{|\boldsymbol{x} - \boldsymbol{\xi}|} \Bigg[\boldsymbol{n} - \\ -3(\boldsymbol{x} - \boldsymbol{\xi}) \frac{(\boldsymbol{x} - \boldsymbol{\xi}) \cdot \boldsymbol{n}}{|\boldsymbol{x} - \boldsymbol{\xi}|^2} \Bigg] \frac{\mathrm{d}a}{|\boldsymbol{x} - \boldsymbol{\xi}|^2} \,, \\ I_2 = \lim_{\eta \to 0} \frac{1}{4\pi} \int\int_{\sigma} \frac{\mu(\boldsymbol{x}) - \mu(\boldsymbol{x}_0)}{|\boldsymbol{x} - \boldsymbol{\xi}|} \Bigg[\boldsymbol{n} - \\ -3(\boldsymbol{x} - \boldsymbol{\xi}) \frac{(\boldsymbol{x} - \boldsymbol{\xi}) \cdot \boldsymbol{n}}{|\boldsymbol{x} - \boldsymbol{\xi}|^2} \Bigg] \frac{\mathrm{d}a}{|\boldsymbol{x} - \boldsymbol{\xi}|^2} \,, \end{aligned} \tag{7.2.21}$$

where $\boldsymbol{\xi} = \boldsymbol{x}_0 + \eta \boldsymbol{n}_0$. The first limit and the integral may be interchanged because the integrand has no singularities on $\Sigma - \sigma$. Hence one obtains

$$\begin{aligned} I_1 = \frac{1}{4\pi} \int\int_{\Sigma - \sigma} \frac{\mu(\boldsymbol{x}) - \mu(\boldsymbol{x}_0)}{|\boldsymbol{x} - \boldsymbol{x}_0|} \Bigg[\boldsymbol{n} - \\ -3(\boldsymbol{x} - \boldsymbol{x}_0) \frac{(\boldsymbol{x} - \boldsymbol{x}_0) \cdot \boldsymbol{n}}{|\boldsymbol{x} - \boldsymbol{x}_0|^2} \Bigg] \frac{\mathrm{d}a}{|\boldsymbol{x} - \boldsymbol{x}_0|^2} \,. \end{aligned} \tag{7.2.22}$$

For every $\varepsilon > 0$ this integral exists if $\mu(\boldsymbol{x})$ satisfies Hölder's condition. So, we deduce

$$\boldsymbol{v}(\boldsymbol{x}_0) = \lim_{\varepsilon \to 0} (I_1 + I_2) = \boldsymbol{w}(\boldsymbol{x}_0) + \lim I_2 \,, \tag{7.2.23}$$

where

$$\boldsymbol{w}(\boldsymbol{x}_0) = \frac{1}{4\pi} \iint_\Sigma' \frac{\mu(\boldsymbol{x}) - \mu(\boldsymbol{x}_0)}{|\boldsymbol{x} - \boldsymbol{x}_0|} \left[\boldsymbol{n} - \right.$$
$$\left. -3(\boldsymbol{x} - \boldsymbol{x}_0) \frac{(\boldsymbol{x} - \boldsymbol{x}_0) \cdot \boldsymbol{n}}{|\boldsymbol{x} - \boldsymbol{x}_0|^2} \right] \frac{\mathrm{d}\,a}{|\boldsymbol{x} - \boldsymbol{x}_0|^2}. \tag{7.2.24}$$

For calculating the last limit from (7.2.23), when ε is small enough, we may replace σ with its projection A on the tangent plane in Q_0 and we shall utilize (7.2.17). It results that we have

$$\boldsymbol{x} - \boldsymbol{\xi} = r(\cos\theta \boldsymbol{i}_0 + \sin\theta \boldsymbol{j}_0) - \eta \boldsymbol{n}_0, \quad |\boldsymbol{x} - \boldsymbol{\xi}|^2 = r^2 + \eta^2, \tag{7.2.25}$$

$$\mu(\boldsymbol{x}) - \mu(\boldsymbol{x}_0) = (\nabla\mu)(\boldsymbol{x}_0) \cdot (\boldsymbol{x} - \boldsymbol{x}_0) + \ldots =$$
$$= r\nabla\mu(\boldsymbol{x}_0) \cdot (\cos\theta \boldsymbol{i}_0 + \sin\theta \boldsymbol{j}_0,$$

where $\nabla\mu(\boldsymbol{x}_0) = (\nabla\mu)(\boldsymbol{x}_0)$, according to the usual notation in Analysis.

Considering the scalar product

$$\nabla\mu \cdot (\cos\theta \boldsymbol{i}_0 + \sin\theta \boldsymbol{j}_0),$$

in the basis $\boldsymbol{i}_0, \boldsymbol{j}_0$ we obtain the identity

$$[\nabla\mu \cdot (\cos\theta \boldsymbol{i}_0 + \sin\theta \boldsymbol{j}_0)](\cos\theta \boldsymbol{i}_0 + \sin\theta \boldsymbol{j}_0) =$$
$$= [(\nabla\mu \cdot \boldsymbol{i}_0)\cos\theta + (\nabla\mu \cdot \boldsymbol{j}_0)\sin\theta](\cos\theta \boldsymbol{i}_0 + \sin\theta \boldsymbol{j}_0) =$$
$$= (\nabla\mu \cdot \boldsymbol{i}_0)(\cos^2\theta \boldsymbol{i}_0 + \cos\theta\sin\theta \boldsymbol{j}_0) +$$
$$+ (\nabla\mu \cdot \boldsymbol{j}_0)(\sin\theta\cos\theta \boldsymbol{i}_0 + \sin^2\theta \boldsymbol{j}_0). \tag{7.2.26}$$

Noticing that some terms vanish after integrating with respect to θ, from I_2 it remains

$$I_2 = \lim_{\eta \to 0} \frac{3}{4\pi} \int_0^\varepsilon \int_0^{2\pi} \frac{r^3 \eta}{(r^2 + \eta^2)^{5/2}} \times$$
$$\times \begin{bmatrix} (\nabla\mu \cdot \boldsymbol{i}_0) \times (\cos^2\theta \boldsymbol{i}_0 + \cos\theta\sin\theta \boldsymbol{j}_0) + \\ + (\nabla\mu \cdot \boldsymbol{j}_0) \times (\sin\theta\cos\theta \boldsymbol{i}_0 + \sin^2\theta \boldsymbol{j}_0) \end{bmatrix} \mathrm{d}r\mathrm{d}\theta,$$

$$I_2 = \lim_{\eta \to 0} \frac{3}{4} \int_0^\varepsilon \frac{r^3 \eta}{(r^2 + \eta^2)^{5/2}} [(\nabla\mu \cdot \boldsymbol{i}_0)\boldsymbol{i}_0 + (\nabla\mu \cdot \boldsymbol{j}_0)\boldsymbol{j}_0] \mathrm{d}\,r.$$

One obtains $(\nabla\mu)(\boldsymbol{x}_0)$ in the square bracket and we take it off from the integral. So,

$$I_2 = \frac{3}{4}(\nabla\mu)(\boldsymbol{x}_0)\lim_{\eta\to 0}\int_0^{\varepsilon}\frac{r^3\eta}{(r^2+\eta^2)^{5/2}}\mathrm{d}\,r = \frac{1}{2}\nabla\mu(\boldsymbol{i}_0)\,.$$

Hence, taking (7.2.23) into account, we obtain the velocity on Σ by means of the formula

$$\boldsymbol{v}(\boldsymbol{x}_0) = \frac{1}{2}(\nabla\mu)(\boldsymbol{x}_0) + \boldsymbol{w}(\boldsymbol{x}_0)\,, \tag{7.2.27}$$

$\boldsymbol{w}(\boldsymbol{x}_0)$ being given in (7.2.24). This formula was obtained by N.Markov in a unpublished paper.

Since $P(\boldsymbol{x})$ and $Q_0(\boldsymbol{x}_0)$ from (7.2.25) belong to the plane which is tangent to Σ in Q_0, we deduce that $\nabla\mu(\boldsymbol{x}_0)$ is in the tangent plane. Hence, the boundary condition

$$(\boldsymbol{i}+\boldsymbol{v})\cdot\boldsymbol{n} = 0 \ \text{ pe } \Sigma$$

determined the following integral equation:

$$\frac{1}{4\pi}\int_{\Sigma}\frac{\mu(\boldsymbol{x})-\mu(\boldsymbol{x}_0)}{|\boldsymbol{x}-\boldsymbol{x}_0|}\left[\boldsymbol{n}\cdot\boldsymbol{n}_0 - \right.$$

$$\left. -3(\boldsymbol{x}-\boldsymbol{x}_0)\cdot\boldsymbol{n}_0\frac{(\boldsymbol{x}-\boldsymbol{x}_0)\cdot\boldsymbol{n}}{|\boldsymbol{x}-\boldsymbol{x}_0|^2}\right]\frac{\mathrm{d}\,a}{|\boldsymbol{x}-\boldsymbol{x}_0|^2} = -\boldsymbol{n}_x^0\,,\ \ \forall\boldsymbol{x}_0\in\Sigma$$

which is an alternative to (7.2.8).

7.3 The Direct Method. The Incompressible Fluid

7.3.1 The Integral Representation Formula

For writing this subsection we used the paper [7.3]. Since we study the same problem like in the previous subsections, we shall utilize the representation (4.6.1), the equations (4.6.2), the boundary condition (4.6.3) and the condition to infinity (4.6.4). The difference is now that the problem is three - dimensional. In (4.6.3), C will be replaced by Σ, the surface of the perturbing body B. For avoiding the singular integrals, we shall replace the equations (4.6.2) by the equations

$$\operatorname{div}(\boldsymbol{v}-\boldsymbol{c}) = 0, \quad \operatorname{rot}(\boldsymbol{v}-\boldsymbol{c}) = 0\,, \tag{7.3.1}$$

c being a vectorial constant.

The system

$$\operatorname{div} \boldsymbol{v}^* = \delta(\boldsymbol{x} - \boldsymbol{x}_0), \quad \operatorname{rot} \boldsymbol{v}^* = 0 \tag{7.3.2}$$

for every point $\boldsymbol{x}_0 \in D + \Sigma$, D representing the domain occupied by the fluid (the exterior of the body B), defines the fundamental solution

$$\boldsymbol{v}^* = \frac{1}{4\pi} \frac{\boldsymbol{x} - \boldsymbol{x}_0}{|\boldsymbol{x} - \boldsymbol{x}_0|^3}. \tag{7.3.3}$$

From the equations (7.3.1) we deduce the identity

$$\int_{D_0} [f \operatorname{div}(\boldsymbol{v} - \boldsymbol{c}) + \boldsymbol{g} \cdot \operatorname{rot}(\boldsymbol{v} - \boldsymbol{c})] \mathrm{d}\, v = 0 \tag{7.3.4}$$

for every two functions, or regular distributions, f and $\boldsymbol{g}$. We denoted by D_0 the exterior of B, bounded by a sphere $S(O, R), R$ being great enough, such that the body B is included into the interior of the sphere. Utilizing the identity $(4.6.9)_1$ and the identity

$$\operatorname{rot}[\boldsymbol{g} \times (\boldsymbol{v} - \boldsymbol{c})] = (\boldsymbol{v} - \boldsymbol{c}) \cdot \operatorname{rot} \boldsymbol{g} - \boldsymbol{g} \cdot \operatorname{rot}(\boldsymbol{v} - \boldsymbol{c}) \tag{7.3.5}$$

and applying Gauss's formula, from (7.3.4) we deduce

$$\begin{aligned} &\int_{D_0} (\boldsymbol{v} - \boldsymbol{c}) \cdot (\operatorname{grad} f - \operatorname{rot} \boldsymbol{g}) \mathrm{d}\, v = \\ &= \int_{\Sigma + \Sigma_R} (\boldsymbol{v} - \boldsymbol{c}) \cdot [f \boldsymbol{n} - (\boldsymbol{n} \times \boldsymbol{g})] \mathrm{d}\, a, \end{aligned} \tag{7.3.6}$$

$\boldsymbol{n}$ being the outward pointing normal on D_0.

Substituting successively

$$\begin{aligned} (f, \boldsymbol{g}) &\to (\boldsymbol{i} \cdot \boldsymbol{v}^*, -\boldsymbol{i} \times \boldsymbol{v}^*) \\ (f, \boldsymbol{g}) &\to (\boldsymbol{j} \cdot \boldsymbol{v}^*, -\boldsymbol{j} \times \boldsymbol{v}^*) \\ (f, \boldsymbol{g}) &\to (\boldsymbol{k} \cdot \boldsymbol{v}^*, -\boldsymbol{k} \times \boldsymbol{v}^*), \end{aligned} \tag{7.3.7}$$

we find the projections on the axes of coordinates of the following identity:

$$\begin{aligned} \int_{D_0} (\boldsymbol{v} - \boldsymbol{c}) \operatorname{div} \boldsymbol{v}^* \mathrm{d}\, v = \int_{\Sigma + \Sigma_R} \{\boldsymbol{n} \cdot (\boldsymbol{v} - \boldsymbol{c}) \boldsymbol{v}^* + [\boldsymbol{n} \times (\boldsymbol{v} - \boldsymbol{c})] \times \boldsymbol{v}^*\} \mathrm{d}\, a, \\ \boldsymbol{x}_0 \in D_0 + \Sigma \end{aligned} \tag{7.3.8}$$

proving in this way that it is correct.

Taking (7.3.2) into account, it results

$$\boldsymbol{v}(\boldsymbol{x}_0) - \boldsymbol{c} = \int_{\Sigma+\Sigma_R} \{\boldsymbol{n}\cdot(\boldsymbol{v}-\boldsymbol{c})\boldsymbol{v}^* + [\boldsymbol{n}\times(\boldsymbol{v}-\boldsymbol{c})]\times\boldsymbol{v}^*\}\mathrm{d}\,a\,. \tag{7.3.9}$$

The integral on Σ_R may be written as follows

$$\int_{\Sigma_R} [(\boldsymbol{n}\cdot\boldsymbol{v})\boldsymbol{v}^* + (\boldsymbol{n}\times\boldsymbol{v})\times\boldsymbol{v}^*]\mathrm{d}\,a - \int_{\Sigma_R} [(\boldsymbol{n}\cdot\boldsymbol{c})\boldsymbol{v}^* + (\boldsymbol{n}\times\boldsymbol{c})\times\boldsymbol{v}^*]\,\mathrm{d}\,a\,.$$

The first term vanishes when $R\to\infty$, because of the condition (4.6.4). For calculating the second term we use the spherical coordinates: R, θ, φ:

$$x = R\sin\theta\cos\varphi$$

$$y = R\sin\theta\sin\varphi$$

$$z = R\cos\theta, \qquad 0\le\theta<\pi\,.$$

We obtain

$$\lim_{R\to\infty}\int_{\Sigma_R} [(\boldsymbol{n}\cdot\boldsymbol{c})\boldsymbol{v}^* + (\boldsymbol{n}\times\boldsymbol{c})\times\boldsymbol{v}^*]\mathrm{d}\,a =$$

$$= \frac{\boldsymbol{c}}{4\pi}\int_0^{2\pi}\int_0^{\pi}\sin\theta\mathrm{d}\,\theta\mathrm{d}\,\varphi = \boldsymbol{c}\,.$$

Now, the representation (7.3.9) becomes

$$\boldsymbol{v}(\boldsymbol{x}_0) = \int_{\Sigma} \{\boldsymbol{n}\cdot(\boldsymbol{v}-\boldsymbol{c})\boldsymbol{v}^* + [\boldsymbol{n}\times(\boldsymbol{v}-\boldsymbol{c})]\times\boldsymbol{v}^*\}\mathrm{d}\,a\,,$$

$$\boldsymbol{x}_0\in D+\Sigma\,.$$

Setting $\boldsymbol{c} = \boldsymbol{v}(\boldsymbol{x}_0)\equiv\boldsymbol{v}_0$ we obtain the representation formula

$$\boldsymbol{v}(\boldsymbol{x}_0) = \int_{\Sigma} \{\boldsymbol{n}\cdot(\boldsymbol{v}-\boldsymbol{v}_0)\boldsymbol{v}^* + [\boldsymbol{n}\times(\boldsymbol{v}-\boldsymbol{v}_0)]\times\boldsymbol{v}^*\}\mathrm{d}\,a\,. \tag{7.3.10}$$

This is a *regularized* representation, valid both for $\boldsymbol{x}_0$ in D and for $\boldsymbol{x}_0$ on Σ. Obviously when $\boldsymbol{x}_0\in D$, the integrand has no singularity. This property is also valid when $\boldsymbol{x}_0\in\Sigma$ because of the factor $\boldsymbol{v}-\boldsymbol{v}_0$ which vanishes for $\boldsymbol{x}\to\boldsymbol{x}_0$.

This representation formula is fundamental [7.1].

Utilizing the boundary condition (4.6.3), the formula (7.3.10) becomes

$$\boldsymbol{v}_0 = \int_{\Sigma} \{-(n_x + \boldsymbol{n}\cdot\boldsymbol{v}_0)\boldsymbol{v}^* + [\boldsymbol{n}\times(\boldsymbol{v}-\boldsymbol{v}_0)]\times\boldsymbol{v}^*\}\mathrm{d}\,a\,. \tag{7.3.11}$$

7.3.2 The Integral Equation

The vector

$$\boldsymbol{F} \equiv \boldsymbol{n} \times \boldsymbol{v} \tag{7.3.12}$$

which intervenes in the *surface circulation*

$$\boldsymbol{C} = \int_{\Sigma} \boldsymbol{n} \times \boldsymbol{v} \mathrm{d}\, a\,. \tag{7.3.13}$$

introduced by *Pascal* and utilized again by *R.von Mises* [1.27], [1.20], has a great importance herein.

We shall deduce the equation for F. To this aim we write the formula (7.3.11) as follows:

$$\boldsymbol{v}_0 = \int_{\Sigma} [-(n_x + \boldsymbol{n}\cdot\boldsymbol{v}_0)\boldsymbol{v}^* + \boldsymbol{F}\times\boldsymbol{v}^* + (\boldsymbol{v}^*\cdot\boldsymbol{v}_0)\boldsymbol{n} - (\boldsymbol{n}\cdot\boldsymbol{v}^*)\boldsymbol{v}_0]\mathrm{d}\, a\,. \tag{7.3.14}$$

After the vectorial multiplication by $\boldsymbol{n}^0$ one obtains

$$\begin{aligned} \boldsymbol{F}_0 = \int_{\Sigma} [&-(n_x + \boldsymbol{n}\cdot\boldsymbol{v}_0)(\boldsymbol{n}^0 \times \boldsymbol{v}^*) + (\boldsymbol{n}^0\cdot\boldsymbol{v}^*)\boldsymbol{F} - \\ &-(\boldsymbol{n}^0\cdot\boldsymbol{F})\boldsymbol{v}^* - (\boldsymbol{n}\cdot\boldsymbol{v}^*)\boldsymbol{F}_0 + (\boldsymbol{v}_0\cdot\boldsymbol{v}^*)(\boldsymbol{n}^0\times\boldsymbol{n})]\mathrm{d}\, a\,. \end{aligned} \tag{7.3.15}$$

But, from the boundary condition $\boldsymbol{n}\cdot\boldsymbol{v} = -n_x$ and from (7.3.12) we deduce $\boldsymbol{v} = -n_x\boldsymbol{n} - \boldsymbol{n}\times\boldsymbol{F}$ such that

$$\begin{aligned} &\boldsymbol{n}\cdot\boldsymbol{v}_0 = -n_x^0\boldsymbol{n}\cdot\boldsymbol{n}^0 - (\boldsymbol{n}, \boldsymbol{n}^0, \boldsymbol{F}_0) \\ &(\boldsymbol{v}^*\cdot\boldsymbol{v}_0)(\boldsymbol{n}^0\times\boldsymbol{n}) = -n_x^0(\boldsymbol{v}^*\cdot\boldsymbol{n}^0)(\boldsymbol{n}^0\times\boldsymbol{n}) - (\boldsymbol{n}^0, \boldsymbol{F}_0, \boldsymbol{v}^*)(\boldsymbol{n}^0\times\boldsymbol{n})\,. \end{aligned} \tag{7.3.16}$$

One obtains therefore

$$\begin{aligned} \boldsymbol{F}_0 = \int_{\Sigma} [&-(n_x - n_x^0\boldsymbol{n}\cdot\boldsymbol{n}^0)(\boldsymbol{n}^0\times\boldsymbol{v}^0) + (\boldsymbol{n}, \boldsymbol{n}^0, \boldsymbol{F}_0)(\boldsymbol{n}^0\times\boldsymbol{v}^0) + \\ &+(\boldsymbol{n}^0\cdot\boldsymbol{v}^*)\boldsymbol{F} - (\boldsymbol{n}^0\cdot\boldsymbol{F})\boldsymbol{v}^* - (\boldsymbol{n}\cdot\boldsymbol{v}^*)\boldsymbol{F}_0 - \\ &-n_x^0(\boldsymbol{v}^*\cdot\boldsymbol{n}^0)(\boldsymbol{n}^0\times\boldsymbol{n}) - (\boldsymbol{n}^0, \boldsymbol{F}_0, \boldsymbol{v}^*)(\boldsymbol{n}^0\times\boldsymbol{n})]\mathrm{d}\, a\,. \end{aligned} \tag{7.3.17}$$

Utilizing the double vectorial product formula

$$(\boldsymbol{U}\times\boldsymbol{V})\times\boldsymbol{W} = (\boldsymbol{U}\cdot\boldsymbol{W})\boldsymbol{V} - (\boldsymbol{V}\cdot\boldsymbol{W})\boldsymbol{U}$$

it results

$$(\boldsymbol{n}, \boldsymbol{n}^0, \boldsymbol{F}_0)(\boldsymbol{n}^0 \times \boldsymbol{v}^0) - (\boldsymbol{n}^0, \boldsymbol{F}_0, \boldsymbol{v}^*)(\boldsymbol{n}^0 \times \boldsymbol{n}) =$$

$$= \boldsymbol{n}^0 \times [(\boldsymbol{n}^0, \boldsymbol{F}_0, \boldsymbol{n})\boldsymbol{v}^* - (\boldsymbol{n}^0, \boldsymbol{F}_0, \boldsymbol{v}^*)\boldsymbol{n}] = \tag{7.3.18}$$

$$= \boldsymbol{n}^0 \times [(\boldsymbol{n} \times \boldsymbol{v}^*) \times (\boldsymbol{n}^0 \times \boldsymbol{F}_0)] = -(\boldsymbol{n}^0, \boldsymbol{n}, \boldsymbol{v}^*)(\boldsymbol{n}^0 \times \boldsymbol{F}_0)\,.$$

Hence, the equation (7.3.10) becomes

$$\boldsymbol{F}_0 + \int_\Sigma \Big[(\boldsymbol{n}^0 \cdot \boldsymbol{F})\boldsymbol{v}^* - (\boldsymbol{n}^0 \cdot \boldsymbol{v}^*)\boldsymbol{F} + (\boldsymbol{n} \cdot \boldsymbol{v}^*)\boldsymbol{F}_0\Big]\mathrm{d}\,a +$$

$$+ \boldsymbol{n}^0 \times \boldsymbol{F}_0) \int_\Sigma (\boldsymbol{n}^0 \times \boldsymbol{n}) \cdot \boldsymbol{v}^* \mathrm{d}\,a = \int_\Sigma \Big\{ n_x^0 (\boldsymbol{v}^* \cdot \boldsymbol{n}^0)(\boldsymbol{n} \times \boldsymbol{n}^0) +$$

$$+ \big[n_x^0 (\boldsymbol{n} \cdot \boldsymbol{n}^0) - n_x\big](\boldsymbol{n}^0 \times \boldsymbol{v}^*)\Big\}\mathrm{d}\,a\,. \tag{7.3.19}$$

this representing the integral equation of the problem. This is a regularized equation. The integrals are not singular. Although the denominators of the distribution $\boldsymbol{v}^*$ vanish for $\boldsymbol{x} \to \boldsymbol{x}_0$, we have to take into account that in these points the numerators vanish too. Indeed,

$$(\boldsymbol{n}^0 \cdot \boldsymbol{F}) \to 0, (\boldsymbol{n} \cdot \boldsymbol{v}^*)\boldsymbol{F}_0 - (\boldsymbol{n}^0 \cdot \boldsymbol{v}^*)\boldsymbol{F} \to 0, \boldsymbol{n} \times \boldsymbol{n}^0 \to 0\,,$$

$$n_x^0 \boldsymbol{n} \cdot \boldsymbol{n}^0 - n_x \to 0\,.$$

7.3.3 Kutta's Condition

In the sequel we shall study the flow around lifting bodies (with smooth surfaces). In order to solve this problem, the classical theory, beginning with the lifting line theory (Prandtl 1918), introduces a vortical surface behind the body. In this way the flow in no longer irrotational in the exterior of the body, hence the above solution cannot be utilized. For defining the fluid flow with the aid of the solution given herein, we shall consider a discrete distribution of vortices located in certain points from the interior of the body [7.3]. In this way, the external flow remains everywhere irrotational and it will be characterized by the solution given herein. Moreover, the circulation C defined by (7.3.13) does not vanish on the trailing edge. It remains to discuss the location of the vortices.

The Kutta-Joukowskicondition concerning the continuity of the pressure on the trailing edge will be imposed, like in the two-dimensional case, writing that the velocity in plane cross-sections, perpendicular to the trailing edge, is continuous when P_s and $P_i \to P_f$ (4.6.21). Considering in figure 7.3.1 the point P_s and the local frame $\boldsymbol{n}$, $\boldsymbol{s}$ and

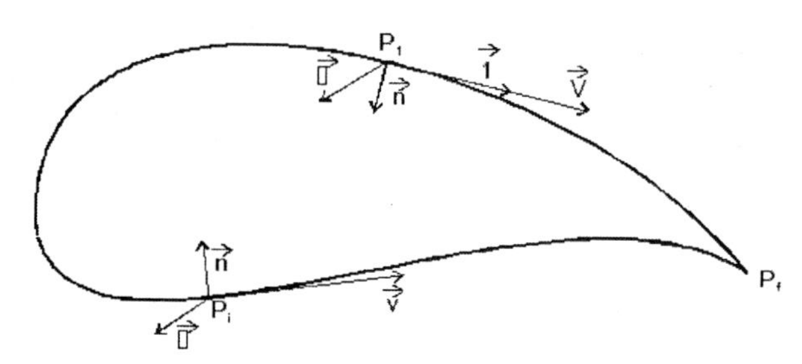

Fig. 7.3.1.

$\boldsymbol{\beta} = \boldsymbol{n} \times \boldsymbol{s}$ and knowing that $\boldsymbol{V} = V_s \boldsymbol{s}$, we deduce

$$\boldsymbol{n} \times \boldsymbol{V} = V_s \boldsymbol{\beta} \,. \tag{7.3.20}$$

Hence, the vectorial form of (4.6.21) is:

$$\begin{gathered} (\boldsymbol{n} \times \boldsymbol{V})(P_s) + (\boldsymbol{n} \times \boldsymbol{V})(P_i) \to 0 \\ P_s, P_i \to P_f \,. \end{gathered} \tag{7.3.21}$$

Taking into account that $\boldsymbol{V}$ has the form (4.6.1) it results the condition

$$< \boldsymbol{F} > + < n_z > \boldsymbol{j} - < n_y > \boldsymbol{k} = 0, \tag{7.3.22}$$

with the notation

$$< \phi >= \phi(P_s) + \phi(P_i) \,. \tag{7.3.23}$$

7.3.4 The Lifting Flow

One knows from the theory of the two - dimensional potential incompressible flow that the lift is generated by a vortex of intensity $\Gamma \boldsymbol{k}$ placed in a point from the interior of the body. For example, in the case of the flow past the circular obstacle with the center in z_0 (see [1.11] p. 117), the potential

$$U_\infty \left[(z - z_0)\mathrm{e}^{-i\alpha} + \frac{R^2}{z - z_0}\mathrm{e}^{i\alpha} \right]$$

characterizes the non-lifting flow and the potential

$$\frac{\Gamma}{2\pi i} \ln(z - z_0)$$

defines the lifting flow, the lift being $-\rho U_\infty \Gamma$. According to this model, we shall try to generate the lift by means of some vortices from the interior of the body, the number of vortices being determined by the number of pairs of panels adjacent to the trailing edge for which we have to write Kutta's condition (7.3.22). To this aim we need at first the velocity field $\boldsymbol{w}^*$ generated by a vortex line of constant intensity Γ. This will be determined by the system

$$\operatorname{div} \boldsymbol{w}^* = 0, \quad \operatorname{rot} \boldsymbol{w}^* = \boldsymbol{\Gamma}\delta(\boldsymbol{x}), \tag{7.3.24}$$

$\boldsymbol{w}^*$ being a distribution (for this reason it was marked by $*$). For determining the solution we apply the Fourier transform, like in the case of the system (7.3.2). Using the formulas (A.6.5) we deduce

$$\boldsymbol{\alpha} \cdot \widehat{\boldsymbol{w}}^* = 0, \quad \boldsymbol{\alpha} \times \widehat{\boldsymbol{w}}^* = i\boldsymbol{\Gamma},$$

whence, utilizing (A.6.9),

$$\widehat{\boldsymbol{w}}^* = \boldsymbol{\Gamma} \times \left(-\frac{\mathrm{i}\boldsymbol{\alpha}}{\alpha^2}\right), \quad \boldsymbol{w}^* = \boldsymbol{\Gamma} \times \nabla F^{-1}\left[\frac{1}{\alpha^2}\right],$$

and then (A.7.10) and (7.3.3)

$$\boldsymbol{w}^* = \boldsymbol{v}^* \times \boldsymbol{\Gamma}. \tag{7.3.25}$$

If the vortex is located in the interior of the body, in a point having the vector of position $\boldsymbol{x}_1$, then

$$\boldsymbol{w}(\boldsymbol{x}) = \boldsymbol{v}^*(\boldsymbol{x} - \boldsymbol{x}_1) \times \boldsymbol{\Gamma}, \tag{7.3.26}$$

and if there are L lines, since the equations (7.3.24) are linear, it results

$$\boldsymbol{w}^*(\boldsymbol{x}) = \sum_{k=1}^{L} \boldsymbol{v}^*(\boldsymbol{x} - \boldsymbol{x}_k) \times \boldsymbol{\Gamma}_k. \tag{7.3.27}$$

Returning to our problem and considering, for the sake of simplicity, a single line, we shall write the formula (4.6.1) as follows:

$$\boldsymbol{V} = U_\infty(\boldsymbol{i} + \boldsymbol{w}^* + \boldsymbol{v}). \tag{7.3.28}$$

From $\operatorname{div} \boldsymbol{v} = 0$ and $\operatorname{rot} \boldsymbol{v} = 0$, taking (7.3.24) into account, it results

$$\operatorname{div} \boldsymbol{v} = 0, \ \operatorname{rot} \boldsymbol{v} = -\boldsymbol{\Gamma}\delta(\boldsymbol{x} - \boldsymbol{x}_1), \tag{7.3.29}$$

and the equations (4.6.5) will be replaced by

$$\operatorname{div}(\boldsymbol{v}-\boldsymbol{c})=0 \quad \operatorname{rot}(\boldsymbol{v}-\boldsymbol{c})=-\boldsymbol{\Gamma}\delta(\boldsymbol{x}-\boldsymbol{x}_1)\,. \tag{7.3.30}$$

The identity (7.3.4) will be written as follows:

$$\int_{D_0}\Big[f\operatorname{div}(\boldsymbol{v}-\boldsymbol{c})+\boldsymbol{g}\cdot\operatorname{rot}(\boldsymbol{v}-\boldsymbol{c})\Big]\mathrm{d}\,v=\begin{cases}-\boldsymbol{\Gamma}\cdot\boldsymbol{g}(\boldsymbol{x}_1) & ,\boldsymbol{x}_1\in D_0\\ 0 & ,\boldsymbol{x}_1\notin D_0\,.\end{cases} \tag{7.3.31}$$

Since the point having the vector of position $\boldsymbol{x}_1$ is in the interior of the body, we shall obtain a second equality identical to (7.3.4), such that all the consequences follow like above. We have therefore (7.3.10).

From the boundary condition $\boldsymbol{v}\cdot\boldsymbol{n}=0$ we deduce

$$\boldsymbol{v}\cdot\boldsymbol{n}=-n_x-w_n^*,\quad w_n^*\equiv\boldsymbol{w}^*\cdot\boldsymbol{n}\,, \tag{7.3.32}$$

such that the formula (7.3.11) becomes

$$\boldsymbol{v}_0=\overset{R}{\int_{\Sigma}}\Big\{-(n_x+w_n^*+\boldsymbol{n}\cdot\boldsymbol{v}_0)\boldsymbol{v}^*+\Big[\boldsymbol{n}\times(\boldsymbol{v}-\boldsymbol{v}_0)\Big]\times\boldsymbol{v}^*\Big\}\mathrm{d}\,a \tag{7.3.33}$$

$x_0\in D+\Sigma\,.$

The relation between $\boldsymbol{v}$ and $\boldsymbol{F}$ is now

$$\boldsymbol{v}=-(n_x+w_n^*)\boldsymbol{n}-\boldsymbol{n}\times\boldsymbol{F}\,, \tag{7.3.34}$$

and the integral equation (7.3.19) becomes

$$\begin{aligned}\boldsymbol{F}_0+&\int_{\Sigma}\Big[(\boldsymbol{n}^0\cdot\boldsymbol{F})\boldsymbol{v}^*-(\boldsymbol{n}^0\cdot\boldsymbol{v}^*)\boldsymbol{F}+(\boldsymbol{n}\cdot\boldsymbol{v}^*)\boldsymbol{F}_0\Big]\mathrm{d}\,a+\\ &+(\boldsymbol{n}^0\times\boldsymbol{F}_0)\int_{\Sigma}(\boldsymbol{n}^0\times\boldsymbol{n})\cdot\boldsymbol{v}^*\mathrm{d}\,a+w_n^*(\boldsymbol{x}_0)\int_{\Sigma}\Big[(\boldsymbol{n}^0\cdot\boldsymbol{v}^*)(\boldsymbol{n}^0\times\boldsymbol{n})-\\ &-(\boldsymbol{n}^0\cdot\boldsymbol{n})(\boldsymbol{n}^0\times\boldsymbol{v}^*)\Big]\mathrm{d}\,a+\int_{\Sigma}w_n^*(\boldsymbol{x})(\boldsymbol{n}^0\times\boldsymbol{v}^*)\mathrm{d}\,a=\\ &=\int_{\Sigma}\Big\{n_x^0(\boldsymbol{n}^0\cdot\boldsymbol{v}^*)(\boldsymbol{n}\times\boldsymbol{n}^0)+\big[n_x^0(\boldsymbol{n}\cdot\boldsymbol{n}^0)-n_x\big](\boldsymbol{n}^0\times\boldsymbol{v}^*)\Big\}\mathrm{d}\,a\,,\end{aligned} \tag{7.3.35}$$

where, taking (7.3.27) into account, we have:

$$\begin{aligned}w_n^*(\boldsymbol{x}_0)&=\Big[\boldsymbol{n}^0\times\boldsymbol{v}^*(\boldsymbol{x}_0-\boldsymbol{x}_1)\Big]\cdot\boldsymbol{\Gamma}\\ w_n^*(\boldsymbol{x})&=\Big[\boldsymbol{n}^0\times\boldsymbol{v}^*(\boldsymbol{x}_0-\boldsymbol{x}_1)\Big]\cdot\boldsymbol{\Gamma}\,.\end{aligned} \tag{7.3.36}$$

In (7.3.35) the dependence of $\boldsymbol{F}$ and $\boldsymbol{\Gamma}$ is linear. In the case of L vortices, the equation (7.3.35) remains unchanged and the expressions (7.3.36) become

$$
\begin{aligned}
w_n^*(\boldsymbol{x}_0) &= \sum_{k=1}^{L} \left[\boldsymbol{n}^0 \times \boldsymbol{v}^*(\boldsymbol{x}_0 - \boldsymbol{x}_k)\right] \cdot \boldsymbol{\Gamma} \\
w_n^*(\boldsymbol{x}) &= \sum_{k=1}^{L} \left[\boldsymbol{n} \times \boldsymbol{v}^*(\boldsymbol{x}_0 - \boldsymbol{x}_k)\right] \cdot \boldsymbol{\Gamma} \,.
\end{aligned}
\tag{7.3.37}
$$

In equation (7.3.34), $\boldsymbol{F}$, $\boldsymbol{\Gamma}_1$, $\boldsymbol{\Gamma}_L$ are unknown.

For specifying the conditions (7.3.21) we shall write the product $\boldsymbol{n} \times \boldsymbol{V}$ as a function of the principal unknown F. From (7.3.28) and (7.3.34) it results

$$
\boldsymbol{n} \times \boldsymbol{V} = U_\infty(\boldsymbol{F} + n_z \boldsymbol{j} - n_y \boldsymbol{k}) \,. \tag{7.3.38}
$$

The condition (7.3.21) will be written as follows

$$
\left[\boldsymbol{F} + \begin{pmatrix} 0 \\ n_z \\ -n_y \end{pmatrix}\right] (P_s) + \left[\boldsymbol{F} + \begin{pmatrix} 0 \\ n_z \\ -n_y \end{pmatrix}\right] (P_i) = 0 \,, \tag{7.3.39}
$$

for every pair of panels from the trailing edge. It will result therefore L equations. These ones, together with the equation (7.3.35) discretized below, will determine the solution $\boldsymbol{F}, \boldsymbol{\Gamma}_1, \ldots, \boldsymbol{\Gamma}_l$. The numerical results are more accurate when the points P_s and P_i are closer from the corresponding point P_f from the trailing edge (see [7.3], p. 364).

7.3.5 The Discretization of the Integral Equation

Like above, one approximates the surface of the body by N panels (triangles and quadrilaterals) and we approximate on every panel Π_j the function $\boldsymbol{F}$ by the constant value $\boldsymbol{F}_j$, that it has in the center $G_j(\boldsymbol{x}_j^0)$ (the collocation method). If we impose to the equation just obtained from (7.3.35) to be verified in all the centers $G_i(\boldsymbol{x}_i^0)$ and we denote

$$
\boldsymbol{X}_{ji} = \int_{\Pi_j} \boldsymbol{v}^*(\boldsymbol{x} - \boldsymbol{x}_i^0) \mathrm{d}\, a \tag{7.3.40}
$$

we deduce, for $i = \overline{1, N}$

$$\begin{aligned}
&\boldsymbol{F}_i + \sum_{j=1}^{N} \Big[(\boldsymbol{n}_i \cdot \boldsymbol{F}_j)\boldsymbol{X}_{ji} - (\boldsymbol{n}_i \cdot \boldsymbol{X}_{ji})\boldsymbol{F}_j + (\boldsymbol{n}_j \cdot \boldsymbol{X}_{ji})\boldsymbol{F}_i \Big] + \\
&+ (\boldsymbol{n}_i \times \boldsymbol{F}_i) \sum_{j=1}^{N} (\boldsymbol{n}_i \times \boldsymbol{n}_j) \cdot \boldsymbol{X}_{ji} + \sum_{k=1}^{L} \Big[(\boldsymbol{n}_i \times \boldsymbol{X}_{ik}) \cdot \boldsymbol{\Gamma}_k \Big] \cdot \\
&\cdot \sum_{j=1}^{N} \Big[(\boldsymbol{n}_i \times \boldsymbol{n}_j)(\boldsymbol{n}_i \cdot \boldsymbol{X}_{ji}) - (\boldsymbol{n}_i \times \boldsymbol{X}_{ji})(\boldsymbol{n}_i \cdot \boldsymbol{n}_j) \Big] + \qquad (7.3.41) \\
&+ \sum_{j=1}^{N} \sum_{k=1}^{L} \Big[(\boldsymbol{n}_j \times \boldsymbol{X}_{jk}) \cdot \boldsymbol{\Gamma}_k \Big] (\boldsymbol{n}_i \times \boldsymbol{X}_{ji}) = \sum_{j=1}^{N} \Big\{ - n_x^i (\boldsymbol{n}_i \cdot \\
&\cdot \boldsymbol{X}_{ji})(\boldsymbol{n}_i \times \boldsymbol{n}_j) + \Big[n_x^i (\boldsymbol{n}_i \cdot \boldsymbol{n}_j) - n_x^j \Big] (\boldsymbol{n}_i \times \boldsymbol{X}_{ji}) \Big\},
\end{aligned}$$

In this system, the singular terms $\boldsymbol{X}_{ii}$ disappear, because the integrals from (7.3.35) are regularized. Taking into account the relations

$$(\boldsymbol{n}_i \cdot \boldsymbol{F}_j)\boldsymbol{X}_{ji} - (\boldsymbol{n}_i \cdot \boldsymbol{X}_{ji})\boldsymbol{F}_j = \boldsymbol{n}_i \times (\boldsymbol{X}_{ji} \times \boldsymbol{F}_j)$$

$$(\boldsymbol{n}_i \cdot \boldsymbol{n}_j)\boldsymbol{X}_{ji} - (\boldsymbol{n}_i \cdot \boldsymbol{X}_{ji})\boldsymbol{n}_j = \boldsymbol{n}_j \times (\boldsymbol{X}_{ji} \times \boldsymbol{n}_i),$$

we deduce that the system (7.3.41) may be written as follows

$$\sum_{j=1}^{N} \boldsymbol{A}_{ij} \boldsymbol{F}_j + \sum_{k=1}^{L} \boldsymbol{B}_k^{(i)} \boldsymbol{\Gamma}_k = \boldsymbol{C}_i, \quad i = 1, \ldots, N \qquad (7.3.42)$$

where

$$\boldsymbol{A}_{ii}\boldsymbol{F}_i = \boldsymbol{F}_i + \boldsymbol{F}_i \sum_{j \neq i} \boldsymbol{n}_j \cdot \boldsymbol{X}_{ji} + (\boldsymbol{n}_i \times \boldsymbol{F}_i) \sum_{j \neq i} (\boldsymbol{n}_i \times \boldsymbol{n}_j) \cdot \boldsymbol{X}_{ji},$$

$$\boldsymbol{A}_{ij}\boldsymbol{F}_j = \boldsymbol{n}_i \times \sum_{j \neq i} (\boldsymbol{X}_{ji} \times \boldsymbol{F}_j),$$

$$\begin{aligned}
\boldsymbol{B}_k^{(i)} \boldsymbol{\Gamma}_k = &\Big[\boldsymbol{\Gamma}_k (\boldsymbol{n}_i \times \boldsymbol{X}_{ik}) \Big] \sum_{j \neq i} (\boldsymbol{n}_i \cdot \boldsymbol{X}_{ji})(\boldsymbol{n}_i \times \boldsymbol{n}_j) + \\
&+ \sum_{j \neq i} \Big\{ \boldsymbol{\Gamma}_k \cdot \Big[- (\boldsymbol{n}_i \times \boldsymbol{X}_{ik})(\boldsymbol{n}_i \cdot \boldsymbol{n}_j) + \boldsymbol{n}_j \times \boldsymbol{X}_{jk} \Big] \Big\} (\boldsymbol{n}_i \times \boldsymbol{X}_{ji}),
\end{aligned}$$

$$\boldsymbol{C}_i = \boldsymbol{n}_i \times \sum_{j \neq i} \left[n_x^i (\boldsymbol{n}_j \times \boldsymbol{X}_{ji}) \times \boldsymbol{n}_i - n_x^i \boldsymbol{X}_{ji} \right]$$

the unknowns being $\boldsymbol{F}_1, \ldots, \boldsymbol{F}_N$, $\boldsymbol{\Gamma}_1, \ldots, \boldsymbol{\Gamma}_l$. To this system of N equations, we add the equations (7.3.39) written for every pair of panels adjacent to the trailing edge. The number L of vortices equals the number of pairs; in this way, the number of $(N+L)$ equations obtained from (7.3.42) and (7.3.39) equals the number of unknowns $\boldsymbol{F}_1, \ldots, \boldsymbol{F}_N$, $\boldsymbol{\Gamma}_1, \ldots, \boldsymbol{\Gamma}_l$.

For calculating the coefficients $\boldsymbol{X}_{ji}$ (7.3.40) one utilizes either the quadrature formulas (for example, the Gauss-type formulas), or the analytical formulas given by Hess and Smith in [7.10]. Denoting by $\boldsymbol{x}_1^j$, $\boldsymbol{x}_2^j$, $\boldsymbol{x}_3^j$, $(\boldsymbol{x}_4^j)$ the vectors of position of the vertices of the panel Π_j which is a triangle (or a quadrilateral), with the definitions

$$d_{k,k+1} = \sqrt{(x_{k+1}^j - x_k^j)^2 + (y_{k+1}^j - y_k^j)^2 + (z_{k+1}^j - z_k^j)^2}\,,$$

$$r_k = \sqrt{(x_i - x_k^j)^2 + (y_i - y_k^j)^2 + (z_i - z_k^j)^2}\,,$$

$$m_{k,k+1} = \frac{y_{k+1}^j - y_k^j}{x_{k+1}^j - x_k^j}\,,\ e_k = (x_i - x_k^j)^2 + (z_i - z_k^j)^2\,,\ h_k = (x_i - x_k^j)^2 (y_i - y_k^j)^2\,,$$

the formulas of Hess and Smith are

$$\boldsymbol{X}_{ji} = (X_{ji}, Y_{ji}, Z_{ji})\,,$$

$$X_{ji} = \frac{1}{4\pi} \sum_{k=1}^{3(4)} \frac{y_{k+1}^j - y_k^j}{d_{k,k+1}} \ln \frac{r_k + r_{k+1} - d_{k,k+1}}{r_k + r_{k+1} + d_{k,k+1}}\,,$$

$$Y_{ji} = -\frac{1}{4\pi} \sum_{k=1}^{3(4)} \frac{x_{k+1}^j - x_k^j}{d_{k,k+1}} \ln \frac{r_k + r_{k+1} - d_{k,k+1}}{r_k + r_{k+1} + d_{k,k+1}}\,,$$

$$Z_{ji} = -\frac{1}{4\pi} \sum_{k=1}^{3(4)} \left(\arctan \frac{m_{k,k+1} e_k - h_k}{(z_i - z_k^j) r_k} - \arctan \frac{m_{k,k+1} + e_{k+1} - h_{k+1}}{(z_i - z_k^j) r_{k+1}} \right).$$

In fact, Hess and Smith deduced the formulas assuming that the panels Π_j are situated in planes parallel to xOy $(z_k^j = 0)$.

An example is presented in [7.3].

Chapter 8

The Supersonic Steady Flow

8.1 The Thin Airfoil of Infinite Span

8.1.1 The Analytical Solution

In this subsection we study the same problem like in 3.1. The only difference is that the unperturbed flow is assumed to be supersonic ($M > 1$). Hence the uniform flow, having the velocity $U\boldsymbol{i}$, the pressure p_∞ and the density ρ_∞, is perturbed by the presence of an infinite cylindrical body. The xOy reference plane represents a cross section and the leading edge is the origin of the axes of coordinates. The length of the projection of the airfoil on the direction of the unperturbed stream (which does not always coincide to the length of the chord) will be taken as reference length (fig 8.1.1). Utilizing the coordinates defined

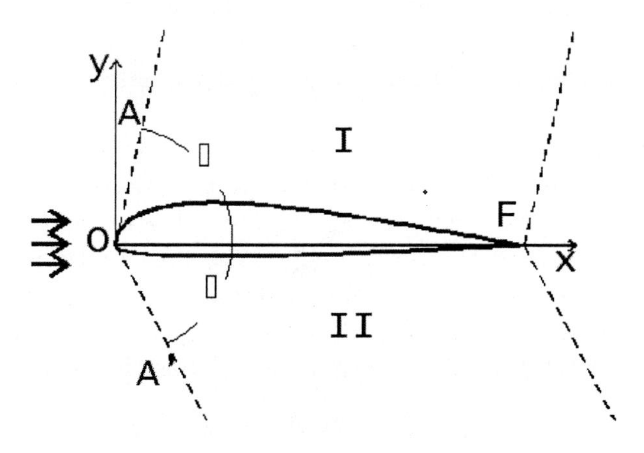

Fig. 8.1.1.

by (2.1.1), we shall denote the equations of the profile determined by the xOy plane

$$y = h_\pm(x)\,, \tag{8.1.1}$$

the functions $h_{\pm}(x)$ being defined on the interval $[0,1]$ and and possessing first order derivatives.

Taking into account the formulas (2.1.3), we deduce that the perturbation (which is obviously steady), will be defined by the system (2.1.32) and by the boundary conditions (2.1.33). From this system, taking into account that the perturbation is plane, it results:

$$u = -p, \quad v_x + p_y = 0, \quad M^2 p_x + u_x + v_y = 0\,, \tag{8.1.2}$$

whence one obtains

$$k^2 v_{xx} - v_{yy} = 0\,, \tag{8.1.3}$$

with the usual notation $k = \sqrt{M^2 - 1}$. The boundary condition (2.1.32) becomes

$$v(x, \pm 0) = h'_{\pm}(x), \quad 0 \le x \le 1\,, \tag{8.1.4}$$

and the damping conditions at infinity

$$\lim_{x \to -\infty} (p, u, v) = 0. \tag{8.1.5}$$

The equation (8.1.3) is hyperbolic and the families of characteristics C_+ and C_- are defined by the equations

$$x - ky = c_+, \quad x + ky = c_-\,. \tag{8.1.6}$$

Obviously, the first family consists of parallel half-lines which make the angle μ with the Ox axis, the second consists of parallel half-lines which make the angle $-\mu$ with the Ox axis, where

$$\tan\mu = \frac{1}{k}\left(\sin\mu = \frac{1}{M}, \quad \cos\mu = \frac{k}{M}\right). \tag{8.1.7}$$

For determining the characteristics passing by a certain point, we write that the coordinates of the point verify the equations (8.1.15) and and then we determine the constants c_+ and c_-. For example, the characteristics detaching from the leading edge have the equations $x - ky = 0$ and $x + ky = 0$. We assume that the profile is in the interior of the angle having the opening 2μ which is determined by these characteristics.

It is well known that the equation (8.1.3) has the general solution

$$v(x, y) = F(x - ky) + G(x + ky), \tag{8.1.8}$$

such that from (8.1.2) and (8.1.5) it results

$$\begin{aligned} kp(x, y) &= F - G \\ ku(x, y) &= -F + G\,. \end{aligned} \tag{8.1.9}$$

The function F is defined on the characteristics from the superior half-plane, and the function G on the characteristics from the inferior half-plane. We have in view the characteristics detaching from the segment $[0, 1]$ which replaces the profile that represents the source of perturbations. The solution from the strip I will be determined by $F(x - ky)$ which will be denoted by $F_+(x - ky)$ and the solution from the strip II will be determined by $G(x + ky)$ which will be denoted by $F_-(x + ky)$. Hence the solution has the shape

$$\begin{aligned} v_\pm(x,y) &= F_\pm(x \mp ky)\,, \\ kp_\pm(x,y) &= \pm F_\pm(x \mp ky), \\ ku_\pm(x,y) &= \mp\, F_\pm(x \mp ky), \end{aligned} \tag{8.1.10}$$

the upper sign corresponding to the solution from the $y > 0$ half-plane and the lower sign to the solution from the $y < 0$ half-plane.

In fact one may demonstrate that if upstream of AOA' the flow is not perturbed, then the solution of the equation (8.1.3) has necessarily the form (8.1.10). Indeed, if the assumption is true, it means that OA and OA' are lines of discontinuity where we have to impose the jump relations (1.3.8), (1.3.9) and (1.3.10)

$$[\![\rho_1 V_{1n}]\!] = 0\,, \quad [\![\rho_1 \boldsymbol{V}_1 V_{1n} + p_1 \boldsymbol{n}]\!] = 0\,, \tag{8.1.11}$$

with the notations (2.1.1) and (2.1.3). Taking into account the order of magnitude of the perturbations, by linearization it results

$$[\![\boldsymbol{v} \cdot \boldsymbol{n} + \rho n_x]\!] = 0\,, \quad [\![\boldsymbol{v} n_x + p\boldsymbol{n}]\!] = 0\,. \tag{8.1.12}$$

On the line OA we have $\boldsymbol{n} = (\sin\mu, -\cos\mu)$. Taking into account that in front of OA and OA' the perturbation vanishes, from (8.1.12) we obtain:

$$\begin{aligned} 0 &= u\sin\alpha - v\cos\alpha + \rho\sin\alpha \\ 0 &= v\sin\alpha - p\cos\alpha\,, \end{aligned} \tag{8.1.13}$$

the relationships being imposed on the line $x = ky$. Replacing here u, v and p given by (8.1.8) and (8.1.9) and taking (8.1.7) into account, from the second condition we find $G = 0$. Since in the linear steady theory we have $\rho = M^2 p$, it follows that the first condition is identically verified. In the same way one demonstrates that in the zone II $v = G(x + ky)$. Hence the solution of the equation (8.1.3) has the form (8.1.10).

Imposing the boundary conditions (8.1.4), it results

$$F_{\pm}(x) = h'_{\pm}(x), \quad 0 \leq x \leq 1 . \tag{8.1.14}$$

These relations determine the functions $F_{\pm}(x)$ on the segment $[0,1]$. In the exterior of this segment we take $F_{\pm}(x) = 0$ because v and p are continuous functions on the Ox axis in the exterior of the segment $[0,1]$ (it results from (8.1.12), setting $\boldsymbol{n} = (0,1)$), such that $F_+ = F_-, F_+ = -F_-$. Hence the perturbation zones are I and II. For $c \in [0,1]$, on the half-line having the equation $x - ky = c$ we deduce:

$$v(x,y) = h'_+(c), \quad kp(x,y) = h'_+(c), \tag{8.1.15}$$

and on the half-line $x + ky = c$

$$v(x,y) = h'_-(c), \quad kp(x,y) = h'_-(c) . \tag{8.1.16}$$

Along a characteristic, the velocity and the pressure take the values that they have in the point where the characteristic intersects the segment $[0,1]$.

In the sequel we shall prove that all these results may be obtained directly with the fundamental solutions method, as we could expect, taking into account that this is a global method.

8.1.2 The Fundamental Solutions Method

We know that the perturbation produced in the uniform stream having the velocity $U\boldsymbol{i}$, the pressure p_∞ and the density ρ_∞ by a force density (f_1, f_2) uniformly distributed along an axis which is parallel to Oz and intersects the xOy plane in the point of coordinates $(\xi, 0)$, is given by the formulas (2.3.30) and (2.3.31). Replacing the profile from figure 8.1.1 with such a density applied on the segment $[0,1]$ from the Ox axis (i.e. on the strip whose cross section is the segment $[0,1]$ from the Ox axis), the perturbation will be given by the formulas

$$\begin{aligned} 2kp(x,y) &= \int_0^1 \big[- f_1(\xi) + kf(\xi)\operatorname{sign} y \big] \delta(x_0 - k|y|)\mathrm{d}\,\xi \\ 2v(x,y) &= \int_0^1 \big[- f_1(\xi)\operatorname{sign} y + kf(\xi) \big] \delta(x_0 - k|y|)\mathrm{d}\,\xi . \end{aligned} \tag{8.1.17}$$

Taking into account the property of the distribution δ:

$$\int_a^b f(\xi)\delta(x-\xi)\mathrm{d}\,\xi = \begin{cases} f(x), & \text{if} x \in (a,b) \\ 0, & \text{if} x \in C(a,b), \end{cases} \tag{8.1.18}$$

we deduce that if $x - k|y| = c \in (0,1)$, then we have:

$$\begin{aligned} 2kp(x,y) &= -f_1(c) + kf(c)\operatorname{sign} y \\ 2v(x,y) &= -f_1(c)\operatorname{sign} y + kf(c)\,. \end{aligned} \tag{8.1.19}$$

Conversely, if $c \in C(0,1)$, then $p = 0, v = 0$. The set of points for which $x - k|y| = c$ determines the characteristic half-lines

$$x \pm ky = c \tag{8.1.20}$$

which detach from the point $x = c$ belonging to the segment $(0,1)$. On these half-lines, the pressure and the velocity have constant values, equal to their values in c.

8.1.3 The Aerodynamic Action

From the representation (8.1.10) we deduce that the jump of the perturbation pressure on the profile

$$[\![p]\!] \equiv p(x,-0) - p(x,+0)$$

is given by the formula

$$k[\![p]\!] = -[F_+(x) + F_-(x)] = -[h'_+(x) + h'_-(x)] \tag{8.1.21}$$

which results from (8.1.14). In this way, the lift and moment coefficients

$$c_L = \frac{L}{(1/2)\rho_\infty U^2 L_0}\,, \quad c_M = \frac{M}{(1/2)\rho_\infty U^2 L_0^2}\,, \tag{8.1.22}$$

are given by the formulas

$$\begin{aligned} c_L &= -\frac{2}{k}\int_0^1 [h'_+(x) + h'_-(x)]\,\mathrm{d}\,x \\ c_M &= -\frac{2}{k}\int_0^1 x[h'_+(x) + h'_-(x)]\,\mathrm{d}\,x\,, \end{aligned} \tag{8.1.23}$$

the moment M being calculated with respect to the origin of the axes of coordinates.

We notice from the formulas (8.1.10) and (8.1.23) that the linear theory cannot be applied in the vicinity of the value $M = 1 (k = 0)$.

As a first example, we shall consider again the case of the flat plate having the angle of attack ε with $\varepsilon < \mu$ (fig. 8.1.2). Since the equation of this profile is $y = -\varepsilon x$, we deduce $h'_+ = h'_- = -\varepsilon$ whence

$$v = -\varepsilon, \quad ku = \pm\varepsilon, \quad kp = \mp\,\varepsilon\,, \tag{8.1.24}$$

the total velocity and pressure being

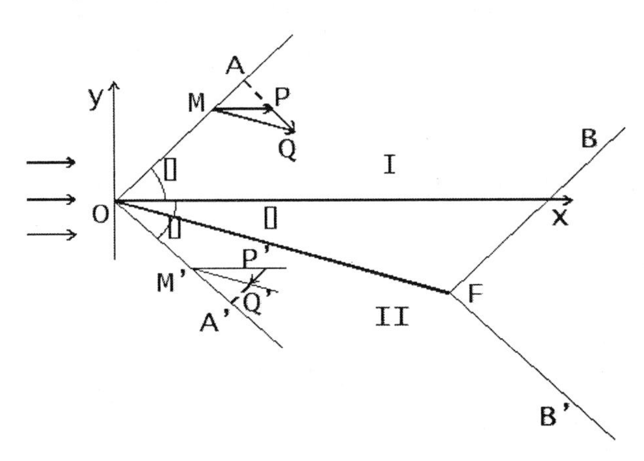

Fig. 8.1.2.

$$\boldsymbol{v}_1 = U\left[\left(1 \pm \frac{\varepsilon}{k}\right)\boldsymbol{i} - \varepsilon\boldsymbol{j}\right], \quad p_1 = p_\infty \mp \rho_\infty U^2 \frac{\varepsilon}{k}\,. \tag{8.1.25}$$

For c_L and c_M we obtain

$$c_L = \frac{4\varepsilon}{k}\,, \quad c_M = \frac{2\varepsilon}{k}\,. \tag{8.1.26}$$

These formulas have been obtained for the first time by Ackeret in 1925. The torsor of the aerodynamic forces reduces to a lifting force L, applied in the middle of the chord of the profile, force which tends to rotate the airfoil in the trigonometric sense.

The graphic representation of the function $c_L(M)$ (fig. 8.1.3) shows that the linear theory is not valid in the vicinity of the value $M = 1$, because the lift cannot be infinitely great. One estimates that the validity of this theory begins from approximately $M = 1,2$ and finishes to approximately $M = 2,5$, because the lift cannot be extremely small. For the zone $M = 1 (0,8 < M < 1,2)$ one elaborates the *theory of transonic flow* (see Chapter IX), and for $M > 2,5$, the *theory of hypersonic flow*.

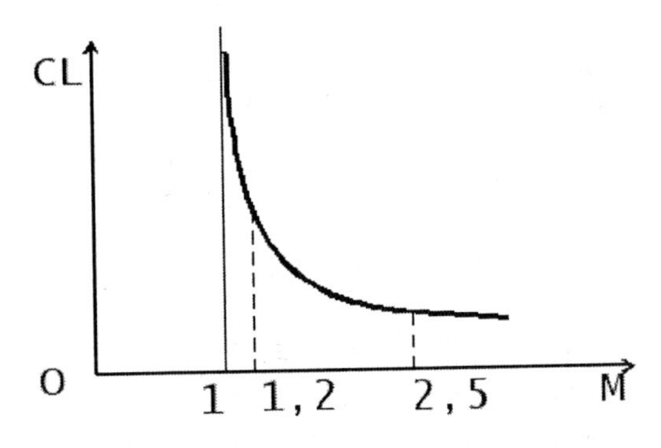

Fig. 8.1.3.

8.1.4 The Graphical Method

This method relies on the following fundamental theorem: *in every point the perturbation velocity is perpendicular to Mach's line (the characteristic line) which is passing through that point.* The demonstration results at once if we take into account that the characteristic lines have the versors $\boldsymbol{i}$ of coordinates $(\cos\mu, \pm\sin\mu)$ and we utilize the representation (8.1.10) and the formulas (8.1.7). Indeed, we have

$$\boldsymbol{v}\cdot\boldsymbol{i}_{\pm} = u\cos\mu \pm v\sin\mu = \mp\frac{1}{k}F_{\pm}\frac{k}{M} \pm F_{\pm}\frac{1}{M} = 0\,. \tag{8.1.27}$$

Let us show for example how we can utilize this theorem for obtaining the graphical representation of the perturbation in the case of the flat plate. We consider at first the region I from the 8.1.2. In an arbitrary point M from OA we draw the vector $\boldsymbol{MP} = U\boldsymbol{i}$ and a vector $\boldsymbol{V}_1$ parallel to the plate and having an arbitrary magnitude; the perpendicular on OA which passes through the vertex P of the vector $U\boldsymbol{i}$ determines the magnitude of the total velocity $\boldsymbol{V}_1 = \vec{\boldsymbol{MQ}}$; the perturbation velocity $U\boldsymbol{v}$ is given by the vector $\vec{\boldsymbol{PQ}}$. For determining the perturbation velocity from the region II, in a certain point M' from OA' ones draw the vectors $\vec{M'P'} = U\boldsymbol{i}$ and $\boldsymbol{V}_1'$, the last being parallel to the plate and having the magnitude unprecised yet; drawing from P' (the vertex of the vector $U\boldsymbol{i}$) the perpendicular line on OA' we determine the vector $\boldsymbol{V}_1' = \vec{\boldsymbol{M'Q'}}$ and then the perturbation $U_\infty\boldsymbol{v} = \vec{P'Q'}$.

Beyond FB and FB' the flow becomes again parallel to Ox.

One knows the non-linear solution of the supersonic flow past a convex dihedron(the Prandtl-Meyer fan). See for example [1.21], [1.34]. In the framework of the non-linear theory, the velocity which is parallel to Ox before OA, becomes at last parallel to the plate, crossing the Prandtl-Meyer fan. In the framework of the linear theory this fan is reduced to the half-line OA.

8.1.5 The Theory of Polygonal Profiles

In the case of polygonal profiles, the solution may be easily determined starting from the solution for the flat plate. The solution, which obviously is piecewise constant, may be analytical, graphical or mixed.

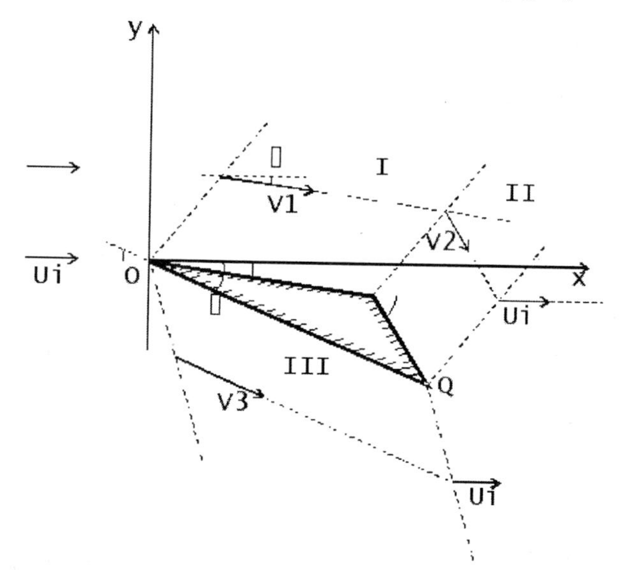

Fig. 8.1.4.

We shall consider at first the profile from figure 8.1.4 ($\varepsilon < \mu$) with $OP = \ell_1$, $PQ = \ell_2$ and $\ell_1 cos\varepsilon_1 + \ell_2 \cos\varepsilon_2 = 1$. According to the linear theory, the last relation will be replaced by $\ell_1 + \ell_2 = 1$. From the formulas (8.1.14) it results

$$F_+ = -\varepsilon_1, \quad 0 < x < \ell_1 \cos\varepsilon_1 \simeq \ell_1\,,$$

$$F_+ = -\varepsilon_2, \quad \ell_1 \cos\varepsilon_1 \simeq \ell_1 < x < 1\,, \tag{8.1.28}$$

$$F_1 = \varepsilon, \quad 0 < x < 1\,.$$

We deduce,

$$k\,[\![p]\!] = \begin{cases} \varepsilon + \varepsilon_1, & 0 < x < \ell_1 \\ \varepsilon + \varepsilon_2, & \ell_1 < x < 1 \end{cases} \tag{8.1.29}$$

whence

$$c_L = \frac{2}{k}(\varepsilon + \varepsilon_1\ell_1 + \varepsilon_2\ell_2),\ c_M = \frac{\varepsilon + \varepsilon_1}{k}\ell_1^2 + \frac{\varepsilon + \varepsilon_2}{k}(1 - \ell_1^2)\,. \qquad (8.1.30)$$

Further we shall determine the pressure in different zones starting from the solution for the flat plate. Taking (2.1.3) and (8.1.24) into account, we deduce in zone I that

$$p_1 = p_\infty - \rho_\infty U^2\frac{\varepsilon_1}{k}\,. \qquad (8.1.31)$$

In zone II we take into account that the deviation with respect to the direction of the uniform stream from zone I is $\varepsilon_2 - \varepsilon_1$. Using again the formula (8.1.24) it results

$$p_2 = p_1 - \rho_1 V_1^2\frac{\varepsilon_2 - \varepsilon_1}{k}\,. \qquad (8.1.32)$$

Using the formulas (2.1.3), we deduce $\rho_1 V_1^2 = \rho_\infty U^2 + \ldots$ Taking into account that the factor $(\varepsilon_2 - \varepsilon_1)$ has the order of magnitude of the perturbations, in the framework of the linear theory we have to retain

$$p_2 = p_1 - \rho_\infty U^2\frac{\varepsilon_2 - \varepsilon_1}{k} \simeq p_\infty - \rho_\infty U^2\frac{\varepsilon_2}{k}\,. \qquad (8.1.33)$$

In zone III we have:

$$p_3 = p_\infty + \rho_\infty U^2\frac{\varepsilon}{k}\,. \qquad (8.1.34)$$

For the jump of the pressure we get the formulas (8.1.29). In fact the formulas (8.1.33) and (8.1.34) coincide with the formulas given by the analytic method.

We shall calculate now, in a different way, the resultant of the pressure on the wing. The perpendicular to the plate OP oriented towards the body has the director cosines

$$\boldsymbol{n}_1 = \left(\cos\left(\varepsilon_1 + \frac{\pi}{2}\right), -\sin\left(\varepsilon_1 + \frac{\pi}{2}\right)\right) \simeq (-\varepsilon_1, -1)\,. \qquad (8.1.35)$$

In the linear approximation, the perpendicular to the plate PQ has the director cosines $(-\varepsilon_2, -1)$, and the perpendicular to OQ pointing towards the body $(\varepsilon, 1)$. Hence the resultant of the pressure is

$$\boldsymbol{R} = (p_1\ell_1\boldsymbol{n}_1 + p_2\ell_2\boldsymbol{n}_2 + p_3\boldsymbol{n}_3)L_0 \qquad (8.1.36)$$

We took into account that the real lengths of the plates are $\ell_1 L_0, l_2 L_0, L_0$. For the lift we obtain the formula

$$L = R_y = \rho_\infty\frac{U^2 L_0}{k}(\varepsilon + \varepsilon_1\ell_1 + \varepsilon_2\ell_2)\,. \qquad (8.1.37)$$

In the framework of the linear theory the drag

$$R_x = (\varepsilon - \ell_1\varepsilon_1 - \ell_2\varepsilon_2)L_0 p_\infty \tag{8.1.38}$$

vanishes because, from the projection of the contour OPQ on the Oy axis we deduce

$$\varepsilon = \ell_1\varepsilon_1 + \ell_2\varepsilon_2$$

In fact this is a known result. The drag has the order of magnitude two

$$c_D = \frac{2}{k}(\varepsilon^2 + \varepsilon_1^2\ell_1 + \varepsilon_2^2\ell_2) \tag{8.1.39}$$

In the sequel we shall consider the polygonal profile from figure 8.1.5, symmetric with respect to the OR axis. We assume that all the ε-s are small. Obviously, the profile is considered to be in the interior of Mach's angle with the vertex in the origin. We deduce like above that in zone I the pressure is given by the formula

$$p_1 = p_\infty + \rho_\infty U^2 \frac{\varepsilon_1}{k}, \tag{8.1.40}$$

in zone II by the formula

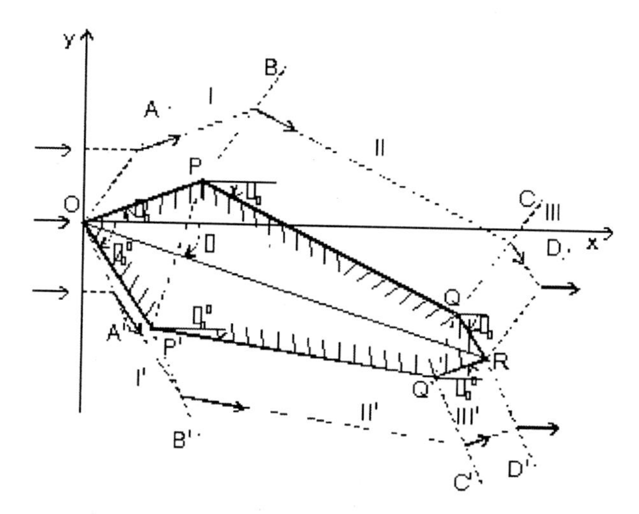

Fig. 8.1.5.

$$p_2 = p_1 - \rho_1 V_1^2 \frac{\varepsilon_1 + \varepsilon_2}{k} \simeq p_\infty - \rho_\infty U^2 \frac{\varepsilon_2}{k}, \tag{8.1.41}$$

in zone III by

$$p_3 = p_2 - \rho_2 V_2^2 \frac{\varepsilon_3 - \varepsilon_2}{k} \simeq p_\infty - \rho_\infty U^2 \frac{\varepsilon_3}{k}, \tag{8.1.42}$$

in zone I',

$$p_1' = p_\infty + \rho_\infty U^2 \frac{\varepsilon_1'}{k}, \quad \varepsilon_1' = 2\varepsilon + \varepsilon_1, \tag{8.1.43}$$

in zone II',

$$p_2' = p_1' - \rho_1' V_1'^2 \frac{\varepsilon_1' - \varepsilon_2'}{k} \simeq p_\infty + \rho_\infty U^2 \frac{\varepsilon_2'}{k}, \varepsilon_2' = 2\varepsilon - \varepsilon_2 \tag{8.1.44}$$

and in zone III',

$$p_3' = p_2' - \rho_2' V_2'^2 \frac{\varepsilon_2' + \varepsilon_3'}{k} \simeq p_\infty - \rho_\infty U^2 \frac{\varepsilon_3'}{k}, \varepsilon_3' = \varepsilon_3 - 2\varepsilon \tag{8.1.45}$$

We denote by $\boldsymbol{n}_1, \boldsymbol{n}_2, \boldsymbol{n}_3, \boldsymbol{n}_1', \boldsymbol{n}_2'$ and $\boldsymbol{n}_3'$ the perpendiculars to OP, respectively $PQ, QR, OP', P'Q'$ and $Q'R$ pointing towards the body. In the framework of the linear theory they will have the coordinates

$$(\varepsilon_1, -1), (-\varepsilon_2, -1), (-\varepsilon_3, -1), (\varepsilon_1', 1), (\varepsilon_2', 1) \text{ and } (-\varepsilon_3', 1).$$

The resultant of the pressure is

$$\boldsymbol{R} = L_0(\ell_1 p_1 \boldsymbol{n}_1 + \ell_2 p_2 \boldsymbol{n}_2 + \ell_3 p_3 \boldsymbol{n}_3 + \ell_1 p_1' \boldsymbol{n}_1' + \ell_2 p_2' \boldsymbol{n}_2' + \ell_3 p_3' \boldsymbol{n}_3'). \tag{8.1.46}$$

The lift is

$$R_y = \frac{\rho_\infty U^2 L_0}{k} \left[\ell_1(\varepsilon_1' - \varepsilon_1) + \ell_2(\varepsilon_2' + \varepsilon_2) + \ell_3(\varepsilon_3 - \varepsilon_3') \right]. \tag{8.1.47}$$

Taking into account $\varepsilon_1', \varepsilon_2', \varepsilon_3'$, and the relation $\ell_1 + \ell_2 + \ell_3 = 1$, we deduce

$$c_L = \frac{4\varepsilon}{k}. \tag{8.1.48}$$

In the framework of the linear theory, the drag

$$R_x = 2p_\infty L_0(\varepsilon + \ell_1\varepsilon_1 - \ell_2\varepsilon_2 - \ell_3\varepsilon_3) \tag{8.1.49}$$

vanishes, because from the projection of the polygonal contour $ORQ'P'$ on the Oy axis we deduce

$$\varepsilon = \ell_1\varepsilon_1' + \ell_2\varepsilon_2' - \ell_3\varepsilon_3' \tag{8.1.50}$$

whence, taking into account the expressions $\varepsilon_1', \varepsilon_2', \varepsilon_3'$ we get

$$\varepsilon + \ell_1\varepsilon_1 - \ell_2\varepsilon_2 - \ell_3\varepsilon_3 = 0.$$

The formula (8.1.39) is very important. It shows that c_L has the same expression like in the case of the wing consisting only of its skeleton OR. The result is natural, because of the symmetry of the wing with respect to OR.

8.1.6 Validity Conditions

We shall investigate, for the flat plate, the limits of validity of the linear theory.

Since the flow is defined by the formulas (8.1.25) in I and II (fig. 8.1.1) and by $V_1 = U\boldsymbol{i}$, $p_1 = p_\infty$ in the regions upstream of AOA' and downstream of BQB', we find that $p_1^I < p_1^{II}$ and $p_1^I < p_\infty$. Hence, in order to have a positive pressure all over the fluid (this is a physical condition), we must have

$$p_1^I > 0\,, \tag{8.1.51}$$

this representing a *first* condition of validity. Taking into account the expression of p_1^I given in (8.1.25), we deduce, with the notation $kX = M^2\varepsilon$,

$$0 < \gamma X < 1\,. \tag{8.1.52}$$

A second condition must ensure us that the flow is everywhere supersonic. Since from (8.1.25) or from Bernoulli's integral it results $|V_1^{II}|^2 < |V_1^I|^2$, we deduce that we must have $(c_1^{II})^2 < (V_1^{II})^2$, whence

$$\gamma p_1^{II}/\rho_1^{II} = (c_1^{II})^2 < U^2\left[\left(1-\frac{\varepsilon}{k}\right)^2 + \varepsilon^2\right]. \tag{8.1.53}$$

Taking into account that in the framework of the steady linear theory we have $\rho = M^2 p$ (it results from (2.1.7)), we deduce the inequality

$$\frac{1+\gamma X}{1+X} < M^2\left[\left(1-\frac{\varepsilon}{k}\right)^2 + \varepsilon^2\right], \tag{8.1.54}$$

equivalent to

$$1+\gamma X < (1+X)(M^2 - 2X + X^2)\,. \tag{8.1.55}$$

In the linear approximation with respect to ε, this inequality becomes

$$1 - M^2 < (M^2 - 2 - \gamma)X\,. \tag{8.1.56}$$

Further we shall deal with the inequalities (8.1.52) and (8.1.56) when the fluid is the air ($\gamma = 1{,}405$). For $M^2 < 3{,}405$, the two inequalities are satisfied in the shaded region from the (M^2, X) plane (fig. 8.1.6), where

$$f(M^2) = \frac{M^2-1}{3{,}405 - M^2}\,. \tag{8.1.57}$$

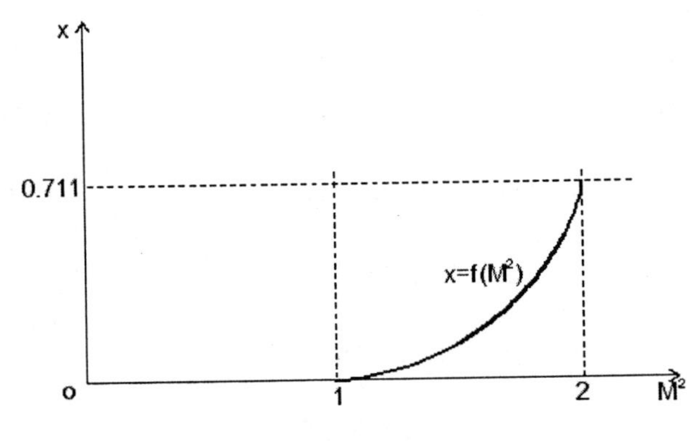

Fig. 8.1.6.

This result shows that the linear theory is valid only in the zone $1 < M^2 < 2$. For a certain value given to M^2 in this zone, ε is subjected to the restriction

$$\varepsilon < \frac{(M^2-1)^{3/2}}{M^2(3,405-M^2)} \cdot \tag{8.1.58}$$

The last ratio increases when M^2 increases, hence the superior limit of the angle of attack is increasing with M^2.

Other results concerning this subject are given in [8.17] [8.32].

8.2 Ground and Tunnel Effects

8.2.1 The General Solution

We assume now that the wing lies between two parallel planes having the equations

$$y = -\frac{a}{2} \quad \text{and} \quad y = \frac{b}{2} \cdot \tag{8.2.1}$$

In this case, the general solution (8.1.8) and (8.1.9)

$$\begin{aligned} ku(x,y) &= -F(x-ky) + G(x+ky) \\ v(x,y) &= F(x-ky) + G(x+ky) \\ p(x,y) &= -u(x,y)\,, \end{aligned} \tag{8.2.2}$$

with the damping condition

$$\lim_{x\to-\infty} F(x-ky) = \lim_{x\to-\infty} G(x+ky) = 0\,, \tag{8.2.3}$$

has to satisfy besides the slipping conditions on the profile,

$$v(x, \pm 0) = h'_{\pm}(0), \quad x \in [0,1], \tag{8.2.4}$$

the slipping conditions on the planes

$$v\left(x, -\frac{a}{2}\right) = 0, \quad v\left(x, \frac{b}{2}\right) = 0, \quad x \in \mathbb{R}. \tag{8.2.5}$$

We assumed that the equations of the profile are

$$y = h_{\pm}(x) \tag{8.2.6}$$

In the sequel we have to calculate the velocity field in the domain

$$D^- = \left\{(x,y) : x + ky \in [0,1], 0 \geq y \geq -\frac{a}{2}\right\}. \tag{8.2.7}$$

We consider the natural number n such that $nka < 1 \leq (n+1)ka$ and the segments $(0, ka)$, $(ka, 2ka), ...,$ $((n-1)ka, nka)$, $(nka, 1)$ on the chord of the profile which is the segment $[0,1]$ from the Ox axis. We consider the characteristic having the equation $\xi - k\eta = C \in (0, ka)$. For $\eta = -\frac{a}{2}$ we have $\xi \in \left(-\frac{ka}{2}, \frac{ka}{2}\right)$.

The characteristic having the equation $x + ky = \xi - \frac{ka}{2}$, $\xi \in$ $\in \left(-\frac{ka}{2}, \frac{ka}{2}\right)$ does not intersect the chord of the profile, i.e. the segment $y = 0$, $x \in [0,1]$. We have, by virtue of condition (8.2.3)

$$G\left(\xi - \frac{ka}{2}\right) = \lim_{x+ky=\xi-\frac{ka}{2}, x \to -\infty} G(x+ky) = 0, \; \xi \in \left(-\frac{ka}{2}, \frac{ka}{2}\right). \tag{8.2.8}$$

From (8.2.5) it results

$$0 = v\left(\xi, -\frac{a}{2}\right) = F\left(\xi + \frac{ka}{2}\right) + G\left(\xi - \frac{ka}{2}\right), \tag{8.2.9}$$

and from (8.2.8) and (8.2.9)

$$F\left(\xi + \frac{ka}{2}\right) = 0, \quad \xi \in \left(-\frac{ka}{2}, \frac{ka}{2}\right). \tag{8.2.10}$$

Utilizing the notations

$$G(x+ky) = G^-(x+ky), F(x-ky) = F^-(x-ky), (x,y) \in D^- \tag{8.2.11}$$

we have on the upper surface of the profile

$$F^{-}(x) = 0, \quad x \in (0, ka). \tag{8.2.12}$$

From the previous equation and from the slipping condition on the profile (8.2.4)

$$h'_{-}(x) = v(x, -0) = F^{-}(x) + G^{-}(x), x \in [0, 1] \tag{8.2.13}$$

it follows that

$$G^{-}(x) = h'_{-}(x) \quad x \in (0, ka). \tag{8.2.14}$$

Further one demonstrates by induction that for $x \in (jka, (j+1)ka)$, $j \in N$, $j \leq n$, we have:

$$G^{-}(x) = h'_{-}(x) + h'_{-}(x - ka) + h'_{-}(x - 2ka) + \ldots$$
$$\ldots + h'_{-}(x - jka), \tag{8.2.15}$$
$$F^{-}(x) = -h'_{-}(x - ka) - h'_{-}(x - 2ka) - \ldots - h'_{-}(x - jka). \tag{8.2.16}$$

Indeed, for $j = 0$ the relations are true. Assuming that the relations are true for j, we shall demonstrate that they are also true for $j + 1$.

Let us consider $x \in \big((j + 1)ka, (j + 2)ka\big)$. From (8.2.5) we obtain:

$$0 = v\left(x - k\frac{a}{2}, -\frac{a}{2}\right) = G^{-}(x - ka) + F^{-}(x).$$

Since $x - ka \in (\ell ka, (\ell + 1)ka)$, by the induction hypothesis we have

$$F^{-}(x) = -G^{-}(x - ka) = -h'_{-}(x - ka) - h'_{-}(x - 2ka) - \ldots$$
$$\ldots - h'_{-}(x - (j + 1)ka). \tag{8.2.17}$$

From the slipping condition on the profile

$$h'_{-}(x) = v(x, -0) = F^{-}(x) + G^{-}(x)$$

and from (8.2.14) it follows:

$$G^{-}(x) = h'_{-}(x) + h'_{-}(x - ka) + h'_{-}(x - 2ka) + \ldots + h'_{-}(x - (j + 1)ka), \tag{8.2.18}$$

whence we deduce by induction the validity of the relations (8.2.15) and (8.2.16).

Taking into account that

$$p(x, y) = -u(x, y) = \frac{1}{k}\left[F(x - ky) - G(x + ky)\right], \tag{8.2.19}$$

we may calculate the velocity field on the lower surface of the profile

$$p(x,-0) = -\frac{1}{k}\left[h'_-(x) + 2h'_-(x-ka) + \ldots + 2h'_-(x-jka)\right], \qquad x \in (jka,(j+1)ka) \cap [0,1]. \tag{8.2.20}$$

Considering the domain

$$D^+ = \left\{(x,y) : x - ky \in [0,1], \quad 0 \le y \le \frac{b}{2}\right\} \tag{8.2.21}$$

and utilizing the notations

$$G(x+ky) = G^+(x+ky), F(x-ky) = F^+(x-ky), \quad (x,y) \in D^+, \tag{8.2.22}$$

we can prove, like in the previous case, that for $x \in (jkb,(j+1))kb) \cap [0,1]$ we have

$$G^+(x) = -h'_+(x-kb) - h'_+(x-2kb) - \ldots - h'_+(x-jkb), \tag{8.2.23}$$

$$F^+(x) = h'_+(x) + h'_+(x-kb) + h'_+(x-2kb) + \ldots + h'_+(x-jkb), \tag{8.2.24}$$

whence it results

$$p(x,+0) = \frac{1}{k}\left[h'_+(x) + 2h'_+(x-kb) + \ldots + 2h'_+(x-jkb)\right], \qquad x \in (jkb,(j+1)kb) \cap [0,1]. \tag{8.2.25}$$

8.2.2 The Aerodynamic Coefficients

As we know from section 8.1, the aerodynamic coefficients are given by the formulas

$$c_L = \int_0^1 [p(x,-0) - p(x,+0)]\,\mathrm{d}\,x$$
$$c_M = \int_0^1 x\,[p(x,-0) - p(x,+0)]\,\mathrm{d}\,x. \tag{8.2.26}$$

In the sequel we shall consider some examples.

For the profile in a *free stream* $(a = \infty, b = \infty)$, it results the already known formulas (8.1.23).

For the thin profile in *ground effects* ($b = \infty$), taking for example $1 \geq ka \geq 1/2$, we deduce

$$c_L = -\frac{1}{k}\int_0^1 \left[h'_+(x) + h'_-(x)\right] \mathrm{d}\,x - \frac{2}{k}\int_0^{1-ka} h'_- \mathrm{d}\,x$$

$$c_M = -\frac{1}{k}\int_0^1 x\left[h'_+(x) + h'_-(x)\right] \mathrm{d}\,x - \frac{2}{k}\int_0^{1-ka} (x+ka)h'_- \mathrm{d}\,x\,. \tag{8.2.27}$$

For the thin profile in a wind tunnel, taking for example $1 \geq ka \geq \frac{1}{2}$, $1 \geq kb \geq 1/2$, we get

$$c_L = -\frac{1}{k}\int_0^1 \left[h'_+(x) + h'_-(x)\right] \mathrm{d}\,x - \frac{2}{k}\int_0^{1-ka} h'_-(x) \mathrm{d}\,x -$$

$$-\frac{2}{k}\int_0^{1-kb} h'_+(x) \mathrm{d}\,x\,,$$

$$c_M = -\frac{1}{k}\int_0^1 x\left[h'_+(x) + h'_-(x)\right] \mathrm{d}\,x - \frac{2}{k}\int_0^{1-ka} (x+ka)h'_-(x) \mathrm{d}\,x -$$

$$-\frac{2}{k}\int_0^{1-kb} (x+kb)h'_+(x) \mathrm{d}\,x\,. \tag{8.2.28}$$

For the flat plate with the angle of attack ε, we have $h_+(x) = h_-(x) = -\varepsilon x$, $x \in [0,1]$, whence

$$c_L = \frac{2\varepsilon}{k}\,, \quad c_M = \frac{\varepsilon}{k}\,, \tag{8.2.29}$$

for the profile in a free stream,

$$c_L = \frac{2\varepsilon}{k}(2-ka), c_M = \frac{2}{k}(2-k^2a^2)\,, \tag{8.2.30}$$

for the profile in ground effects ($1 \geq k \geq 1/2$) and

$$c_L = \frac{2\varepsilon}{k}(3-ka-kb), \quad c_M = \frac{\varepsilon}{k}(3-k^2a^2-k^2b^2)\,, \tag{8.2.31}$$

for the profile in tunnel effects ($1 \geq ka \geq 1/2$, $1 \geq kg \geq 1/2$).

This subsection was written following the paper [8.8].

8.3 The Three-Dimensional Wing

8.3.1 Subsonic and Supersonic Edges

In this subsection we present the general theory of the thin wing in a supersonic stream. We shall utilize the coordinates (2.1.1) and the fields (2.1.3). The free flow is by hypothesis supersonic. Like in the subsonic case, we shall denote by

$$z = h(x,y) \pm h_1(x,y)$$
$$= \varepsilon[\overline{h}(x,y) \pm \overline{h}_1(x,y)] \tag{8.3.1}$$

the equations of the upper and lower surfaces of the wing. The projection of the wing on the xOy plane will be the domain D, assumed to be simple connected. On the boundary Γ of this domain we have:

$$h_1(x,y) = 0\,.$$

We assume that Γ is smooth. Then there exist a point F where the tangent to Γ makes with the direction of the stream at infinity the angle of Mach μ defined by the formulas (8.1.7) and a point A, where the tangent to Γ makes with the direction of the stream at infinity the angle $-\mu$ (fig. 8.3.1). The point of intersection of these tangents will be considered the origin of the frame of reference. There also exist two

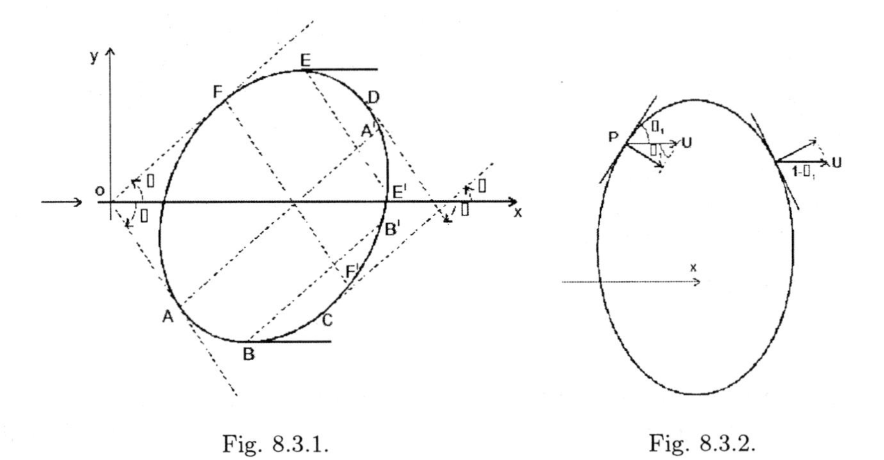

Fig. 8.3.1. Fig. 8.3.2.

points B and E where the tangents are parallel to the direction of the stream at infinity. As we know from the subsonic case, the points B and E separate the boundary Γ in two portions: the leading edge

$EFAB$ and the trailing edge $BCDE$ (C and D are the points where the tangents make the angles μ respectively $-\mu$ with the direction of the free stream). Obviously, we assume here again that every parallel to the direction of the stream at infinity intersects the edge Γ in at most two points at a finite distance.

Definition. *We name supersonic (subsonic) part of the leading or trailing edge, the part for which the absolute value of the component normal to the edge of the velocity of the free stream is greater (smaller) then the sound velocity.*

We shall prove that this definition is equivalent to the following one: *If in a certain point of the leading or trailing edge the angle of the tangent to the edge with the direction of the unperturbed stream is greater (respectively smaller) than Mach's angle, then in that point the edge is supersonic (respectively subsonic).*

Indeed, from figure 8.3.2 it results that the component normal to the edge of the velocity of the free stream in the generic point P has the magnitude $U \sin \mu_1$. If this is greater than the velocity of the sound in the unperturbed flow we have $U \sin \mu > c$, whence

$$\sin \mu_1 > \frac{1}{M} = \sin \mu$$

Utilizing the second definition, it results that the edge FA_1A from figure 8.3.1 is a supersonic leading edge, the edges AB and FE are subsonic leading edges, the edges BC and DE are subsonic trailing edges, and the edge $CB'E'D$ is a supersonic trailing edge.

It is known from the theory of hyperbolic partial differential equations (see also the plane problem from 8.1 and 8.2) that the zones of influence are the zones delimited by the characteristic lines. For example, in figure 8.3.1, the zone of influence of the subsonic leading edge FE is $FF'E'E$, FF' and EE' being parallel to OA.

Definition. *We name wing with independent subsonic leading or trailing edges, a wing for which the zones of influence of these edges are disjoint.*

It results therefore that a wing has independent subsonic leading edges if the Mach lines AA' and FF' do not intersect in the domain D and independent subsonic trailing edges if BB' and EE' do not intersect in D. For example the wing from figure 8.3.1 has dependent subsonic leading edges and independent subsonic trailing edges and the wing from figure 8.3.3 has only independent subsonic edges.

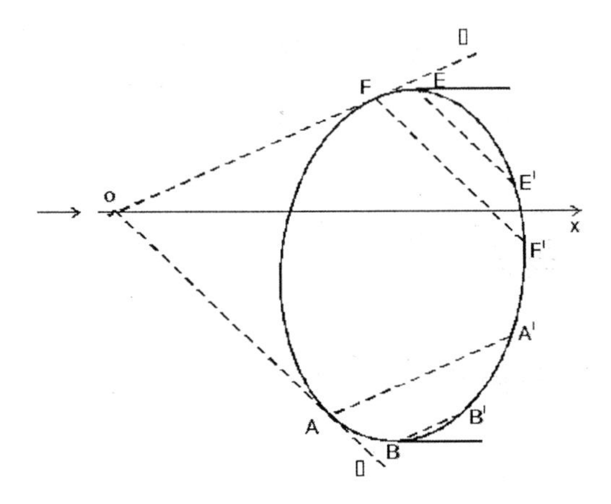

Fig. 8.3.3.

8.3.2 The Representation of the General Solution

Like in the subsonic case, we shall replace the wing with a continuous distribution of forces having the form $f = (f_1, 0, f)$ defined on D. We shall see that we may determine such a structure of $\boldsymbol{f}$, in order to satisfy the boundary conditions. The perturbation of the pressure determined in the uniform stream by the distribution $\boldsymbol{f}$ will be, according to the formula (2.3.32),

$$p(x, y, z) = -\frac{1}{2\pi} \iint_D \left[f_1(\xi, \eta) \frac{\partial}{\partial x} + f(\xi, \eta) \frac{\partial}{\partial x} \right] G(x_0, y_0, z) \mathrm{d}\xi \, \mathrm{d}\eta \,, \tag{8.3.2}$$

where

$$G(x_0, y_0, z) = \frac{H(x_0 - s)}{\sqrt{x_0^2 - s^2}} \,, \quad s = k\sqrt{y_0^2 + z^2} \,, \quad x_0 = x - \xi \,, \quad y_0 = y - \eta \,. \tag{8.3.3}$$

For the velocity field, from (2.3.12) and (2.3.34), it results

$$v(x, y, z) = \delta(z) \iint_D f(\xi, \eta) H(x_0) \delta(y_0) \mathrm{d}\xi \, \mathrm{d}\eta + \nabla \varphi \,, \tag{8.3.4}$$

where

$$\varphi(x, y, z) = \frac{1}{2\pi} \iint_D \left[f_1(\xi, \eta) - f(\xi, \eta) \frac{x_0 z}{y_0^2 + z^2} \right] G(x_0, y_0, z) \mathrm{d}\xi \, \mathrm{d}\eta \,, \tag{8.3.5}$$

Obviously, the perturbation is potential excepting the trace of the domain D in the uniform stream, where the first term from the expression of $\boldsymbol{v}$ does not vanish.

For the component w from (2.3.12) and (2.3.34) it results:

$$w(x,y,z) = \delta(z)\iint_D f(\xi,\eta)H(x_0)\delta(y_0)\mathrm{d}\xi\ \mathrm{d}\eta +$$

$$+\frac{1}{2\pi}\frac{\partial}{\partial z}\iint_D \left[f_1(\xi,\eta) - f(\xi,\eta)\frac{x_0 z}{y_0^2+z^2}\right] G(x_0,y_0,z)\mathrm{d}\xi\ \mathrm{d}\eta\,. \tag{8.3.6}$$

For w we also have the representation:

$$\begin{aligned} w(x,y,z) &= \frac{1}{2\pi}\iint_D f_1(\xi,\eta)\frac{\partial}{\partial z}G(x_0,y,z)\mathrm{d}\xi\ \mathrm{d}\eta + \\ &+\frac{1}{2\pi}\iint_D f(\xi,\eta)N(x_0,y_0,z)\mathrm{d}\xi\ \mathrm{d}\eta\,, \end{aligned} \tag{8.3.7}$$

where

$$N(x_0,y_0,z) = k^2\frac{\partial}{\partial x}G(x_0,y_0,z) + \frac{\partial}{\partial y}\left[\frac{x_0 y_0}{y_0^2+z^2}G(x_0,y_0,z)\right] \tag{8.3.8}$$

which results from (2.3.37) and the representation

$$w(x,y,z) = \frac{1}{2\pi}\iint_D f_1(\xi,\eta)\frac{\partial}{\partial z}G(x_0,y_0,z)\mathrm{d}\xi\ \mathrm{d}\eta +$$

$$+\frac{1}{2\pi}\left(k^2\frac{\partial^2}{\partial x^2} - \frac{\partial^2}{\partial y^2}\right)\iint_D f(\xi,\eta)\left[\int_{-\infty}^{x_0}\frac{H(\tau-s)}{\sqrt{\tau^2-s^2}}\mathrm{d}\tau\right]\mathrm{d}\xi\ \mathrm{d}\eta\,, \tag{8.3.9}$$

which results from (2.3.38). Each of these representations determine an integral equation for the function $f(x,y)$. All the known representations (Evvard [8.9], Ward [8.34], Krasilscicova [8.20], Heaslet and Lomax [8.15], Homentcovschi [8.16] and Dragoş [8.7]) are found in the formulas (8.3.6)-(8.3.9). Prolonging the functions f_1 and f with 0 in $\mathbb{R}^2\backslash\overline{D}$, the above representations may be written as convolutions. For example, $p(x,y,z)$ and $\varphi(x,y,z)$ have the following form:

$$\begin{aligned} p &= -\frac{1}{2\pi}\left(\frac{\partial}{\partial x}f_1 * G + \frac{\partial}{\partial z}f * G\right) \\ \varphi &= \frac{1}{2\pi}\left(f_1 * G - f * \frac{xz}{y^2+z^2}G\right)\,, \end{aligned} \tag{8.3.10}$$

where the sign $*$ indicates the convolution relative to the variables x, y.

8.3.3 The Influence Zones. The Domain D_1

First of all we must notice that the perturbation my be represented by integrals whose integrand contains the factor $H(x_0 - s)$. Indeed, for (8.3.5) this is obvious. Taking into account the formulas (2.3.35) and (2.3.36), it results that the assertion is also valid for (8.3.2) and (8.3.7). In (8.3.9) we have

$$\int_{-\infty}^{x_0} \frac{H(\tau - s)}{\sqrt{\tau^2 - s^2}} \mathrm{d}\tau = H(x_0 - s) \int_s^{x_0} \frac{\mathrm{d}\tau}{\sqrt{\tau^2 - s^2}} \,. \tag{8.3.11}$$

Since the above integrands contain the factor $H(x_0 - s)$ we deduce that for a point $M(x, y, z)$ from the domain occupied by the fluid, the integrals on D are in fact calculated only on the domain D_1 where we have:

$$x_0 > s \tag{8.3.12}$$

This inequality implies

$$\xi < x \quad (x - \xi)^2 \geq k^2[(y - \eta)^2 + z^2] \,. \tag{8.3.13}$$

The points from D verifying these inequalities are situated between the leading edge and the hyperbola C which has the equation

$$(x - \xi)^2 = k^2[(y - \eta)^2 + z^2] \tag{8.3.14}$$

and the branches to $-\infty$ because $\xi < x$ (fig. 8.3.4). The hyperbola C (the variables are ξ and η) has the axis parallel to Ox. In fact, C represents the intersection of the cone having the equation

$$(x - \xi)^2 = k^2 \left[(y - \eta)^2 + (z - \zeta)^2\right]$$

with the plane $\zeta = 0$. This is Mach's cone. It has the vertex in M and the axis parallel to Ox. From the mechanical point of view, this result represents a consequence of a fact known from the hyperbolic partial differential equations theory [1.6] namely the fact that in M one can receive only the perturbations produced in the points belonging to the interior of Mach's cone with the vertex in M. When M will be on the wing $(z = 0)$, the hyperbola C will be reduced to the half-lines

$$x - \xi = \pm k(y - \eta) \,. \tag{8.3.15}$$

These are the characteristics issuing from M (fig. 8.3.5).

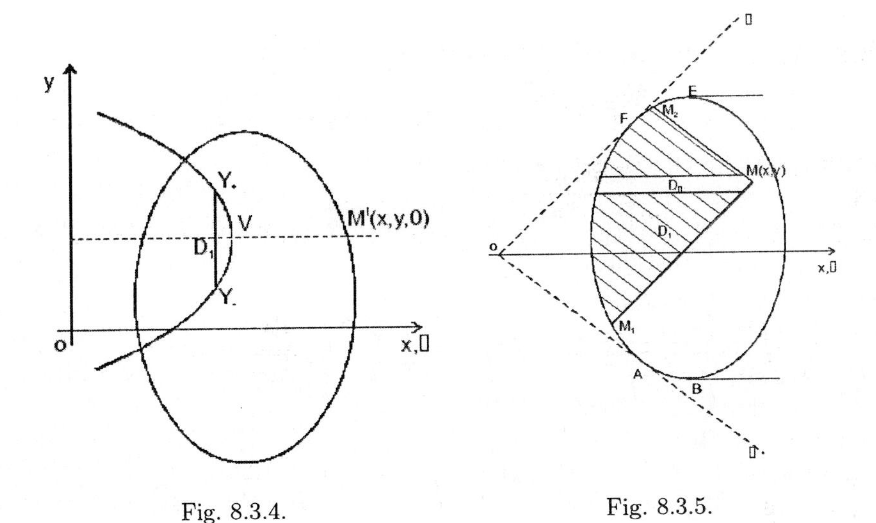

Fig. 8.3.4. Fig. 8.3.5.

We may easily explain why the points from $D - D_1$ do not affect the perturbation in M if we have in view (the significance of the fundamental solution) that the perturbation produced in a point $Q \in D$ propagates only in the interior of Mach's cone with the vertex in Q. The point M is in the interior of all the cones with the vertices in D_1 and in the exterior of all the cones with the vertices in $D - D_1$.

It also results that in the fluid exterior to the envelope of the posterior cones with the vertices on D, the perturbation is zero.

Hence we can give up the factor $H(x_0 - s)$ in the integrals expressing the perturbation if we replace the domain D by D_1. Prolonging the functions f_1 and f in the exterior of D with the value zero, for a given ξ , η will vary between Y_- and Y_+ defined by (8.3.14) through

$$kY_{\pm} = ky \pm \sqrt{x_0^2 - k^2 z^2} \,. \tag{8.3.16}$$

The vertex of the hyperbola C has the coordinates $\eta = y, \xi = x - k|z|$ (obtained for $Y_+ = Y_-$).

8.3.4 The Boundary Values of the Pressure

For the integrals having the form:

$$I(x, y, z) = \iint_D f(\xi, \eta) \frac{\partial}{\partial z} G(x_0, y_0, z) \mathrm{d}\xi \, \mathrm{d}\eta \tag{8.3.17}$$

we have:

$$I = \frac{\partial}{\partial z}\int_{-\infty}^{x-k|z|} \mathrm{d}\xi \int_{Y_-}^{Y_+} \frac{f(\xi,\eta)}{\sqrt{x_0^2 - s^2}}\mathrm{d}\eta\,. \tag{8.3.18}$$

Performing the change of variables $\eta \rightarrow \theta$:

$$k\eta = ky - \sqrt{x_0^2 - k^2 z^2}\cos\theta \tag{8.3.19}$$

we obtain

$$I(x,y,z) = \frac{1}{k}\frac{\partial}{\partial z}\int_{-\infty}^{x-k|z|} \mathrm{d}\xi \int_0^\pi f\left(\xi, y - \frac{1}{k}\sqrt{x_0^2 - k^2 z^2}\cos\theta\right)\mathrm{d}\theta =$$

$$= -\mathrm{sign}\ z\int_0^\pi f(x-k|z|,y)\mathrm{d}\theta + \int_{-\infty}^{x-k|z|}\mathrm{d}\xi\int_0^\pi \frac{\partial f}{\partial \eta}\frac{z\cos\theta}{\sqrt{x_0^2 - k^2 z^2}}\mathrm{d}\theta\,.$$

Hence,

$$I(x,y,\pm 0) = \mp\ \pi f(x,y)\,. \tag{8.3.20}$$

Using this formula, from (8.3.2) we obtain

$$p(x,y,\pm) = -\frac{1}{2\pi}\iint_D f_1(\xi,\eta)\frac{\partial}{\partial x}G(x_0,y_0,0)\mathrm{d}\xi\ \mathrm{d}\eta \pm \frac{1}{2}f(x,y)$$

whence,

$$f(x,y) = p(x,y,+0) - p(x,y,-0)\,. \tag{8.3.21}$$

This result puts into evidence the significance of the function f.

8.3.5 The First Form of the Integral Equation

The simplest way to obtain the lifting surface equation relies on the representation (8.3.7). Taking into account the derivation formula (2.3.35), we may write the kernel (8.3.8) as follows:

$$N(x_0,y_0,z) = -k^2\frac{x_0 z^2}{y_0^2+z^2}\frac{H(x_0-s)}{(x_0^2-s^2)^{3/2}} - \frac{x_0(y_0^2-z^2)}{(y_0^2+z^2)^2}\frac{H(x_0-s)}{\sqrt{x_0^2-s^2}}\,. \tag{8.3.22}$$

We notice that for $z = \pm 0$ it appears the singular line $\eta = y$. Detaching from D the domain D_ε defined by $y-\varepsilon < \eta < y+\varepsilon$ in $D_1 - D_\varepsilon$ (fig. 8.3.5), it is possible to simplify by y_0 after putting $z = \pm 0$. Performing this operation we deduce

$$N(x_0,y_0,\pm 0) = -\frac{x_0}{y_0^2}\frac{H(x_0-s_0)}{\sqrt{x_0^2-s_0^2}} \equiv N(x_0,y_0)\,. \tag{8.3.23}$$

Adopting the definition

$$\overset{=}{\iint}_{D_1} = \lim_{\varepsilon\to 0} \iint_{D_1 - D_\varepsilon}, \tag{8.3.24}$$

we shall prove that

$$J = \lim_{\varepsilon\to 0} \iint_{D_\varepsilon} f(\xi,\eta) N(x_0, y_0, z) \mathrm{d}\xi \, \mathrm{d}\eta = 0 . \tag{8.3.25}$$

For ε small enough, we may perform the replacement $\eta = y$ in the integrand. Hence,

$$J = \lim_{\varepsilon\to 0} \int_{-\infty}^{x-k|z|} \left[\int_{y-\varepsilon}^{y-\varepsilon} f(\xi,\eta) N(x_0, y_0, z) \mathrm{d}\eta \right] \mathrm{d}\xi =$$

$$= \lim_{\varepsilon\to 0} \int_{-\infty}^{x-k|z|} \left[f(\xi,\eta) N(x_0, y_0, z) \int_{y-\varepsilon}^{y-\varepsilon} \mathrm{d}\eta \right] \mathrm{d}\xi = 0 .$$

Using the formula (8.3.20) from (8.3.7) we deduce:

$$w(x, y, \pm 0) = \mp \, \frac{1}{2} f_1(x,y) + \frac{1}{2\pi} \overset{=}{\iint}_{D} f(\xi,\eta) N(x_0, y_0) \mathrm{d}\xi \, \mathrm{d}\eta . \tag{8.3.26}$$

Adding and subtracting the boundary conditions

$$w(x, y, \pm 0) = h_x(x,y) \pm h_{1x}(x,y) \quad (x,y) \in D , \tag{8.3.27}$$

one obtains:

$$\frac{1}{2\pi} \overset{=}{\iint}_{D} f(\xi,\eta) N(x_0, y_0) \mathrm{d}\xi \, \mathrm{d}\eta = h_x(x,y) , \tag{8.3.28}$$

$$f_1(x,y) = -2 h_{1x}(x,y) , \qquad (x,y) \in D . \tag{8.3.29}$$

The equation (8.3.28) is *the lifting surface equation* in the supersonic stream. It can be also written as follows:

$$-\frac{1}{2\pi} \overset{=}{\iint}_{D_1} \frac{f(\xi,\eta)}{y_0^2} \frac{x_0}{\sqrt{x_0^2 - k^2 y_0^2}} \mathrm{d}\xi \, \mathrm{d}\eta = h_x(x,y) , \tag{8.3.30}$$

D_1 representing the shaded domain from figure 8.3.5.

The analogy of this equation with equation (5.1.28) is obvious. The equation (8.3.30) was given in [8.7]. For the sake of simplicity we shall name it *the equation D*.

8.3.6 The Equation D in Coordinates on Characteristics

We know from (8.3.15) that if the current point M is on the wing $(x, y) \in D$, $z = 0$, the hyperbola C from (8.3.4) degenerates in the characteristics MM_1 and MM_2 (fig. 8.3.5) having the equations $\xi - k\eta = x - ky$ respectively $\xi + k\eta = x + ky$ (ξ and η are the variables, x and y are the coordinates of the point M). Performing the change of variables $\xi, \eta \to \alpha, \beta$, $x, y \to a, b$ defined by the formulas

$$\begin{aligned} \xi - k\eta &= \alpha \qquad & x - ky &= a \\ \xi + k\eta &= \beta \qquad & x + ky &= b\,, \end{aligned} \tag{8.3.31}$$

we deduce

$$2k\mathrm{d}\xi\,\mathrm{d}\eta = \mathrm{d}\alpha\,\mathrm{d}\beta\,.$$

The characteristic MM_1 has the equation $\alpha = a$ and the characteristic MM_2, the equation $\beta = b$. The domain D_1 given in the old variables by the inequalities

$$\xi - k\eta < x - ky, \quad \xi + k\eta < x + ky \tag{8.3.32}$$

will be characterized in the new variables by

$$\alpha < a; \quad \beta < b\,. \tag{8.3.33}$$

The axes OA and OF will be the new axes of coordinates because on OA we have $\beta = 0$ and α is variable, and on OF, $\alpha = 0$ and β is variable. The coordinates of M with respect to the new axes are (a, b) (fig. 8.3.6). Since the new axes are characteristic lines, the coordinates α and β will be named *coordinates on characteristics.* In these coordinates the equation (8.3.30) becomes

$$-\frac{k}{2\pi} \overline{\iint_{D_1}} f(\alpha, \beta) \frac{a - \alpha + b - \beta}{[a - \alpha - (b - \beta)]^2} \frac{\mathrm{d}\alpha \mathrm{d}\beta}{\sqrt{(a - \alpha)(b - \beta)}} = H(a, b)\,, \tag{8.3.34}$$

where we denoted by $f(\alpha, \beta)$ the unknown f in the new variables, $H(a, b)$ the function given by $h_x(x, y)$ in the new variables and D_1, the domain D in the new variables, i.e. the shaded domain from figure 8.3.6 ($\alpha < a, \beta < b$).

Obviously, the integrand has in M ($\alpha = a, \beta = b$) a strong singularity. We shall isolate therefore this point drawing the parallel $\beta = b - \varepsilon$

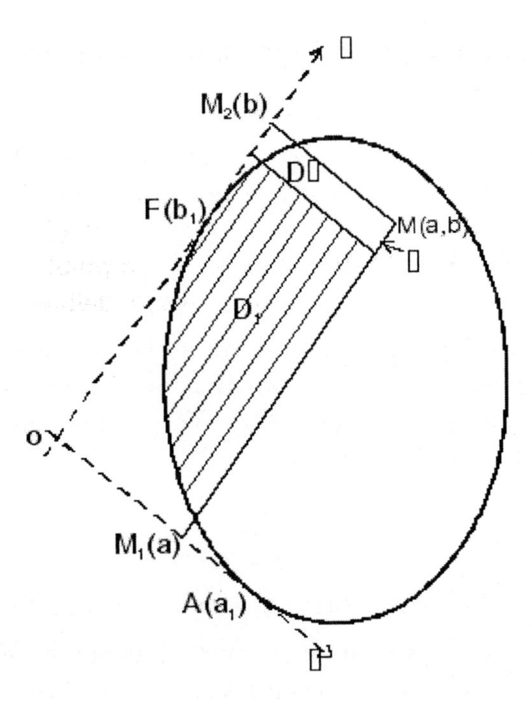

Fig. 8.3.6.

and denoting by D_ε the parallelogram indicated on figure 8.3.6. We adopt in (8.3.34) the definition

$$\overline{\overline{\iint}}_{D_1} = \lim_{\varepsilon \to 0} \iint_{D_1 - D_\varepsilon} . \tag{8.3.35}$$

In this way, the non - integrable denominator from (8.3.34) does not vanish (even if $\alpha = a$). In $D_1 - D_\varepsilon$ we may utilize the identities

$$\frac{a-\alpha+b-\beta}{[a-\alpha-(b-\beta)]^2}\frac{1}{\sqrt{(a-\alpha)(b-\beta)}} = -\frac{2}{\sqrt{b-\beta}}\frac{\partial}{\partial a}\left[\frac{\sqrt{a-\alpha}}{a-\alpha-(b-\beta)}\right] =$$

$$= -2\frac{\partial^2}{\partial a \partial b}\ln\frac{\sqrt{a-\alpha}+\sqrt{b-\beta}}{|\sqrt{a-\alpha}-\sqrt{b-\beta}|}, \tag{8.3.36}$$

such that the equation (8.3.34) may be also written as follows

$$\frac{k}{\pi}\frac{\partial}{\partial a}\iint'_{D_1} f(\alpha,\beta)\sqrt{\frac{a-\alpha}{b-\beta}}\,\frac{\mathrm{d}\alpha \mathrm{d}\beta}{a-\alpha-(b-\beta)} = H(a,b)\,, \tag{8.3.37}$$

$$\frac{k}{\pi}\frac{\partial^2}{\partial a \partial b}\iint_{D_1} f(\alpha,\beta)\ln\frac{\sqrt{a-\alpha}+\sqrt{b-\beta}}{|\sqrt{a-\alpha}-\sqrt{b-\beta}|}\mathrm{d}\alpha\,\mathrm{d}\beta = H(a,b) \tag{8.3.38}$$

in order to have weaker singularities. In [8.7] one gives another form for equation (8.3.37).

8.3.7 The Plane Problem

We act like in 5.1.7. In order to have a plane-parallel flow, we must consider the case when the conditions that determine the flow are identical in every plane parallel to xOz. We assume therefore that the equations (8.3.1) have the form

$$z = h(x) \pm h_1(x)\,, \tag{8.3.39}$$

and the domain D is rectangular, such that the span $2b$ tends to infinity ($b \to \infty$). Hence we assume that D is characterized by the inequalities $0 < x < 1$, $-b < y < b$. These conditions imply $f_1 = f_1(\xi)$ and $f = f(\xi)$. Noticing that

$$\begin{aligned}\int_{-\infty}^{+\infty} E\mathrm{d}\eta &= \frac{2}{k}\int_0^{\infty} \frac{H(x_0 - \sqrt{u^2 + k^2z^2})}{\sqrt{x_0^2 - k^2z^2 - u^2}}\mathrm{d}u = \\ &= \frac{2}{k}H(x_0 - k|z|)\int_0^{\sqrt{x_0^2 - k^2z^2}} \frac{\mathrm{d}u}{\sqrt{x_0^2 - k^2z^2 - u^2}} = \\ &= \frac{2}{k}H(x_0 - k|z|)\int_0^1 \frac{\mathrm{d}v}{\sqrt{1 - v^2}} = \frac{\pi}{k}H(x_0 - k|z|)\,,\end{aligned} \tag{8.3.40}$$

$$\begin{aligned}&\iint_D f(\xi)\frac{\partial}{\partial y}\left(\frac{x_0y_0}{y_0^2 + z^2}E\right)\mathrm{d}\xi\,\mathrm{d}\eta = \\ &= -\int_0^1 f(\xi)\left[\int_{-\infty}^{\infty}\frac{\partial}{\partial \eta}\left(\frac{x_0y_0}{y_0^2 + z^2}E\right)\mathrm{d}\eta\right]\mathrm{d}\xi = 0\,,\end{aligned} \tag{8.3.41}$$

we deduce the representation (8.1.17).

8.3.8 The Equation of Heaslet and Lomax (the HL Equation)

This equation may be deduced from the representation (8.3.6). To this aim we denote

$$L(x,y,z) = \frac{\partial}{\partial z}\iint_{D_1} \frac{f(\xi,\eta)}{\sqrt{x_0^2 - s^2}}\,\frac{x_0 z}{y_0^2 + s^2}\mathrm{d}\xi\,\mathrm{d}\eta =$$
$$= \frac{\partial}{\partial z}\int_{-\infty}^{x-k|z|}\mathrm{d}\xi \int_{Y_-}^{Y_+} \frac{f(\xi,\eta)}{\sqrt{x_0^2 - s^2}}\,\frac{x_0 z}{y_0^2 + z^2}\mathrm{d}\xi\,\mathrm{d}\eta \qquad (8.3.42)$$

the functions Y_+, Y_- being defined in (8.3.16). Here the change of variable (8.3.19) is not indicated, because the term that one obtains by deriving the superior limit of the first integral becomes ∞ for $z \to 0$. We act therefore as follows:

$$L = \lim_{\varepsilon\to 0}\frac{\partial}{\partial z}\int_{-\infty}^{x-k\sqrt{z^2+\varepsilon^2}}\mathrm{d}\xi \int_{Y_-}^{Y_+} \frac{f(\xi,\eta)}{\sqrt{x_0^2 - s^2}}\,\frac{x_0 z}{y_0^2 + z^2}\mathrm{d}\eta \equiv$$
$$\equiv L_1(x,y,z) + \int_{-\infty}^{x-k|z|} x_0 \ell(x_0,y,z)\mathrm{d}\xi\,, \qquad (8.3.43)$$

where, after performing the change of variable $\eta \to u : \eta - y = u$, we have:

$$L_1 = -kz^2 \lim_{\varepsilon\to 0}\int_{y-\varepsilon}^{y+\varepsilon} \frac{f(x - k\sqrt{z^2+\varepsilon^2},\eta)}{\sqrt{\varepsilon^2 - y_0^2}}\,\frac{\mathrm{d}\eta}{y_0^2 + z^2}$$
$$= -kz^2 f(x - k|z|, y)\lim_{\varepsilon\to 0}\int_{-\varepsilon}^{+\varepsilon}\frac{\mathrm{d}u}{\sqrt{\varepsilon^2 - u^2}(u^2+z^2)} = -k\pi f(x-k|z|,y)\,, \qquad (8.3.44)$$

$$\ell(x_0,y,z) = \frac{\partial}{\partial z}\int_{Y_-}^{Y_+}\frac{f(\xi,\eta)}{\sqrt{x_0^2 - s^2}}\,\frac{z}{y_0^2+z^2}\mathrm{d}\eta = \frac{\partial}{\partial z}\int_{-a(z)}^{+a(z)}\frac{F(u,z)z}{u^2+z^2}\mathrm{d}u\,, \qquad (8.3.45)$$

with the notations

$$a(z) = \frac{1}{k}\sqrt{x_0^2 - k^2 z^2}, \quad F(u,z) = \frac{f(\xi, u+y)}{\sqrt{x_0^2 - k^2(u^2+z^2)}}\,. \qquad (8.3.46)$$

We deduce therefore

$$L(x,y,\pm 0) = -k\pi f(x,y) + \int_{-\infty}^{x} x_0\ell(x_0,y,\pm 0)\mathrm{d}\xi\,. \qquad (8.3.47)$$

For determining the function $\ell(x_0, y, \pm 0)$ we notice that $a'(\pm 0) = 0$. It results that the derivative from (8.3.45) and the integral interchanges. We have therefore:

$$\ell(x_0, y, \pm 0) = \lim_{z \to \pm 0} \frac{\partial}{\partial z} \int_{-a}^{+a} F(u, z) \mathrm{d} \arctan \frac{u}{z} =$$

$$= \lim_{z \to \pm 0} \frac{\partial}{\partial z} \{ [F(a, z) + F(-a, z)] \arctan \frac{a}{z} \} -$$

$$- \lim_{z \to \pm 0} \int_{-a}^{+a} \frac{\partial^2 F}{\partial u \partial z} \arctan \frac{u}{z} \mathrm{d}\, u + \lim_{z \to \pm 0} \int_{-a}^{+a} \frac{\partial F}{\partial u} \frac{u}{u^2 + z^2} \mathrm{d}\, u =$$

$$= - \lim_{z \to \pm 0} [F(a, z) + F(-a, z)] \frac{a}{z^2 + a^2} + \lim_{z \to \pm 0} \int_{-a}^{+a} \frac{\partial F}{\partial z} \frac{z}{u^2 + z^2} \mathrm{d}\, u +$$

$$+ \lim_{z \to \pm 0} \int_{-a}^{+a} \frac{\partial F}{\partial u} \frac{u}{u^2 + z^2} \mathrm{d}\, u \, .$$

But

$$\lim_{z \to \pm 0} \int_{-a}^{+a} \frac{\partial F}{\partial z} \frac{z}{u^2 + z^2} \mathrm{d}\, u = \lim_{z \to \pm 0} \frac{\partial F}{\partial z}(0, z) z \int_{-\varepsilon}^{+\varepsilon} \frac{\mathrm{d}\, u}{u^2 + z^2} = \pm \pi \frac{\partial F}{\partial z}(0, 0) \, ,$$

$$\lim_{z \to \pm 0} \int_{-a}^{+a} \frac{\partial F}{\partial u} \frac{u}{u^2 + z^2} \mathrm{d}\, u = \lim_{\varepsilon \to 0} \lim_{z \to \pm 0} \left(\int_{-a}^{-\varepsilon} + \int_{a}^{+\varepsilon} \right) \frac{\partial F}{\partial u} \frac{u}{u^2 + z^2} \mathrm{d}\, u =$$

$$= {\int_{-a_0}^{+a_0}}' \frac{\partial F}{\partial u}(u, 0) \frac{\mathrm{d}\, u}{u} \, ,$$

where $a_0 = a(0)$. Finally, taking the formula (D.3.7) into account, we obtain:

$$\begin{aligned} \ell(x_0, y, \pm 0) &= - \frac{F(a_0, 0)}{a_0} - \frac{F(-a_0, 0)}{a_0} + {\int_{-a_0}^{+a_0}}' \frac{\partial F}{\partial u}(u, 0) \frac{\mathrm{d}\, u}{u} \pm \\ \pm \pi \frac{\partial F}{\partial z}(0, 0) &= {\int_{-a_0}^{+a_0}}^{*} \frac{F(u, 0)}{u^2} \mathrm{d}\, u \pm \pi \frac{\partial F}{\partial z}(0, 0) \, . \end{aligned} \tag{8.3.48}$$

Having in view the expression of $F(u, z)$ (8.3.46), we deduce:

$$L(x, y, \pm 0) = -k \pi f(x, y) + \int_{-\infty}^{x} x_0 \left[{\int_{y - \frac{x_0}{k}}^{y + \frac{x_0}{k}}}^{*} \frac{f(\xi, \eta)}{\sqrt{x_0^2 - k^2 y_0^2}} \frac{\mathrm{d}\, \eta}{y_0^2} \right] \mathrm{d}\, \xi \, . \tag{8.3.49}$$

Hence,

$$w(x,y,\pm 0) = \mp\ \frac{1}{2}f_1(x,y) + \frac{k}{2}f(x,y) - \\ -\frac{1}{2\pi}\int_{-\infty}^{x} x_0\left[{}^{*}\!\!\int_{y-\frac{x_0}{k}}^{y+\frac{x_0}{k}} \frac{f(\xi,\eta)}{\sqrt{x_0^2-k^2y_0^2}}\,\frac{\mathrm{d}\,\eta}{y_0^2}\right]\mathrm{d}\,\xi\,. \tag{8.3.50}$$

such that, imposing the boundary conditions (8.3.27), we find:

$$kf(x,y) - \frac{1}{\pi}\int_{-\infty}^{x} x_0\left[{}^{*}\!\!\int_{y-\frac{x_0}{k}}^{y+\frac{x_0}{k}} \frac{f(\xi,\eta)}{\sqrt{x_0^2-k^2y_0^2}}\,\frac{\mathrm{d}\,\eta}{y_0^2}\right]\mathrm{d}\,\xi = 2h_x(x,y)\,, \tag{8.3.51}$$

$$f_1(x,y) = (h'_- - h'_+)(x,y)\,. \tag{8.3.52}$$

(8.3.51) is the HL equation.

Integrating in (8.3.42), at first with respect to ξ and then with respect to η, we obtain

$$w(x,y,\pm 0) = \mp\ \frac{1}{2}f_1(x,y) - \frac{1}{2\pi}\,{}^{*}\!\!\int_{-\infty}^{+\infty}\frac{\mathrm{d}\,\eta}{y_0^2}\left[\int_{\sqrt{x_0^2-k^2y_0^2}}^{x-k|y_0|}\mathrm{d}\,\xi\right].$$

8.3.9 The Deduction of HL Equation from D Equation

This deduction was performed by V. Iftimie. To this aim, we shall extend the integral from (8.3.30) from the shaded domain D_1, in figure 8.3.5 to the infinite domain $D_{\xi\eta}$ situated between the characteristics MM_1 and MM_2 extended to infinity, putting $f=0$ in the exterior of the arc M_1M_2. We denote by I the integral obtained in this way and we perform the change of variables $\xi,\eta \to u,v$:

$$u = x - \xi\,, \quad v = y - \eta\,. \tag{8.3.53}$$

In the new variables, the domain $D_{\xi\eta}$ will be transformed into the domain D_{uv} (fig. 8.3.7) determined by the angle M_1MM_2. Denoting $g(u,v) = f(x-u,y-v)$, we deduce

$$f(x,y) = g(0,0)\,. \tag{8.3.54}$$

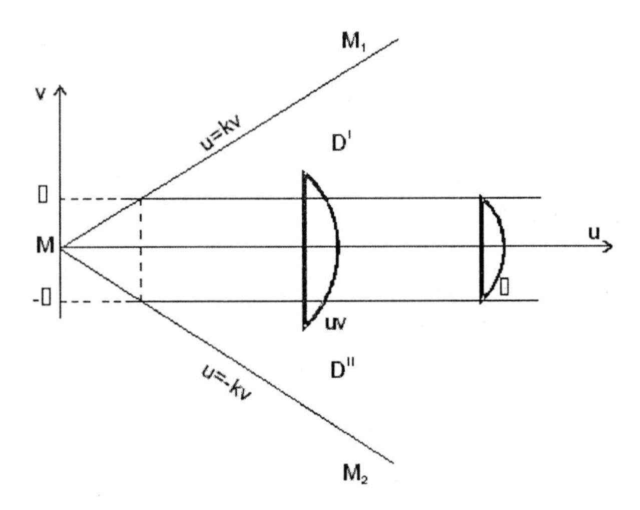

Fig. 8.3.7.

The HL equations result from the formulas

$$
\begin{aligned}
I &= \overset{*}{\iint}_{D_{uv}} g(u,v)\frac{u}{v^2}\,\frac{\mathrm{d}\,u\,\mathrm{d}\,v}{\sqrt{u^2-k^2v^2}} = \\
&= -k\pi g(0,0) + \int_0^{\infty}\mathrm{d}\,u \overset{*}{\int}_{-u/k}^{u/k} g(u,v)\frac{u}{v^2}\,\frac{\mathrm{d}\,v}{\sqrt{u^2-k^2v^2}} = \qquad (8.3.55)\\
&= \overset{*}{\int}_{-\infty}^{+\infty}\mathrm{d}\,v\int_{k|v|}^{\infty} g(u,v)\frac{u}{v^2}\,\frac{\mathrm{d}\,u}{\sqrt{u^2-k^2v^2}}
\end{aligned}
$$

which have to be demonstrated (the demonstration was given by V.Iftimie). For proving the first formula, we isolate the singularity by two parallels to the Mu axis (fig. 8.3.7) at the distance ε from this axis. We denote by D' and D'' the domains composing $D_{uv} - D_\varepsilon$ and by I' and I'' the integrals I corresponding to these domains. Using the definition (8.3.24) we get

$$
I = \lim_{\varepsilon\to 0}(I' + I'')\,. \qquad (8.3.56)
$$

Putting $v = tu$, we deduce:

$$
\begin{aligned}
I' &= \iint_{D'} g(u,v)\frac{u}{v^2}\,\frac{\mathrm{d}\,u\,\mathrm{d}\,v}{\sqrt{u^2-k^2v^2}} = \int_{k\varepsilon}^{\infty} u\left[\int_{\varepsilon}^{u/k}\frac{g(u,v)}{\sqrt{u^2-k^2v^2}}\mathrm{d}\,v\right]\mathrm{d}\,u = \\
&= \int_{k\varepsilon}^{\infty}\frac{1}{u}\left[\int_{\frac{\varepsilon}{k}}^{\frac{1}{k}}\frac{g(u,tu)}{t^2\sqrt{1-k^2t^2}}\mathrm{d}\,t\right]\mathrm{d}\,u\,.
\end{aligned}
$$

We expand g in a Taylor series:

$$g(u,tu) = g(u,0) + tug_v(u,0) + t^2u^2m(u,tu)\,.$$

and we introduce this series in the above integral.

Acting analogously for I'', we deduce:

$$I'' = \int_{k\varepsilon}^{\infty} u\left[\int_{-\frac{u}{k}}^{-\varepsilon} \frac{g(u,v)}{v^2\sqrt{u^2-k^2v^2}}\mathrm{d}\,v\right]\mathrm{d}\,u =$$

$$= \int_{k\varepsilon}^{\infty} \frac{1}{u}\left[\int_{-\frac{1}{k}}^{-\frac{\varepsilon}{u}} \frac{g(u,tu)}{t^2\sqrt{1-k^2t^2}}\mathrm{d}\,t\right]\mathrm{d}\,u$$

whence,

$$I' + I'' = \int_{k\varepsilon}^{\infty} \frac{g(u,0)}{u}\left[\int_{\frac{\varepsilon}{u}}^{\frac{1}{k}} \frac{\mathrm{d}\,t}{t^2\sqrt{1-k^2t^2}} + \int_{-\frac{1}{k}}^{-\frac{\varepsilon}{u}} \frac{\mathrm{d}\,t}{t^2\sqrt{1-k^2t^2}}\right]\mathrm{d}\,u +$$

$$+ \int_{k\varepsilon}^{\infty} g_v(u,0)\left[\int_{\frac{\varepsilon}{u}}^{\frac{1}{k}} \frac{\mathrm{d}\,t}{t\sqrt{1-k^2t^2}} + \int_{-\frac{1}{k}}^{-\frac{\varepsilon}{u}} \frac{\mathrm{d}\,t}{t\sqrt{1-k^2t^2}}\right]\mathrm{d}\,u +$$

$$+ \int_{k\varepsilon}^{\infty} u\left[\int_{\frac{\varepsilon}{u}}^{\frac{1}{k}} \frac{m(u,tu)}{\sqrt{1-k^2t^2}}\mathrm{d}\,t + \int_{-\frac{1}{k}}^{-\frac{\varepsilon}{u}} \frac{m(u,tu)}{\sqrt{1-k^2t^2}}\mathrm{d}\,t\right]\mathrm{d}\,u\,.$$

The factor which multiplies $g_v(u,0)$ is zero if we replace in the second integral t by $-t$. An elementary calculus gives

$$\int_{\frac{\varepsilon}{u}}^{\frac{1}{k}} \frac{\mathrm{d}\,t}{t^2\sqrt{1-k^2t^2}} = \int_{-\frac{1}{k}}^{-\frac{\varepsilon}{u}} \frac{\mathrm{d}\,t}{t^2\sqrt{1-k^2t^2}} = \frac{\sqrt{u^2-k^2\varepsilon^2}}{\varepsilon}\,.$$

whence,

$$I' + I'' = \frac{2}{\varepsilon}\int_{k\varepsilon}^{\infty} \frac{g(u,0)}{u}\sqrt{u^2-k^2\varepsilon^2}\mathrm{d}\,u +$$
$$+ \int_{k\varepsilon}^{\infty} u\left(\int_{\frac{\varepsilon}{u}}^{\frac{1}{k}} + \int_{-\frac{1}{k}}^{-\frac{\varepsilon}{u}}\right) \frac{m(u,tu)}{\sqrt{1-k^2t^2}}\mathrm{d}\,t\,\mathrm{d}\,u \qquad (8.3.57)$$

When $\varepsilon \to 0$, the first integral is a Finite Part. Hence

$$\frac{2}{\varepsilon}\int_{k\varepsilon}^{\infty} \frac{g(u,0)}{u}\sqrt{u^2-k^2\varepsilon^2}\mathrm{d}\,u = \frac{2}{\varepsilon}\int_{k\varepsilon}^{1} \frac{g(u,0)-g(0,0)}{u}\sqrt{u^2-k^2\varepsilon^2} +$$

$$+\frac{2}{\varepsilon}g(0,0)\int_{k\varepsilon}^{1}\frac{\sqrt{u^2-k^2\varepsilon^2}}{u}\mathrm{d}\,u+\frac{2}{\varepsilon}\int_{1}^{\infty}\frac{g(u,0)}{u}\sqrt{u^2-k^2\varepsilon^2}\,\mathrm{d}\,u\,.$$

The first and the last integrals from the right hand member are finite. When we consider the Finite Part, we neglect these integrals because they have the factor ε^{-1} . One calculates the integral from the middle and one obtains

$$FP\frac{2}{\varepsilon}\int_{k\varepsilon}^{\infty}\frac{g(u,0)}{u}\sqrt{u^2-k^2\varepsilon^2}\mathrm{d}\,u=-k\pi g(0,0)\,.$$

Hence, from (8.3.57) we deduce:

$$I=\lim_{\varepsilon\to0}(I'+I'')=-k\pi g(0,0)+\int_0^{\infty}u\left[\int_{-\frac{1}{k}}^{\frac{1}{k}}\frac{m(u,tu)}{\sqrt{1-k^2t^2}}\mathrm{d}\,t\right]\mathrm{d}\,u\,. \quad (8.3.58)$$

We set now

$$J=\int_0^{\infty}\left[\int_{-\frac{u}{k}}^{\frac{u}{k}}g(u,v)\frac{u}{v^2}\frac{\mathrm{d}\,v}{\sqrt{u^2-k^2v^2}}\right]\mathrm{d}\,u\,. \quad (8.3.59)$$

Using the same Taylor expansion we deduce:

$$J'=\int_{\varepsilon}^{\frac{u}{k}}g(u,v)\frac{u}{v^2\sqrt{u^2-k^2v^2}}\mathrm{d}\,v=\int_{\frac{\varepsilon}{u}}^{1/k}\frac{g(u,tu)}{u}\frac{\mathrm{d}\,t}{t^2\sqrt{1-k^2t^2}}=$$

$$=\frac{g(u,0)}{u}\int_{\frac{\varepsilon}{u}}^{\frac{1}{n}}\frac{\mathrm{d}\,t}{t^2\sqrt{1-k^2t^2}}+g_v(u,0)\int_{\frac{\varepsilon}{u}}^{\frac{1}{k}}\frac{\mathrm{d}\,t}{t\sqrt{1-k^2t^2}}+$$

$$+u\int_{\frac{\varepsilon}{u}}^{\frac{1}{k}}\frac{m(u,tu)}{\sqrt{1-k^2t^2}}\mathrm{d}\,t\,,$$

$$J''=\int_{\frac{u}{k}}^{-\varepsilon}g(u,v)\frac{u}{v^2\sqrt{u^2-k^2v^2}}\mathrm{d}\,v=\frac{g(u,0)}{u}\int_{-\frac{1}{k}}^{-\frac{\varepsilon}{u}}\frac{\mathrm{d}\,t}{t^2\sqrt{1-k^2t^2}}+$$

$$+g_v(u,0)\int_{-\frac{1}{k}}^{-\frac{\varepsilon}{u}}\frac{\mathrm{d}\,t}{t\sqrt{1-k^2t^2}}+\dots$$

whence,

$$J'+J''=\frac{2g(u,0)}{u}\frac{\sqrt{u^2-k^2\varepsilon^2}}{\varepsilon}+u\left(\int_{-\frac{\varepsilon}{u}}^{\frac{1}{k}}+\int_{-\frac{1}{k}}^{-\frac{\varepsilon}{u}}\right)\frac{m(u,tu)}{\sqrt{1-k^2t^2}}\mathrm{d}\,t\,.$$

When we consider the Finite Part (FP) we neglect the first term and we get

$$\lim_{\varepsilon\to 0}(J'+J'') = u\int_{-\frac{1}{k}}^{\frac{1}{k}} \frac{m(u,tu)}{\sqrt{1-k^2t^2}} \Rightarrow J = \int_0^\infty u\left[\int_{-\frac{1}{k}}^{\frac{1}{k}} \frac{m(u,tu)}{\sqrt{1-k^2t^2}}\mathrm{d}\,t\right]\mathrm{d}\,u\,. \tag{8.3.60}$$

From (8.3.58) and (8.3.60) it results

$$I = -k\pi g(0,0) + J\,, \tag{8.3.61}$$

which demonstrates the first equality from (8.3.55).

For proving the second equality, we denote:

$$K = {}^{*}\!\!\int_{-\infty}^{+\infty}\left[\int_{k|v|}^{\infty} g(u,v)\frac{u}{v^2\sqrt{u^2-k^2v^2}}\mathrm{d}\,u\right]\mathrm{d}\,v = \lim_{\varepsilon\to 0}(K'+K'')\,,$$

where (we may change the order of integration)

$$\begin{aligned} K' &= \int_{-\infty}^{-\varepsilon}\left[\int_{-kv}^{\infty}\frac{ug(u,v)}{v^2\sqrt{u^2-k^2v^2}}\mathrm{d}\,u\right]\mathrm{d}\,v = \\ &= \int_{k\varepsilon}^{\infty} u\left[\int_{\varepsilon}^{\frac{u}{k}}\frac{g(u,v)}{v^2\sqrt{u^2-k^2v^2}}\mathrm{d}\,v\right]\mathrm{d}\,u = I', \\ K'' &= \int_{\varepsilon}^{\infty}\left[\int_{kv}^{\infty}\frac{ug(u,v)}{v^2\sqrt{u^2-u^2v^2}}\mathrm{d}\,u\right]\mathrm{d}\,v = \\ &= \int_{k\varepsilon}^{\infty} u\left[\int_{-\frac{u}{k}}^{-\varepsilon}\frac{g(u,v)}{v^2\sqrt{u^2-k^2v^2}}\mathrm{d}\,v\right]\mathrm{d}\,u = I''\,. \end{aligned} \tag{8.3.62}$$

Hence $K = I$ and the demonstration is finished. The HL equations resulting from (8.3.27) are

$$kf(x,y) - \frac{1}{\pi}\int_{-\infty}^{x} x_0\left[{}^{*}\!\!\int_{y-\frac{x_0}{k}}^{y+\frac{x_0}{k}}\frac{f(\xi,\eta)}{\sqrt{x_0^2-k^2y_0^2}}\frac{\mathrm{d}\,\eta}{y_0^2}\right]\mathrm{d}\,\xi = 2h_x(x,y), \tag{8.3.63}$$

$$-\frac{1}{\pi}\,{}^{*}\!\!\int_{-\infty}^{+\infty}\frac{\mathrm{d}\,\eta}{y_0^2}\left[\int_{-\infty}^{x-k|y_0|}\frac{f(\xi,\eta)x_0}{\sqrt{x_0^2-k^2y_0^2}}\mathrm{d}\,\xi\right] = 2h_x(x,y)\,. \tag{8.3.64}$$

These are the equations (10-21) and (10-22) from [8.15].

8.3.10 The Equation of Homentcovschi (H Equation)

This equation results from the representation (8.3.9). Taking into account (8.3.20), we deduce:

$$w(x,y,\pm 0) = \mp\ \frac{1}{2}f_1(x,y)+$$

$$+\frac{1}{2\pi}\left(k^2\frac{\partial^2}{\partial x^2}-\frac{\partial^2}{\partial y^2}\right)\iint_D f(\xi,\eta)\left[\int_{-\infty}^{x_0}\frac{H(\tau-s_0)}{\sqrt{\tau^2-s_0^2}}\mathrm{d}\,\tau\right]\mathrm{d}\,\xi\,\mathrm{d}\,\eta\,, \tag{8.3.65}$$

such that imposing the boundary conditions (8.3.27), we obtain the equation

$$\frac{1}{\pi}\left(k^2\frac{\partial^2}{\partial x^2}-\frac{\partial^2}{\partial y^2}\right)\iint_D f(\xi,\eta)\left[\int_{-\infty}^{x_0}\frac{H(\tau-s_0)}{\sqrt{\tau^2-s_0^2}}\mathrm{d}\,\tau\right]\mathrm{d}\,\xi\,\mathrm{d}\,\eta = 2h_x(x,y)\,, \tag{8.3.66}$$

which was given for the first time (in a different way) by D.Homentcovschi [8.16].

We have to notice that after evaluating the interior integral with the formula (8.3.11) one obtains the equation

$$\frac{1}{\pi}\left(k^2\frac{\partial^2}{\partial x^2}-\frac{\partial^2}{\partial y^2}\right)\iint_{D_1} f(\xi,\eta)\ln\frac{x_0+\sqrt{x_0^2-s_0^2}}{s_0}\mathrm{d}\,\xi\,\mathrm{d}\,\eta = 2h_x(x,y)\,. \tag{8.3.67}$$

In the variables on characteristics this equation becomes (8.3.38). However, Homentcovschi does not follow this way. Inspired probably by the papers of Evvard [8.9], Ward [8.33] and Krasilscicova [8.20], he introduces the unknown N, related to f by the equality

$$N(x,y) = \int_{-\infty}^{x} f(\xi,y)\mathrm{d}\,\xi\,, \tag{8.3.68}$$

whence it results

$$f(x,y) = N_x(x,y)\,. \tag{8.3.69}$$

We have seen in (5.4.5) that o function similar to N also intervenes in the theory of low aspect wings in a subsonic stream.

Taking (8.3.69) into account, the integrand from (8.3.66) becomes:

$$N_\xi\int_{-\infty}^{x_0}\frac{H(\tau-s_0)}{\sqrt{\tau^2-s_0^2}}\mathrm{d}\,\tau = \frac{\partial}{\partial\xi}\left[N\int_{-\infty}^{x_0}\frac{H(\tau-s_0)}{\sqrt{\tau^2-s_0^2}}\mathrm{d}\,\tau\right]+N\frac{H(x_0-s_0)}{\sqrt{x_0^2-s_0^2}} =$$

$$= \frac{\partial}{\partial\xi}\left[NH(x_0-s_0)\int_{s_0}^{x_0}\frac{\mathrm{d}\,\tau}{\sqrt{\tau^2-s_0^2}}\right]+N\frac{H(x_0-s_0)}{\sqrt{x_0^2-s_0^2}} =$$

$$= H(x_0 - s_0)\left\{\frac{\partial}{\partial \xi}\left[N(\xi,\eta)\ln\frac{x_0 + \sqrt{x_0^2 - s_0^2}}{s_0}\right] + \frac{F(\xi,\eta)}{\sqrt{x_0^2 - s_0^2}}\right\}.$$

In this way, after applying Green's formula, the equation (8.3.66) becomes

$$\frac{1}{\pi}\left(k^2\frac{\partial^2}{\partial x^2} - \frac{\partial^2}{\partial y^2}\right)\left\{\iint_{D_1}\frac{N(\xi,\eta)}{\sqrt{x_0^2 - s_0^2}}\mathrm{d}\,\xi\,\mathrm{d}\,\eta + \right.$$
$$\left. + \oint_{\partial D_1} N(\xi,\eta)\ln\frac{x_0 + \sqrt{x_0^2 - s_0^2}}{s_0}\mathrm{d}\,\eta\right\} = 2h_x(x,y)\,. \tag{8.3.70}$$

In variables on characteristics this equation becomes

$$\frac{k}{\pi}\frac{\partial^2}{\partial a \partial b}\left\{\iint_{D_1}\frac{N(\alpha,\beta)\mathrm{d}\,\alpha\,\mathrm{d}\,\beta}{\sqrt{(a-\alpha)(b-\beta)}} + \right.$$
$$\left. + \oint_{\partial D_1} N(\alpha,\beta)\ln\frac{\sqrt{a-\alpha} + \sqrt{b-\beta}}{|\sqrt{a-\alpha} - \sqrt{b-\beta}|}\mathrm{d}\,(\beta-\alpha)\right\} = H(a,b)\,. \tag{8.3.71}$$

From (8.3.68) it follows that $N(\xi,\eta)$ vanishes on the leading edge, hence

$$N = 0, \quad \text{on } EFAB\,. \tag{8.3.72}$$

Indeed, for every point $(\xi,\eta) \in EFAB$, the integral (8.3.68) is calculated on the half -line having the applicate η upstream of the wing, where $f = 0$. We also notice that the curvilinear integral from (8.3.71) vanishes on the characteristics $\alpha = a$ and $\beta = b$ (fig. 8.3.6), because the logarithm vanishes. Hence, for a given $M(a,b)$, the curvilinear integral from (8.3.71) will differ from zero only in those points of the trailing edge belonging to ∂D_1. For example, for the wing from 8.3.1, the curvilinear integral will differ from zero only for M situated in the zones of influence of the subsonic trailing edges BC and DE, respectively in the domains BCB' and $EE'D$. Utilizing the definition (D.3.6) for (8.3.71) we also have the equivalent forms:

$$\frac{k}{4\pi}\,{}^{*}\!\!\int {}^{*}\!\!\int_{D_1}\frac{N(\alpha,\beta)\mathrm{d}\,\alpha\,\mathrm{d}\,\beta}{(a-\alpha)^{3/2}(b-\beta)^{3/2}} +$$
$$+\frac{k}{\pi}\frac{\partial}{\partial a}\int_{\partial D_1}\frac{N(\alpha,\beta)}{a-\alpha-(b-\beta)}\frac{\sqrt{a-\alpha}}{\sqrt{b-\beta}}\mathrm{d}\,(\beta-\alpha) = H(a,b)\,, \tag{8.3.73}$$

$$\frac{k}{4\pi} {}^{*}\!\!\int {}^{*}\!\!\int_{D_1} \frac{N(\alpha,\beta)\mathrm{d}\,\alpha\,\mathrm{d}\,\beta}{(a-\alpha)^{3/2}(b-\beta)^{3/2}} +$$

$$+\frac{k}{\pi}\frac{\partial}{\partial b}\int_{\partial D_1} \frac{N(\alpha,\beta)}{b-\beta-(a-\alpha)}\frac{\sqrt{b-\beta}}{\sqrt{a-\alpha}}\mathrm{d}\,(\beta-\alpha) = H(a,b)\,, \tag{8.3.74}$$

the curvilinear integral being different from zero only in those points of the trailing edge where $\alpha < a$ and $\beta < b$ (see the definition of D_1 in coordinates on characteristics (fig. 8.3.6).

The equation (8.3.71) is *the equation of Homentcovschi* (H equation). In the following subsection we shall present the solutions of this equations as they were determined by the author himself.

8.4 The Theory of Integration of the H Equation

8.4.1 Abel's Equation

Abel's equation has the form

$$\int_a^y \frac{f(x)}{\sqrt{y-x}}\mathrm{d}\,x = h(y)\,, \qquad a < y\,. \tag{8.4.1}$$

This was the first equation encountered in applications. Multiplying (8.4.1) by $(\lambda - y)^{-1/2}$ and integrating with respect to y on the interval (a, λ) we obtain:

$$\int_a^\lambda \frac{\mathrm{d}\,y}{\sqrt{\lambda-y}}\int_a^y \frac{f(x)}{\sqrt{y-x}}\mathrm{d}\,x = \int_a^\lambda \frac{h(y)}{\sqrt{\lambda-y}}\mathrm{d}\,y\,.$$

Changing the order of integration in the first member (fig. 8.4.1), we have:

$$\int_a^\lambda f(x)\left(\int_x^\lambda \frac{\mathrm{d}\,y}{\sqrt{(\lambda-y)(y-x)}}\right)\mathrm{d}\,x = \int_a^\lambda \frac{h(y)}{\sqrt{\lambda-y}}\mathrm{d}\,y$$

and then

$$\int_a^\lambda f(x)\mathrm{d}\,x = \frac{1}{\pi}\int_a^\lambda \frac{h(y)}{\sqrt{\lambda-y}}\mathrm{d}\,y\,.$$

Deriving this relation and utilizing the definition of the Finite Part (D.4.3), we deduce:

$$f(\lambda) = \frac{1}{\pi}\frac{\mathrm{d}}{\mathrm{d}\,\lambda}\int_a^\lambda \frac{h(y)}{\sqrt{\lambda-y}}\mathrm{d}\,y = -\frac{1}{2\pi} {}^{*}\!\!\int_a^\lambda \frac{h(y)}{(\lambda-y)^{3/2}}\mathrm{d}\,y\,. \tag{8.4.2}$$

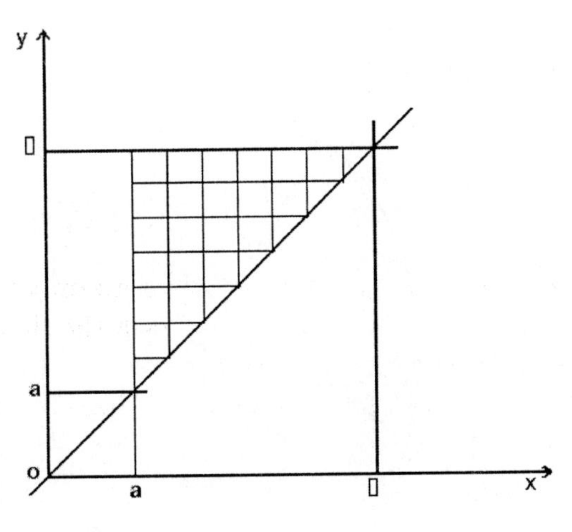

Fig. 8.4.1.

It results that the solution of the equation (8.4.1) is (8.4.2) and conversely, the solution of the integral equation (8.4.2), where h is the unknown, is given by the formula (8.4.1).

8.4.2 The Solution of the H Equation in the Domain of Influence of the Supersonic Trailing Edge

If M is in the zone of influence of the supersonic trailing edge (fig. 8.4.2), as we noticed at the end of the section 8.3., the curvilinear integral from the equations (8.3.73) and (8.3.74) disappears. We see in fact on the figure 8.4.2 that on the arc BM_2 we have $\alpha > a$ and on the arc M_1E, $\beta > b$. It remains to integrate the equation:

$$\frac{k}{4\pi} {}^{*}\!\!\int\!\!\int^{*}_{D_1} \frac{N(\alpha,\beta)\mathrm{d}\,\alpha\,\mathrm{d}\,\beta}{(a-\alpha)^{3/2}(b-\beta)^{3/2}} = H(a,b)\,. \tag{8.4.3}$$

Denoting:

$$\begin{aligned} &\alpha = A(\beta), \text{ the equation of the arc } DEFA \text{ and} \\ &\beta = B(\alpha), \text{ the equation of the contour } FABC, \end{aligned} \tag{8.4.4}$$

the equation (8.4.3) may be written as follows

$$\frac{k}{4\pi} {}^{*}\!\!\int_{B(a)}^{b} \frac{N_1(a,\beta)}{(b-\beta)^{3/2}}\mathrm{d}\,\beta = H(a,b)\,, \tag{8.4.5}$$

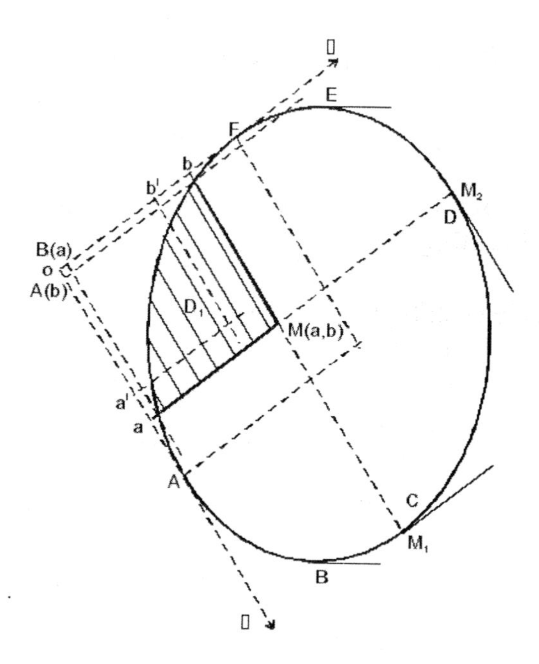

Fig. 8.4.2.

where

$$N_1(a,\beta) = \int_{A(\beta)}^{a} \frac{N(\alpha,\beta)}{(a-\alpha)^{3/2}}\mathrm{d}\,\alpha\,. \tag{8.4.6}$$

Utilizing (8.4.2) and (8.4.1), we deduce from (8.4.5)

$$N_1(a,\beta) = -\frac{2}{k}\int_{B(a)}^{\beta} \frac{H(a,b')}{\sqrt{\beta-b'}}\mathrm{d}\,b'\,. \tag{8.4.7}$$

In this way, the equation (8.4.6) becomes

$$-\frac{1}{2\pi}\int_{A(\beta)}^{a} \frac{N(\alpha,\beta)}{(a-\alpha)^{3/2}}\mathrm{d}\,\alpha = \frac{1}{\pi k}\int_{B(a)}^{\beta} \frac{H(a,b')}{\sqrt{\beta-b'}}\mathrm{d}\,b'\,. \tag{8.4.8}$$

The equation (8.4.8) is similar to (8.4.2), so that

$$N(\alpha,\beta) = \int_{A(\beta)}^{\alpha} \frac{\mathrm{d}\,a'}{\sqrt{\alpha-a'}}\frac{1}{\pi k}\int_{B(a')}^{\beta} \frac{H(a',b')}{\sqrt{\beta-b'}}\mathrm{d}\,b'\,.$$

For $\alpha = a,\ \beta = b$ we obtain

$$\begin{aligned} N(a,b) &= \frac{1}{\pi k}\int_{A(b)}^{a} \frac{\mathrm{d}\,a'}{\sqrt{a-a'}}\int_{B(a')}^{b} \frac{H(a',b')}{\sqrt{b-b'}}\mathrm{d}\,b' \\ &= \frac{1}{\pi k}\iint_{D_1(a,b)} \frac{H(a',b')\mathrm{d}\,a'\,\mathrm{d}\,b'}{\sqrt{(a-a')(b-b')}}\,. \end{aligned} \tag{8.4.9}$$

This is the solution in the zone of influence of the supersonic leading edge.

8.4.3 The Solution in the Domains of Influence of the Subsonic Leading Edge

We assume that $M(a,b)$ is situated in the zone of influence of the subsonic leading edge AB (fig. 8.4.3). In this case, the curvilinear integral also vanishes, because on BM_2 we have $\alpha > a$, and on M_1E, $\beta > b$. The equation which has to be integrated is also (8.4.3). Obviously, $A(b) \le a_1 \le a$, $B(a) \le b \le b_1$.

The equation (8.4.3) may be written as follows

$$\frac{k}{4\pi} {}^{*}\!\!\int_{A(b)}^{a} \frac{\mathrm{d}\,\alpha}{(a-\alpha)^{3/2}} \int_{B(\alpha)}^{b} \frac{N(\alpha,\beta)}{(b-\beta)^{3/2}} \mathrm{d}\,\beta = H(a,b)\,, \tag{8.4.10}$$

such that we denote

$$N_2(\alpha,b) = \frac{1}{2}\int_{B(\alpha)}^{b} \frac{N(\alpha,\beta)}{(b-\beta)^{3/2}} \mathrm{d}\,\beta \tag{8.4.11}$$

we obtain the integral equation

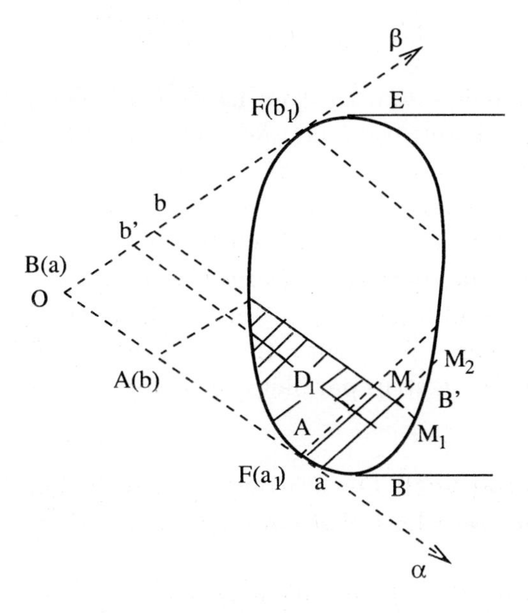

Fig. 8.4.3.

$$\frac{k}{2\pi}\int_{A(b)}^{a}\frac{N_2(\alpha,b)}{(a-\alpha)^{3/2}}\mathrm{d}\,\alpha=H(a,b) \tag{8.4.12}$$

whose solution is

$$N_2(\alpha,b)=-\frac{1}{k}\int_{A(b)}^{\alpha}\frac{H(a',b)}{\sqrt{\alpha-a'}}\mathrm{d}\,a' \tag{8.4.13}$$

for $\alpha > A(b)$. From (8.4.11) we obtain for $N(\alpha,\beta)$ the following integral equation

$$-\frac{1}{2\pi}\int_{B(\alpha)}^{b}\frac{N(\alpha,\beta)}{(b-\beta)^{3/2}}\mathrm{d}\,\beta=\frac{1}{\pi k}\int_{A(b)}^{\alpha}\frac{H(a',b)}{\sqrt{\alpha-a'}}\mathrm{d}\,a'\,. \tag{8.4.14}$$

Utilizing again the solution of the equation (8.4.2), we deduce:

$$N(\alpha,\beta)=\frac{1}{\pi k}\int_{B(\alpha)}^{\beta}\frac{\mathrm{d}\,b'}{\sqrt{\beta-b'}}\int_{A(b')}^{\alpha}\frac{H(a',b')}{\sqrt{\alpha-a'}}\mathrm{d}\,a'\,,$$

for $\alpha > A(b)$ and $\beta > B(\alpha)$. Putting $\alpha = a$, $\beta = b$, we obtain:

$$\begin{aligned}N(a,b)&=\frac{1}{\pi k}\int_{B(a)}^{b}\frac{\mathrm{d}\,b'}{\sqrt{b-b'}}\int_{A(b')}^{\alpha}\frac{H(a',b')}{\sqrt{\alpha-a'}}\mathrm{d}\,a'\\&=\frac{1}{\pi k}\iint_{D_1(a,b)}\frac{H(a',b')\mathrm{d}\,a'\,\mathrm{d}\,b'}{\sqrt{(a-a')(b-b')}}\,,\end{aligned} \tag{8.4.15}$$

D_1 being the shaded domain from figure 8.4.3 We notice that in this case D_1 is not the entire domain determined by the leading edge and the characteristics issuing from M. From this domain one eliminates the strip where $b' < B(a)$. This result was obtained for the first time in 1949, independently, by Evvard and Krasilscicova and it is called in some books *the theorem of Evvard and Krasilscicova.*

The solution is obtained analogously when M is in the zone of influence of the edge FE, with the difference that in this case one eliminates from D_1 a strip parallel to the $O\beta$ axis.

8.4.4 The Wing with Dependent Subsonic Leading Edges and Independent Subsonic Trailing Edges

For a wing with dependent subsonic leading edges and independent subsonic trailing edges (fig. 8.4.4) the solution in the domain bounded by the curve $AHFA$ is given by the formula (8.4.9), the solution in the domain bounded by $ABH'HA$ – by the formula (8.4.15), and the solution in the domain bounded by $FHH''EF$ – by a formula analogous to (8.4.15). The case when M is in the common zone of influence $HH'B'E'H''H$ of the subsonic leading edges is presented in the sequel.

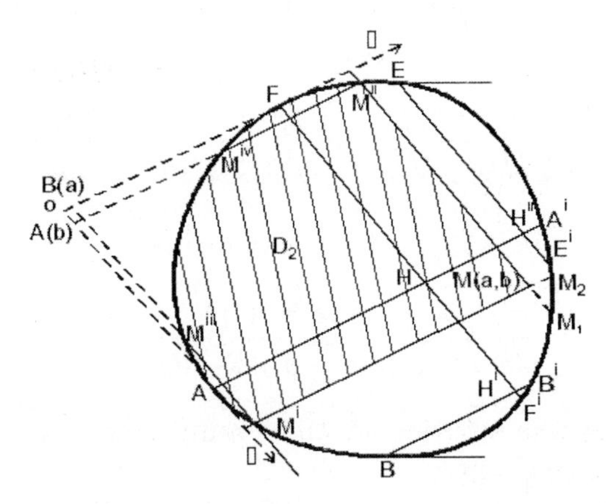

Fig. 8.4.4.

We notice at first that in this case the curvilinear integral from (8.3.73) also vanishes, because on BM_2 we have $\alpha > a$, and on $M_1E\beta > b$. Hence, we have to integrate the equation (8.4.3). Denoting by R_1, R_2 the domains bounded by the curves $FM''RF$, respectively $FM^{IV}M''F$ and by Q_1, Q_2 the domains bounded by the curves AQM', respectively $AM'M'''A$, we notice that on R_1 and Q_1 we have $N = 0$ (it results from (8.3.68)). Hence, the domain of integration from (8.4.3) may be prolonged to the domain bounded by $MRFAQM$, i.e. to the domain

$$D_2 + R_1 + R_2 + Q_1 + Q_2\,,$$

where D_2 is the shaded region from figure 8.4.4, i.e. the region bounded by the contour $M'''M'MM''M^{IV}$. The leading edge of this domain is entirely supersonic, such that the solution of the equation (8.4.3) is given by the formula (8.4.9)

$$N(a,b) = \frac{1}{\pi k}\left[\iint_{D_2} + \iint_{R_1} + \iint_{R_2} + \iint_{Q_1} + \iint_{Q_2}\right] \cdot \\ \cdot \frac{H(a',b')}{\sqrt{(a-a')(b-b')}} \mathrm{d}\,a'\,\mathrm{d}\,b'\,. \tag{8.4.16}$$

Let us consider now that M belongs to the zone of influence of the subsonic leading edge AB. From formula (8.4.15) we deduce:

$$N(a,b) = \frac{1}{\pi k}\left[\iint_{D_2} + \iint_{R_1} + \iint_{R_2}\right] \frac{H(a',b')}{\sqrt{(a-a')(b-b')}} \mathrm{d}\,a'\,\mathrm{d}\,b' . \tag{8.4.17}$$

Similarly, taking into account that M belongs to the zone of influence of the leading edge FE, we deduce:

$$N(a,b) = \frac{1}{\pi k}\left[\iint_{D_2} + \iint_{Q_1} + \iint_{Q_2}\right] \frac{H(a',b')}{\sqrt{(a-a')(b-b')}} \mathrm{d}\,a'\,\mathrm{d}\,b' . \tag{8.4.18}$$

From the formulas (8.4.16) – (8.4.18) it results

$$N(a,b) = \frac{1}{\pi k}\iint_{D_2} \frac{H(a',b')}{\sqrt{(a-a')(b-b')}} \mathrm{d}\,a'\,\mathrm{d}\,b' , \tag{8.4.19}$$

this representing the solution in the common zone of influence of the two subsonic leading edges.

We notice that the solutions from the previously considered domains have the same form, differing only the domains of integration. But we may establish a common rule for determining the domains of integration. They are bounded by the parallels to the characteristics issuing from the points where the first parallels intersect the leading edge and by the remaining portion of the leading edge. In the first case, when M is in the zone of influence of the supersonic edge, the parallels to the characteristics issuing from the points where the parallels from M towards infinity upstream intersect the leading edge, do not intersect any longer this edge.

8.4.5 The Wing with Dependent Subsonic Trailing Edges

We consider now a wing for which the parallels $M'M'''$, $M''M^{IV}$ intersect in a point P belonging to the interior of the domain D (fig. 8.4.5). In this case, denoting by R_2 the domain bounded by $FM'''PM''F$ and by Q_2 the domain bounded by $AM'PM^{IV}A$, (the domains R_1 and Q_1 keeping the same definition) and performing the same reasoning like in the previous subsection, we obtain successively

$$N(a,b) = I(D_3 + R_1 + R_2 + D_4 + Q_2 + Q_1)$$

$$N(a,b) = I(D_3 + R_1 + R_2)\,, \quad N(a,b) = I(D_3 + Q_2 + Q_1)\,,$$

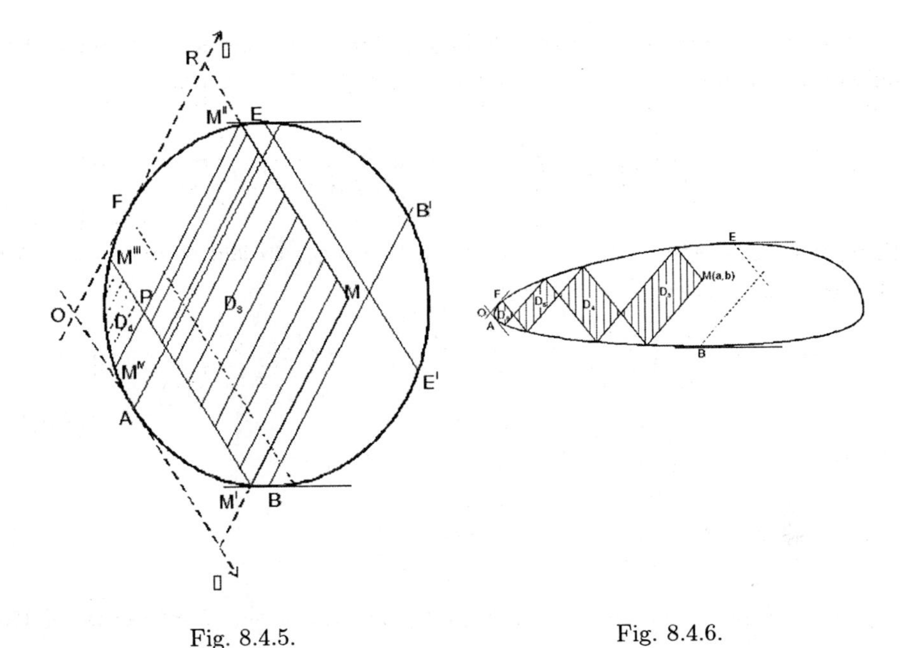

Fig. 8.4.5. Fig. 8.4.6.

I representing the symbol for the integral appearing in (8.4.19).

It results the solution

$$
\begin{aligned}
N(a,b) = \frac{1}{\pi k}\iint_{D_3} \frac{H(a',b')}{\sqrt{(a-a')(b-b')}}\mathrm{d}\,a'\,\mathrm{d}\,b' - \\
-\frac{1}{\pi k}\iint_{D_4} \frac{H(a',b')}{\sqrt{(a-a')(b-b')}}\mathrm{d}\,a'\,\mathrm{d}\,b'
\end{aligned}
\tag{8.4.20}
$$

The result may be generalized for every wing having finite dimensions. For example, for the wing from figure 8.4.6 the solution in the point $M(a,b)$ is

$$N(a,b) = I(D_3 - D_4 + D_5 - D_6)\,. \tag{8.4.21}$$

We stop here the presentation of the solutions in the zones of influence of the leading edge.

8.4.6 The Solution in the Zone of Influence of the Subsonic Edges under the Hypothesis that the Subsonic Leading Edges are Independent

For the wing from figure 8.4.7 the solution is determined in the domain bounded by the curve $ABB'E'EFA$. It remains to determine the solution in the zones $BCB'B$ and $E'DEE'$. To this aim we shall use the method of Homentcovschi [8.16].

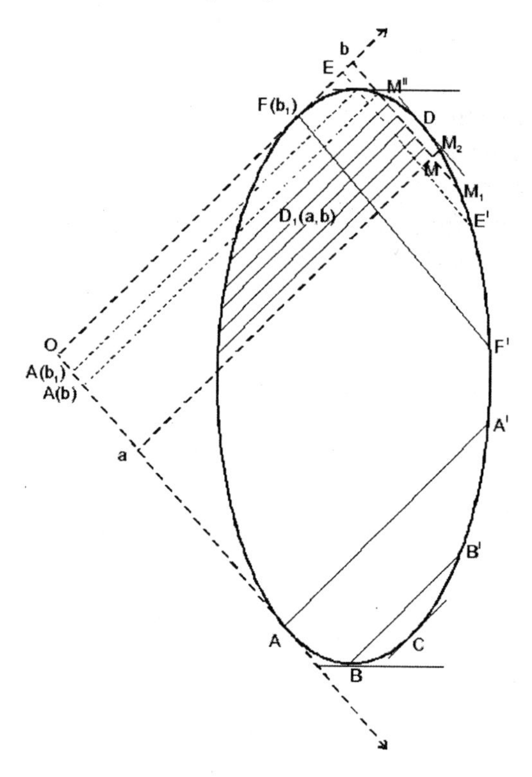

Fig. 8.4.7.

Let $M(a,b)$ be situated in the last zone. In this case, the curvilinear integral from (8.3.73) does not vanish. More precisely, it is zero on BM_2, where $\alpha > a$, it is zero on M_1M'', where $\beta > b$, but it is not zero on $M''E$. We assume at first that N is known on the trailing edge. Since M is in the zone of influence of the subsonic leading edge FE, the double integral may be inverted with a formula similar to (8.4.15). We have therefore

$$N(a,b) = \frac{1}{\pi k}\iint_{D_1(a,b)} \frac{H(a',b')\mathrm{d}\,a'\,\mathrm{d}\,b'}{\sqrt{(a-a')(b-b')}} + \frac{1}{\pi^2}\iint_{D_1} \frac{\partial}{\partial a'}\cdot$$

$$\cdot\left[\int_{EM''}\sqrt{\frac{a'-\alpha}{b'-\beta}}\,\frac{N(\alpha,\beta)}{a'-\alpha-(b'-\beta)}\mathrm{d}\,(\beta-\alpha)\right]\frac{\mathrm{d}\,a'\,\mathrm{d}\,b'}{\sqrt{(a-a')(b-b')}}, \tag{8.4.22}$$

where $D_1(a,b)$ is in the shaded domain from figure 8.4.7, i.e. the domain $A(b) < a' < a$, $b' < b$. In front of the last integral we have the sign $+$ because we have changed the sense on $M''E$ to EM''. On $M''E$ we shall denote

$$N(\alpha, \beta) = N(A(\beta), \beta) \equiv N(\beta)\,, \tag{8.4.23}$$

because the equation of the edge $DEFA$ is $\alpha = A(\beta)$.

The curvilinear integral imposes to eliminate from $D_1(a,b)$ the points where $b' < \beta$. D_1 from the second integral (8.4.22)is therefore the shaded domain from figure 8.4.8. Denoting this integral by T and interchanging the curvilinear integral and the integral on the domain we get:

$$T = \frac{1}{\pi^2}\iint_{D_1}\frac{\partial}{\partial a'}\left[\int_{M'E} N(\beta)\sqrt{\frac{a'-A(\beta)}{b'-\beta}}\,\frac{[1-A'(\beta)]\mathrm{d}\,\beta}{a'-A(\beta)-(b'-\beta)}\right]\cdot$$

$$\cdot\frac{\mathrm{d}\,a'\,\mathrm{d}\,b'}{\sqrt{(a-a')(b-b')}} = \frac{1}{\pi^2}\int_{b_2}^{b} N(\beta)[1-A'(\beta)]\mathrm{d}\,\beta\int_{A(b)}^{a}\frac{\mathrm{d}\,a'}{\sqrt{a-a'}}\,\frac{\partial}{\partial a'}\cdot$$

$$\cdot\left[\sqrt{a'-A(\beta)}\int_{\beta}^{b}\frac{1}{\sqrt{(b'-\beta)(b-b')}}\,\frac{\mathrm{d}\,b'}{u-b'}\right], \tag{8.4.24}$$

where $u = a' + \beta - A(\beta)$.

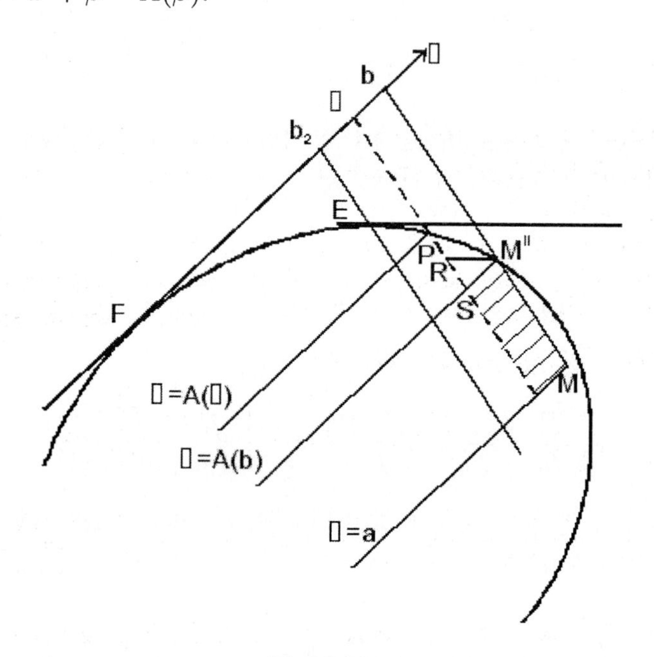

Fig. 8.4.8.

For calculating the interior integral we shall prove first that $b < u$, i.e. that

$$b < a' + \beta - A(\beta) . \tag{8.4.25}$$

Denoting by P, R, S the points having the ordinates β namely, P on the trailing edge, R on the la intersection with the direction of the unperturbed stream and S on the boundary of the domain D_1 (fig. 8.4.8), we have $R(A(\beta),\beta)$, $S(A(b),\beta)$, $\sphericalangle\, RM''S = \sphericalangle\, SRM'' = \mu$, μ representing Mach's angle (the angle made by the characteristic lines with the direction of the unperturbed stream). It results

$$RS = SM'' = b - \beta, \quad PS = A(b) - A(\beta) .$$

Since $M''P$ is an arc on the trailing edge we have $RS < PS$ whence

$$b \leq A(b) + \beta - A(\beta) \leq a' + \beta - A(\beta) , \tag{8.4.26}$$

because $a' > A(b)$.

Employing the substitution

$$b' = \frac{b+\beta}{2} + \frac{b-\beta}{2}\cos\theta , \quad u = \frac{b+\beta}{2} + \frac{b-\beta}{2}s$$

we obtain

$$I \equiv \int_\beta^b \frac{1}{\sqrt{(b'-\beta)(b-b')}}\frac{db'}{u-b'} = -\frac{2}{b-\beta}\int_0^\pi \frac{d\,\theta}{\cos\theta - s} ,$$

the inequality $b < u$ implying $1 < s$. This integral has the form (B.6.1) and the solution is given by (B.6.4).

One obtains:

$$I_1 = \frac{\pi}{\sqrt{(u-b)(u-\beta)}} = \frac{\pi}{\sqrt{[a' - A(\beta)][a' + \beta - A(\beta) - b]}} . \tag{8.4.27}$$

The following integral has the form

$$\int_{A(b)}^a \frac{1}{\sqrt{a-a'}}\frac{\partial}{\partial a'}\left(\frac{1}{\sqrt{a'-v}}\right) d\,a' = -\frac{1}{2}\int_{A(b)}^a \frac{1}{(a'-v)^{3/2}}\frac{d\,a'}{\sqrt{a-a'}} ,$$

where $v = b + A(\beta) - \beta < A(b)$. This inequality results from the first inequality (8.4.26). Noticing that

$$\int_{A(b)}^a \frac{1}{(a'-v)^{3/2}}\frac{d\,a'}{\sqrt{a-a'}} = \frac{2}{a-v}\sqrt{\frac{a-A(b)}{A(b)-v}} ,$$

we deduce

$$T = -\frac{1}{\pi}\sqrt{a - A(b)} \int_{b_2}^{b} \frac{N(\beta)}{\sqrt{\beta - A(\beta) + A(b) - b}} \frac{1 - A'(\beta)}{\beta - A(\beta) + a - b} \tag{8.4.28}$$

whence

$$N(a,b) = \frac{1}{\pi k} \iint_{D_1(a,b)} \frac{H(a',b')\mathrm{d}\,a'\,\mathrm{d}\,b'}{\sqrt{(a-a')(b-b')}} + T \tag{8.4.29}$$

in T, $N(\beta)$ being unknown. The form (8.4.29) is given in [8.16].

We consider that M tends to the position M'' on the boundary. It means that we make in (8.4.29) the substitution, $a \to A(b)$. The first integral is

$$I \equiv \frac{1}{\pi k} \iint_{D_1(a,b)} \frac{H(a',b')\mathrm{d}\,a'\,\mathrm{d}\,b'}{\sqrt{(a-a')(b-b')}} = \frac{1}{\pi k} \int_{A(b)}^{a} \frac{L(a')\mathrm{d}\,a'}{\sqrt{a-a'}}, \tag{8.4.30}$$

where

$$L(a') = \int_{B(a)}^{b} \frac{H(a',b')\mathrm{d}\,b'}{\sqrt{b-b'}}, \tag{8.4.31}$$

$b = B(a)$ representing the equation of the edge $FABC$. Obviously, this integral vanishes when $a \to A(b)$.

We shall perform in the expression of T the substitution $\beta \to t$: $\beta - A(\beta) = t$ and we shall denote $N(\beta) = N_1(t)$. We also denote

$$a - A(b) = \varepsilon, \quad b - A(b) = c, \quad b_2 - A(b_2) = c_2\,. \tag{8.4.32}$$

From the first inequality (8.4.26) it results $\beta - A(\beta) > b - A(b)$ whence $c_2 > c$.

Hence,

$$T = \frac{\sqrt{\varepsilon}}{\pi} \int_{c}^{c_2} \frac{N_1(t)}{\sqrt{t-c}} \frac{\mathrm{d}\,t}{t - c + \varepsilon}\,.$$

We shall integrate by parts setting

$$u = N_1(t)\,, \quad \mathrm{d}v = \frac{1}{\sqrt{t-c}} \frac{\mathrm{d}(t-c)}{t-c+\varepsilon}\,.$$

It results

$$v = \frac{2}{\sqrt{\varepsilon}} \arctan \sqrt{\frac{t-c}{\varepsilon}}\,.$$

The integrated term vanishes because $t = c_2$ implies $\beta = b_2$ and

$$N_1(c_2) = N(b_2) = 0\,.$$

We obtain therefore:

$$T = -\frac{2}{\pi}\int_c^{c_2} N_1'(t)\arctan\sqrt{\frac{t-c}{\varepsilon}}\mathrm{d}\,t \tag{8.4.33}$$

and

$$\lim_{\varepsilon\to 0} T = -\frac{2}{\pi}\,\frac{\pi}{2}\int_c^{c_2} N_1'(t)\mathrm{d}\,t = N_1(c) = N(b)\,.$$

In this way, passing to the limit in (8.4.29) we obtain

$$N(A(b), b) \equiv N(b)\,,$$

i.e. an identity. This means that for an arbitrary given $N(b)$, the function $N(a,b)$ given by (8.4.25) is a solution of the integral equation (8.3.73). One obtains an indetermination like in the subsonic case). This indetermination exists in the zones of influence of the subsonic edges $EE'DE$ and $BCB'B$. In these zones the integral equation of the problem is not sufficient for determining the solution.

Like in the subsonic case we remove this indetermination imposing the Kutta-Joukovsky condition. Imposing a finite velocity on EE', it results that the jump of the velocity on EE' is finite. Since the jump of the component u is given by the jump of the pressure with the changed sign, and the jump of the pressure by $f(x,y)$, it results that it is sufficient to impose for f to have finite values on the subsonic trailing edge.

We have

$$f(x,y) = \frac{\partial}{\partial x}N(x,y) = \left(\frac{\partial}{\partial a}+\frac{\partial}{\partial b}\right)N(a,b) = \left(\frac{\partial}{\partial a}+\frac{\partial}{\partial b}\right)(I+T)\,. \tag{8.4.34}$$

Performing in (8.4.30) the change of variable $a' \to s : a - a' = = [a - A(b)]s$ and keeping the notation $a - A(b) = \varepsilon$ we deduce

$$I = \frac{\sqrt{\varepsilon}}{\pi k}\int_0^1 L(a-\varepsilon s)\frac{\mathrm{d}\,s}{\sqrt{s}} = \frac{2\sqrt{\varepsilon}}{\pi k}L(a) - O(\varepsilon^{3/2}) =$$

$$= \frac{2}{\pi k}\sqrt{a - A(b)}L(a) - O(\varepsilon^{3/2})\,,$$

whence

$$\begin{aligned}\left(\frac{\partial}{\partial a}+\frac{\partial}{\partial b}\right)I &= \frac{1}{\pi k}\left(\frac{1}{\sqrt{\varepsilon}} - \frac{A'(b)}{\sqrt{\varepsilon}}\right)L(a) + O(\sqrt{\varepsilon}) \\ &= \frac{1-A'(b)}{\pi\sqrt{\varepsilon}k}\int_{B(a)}^{b}\frac{H(a,b')\mathrm{d}\,b'}{\sqrt{b-b'}} + O(\sqrt{\varepsilon})\,.\end{aligned} \tag{8.4.35}$$

From (8.4.32) we deduce

$$T = -\frac{2}{\pi}\int_{b-A(b)}^{b_2-A(b_2)} N_1'(t)\,\arctan\frac{\sqrt{t-b+A(b)}}{\sqrt{a-A(b)}}\,\mathrm{d}\,t$$

whence,

$$\left(\frac{\partial}{\partial a}+\frac{\partial}{\partial b}\right)T = \frac{1-A'(b)}{\pi\sqrt{\varepsilon}}\int_{b-A(b)}^{b_2-A(b_2)} N_1'(t)\frac{\sqrt{t-b+A(b)}}{t-b+a}\,\mathrm{d}\,t + O(\sqrt{\varepsilon})\,.$$

But

$$\frac{\sqrt{t-b+A(b)}}{t-b+a} = \frac{\sqrt{t-b+A(b)}}{t-b+A(b)+\varepsilon} = \frac{1}{\sqrt{t-b+A(b)}}[1+O(\varepsilon)]\,,$$

such that finally we have:

$$\left(\frac{\partial}{\partial a}+\frac{\partial}{\partial b}\right)T = \frac{1-A'(b)}{\pi\sqrt{\varepsilon}}\int_{b-A(b)}^{b_2-A(b_2)} \frac{N_1'(t)\mathrm{d}\,t}{\sqrt{t-b+A(b)}} + O(\sqrt{\varepsilon})\,. \quad (8.4.36)$$

We obtain therefore

$$f(x,y) = \frac{1-A'(b)}{\pi\sqrt{\varepsilon}}\left[\frac{1}{k}\int_{B(a)}^{b}\frac{H(a,b')}{\sqrt{b-b'}}\mathrm{d}\,b' + \int_{b-A(b)}^{b_2-A(b_2)}\frac{N_1'(t)\mathrm{d}\,t}{\sqrt{t-b+A(b)}}\right] + O(\sqrt{\varepsilon})\,. \quad (8.4.37)$$

The function $f(x,y)$ has finite values on the trailing edge $(a \to A(b),\ \varepsilon \to 0)$ when the square bracket vanishes i.e. when

$$\int_{b-A(b)}^{b_2-A(b_2)}\frac{N_1'(t)\mathrm{d}\,t}{\sqrt{t-b+A(b)}} = -\frac{1}{k}\int_{B(A(b))}^{b}\frac{H(A(b),b')}{\sqrt{b-b'}}\mathrm{d}\,b' \equiv -G(b)\,. \quad (8.4.38)$$

This condition is an integral equation (of Abel type) for determining the unknown $N_1'(t)$. Using the notations (8.4.27) we may write this equation as follows

$$\int_{c}^{c_2}\frac{N_1'(t)\mathrm{d}\,t}{\sqrt{t-c}} = -G_1(c)\,, \quad (8.4.39)$$

where $G_1(c) = G_1(b-A(b)) = G(b)$. We deduce

$$\int_{x}^{c_2}\frac{\mathrm{d}\,c}{\sqrt{c-x}}\int_{c}^{c_2}\frac{N_1'(t)\mathrm{d}\,t}{\sqrt{t-c}} = -\int_{x}^{c_2}\frac{G_1(c)}{\sqrt{c-x}}\mathrm{d}\,c\,.$$

Changing the order of integration in the left hand member we get:

$$\int_x^{c_2} N_1'(t)\mathrm{d}\,t \int_x^t \frac{\mathrm{d}\,c}{\sqrt{(c-x)(t-c)}} = -\int_x^{c_2} \frac{G_1(c)}{\sqrt{c-x}}\mathrm{d}\,c\,.$$

Since $N_1(c_2) = N_1(b_2 - A(b_2)) = N(b_2) = 0$ (on the leading edge N vanishes), we obtain

$$N_1(x) = \frac{1}{\pi}\int_x^{c_2} \frac{G_1(t)}{\sqrt{t-x}}\mathrm{d}\,t\,.$$

For $x = b - A(b)$ we deduce

$$N(b) = \frac{1}{\pi}\int_{b-A(b)}^{b_2-A(b_2)} \frac{G_1(t)}{\sqrt{t-b+A(b)}}\mathrm{d}\,t\,,$$

or, using the change of variable $t \to \beta : \beta - A(\beta) = t$,

$$\begin{aligned} N(b) &= \frac{1}{\pi}\int_b^{b_2} \frac{G(\beta)[1-A'(\beta)]\mathrm{d}\,\beta}{\sqrt{\beta - A(\beta) - b + A(b)}} \\ &= -\frac{1}{\pi}\int_{b_2}^{b} \frac{K(\beta)[1-A'(\beta)]\mathrm{d}\,\beta}{\sqrt{\beta - b + A(b) - A(\beta)}}\,, \end{aligned} \tag{8.4.40}$$

where

$$K(\beta) = \frac{1}{k}\int_{B(A(\beta))}^{\beta} \frac{H(A(\beta), b')}{\sqrt{\beta - b'}}\mathrm{d}\,b'\,. \tag{8.4.41}$$

This formula determines N in every point of the subsonic trailing edge ED. Replacing this expression in (8.4.28), we may find out T.

In the sequel we shall give an explicit expression of T. Changing the order of integration (fig. 8.4.9), we deduce:

$$\begin{aligned} T &= \frac{1}{\pi^2}\sqrt{a - A(b)}\int_{b_2}^{b} \frac{1 - A'(\beta)}{\sqrt{\beta - A(\beta) + A(b) - b}}\cdot \\ &\quad\cdot \frac{\mathrm{d}\,\beta}{\beta - A(\beta) + a - b}\cdot\left[\int_{b_2}^{\beta} \frac{K(\beta')[1-A'(\beta')]\mathrm{d}\,\beta'}{\sqrt{\beta - b + A(b) - A(\beta)}}\right] = \\ &= \frac{1}{\pi^2}\sqrt{a - A(b)}\int_{b_2}^{b} K(\beta')I(\beta')[1-A'(\beta')]\mathrm{d}\,\beta'\,, \end{aligned} \tag{8.4.42}$$

where, using the substitution $\beta \to t : \beta - A(\beta) = t, \beta' - A(\beta') = t'$,

$$I(\beta') = -\int_c^{t'} \frac{1}{\sqrt{(t-c)(t'-t)}}\,\frac{\mathrm{d}\,t}{t-(b-a)}\,. \tag{8.4.43}$$

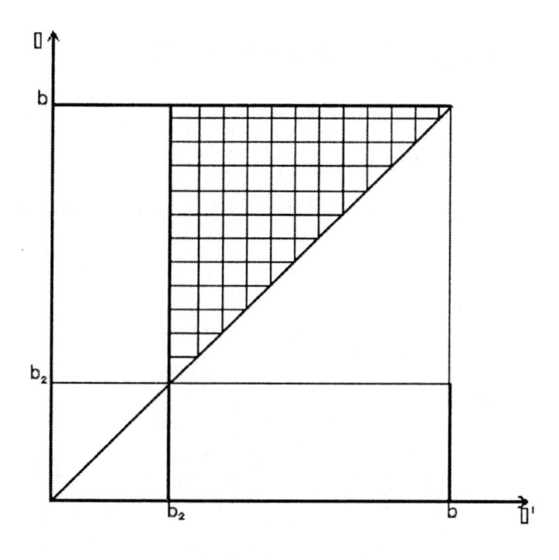

Fig. 8.4.9.

From (8.4.26) it results $b - A(b) < \beta - A(\beta)$ for every β, hence for $\beta = \beta'$. This implies $c < t'$. The integral from (8.4.43) has the form (B.6.11). It results

$$I = -\frac{\pi}{\sqrt{[a - A(b)][\beta' - A(\beta') - b + a]}},$$

whence

$$T = -\frac{1}{\pi}\int_{b_2}^{b} \frac{K(\beta)[1 - A'(\beta)]\mathrm{d}\,\beta}{\sqrt{\beta - A(\beta) - b + a}}. \tag{8.4.44}$$

Using this form of T, we express the solution in the domain bounded by the curve $EE'DE$ by means of the formula (8.4.29).

2. If $M(a,b)$ is in the domain bounded by the curve $BCB'B$ (fig. 8.4.10), then we shall utilize the integral equation (8.3.74). Obviously, the curvilinear integral does not vanish on the arc BM'. Utilizing the solution (8.4.15), we deduce:

$$N(a,b) = \frac{1}{\pi k}\iint_{D_1(a,b)} \frac{H(a',b')}{\sqrt{(a-a')(b-b')}}\mathrm{d}\,a'\,\mathrm{d}\,b' + T_1 \tag{8.4.45}$$

where

$$T_1 = -\frac{1}{\pi^2}\iint_{D_1} \frac{\mathrm{d}\,a'\,\mathrm{d}\,b'}{\sqrt{(a-a')(b-b')}}\cdot$$

$$\cdot\frac{\partial}{\partial b'}\int_{a_2}^{a'} N(\alpha)\sqrt{\frac{b' - B(\alpha)}{a' - \alpha}}\,\frac{[B'(\alpha) - 1]\mathrm{d}\,\alpha}{b' - a' - B(\alpha) + \alpha}. \tag{8.4.46}$$

Taking into account that the equation of the edge BC is $\beta = B(\alpha)$,

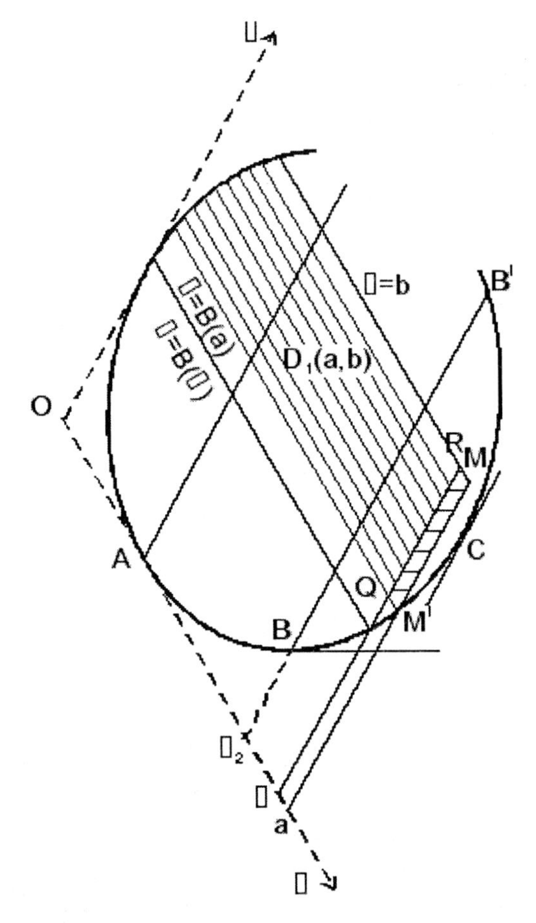

Fig. 8.4.10.

we denoted $N(\alpha, B(\alpha)) = N(\alpha)$. $D_1(a,b)$ is the shaded from figure (8.4.10), i.e. the domain bounded by MM', the parallels $\beta = B(a)$ and $\beta = b$ and the leading edge included between these parallels. D_1 is the domain bounded by the contour $MPQM'$ imposed by the condition $\alpha < a'$.

Interchanging in (8.4.45) the curvilinear integral and the double integral we obtain

$$T_1 = \frac{1}{\pi^2}\int_{a_2}^{a} N(\alpha)[1 - B'(\alpha)]\Big\{\int_{B(a)}^{b} \frac{1}{\sqrt{b-b'}}\cdot$$

$$\cdot\frac{\partial}{\partial b'}\left[\sqrt{b'-B(\alpha)}\int_{\alpha}^{a}\frac{1}{\sqrt{(a-a')(a'-\alpha)}}\frac{\mathrm{d}\,a'}{w-a'}\right]\mathrm{d}\,b'\Big\}\mathrm{d}\,\alpha \qquad (8.4.47)$$

where

$$w = b' + \alpha - B(\alpha)\,.$$

The similarity of the expressions (8.4.47) and (8.4.24) is obvious. T_1 may be obtained from T replacing b, β and A by a, α and B and conversely. It results therefore

$$T_1 = -\frac{1}{\pi}\sqrt{b - B(a)}\int_{a_2}^{a}\frac{N(\alpha)}{\sqrt{\alpha - B(\alpha) + B(a) - a}}\frac{1 - B'(\alpha)}{\alpha - B(\alpha) + b - a}\mathrm{d}\,\alpha \qquad (8.4.48)$$

and then

$$T_1 = -\frac{1}{\pi}\int_{a_2}^{a}\frac{K_1(\alpha)[1 - B'(\alpha)]\mathrm{d}\,\alpha}{\sqrt{\alpha - B(\alpha) - a + b}} \qquad (8.4.49)$$

where

$$K_1(\alpha) = \frac{1}{k}\int_{A(B(\alpha))}^{\alpha}\frac{H(a', B(\alpha))}{\sqrt{\alpha - a'}}\mathrm{d}\,a' \qquad (8.4.50)$$

The solution in the domain bounded by $BCB'B$ is (8.4.45) where T_1 is given by (8.4.49).

8.4.7 The Wing with Dependent Subsonic Trailing Edges

For a wing with subsonic dependent trailing edges (fig. 8.4.11) the solution in the zone AHF is determined by the formula (8.4.9), the solution in the zone $ABF'H$ by the formula (8.4.15), the solution in the zone $FHA'E$ by a formula analogous to (8.4.15), the solution in the zone $HF'IA'$ by the formula (8.4.19), the solution in the zone $BCE'IB$ by the formula (8.4.45), where T_1 is (8.4.48), and the solution in the zone $EIB'DE$ by the formula (8.4.29), where T is (8.4.44). It remains to determine the solution in the zone $IE'B'I$, i.e. in the common zone of influence of the subsonic trailing edges.

Noticing that the curvilinear integral does not vanish on BM' and $M''E$ and utilizing in the first case the expression from (8.3.74), and in the second case the expression from (8.3.73), we deduce that the integral equation has the following form:

$$\frac{k}{4\pi}\, {}^{**}\!\!\iint_{D_1} \frac{N(\alpha,\beta)}{(a-\alpha)^{3/2}(b-\beta)^{3/2}}\,\mathrm{d}\,\alpha\,\mathrm{d}\,\beta+$$

$$+\frac{k}{\pi}\frac{\partial}{\partial b}\int_{BM'} N(\alpha)\sqrt{\frac{b-B(\alpha)}{a-\alpha}}\;\frac{[B'(\alpha)-1]\mathrm{d}\,a}{b-a-B(\alpha)+\alpha}+ \tag{8.4.51}$$

$$+\frac{k}{\pi}\frac{\partial}{\partial a}\int_{M''E}\sqrt{\frac{a-A(\beta)}{b-\beta}}\;\frac{[1-A'(\beta)]\mathrm{d}\,\beta}{a-b-A(\beta)+\beta}=H(a,b)\,.$$

the integral on BM' representing in fact the integral with respect to α on the interval (a_2,a) and the integral on $M''E$ representing the integral with respect to β on the interval (b,b_2).

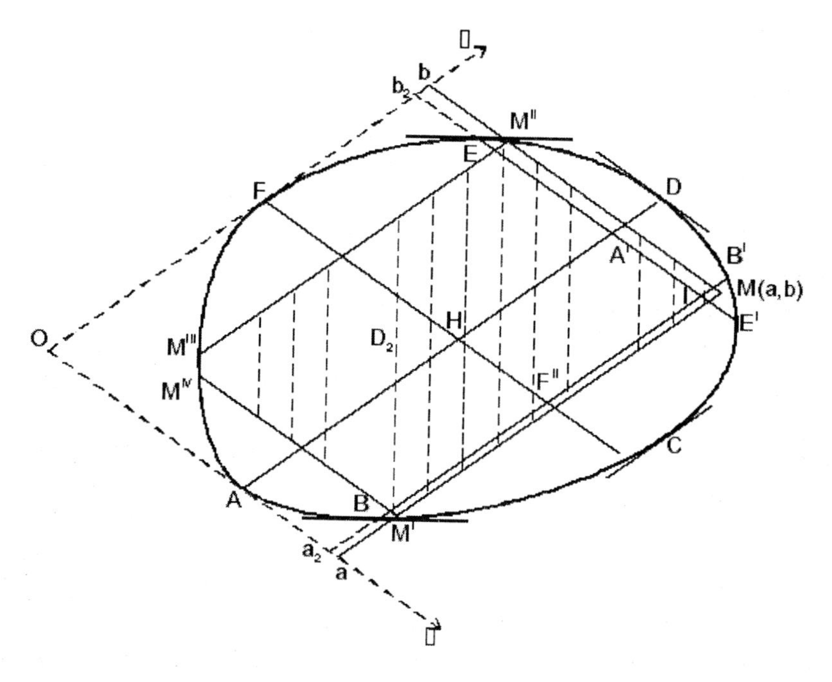

Fig. 8.4.11.

In this case, the point $M(a,b)$ is in the common zone of influence of the subsonic leading edges AB and FE. Hence the double integral may be inverted according to the formula (8.4.19), D_2 representing the shaded domain from 8.4.11, i.e. the domain bounded by the curve

$MM''M'''M^{IV}M'M$.

$$N(a,b) = \frac{1}{\pi k} \iint_{D_2} \frac{H(a',b')}{\sqrt{(a-a')(b-b')}} \mathrm{d}\,a'\,\mathrm{d}\,b' + T_1 + T\,, \qquad (8.4.52)$$

where T_1 has the expression (8.4.43), and T, (8.4.27).

Setting $M \to M''(a \to A(b))$, the integral on BM' vanishes (as we can see on the figure), such that from (8.4.52) one obtains $N(A(b),b) = N(b)$. Imposing the Kutta-Joukovsky condition, we deduce that T has the form (8.4.43). Similarly we deduce that T_1 has the expression (8.4.49).

Now the problem is completely solved.

In the end, it is a pleasant duty for me to mention that for elaborating this section I utilized especially Homentcovschi's paper [8.16] and the license thesis of my former student Luminiţa Berechet [8.2].

8.5 The Theory of Conical Motions

8.5.1 Introduction

The theory of conical motions was initiated by Busemann in 1943, [8.3]. It refers to wings bounded by conical surfaces with the vertex in the origin of the system of coordinates, the body being placed downstream. The surface of such a body is a smooth surface consisting of half-lines issuing from the origin and leaning on a closed curve situated in the plane $x = 1(x_1 = L_1)$. According to the boundary conditions the velocity is constant along every half-line passing through the origin and belonging to the boundary of the body. The hypothesis of conical flow leads to the assumption that the velocity has everywhere in the fluid this property. We have therefore

$$\boldsymbol{v}(mx, my, mz) = \boldsymbol{v}(x, y, z) \qquad (8.5.1)$$

for every m, real and positive. It means that the velocity is a homogeneous function having the zero degree. Under this assumption the equation of the potential becomes simpler, the unknowns depending not on three but on two variables. After Busemann, many authors (Langerstrom [8.22], Germain [8.11], Poritsky [8.28], Ward [8.34], Heaslet and Lomax [8.15], Iacob [8.18], Carafoli [1.5] ş.a.) have contributed decisively to the development of this theory. In all this theory, which will be called the classical theory, we make the hypothesis that the motion is conical.

Starting from the lifting surface equation in a supersonic stream, one may prove that if the wing is conical, then the solution of the integral equation is conical. For the equation (8.3.30) this thing is done in [8.5], and for equation (8.3.71), in [8.16]. In the present subsection, utilizing the solution from the previous subsection, we shall give the solution of the conical motions by particularization. We shall also give the basic elements of the classical method, because they may be obtained directly, without knowing the solution of the lifting surface equation.

8.5.2 The Wing with Supersonic Leading Edges

We assume that the surface of the wing is a conical surface. From $z = h(x, y)$ it results that $\lambda z = h(\lambda x, \lambda y)$ and, with $\lambda = (1/x)$,

$$h(x, y) = xh\left(1, \frac{y}{x}\right) \equiv xg\left(\frac{y}{x}\right) . \tag{8.5.2}$$

We deduce therefore

$$h_x = g\left(\frac{y}{x}\right) - \frac{y}{x}g'\left(\frac{y}{x}\right)$$

and then

$$H(a, b) = F\left(\frac{b}{a}\right) . \tag{8.5.3}$$

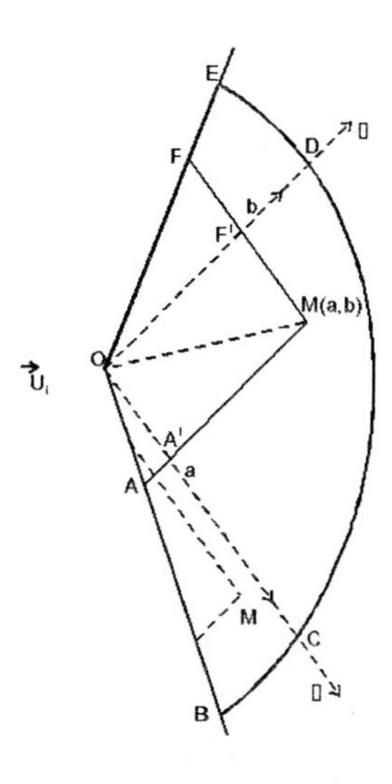

Fig. 8.5.1.

We shall consider now a wing with supersonic leading edges (fig. 8.5.1). Denoting by $b = m_1 a$ the equation of the edge OA and by $b = m_2 a$ the equation of the edge OF, it is obvious that $m_1 < 0$, $m_2 < 0$, because on OA we have $a > 0$, $b < 0$, and on OF, $a < 0$, $b > 0$. Since the entire domain D is only in the zone of influence of the supersonic leading edge, for every $M(a,b)$ the solution is given by the formula (8.4.9) where $D_1(a,b)$ is the domain limited by the curve $OAMFO$, and $H(a',b')$ will be replaced by $F(\mu)$, where $b' = \mu a'$. We have therefore to put $b = ma$ and to replace the variables b and b' by m and μ.

For M in the zone OCD we shall denote $N = N_2$

$$N_2(a, ma) = N_{21} + N_{22} + N_{23}\,, \tag{8.5.4}$$

where

$$N_{21}(a,m) = \frac{1}{\pi k}\int_{m_1}^{0} \frac{F(\mu)}{\sqrt{-\mu}}\left(\int_0^a \frac{a'\mathrm{d}\,a'}{\sqrt{(a-a')(a'-c)}}\right)\mathrm{d}\,\mu,\ c = \frac{m}{\mu}a,.$$

$$N_{22}(a,m) = \frac{1}{\pi k}\int_{0}^{m} \frac{F(\mu)}{\sqrt{\mu}}\left(\int_0^a \frac{a'\mathrm{d}\,a'}{\sqrt{(a-a')(c-a')}}\right)\mathrm{d}\,\mu +$$

$$+\frac{1}{\pi k}\int_{m}^{\infty} \frac{F(\mu)}{\sqrt{\mu}}\left(\int_0^c \frac{a'\mathrm{d}\,a'}{\sqrt{(a-a')(c-a')}}\right)\mathrm{d}\,\mu\,,$$

$$N_{23}(a,m) = \frac{1}{\pi k}\int_{-\infty}^{m_2} \frac{F(\mu)}{\sqrt{-\mu}}\left(\int_c^0 \frac{a'\mathrm{d}\,a'}{\sqrt{(a-a')(a'-c)}}\right)\mathrm{d}\,\mu\,, \tag{8.5.5}$$

N_{21} representing the integral on OAA', N_{22} on $OA'MF'$ and N_{23} the integral on $OF'F$. Performing the calculations we find:

$$N_{21} = \frac{a}{\pi k}\int_{m_1}^{0}\left(\frac{m+\mu}{\mu\sqrt{-\mu}}\arctan\sqrt{\frac{-\mu}{m}} - \frac{\sqrt{m}}{\mu}\right)F(\mu)\mathrm{d}\,\mu\,,$$

$$N_{23} = -\frac{a}{\pi k}\int_{-\infty}^{m_2}\left(\frac{m+\mu}{\mu\sqrt{-\mu}}\operatorname{arccot}\sqrt{\frac{-\mu}{m}} - \frac{\sqrt{m}}{\mu}\right)F(\mu)\mathrm{d}\,\mu\,, \tag{8.5.6}$$

$$N_{22} = \frac{a}{\pi k}\int_{0}^{\infty}\left(\frac{m+\mu}{2\mu\sqrt{\mu}}\ln\frac{\sqrt{\mu}+\sqrt{m}}{|\sqrt{\mu}-\sqrt{m}|} - \frac{\sqrt{m}}{\mu}\right)F(\mu)\mathrm{d}\,\mu_0\,.$$

It is obvious that

$$f(x,y) = \left(\frac{\partial}{\partial a} + \frac{\partial}{\partial b}\right)N_2 \tag{8.5.7}$$

is constant on every half-line issuing from the origin. The flow is conical.

If $M(a,b)$ is in the zone OBC, then the solution is

$$N_2(a,m) = \frac{1}{\pi k}\int_{m_1}^{m}\frac{F(\mu)}{\sqrt{-\mu}}\left(\int_c^a \frac{a'\,\mathrm{d}\,a'}{\sqrt{(a-a')(a'-c)}}\right)\mathrm{d}\,\mu = \\ = \frac{a}{2k}\int_{m_1}^{m}\frac{F(\mu)}{\sqrt{-\mu}}\,\frac{m+\mu}{\mu}\mathrm{d}\,\mu\,, \tag{8.5.8}$$

and if M is in the zone ODE, then

$$N_3(a,m) = \frac{1}{\pi k}\int_{m}^{m_2}\frac{F(\mu)}{\sqrt{-\mu}}\left(\int_a^c \frac{a'\,\mathrm{d}\,a'}{\sqrt{(a-a')(a'-c)}}\right)\mathrm{d}\,\mu = \\ = \frac{a}{2k}\int_{m}^{m_2}\frac{F(\mu)}{\sqrt{-\mu}}\,\frac{m+\mu}{\mu}\mathrm{d}\,\mu\,, \tag{8.5.9}$$

8.5.3 The Wing with a Supersonic Leading Edge and with Another Subsonic Leading or Trailing Edge

Further we shall consider a wing having a supersonic leading edge (the edge OB from figures 8.5.2) and another subsonic leading edge (fig. 8.5.2a)), or a subsonic trailing edge (the edge OE from fig. 8.5.2b)).

In this case, the solution is obtained with the formula (8.4.15) where D_1 is the domain limited by $MFF'AM$. For M belonging to the interior of Mach's cone i.e. M in the zone $OCEO$, noticing that the equation of the line FF' is $a' = d$ (it is obtained from the intersection of $b' = b$ with $b' = m_2 a'$) where $d = \dfrac{m}{m_2}a$, we deduce

$$N_2(a,m) = \frac{1}{\pi k}\int_{m_1}^{0}\frac{F(\mu)}{\sqrt{-\mu}}\mathrm{d}\,\mu\int_d^a \frac{a'\,\mathrm{d}\,a'}{\sqrt{(a-a')(a'-c)}} + \\ + \frac{1}{\pi k}\int_0^m \frac{F(\mu)}{\sqrt{\mu}}\mathrm{d}\,\mu\int_d^a \frac{a'\,\mathrm{d}\,a'}{\sqrt{(a-a')(c-a')}} + \\ + \frac{1}{\pi k}\int_m^{m_2}\frac{F(\mu)\mathrm{d}\,\mu}{\sqrt{\mu}}\int_d^c \frac{a'\,\mathrm{d}\,a'}{\sqrt{(a-a')(c-a')}}\,, \tag{8.5.10}$$

c being defined in the previous section. The interior integrals are elementary.

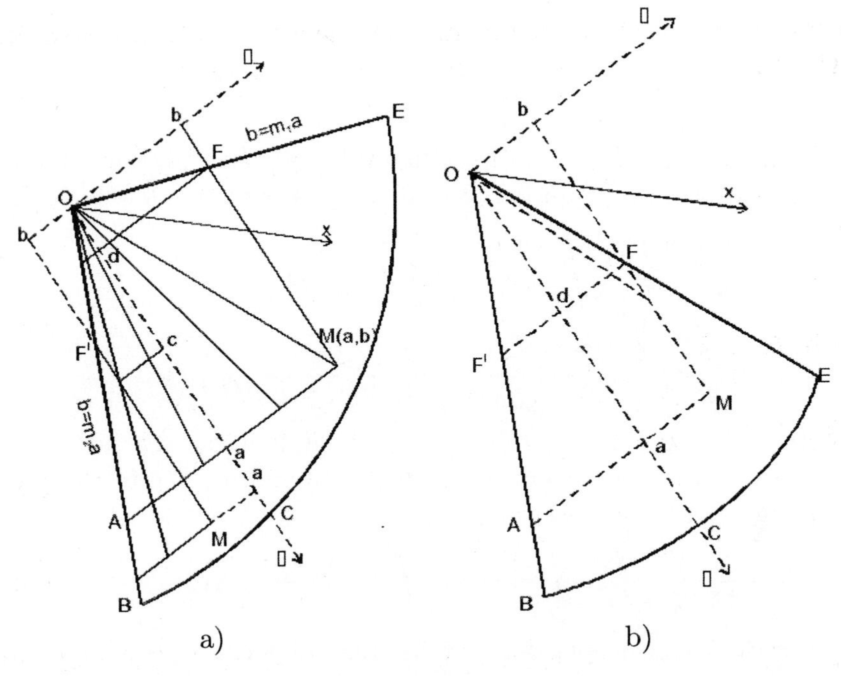

Fig. 8.5.2.

For M in the zone OBC noticing that the intersection of the line $b' = \mu a'$ with $b' = b$ has the abscissa c, we obtain

$$N_1(a, m) = \frac{1}{\pi k} \int_{m_1}^{m} \frac{F(\mu)}{\sqrt{-\mu}} \mathrm{d}\,\mu \int_{c}^{a} \frac{a' \mathrm{d}\, a'}{\sqrt{(a - a')(a' - c)}} = \tag{8.5.11}$$

$$= \frac{a}{2k} \int_{m_1}^{m} \frac{F(\mu)}{\sqrt{-\mu}} \frac{m + \mu}{\mu} \mathrm{d}\,\mu\,.$$

8.5.4 The Wing with Subsonic Leading Edges

When the two leading edges are subsonic, it is difficult to utilize the solution from the previous subsection. We shall use therefore Homentcovschi's idea concerning the direct integration of the equation (8.3.73). Assuming that $N(\alpha, \mu\alpha)$ has the form

$$N(\alpha, \mu\alpha) = \alpha N(\mu)\,, \tag{8.5.12}$$

the equation we have in view reduces to

$$\frac{k}{4\pi}\iint_D \frac{\alpha^2 N(\mu)\mathrm{d}\,\alpha\,\mathrm{d}\,\mu}{(a-\alpha)^{3/2}(ma-\mu\alpha)^{3/2}} = F(m)\,, \tag{8.5.13}$$

D being the shaded domain from figure 8.5.3. With the same notation for $c\,(=ma/\mu)$, the equation (8.5.13) may be written as follows

$$\begin{aligned}&\frac{k}{4\pi}\int_{m_1}^{m}\frac{N(\mu)}{\mu^{3/2}}\left(\int_0^a \frac{\alpha^2\mathrm{d}\,\alpha}{(a-\alpha)^{3/2}(c-\alpha)^{3/2}}\right)\mathrm{d}\,\mu+\\ &+\frac{k}{4\pi}\int_{m_1}^{m}\frac{N(\mu)}{\mu^{3/2}}\left(\int_0^c \frac{\alpha^2\mathrm{d}\,\alpha}{(a-\alpha)^{3/2}(c-\alpha)^{3/2}}\right)\mathrm{d}\,\mu\,.\end{aligned} \tag{8.5.14}$$

Obviously, in the first integral $a < c$, and in the second, $c < a$. The interior integrals are considered in Hadamard's Finite Part sense. Taking into account the formula

$$\frac{\mathrm{d}}{\mathrm{d}\,a}\int_0^a \frac{f(\alpha)}{\sqrt{a-\alpha}}\mathrm{d}\,\alpha = -\frac{1}{2}\,{}^*\!\!\int_0^a \frac{f(\alpha)}{(a-\alpha)^{3/2}}\mathrm{d}\,\alpha\,, \tag{8.5.15}$$

given in (D.4.3), the equation (8.5.14) may be written as follows

$$\begin{aligned}&\frac{k}{\pi}\int_{m_1}^{m}\frac{N(\mu)}{\mu^{3/2}}\left(\frac{\partial^2}{\partial a\partial c}\int_0^a \frac{\alpha^2\mathrm{d}\,\alpha}{\sqrt{(a-\alpha)(c-\alpha)}}\right)\mathrm{d}\,\mu+\\ &+\frac{k}{\pi}\int_{m_1}^{m}\frac{N(\mu)}{\mu^{3/2}}\left(\frac{\partial^2}{\partial a\partial c}\int_0^c \frac{\alpha^2\mathrm{d}\,\alpha}{\sqrt{(a-\alpha)(c-\alpha)}}\right)\mathrm{d}\,\mu\end{aligned} \tag{8.5.16}$$

We notice that in the first case $(c > a)$, we have

$$\int_0^a \frac{\alpha^2\mathrm{d}\,\alpha}{\sqrt{(a-\alpha)(c-\alpha)}} = -\frac{3}{4}(a+c)\sqrt{ac}+\frac{3(a+c)^2-4ac}{8}\ln\frac{\sqrt{c}+\sqrt{a}}{\sqrt{c}-\sqrt{a}}\,,$$

and in the second $(c < a)$,

$$\int_0^c \frac{\alpha^2\mathrm{d}\,\alpha}{\sqrt{(a-\alpha)(c-\alpha)}} = -\frac{3}{4}(a+c)\sqrt{ac}+\frac{3(a+c)^2-4ac}{8}\ln\frac{\sqrt{c}+\sqrt{a}}{\sqrt{c}-\sqrt{a}}\,. \tag{8.5.17}$$

The results are the same if we put under logarithm $(\sqrt{a}-\sqrt{c})$. Performing the calculations, it results that the integral equation (8.5.16) may be written as follows

$$\frac{k}{4\pi}\,{}^*\!\!\int_{m_1}^{m_2}\frac{N_1(\mu)}{\sqrt{\mu}}\left\{\ln\frac{\sqrt{m}+\sqrt{\mu}}{|\sqrt{m}-\sqrt{\mu}|}-2\sqrt{m\mu}\frac{m+\mu}{(m-\mu)^2}\right\}\mathrm{d}\,\mu = F(m) \tag{8.5.18}$$

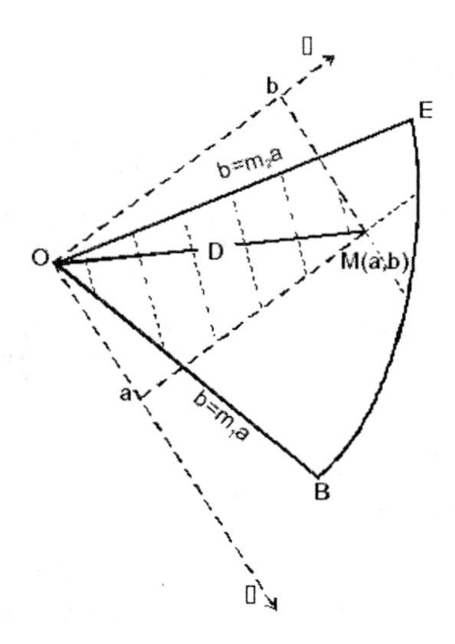

Fig. 8.5.3.

for $m_1 < m < m_2$. We put $N(\mu) = \mu N_1(\mu)$.

Denoting

$$H(m) = \frac{k}{\pi} \int_{m_1}^{m_2} \frac{N_1(\mu)}{\sqrt{\mu}} \ln \frac{\sqrt{m} + \sqrt{\mu}}{|\sqrt{m} - \sqrt{\mu}|} \mathrm{d}\,\mu\,, \tag{8.5.19}$$

on the basis of the equation (8.5.18) we obtain the following differential equation

$$m^2 H''(m) + mH'(m) - \frac{1}{4} H(m) = -F(m)\,. \tag{8.5.20}$$

The homogeneous equation has the linearly independent solutions $\sqrt{m}$ and $1/\sqrt{m}$. Hence, the general solution of the equation (8.5.20) is:

$$H(m) = 2c_1\sqrt{m} - \frac{2c_2}{\sqrt{m}} + F_0(m)\,, \tag{8.5.21}$$

c_1 and c_2 representing constant which have to be determined, and F_0 representing a particular solution of the non-homogeneous equation.

From (8.5.19) and (8.5.21) we deduce the following integral equation for N_1

$$\frac{k}{\pi} \int_{m_1}^{m_2} \frac{N_1(\mu)}{\sqrt{\mu}} \ln \frac{\sqrt{m} + \sqrt{\mu}}{|\sqrt{m} - \sqrt{\mu}|} \mathrm{d}\,\mu = 2c_1\sqrt{m} - \frac{2c_2}{\sqrt{m}} + F_0(m)\,. \tag{8.5.22}$$

Deriving with respect to m we obtain

$$\frac{k}{\pi} {}' \!\! \int_{m_1}^{m_2} \frac{N_1(\mu)}{\mu - m} \mathrm{d}\,\mu = c_1 + \frac{c_2}{m} + \sqrt{m} F_0'(m)\,, \tag{8.5.23}$$

which is the classical equation of the thin profiles. As we have already observed, from the definition of N it results that $N_1(\mu)$ vanishes on the leading edge. The solution of the equation (8.5.23) which vanishes for $m = m_1$ and $m = m_2$ has the form (C.1.14) with a condition having the form (C.1.13). Taking also into account (B.5.8) it results

$$N_1(m) = -\frac{1}{\pi}\sqrt{(m-m_1)(m_2-m)}\,{}'\!\!\int_{m_1}^{m_2} \frac{\sqrt{\mu}F_0'(\mu)}{\sqrt{(\mu-m_1)(m_2-\mu)}}\frac{\mathrm{d}\,\mu}{\mu-m} + \\ + c_2\frac{\sqrt{(m-m_1)(m_2-m)}}{m\sqrt{m_1 m_2}}. \tag{8.5.24}$$

The condition (C.1.13) will give

$$\pi\left(c_1 + \frac{c_2}{\sqrt{m_1 m_2}}\right) + \int_{m_1}^{m_2} \frac{\sqrt{\mu}F_0'(\mu)}{\sqrt{(\mu-m_1)(m_2-\mu)}}\mathrm{d}\,\mu = 0 \tag{8.5.25}$$

and will be useful for the determination of the constant c_1, after determining the constant c_2. In fact, the constant c_1 is of no interest.

The constant c_2 which intervenes effectively in the solution (8.5.24) will be determined imposing for the solution to verify the integral equation (8.5.18). This condition is necessary because the solution was determined after some derivations.

Writing the equation (8.5.18) as follows

$$\,^{*}\!\!\int_{m_1}^{m_2} \frac{N_1(\mu)}{(\mu-m^2)}\mathrm{d}\,\mu + \frac{1}{2m}\,{}'\!\!\int_{m_1}^{m_2} \frac{N_1(\mu)}{\mu-m}\mathrm{d}\,\mu - \\ -\frac{1}{4m}\int_{m_1}^{m_2} N_1(\mu)K(m,\mu)\mathrm{d}\,\mu = -\frac{\pi}{k}\frac{F(m)}{m^{3/2}}, \tag{8.5.26}$$

where $K(m,\mu)$ is the symmetric kernel

$$K(m,\mu) = \frac{1}{\sqrt{m\mu}}\ln\left|\frac{\sqrt{m}+\sqrt{\mu}}{\sqrt{m}-\sqrt{\mu}}\right|, \tag{8.5.27}$$

we notice that the equations (8.5.24) and (8.5.26) will determine the unknowns N_1 and c_2. The replacement of N_1 from (8.5.24) in equation (8.5.26) leads to difficult calculations. It is necessary, for example, to know the formulas for interchanging the FP (Finite Part) with PV (Principal Value) and PV with FP (the formula of Poincaré-Bertrand [A.27]).

The equation (8.5.26) may be also solved numerically using the Gauss-type quadrature formulas (because N has the form

$$N_1(\mu) = \sqrt{(\mu - m_1)(m_2 - \mu)}n(\mu)\,, \tag{8.5.28}$$

$n(\mu)$ representing the new unknown).

8.6 Flat Wings

8.6.1 The Angular Wing with Supersonic Leading Edges

For the flat wings having the angle of attack ε we have $F = -\varepsilon$. The theory of the angular airfoil with supersonic leading edges may be obtained from 8.5.2, putting $F = -2$. Since

$$m = \frac{b}{a}, \quad \mu = \frac{b'}{a'}, \quad c = \frac{m}{\mu}a\,, \tag{8.6.1}$$

it results

$$\int_0^a \frac{a'\mathrm{d}\,a'}{\sqrt{(a-a')(a'-c)}} = a\left(1+\frac{m}{\mu}\right)\arctan\sqrt{-\frac{\mu}{m}} + a\sqrt{-\frac{m}{\mu}}\,, \tag{8.6.2}$$

and then, after elementary calculations,

$$N_{21}(a,m) = -\frac{2a\varepsilon}{k\pi}\left(\frac{m-m_1}{\sqrt{-m_1}}\arctan\sqrt{-\frac{m_1}{m}} - \sqrt{m}\right). \tag{8.6.3}$$

Similarly one obtains

$$N_{23}(a,m) = -\frac{2a\varepsilon}{k\pi}\left(\frac{m-m_2}{\sqrt{-m_2}}\arctan\sqrt{-\frac{m}{m_2}} - \sqrt{m}\right). \tag{8.6.4}$$

We have also

$$\int_0^a \frac{a'\mathrm{d}\,a'}{\sqrt{(a-a')(c-a')}} = -a\sqrt{\frac{m}{\mu}} + \frac{a}{2}\left(1+\frac{m}{\mu}\right)\ln\frac{\sqrt{m}+\sqrt{\mu}}{|\sqrt{\mu}-\sqrt{m}|}\,,$$

and then, from (8.5.5) or directly from (8.5.6),

$$N_{22}(a,m) = \frac{a\varepsilon}{k\pi}I,\; I = \int_0^\infty \left[\sqrt{\frac{m}{\mu}} - \frac{1}{2}\left(1+\frac{m}{\mu}\right)\ln\frac{\sqrt{m}+\sqrt{\mu}}{|\sqrt{\mu}-\sqrt{m}|}\right]\frac{\mathrm{d}\,\mu}{\sqrt{\mu}}. \tag{8.6.5}$$

In I we make the change of variable $\sqrt{\mu} = x$ and we denote $\sqrt{m} = q$. In this way one obtains

$$I = I_1 + q^2 I_2 - 2q I_3 , \tag{8.6.6}$$

where

$$I_1 = \int_0^q \ln \frac{x+q}{|x-q|} \mathrm{d}\, x + \int_q^\infty \ln \frac{x+q}{|x-q|} \mathrm{d}\, x = I_{11} + I_{12} ,$$

$$I_2 = \int_0^q \frac{1}{x^2} \ln \frac{x+q}{|x-q|} \mathrm{d}\, x + \int_q^\infty \frac{1}{x^2} \ln \frac{x+q}{|x-q|} \mathrm{d}\, x = I_{21} + I_{22} , \tag{8.6.7}$$

$$I_3 = \int_0^q \frac{\mathrm{d}\, x}{x} + \int_q^\infty \frac{\mathrm{d}\, x}{x} = I_{31} + I_{32} .$$

The integral I_{11} is elementary (it has an integrable singularity). One obtains

$$I_{11} = 2q \ln 2 . \tag{8.6.8}$$

The integral I_{21} has a strong singularity in $x = 0$. It must be considered in the Finite Part sense. On the basis of the formula (D.2.2) we have

$$I_{21} = \int_0^q \frac{\ln(x+q) - \ln(q-x) - 2x/q}{x^2} + \frac{2}{q} \ln q .$$

Integrating by parts, one obtains

$$I_{21} = \frac{2}{q} (\ln 2q + 1) . \tag{8.6.9}$$

Using (D.2.3) we deduce

$$I_{31} = \ln q . \tag{8.6.10}$$

For calculating I_{12}, I_{22} and I_{32} we make the substitution $x = 1/y$ and we utilize the results (8.6.8) – (8.6.10). One obtains $I_3 = 0$,

$$I_1 = 2q \ln 2 + \frac{2}{q} \left(\ln \frac{2}{q} + 1 \right) ,$$

$$I_2 = 2q \left(\ln \frac{2}{q} + 1 \right) + \frac{2}{q} \ln 2 , \tag{8.6.11}$$

such that

$$N_{22}(a, m) = \frac{4a\varepsilon}{k\pi} \sqrt{m} (2 \ln 2 + 1) . \tag{8.6.12}$$

In this way, taking into account that $m = b/a$, the formula (8.5.4), together with (8.6.3), (8.6.4) and (8.6.12) give

$$N_2(a,b) = -\frac{2\varepsilon}{k\pi}\left(-\frac{b-m_1a}{\sqrt{-m_1}}\arctan\sqrt{-\frac{m_1a}{b}} - \sqrt{ab}\right) -$$

$$-\frac{2\varepsilon}{k\pi}\left(-\frac{b-m_2a}{\sqrt{-m_2}}\arctan\sqrt{-\frac{b}{am_2}} - \sqrt{ab}\right) - \frac{4\varepsilon}{k\pi}\sqrt{ab}(2\ln 2+1)\,. \tag{8.6.13}$$

This is the solution when M is in the zone limited by the characteristics OC and OD.

If $M(a,b)$ is in the domain limited by OD and OE, then we use (8.5.9). We deduce that

$$N_3(a,b) = -\frac{\varepsilon}{k\sqrt{-m_2}}(m_2a-b)\,, \tag{8.6.14}$$

and if M is in the zone OB, OC (fig. 8.5.1)

$$N_1(a,b) = -\frac{\varepsilon}{k}\,\frac{b-m_1a}{\sqrt{-m_1}}\,. \tag{8.6.15}$$

8.6.2 The Triangular Wing. The Calculation of the Aerodynamic Action

In order to obtain a finite action, it is necessary to consider a wing having a finite area. We assume that in the physical plane it has the triangular form from figure 8.6.1. In order to obtain a well determined wing we must give the coordinates of the points A and F. Let m_1 be the inclination of the line OA and a_1 the ordinate of the point A in the frame of reference $O\alpha\beta$ and m_2 the inclination of the line OF and a_2 the ordinate of the point F in the same frame of reference. Then the equations of the edges OA and OF will be $b = m_1a$ respectively $b = m_2a$, and the coordinates of the points A and F respectively $(a_1, b_1 = m_1a_1)$ and $(a_2, b_2 = m_2a_2)$. The equation of the line is

$$b = (m_1 - m_3)a_1 + m_3a\,, \tag{8.6.16}$$

where

$$m_3 \equiv \frac{m_2a_2 - m_1a_1}{a_2 - a_1}\,. \tag{8.6.17}$$

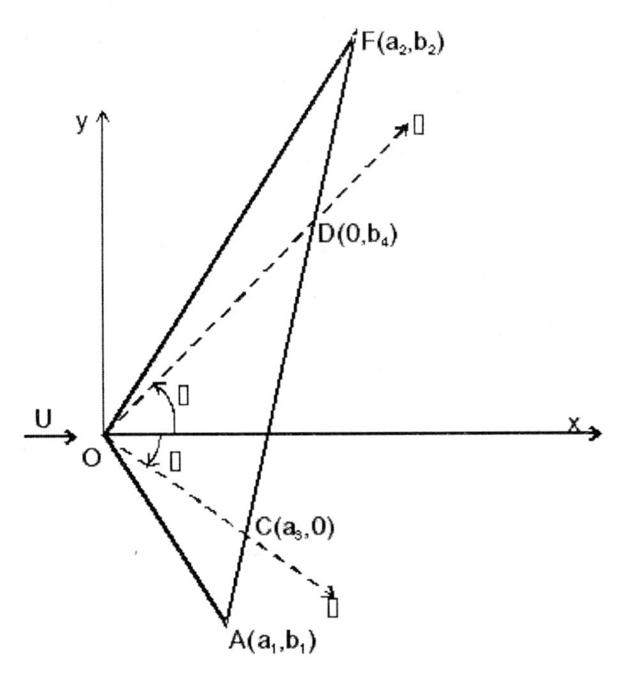

Fig. 8.6.1.

Denoting by (a_3, b_3) the coordinates of the point C and by (a_4, b_4) the coordinates of the point D, we deduce

$$a_3 = 1 - \frac{m_1}{m_3}, \quad b_3 = 0; \quad a_4 = 0, \quad b_4 = (m_1 - m_3)a_1 . \tag{8.6.18}$$

As we already know, the lift is given by the formula

$$L = -\iint_{D(x,y)} [\![p_1]\!] \, \mathrm{d}\, x_1 \, \mathrm{d}\, y_1 = -\rho_\infty U_\infty^2 L_0^2 \iint_{D(x,y)} [\![p]\!] \, \mathrm{d}\, x \, \mathrm{d}\, y ,$$

and the lift coefficient c_L by the formula

$$c_L = \frac{L}{\frac{1}{2}\rho_\infty U_\infty^2 A}, \tag{8.6.19}$$

where A is the area of the wing and L_0, the reference wing.

Taking (8.3.69) into account, passing to coordinates on characteristics and applying Green's formula, it results

$$
\begin{aligned}
c_L &= -\frac{2L_0^2}{A}\iint_{D(x,y)} f(x,y)\mathrm{d}\,x\,\mathrm{d}\,y = \\
&= -\frac{2L_0^2}{A}\iint_{D(x,y)} \frac{\partial}{\partial x}N(x,y)\mathrm{d}\,x\,\mathrm{d}\,y = \\
&= -\frac{L_0^2}{A}\iint_{D(a,b)} \left(\frac{\partial}{\partial a}+\frac{\partial}{\partial b}\right)N(a,b)\mathrm{d}\,a\,\mathrm{d}\,b = \\
&= -\frac{L_0^2}{A}\oint_{\partial D(a,b)} N\,\mathrm{d}(b-a) = -\frac{L_0^2}{A}I\,,
\end{aligned}
\tag{8.6.20}
$$

where

$$
\begin{aligned}
&I = I_1 + I_2 + I_3, \quad I_1 = \oint_{OA+AC+CO} N_1\,\mathrm{d}(b-a)\,, \\
&I_2 = \oint_{OC+CD+DO} N_2\,\mathrm{d}\,(b-a), \quad I_3 = \oint_{OD+DF+FO} N_3\,\mathrm{d}\,(b-a)\,.
\end{aligned}
\tag{8.6.21}
$$

Taking (8.6.15) into account, it results

$$
N_1\mathrm{d}(b-a)\Big|_{OA} = 0
$$

$$
\begin{aligned}
N_1\mathrm{d}(b-a)\Big|_{AC} &= N_1\mathrm{d}(b-a)\Big|_{b=(m_1-m_3)a_1+m_3a} \\
&= \frac{-\varepsilon(m_3-m_1)(m_3-1)}{k\sqrt{-m_1}}(a-a_1)\mathrm{d}a
\end{aligned}
$$

$$
N_1\mathrm{d}(b-a)\Big|_{CO} = N_1\mathrm{d}(b-a)\Big|_{b=0} = \frac{\varepsilon\sqrt{-m_1}}{k}a\mathrm{d}a\,,
$$

such that

$$
\begin{aligned}
I_1 &= -\frac{\varepsilon(m_3-m_1)(m_3-1)}{k\sqrt{-m_1}}\int_{a_1}^{a_3}(a-a_1)\mathrm{d}a - \frac{\varepsilon\sqrt{-m_1}}{k}\int_0^{a_3} a\mathrm{d}a = \\
&= -\frac{\varepsilon(m_3-m_1)(m_3-1)}{k\sqrt{-m_1}}\frac{(a_3-a_1)^2}{2} - \frac{\varepsilon\sqrt{-m_1}}{k}\frac{a_3^2}{2}\,.
\end{aligned}
\tag{8.6.22}
$$

In the same way one calculates I_3. Taking into account (8.6.14), one obtains

$$I_3 = \frac{\varepsilon}{k\sqrt{-m_2}} \frac{b_4^3}{2} - \frac{\varepsilon(m_3 - 1)(m_2 - m_3)}{k\sqrt{-m_2}} \frac{a_2^2}{2} + \\ + \frac{\varepsilon a_1 a_2 (m_3 - 1)(m_1 - m_3)}{k\sqrt{-m_2}} \tag{8.6.23}$$

and the problem of calculating c_L is solved.

8.6.3 The Trapezoidal Wing with Subsonic Lateral Edges

We assume that the projection of the wing (which is flat and has the angle of attack ε) on the plane xOy is the isosceles trapezoid $ABEF$ from figure 8.6.2, having the bases $2l$, $2L$ and the height h (dimensionless quantities). The direction of the unperturbed stream is perpendicular to the bases.

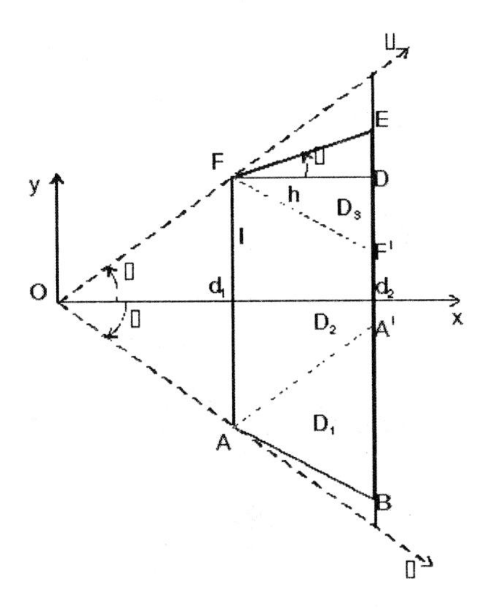

Fig. 8.6.2.

We consider the case $\alpha < \mu$. The leading edges AB and FE are subsonic and the edge AF is supersonic. We also assume that h is such that the subsonic edges are independent (for the sake of simplicity). Obviously, in this case the fluid motion is not conical. In the domain D_2 the solution may be obtained from (8.4.9), and in D_1 and D_3 from (8.4.15). For developing these solutions, we must characterize the domains in coordinates on characteristics. We must specify at first the physical coordinates. We introduce the parameter m by the formula

$$0 < \frac{m}{k} \equiv \tan \alpha = \frac{L - \ell}{h} < \tan \mu \equiv \frac{1}{k}. \tag{8.6.24}$$

It results therefore $0 < m < 1$. The distances d_1 and d_2 are defined bu the formulas

$$d_1 = \ell k, d_2 = d_1 + h. \tag{8.6.25}$$

The equations of the straight lines AB and FE are respectively

$$-y = \ell + \frac{m}{k}(x - d_1), y = \ell + \frac{m}{k}(x - d_1). \tag{8.6.26}$$

The equation of the wing (see figure 8.6.3) is

$$z = -\varepsilon \left(x - \frac{d_1 + d_2}{2} \right), \tag{8.6.27}$$

the function being defined on the domain $D + D_1 + D_2 + D_3$ from the xOy plane:

$$\begin{gathered} d_1 < x < d_2 \\ -y_1(x) < y < y_1(x), \end{gathered} \tag{8.6.28}$$

where

$$y_1(x) = \ell_1 \frac{m}{k}(x - d_1). \tag{8.6.29}$$

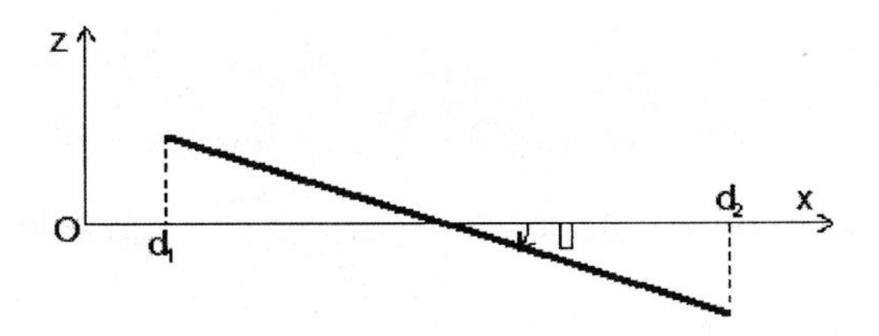

Fig. 8.6.3.

It obviously results

$$H(x, y) = -\varepsilon. \tag{8.6.30}$$

Passing to coordinates on characteristics we put

$$x = \frac{a + b}{2}, y = \frac{b - a}{2k}. \tag{8.6.31}$$

It results

$$H(a, b) = -\varepsilon \tag{8.6.32}$$

and the following equations for the sides of the domain D

$$AF : a+b=2d_1, \qquad BE : a+b=2d_2$$
$$AB : a=2d_1+m_0 b, FE : b=2d_1+m_0 a, \tag{8.6.33}$$

where we denoted

$$m_0=\frac{1+m}{1-m}. \tag{8.6.34}$$

For the vertices we deduce the following coordinates:

$$A=(2d_1,0), B=(d_2+kL, d_2-kL),$$
$$F=(0,2d_1), E=(d_2-kL, d_2+kL). \tag{8.6.35}$$

Taking (8.4.9) into account, we deduce for the solution in D_2

$$N_2(a,b)=-\frac{\varepsilon}{k\pi}\int_{2d_1-b}^{a}\frac{\mathrm{d}\,a'}{\sqrt{a-a'}}\int_{2d_1-a'}^{b}\frac{\mathrm{d}\,b'}{\sqrt{b-b'}}$$
$$=-\frac{\varepsilon}{\pi}(a+b-2d_1). \tag{8.6.36}$$

The solution in D_1 is given by (8.4.15). Noticing that

$$A(b)=2d_1-b, \quad B(a)=2d_1-a, \tag{8.6.37}$$

it results

$$N_1(a,b)=-\frac{\varepsilon}{k\pi}\int_{2d_1-a}^{b}\frac{\mathrm{d}\,b'}{\sqrt{b-b'}}\int_{2d_1-b'}^{a}\frac{\mathrm{d}\,a'}{\sqrt{a-a'}}. \tag{8.6.38}$$

Obviously, this expression may be obtained from (8.6.36), changing a with b. Hence

$$N_1(a,b)=-\frac{\varepsilon}{k}(a+b-2d_1) \tag{8.6.39}$$

and an identical expression for $N_3(a,b)$ (because of the symmetry).

According to (8.6.20), for c_L we have the expression

$$c_L=-\frac{\varepsilon L_0^2}{kA}I, \tag{8.6.40}$$

where

$$I=\oint_{\partial D}N(a+b-2d_1)\mathrm{d}\,(b-a)=$$
$$=\int_{AB+BE+EF}(a+b-2d_1)\mathrm{d}\,(b-a)=(1-m_0^2)(d_2-kL)^2.$$

8.6.4 The Trapezoidal Wing with Lateral Supersonic Edges

We consider an isosceles trapezium with the bases having the length $2L$ respectively 2ℓ perpendicular on the direction of the stream at infinity (fig. 8.6.4). We introduce here again the parameter m defined by the relation:

$$\frac{m}{k} \equiv \tan\alpha = \frac{L-\ell}{h} > \tan\mu = \frac{1}{k} \tag{8.6.41}$$

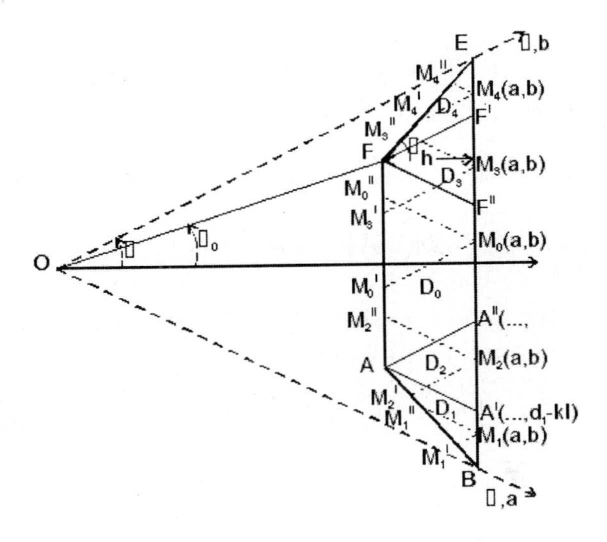

Fig. 8.6.4.

Obviously, $m > 1$. Denoting by d_1 the distance from the small basis to O and by d_2 the distance corresponding to the great basis, we obviously have $d_2 = d_1 + h$, and for the cartesian coordinates of the vertices of the trapezium

$$A = (d_1, \ell), B = (d_2, -L), F = (d_1, \ell), E = (d_2, L) \tag{8.6.42}$$

The coordinates on characteristics are obtained from the formulas

$$a = x - ky, \quad b = x + ky\,. \tag{8.6.43}$$

Noticing that $d_2 = kL$, we deduce

$$A = (d_1 + k\ell,\, d_1 - k\ell), \quad B = (2kL, 0)$$

$$F = (d_1 - k\ell, d_1 + k\ell), \quad E = (0, 2kL) \tag{8.6.44}$$

Using the notation

$$k = \frac{d_1 - k\ell}{d_1 + 2h - k\ell}\left(k - 1 = -\frac{2}{m+1}\right), \tag{8.6.45}$$

and observing that from figure 8.6.4 we have the compatibility condition $\alpha_0 < \mu$, which implies

$$\frac{\ell}{d_1} = \tan\alpha_0 < \tan\mu = \frac{1}{k} \Longrightarrow d_1 > k\ell\,, \tag{8.6.46}$$

we deduce $0 < k < 1$. Now, the equations of the sides of the trapezium may be written as follows

$$\begin{aligned} &AB: a + kb = 2d_2\,, FE: k^{-1}a + b = 2d_2\,, \\ &BE: a + b = 2d_2\,, AF: a + b = 2d_1\,. \end{aligned} \tag{8.6.47}$$

The entire leading edge is supersonic. The solution may be expressed by means of the formula (8.4.9). To this aim it is necessary to specify the functions $\alpha = A(\beta)$ and $\beta = B(\alpha)$. We have:

$$\begin{aligned} &-\text{ on the edge } BA\,, \\ &\qquad b = B_1(a) = \frac{2d_2 - a}{k}, \quad a = B_1^{-1}(b) = 2d_2 - kb\,, \\ &-\text{ on the edge } AF\,, \\ &\qquad b = B_2(a) = 2d_1 - a, \quad a = \begin{Bmatrix} B_2^{-1}(b) \\ A_2(b) \end{Bmatrix} = 2d_1 - b\,. \end{aligned} \tag{8.6.48}$$

We must also observe that for determining the lift coefficient we do not need $N(a, b)$ on the entire wing, but only on the trailing edge BE. Indeed, this may be expressed with the formula (8.6.20), and N on the leading edge $BAFE$ vanishes as we have already mentioned in formula (8.3.68). The domains of influence are (fig. 8.6.3) :

$$D_1 = ABA'A, \quad D_2 = AA'A''A, \quad D_0 = AA''F''A\,,$$

$$D_3 = FF''F'F, \quad D_4 = FF'EF$$

Hence, we shall put

$$I = I_1 + I_2 + I_3 + I_4\,, \tag{8.6.49}$$

where

$$I_1 = \int_{BA'} N_1(a,b)\mathrm{d}\,(b-a), \quad I_2 = \int_{A'A''} N_2(a,b)\mathrm{d}\,(b-a)\,,$$

$$I_0 = \int_{A''F''} N_0(a,b)\mathrm{d}\,(b-a), \; I_3 = \int_{F''F'} N_3(a,b)\mathrm{d}\,(b-a), \qquad (8.6.50)$$

$$I_4 = \int_{F''E} N_4(a,b)\mathrm{d}\,(b-a)\,.$$

Using the formula (8.4.9) and the equations (8.6.47), we deduce

$$N_1(a,b)\Big|_{BA'} = -\frac{\varepsilon}{k\pi}\int_{B_1^{-1}(b)}^{a}\frac{\mathrm{d}\,a'}{\sqrt{a-a'}}\int_{B_1(a')}^{b}\frac{\mathrm{d}\,b'}{\sqrt{b-b'}}\Big|_{BA'} = \frac{\varepsilon b(1-k)}{k\sqrt{k}}\,,$$

$$N_2(a,b)\Big|_{A'A''} = -\frac{\varepsilon}{k\pi}\int_{B_1^{-1}(b)}^{a}\frac{\mathrm{d}\,a'}{\sqrt{a-a'}}\int_{B_2(a')}^{b}\frac{\mathrm{d}\,b'}{\sqrt{b-b'}}\Big|_{A'A''} = \ldots\,,$$

$$N_0 = -\frac{2\varepsilon}{k}h\,, \quad N_4 = -\frac{\varepsilon(1-k)a}{k\sqrt{k}}\,.$$

After elementary calculations we deduce:

$$I_1 = \frac{\varepsilon(1-k)}{k^2\sqrt{k}}\int_0^{d_2-kl-h} b\mathrm{d}\,b = \frac{\varepsilon(1-k)}{k^2\sqrt{k}}\frac{(d_1-k\ell)}{2}\,,$$

$$I_4 = -\frac{\varepsilon(1-k)}{k^2\sqrt{k}}\frac{(d_1-k\ell)^2}{2}\,, \quad I_0 = -\frac{4\varepsilon h}{k^2}(k\ell-h)\,,$$

and finally,

$$c_L = \left(2\ell - \frac{h}{k}\right)\frac{h}{2}\,.$$

Chapter 9

The Steady Transonic Flow

9.1 The Equations of the Transonic Flow

9.1.1 The Presence of the Transonic Flow

We call *transonic flow* the flow which is subsonic in a domain of the space and supersonic in the adjacent domain. One demonstrates (for the potential flow – see [1.21] pp 517, 518) that the equality $v = c$ comes true in E_2 only on curves separating the domains where the flow is subsonic from the domains where the flow is supersonic, and in E_3 on the surfaces which separate such domains. The name of transonic flow was introduced by Th. von Kármán in 1947.

In the present paper the transonic flow has been encountered in several situations.

At first, we have to mention the one-dimensional flow [1.11] §4.5. The formula (4.5.8) which gives the variation of the velocity against the variation of the cross section indicates that, in the subsonic flow ($M < 1$), the velocity increases when the area decreases and decreases when area increases (like ca in the incompressible fluid), while in the supersonic flow ($M > 1$) the variations are produced in the same sense. This circumstance leads to the conclusion that in a tube having the shape from figure 9.1.1 the flow may become transonic. To this aim it is sufficient for the upstream subsonic velocity to have the critical value in the section of minimum area. Further since the area of the section increases, the velocity also increases, remaining supersonic.

In the linearized theory we deduced for the aerodynamic action the formulas (3.1.33) and (3.1.34) in the subsonic case and (8.1.9) in the supersonic case. It is obvious that these formulas are not valid in the vicinity of $M = 1$. For the flat plate these formulas become (3.1.35) and (8.1.22). The figures (3.1.3) and (8.1.3) are very suggestive.

In the case of the subsonic flow with great velocity past thick bodies like in figure 9.1.2, the flow may become transonic. Indeed, considering

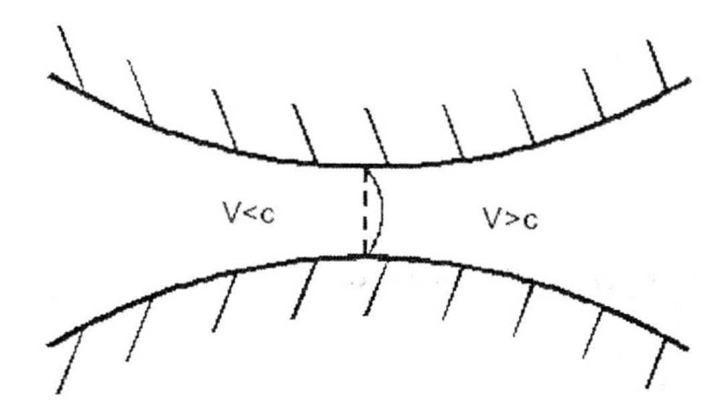

Fig. 9.1.1.

the flow between the streamline which includes the boundary and a neighbor streamline L, we shall find that the flow is like in a tube. Since the domain between these lines narrows because of the body, it

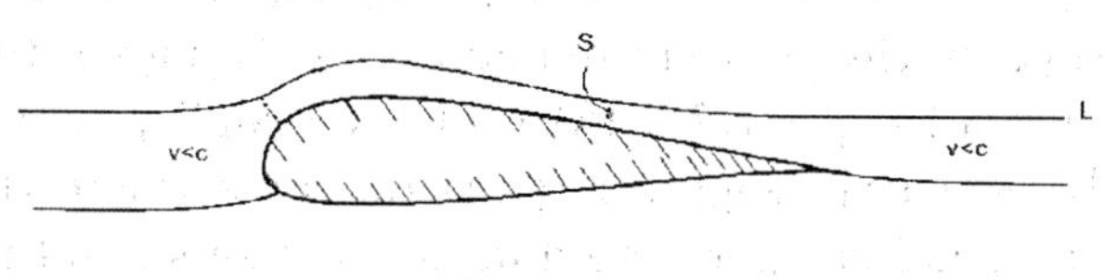

Fig. 9.1.2.

follows that in the vicinity of the body the flow may become supersonic. The transition from the supersonic flow to the subsonic flow is performed by a shock wave S according to the scheme described in 1.3.6. Until S the flow is transonic. We shall deduce in the sequel the equations which describe this flow. The flow with great subsonic velocity past thick bodies is described by the scheme from 9.1.3.

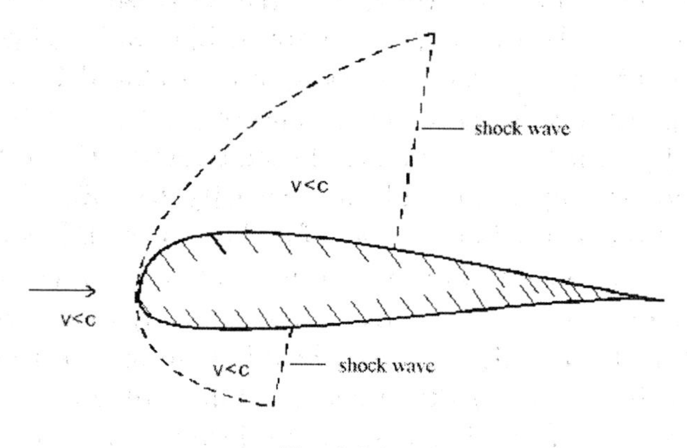

Fig. 9.1.3.

Finally, in the supersonic flow, for great velocities, practically in the hypersonic regime, it appears, as we noticed in 1.3.6, detachedor attached shock waves (figure 1.3.5). Behind these waves the flow is transonic (it passes from the subsonic regime $(M_2 < 1)$ to the supersonic one $(M_2 > 1)$).

As we could see, in modern aerodynamics the transonic regime is frequent. So one explains the great number of papers devoted to this subject in the last years. We mention especially the papers of Bauer, Garabedian and Korn [9.1] devoted to the theory of minimum drag wings. There are three dominant methods for studying the transonic flow, namely:

1° *the hodograph method*, suitable only for the plane steady jet flow (see for example Ferrari and Tricomi [9.11], Manwell [9.30], [9.31] etc.);

2° *direct analytical methods*, based on the semi-linearized equation of the potential. They lead to integral equations which may be solved numerically;

3° *numerical methods* applied directly to the system of equations which describes the fluid flow (we mention especially the finite elements method).

In this chapter we present some direct analytic methods.

9.1.2 The Equation of the Potential

The reasoning based on the assumption that the independent variables x, y, z have the same role in the structure of φ, (utilized for deducing the equation (2.1.39)), is not valid for the flow in the vicinity of $M = 1$. Indeed, in this vicinity $M^2 - 1$ becomes itself a small parameter. If, for example $M^2 - 1 = O(\varepsilon)$, then for $\varphi_{xx} = O(\varphi)$ it results φ_{yy} and $\varphi_{zz} = O(\varepsilon^2)$. One imposes an analysis of the order of magnitude of the perturbations depending on the geometry of the body and the conditions which determine the flow (Mach's number M, the thickness and length parameters, the angle of attack, etc.). In fact, the idea that the variables y and z do not behave like the variable x results from the special property of the Ox axis (which is parallel to the direction on the unperturbed stream). We shall introduce therefore the variables

$$\overline{y} = \nu(\varepsilon)y, \quad \overline{z} = \nu(\varepsilon)z\,, \tag{9.1.1}$$

expecting for $\nu(\varepsilon)$, like for $\eta(\varepsilon)$ from the expansion

$$\phi(x, y, z, \varepsilon) = U\left[x + \eta(\varepsilon)\varphi(x, \overline{y}, \overline{z}) + \dots\right] \tag{9.1.2}$$

to be determined by comparing the orders of magnitude.

Coming back to the linearized theory 2.1, we notice that for $M =$ $= 1$ a catastrophe is produced (it disappears terms from the equations). The lift and moment coefficients become therefore infinite. But this catastrophe has only a mathematical nature, not a physical one. It is determined by the fact that in the vicinity of the value $M = 1$, the order of magnitude of all the first order derivatives is not the same (ε). One imposes therefore (9.1.1).

It results

$$\begin{aligned} &\phi_x = U(1+\eta\varphi_x+\ldots), \phi_y = U\eta\nu\varphi_{\overline{y}}+\ldots, \phi_z = U\eta\nu\varphi_{\overline{z}} \\ &\phi_{xx} = U\eta\varphi_{xx}, \qquad \phi_{xy} = U\eta\nu\varphi_{x\overline{y}}, \qquad \phi_{yy} = U\eta\nu^2\varphi_{\overline{y}\overline{y}}\,. \end{aligned} \tag{9.1.3}$$

From (1.2.17) we deduce

$$c^2 = c_\infty^2 - (\gamma-1)U^2\eta\varphi_x + O(\eta^2)\,, \tag{9.1.4}$$

and from (1.2.16) written explicitly as follows

$$(c^2-\phi_x^2)\phi_{xx} + (c^2-\phi_y^2)\phi_{yy} - 2\phi_x\phi_y\phi_{xy} + \ldots = 0\,, \tag{9.1.5}$$

we deduce

$$\begin{aligned} &[1-M^2-(\gamma+1)(M^2-1)\eta\varphi_x - (\gamma+1)\eta\varphi_x + O(\eta^2)]\varphi_{xx} + \\ &\quad + [1-(\gamma-1)(M^2-1)\eta\varphi_x - (\gamma-1)\eta\varphi_x]\nu^2\varphi_{\overline{y}\overline{y}} - \\ &\quad - 2M^2\nu\eta^2\varphi_{\overline{y}}\varphi_{x\overline{y}} + O(\eta^2\nu^2) + \ldots = 0\,. \end{aligned} \tag{9.1.6}$$

For a fixed M, we see that the equation is consistent if $\nu \to 0$ when $\eta \to 0$, so

$$\nu^2 \sim \eta, 1 - M^2 \sim \eta\,. \tag{9.1.7}$$

We introduce now the boundary condition. If

$$z = \varepsilon h(x,y) \tag{9.1.8}$$

is the equation of the perturbing surface, imposing the condition to be a material surface i.e.

$$\varepsilon h'_x\phi_x + \varepsilon h'_y\phi_y = \phi_z \tag{9.1.9}$$

which implies, taking into account (9.1.2)

$$\varepsilon h'_x = \eta\nu\varphi_{\overline{z}} + \ldots$$

whence

$$\varepsilon = \eta\nu\,. \tag{9.1.10}$$

Taking (9.1.7) into account, we deduce

$$\nu = \varepsilon^{1/3}\,, \qquad \eta = \varepsilon^{2/3}\,. \tag{9.1.11}$$

When $M \to 1$ we have to compare in (9.1.6) the terms of order immediately superior to those which gave (9.1.7). It results $1 - M^2 = = K\nu^2$ whence

$$K = \frac{1 - M^2}{\varepsilon^{2/3}}\,. \tag{9.1.12}$$

K is called *the parameter of the transonic similitude.* In this way, the first approximation from (9.1.6) (*the dominant equation*) is

$$[K - (\gamma + 1)\varphi_x]\varphi_{xx} + \varphi_{\overline{y}\,\overline{y}} + \varphi_{\overline{z}\,\overline{z}} = 0\,. \tag{9.1.13}$$

This is the equation of the *transonic flow* (the equation of the potential). It is elliptic if $\varphi_1 < K/(\gamma + 1)$ and hyperbolic if $\varphi_x > K/(\gamma + 1)$.

The relation $\varphi_x = K/(\gamma + 1)$ is verified on the surface where $V^2 = c^2$. Indeed, using the notations (2.1.3), and taking (1.3.32) into account, the condition $V^2 = c^2$ becomes

$$V_1^2 = c_1^2 = c_0^2 - \frac{\gamma - 1}{2}V_1^2 = c_\infty^2 - \frac{\gamma - 1}{2}(V_1^2 - U^2)\,. \tag{9.1.14}$$

Here, the dominant relation is

$$U^2\left(1 + 2\frac{\varepsilon}{\nu}\widehat{u}\,,\right) = c_\infty^2 - (\gamma - 1)U^2\frac{\varepsilon}{\nu}\widehat{u}\,, \tag{9.1.15}$$

whence $\widehat{u} = K/(\gamma + 1)$.

Now it is clear that the non - linearity is necessary for making this transition possible.

The first study of the transonic flow has been performed by von Kármán [9.26]. By various methods the problem was investigated by Ovsiannikov [9.53], Guderley [9.15], Cole & Messiter [9.6] etc. Cole's study from 1975 relying on the method of perturbations was continued by the same author in 1978. In the last study one proves that if we denote by ε the thickness parameter and we set for the cross sections

$$\overline{y} = \varepsilon^{1/3}y\,, \quad \overline{z} = \varepsilon^{1/3}z\,,$$

then the potential ϕ has the following structure [9.54]:

$$\begin{aligned}\phi(x, y, z; M\,, \alpha, \delta, b) = U\big[x + \varepsilon^{2/3}\varphi(x, \overline{y}, \overline{z}; K, A, B) +\\ +\varepsilon^{4/3}\varphi_2(x, \overline{y}, \overline{z}; K, A, B) + \ldots\big]\,,\end{aligned} \tag{9.1.16}$$

where K is the transonic parameter (9.1.12), A, the parameter of the angle of attack $= \alpha/\varepsilon$, and β, the span parameter $= b\varepsilon^{1/3}$.

For φ one obtains the equation (9.1.13).

9.1.3 The System of Transonic Flow

It is rigorous to perform the asymptotic analysis on the system of equations and not on the equation of the potential which has been obtained from the system by derivation with respect to the x, y, z coordinates. We present here such an analysis which was performed together with professor A. Halanay in the years '80. We utilize the coordinates $\overline{y}$ and $\overline{z}$ in the form (9.1.1) and we denote

$$\overline{v}(x,\overline{y},\overline{z},\varepsilon) = v\left(x, \frac{\overline{y}}{\nu(\varepsilon)}, \frac{\overline{z}}{\nu(\varepsilon)}\right), \ \overline{h}(x,\overline{y},\varepsilon) = h\left(x, \frac{\overline{y}}{\nu(\varepsilon)}\right). \quad (9.1.17)$$

It results

$$\overline{h}'_x(x,\overline{y},\varepsilon) = h'_x\left(x, \frac{\overline{y}}{\nu(\varepsilon)}\right), \ \overline{h}'_{\overline{y}}(x,\overline{y},\varepsilon) = h'_y\left(x, \frac{\overline{y}}{\nu(\varepsilon)}\right)\frac{1}{v(\varepsilon)},$$

and the boundary condition

$$\varepsilon h'_x(x,y)[1+u(x,y,\varepsilon h(x,y))] + \varepsilon h'_y(x,y)v(x,y,\varepsilon h(x,y)) = \\ = w(x,y,\varepsilon h(x,y))$$

becomes

$$\varepsilon\overline{h}'_x(x,\overline{y},\varepsilon)\left[1+\overline{u}(x,\overline{y},\frac{\varepsilon}{\nu(\varepsilon)}\overline{h}(x,\overline{y},\varepsilon),\varepsilon)\right] + \\ +\varepsilon\nu(\varepsilon)\overline{h}'_{\overline{y}}(x,\overline{y},\varepsilon)v\left(x,\overline{y},\frac{\varepsilon}{\nu(\varepsilon)}\overline{h}(x,\overline{y},\varepsilon)\right) = \\ = \overline{w}\left(x,\overline{y},\frac{\varepsilon}{\nu(\varepsilon)}\overline{h}(x,\overline{y},\varepsilon),\varepsilon\right). \quad (9.1.18)$$

The dominant term in the first member would be $\varepsilon\overline{h}'_x(x,\overline{y},\varepsilon)$ if $\overline{y}$ would not disturb. But for a small $\overline{y}$ we have

$$\overline{h}'_x(x,\overline{y},\varepsilon) = h'_x\left(x, \frac{\overline{y}}{\nu(\varepsilon)}\right) = h'_x(x,0) + \frac{\overline{y}}{\nu(\varepsilon)}h''_{xy}(x,0) + \ldots.$$

From the physical conditions of the problem it results that $h'_x(x,0) \neq 0$.

The condition (9.1.18) suggests that the right hand and member has the order of ε. Hence,

$$\overline{w}\left(x,\overline{y},\frac{\varepsilon}{\nu(\varepsilon)}\overline{h}(x,\overline{y},\varepsilon)\right)=\varepsilon\widehat{w}\left(x,\overline{y},\frac{\varepsilon}{\nu(\varepsilon)}\overline{h}(x,\overline{y},\varepsilon),\varepsilon\right). \tag{9.1.19}$$

We assume that this is valid in the entire domain occupied by the fluid, i.e.:

$$\overline{w}(x,\overline{y},\overline{z},\varepsilon)=\varepsilon\widehat{w}(x,\overline{y},\overline{z},\varepsilon). \tag{9.1.20}$$

Taking (9.1.19) into account, from (9.1.18) we retain in the first approximation, under the hypothesis that $\dfrac{\varepsilon}{\nu(\varepsilon)}\to 0$

$$\overline{h}'_x(x,\overline{y},\varepsilon)\left[1+\overline{u}(x,\overline{y},0,\varepsilon)\right]=\widehat{w}(x,\overline{y},0,\varepsilon). \tag{9.1.21}$$

Using the notations (2.1.3) the system which determines the perturbation produced by a fixed body in the uniform flow of a compressible fluid characterized by M is determined (see (2.1.10) - (2.1.13)) by the system

$$(1+\rho)M^2\dot{p}=(1+\gamma M^2p)\dot{\rho} \tag{9.1.22}$$

$$M^2\dot{p}+(1+\gamma M^2)\operatorname{div}\boldsymbol{v}=0 \tag{9.1.23}$$

$$(1+\rho)\dot{\boldsymbol{v}}+\operatorname{grad}p=0 \tag{9.1.24}$$

where

$$\dot{p}=\left[(1+u)\frac{\partial}{\partial x}+v\frac{\partial}{\partial y}+w\frac{\partial}{\partial y}\right]p,\ldots. \tag{9.1.25}$$

We notice now that from the structure

$$\overline{p}(x,\overline{y},\overline{z},\varepsilon)=p\left(x,\frac{\overline{y}}{\nu(\varepsilon)},\frac{\overline{z}}{\nu(\varepsilon)}\right). \tag{9.1.26}$$

it results the formulas

$$\frac{\partial\overline{p}}{\partial x}=\frac{\partial p}{\partial x},\ \frac{\partial\overline{p}}{\partial\overline{y}}=\frac{1}{\nu(\varepsilon)}\frac{\partial p}{\partial y},\ \frac{\partial\overline{p}}{\partial\overline{z}}=\frac{1}{\nu(\varepsilon)}\frac{\partial p}{\partial z}, \tag{9.1.27}$$

which will be replaced in the projections of the equation (9.1.24) on the axes of coordinates. In this way, the projection on Oz gives

$$(1+\overline{\rho})\left[(1+\overline{u})\varepsilon\frac{\partial\widehat{w}}{\partial x}+\varepsilon\nu\overline{v}\frac{\partial\widehat{w}}{\partial\overline{y}}+\varepsilon^2\nu\widehat{w}\frac{\partial\widehat{w}}{\partial\overline{z}}\right]+\nu\frac{\partial\overline{p}}{\partial\overline{z}}=0. \tag{9.1.28}$$

Comparing here the dominant terms we deduce that

$$\overline{p}(x,\overline{y},\overline{z},\varepsilon)=\frac{\varepsilon}{\nu(\varepsilon)}\widehat{p}(x,\overline{y},\overline{z},\varepsilon), \tag{9.1.29}$$

and from (9.1.28) one retains

$$(1+\overline{\rho})(1+\overline{u})\frac{\partial\widehat{w}}{\partial x}+\frac{\partial\widehat{p}}{\partial\overline{z}}=0\,. \tag{9.1.30}$$

Analogously, from the projection of the equation (9.1.24) on the Oy axis, it results

$$(1+\overline{\rho})\left[(1+\overline{u})\frac{\partial\overline{v}}{\partial x}+\nu(\varepsilon)\overline{v}\frac{\partial\overline{v}}{\partial\overline{y}}+\varepsilon\nu(\varepsilon)\widehat{w}\frac{\partial\overline{v}}{\partial\overline{z}}\right]+\varepsilon\frac{\partial\widehat{p}}{\partial y}=0\,, \tag{9.1.31}$$

From this equation it follows

$$\overline{v}(x,\overline{y},\overline{z},\varepsilon)=\varepsilon\widehat{v}(x,\overline{y},\overline{z},\varepsilon) \tag{9.1.32}$$

and then

$$(1+\overline{\rho})(1+\overline{u})\frac{\partial\widehat{v}}{\partial x}+\frac{\partial\widehat{p}}{\partial\overline{y}}=0\,. \tag{9.1.33}$$

At last, the projection of the equation (9.1.24) on the Ox axis gives

$$(1+\overline{\rho})\left[(1+\overline{u})\frac{\partial\overline{u}}{\partial x}+\varepsilon\nu\widehat{v}\frac{\partial\overline{u}}{\partial\overline{y}}+\varepsilon\nu\widehat{w}\frac{\partial\overline{u}}{\partial\overline{z}}\right]+\frac{\varepsilon}{\nu(\varepsilon)}\frac{\partial\widehat{p}}{\partial x}=0\,,$$

whence we obtain

$$\overline{u}(x,\overline{y},\overline{z},\varepsilon)=\frac{\varepsilon}{\nu(\varepsilon)}\widehat{u}(x,\overline{y},\overline{z},\varepsilon)\,, \tag{9.1.34}$$

and then

$$(1+\overline{\rho})\frac{\partial\widehat{u}}{\partial x}+\frac{\partial\widehat{p}}{\partial x}=0\,. \tag{9.1.35}$$

The behaviour (9.1.34) determines for (9.1.33) and (9.1.30) the forms:

$$(1+\overline{\rho})\frac{\partial\widehat{v}}{\partial x}+\frac{\partial\widehat{p}}{\partial\overline{y}}=0\,,\quad (1+\overline{\rho})\frac{\partial\widehat{w}}{\partial x}+\frac{\partial\widehat{p}}{\partial\overline{z}}=0\,, \tag{9.1.36}$$

and the boundary condition (9.1.21) determines the equality

$$\overline{h}'_x(x,\overline{y},\varepsilon)=\widehat{w}(x,\overline{y},0,\varepsilon)$$

which implies

$$h'_x(x,y)=w(x,y,0)\,. \tag{9.1.37}$$

Knowing that $M^2=1$ constitutes a singularity, we shall consider in (9.1.22) and (9.1.23) $M^2=1+\mu$ and we shall keep the dominant terms for a small p. Utilizing the previous results, the equation (9.1.22) becomes

$$(1+\mu)(1+\overline{\rho})\left[\left(1+\frac{\varepsilon}{\nu}\widehat{u}\right)\frac{\partial}{\partial x}+\varepsilon\nu\widehat{v}\frac{\partial}{\partial \overline{y}}+\varepsilon\nu\widehat{w}\frac{\partial}{\partial \overline{z}}\right]\frac{\varepsilon}{\nu}\widehat{p}=$$

$$=\left[1+\gamma(1+\mu)\frac{\varepsilon}{\nu}\widehat{p}\right]\left[\left(1+\frac{\varepsilon}{\nu}\widehat{u}\right)\frac{\partial}{\partial x}+\varepsilon\nu\widehat{v}\frac{\partial}{\partial \overline{y}}+\varepsilon\nu\widehat{w}\frac{\partial}{\partial \overline{z}}\right]\overline{\rho},$$

whence we deduce

$$\overline{\rho}(x,\overline{y},\overline{z},\varepsilon)=\frac{\varepsilon}{\nu(\varepsilon)}\widehat{\rho}(x,\overline{y},\overline{z},\varepsilon), \tag{9.1.38}$$

and then

$$\frac{\partial\widehat{\rho}}{\partial x}=\frac{\partial\widehat{p}}{\partial x}. \tag{9.1.39}$$

Having in view the damping condition at infinity for the perturbation, from the last equation we deduce

$$\widehat{\rho}=\widehat{p}. \tag{9.1.40}$$

Taking the relation (9.1.38) into account, it results that the dominant parts in the equations (9.1.35) and (9.1.36) are

$$\frac{\partial\widehat{u}}{\partial x}+\frac{\partial\widehat{p}}{\partial x}=0,\quad \frac{\partial\widehat{v}}{\partial x}+\frac{\partial\widehat{p}}{\partial \overline{y}}=0,\quad \frac{\partial\widehat{w}}{\partial x}+\frac{\partial\widehat{p}}{\partial \overline{z}}=0, \tag{9.1.41}$$

whence it results

$$\widehat{u}=-\widehat{p},\quad \frac{\partial\widehat{v}}{\partial x}-\frac{\partial\widehat{u}}{\partial \overline{y}}=0,\quad \frac{\partial\widehat{w}}{\partial x}-\frac{\partial\widehat{u}}{\partial \overline{z}}=0. \tag{9.1.42}$$

Finally, from the equation (9.1.23) written as follows

$$(1+\mu)\left[\left(1+\frac{\varepsilon}{\nu}\widehat{u}\right)\frac{\partial\widehat{p}}{\partial x}+\varepsilon\nu\widehat{v}\frac{\partial\widehat{p}}{\partial \overline{y}}+\varepsilon\nu\widehat{w}\frac{\partial\widehat{p}}{\partial \overline{z}}\right]+$$

$$+\left[1+\gamma(1+\mu)\frac{\varepsilon}{\nu}\widehat{p}\right]\left[\frac{\partial\widehat{u}}{\partial x}+\nu^2\frac{\partial\widehat{v}}{\partial \overline{y}}+\nu^2\frac{\partial\widehat{w}}{\partial \overline{z}}\right]=0$$

we obtain, if we have in view $(9.1.41)_1$,

$$\left(\mu+\frac{\varepsilon}{\nu}\widehat{u}+\mu\frac{\varepsilon}{\nu}\widehat{u}\right)\frac{\partial\widehat{p}}{\partial x}+(1+\mu)\varepsilon\nu\left(\widehat{v}\frac{\partial\widehat{p}}{\partial \overline{y}}+\widehat{w}\frac{\partial\widehat{p}}{\partial \overline{z}}\right)+$$

$$+\gamma(1+\mu)\frac{\varepsilon}{\nu}\widehat{p}\frac{\partial\widehat{u}}{\partial x}+\left[1+\gamma(1+\mu)\frac{\varepsilon}{\nu}\widehat{p}\right]\nu^2\left(\frac{\partial\widehat{v}}{\partial \overline{y}}+\frac{\partial\widehat{w}}{\partial \overline{z}}\right)=0.$$

The dominant part is obtained from the linear terms. We may write therefore

$$\left(\mu + \frac{\varepsilon}{\nu}\widehat{u}\right)\frac{\partial \widehat{p}}{\partial x} + \gamma\frac{\varepsilon}{\nu}\widehat{p}\frac{\partial \widehat{u}}{\partial x} + \nu^2\left(\frac{\partial \widehat{v}}{\partial \overline{y}} + \frac{\partial \widehat{w}}{\partial \overline{z}}\right) = 0\,,$$

whence

$$\mu = -k\frac{\varepsilon}{\nu}\,, \nu^2 = \frac{\varepsilon}{\nu} \tag{9.1.43}$$

and the residual equation

$$(-K + \widehat{u})\frac{\partial \widehat{p}}{\partial x} + \gamma\widehat{p}\frac{\partial \widehat{u}}{\partial x} + \frac{\partial \widehat{v}}{\partial \overline{y}} + \frac{\partial \widehat{w}}{\partial \overline{z}} = 0\,. \tag{9.1.44}$$

At last, from (9.1.43), (9.1.44) and (9.1.42) one obtains

$$\nu(\varepsilon) = \varepsilon^{1/3}, \quad K = \frac{1 - M^2}{\varepsilon^{2/3}} \tag{9.1.45}$$

$$[K - (1+\gamma)\widehat{u}]\,\frac{\partial \widehat{u}}{\partial x} + \frac{\partial \widehat{v}}{\partial \overline{y}} + \frac{\partial \widehat{w}}{\partial \overline{z}} = 0\,. \tag{9.1.46}$$

This equation, together with the equations (9.1.42) constitutes the general system of equations of the *steady transonic flow*. In the $x, \overline{y}, \overline{z}$ space the equations (9.1.42) give the irrotational conditionof the velocity of coordinates $(\widehat{u}, \widehat{v}, \widehat{w})$. Introducing the potential $\varphi(x, \overline{y}, \overline{z})$ by means of the formulas

$$\widehat{u} = \varphi_x, \quad \widehat{v} = \varphi_{\overline{y}}, \quad \widehat{w} = \varphi_{\overline{z}}\,, \tag{9.1.47}$$

one obtains (9.1.13) from (9.1.46).

9.1.4 The Shock Equations

In the case of the flow with shock waves, from the integral form of the equations of motion (9.1.42) and (9.1.46), written in the conservative form (by means of the div operator),

$$\begin{gathered}\widehat{v}_x + (-\widehat{u})_{\overline{y}} = 0, \quad \widehat{w}_x + (-\widehat{u})_{\overline{z}} = 0 \\ \left[K\widehat{u} - \frac{\gamma+1}{2}\widehat{u}^2\right]_x + \widehat{v}_{\overline{y}} + \widehat{w}_{\overline{z}} = 0\,,\end{gathered} \tag{9.1.48}$$

integrating on every domain which contains the shock surface and passing to the limit as usually, it results

$$[\![\widehat{v}]\!]n_x - [\![\widehat{u}]\!]n_{\overline{y}} = 0, \ [\![\widehat{w}]\!]n_x - [\![\widehat{u}]\!]n_{\overline{z}} = 0\,,$$

$$[\![K\widehat{u} - \frac{\gamma+1}{2}\widehat{u}^2]\!]n_x + [\![\widehat{v}]\!]n_{\overline{y}} + [\![\widehat{w}]\!]n_{\overline{z}} = 0\,, \tag{9.1.49}$$

where $n_x, n_{\overline{y}}, n_{\overline{z}}$ are the coordinates of the normal to the shock surface, i.e.

$$n_x = (\mathrm{d}\overline{y}\mathrm{d}\overline{z})_s, \quad n_{\overline{y}} = (\mathrm{d}\overline{z}\mathrm{d}x)_s, \quad n_{\overline{z}} = (\mathrm{d}x\mathrm{d}\overline{y})_s\,. \tag{9.1.50}$$

If, for example, the parametric equations of the shock surface are

$$x = x(\lambda_1, \lambda_2), \quad \overline{y} = \overline{y}(\lambda_1, \lambda_2), \quad \overline{z} = \overline{z}(\lambda_1, \lambda_2)\,,$$

then from $\boldsymbol{n} = \mathrm{d}_{\lambda_1}\boldsymbol{x} \times \mathrm{d}_{\lambda_2}\boldsymbol{x}$, it results

$$n_x = \left(\frac{\partial \overline{y}}{\partial \lambda_1}\frac{\partial \overline{z}}{\partial \lambda_2} - \frac{\partial \overline{z}}{\partial \lambda_1}\frac{\partial \overline{y}}{\partial \lambda_2}\right)\mathrm{d}\lambda_1\mathrm{d}\lambda_2, \ldots\,. \tag{9.1.51}$$

9.2 The Plane Flow

9.2.1 The Fundamental Solution

We consider, like in Chapter 3, that an uniform stream, having the Mach number M is perturbed by the presence of an infinite cylindrical body, with the generatrices perpendicular on the direction of the stream which coincides with the Ox axis. The Oy axis is in the section perpendicular to the generatrix. Let

$$y = h_{\pm}(x), \quad |x| < 1 \tag{9.2.1}$$

be the equations of the profile determined by the cross section. Our aim is to determine the perturbation and the action of the fluid against the profile.

The flow is obviously plane. The velocity will lie in the xOy plane and will not depend on the variable z. Taking into account the orientation of the axes, we deduce from (9.1.37) the boundary condition

$$v(x, \pm 0) = h'_{\pm}(x), \quad |x| < 1\,. \tag{9.2.2}$$

According to (9.1.42) and (9.1.46), the perturbation will be determined by the equations

$$\widehat{u} = -\widehat{p} \tag{9.2.3}$$

$$\widehat{u}_{\overline{y}} - \widehat{v}_x = 0, \quad K\widehat{u}_x + \widehat{v}_{\overline{y}} = (\gamma+1)\widehat{u}\widehat{u}_x . \tag{9.2.4}$$

Using the change of variables

$$y^* = \sqrt{K}\,\overline{y}, \quad u^* = \sqrt{K}\,\widehat{u}\,, \tag{9.2.5}$$

we reduce the system (9.2.4) to

$$\frac{\partial u^*}{\partial y^*} - \frac{\partial \widehat{v}}{\partial x} = 0, \quad \frac{\partial u^*}{\partial x} + \frac{\partial \widehat{v}}{\partial y^*} = k\frac{\partial}{\partial x}u^{*2}\,, \tag{9.2.6}$$

where we denote $\gamma + 1 = 2kK^{3/2}$. In the sequel we shall integrate the system (9.2.6) without writing the marks $*, \widehat{}$ any longer.

The fundamental solution of this system is determined by the equations

$$\frac{\partial u}{\partial y} - \frac{\partial v}{\partial x} = \ell\delta(x,y)\,, \quad \frac{\partial u}{\partial x} + \frac{\partial v}{\partial y} = k\frac{\partial}{\partial x}u^2 + m\delta(x,y)\,. \tag{9.2.7}$$

For the Fourier transforms $\overline{u}$ and $\overline{v}$ one obtains the formulas

$$\overline{u} = \frac{\mathrm{i}\,\alpha_2\ell + \mathrm{i}\,\alpha_1 m + k\alpha_1^2 F}{\alpha^2}\,, \quad \overline{v} = \frac{-\mathrm{i}\,\alpha_1\ell + \mathrm{i}\,\alpha_2 m + k\alpha_1\alpha_2 F}{\alpha^2}\,, \tag{9.2.8}$$

where $F = \mathcal{F}[u^2], \alpha^2 = \alpha_1^2 + \alpha_2^2$. We take into account the formulas

$$\begin{aligned} &\mathcal{F}^{-1}\left[\frac{1}{\alpha^2}\right] = -\frac{1}{2\pi}\ln r_0 \overset{\text{not}}{=} e\,, \quad r_0 = \sqrt{x^2+y^2} \\ &\mathcal{F}^{-1}\left[\frac{F}{\alpha^2}\right] = u^2 * e = -\frac{1}{4\pi}\iint_{\mathbb{R}^2} u^2(\xi,\eta) r^2 \mathrm{d}\,\xi\,\mathrm{d}\,\eta\,, \end{aligned} \tag{9.2.9}$$

where $u^2 * e$ is the convolution product and

$$r = \sqrt{x_0^2 + y_0^2}, \quad x_0 = x - \xi, \quad y_0 = y - \eta\,. \tag{9.2.10}$$

From (9.2.8) it results

$$\begin{aligned} u(x,y) &= \frac{1}{2\pi}\frac{mx + \ell y}{x^2+y^2} + \frac{k}{2\pi}\frac{\partial}{\partial x}\iint_{\mathbb{R}^2} u^2(\xi,\eta)\frac{x_0}{r^2}\mathrm{d}\,\xi\,\mathrm{d}\,\eta\,, \\ v(x,y) &= \frac{1}{2\pi}\frac{my - \ell x}{x^2+y^2} + \frac{k}{2\pi}\frac{\partial}{\partial x}\iint_{\mathbb{R}^2} u^2(\xi,\eta)\frac{y_0}{r^2}\mathrm{d}\,\xi\,\mathrm{d}\,\eta\,. \end{aligned} \tag{9.2.11}$$

Obviously, the last integrals are singular. They have the shape (E.3). Isolating the singular point (x, y) with a circle having the radius ε and setting $f = x_0/r$ and respectively y_0/r we deduce that the condition (E.5) is satisfied and the integrals are singular. Also, from the formula (E.10) we get

$$I \equiv \frac{\partial}{\partial x} \iint_{\mathbb{R}^2} u^2(\xi,\eta) \frac{x_0}{r^2} \mathrm{d}\,\xi\,\mathrm{d}\,\eta =$$

$$= \iint_{\mathbb{R}^2} u^2(\xi,\eta) \frac{\partial}{\partial x}\left(\frac{x_0}{r^2}\right) \mathrm{d}\,\xi\,\mathrm{d}\,\eta - \pi u^2(x,y),$$

$$J \equiv \frac{\partial}{\partial x} \iint_{\mathbb{R}^2} u^2(\xi,\eta) \frac{y_0}{r^2} \mathrm{d}\,\xi\,\mathrm{d}\,\eta = \iint_{\mathbb{R}^2} u^2(\xi,\eta) \frac{\partial}{\partial x}\left(\frac{y_0}{r_2}\right) \mathrm{d}\,\xi\,\mathrm{d}\,\eta, \tag{9.2.12}$$

or, performing the calculations,

$$I = \iint_{\mathbb{R}^2} u^2(\xi,\eta) \frac{y_0^2 - x_0^2}{r^4} \mathrm{d}\,x\,\mathrm{d}\,\eta - \pi u^2(x,y)$$

$$J = -\iint_{\mathbb{R}^2} u^2(\xi,\eta) \frac{x_0^2 y_0^2}{r^4} \mathrm{d}\,x\,\mathrm{d}\,\eta. \tag{9.2.13}$$

9.2.2 The General Solution

Replacing the profile by a perturbing distribution defined on the segment $(-1,+1)$ (the chord of the profile), it results the following general representation of the perturbation:

$$u(x,y) = \frac{1}{2\pi} \int_{-1}^{+1} \frac{\ell(\xi)y + m(\xi)x_0}{x_0^2 + y^2} \mathrm{d}\,\xi + \frac{k}{2\pi} I,$$

$$v(x,y) = \frac{1}{2\pi} \int_{-1}^{+1} \frac{m(\xi)y - \ell(\xi)x_0}{x_0^2 + y^2} \mathrm{d}\,\xi + \frac{k}{2\pi} J. \tag{9.2.14}$$

Taking into account that

$$x_0^2 + y^2 = (x_0 - \mathrm{i}\,y)(x_0 + \mathrm{i}\,y), \quad x_0^2 + y_0^2 = (x_0 - \mathrm{i}\,y_0)(x_0 + \mathrm{i}\,y_0),$$

for complex velocity

$$w(z) = u(x,y) - \mathrm{i}\,v(x,y),$$

it results the following formula:

$$w(z) = \frac{1}{2\pi} \int_{-1}^{+1} \frac{m(\xi) + \mathrm{i}\,\ell(\xi)}{z - \xi} \mathrm{d}\,\xi + \frac{k}{2\pi} L(z)\,, \tag{9.2.15}$$

where

$$L(z) = \frac{\partial}{\partial x} \iint_{\mathbb{R}^2} \frac{u^2(\xi, \eta)}{z - \zeta} \mathrm{d}\,\xi\,\mathrm{d}\,\eta\,,$$

and

$$z = x + \mathrm{i}\,\eta, \quad \zeta = \xi + \mathrm{i}\,\eta\,.$$

In order to impose the conditions (9.2.2) we shall pass to the limit on the segment $[-1, +1]$. Using Plemelj's formulas we obtain from (9.2.15)

$$\begin{aligned} u(x, \pm 0) - \mathrm{i}\,v(x, \pm 0) &= \mp \frac{\mathrm{i}}{2} \left[m(x) + \mathrm{i}\,\ell(x) \right] + \\ &+ \frac{1}{2\pi} {\int_{-1}^{+1}}' \frac{m(\xi) + \mathrm{i}\,\ell(\xi)}{x_0} \mathrm{d}\,\xi + \frac{k}{2\pi} \frac{\partial}{\partial x} \iint_{\mathbb{R}^2} \frac{u^2(\xi, \eta)}{x - \zeta} \mathrm{d}\,\xi\,\mathrm{d}\,\eta\,, \end{aligned} \tag{9.2.16}$$

whence

$$\begin{aligned} u(x, \pm 0) &= \pm \frac{1}{2} \ell(x) + \frac{1}{2\pi} {\int_{-1}^{+1}}' \frac{m(\xi)}{x_0} \mathrm{d}\,\xi + \\ &+ \frac{k}{2\pi} \frac{\partial}{\partial x} \iint_{\mathbb{R}^2} u^2(\xi, \eta) \frac{x_0}{x_0^2 + \eta^2} \mathrm{d}\,\xi\,\mathrm{d}\,\eta\,, \\ v(x, \pm 0) &= \pm \frac{1}{2} m(x) - \frac{1}{2\pi} {\int_{-1}^{+1}}' \frac{\ell(\xi)}{x_0} \mathrm{d}\,\xi - \\ &- \frac{k}{2\pi} \frac{\partial}{\partial x} \iint_{\mathbb{R}^2} u^2(\xi, \eta) \frac{\eta}{x_0^2 + \eta^2} \mathrm{d}\,\xi\,\mathrm{d}\,\eta\,. \end{aligned} \tag{9.2.17}$$

It results

$$\ell(x) = u(x, +0) - u(x, -0) = \sqrt{K} \left[\widehat{p}(x, -0) - \widehat{p}(x, +0) \right]\,. \tag{9.2.18}$$

The function $\ell(x)$ will determine therefore the *lift*.

Imposing the conditions (9.2.2) we shall obtain

$$m(x) = h_+'(x) - h_-'(x) \overset{\text{not}}{=} h^-(x)\,, \quad |x| < 1 \tag{9.2.19}$$

$$\frac{1}{\pi}\int_{-1}^{+1}{}' \frac{\ell(\xi)}{\xi - x_0}\mathrm{d}\,\xi = \frac{k}{\pi}\frac{\partial}{\partial x}\iint_{\mathbb{R}^2} u^2(\xi,\eta)\frac{\eta}{x_0^2+\eta^2}\mathrm{d}\,\xi\,\mathrm{d}\,\eta + \tag{9.2.20}$$

$$+h^+(x) \overset{\text{not}}{=} G(x)\,, |x| < 1\,,$$

where we have utilized the notation $h^+(x) = h'_+(x) + h'_-(x)$.

The relation (9.2.19) determines the unknown $m(x)$. Considering that $G(x)$ is known, the equation (9.2.20) determines f with the aid of the formula (C.1.9)

$$\ell(x) = -\frac{1}{\pi}\sqrt{\frac{1-x}{1+x}}\int_{-1}^{+1}{}'\sqrt{\frac{1+t}{1-t}}\frac{h^+(t)}{t-x}\mathrm{d}\,t -$$

$$-\frac{1}{\pi}\sqrt{\frac{1-x}{1+x}}\int_{-1}^{+1}\sqrt{\frac{1+t}{1-t}}\frac{\partial}{\partial t}\left(\frac{k}{\pi}\iint_{\mathbb{R}_2} u^2(\xi,\eta)\frac{\eta\mathrm{d}\,\xi\,\mathrm{d}\,\eta}{(t-\xi)^2+\eta^2}\right)\frac{\mathrm{d}\,t}{t-x}\,. \tag{9.2.21}$$

Noticing that according to the formulas (B.5.3) and (B.5.4) we have

$$\int_{-1}^{+1}{}'\sqrt{\frac{1-\xi}{1+\xi}}\frac{\mathrm{d}\,\xi}{(\xi-z)(t-\xi)} = \frac{1}{t-z}\left(\int_{-1}^{+1}\sqrt{\frac{1-\xi}{1+\xi}}\frac{\mathrm{d}\,\xi}{\xi-z} - \right. \tag{9.2.22}$$

$$\left. - \int_{-1}^{+1}\sqrt{\frac{1-\xi}{1+\xi}}\frac{\mathrm{d}\,\xi}{\xi-t}\right) = \frac{\pi}{t-z}\sqrt{\frac{z-1}{z+1}}$$

and, after replacing (9.2.21) in (9.2.15), it results

$$w(z) = -\frac{1}{2\pi}\int_{-1}^{+1}\frac{h^-(\xi)}{\xi-z}\mathrm{d}\,\xi - \tag{9.2.23}$$

$$-\frac{1}{2\pi\mathrm{i}}\sqrt{\frac{z-1}{z+1}}\int_{-1}^{+1}\sqrt{\frac{1+t}{1-t}}\frac{h^+(t)}{t-z}\mathrm{d}\,t + \frac{k}{2\pi}L(z) - \frac{k}{2\pi}M(z)\,,$$

where we denoted

$$M(z) = \frac{1}{\pi\mathrm{i}}\sqrt{\frac{z-1}{z+1}}\int_{-1}^{+1}\sqrt{\frac{1+t}{1-t}}\left[\frac{\mathrm{d}}{\mathrm{d}\,t}\iint_{\mathbb{R}^2} u^2(\xi,\eta)\cdot\right. \tag{9.2.24}$$

$$\left.\cdot\frac{\eta}{(\xi-t)^2+\eta^2}\mathrm{d}\,\xi\,\mathrm{d}\,\eta\right]\frac{\mathrm{d}\,t}{t-z}\,,$$

the formula coinciding with (4.11) from [9.21].

In the sequel we shall deal with $M(z)$. We may write

$$M(z) = \frac{1}{\pi \mathrm{i}}\sqrt{\frac{z-1}{z+1}} \iint_{\mathbb{R}^2} u^2(\xi,\eta)\cdot$$

$$\cdot\left\{\int_{-1}^{+1}\sqrt{\frac{1+t}{1-t}}\frac{\mathrm{d}}{\partial t}\left[\frac{\eta}{(\xi-t)^2+\eta^2}\right]\frac{\mathrm{d}t}{t-z}\right\}\mathrm{d}\xi\,\mathrm{d}\eta = -\frac{1}{\pi \mathrm{i}}\sqrt{\frac{z-1}{z+1}}\cdot$$

$$\cdot\iint_{\mathbb{R}^2} u^2(\xi,\eta)\cdot\left[\frac{\partial}{\partial \xi}\int_{-1}^{+1}\sqrt{\frac{1+t}{1-t}}\frac{\eta}{(\xi-t)^2+\eta^2}\frac{\mathrm{d}t}{t-z}\right]\mathrm{d}\xi\,\mathrm{d}\eta\,.$$

The last integral is calculated by Homentcovschi by means of the residue theorem. It may be calculated elementary noticing that we have

$$\frac{\eta}{(t-\xi)^2+\eta^2}\,\frac{1}{t-z} = \frac{1}{2\mathrm{i}\,(\zeta-z)}\,\frac{1}{t-\zeta}-$$

$$-\frac{1}{2\mathrm{i}\,(\overline{\zeta}-z)}\,\frac{1}{t-\overline{\zeta}}+\frac{\eta}{(\xi-z)^2+\eta^2}\,\frac{1}{t-z}$$

and taking (B.5.1) into account. It results

$$M(z) = -\frac{1}{2}\sqrt{\frac{z-1}{z+1}}\iint_{\mathbb{R}^2} u^2(\xi,\eta)\frac{\partial}{\partial\xi}\left[\frac{1}{\zeta-z}\sqrt{\frac{\zeta+1}{\zeta-1}}-\right.$$

$$\left.-\frac{1}{\overline{\zeta}-z}\sqrt{\frac{\overline{\zeta}+1}{\overline{\zeta}-1}}\right]\mathrm{d}\xi\,\mathrm{d}\eta + \frac{\partial}{\partial z}\iint_{\mathbb{R}^2} u^2(\xi,\eta)\frac{\mathrm{i}\eta}{(\xi-z)^2+\eta^2}\mathrm{d}\xi\,\mathrm{d}\eta\,.$$

We write the first integral as follows

$$\iint_{\mathbb{R}^2} u^2(\xi,\eta)\frac{\partial}{\partial\xi}\left(\frac{1}{\zeta-z}\sqrt{\frac{\zeta+1}{\zeta-1}}\right)\mathrm{d}\xi\,\mathrm{d}\eta-$$

$$-\iint_{\mathbb{R}^2} u^2(\xi,-\eta)\frac{\partial}{\partial\xi}\left(\frac{1}{\zeta-z}\sqrt{\frac{\zeta+1}{\zeta-1}}\right)\mathrm{d}\xi\,\mathrm{d}\eta =$$

$$= -\iint_{\mathbb{R}^2}\frac{u^2(\xi,\eta)-u^2(\xi,-\eta)}{\zeta-z}\sqrt{\frac{\zeta+1}{\zeta-1}}\left(\frac{1}{\zeta-z}-\frac{1}{\zeta^2-1}\right)\mathrm{d}\xi\,\mathrm{d}\eta\,,$$

and the second

$$\frac{\partial}{\partial z}\iint_{\mathbb{R}^2} u^2(\xi,\eta)\left(\frac{1}{\overline{\zeta}-z}-\frac{1}{\zeta-z}\right)\mathrm{d}\xi\,\mathrm{d}\eta =$$

$$= \iint_{\mathbb{R}^2} \frac{u^2(\xi,-\eta) - u^2(\xi,\eta)}{(\zeta - z)^2} \mathrm{d}\,\xi\,\mathrm{d}\,\eta\,.$$

We have therefore

$$\begin{aligned} M(z) = \frac{1}{2} \iint_{-\infty}^{+\infty} \left[u^2(\xi,\eta) - u^2(\xi,-\eta)\right] K(z,\zeta)\mathrm{d}\,\xi\,\mathrm{d}\,\eta - \\ -\frac{1}{2} \iint_{-\infty}^{+\infty} \frac{u^2(\xi,\eta) - u^2(\xi,-\eta)}{(\zeta - z)^2} \mathrm{d}\,\xi\,\mathrm{d}\,\eta\,, \end{aligned} \tag{9.2.25}$$

where

$$K(z,\zeta) = \sqrt{\frac{z-1}{z+1}}\sqrt{\frac{\zeta+1}{\zeta-1}}\left[\frac{1}{(\zeta-z)^2} - \frac{1}{(\zeta-z)(\zeta^2-1)}\right]. \tag{9.2.26}$$

Taking also (9.2.13) and (9.2.14) into account, we have that

$$\frac{y_0^2 - x_0^2 + 2\mathrm{i}\,x_0 y_0}{r^4} = -\frac{1}{(\zeta - z)^2}\,, \tag{9.2.27}$$

$$L(z) = I - \mathrm{i}\,J = -\pi u^2(x,y) - \iint_{\mathbb{R}^2} \frac{u^2(\xi,\eta)}{(\zeta - z)^2} \mathrm{d}\,\xi\,\mathrm{d}\,\eta\,. \tag{9.2.28}$$

We replace the expressions (9.2.25) and (9.2.27) by (9.2.23) and we obtain the complex velocity. Separating the real part of the complex velocity, we obtain the integral equation of the problem. We have:

$$\begin{aligned} u(x,y) = \frac{1}{2\pi} \int_{-1}^{+1} h^-(\xi) \frac{x_0}{x_0^2 + y^2} \mathrm{d}\,\xi - \\ -\frac{1}{4\pi} \int_{-1}^{+1} \sqrt{\frac{1+t}{1-t}} h^+(t) \left[\sqrt{\frac{z-1}{z+1}}\,\frac{1}{t-z} - \sqrt{\frac{\overline{z}-1}{\overline{z}+1}}\,\frac{1}{t-\overline{z}}\right] \mathrm{d}\,t + \\ +\frac{k}{2\pi} I(x,y) - \frac{k}{8\pi} \iint_{-\infty}^{+\infty} \left[u^2(\xi,\eta) - u^2(\xi,-\eta)\right] \left[K(z,\zeta) + \right. \\ \left. + K(\overline{z},\overline{\zeta})\right] \mathrm{d}\xi\,\mathrm{d}\eta + \frac{1}{2} \iint_{-\infty}^{+\infty} \left[u^2(\xi,\eta) - u^2(\xi,-\eta)\right] \frac{y_0^2 - x_0^2}{r^4} \mathrm{d}\,\xi\,\mathrm{d}\,\eta\,. \end{aligned} \tag{9.2.29}$$

This equation, was given in a slightly different form by Homentcovschi in [9.21].

9.2.3 The Lift Coefficient

From (9.2.18) we deduce the following formula for the lift coefficient

$$c_p = \frac{1}{\sqrt{K}}\int_{-1}^{+1}\ell(x)\mathrm{d}\,x = -\frac{1}{\pi K}\int_{-1}^{+1}\sqrt{\frac{1-x}{1+x}}\left[\int_{-1}^{+1}\sqrt{\frac{1+t}{1-t}}\cdot\frac{h^{+}(t)}{t-x}\mathrm{d}\,t\right]\mathrm{d}\,x - \frac{k}{\pi^2\sqrt{K}}\int_{-1}^{+1}\sqrt{\frac{1-x}{1+x}}\left[\int_{-1}^{+1}\sqrt{\frac{1+t}{1-t}}\cdot\frac{\partial}{\partial t}\left[\iint_{\mathbb{R}^2} u^2(\xi,\eta)\frac{\eta}{(t-\xi)^2+\eta^2}\mathrm{d}\,\xi\,\mathrm{d}\,\eta\right]\frac{\mathrm{d}\,t}{t-x}\right]\mathrm{d}\,x\,. \tag{9.2.30}$$

We have

$$\iint_{\mathbb{R}^2} u^2(\xi,\eta)\frac{\partial}{\partial t}\left[\frac{\eta}{(t-\xi)^2+\eta^2}\right]\mathrm{d}\,\xi\,\mathrm{d}\,\eta = \frac{1}{2\mathrm{i}}\iint_{\mathbb{R}^2} u^2(\xi,\eta)\frac{\partial}{\partial t}\left(\frac{1}{t-\zeta}-\frac{1}{t-\overline{\zeta}}\right)\mathrm{d}\,\xi\,\mathrm{d}\,\eta = -\frac{1}{2\mathrm{i}}\int_{\mathbb{R}^2}\frac{u^2(\xi,\eta)-u^2(\xi,-\eta)}{(t-\zeta)^2}\mathrm{d}\,\xi\,\mathrm{d}\,\eta\,. \tag{9.2.31}$$

From (B.5.1), derivating, it results:

$$\frac{1}{\pi}\int_{-1}^{+1}\sqrt{\frac{1+t}{1-t}}\frac{\mathrm{d}\,t}{(t-\zeta)^2} = -\frac{1}{\zeta-1}\frac{1}{\sqrt{\zeta^2-1}}\,. \tag{9.2.32}$$

Using these results we obtain

$$c_p = -\frac{1}{\sqrt{K}}\int_{-1}^{+1}\sqrt{\frac{1+t}{1-t}}h^{+}(t)\mathrm{d}\,t - \frac{1}{2\mathrm{i}}\frac{k}{\sqrt{K}}\iint_{\mathbb{R}^2}\frac{u^2(\xi,\eta)-u^2(\xi,-\eta)}{\zeta-1}\frac{\mathrm{d}\,\xi\,\mathrm{d}\,\eta}{\sqrt{\zeta^2-1}}\,. \tag{9.2.33}$$

9.2.4 The Symmetric Wing

If the wing is symmetric, then the equations (9.2.1) have the form

$$y = \pm h(x) \tag{9.2.34}$$

$$= \iint_{\mathbb{R}^2} \frac{u^2(\xi,-\eta) - u^2(\xi,\eta)}{(\zeta - z)^2} \mathrm{d}\,\xi\,\mathrm{d}\,\eta\,.$$

We have therefore

$$M(z) = \frac{1}{2} \iint\limits_{-\infty}^{+\infty} \Big[u^2(\xi,\eta) - u^2(\xi,-\eta)\Big] K(z,\zeta) \mathrm{d}\,\xi\,\mathrm{d}\,\eta - \\ - \frac{1}{2} \iint\limits_{-\infty}^{+\infty} \frac{u^2(\xi,\eta) - u^2(\xi,-\eta)}{(\zeta - z)^2} \mathrm{d}\,\xi\,\mathrm{d}\,\eta\,, \tag{9.2.25}$$

where

$$K(z,\zeta) = \sqrt{\frac{z-1}{z+1}} \sqrt{\frac{\zeta+1}{\zeta-1}} \left[\frac{1}{(\zeta-z)^2} - \frac{1}{(\zeta-z)(\zeta^2-1)}\right]. \tag{9.2.26}$$

Taking also (9.2.13) and (9.2.14) into account, we have that

$$\frac{y_0^2 - x_0^2 + 2\mathrm{i}\,x_0 y_0}{r^4} = -\frac{1}{(\zeta - z)^2}\,, \tag{9.2.27}$$

$$L(z) = I - \mathrm{i}\,J = -\pi u^2(x,y) - \iint_{\mathbb{R}^2} \frac{u^2(\xi,\eta)}{(\zeta - z)^2} \mathrm{d}\,\xi\,\mathrm{d}\,\eta\,. \tag{9.2.28}$$

We replace the expressions (9.2.25) and (9.2.27) by (9.2.23) and we obtain the complex velocity. Separating the real part of the complex velocity, we obtain the integral equation of the problem. We have:

$$u(x,y) = \frac{1}{2\pi} \int_{-1}^{+1} h^-(\xi) \frac{x_0}{x_0^2 + y^2} \mathrm{d}\,\xi - \\ - \frac{1}{4\pi} \int_{-1}^{+1} \sqrt{\frac{1+t}{1-t}} h^+(t) \left[\sqrt{\frac{z-1}{z+1}} \frac{1}{t-z} - \sqrt{\frac{\overline{z}-1}{\overline{z}+1}} \frac{1}{t-\overline{z}}\right] \mathrm{d}\,t + \\ + \frac{k}{2\pi} I(x,y) - \frac{k}{8\pi} \iint\limits_{-\infty}^{+\infty} \Big[u^2(\xi,\eta) - u^2(\xi,-\eta)\Big] \Big[K(z,\zeta) + \\ + K(\overline{z},\overline{\zeta})\Big] \mathrm{d}\xi\,\mathrm{d}\eta + \frac{1}{2} \iint\limits_{-\infty}^{+\infty} \Big[u^2(\xi,\eta) - u^2(\xi,-\eta)\Big] \frac{y_0^2 - x_0^2}{r^4} \mathrm{d}\,\xi\,\mathrm{d}\,\eta\,. \tag{9.2.29}$$

This equation, was given in a slightly different form by Homentcovschi in [9.21].

9.2.3 The Lift Coefficient

From (9.2.18) we deduce the following formula for the lift coefficient

$$c_p = \frac{1}{\sqrt{K}}\int_{-1}^{+1} \ell(x)\mathrm{d}\,x = -\frac{1}{\pi K}\int_{-1}^{+1}\sqrt{\frac{1-x}{1+x}}\left[\int_{-1}^{+1}\sqrt{\frac{1+t}{1-t}}\cdot\right.$$

$$\left.\cdot\frac{h^+(t)}{t-x}\mathrm{d}\,t\right]\mathrm{d}\,x - \frac{k}{\pi^2\sqrt{K}}\int_{-1}^{+1}\sqrt{\frac{1-x}{1+x}}\left[\int_{-1}^{+1}\sqrt{\frac{1+t}{1-t}}\cdot\right. \tag{9.2.30}$$

$$\left.\cdot\frac{\partial}{\partial t}\left[\iint_{\mathbb{R}^2} u^2(\xi,\eta)\frac{\eta}{(t-\xi)^2+\eta^2}\mathrm{d}\,\xi\,\mathrm{d}\,\eta\right]\frac{\mathrm{d}\,t}{t-x}\right]\mathrm{d}\,x\,.$$

We have

$$\iint_{\mathbb{R}^2} u^2(\xi,\eta)\frac{\partial}{\partial t}\left[\frac{\eta}{(t-\xi)^2+\eta^2}\right]\mathrm{d}\,\xi\,\mathrm{d}\,\eta =$$

$$= \frac{1}{2\mathrm{i}}\iint_{\mathbb{R}^2} u^2(\xi,\eta)\frac{\partial}{\partial t}\left(\frac{1}{t-\zeta}-\frac{1}{t-\overline{\zeta}}\right)\mathrm{d}\,\xi\,\mathrm{d}\,\eta \tag{9.2.31}$$

$$= -\frac{1}{2\mathrm{i}}\int_{\mathbb{R}^2}\frac{u^2(\xi,\eta)-u^2(\xi,-\eta)}{(t-\zeta)^2}\mathrm{d}\,\xi\,\mathrm{d}\,\eta\,.$$

From (B.5.1), derivating, it results:

$$\frac{1}{\pi}\int_{-1}^{+1}\sqrt{\frac{1+t}{1-t}}\,\frac{\mathrm{d}\,t}{(t-\zeta)^2} = -\frac{1}{\zeta-1}\,\frac{1}{\sqrt{\zeta^2-1}}\,. \tag{9.2.32}$$

Using these results we obtain

$$c_p = -\frac{1}{\sqrt{K}}\int_{-1}^{+1}\sqrt{\frac{1+t}{1-t}}h^+(t)\mathrm{d}\,t -$$

$$-\frac{1}{2\mathrm{i}}\,\frac{k}{\sqrt{K}}\iint_{\mathbb{R}^2}\frac{u^2(\xi,\eta)-u^2(\xi,-\eta)}{\zeta-1}\,\frac{\mathrm{d}\,\xi\,\mathrm{d}\,\eta}{\sqrt{\zeta^2-1}}\,. \tag{9.2.33}$$

9.2.4 The Symmetric Wing

If the wing is symmetric, then the equations (9.2.1) have the form

$$y = \pm h(x) \tag{9.2.34}$$

From the boundary conditions (9.2.2) we have the relation

$$v(x,+0) = -v(x,-0)$$

which suggests that the solution of the system (9.2.6) has the property

$$u(x,y) = u(x,-y), \quad v(x,y) = -v(x,-y)\,. \tag{9.2.35}$$

We easily check that if $u(x,y),\ v(x,y)$ is a solution of the system (9.2.6), then $u(x,-y), -v(x,-y)$ also have this property. By virtue of the uniqueness theorem it results (9.2.35) whence $u^2(x,\eta) = u^2(\xi,-\eta)$. The integral equation (9.2.29) receives the form

$$\begin{aligned} u(x,y) &= \frac{1}{\pi}\int_{-1}^{+1} h'(\xi)\frac{x_0}{x_0^2+y^2}\,\mathrm{d}\,\xi + \\ &+\frac{k}{2\pi}\iint\limits_{-\infty}^{+\infty} u^2(\xi,\eta)\frac{y_0^2-x_0^2}{r^4}\,\mathrm{d}\,\xi\,\mathrm{d}\,\eta - \frac{k}{2}u^2(x,y)\,, \end{aligned} \tag{9.2.36}$$

given for the first time by Oswatitsch [9.49].

Obviously, the lift coefficient vanishes. The result is natural because the angle of attack of the wing is zero.

9.2.5 The Solution in Real

We shall present a solution which is different from Homentcovschi's solution which utilizes the complex velocity. Using the formulas (3.1.19) and (3.1.20), one obtains from the general solution (9.2.14):

$$\begin{aligned} u(x,\pm 0) &= \pm\frac{1}{2}\ell(x) + \frac{1}{2\pi}\int_{-1}^{+1}\frac{m(s)}{x-s}\,\mathrm{d}\,s + \frac{k}{2\pi}I(x,\pm 0)\,, \\ v(x,\pm 0) &= \pm\frac{1}{2}m(x) - \frac{1}{2\pi}\int_{-1}^{+1}\frac{\ell(s)}{x-s}\,\mathrm{d}\,s + \frac{k}{2\pi}J(x,\pm 0)\,, \\ &\qquad -1 < s < 1\,, \end{aligned} \tag{9.2.37}$$

where

$$\begin{aligned} I(x,\pm 0) &= -\iint_{\mathbb{R}^2} u^2(\xi,\eta)\frac{x_0^2-\eta^2}{(x_0^2+\eta^2)^2}\,\mathrm{d}\,\xi\,\mathrm{d}\,\eta - \pi u^2(x,\pm 0)\,, \\ J(x,\pm 0) &= 2\iint_{\mathbb{R}^2} u^2(\xi,\eta)\frac{\eta x_0}{(x_0^2+\eta^2)^2}\,\mathrm{d}\,\xi\,\mathrm{d}\,\eta, \quad -\infty < \xi,\eta < \infty. \end{aligned} \tag{9.2.38}$$

We obtain the formula (9.2.18), and from the boundary conditions (9.2.2)

$$m(x) = h^-(x) \tag{9.2.39}$$

$$\frac{1}{\pi}{\int_{-1}^{+1}}' \frac{\ell(s)}{s-x}\mathrm{d}\,s = h^+(x) - \frac{k}{\pi}J(x, \pm 0)\,, \quad |x| < 1\,. \tag{9.2.40}$$

By means of the formula (C.1.9) we deduce

$$\begin{aligned}\ell(x) = &-\frac{1}{\pi}\sqrt{\frac{1-x}{1+x}}{\int_{-1}^{+1}}' \sqrt{\frac{1+t}{1-t}}\frac{h^+(t)}{t-x}\mathrm{d}\,t+ \\ &+\frac{k}{\pi^2}\sqrt{\frac{1-x}{1+x}}{\int_{-1}^{+1}}' \sqrt{\frac{1+t}{1-t}}\frac{J(t,\pm 0)}{t-x}\mathrm{d}\,t\,.\end{aligned} \tag{9.2.41}$$

For obtaining the integral equation of the problem, we shall replace m and ℓ in $(9.2.14)_1$. We shall denote by s the superposition variable ξ from the first integrals (9.2.14) and $x_1 = x - s$. Changing the order of integration, we deduce

$$\begin{aligned}u(x,y) = &\frac{y}{2\pi^2}\int_{-1}^{+1}\sqrt{\frac{1+t}{1-t}}h^+(t)N(y,t)\mathrm{d}\,t- \\ &-\frac{y}{2\pi}\frac{k}{\pi^2}\int_{-1}^{+1}\sqrt{\frac{1+t}{1-t}}J(t,\pm 0)N(x,y,t)\mathrm{d}\,t+ \\ &+\frac{1}{2\pi}\int_{-1}^{+1}\frac{x_1 h^-(s)}{x_1^2+y^2}\mathrm{d}\,s + \frac{k}{2\pi}I(x,y), \quad x_1 \equiv x - s\,,\end{aligned} \tag{9.2.42}$$

where we denoted

$$N(x,y,t) = \int_{-1}^{+1}\sqrt{\frac{1-s}{1+s}}\frac{1}{x_1^2+y^2}\frac{\mathrm{d}\,s}{s-t}\,. \tag{9.2.43}$$

Since

$$\begin{aligned}\frac{1}{x_1^2+y^2}\frac{1}{s-t} = &\frac{1}{z-\overline{z}}\left(\frac{1}{t-\overline{z}}\frac{1}{s-\overline{z}} - \frac{1}{t-z}\frac{1}{s-z}\right)+ \\ &+\frac{1}{(x-t)^2+y^2}\frac{1}{s-t}\,,\end{aligned}$$

taking the formulas (B.5.3) and (B.5.4) into account, we deduce

$$N = \frac{\pi}{z-\overline{z}}\left(\frac{1}{t-\overline{z}}\sqrt{\frac{\overline{z}-1}{\overline{z}+1}} - \frac{1}{t-z}\sqrt{\frac{z-1}{z+1}}\right)\,. \tag{9.2.44}$$

Let us calculate now the term

$$T=\int_{-1}^{+1}\sqrt{\frac{1+t}{1-t}}J(t,\pm 0)N(x,y,t)\mathrm{d}\,t\,. \tag{9.2.45}$$

To this aim we shall specify $J(t,\pm 0)$. Taking into account (9.2.13) and the identity

$$\frac{2\mathrm{i}\,\eta}{(t-\xi)^2+\eta^2}=\frac{1}{t-\zeta}-\frac{1}{t-\overline{\zeta}}\,,$$

we deduce

$$\begin{aligned}J(x,\pm 0)&=-\iint_{\mathbb{R}^2}\eta u^2(\xi,\eta)\frac{\partial}{\partial\xi}\left[\frac{1}{(t-\xi)^2+\eta^2}\right]\mathrm{d}\,\xi\,\mathrm{d}\,\eta=\\&=\frac{1}{2\mathrm{i}}\frac{\partial}{\partial t}\iint_{\mathbb{R}^2}u^2(\xi,\eta)\left(\frac{1}{t-\zeta}-\frac{1}{t-\overline{\zeta}}\right)\mathrm{d}\,\xi\,\mathrm{d}\,\eta=\\&=\frac{1}{2\mathrm{i}}\frac{\partial}{\partial t}\iint_{\mathbb{R}^2}\frac{u^2(\xi,\eta)-u^2(\xi,-\eta)}{t-\zeta}\mathrm{d}\,\xi\,\mathrm{d}\,\eta=\\&=\frac{1}{2\mathrm{i}}\iint_{\mathbb{R}^2}\frac{u^2(\xi,\eta)-u^2(\xi,-\eta)}{(t-\zeta)^2}\mathrm{d}\,\xi\,\mathrm{d}\,\eta\end{aligned} \tag{9.2.46}$$

and then

$$\begin{aligned}T=-\frac{1}{2\mathrm{i}}\int_{-1}^{+1}\sqrt{\frac{1+t}{1-t}}\left[\iint_{\mathbb{R}^2}\frac{u^2(\xi,\eta)-u^2(\xi,-\eta)}{(t-\zeta)^2}\mathrm{d}\,\xi\,\mathrm{d}\,\eta\right]\frac{\pi}{2\mathrm{i}\,y}\cdot\\\cdot\left(\frac{1}{t-z}\sqrt{\frac{z-1}{z+1}}-\frac{1}{t-\overline{z}}\sqrt{\frac{\overline{z}-1}{\overline{z}+1}}\right)\mathrm{d}\,t\,.\end{aligned} \tag{9.2.47}$$

Changing the order of integration and denoting

$$T_0(\zeta,z)=\int_{-1}^{+1}\sqrt{\frac{1+t}{1-t}}\frac{1}{(t-\zeta)^2}\frac{1}{t-z}\mathrm{d}\,t\,, \tag{9.2.48}$$

we deduce

$$\begin{aligned}T=\frac{\pi}{4y}\iint_{\mathbb{R}^2}\left[u^2(\xi,\eta)-u^2(\xi,-\eta)\right]\Bigg[T_0(\zeta,z)\sqrt{\frac{z-1}{z+1}}-\\-T_0(\zeta,\overline{z})\sqrt{\frac{\overline{z}-1}{\overline{z}+1}}\Bigg]\mathrm{d}\,\xi\,\mathrm{d}\,\eta\end{aligned} \tag{9.2.49}$$

for specifying the function T_0 we notice that

$$\frac{1}{(t-\zeta)^2}\frac{1}{t-z} = -\frac{t}{\zeta(z-\zeta)}\frac{1}{(t-\zeta)^2} +$$

$$+\left[\frac{1}{\zeta}\frac{1}{z-\zeta} - \frac{1}{(z-\zeta)^2}\right]\frac{1}{t-\zeta} + \frac{1}{(z-\zeta)^2}\frac{1}{t-z}.$$

Taking (B.5.1) into account , one obtains

$$T_0(\zeta, z) = \frac{\pi}{(z-\zeta)^2}\left(\sqrt{\frac{1+\zeta}{\zeta-1}} - 1\right) - \frac{\pi}{(z-\zeta)^2}\left(\sqrt{\frac{z+1}{z-1}} - 1\right) -$$

$$-\frac{\pi}{z-\zeta}\frac{\mathrm{d}}{\mathrm{d}\zeta}\sqrt{\frac{1+\zeta}{\zeta-1}}. \tag{9.2.50}$$

The term T which intervenes in the integral equation (9.2.42) has therefore the form (9.2.49) where T_0 It is given by the formula (9.2.50). Using the expression (9.2.41) for the lift coefficient

$$c_L = \frac{1}{\sqrt{K}}\int_{-1}^{+1} \ell(x)\mathrm{d}\,x \tag{9.2.51}$$

one obtains the formula

$$c_L = -\frac{1}{\sqrt{K}}\int_{-1}^{+1}\sqrt{\frac{1+t}{1-t}}h^+(t)\mathrm{d}\,t + \frac{k}{\pi\sqrt{K}}\int_{-1}^{+1}\sqrt{\frac{1+t}{1-t}}J(t,\pm 0)\mathrm{d}\,t. \tag{9.2.52}$$

The function $J(t, \pm 0)$ is given in (9.2.46). For calculating the last term we change the order of integration and we take the formula (9.2.32) into account. One obtains the formula (9.2.33).

9.2.6 The Symmetric Wing

We gave in (9.2.36) the equation of Oswatitsch for the symmetric wing. We obtained this equation from the general theory presented in (9.2.2). We present here the direct deduction based on the equation of the potential

$$\varphi_{xx} + \varphi_{yy} = k\varphi_x^2 \tag{9.2.53}$$

which may be obtained from (9.1.13) or (9.2.6).

The fundamental solution is defined by the equation

$$\mathcal{E}_{xx} + \mathcal{E}_{yy} = k\frac{\partial}{\partial x}\mathcal{E}_x^2 + m\delta(x,y)\,. \tag{9.2.54}$$

Applying the Fourier transform it results

$$(\alpha_1^2 + \alpha_2^2)\widehat{\mathcal{E}} = \mathrm{i}\,\alpha_1 kF - m\,,\quad F \equiv \mathcal{F}[\mathcal{E}_x^2]\,,$$

whence

$$\widehat{\mathcal{E}} = k\frac{\mathrm{i}\,\alpha_1 F}{\alpha_1^2 + \alpha_2^2} - \frac{m}{\alpha_1^2 + \alpha_2^2}$$

and then

$$\mathcal{E} = -k\frac{\partial}{\partial x}\mathcal{F}^{-1}\left[\frac{F}{\alpha_1^2 + \alpha_2^2}\right] - m\mathcal{F}^{-1}\left[\frac{1}{\alpha_1^2 + \alpha_2^2}\right]. \tag{9.2.55}$$

Applying the convolution theorem (A.6.13) and taking (A.7.11) into account, we obtain

$$\mathcal{F}^{-1}\left[\frac{F}{\alpha_1^2 + \alpha_2^2}\right] = \mathcal{E}_x^2 * \left(-\frac{1}{4\pi}\ln(x^2 + y^2)\right) =$$

$$= -\frac{1}{4\pi}\iint\limits_{-\infty}^{+\infty} \mathcal{E}_x^2(\xi,\eta)\ln(x_0^2 + y_0^2)\mathrm{d}\,\xi\,\mathrm{d}\,\eta\,,$$

such that

$$\mathcal{E} = +\frac{k}{2\pi}\iint\limits_{-\infty}^{+\infty} \mathcal{E}_x^2(\xi,\eta)\frac{x_0}{x_0^2 + y_0^2}\mathrm{d}\,\xi\,\mathrm{d}\,\eta + \frac{m}{4\pi}\ln(x^2 + y^2)\,. \tag{9.2.56}$$

Replacing the wing with a continuous distribution of perturbation sources defined on the chord of the profile (only in the symmetric case the sources distribution is sufficient), from (9.2.56) it results, in the domain occupied by the fluid, the following general representation of the solution of the equation (9.2.53):

$$\varphi(x,y) = \frac{1}{4\pi}\int_{-1}^{+1} m(\xi)\ln(x_0^2 + y^2)\mathrm{d}\,\xi + \frac{k}{2\pi}\iint_{\mathbb{R}^2} u^2(\xi,\eta)\frac{x_0}{x_0^2 + y_0^2}\mathrm{d}\,\xi\,\mathrm{d}\,\eta\,, \tag{9.2.57}$$

where $m(\xi)$ is a function which has to be determined by the shape of the profile, i.e. by the boundary condition

$$v(x,\pm 0) = \pm h'(x),\quad x \in [0,1]\,, \tag{9.2.58}$$

which is imposed by (9.2.34).

From (9.2.57) we deduce

$$
\begin{aligned}
u = \varphi_x &= \frac{1}{2\pi}\int_{-1}^{+1} m(\xi)\frac{x_0}{x_0^2+y^2}\,\mathrm{d}\,\xi + \frac{k}{2\pi}\frac{\partial}{\partial x}I \\
v = \varphi_y &= \frac{1}{2\pi}\int_{-1}^{+1} m(\xi)\frac{y}{x_0^2+y^2}\,\mathrm{d}\,\xi + \frac{k}{2\pi}\frac{\partial}{\partial y}I\,,
\end{aligned}
\tag{9.2.59}
$$

where

$$
I = \iint_{\mathbb{R}^2} u^2(\xi,\eta)\frac{x_0}{x_0^2+y_0^2}\,\mathrm{d}\,\xi\,\mathrm{d}\,\eta\,. \tag{9.2.60}
$$

The singular integral I has the shape (E.9). Applying the formula (E.10) we deduce

$$
\begin{aligned}
\frac{\partial}{\partial x}I &= \iint_{\mathbb{R}^2} u^2(\xi,\eta)\frac{\partial}{\partial x}\left(\frac{x_0}{x_0^2+y_0^2}\right)\mathrm{d}\,\xi\,\mathrm{d}\,\eta - \pi u^2(x,y) = \\
&= -\iint_{\mathbb{R}^2} u^2(\xi,\eta)\frac{x_0^2-y_0^2}{(x_0^2+y_0^2)^2}\,\mathrm{d}\,\xi\,\mathrm{d}\,\eta - \pi u^2(x,y)\,, \qquad (9.2.61)\\
\frac{\partial}{\partial y}I &= -2\iint_{\mathbb{R}^2} u^2(\xi,\eta)\frac{x_0y_0}{(x_0^2+y_0^2)^2}\,\mathrm{d}\,\xi\,\mathrm{d}\,\eta\,.
\end{aligned}
$$

Hence we have the following representation of the solution

$$
\begin{aligned}
u(x,y) &= \frac{1}{2\pi}\int_{-1}^{+1} m(\xi)\frac{x_0}{x_0^2+y^2}\,\mathrm{d}\,\xi - \\
&\quad -\frac{k}{2\pi}\iint_{\mathbb{R}^2} u^2(\xi,\eta)\frac{x_0^2-y_0^2}{(x_0^2+y_0^2)^2}\,\mathrm{d}\,\xi\,\mathrm{d}\,\eta - \frac{k}{2\mathrm{i}}u^2(x,y)\,,\\
v(x,y) &= \frac{1}{2\pi}\int_{-1}^{+1} m(\xi)\frac{y}{x_0^2+y^2}\,\mathrm{d}\,\xi - \frac{k}{\pi}\iint_{\mathbb{R}^2} u^2(\xi,\eta)\frac{x_0^2y_0^2}{(x_0^2+y_0^2)^2}\,\mathrm{d}\,\xi\,\mathrm{d}\,\eta
\end{aligned}
\tag{9.2.62}
$$

which is also given in (9.2.14).

The boundary values of the first integrals are given in (3.1.19) and (3.1.20). We obtain therefore

$$
v(x_1 \pm 0) = \pm\frac{1}{2}m(x) + \frac{k}{\pi}\iint_{\mathbb{R}^2} u^2(\xi,\eta)\frac{x_0\eta}{(x_0^2+\eta^2)^2}\,\mathrm{d}\,\xi\,\mathrm{d}\,\eta\,. \tag{9.2.63}
$$

Imposing the boundary conditions (9.2.58) we find

$$\iint_{-\infty}^{+\infty} u^2(\xi,\eta)\frac{x_0\eta}{(x_0^2+\eta^2)^2}\mathrm{d}\,\xi\,\mathrm{d}\,\eta\,, \tag{9.2.64}$$

whence we deduce

$$u^2(\xi,\eta) = u^2(\xi,-\eta) \Longrightarrow u^2(x,y) = u^2(x,-y)\,. \tag{9.2.65}$$

Taking into account (9.2.63) and the previous relation, it results

$$m(x) = 2h'(x)\,. \tag{9.2.66}$$

In this way we determine the distribution m. Coming back to (9.2.62), we obtain the equation:

$$u(x,y) + \frac{k}{2}\,u^2(x,y) + \frac{k}{2\pi}\iint_{\mathbb{R}^2} u^2(\xi,\eta)\frac{x_0^2-y_0^2}{(x_0^2+y_0^2)^2}\mathrm{d}\,\xi\,\mathrm{d}\,\eta = \tag{9.2.67}$$
$$= \frac{1}{\pi}\int_{-1}^{+1} h'(\xi)\frac{x_0}{x_0^2+y^2}\mathrm{d}\,\xi\,,$$

which coincides with (9.2.36).

9.3 The Three-Dimensional Flow

9.3.1 The Fundamental Solution

In the last 40 years, a great number of papers was devoted to the steady transonic flow past thin bodies. Usually one assumes that the flow is irrotational, the potential satisfying a non-linear equation having the form (9.1.13). For deducing the integral equations of the problem, we apply Green's formula to the equation of Poisson and we assume that a vortices layer is present downstream the wing. Derivating, we obtain the non-linear integral system for the components of the velocity (see for example [9.36]).

In the case of the symmetric profiles the system reduces to a single equation for the component $u(x,y,z)$. In this sense, after the initial paper of Oswatitsch [9.50] where one defines a principal value for the singular integral which intervenes in the representation, it followed the paper of Heaslet and Spreiter [9.17] where one gives a general representation which in the symmetric case reduces to an equation. The representation is valid both for the flow with shock waves and the flow without shock waves.

For the lifting wings the forms of Norstrud [9.41] and Nixon [9.39] are available. In this case, the problem reduces to a system of two non-linear integral equations. At last, we mention the paper of Ogana [9.47] where one shows how the integral equations depend on the definition given to the principal value of the singular integrals.

A new point of view, belonging to D. Homentcovschi [9.20], [9.21] and L. Dragoş [9.8] [9.9] does not assume that the flow is potential. Utilizing the system of equations of motion it is necessary to assume the existence of the vortices layer downstream. In the sequel we shall utilize the method of fundamental solutions [9.8].

The system which determines the perturbation is (9.1.42) and (9.1.46). Performing the change of variables

$$y^* = \sqrt{K}\overline{y}, z^* = \sqrt{K}\overline{z}, \qquad u^* = \sqrt{K}\overline{u}\,, \tag{9.3.1}$$

and omitting the marks $*$ and $\wedge$, the system becomes

$$\begin{aligned} &u_y - v_x = 0 \quad u_z - w_x = 0 \\ &u_x + v_y + w_z = k(u^2)_x\,, \end{aligned} \tag{9.3.2}$$

where k has the same significance like in (9.2.6). We shall see further that employing a fundamental solution similar to the fundamental solution of the system

$$\begin{aligned} &u_y - v_x = \ell\delta(x, y, z), \quad u_z - w_x = 0 \\ &u_x + v_y + w_z = k(u^2)_x + m\delta(x, y, z) \end{aligned} \tag{9.3.3}$$

we may satisfy all the conditions of the problem. This solution will be determined in the manner described in 2.3. Applying the Fourier transform, solving the algebraic system just obtained and considering the inverse Fourier transform, on the basis of the formulas from appendix

A, we obtain:

$$u(x,y,z) = -\frac{1}{4\pi}\left(m\frac{\partial}{\partial x} + \ell\frac{\partial}{\partial y}\right)\frac{1}{r} - \frac{k}{4\pi}\frac{\partial^2}{\partial x^2}u^2 * \frac{1}{r},$$

$$v(x,y,z) = \frac{\ell}{4\pi}\frac{\partial}{\partial x}\left(\frac{1}{r}\right) - \frac{m}{4\pi}\frac{\partial}{\partial y}\left(\frac{1}{r}\right) + \ell\frac{\partial}{\partial z}\mathcal{F}^{-1}\left[\frac{\alpha_3}{\alpha_1\alpha^2}\right] -$$

$$-\frac{k}{4\pi}\frac{\partial^2}{\partial x\partial y}u^2 * \frac{1}{r},$$

$$w(x,y,z) = -\frac{m}{4\pi}\frac{\partial}{\partial z}\left(\frac{1}{r}\right) - \ell\frac{\partial}{\partial z}\mathcal{F}^{-1}\left[\frac{\alpha_2}{\alpha_1\alpha^2}\right] - \frac{k}{4\pi}\frac{\partial^2}{\partial x\partial z}u^2 * \frac{1}{r}, \tag{9.3.4}$$

where $r = \sqrt{x^2+y^2+z^2}$ and,

$$u^2 * \frac{1}{r} = \iiint_{\mathbb{R}^3}\frac{u^2(\xi,\eta,\zeta)}{|\boldsymbol{x}-\boldsymbol{\xi}|}\mathrm{d}v\,, \tag{9.3.5}$$

with the notation $\mathrm{d}v = \mathrm{d}\xi\,\mathrm{d}\eta\,\mathrm{d}\zeta$. This integral is called *the convolution* of the functions u^2 and $1/r$. Taking $M = 0$, from the formulas (2.3.11), (2.2.6) and (2.3.27) it results

$$\mathcal{F}^{-1}\left[\frac{\alpha_1}{\alpha_1\alpha^2}\right] = -\frac{\partial}{\partial y}\mathcal{F}^{-1}\left[\frac{1}{\mathrm{i}\,\alpha_1\alpha^2}\right] =$$

$$= \frac{1}{4\pi}\frac{\partial}{\partial y}\int_{-\infty}^{x}\frac{\mathrm{d}x}{r}\mathrm{d}^2x = -\frac{1}{4\pi}\frac{y}{y^2+z^2}\left(1+\frac{x}{r}\right) \tag{9.3.6}$$

and a similar formula. In fact one obtains the following form of the fundamental solution:

$$u(x,y,z) = -\frac{1}{4\pi}\left(m\frac{\partial}{\partial x} + \ell\frac{\partial}{\partial y}\right)\frac{1}{r} - \frac{k}{4\pi}\frac{\partial}{\partial x}I_0\,,$$

$$v(x,y,z) = \frac{1}{4\pi}\left(\ell\frac{\partial}{\partial x} - m\frac{\partial}{\partial y}\right)\frac{1}{r} -$$

$$-\frac{\ell}{4\pi}\frac{\partial}{\partial z}\left[\frac{z}{y^2+z^2}\left(1+\frac{x}{r}\right)\right] - \frac{k}{4\pi}\frac{\partial}{\partial x}J_0\,,$$

$$w(x,y,z) = -\frac{m}{4\pi}\frac{\partial}{\partial z}\left(\frac{1}{r}\right) + \frac{\ell}{4\pi}\frac{\partial}{\partial z}\left[\frac{y}{y^2+z^2}\left(1+\frac{x}{\pi}\right)\right] - \frac{k}{4\pi}\frac{\partial}{\partial x}K_0\,, \tag{9.3.7}$$

where we denoted

$$I_0 = \frac{\partial}{\partial x} u^2 * \frac{1}{r}, \quad J_0 = \frac{\partial}{\partial y} u^2 * \frac{1}{r}, \quad K_0 = \frac{\partial}{\partial z} u^2 * \frac{1}{r}, \tag{9.3.8}$$

ℓ and m being constants.

9.3.2 The Study of the Singular Integrals

The integral (9.3.5) has a weak (integrable) singularity. The integral exists, (it is convergent) (u^2 is zero far away) and it may be derived (the convolution, if it exists may be derived (A.3.7)), such that we have

$$I_0 = u^2 * \frac{\partial}{\partial x}\left(\frac{1}{r}\right) = -u^2 * \frac{x}{|\boldsymbol{x}|^3} = \int_{\mathbb{R}^3} u^2(\boldsymbol{\xi}) \frac{\xi - x}{|\boldsymbol{\xi} - \boldsymbol{x}|^3} \mathrm{d}\, v$$

and similar expressions for J_0 and K_0. Since the integral has the form (E.3), it is convergent. With the notation

$$f = \frac{\xi - x}{|\boldsymbol{\xi} - \boldsymbol{x}|},$$

I_0 has the form (E.9) and may be derived according to the formula (E.10). For calculating the last term one utilizes the spherical coordinates with the center in the point having the vector of position $\boldsymbol{x}$:

$$\xi - x = \sin\theta \cos\varphi$$

$$\eta - y = \sin\theta \sin\varphi$$

$$\zeta - z = \cos\theta\,.$$

One obtains

$$\int_S f \cos(\boldsymbol{n}, \boldsymbol{x}) \mathrm{d}\, a = \frac{4}{3}\pi$$

whence it follows the formula

$$I = \frac{\partial}{\partial x} I_0 = \int_{\mathbb{R}^3} u^2(\boldsymbol{\xi}) \frac{\partial}{\partial x}\left(\frac{\xi - x}{|\boldsymbol{\xi} - \boldsymbol{x}|^3}\right) \mathrm{d}v - \frac{4\pi}{3} u^2(\boldsymbol{x}) =$$

$$= \int_{\mathbb{R}^3} u^2(\boldsymbol{\xi}) \frac{2x_0^2 - y_0^2 - z_0^2}{|\boldsymbol{x} - \boldsymbol{\xi}|^5} \mathrm{d}v - \frac{4\pi}{3} u^2(\boldsymbol{x})\,. \tag{9.3.9}$$

Analogously one demonstrates that

$$J = \frac{\partial}{\partial x} J_0 = \int_{\mathbb{R}^3} u^2(\boldsymbol{\xi}) \frac{3x_0y_0}{|\boldsymbol{x}-\boldsymbol{\xi}|^5} \mathrm{d}v,$$
$$K = \frac{\partial}{\partial x} K_0 = \int_{\mathbb{R}^3} u^2(\boldsymbol{\xi}) \frac{3x_0z_0}{|\boldsymbol{x}-\boldsymbol{\xi}|^5} \mathrm{d}v, \tag{9.3.10}$$

where $x_0 = x-\xi, y_0 = y-\eta, z_0 = z-\zeta$. The integrals we have obtained are convergent if u^2 satisfies Hölder's condition and if its behaviour at infinity is $u^2(\boldsymbol{\xi}) = O(|\boldsymbol{\xi}|^{-\ell})$ with $\ell > 1$.

9.3.3 The General Solution

Denoting by D the projection of the wing on the xOz plane and by

$$y = h(x,z) \pm h_1(x,z), \quad (x,z) \in D \tag{9.3.11}$$

the equations of the wing (which is assumed to be thin), we shall be able to satisfy the conditions of the problem with a continuous superposition of solutions having the form (9.3.7), defined on D. It results the following general representation:

$$u(x,y,z) = -\frac{1}{4\pi} \iint_D \left[m(\xi,\zeta)\frac{\partial}{\partial x} + \ell(\xi,\zeta)\frac{\partial}{\partial y} \right] \left(\frac{1}{R}\right) \mathrm{d}\xi\, \mathrm{d}\zeta - $$
$$ - \frac{k}{4\pi} I(x,y,z), \tag{9.3.12}$$

$$v(x,y,z) = \frac{1}{4\pi} \iint_D \left[\ell(\xi,\zeta)\frac{\partial}{\partial x} - m(\xi,\zeta)\frac{\partial}{\partial y} \right] \left(\frac{1}{R}\right) \mathrm{d}\xi\, \mathrm{d}\zeta - $$
$$ - \frac{1}{4\pi} \iint_D \ell(\xi,\zeta)\frac{\partial}{\partial z} \left[\frac{z_0}{y^2+z_0^2}\left(1+\frac{x_0}{R}\right)\right] \mathrm{d}\xi\, \mathrm{d}\zeta - \tag{9.3.13}$$
$$ - \frac{k}{4\pi} J(x,y,z),$$

$$w(x,y,z) = -\frac{1}{4\pi} \iint_D m(\xi,\zeta)\frac{\partial}{\partial z}\left(\frac{1}{R}\right) + $$
$$ + \frac{1}{4\pi} \iint_D \ell(\xi,\zeta)\frac{\partial}{\partial z} \left[\frac{y}{y^2+z_0^2}\left(1+\frac{x_0}{R}\right)\right] \mathrm{d}\xi\, \mathrm{d}\zeta, \tag{9.3.14}$$

where we denoted

$$x_0 = x - \xi, \quad z_0 = z - \zeta, \quad R = \sqrt{x_0^2 + y^2 + z_0^2}\,. \tag{9.3.15}$$

Taking the formulas (5.1.16), (5.1.18) and (5.1.24) into account, it results

$$u(x, \pm 0, z) = \pm\frac{1}{2}\ell(x, z) + \frac{1}{4\pi}\overset{\odot}{\iint_D} m(\xi, \zeta)\frac{x_0}{R_0^3}\mathrm{d}\,\xi\,\mathrm{d}\,\zeta - \frac{k}{4\pi}I(x, \pm 0, z) \tag{9.3.16}$$

$$v(x, \pm 0, z) = \pm\frac{1}{2}m(x, z) + \frac{1}{4\pi}\overset{*}{\iint_D} \frac{\ell(\xi, \zeta)}{(z - \zeta)^2}\left(1 + \frac{x_0}{R_0}\right)\mathrm{d}\,\xi\,\mathrm{d}\,\zeta - \\ - \frac{k}{4\pi}J(x, \pm 0, z) \tag{9.3.17}$$

where

$$R_0 = \sqrt{x_0^2 + z_0^2} \tag{9.3.18}$$

and the mark $*$ indicates the Finite Part like in (5.1.24). From (9.1.42) and (9.3.16) we deduce the significance of the function

$$\ell(x, z) : \ell(x, z) = p(x, -0, z) - p(x, +0, z)\,. \tag{9.3.19}$$

Hence, $\ell(x, z)$ gives the jump of the pressure. This function will be utilized for calculating the aerodynamic action.

From the expression of $v(x, \pm 0, z)$ and from the boundary condition

$$v(x, \pm 0, z) = h'(x, z) \pm h_1'(x, z) \quad (x, z) \in D\,, \tag{9.3.20}$$

where the mark "prime" indicates the derivative with respect to the x variable, it results after subtracting and adding

$$m(x, z) = 2h_1'(x, z)\,, \tag{9.3.21}$$

$$\frac{1}{4\pi}\overset{*}{\iint_D} \frac{\ell(\xi, \zeta)}{z_0^2}\left(1 + \frac{x_0}{R_0}\right)\mathrm{d}\,\xi\,\mathrm{d}\,\zeta + \\ + \frac{k}{2\pi}\int_{\mathbb{R}^3} u^2(\boldsymbol{\xi})\frac{3x_0\eta}{(x_0^2 + \eta^2 + z_0^2)^{5/2}}\mathrm{d}\,v = h'(x, z)\,,\ (x, z) \in D\,. \tag{9.3.22}$$

The formula (9.3.21) determines directly the unknown $m(x,z)$. In the equation for $\ell(x,z)$ it intervenes the values of u^2 in $\mathbb{R}^3$. They are obtained from (9.3.12) after replacing m by (9.3.21).

We deduce

$$\begin{aligned} u(\boldsymbol{x}) - \frac{k}{3}u^2(\boldsymbol{x}) + \frac{k}{4\pi}\int_{\mathbb{R}^3} u^2(\boldsymbol{\xi})\frac{2x_0^2 - y_0^2 - z_0^2}{|\boldsymbol{x}-\boldsymbol{\xi}|^5}\mathrm{d}\,v - \\ -\frac{1}{4\pi}\int_D \ell(\xi,\zeta)\frac{y}{R^3}\mathrm{d}\,\xi\,\mathrm{d}\,\eta = \frac{1}{2\pi}\int_D h_1'(\xi,\zeta)\frac{x_0}{R^3}\mathrm{d}\,\xi\,\mathrm{d}\,\zeta\,. \end{aligned} \tag{9.3.23}$$

Hence, for determining the unknown $\ell(x,z)$ on D we have to solve the system consisting of the equations (9.3.22) and (9.3.23) where $u(\boldsymbol{x})$ is defined on $\mathbb{R}^3$. For $u(x,\pm 0,z)$ we shall utilize the values (9.3.16). The mathematical problem is extremely difficult and there are not known any attempts for solving it.

For the symmetric wing $(h=0)$, the solution is obtained for $\ell=0$ and $u(x,y,z)=u(x,-y,z)$ if

$$\begin{aligned} u(x^2) - \frac{k}{3}u^2(\boldsymbol{x}) + \frac{k}{4\pi}\int_{\mathbb{R}^3} u^2(\boldsymbol{\xi})\frac{2x_0^2 - y_0^2 - z_0^2}{|\boldsymbol{x}-\boldsymbol{\xi}|^5}\mathrm{d}\,v = \\ = \frac{1}{2\pi}\int_D h_1'(\xi,\zeta)\frac{x_0}{R^3}\mathrm{d}\,\xi\,\mathrm{d}\,\zeta\,. \end{aligned} \tag{9.3.24}$$

9.3.4 Flows with Shock Waves

In the case of the flow with shock waves, the general solution has also the form (9.3.12)–(9.3.14). We can see it in the simplest way if we utilise the notion of Fourier transform for bounded domains, introduced by D.Homentcovschi [9.19]. Indeed, in the fluid domain D the equations

$$\begin{aligned} v_x - u_y = 0\,, \quad w_x - u_z = 0\,, \\ u_z + v_y + w_z = k(u^2)_x\,, \end{aligned} \tag{9.3.25}$$

with the notations from (9.3.2) have to be satisfied. On the shock waves Σ one imposes the relations

$$\begin{aligned} [\![v]\!]n_x - [\![u]\!]n_y = 0\,, \quad [\![w]\!]n_x - [\![u]\!]n_z = 0\,, \\ [\![u]\!]n_x + [\![v]\!]n_y + [\![w]\!]n_z = k[\![u^2]\!]n_x\,, \end{aligned} \tag{9.3.26}$$

deduced from (9.1.49), and on the borders S_+ (upper surface) and S_- (lower surface), the conditions (9.3.20).

Applying the Fourier transform for bounded domains, we shall utilize the formulas of the type (A.8.1). From (9.3.25) we deduce

$$\begin{aligned} -\mathrm{i}\,\alpha_1\widehat{v} + \mathrm{i}\,\alpha_2\widehat{u} &= S_1 + T_1\,, \\ -\mathrm{i}\,\alpha_1\widehat{w} + \mathrm{i}\,\alpha_3\widehat{u} &= S_2 + T_2\,, \\ -\mathrm{i}\,\alpha_1\widehat{u} - \mathrm{i}\,\alpha_2\widehat{v} - \mathrm{i}\,\alpha_3\widehat{w} + k\mathrm{i}\,\alpha_1\widehat{u^2} &= S_3 + T_3\,, \end{aligned} \tag{9.3.27}$$

where, taking into account that on S_+ we have $\boldsymbol{n} = (0,1,0)$, and on S_-, $\boldsymbol{n} = (0,-1,0)$

$$\begin{aligned} S_1 &= \int_{S_+ + S_-} (v n_x - u n_y) \mathrm{e}^{\mathrm{i}\,\boldsymbol{\alpha}\cdot\boldsymbol{x}} \mathrm{d}\,a = \\ &= -\int_D [\![u]\!] \mathrm{e}^{\mathrm{i}\boldsymbol{\alpha}\boldsymbol{x}} \mathrm{d}\,a = -\int_D \ell(x,z) \mathrm{e}^{\mathrm{i}(\alpha_1 x + \alpha_3 z)} \mathrm{d}\,a\,, \\ S_2 &= \int_{S_+ + S_-} (w n_x - u n_z) \mathrm{e}^{\mathrm{i}\,\boldsymbol{\alpha}\cdot\boldsymbol{x}} \mathrm{d}\,a = 0\,, \\ S_3 &= \int_{S_+ + S_-} (u n_x + v n_y + w n_z - k u^2 n_x) \mathrm{e}^{\mathrm{i}\,\boldsymbol{\alpha}\cdot\boldsymbol{x}} \mathrm{d}\,a = \\ &= \int_D [\![v]\!] \mathrm{e}^{\mathrm{i}\,\boldsymbol{\alpha}\cdot\boldsymbol{x}} \mathrm{d}\,a = \int_D m(x,z) \mathrm{e}^{\mathrm{i}(\alpha_1 x + \alpha_3 z)} \mathrm{d}\,a\,. \end{aligned} \tag{9.3.28}$$

$\ell(x,z)$ and $m(x,z)$ having the significations from (9.3.19) and (9.3.21).

The integrals

$$T_1 = \int_\Sigma \left([\![v]\!] n_x - [\![u]\!] n_y\right) \mathrm{e}^{\mathrm{i}\,\boldsymbol{\alpha}\cdot\boldsymbol{x}} \mathrm{d}\,a\,,\quad T_2 = \int_\Sigma \left([\![w]\!] n_x - [\![u]\!] n_z\right) \mathrm{e}^{\mathrm{i}\,\boldsymbol{\alpha}\cdot\boldsymbol{x}} \mathrm{d}\,a$$

$$T_3 = \int_\Sigma \left([\![u]\!] n_x + [\![v]\!] n_y + [\![w]\!] n_z - k[\![u^2]\!] n_x\right) \mathrm{e}^{\mathrm{i}\,\boldsymbol{\alpha}\cdot\boldsymbol{x}} \mathrm{d}\,a\,,$$

vanish because of the relations (9.3.26). Hence, the system (9.3.27) reduces to

$$\begin{aligned} -\mathrm{i}\,\alpha_1\widehat{v} + \mathrm{i}\,\alpha_2\widehat{u} &= S_1 \\ -\mathrm{i}\,\alpha_1\widehat{w} + \mathrm{i}\,\alpha_3\widehat{u} &= 0 \\ -\mathrm{i}\,\alpha_1\widehat{u} - \mathrm{i}\,\alpha_2\widehat{v} - \mathrm{i}\,\alpha_3\widehat{w} &= S_3 - k\mathrm{i}\,\alpha_1\widehat{u^2}\,, \end{aligned} \tag{9.3.29}$$

which has the solution

$$\widehat{u} = \frac{-\mathrm{i}\,\alpha_2 S_1 + \mathrm{i}\,\alpha_1 S_3}{\alpha^2} + k\frac{\alpha_1^2}{\alpha^2}\widehat{u^2},$$

$$\widehat{v} = \frac{\mathrm{i}\,\alpha_2}{\alpha^2}S_3 - \frac{\alpha_1^2 + \alpha_3^2}{\mathrm{i}\,\alpha_1\alpha^2}S_1 + k\frac{\alpha_1\alpha_2}{\alpha^2}\widehat{u^2}, \tag{9.3.30}$$

$$\widehat{w} = \frac{\mathrm{i}\,\alpha_3}{\alpha^2}S_3 - \frac{\mathrm{i}\,\alpha_3}{\mathrm{i}\,\alpha_1\alpha^2}S_1 + k\frac{\alpha_1\alpha_3}{\alpha^2}\widehat{u^2},$$

where $\alpha^2 = \alpha_1^2 + \alpha_2^2 + \alpha_3^2$.

Considering the inverse Fourier transform and utilizing the formulas (A.6.9) we obtain

$$u = \frac{\partial}{\partial y}\mathcal{F}^{-1}\left[\frac{S_1}{\alpha^2}\right] - \frac{\partial}{\partial x}\mathcal{F}^{-1}\left[\frac{S_3}{\alpha^2}\right] - k\frac{\partial^2}{\partial x^2}\mathcal{F}^{-1}\left[\frac{\alpha_2}{\alpha^2}\right],$$

$$v = -\frac{\partial}{\partial y}\mathcal{F}^{-1}\left[\frac{S_3}{\alpha^2}\right] - \frac{\partial}{\partial x}\mathcal{F}^{-1}\left[\frac{S_1}{\alpha^2}\right] + \frac{\partial^2}{\partial z^2}\mathcal{F}^{-1}\left[\frac{S_1}{\mathrm{i}\,\alpha_1\alpha^2}\right], \tag{9.3.31}$$

$$w = -\frac{\partial}{\partial z}\mathcal{F}^{-1}\left[\frac{S_3}{\alpha^2}\right] + \frac{\alpha^2}{\partial z\partial y}\mathcal{F}^{-1}\left[\frac{S_1}{\mathrm{i}\,\alpha_1\alpha^2}\right] -$$

$$-k\frac{\partial^2}{\partial z\partial x}\mathcal{F}^{-1}\left[\frac{\widehat{u^2}}{\alpha^2}\right] - k\frac{\partial^2}{\partial x\partial y}\mathcal{F}^{-1}\left[\frac{\widehat{u^2}}{\alpha^2}\right].$$

By direct calculations, we deduce

$$\mathcal{F}^{-1}\left[\frac{S_1}{\alpha^2}\right] =$$

$$= \frac{1}{(2\pi)^3}\frac{\partial}{\partial y}\int_{\mathbb{R}^3}\frac{1}{\alpha^2}\left[-\int_D \ell(\xi,\zeta)\mathrm{e}^{\mathrm{i}\,(\alpha_1\xi+\alpha_3\zeta)}\mathrm{d}\,\xi\,\mathrm{d}\,\zeta\right]\mathrm{e}^{-\mathrm{i}(\alpha_1 x+\alpha_2 y+\alpha_3 z)}\mathrm{d}\,\boldsymbol{\alpha} =$$

$$= -\int_D \ell(\xi,\zeta)\left[\frac{1}{(2\pi)^3}\frac{\partial}{\partial y}\int_{\mathbb{R}^3}\frac{\mathrm{e}^{-\mathrm{i}\,[\alpha_1(x-\xi)+\alpha_2 y+\alpha_3(z-\zeta)]}}{\alpha^2}\mathrm{d}\,\boldsymbol{\alpha}\right]\mathrm{d}\xi\,\mathrm{d}\zeta =$$

$$= -\frac{1}{4\pi}\int_D \ell(\xi,\zeta)\frac{\partial}{\partial y}\left(\frac{1}{R}\right)\mathrm{d}\,\xi\,\mathrm{d}\,\zeta. \tag{9.3.32}$$

where, with the notation

$$R = \sqrt{x_0^2 + y^2 + z_0^2}, \tag{9.3.33}$$

we utilized the formula (A.7.10). From (2.3.11) and (9.3.27), it also results

$$\frac{\partial}{\partial z}\mathcal{F}^{-1}\left[\frac{1}{\mathrm{i}\,\alpha_1\alpha^2}\right] = \frac{1}{4\pi}\frac{\partial}{\partial z}\int_{-\infty}^{x}\frac{\mathrm{d}\,x}{\sqrt{x^2+y^2+z^2}} =$$

$$= -\frac{1}{4\pi}\int_{-\infty}^{x}\frac{z\mathrm{d}\,x}{(x^2+y^2+z^2)^{3/2}} = -\frac{1}{4\pi}\frac{z}{y^2+z^2}\left(1+\frac{x}{r}\right). \tag{9.3.34}$$

On the basis of this formula we deduce

$$\frac{\partial}{\partial z}\mathcal{F}^{-1}\left[\frac{S_1}{\mathrm{i}\,\alpha_1\alpha^2}\right] =$$

$$= \frac{1}{(2\pi)^3}\frac{\partial}{\partial z}\int_{\mathbb{R}^3}\frac{1}{\mathrm{i}\,\alpha_1\alpha^2}\left[-\int_D \ell(\xi,\zeta)\mathrm{e}^{\mathrm{i}\,(\alpha_1\xi+\alpha_3\zeta)}\mathrm{d}\,a\right]\mathrm{e}^{-\mathrm{i}\,(\alpha_1 x+\alpha_2 y+\alpha_3 z)}\mathrm{d}\,\boldsymbol{\alpha} =$$

$$= -\frac{1}{(2\pi)^3}\int_D \ell(\xi,\zeta)\left[\frac{\partial}{\partial z}\int_{\mathbb{R}^3}\frac{\mathrm{e}^{-\mathrm{i}\,[\alpha_1(x-\xi)+\alpha_2 y+\alpha_3(z-\xi)]}}{\mathrm{i}\,\alpha_1\alpha^2}\mathrm{d}\,\boldsymbol{\alpha}\right]\mathrm{d}\,\xi\,\mathrm{d}\,\zeta =$$

$$= \frac{1}{4\pi}\int_D \ell(\xi,\zeta)\frac{z_0}{y^2+z_0^2}\left(1+\frac{x_0}{R}\right)\mathrm{d}\,\xi\,\mathrm{d}\,\zeta\,. \tag{9.3.35}$$

At last, taking into account the definition of the convolution product, it results

$$\mathcal{F}^{-1}\left[\frac{\widehat{u^2}}{\alpha^2}\right] = \frac{1}{4\pi}\int_{\mathbb{R}^3}\frac{u^2(\boldsymbol{\xi})}{|\boldsymbol{x}-\boldsymbol{\xi}|}\mathrm{d}\,v\,. \tag{9.3.36}$$

With these formulas and with the similar ones it is not difficult to see that in (9.3.31) we have just the solution (9.3.12)–(9.3.14).

9.4 The Lifting Line Theory

9.4.1 The Velocity Field

The lifting line theory in the transonic flow is studied in [9.55] and [9.8]. In the last reference, it is obtained, as it is natural to do, from the lifting surface theory. This method is also utilized herein.

We shall deduce the equations of the lifting line theory from the lifting surface equations using the assumptions $1^0, 2^0, 3^0$ (Prandtl's hypotheses) from 6.1. Hence we shall take $h_1 = 0$ and we shall consider that the unknown is the circulation

$$C(\zeta) = + \int_{x_-(\zeta)}^{x_+(\zeta)} \ell(\xi, \zeta) \mathrm{d}\,\xi\,, \tag{9.4.1}$$

$$C(\pm c) = 0 \tag{9.4.2}$$

and we shall utilize the formula

$$\begin{aligned} \lim_{x_-(\zeta)\to 0 \leftarrow x_+(\zeta)} \iint_D \ell(\xi, \zeta) k(x, y, z, \xi, \zeta) \mathrm{d}\,\xi\, \mathrm{d}\,\zeta = \\ = \int_{-c}^{+c} C(\zeta) k(x, y, z, 0, \zeta) \mathrm{d}\,\zeta\,, \end{aligned} \tag{9.4.3}$$

$2c$ representing the span of the wing on the direction of the Oz axis. On the basis of the first hypothesis $(h_1 = 0)$, from (9.3.21) it results $m = 0$ and the representation

$$\begin{aligned} u(x, y, z) &= \frac{1}{4\pi} \iint_D \ell(\xi, \zeta) \frac{y}{R^3} \mathrm{d}\,\xi\, \mathrm{d}\,\zeta - \frac{k}{4\pi} I(x, y, z) \\ v(x, y, z) &= -\frac{1}{4\pi} \iint_D \ell(\xi, \zeta) \frac{x_0}{R^3} \mathrm{d}\,\xi\, \mathrm{d}\,\zeta - \\ &\quad - \frac{1}{4\pi} \iint_D \ell(\xi, \zeta) \frac{\partial}{\partial z} \left[\frac{z_0}{y^2 + z_0^2} \left(1 + \frac{x_0}{R}\right) \right] \mathrm{d}\,\xi\, \mathrm{d}\,\zeta - \frac{k}{4\pi} J(x, y, z)\,, \\ w(x, y, z) &= \frac{1}{4\pi} \iint_D \ell(\xi, \zeta) \frac{\partial}{\partial z} \left[\frac{y}{y^2 + z_0^2} \left(1 + \frac{x_0}{R}\right) \right] \mathrm{d}\,\xi\, \mathrm{d}\,\zeta - \frac{k}{4\pi} I(x, y, z) \end{aligned} \tag{9.4.4}$$

or, using also the formulas (9.4.1) and (9.4.3):

$$\begin{aligned} u(x, y, z) &= \frac{1}{4\pi} \int_{-c}^{+c} C(\zeta) \frac{y}{R_1^3} \mathrm{d}\,\zeta - \frac{k}{4\pi} I(x, y, z)\,, \\ v(x, y, z) &= -\frac{1}{4\pi} \int_{-c}^{+c} C(\zeta) \frac{x}{R_1^3} \mathrm{d}\,\zeta - \\ &\quad - \frac{1}{4\pi} \int_{-c}^{+c} C'(\zeta) \frac{z_0}{y^2 + z_0^2} \left(1 + \frac{x}{R_1}\right) \mathrm{d}\,\zeta - \frac{k}{4\pi} J(x, y, z)\,, \\ w(x, y, z) &= \frac{1}{4\pi} \int_{-c}^{+c} C(\zeta) \frac{y}{y^2 + z_0^2} \left(1 + \frac{x_0}{R_1}\right) \mathrm{d}\,\zeta - \frac{k}{4\pi} K(x, y, z)\,, \end{aligned} \tag{9.4.5}$$

where we denoted

$$R_1 = \sqrt{x^2 + y^2 + z_0^2}\,. \tag{9.4.6}$$

9.4.2 The Integral Equations

Using the identity

$$\frac{1}{z_0^2}\left(1 + \frac{x_0}{R_0}\right) = -\frac{\partial}{\partial z}\,\frac{x_0 + R_0}{x_0 z_0}\,, \tag{9.4.7}$$

where R_0 is (9.3.18), we deduce for (9.3.22)

$$T \equiv {}^{*}\!\!\iint_D \frac{\ell(\xi,\zeta)}{z_0^2}\left(1 + \frac{x_0}{R_0}\right) \mathrm{d}\,\xi\,\mathrm{d}\,\zeta = -\frac{\partial}{\partial z}\,{}'\!\!\iint_D \ell(\xi,\zeta)\frac{x_0 + R_0}{x_0 z_0}\mathrm{d}\,\xi\,\mathrm{d}\,\zeta =$$

$$= -\frac{\mathrm{d}}{\mathrm{d}\,z}\,{}'\!\!\int_{-c}^{+c} \frac{C(\zeta)}{z_0}\mathrm{d}\,\zeta - \frac{\partial}{\partial z}\,{}'\!\!\iint \ell(\xi,\zeta)\frac{R_0}{x_0 z_0}\mathrm{d}\,\xi\,\mathrm{d}\,\zeta\,.$$

Using the calculations (6.1.11) - (6.1.13) we deduce in the sequel

$$T = -\,{}'\!\!\int_{-c}^{+c} \frac{C'(\zeta)}{z_0}\mathrm{d}\,\zeta - \frac{\partial}{\partial z}\,{}'\!\!\iint \ell(\xi,\zeta)\frac{\operatorname{sign} z_0}{x_0}\mathrm{d}\,\xi\,\mathrm{d}\,\zeta =$$

$$= -\,{}'\!\!\int_{-c}^{+c} \frac{C'(\zeta)}{z_0}\mathrm{d}\zeta - 2\int_{-c}^{+c}\left[\int_{x_-(\zeta)}^{x_+(\zeta)} \frac{\ell(\xi,\zeta)}{x_0}\mathrm{d}\,\xi\right]\delta(z-\zeta)\mathrm{d}\,\zeta = \tag{9.4.8}$$

$$= -\,{}'\!\!\int_{-c}^{+c} \frac{C'(\zeta)}{z_0}\mathrm{d}\,\zeta - 2\,{}'\!\!\int_{x_-(z)}^{x_+(z)} \frac{\ell(\xi,z)}{x_0}\mathrm{d}\,\xi\,.$$

Substituting this in (9.3.22), multiplying the obtained equation by

$$\sqrt{\frac{x - x_-(z)}{x_+(z) - x}}$$

and, integrating with respect to x on the interval $(x_-(z), x_+(z))$, one obtains the following equation:

$$C(z) = \frac{a(z)}{2}\,{}'\!\!\int_{-c}^{+c} \frac{C'(\zeta)}{\zeta - z}\mathrm{d}\,\zeta + \frac{k}{\pi}\int_{\mathbb{R}^3} u^2(\boldsymbol{\xi})S(z,\xi,\eta,\zeta)\mathrm{d}\,v + H(z)\,, \tag{9.4.9}$$

where

$$S(z,\xi,\eta,\zeta) = \int_{x_-(z)}^{x_+(z)} \sqrt{\frac{x - x_-(z)}{x_+(z) - x}} \frac{3x_0\eta \mathrm{d}\, x}{(x_0^2 + \eta^2 + z_0^2)^{5/2}}\,,$$

$$H(z) = -2\int_{x_-(z)}^{x_+(z)} \sqrt{\frac{x - x_-(z)}{x_+(z) - x}} h'(x,z)\mathrm{d}\, x\,. \tag{9.4.10}$$

The equation (9.3.23) for the unknown $C(z)$ is

$$u(\boldsymbol{x}) - \frac{k}{3}u^2(\boldsymbol{x}) + \frac{k}{4\pi}\int_{\mathbb{R}^3} u^2(\boldsymbol{\xi})\frac{2x_0^2 - y_0^2 - z_0^2}{|\boldsymbol{x} - \boldsymbol{\xi}|^5}\mathrm{d}\, v =$$

$$= \frac{y}{4\pi}\int_{-c}^{+c} \frac{C(\zeta)\mathrm{d}\,\zeta}{(x^2 + y^2 + z_0^2)^{3/2}}\,. \tag{9.4.11}$$

We have therefore the equations (9.4.9) and (9.4.11) for the unknowns $C(z)$ and $u(x,y,z)$, C being defined on $[-c,+c]$, and u, on $\mathbb{R}^3$. The continuation of this reasoning may be found in [9.8]. Obviously, we have to consider only numerical solutions.

Chapter 10

The Unsteady Flow

10.1 The Oscillatory Profile in a Subsonic Stream

10.1.1 The Statement of the Problem

As we have already mentioned (see Chapter 2), the general problem of aerodynamics, i.e. the problem concerning the determination of the perturbation produced by an arbitrary moving body, in a fluid whose state is known, is very difficult. A presentation of this subject can be found in the papers of Küssner [10.37], [10.38], [10.39]. We consider in this chapter, the particular case of the wing which is oscillating harmonically in an uniform stream. The problem is important in the flutter theory. For the incompressible fluid the plane problem was investigated by Theodorsen [10.75]. For the compressible, subsonic fluid in a plane flow, the problem was studied at first by Possio [10.59] and then by Dietze [10.14] and Haskind [10.30]. These authors used the potential of accelerations ψ, replacing the body by a doublets distribution defined on the chord of the profile. The integral equation which was obtained made the object of many researches [10.22], [10.23]. In [1.1], [1.3], [1.8], [1.18] one may find syntheses and references.

Considering that the replacement of the wing by a doublets distribution has no physical justification, in [10.15] we deduced the integral equation starting from the idea that the wing must be replaced according to Cauchy's principle by a distribution of forces. In the same paper, we gave a solution for the integral equation for small values of the frequency ($\omega \ll 1$). The present section is written on the basis of this paper.

10.1.2 The Fundamental Solution

We utilize the dimensionless variables x, y, z, t introduced by (2.1.1), and for the velocity field $\boldsymbol{V}_1$ and pressure $\boldsymbol{P}_1$ resulting from the superposition of the perturbation over the basic flow, we set

$$V_1 = U_\infty(i + V), \quad p_1 = P_\infty + \rho_\infty U_\infty^2 P. \tag{10.1.1}$$

Like in the previous sections $U_\infty i$ is the velocity of the unperturbed stream, and p_∞ and ρ_∞ the pressure and the density in that stream. Imposing the condition that these fields verify the equations of motion, it results (see the equations (2.1.26) and (2.1.27)) the system

$$\begin{aligned} &\partial V/\partial t + \partial V/\partial x + \operatorname{grad} P = F, \\ &M^2(\partial P/\partial t + \partial P/\partial x) + \operatorname{div} V = 0, \\ &\lim_{|x|\to\infty} (V, P) = 0, \end{aligned} \tag{10.1.2}$$

where F is the force density assumed to be small (the system is the result of a linearization), and M is Mach's number for the unperturbed stream.

If the uniform flow of the fluid is perturbed by harmonic forces having the form

$$f \cos(\mathrm{i}\omega t), \quad f \sin(\mathrm{i}\omega t),$$

applied in the origin of the axes of coordinates, then we shall put in (10.1.2)

$$F = f\delta(x)\mathrm{e}^{\mathrm{i}\omega t}. \tag{10.1.3}$$

Solving the system, we shall obtain that the real part of the solution is determined by the force $f\cos(\mathrm{i}\omega t)$, and the imaginary part by $f\sin(\mathrm{i}\omega t)$.

The perturbation produced by (10.1.3) will obviously have the shape

$$V = v(x)\mathrm{e}^{\mathrm{i}\omega t}, \quad P = p(x)\mathrm{e}^{\mathrm{i}\omega t}. \tag{10.1.4}$$

Replacing in 10.1.2 one obtains the system

$$\begin{aligned} &\mathrm{i}\omega v + \partial v/\partial x + \operatorname{grad} p = f\delta(x), \\ &M^2(\mathrm{i}\omega p + \partial p/\partial x) + \operatorname{div} v = 0, \\ &\lim_{|x|\to\infty} (v, p) = 0. \end{aligned} \tag{10.1.5}$$

The solutions of this system are given in 2.4 (they are the solutions of the system (2.4.3)).

In the case of the subsonic, two-dimensional flow, assuming that f has the form $(0, f)$, it results from (2.4.6)

$$p(x,y) = -f\frac{\partial}{\partial y}G_0(x,y), \tag{10.1.6}$$

and from (2.4.16) and (2.4.21)

$$v(x,y) = fe^{-i\omega x}\left[2i\omega G - \beta^2 G_x + \omega^2\int_{-\infty}^{x} G(\tau,y)\,d\tau\right]. \tag{10.1.7}$$

We denoted

$$\begin{aligned} G_0(x,y) &= \frac{1}{4i\beta}H_0^{(2)}\left(k\sqrt{x^2+\beta^2y^2}\right)e^{iax} \\ G(x,y) &= \frac{1}{4i\beta}H_0^{(2)}\left(k\sqrt{x^2+\beta^2y^2}\right)e^{i\overline{\omega}x} \end{aligned} \tag{10.1.8}$$

where

$$k = \overline{\omega}M, \quad \overline{\omega} = \omega/\beta^2, \quad a = kM \tag{10.1.9}$$

$H_0^{(2)}$ being Hankel's function.

10.1.3 The Integral Equation

Assuming that the perturbation of the fluid is determined by an oscillatory wing having the equation

$$y = h(x)e^{i\omega t}, \quad |x| \le 1, \tag{10.1.10}$$

from (2.1.29) it results the following boundary condition

$$v(x,0) = h'(x) + i\omega h(x) \equiv H(x), \quad |x| \le 1, \tag{10.1.11}$$

Let us replace the action of this wing against the fluid by the action of a continuous forces distribution having the form $(0, f)$ defined on the interval $[-1, +1]$. From (10.1.6) and (10.1.7) it results the following general representation of the perturbation

$$p(x,y) = -\int_{-1}^{+1} f(\xi)\frac{\partial}{\partial y}G_0(x_0,y)\,d\xi, \tag{10.1.12}$$

$$\begin{aligned} v(x,y) = \int_{-1}^{+1} f(\xi)e^{-i\omega x_0}[2i\omega G(x_0,y) - \beta^2 G_x(x_0,y) + \\ +\omega^2\int_{-\infty}^{x_0} G(\tau,y)d\tau]d\xi, \end{aligned} \tag{10.1.13}$$

where $x_0 = x - \xi$. The function f has to be determined from the condition (10.1.11).

For obtaining the limit values $p(x,0)$ and $v(x,0)$, we shall take into account that Hankel's functions $H_0^{(2)}(u)$ and $H_1^{(2)}(u)$ satisfy the relation

$$\frac{\mathrm{d}}{\mathrm{d}u}H_0^{(2)}(u) = -H_1^{(2)}(u)\,, \tag{10.1.14}$$

and have the following asymptotic behaviour for small values of the argument [1.16], [1.40]:

$$H_0^{(2)}(u) \equiv \Gamma - \frac{2\mathrm{i}}{\pi}\ln u\,,\ \Gamma = 1 - \frac{2\mathrm{i}}{\pi}(C - \ln 2)\,, \tag{10.1.15}$$

$$H_1^{(2)}(u) = \frac{2\mathrm{i}}{\pi}\frac{1}{u}\,, \tag{10.1.16}$$

where $C(= \ln\gamma)$ is Euler's constant $(= 0.577215)$. We deduce

$$p(x,y) = \frac{\beta k}{4\mathrm{i}}\,y\int_{-1}^{+1} f(\xi)H_1^{(2)}(k\sqrt{x_0^2+\beta^2y^2})\frac{\exp(\mathrm{i}ax_0)}{\sqrt{x_0^2+\beta^2y^2}}\,\mathrm{d}\,\xi\,.$$

When we calculate $p(x,0)$ we observe that, because of the presence of the factor y, the product $y\int_{-1}^{+1}$ vanishes excepting the vicinity $(x-\epsilon, x+\epsilon)$ where the integrand becomes infinite (for $\xi = x$). If ϵ is small enough, in this vicinity the function $f(\xi)\exp(\mathrm{i}ax_0)$ may be approximated by its value in the middle of the interval (we assume that $f(\xi)$ is continuous), i.e. with $f(x)$. It results

$$p(x,\pm 0) = \frac{\beta k}{4\mathrm{i}}\,f(x)\lim_{y\to\pm 0} y\int_{x-\epsilon}^{x+\epsilon} H_1^{(2)}(k\sqrt{x_0^2+\beta^2y^2})\frac{\mathrm{d}\,\xi}{\sqrt{x_0^2+\beta^2y^2}}\,,$$

and, performing the substitution $\xi \to t:\ \xi - x = t$, and taking (10.1.16) into account,

$$p(x,\pm 0) = \frac{\beta}{2\pi}\,f(x)\lim_{y\to\pm 0} y\int_{-\epsilon}^{+\epsilon}\frac{\mathrm{d}\,t}{t^2+\beta^2y^2} = \pm\frac{1}{2}\,f(x)\,. \tag{10.1.17}$$

Hence we obtain the significance of the function $f(x)$. We have

$$p(x,+0) - p(x,-0) = f(x)\,. \tag{10.1.18}$$

The component $\mathrm{v}(x,y)$ is the sum of three terms $\mathrm{v}_1, \mathrm{v}_2, \mathrm{v}_3$, which are represented as follows

$$\mathrm{v}_1(x,y) \equiv \int_{-1}^{+1} f(\xi)\mathrm{e}^{-\mathrm{i}\omega x_0}\,G(x_0,y)\mathrm{d}\,\xi = $$
$$= \frac{1}{4\mathrm{i}\beta}\int_{-1}^{+1} f(\xi)\mathrm{e}^{-\,\mathrm{i}\omega x_0}\,H_0^{(2)}(k\sqrt{x_0^2+\beta^2y^2})\mathrm{e}^{\mathrm{i}\overline{\omega}x_0}\,\mathrm{d}\,\xi\,,$$

$$\mathrm{v}_1(x,0)=\frac{1}{4\mathrm{i}\beta}\int_{-1}^{+1} f(\xi)\mathrm{e}^{-\mathrm{i}\omega x_0} H_0^{(2)}(k|x_0|)\mathrm{e}^{\mathrm{i}\overline{\omega}x_0}\,\mathrm{d}\,\xi\,, \tag{10.1.19}$$

the singularity from $H_0^{(2)}$ being integrable. We have also,

$$\mathrm{v}_2(x,y)\equiv\int_{-1}^{+1} f(\xi)\mathrm{e}^{-\,\mathrm{i}\omega x_0}\frac{\partial}{\partial x}G(x_0,y)\,\mathrm{d}\,\xi=$$

$$=-\frac{k}{4\mathrm{i}\beta}\int_{-1}^{+1} f(\xi)\mathrm{e}^{-\mathrm{i}\omega x_0}H_1^{(2)}(k\sqrt{x_0^2+\beta^2y^2})\frac{x_0}{\sqrt{x_0^2+\beta^2y^2}}\mathrm{e}^{\mathrm{i}\overline{\omega}x_0}\mathrm{d}\,\xi+$$

$$+\frac{\overline{\omega}}{4\beta^3}\int_{-1}^{+1} f(\xi)\mathrm{e}^{-\mathrm{i}\omega x_0}H_0^{(2)}(k\sqrt{x_0^2+\beta^2y^2})\mathrm{e}^{\mathrm{i}\overline{\omega}x_0}\,\mathrm{d}\,\xi\,,$$

and taking (10.1.16) into account ,

$$\begin{aligned}\mathrm{v}_2(x,0)=&-\frac{k}{4\mathrm{i}\beta}{\int_{-1}^{+1}}' f(\xi)\mathrm{e}^{-\mathrm{i}\omega x_0} H_1^{(2)}(k|x_0|)\frac{x_0}{|x_0|}\,\mathrm{e}^{\mathrm{i}\overline{\omega}x_0}\,\mathrm{d}\,\xi\\ &+\frac{\omega}{4\beta^3}\int_{-1}^{+1} f(\xi)\mathrm{e}^{-\mathrm{i}\omega x_0} H_0^{(2)}(k|x_0|)\,\mathrm{e}^{\mathrm{i}\overline{\omega}x_0}\,\mathrm{d}\,\xi\,,\end{aligned} \tag{10.1.20}$$

From

$$\mathrm{v}_3(x,y)\equiv\int_{-1}^{+1} f(\xi)\mathrm{e}^{-\,\mathrm{i}\omega x_0}\left[\int_{-\infty}^{x_0} G(\tau,y)\,\mathrm{d}\,\tau\right]\,\mathrm{d}\,\xi\,,$$

one obtains

$$\mathrm{v}_3(x,0)=\frac{1}{4\mathrm{i}\beta}\int_{-1}^{+1} f(\xi)\mathrm{e}^{-\mathrm{i}\omega x_0}\left[\int_{-\infty}^{x_0} H_0^{(2)}(k|\tau|)\,\mathrm{e}^{\mathrm{i}\overline{\omega}\tau}\,\mathrm{d}\,\tau\right]\,\mathrm{d}\,\xi\,,$$

where one performs the change of variable $\tau\to u:\ \overline{\omega}\tau=u$ and one takes into account the formula

$$\int_{-\infty}^{0} H_0^{(2)}(M|u|)\,\mathrm{e}^{\mathrm{i}u}\mathrm{d}\,u=\frac{2}{\pi\beta}\ln\frac{1+\beta}{M}\,, \tag{10.1.21}$$

given in [1.2]. One obtains

$$\mathrm{v}_3(x,0)=\frac{1}{2\mathrm{i}\omega}\int_{-1}^{+1} f(\xi)\mathrm{e}^{-\mathrm{i}\omega x_0}\left[\ln\frac{1+\beta}{M}+\frac{\beta}{2}\int_{0}^{\overline{\omega}x_0} H_0^{(2)}(M|u|)\,\mathrm{e}^{\mathrm{i}u}\,\mathrm{d}\,u\right]\mathrm{d}\,\xi\,, \tag{10.1.22}$$

After all, from (10.1.13) it results

$$\mathrm{v}(x,0)={\int_{-1}^{+1}}' f(\xi)\mathcal{N}_1(x_0)\,\mathrm{d}\,\xi+\int_{-1}^{+1} f(\xi)\mathcal{N}_2(x_0)\,\mathrm{d}\,\xi\,, \tag{10.1.23}$$

where we denoted

$$\mathcal{N}_k(x_0) = n_k(x_0)\,\mathrm{e}^{-\,\mathrm{i}\omega x_0}\,, \quad k = 1,\,2 \tag{10.1.24}$$

$$n_1(x_0) = \frac{\omega M}{4\mathrm{i}\beta} H_1^{(2)}(k|x_0|)\,\frac{x_0}{|x_0|}\,\mathrm{e}^{\mathrm{i}\overline{\omega}x_0}\,,$$

$$n_2(x_0) = \frac{\omega}{4\beta} H_0^{(2)}(k|x_0|)\,\mathrm{e}^{\mathrm{i}\overline{\omega}x_0} \;-\; \frac{\mathrm{i}\omega}{2\pi}\,\ln\frac{1+\beta}{M} - \tag{10.1.25}$$

$$-\frac{\mathrm{i}\omega\beta}{4}\int_0^{\overline{\omega}x_0} H_0^{(2)}(M|u|)\,\mathrm{e}^{\mathrm{i}u}\,\mathrm{d}\,u\,.$$

Imposing the boundary condition 10.1.11 it results the following singular integral equation

$$\int_{-1}^{\prime\,+1} f(\xi)\mathcal{N}(x_0)\,\mathrm{d}\,\xi = H(x)\,, \quad |x| < 1 \tag{10.1.26}$$

where

$$\mathcal{N}(x_0) = \mathcal{N}_1(x_0) + \mathcal{N}_2(x_0) \tag{10.1.27}$$

In fact the kernel $\mathcal{N}$ depends on the variable x_0 and on the parameters M and ω. We deduce therefore that the kernel has the following explicit expression ($a = kM$)

$$\mathcal{N}(x_0, M, \omega) = -\frac{\mathrm{i}\omega M}{4\beta} H_1^{(2)}(k|x_0|)\,\frac{x_0}{|x_0|}\,\mathrm{e}^{\mathrm{i}ax_0} + \frac{\omega}{4\beta} H_0^{(2)}(k|x_0|)\mathrm{e}^{\mathrm{i}ax_0} -$$

$$-\frac{\mathrm{i}\omega\beta}{4}\,\mathrm{e}^{-\mathrm{i}\omega x_0}\int_0^{\overline{\omega}x_0} H_0^{(2)}(M|u|)\,\mathrm{e}^{\mathrm{i}u}\,\mathrm{d}\,u \;-\; \frac{\mathrm{i}\omega}{2\pi}\,\mathrm{e}^{-\mathrm{i}\omega x_0}\,\ln\frac{1+\beta}{M}\,. \tag{10.1.28}$$

The equation (10.1.26) is *Possio's equation* [10.59]. There are a lot of papers devoted to the kernel (10.1.28) in the literature [10.5], [10.22], [10.30].

10.1.4 Considerations on the Kernel

Taking (10.1.16) and (10.1.16) into account, we deduce

$$\mathcal{N}(x_0, M, 0) = \lim_{\omega\to 0} \mathcal{N}(x_0, M, \omega) \;=\; \frac{\beta}{2\pi}\frac{1}{x_0}\,,$$

which is the kernel for *the steady flow.* With this kernel the integral equation (10.1.26) reduces to (3.1.15).

The kernel for the *incompressible* fluid is obtained from (10.1.28) calculating

$$\mathcal{N}(x_0,0,\omega) = \lim_{M\to 0} \mathcal{N}(x_0,M,\omega)\,. \tag{10.1.29}$$

Integrating by parts we transform the formula (10.1.28) into

$$\mathcal{N}(x_0,M,\omega) = \frac{\omega M}{4\beta}\,\mathrm{e}^{\mathrm{i}ax_0}\left[MH_0^{(2)}(k|x_0|) - \mathrm{i}H_1^{(2)}(k|x_0|)\,\frac{x_0}{|x_0|}\right] +$$

$$+\frac{\omega\beta}{4}\,\mathrm{e}^{-\mathrm{i}\omega x_0}\lim_{u\to 0} H_0^{(2)}(M|u|) - \frac{\omega\beta M}{4}\,\mathrm{e}^{-\mathrm{i}\omega x_0}.$$

$$\cdot\int_0^{\overline{\omega}x_0}\frac{u}{|u|}H_1^{(2)}(M|u|)\,\mathrm{e}^{\mathrm{i}u}\,\mathrm{d}\,u - \frac{\mathrm{i}\omega}{2\pi}\,\mathrm{e}^{-\mathrm{i}\omega x_0}\,\ln\frac{1+\beta}{\mathrm{M}}\,.$$

Utilizing now the relations (10.1.15) and (10.1.16) the notations [1.16]

$$\mathrm{Ci}(z) = \ln\gamma z + \int_0^z \frac{\cos u - 1}{u}\,\mathrm{d}\,u = \ln\gamma z + \sum_{n=1}^{\infty}(-1)^n\frac{z^{2n}}{(2n)(2n)!}\,,$$

$$\mathrm{Si}(z) = \int_0^z \frac{\sin u}{u}\,\mathrm{d}\,u = \sum_{n=0}^{\infty}(-1)^n\frac{z^{2n+1}}{(2n+1)(2n+1)!}\,, \tag{10.1.30}$$

called, the first, *integral cosine*, and the second *integral sine*, we obtain

$$\mathcal{N}(x_0,0,\omega) = \frac{1}{2\pi}\frac{1}{x_0} + \frac{\omega}{2\pi\mathrm{i}}\mathrm{e}^{-\mathrm{i}\omega x_0}\left\{\mathrm{Ci}(\omega x_0) + \mathrm{i}\left[\frac{\pi}{2} + \mathrm{Si}(\omega x_0)\right]\right\}. \tag{10.1.31}$$

This is the kernel for the incompressible fluid.

One demonstrates in [10.15] that for small values of the frequency ($\omega \ll 1$), the integral equation (10.1.26) has the form

$$\frac{\beta}{\pi}{\int_{-1}^{+1}}' \frac{f(\xi)}{x-\xi}\,\mathrm{d}\,\xi + \frac{\omega}{\pi\mathrm{i}\beta}\int_{-1}^{+1} f(\xi)(\ln(|x-\xi| + \Gamma))\,\mathrm{d}\,\xi = 2H(x)\,, \tag{10.1.32}$$

where Γ is a constant. This kind of equations are solved in [A.16]. We leave to the reader the task of writing explicitly the solution.

In [10.22] one shows that the general kernel (10.1.28) has the form

$$\mathcal{N}(x_0,M,\omega) = A_0(x_0) + A_1(x_0,M,\omega)\ln(|x_0|) + A_2(x_0,M,\omega)\,, \tag{10.1.33}$$

where

$$A_0 = \frac{\beta}{2\pi}\frac{1}{x_0}\,,\ A_1 = -\frac{\mathrm{i}\omega}{2\pi\beta}B_1(x_0, M, \omega)\mathrm{e}^{-\mathrm{i}\omega x_0}\,,$$
$$A_2 = -\frac{\beta}{2\pi}B_2(x_0, M, \omega)\mathrm{e}^{-\mathrm{i}\omega x_0}\,, \tag{10.1.34}$$

A_1 and A_2 being analytic functions with respect to x_0.

10.2 The Oscillatory Surface in a Subsonic Stream

10.2.1 The General Solution

The problem presented in this subsection was studied in many papers [10.86], [10.87], [10.45], [10.35], [10.83] where the integral equation was obtained by means of the potential of accelerations, replacing the wing by a distribution of doublets. A slightly different investigation was given in [10.12]. We studied this problem in [10.6] utilizing the fundamental solutions method which will be presented in the sequel.

The problem is the following: an uniform stream having the velocity $U_\infty \boldsymbol{i}$, the pressure p_∞ and density ρ_∞, is perturbed by a surface, oscillating according to one of the laws

$$z = h_0(x, y)\cos(\omega t)\,,\ z = h_0 \sin(\omega t)\,,\ (x, y) \in D\,. \tag{10.2.1}$$

One requires to determine the perturbation. One utilizes the dimensionless variables introduced in (2.1.1) and the notations (10.1.1). The problem is simplified if we replace the laws (10.2.1) by

$$z = h_0(x, y)\mathrm{e}^{\mathrm{i}\omega t}\,, \quad (x, y) \in D\,. \tag{10.2.2}$$

In this case the real part of the solution will give the perturbation produced by (10.2.1) and the imaginary part the perturbation produced by (10.2.1b). The boundary condition (2.1.20) and the linearized system (10.1.2) lead to solutions having the form (10.1.4) where the functions v and p are determined by the system (10.1.5) and by the boundary conditions

$$\mathrm{w}(x, y, 0) = \frac{\partial}{\partial x}h_0(x, y) + \mathrm{i}\omega h_0(x, y) \equiv H(x, y)\,,\ (x, y) \in D\,. \tag{10.2.3}$$

The solution of the system (10.1.5) under the assumptions that $\boldsymbol{f} = (0, 0, f)$ and the unperturbed stream is subsonic ($M < 1$) is obtained from (2.4.7) and (2.4.17) as follows

$$p(x, y, z) = -f\frac{\partial}{\partial z}G_0(x, y, z)\,, \tag{10.2.4}$$

$$\mathrm{w}(x,y,z) = f\mathrm{e}^{-\mathrm{i}\omega \mathrm{x}}\left[(2\mathrm{i}\omega - \beta^2\frac{\partial}{\partial \mathrm{x}})\,\mathrm{G} + (\omega^2 - \frac{\partial^2}{\partial \mathrm{y}^2})\int_{-\infty}^{\mathrm{x}} \mathrm{G}(\tau,\mathrm{y},\mathrm{z})\,\mathrm{d}\,\tau\right],$$

where

$$G_0(x,y,z) = \frac{1}{4\pi}\,\frac{\exp\left[\mathrm{i}k(Mx - R_1)\right]}{R_1},$$

$$G(x,y,z) = \frac{1}{4\pi}\,\frac{\exp\left[\mathrm{i}\overline{\omega}(x - M\,R_1)\right]}{R_1}, \tag{10.2.5}$$

$$R_1(x,y,z) = \sqrt{x^2 + \beta^2(y^2 + z^2)}\,.$$

As we already know, the formulas (10.2.4) define the perturbation produced in the uniform stream by the force $(0,0,f)\exp(\mathrm{i}\omega t)$ applied in the origin of the axes of coordinates. Replacing the wing with a continuous distribution of such forces, defined on the domain D, we obtain the following general representation of the perturbation

$$p(x,y,z) = -\iint_D f(\xi,\eta)\frac{\partial}{\partial z}\,G_0(x_0,y_0,z)\,\mathrm{d}\,\xi\,\mathrm{d}\,\eta\,, \tag{10.2.6}$$

$$\begin{aligned}\mathrm{w}(x,y,z) = \iint_D f(\xi,\eta)\,\mathrm{e}^{-\mathrm{i}\omega \mathrm{x}_0}\,[(2\mathrm{i}\omega - \beta^2\frac{\partial}{\partial \mathrm{x}})\,\mathrm{G}(\mathrm{x}_0,\mathrm{y}_0,\mathrm{z}) + \\ +(\omega^2 - \frac{\partial^2}{\partial y^2})\int_{-\infty}^{x} G(\tau,y_0,z)\,\mathrm{d}\,\tau]\,\mathrm{d}\,\xi\,\mathrm{d}\,\eta\,,\end{aligned} \tag{10.2.7}$$

where, as usually, $x_0 = x-\xi\,, y_0 = y-\eta\,$.The function f is the unknown.

10.2.2 The Integral Equation

In order to determine the unknown f, we shall impose the conditions (10.2.3). At first we shall prove that if $f(x,y)$ is a continuous function, then

$$\lim_{z\to\pm 0}\iint_D f(\xi,\eta)\frac{\partial}{\partial z}\,G_0(x_0,y_0,z)\,\mathrm{d}\,\xi\,\mathrm{d}\,\eta = \mp\frac{1}{2}f(x,y)\,, \tag{10.2.8}$$

Indeed, we have

$$p(x,y,\pm 0) = \frac{1}{4\pi}\lim_{z\to\pm 0}\iint_D f(\xi,\eta)\left(\mathrm{i}\omega M + \frac{\beta^2}{R_0}\right)\frac{x\,z}{R_0^2}\,\mathrm{e}^{\mathrm{i}k(\mathrm{M}x_0 - \mathrm{R}_0)}\,\mathrm{d}\,\xi\,\mathrm{d}\,\eta$$

where $R_0 = R_1(x_0,y_0,z)$. We notice that if we set $z = 0$, the integrand will be zero excepting the point $Q(x,y) \in D$. Denoting by D_ϵ the disk having the center Q and the radius ϵ and assuming that ϵ is small enough in order to approximate $f(\xi,\eta)$ with $f(x,y)$ (this is possible if f is continuous) and the exponential with the unity, it results

$$p(x,y,\pm 0) = \frac{1}{4\pi}\lim_{z\to\pm 0}\iint_{D_\epsilon}\frac{z}{R_0^2}\left(\mathrm{i}\omega M + \frac{\beta^2}{R_0}\right)\mathrm{d}\xi\,\mathrm{d}\eta\,.$$

Performing the change of variables $\xi,\eta \to r,\theta$:

$$\xi - x = \beta r\cos\theta\,,\ \ \eta - y = r\sin\theta\,,$$

$$0 < r < \epsilon\,,\ 0 < \theta < 2\pi\,,$$

we deduce

$$\begin{aligned} p(x,y,\pm 0) &= \frac{1}{2\beta}f(x,y)\lim_{z\to\pm 0} z\int_0^\epsilon\left(\mathrm{i}\omega M + \frac{\beta}{\sqrt{r^2+z^2}}\right)\frac{r\,\mathrm{d}r}{r^2+z^2} = \\ &= \pm\frac{1}{2}f(x,y)\,, \end{aligned} \tag{10.2.9}$$

and then

$$p(x,y,+0) - p(x,y,-0) = f(x,y)\,, \tag{10.2.10}$$

The formula (10.2.9) proves (10.2.8) and (10.2.10) gives the significance of the function $f(x,y)$.

In (10.2.7) we may interchange the limit and the derivations (with respect to x and y). It results therefore

$$\mathrm{w}(x,y,0) = \frac{1}{4\pi}\iint_{\mathrm{D}} f(\xi,\eta)\mathrm{e}^{-\mathrm{i}\omega \mathrm{x}_0}\left[n_1 + n_2\right]\mathrm{d}\xi\,\mathrm{d}\eta\,, \tag{10.2.11}$$

where

$$\begin{aligned} n_1 &= \left(2\mathrm{i}\omega - \beta^2\frac{\partial}{\partial x}\right)\frac{\mathrm{e}^{\mathrm{i}\omega \mathrm{x}}}{R}\,, \\ n_2 &= \left(\omega^2 - \frac{\partial^2}{\partial y^2}\right)(E_1 + E_2)\,, \\ E_1 &= \int_{-\infty}^{0}\frac{\mathrm{e}^{\mathrm{i}\overline{\omega}(\tau - \mathrm{M}R_\tau)}}{R_\tau}\,\mathrm{d}\tau = \int_0^\infty\frac{\mathrm{e}^{-\mathrm{i}\overline{\omega}(\tau+\mathrm{M}R_\tau)}}{R_\tau}\,\mathrm{d}\tau\,, \\ E_2 &= \int_0^{x_0}\frac{\mathrm{e}^{\mathrm{i}\overline{\omega}(\tau-\mathrm{M}R_\tau)}}{R_\tau}\,\mathrm{d}\tau\,, \end{aligned} \tag{10.2.12}$$

$$\begin{aligned} &\beta^2 X = (x_0 - MR)\,, \quad R = \sqrt{x_0^2 + \beta^2 y_0^2} \\ &R_\tau = \sqrt{\tau^2 + \beta^2 y_0^2} \end{aligned} \tag{10.2.13}$$

In E_1 we perform the substitution $\tau \to \lambda$:

$$\tau + MR_\tau = \beta^2 \lambda \,, \tag{10.2.14}$$

Using the notation $|y_0| = r$, we deduce

$$\frac{\mathrm{d}\tau}{\sqrt{\tau^2 + \beta^2 r^2}} = \frac{\mathrm{d}\lambda}{\sqrt{\lambda^2 + r^2}} \,, \tag{10.2.15}$$

and then

$$E_1 = \int_{Mr/\beta}^{\infty} \frac{\mathrm{e}^{-\mathrm{i}\omega\lambda}}{\sqrt{\lambda^2 + r^2}} \mathrm{d}\lambda = \int_0^{\infty} \frac{\mathrm{e}^{-\mathrm{i}\omega\lambda}}{\sqrt{\lambda^2 + r^2}} \mathrm{d}\lambda - \int_0^{\frac{Mr}{\beta}} \frac{\mathrm{e}^{-\mathrm{i}\omega\lambda}}{\sqrt{\lambda^2 + r^2}} \mathrm{d}\lambda =$$

$$= K_0(\omega r) - \frac{\mathrm{i}\pi}{2} \left[I_0(\omega r) - L_0(\omega r) \right] - \int_0^{\frac{M}{\beta}} \frac{\mathrm{e}^{-\mathrm{i}\omega r \mu}}{\sqrt{1 + \mu^2}} \mathrm{d}\mu \,, \tag{10.2.16}$$

K_0 and I_0 being Bessel functions, and L_0 the Struve function [1.16]. The expression of E_1 may be derived taking into account that $\partial^2/\partial y^2 = \partial^2/\partial r^2$. Using the relations between Bessel's functions and their derivatives, we deduce

$$\left(\omega^2 - \frac{\partial^2}{\partial y^2} \right) E_1 - = -\frac{\omega}{r} K_1(\omega r) - \frac{\mathrm{i}\pi}{2} \frac{\omega}{r} \left[I_1(\omega r) - L_1(\omega r) \right] -$$

$$-\omega^2 \int_0^{M/\beta} \sqrt{1 + \mu^2} e^{-\mathrm{i}\omega r \mu} \mathrm{d}\mu \,.$$

In E_2 we make the notation $\beta |y_0| = u$ and then the change of variables

$$\tau \to \theta \, : \, \tau = u \sinh \theta \,. \tag{10.2.17}$$

Observing that $\partial^2/\partial y^2 = \beta^2 \partial^2/\partial u^2$, we deduce

$$\frac{\partial^2}{\partial y^2} E_2 = \frac{x_0}{y_0^2} \frac{\mathrm{e}^{\mathrm{i}\omega \mathrm{X}}}{R} + \beta^2 \frac{x_0}{R^3} \mathrm{e}^{\mathrm{i}\omega \mathrm{X}} + \mathrm{i}\omega \mathrm{M} \frac{\mathrm{x}_0}{\mathrm{R}^2} \mathrm{e}^{\mathrm{i}\omega \mathrm{X}} - \frac{\mathrm{i}\omega \mathrm{X}}{\mathrm{y}_0^2} \frac{\mathrm{x}_0}{\mathrm{R}^2} \mathrm{e}^{\mathrm{i}\omega \mathrm{X}} -$$

$$-\frac{\omega^2}{\beta^2} \frac{1}{u^2} \int_0^{x_0} \frac{(\tau - MR_\tau)^2}{R_\tau} \mathrm{e}^{\mathrm{i}\overline{\omega}(\tau - \mathrm{M}\mathrm{R}_\tau)} \, \mathrm{d}\tau \,,$$

where

$$\beta^2 X = x_0 - MR_\tau \,, \; R = \sqrt{x_0^2 + u^2} \,, \; R_\tau = \sqrt{\tau^2 + u^2} \,.$$

Performing inside the integral the change of variable $\tau \to \lambda$:

$$\tau - MR_\tau = \beta^2 \lambda \,, \tag{10.2.18}$$

we notice that the formula (10.2.15) remains valid because it depends only on M^2. One obtains the identity

$$\begin{aligned} E_2 + \frac{1}{\beta^4}\frac{1}{y_0^2}\int_0^{x_0} \frac{(\tau - MR_\tau)^2}{R_\tau} \mathrm{e}^{\mathrm{i}\overline{\omega}(\tau - \mathrm{M}\mathrm{R}_\tau)}\, \mathrm{d}\,\tau = \\ = \frac{1}{y_0^2}\int_{-M|y_0|/\beta}^{X/\beta^2} \sqrt{\lambda^2 + y_0^2}\mathrm{e}^{\mathrm{i}\omega\lambda}\,\mathrm{d}\,\lambda \,, \end{aligned} \tag{10.2.19}$$

Taking these results into account we obtain

$$\begin{aligned} n(x_0, y_0) = n_1 + n_2 = -\frac{x_0}{y_0^2}\frac{\mathrm{e}^{\mathrm{i}\omega \mathrm{X}}}{R} + \mathrm{i}\omega\frac{R - x_0 M}{\beta^2 y_0^2}\mathrm{e}^{\mathrm{i}\omega \mathrm{X}} - \\ -\frac{\omega}{R}K_1(\omega r) - \frac{\mathrm{i}\pi}{2}\frac{\omega}{r}\left[I_1(\omega r) - L_1(\omega r)\right] - \\ -\omega^2 \int_0^{M/\beta} \sqrt{1 + \mu^2}\mathrm{e}^{\mathrm{i}\omega r\mu}\mathrm{d}\mu + \frac{\omega^2}{y_0^2}\int_{-Mr/\beta}^{X/\beta^2} \sqrt{\lambda^2 + r^2}\mathrm{e}^{\mathrm{i}\omega\lambda}\,\mathrm{d}\,\lambda \,, \end{aligned} \tag{10.2.20}$$

where, for the sake of simplicity we maintain the notation $r = |y_0|$, and X and R are given in (10.2.13). Obviously the line $y_0 = 0$ is singular.

Employing this form of the expression $n_1 + n_2$ in (10.2.11) and imposing the boundary condition (10.2.3) we get the following integral equation

$$\frac{1}{4\pi}\iint^{*} f(\xi, \eta)\mathcal{N}(x_0, y_0)\mathrm{d}\,\xi\,\mathrm{d}\,\eta = H(x, y)\,, \quad (x, y) \in D\,, \tag{10.2.21}$$

where

$$\mathcal{N}(x_0, y_0) = \mathrm{e}^{-\mathrm{i}\omega \mathrm{x}_0}\mathrm{n}(\mathrm{x}_0, \mathrm{y}_0) \tag{10.2.22}$$

is the kernel.

Since $\omega \ll 1$ we have the asymptotic expansions

$$K_1(\omega r) = \frac{1}{\omega r} + O(\omega)\,, \; I_1(\omega r) = O(\omega)\,, \; L_1(\omega r) = \frac{2}{\pi} + O(\omega) \tag{10.2.23}$$

we deduce

$$\lim_{\omega \to 0} n = -\frac{1}{y_0^2}\left(1 + \frac{x_0}{R}\right)\,, \tag{10.2.24}$$

i.e. just the kernel from the steady case.

In the case of the incompressible fluid one obtains the kernel setting $M = 0$ in (10.2.20). From (10.2.13) it results

$$X = x_0\,, \; R = \sqrt{x_0^2 + y_0^2}\,, \qquad (10.2.25)$$

whence,

$$n(x_0, y_0) = -\frac{x_0}{y_0^2}\frac{\mathrm{e}^{\mathrm{i}\omega x_0}}{R} + \mathrm{i}\omega\frac{R}{y_0^2}\mathrm{e}^{\mathrm{i}\omega x_0} - \frac{\omega}{\mathrm{r}}\mathrm{K}_1(\omega \mathrm{r}) -$$
$$-\frac{\mathrm{i}\pi}{2}\frac{\omega}{r}\left[I_1(\omega r) - L_1(\omega r)\right] + \frac{\omega^2}{y_0^2}\int_0^{x_0}\sqrt{\lambda^2 + r^2}\mathrm{e}^{\mathrm{i}\omega\lambda}\,\mathrm{d}\,\lambda\,, \qquad (10.2.26)$$

This is the kernel for the *incompressible fluid.*

10.2.3 Other Expressions of the Kernel Function

Because of the Bessel functions, the expression (10.2.20) is considered to be complicated. However it does not contain divergent integrals. The idea concerning the introduction of these functions may be found in [10.86] and it was also taken into consideration in [10.17]. We obtain another expression of the kernel function, starting from the fundamental solution (2.4.13). Indeed, with this one, the component $\mathrm{w}(x, y, z)$ is written as follows

$$\mathrm{w}(x, y, z) = \frac{1}{4\pi}\int\!\!\int_D f(\xi, \eta) n(x_0, y_0, z)\,\mathrm{d}\,\xi\,\mathrm{d}\,\eta\,, \qquad (10.2.27)$$

where

$$n(x_0, y_0, z) = \frac{\partial^2}{\partial z^2}\int_{-\infty}^{x_0}\frac{\mathrm{e}^{\mathrm{i}\overline{\omega}(\tau - \mathrm{M}\mathrm{R}_{1\tau})}}{R_{1\tau}}\mathrm{d}\,\tau \qquad (10.2.28)$$

with the notation

$$R_{1\tau} = \sqrt{\tau^2 + \beta^2(y_0^2 + z^2)}\,. \qquad (10.2.29)$$

The passage to the limit and the derivation with respect to z do not interchange (only the passage to the limit and the derivations with respect to y interchange; for this reason we gave the expression (2.4.23)). Therefore we shall derivate at first with respect to z. The expression

(10.2.28) may be found at Watkins [10.87], Williams [10.89], Dowell and others. Performing the change of variable $\tau \to \lambda$:

$$\beta^2 \lambda = \tau - M R_{1\tau}\,, \tag{10.2.30}$$

we deduce as we have already seen,

$$\frac{\mathrm{d}\tau}{R_\tau} = \frac{\mathrm{d}\lambda}{\sqrt{\lambda^2 + s^2}}\,, \quad s^2 = y_0^2 + z^2\,. \tag{10.2.31}$$

Hence,

$$E(x_0, y_0, z) = \int_{-\infty}^{x_0} \frac{\mathrm{e}^{\mathrm{i}\overline{\omega}(\tau - \mathrm{M R}_{1\tau})}}{R_{1\tau}}\,\mathrm{d}\,\tau = \int_{-\infty}^{X_1} \mathrm{e}^{\mathrm{i}\omega\lambda} \frac{\mathrm{d}\lambda}{\sqrt{\lambda^2 + s^2}}\,, \tag{10.2.32}$$

where

$$\beta^2 X_1 = x_0 - M\sqrt{x_0^2 + \beta^2 s^2}\,. \tag{10.2.33}$$

Derivating we obtain

$$\begin{aligned} \frac{\partial E}{\partial z} &= -\frac{Mz}{\sqrt{x_0^2 + \beta^2 s^2}} \frac{\mathrm{e}^{\mathrm{i}\omega \mathrm{X}_1}}{\sqrt{X_1^2 + s^2}} - z \int_{-\infty}^{X_1} \frac{\mathrm{e}^{\mathrm{i}\omega\lambda}}{\sqrt{\lambda^2 + s^2}}\,\mathrm{d}\,\lambda \\ \lim_{z \to 0} \frac{\partial^2 E}{\partial z^2} &= -\frac{M}{R} \frac{\mathrm{e}^{\mathrm{i}\omega \mathrm{X}_1}}{\sqrt{X_1^2 + y_0^2}} - \int_{-\infty}^{X_1} \frac{\mathrm{e}^{\mathrm{i}\omega\lambda}}{(\lambda^2 + y_0^2)^{3/2}}\,\mathrm{d}\,\lambda\,, \end{aligned} \tag{10.2.34}$$

where

$$R = \sqrt{x_0^2 + \beta^2 y_0^2}\,, \; \beta^2 X_1 = x_0 - MR\,. \tag{10.2.35}$$

Imposing the boundary condition (10.2.3), we get the integral equation
(10.2.21), where

$$n(x_0, y_0) = -\frac{M}{R} \frac{\mathrm{e}^{\mathrm{i}\omega \mathrm{X}_1}}{\sqrt{X_1^2 + y_0^2}} - \int_{-\infty}^{X_1} \frac{\mathrm{e}^{\mathrm{i}\omega\lambda}}{(\lambda^2 + y_0^2)^{3/2}}\,\mathrm{d}\,\lambda\,, \tag{10.2.36}$$

Obviously the line $y_0 = 0$ is singular. This expression of $n(x_0, y_0)$ is simpler than the expression (10.2.20), but here the integral is no longer convergent. We must therefore consider the Finite Part. The expression (10.2.36) is utilized by Ueda and Dowell [10.80], Ando [10.3], etc. Utilizing the identity

$$\frac{1}{\sqrt{X_1^2 + y_0^2}} = \frac{M x_0 + R}{\sqrt{x_0^2 + y_0^2}} = \frac{M}{x_0 - X_1}\,, \tag{10.2.37}$$

one obtains two other expressions for $n(x_0, y_0)$.

For the incompressible fluid ($M = 0$), we deduce

$$n(x_0, y_0) = -\int_{-\infty}^{x_0} \frac{e^{i\omega\lambda}}{(\lambda^2 + y_0^2)^{3/2}}\, d\lambda. \tag{10.2.38}$$

If the unperturbed flow has the sound velocity, we obtain the kernel considering $M \to 0$. Since

$$\lim_{M\to 1} X = \lim_{M\to 1} \frac{x_0 - M\sqrt{x_0^2 + \beta^2 y_0^2}}{\beta^2} = \frac{x_0^2 - y_0^2}{2x_0} \equiv X_1, \tag{10.2.39}$$

it results

$$\mathcal{N}(x_0, y_0) = -e^{-i\omega x_0}\left[\frac{2}{x_0^2 + y_0^2} e^{i\omega X_1} + \int_{-\infty}^{X_1} \frac{e^{i\omega\lambda}}{(\lambda^2 + y_0^2)^{3/2}}\, d\lambda\right]. \tag{10.2.40}$$

This is the kernel for the *sonic flow.*

One may verify the similarity of the representations (10.2.20) and (10.2.40), if we employ the relation

$$\int_{-\infty}^{X} \frac{e^{i\omega\lambda}}{(\lambda^2 + y_0^2)^{3/2}}\, d\lambda = \int_{0}^{X} \frac{e^{i\omega\lambda}}{(\lambda^2 + y_0^2)^{3/2}}\, d\lambda + \frac{\omega}{r} K_1(\omega r) + i\frac{\pi}{2}\frac{\omega}{r}\left[I_1(\omega r) - L_1(\omega r)\right], \tag{10.2.41}$$

which can be easily proved if we take into account the formulas

$$\int_0^\infty \frac{\cos\omega t}{(t^2 + r^2)^{3/2}}\, dt = \frac{\omega}{r} K_1(\omega r),$$
$$\int_0^\infty \frac{\sin\omega t}{(t^2 + r^2)^{3/2}}\, dt = -\frac{\pi}{2}\frac{\omega}{r}\left[I_1(\omega r) - L_1(\omega r)\right] \tag{10.2.42}$$

given in the tables dedicated to Fourier transforms [1.16].

From the integral representation of the function K_0 we deduce the identity

$$\int_{-\infty}^{x_0} \frac{e^{i\overline{\omega}(\tau - MR_{1\tau})}}{(t^2 + r^2)^{3/2}}\, d\tau = 2K_0(\omega\sqrt{y_0^2 + z^2}) - \int_{x_0}^{\infty} \frac{e^{i\overline{\omega}(\tau - MR_{1\tau})}}{R_{1\tau}}\, d\tau, \tag{10.2.43}$$

with the notation (10.2.25). Putting this in 10.2.24) and deriving according to the formula

$$\frac{\partial^2}{\partial z^2} = \frac{\beta^2}{s}\frac{\partial}{\partial s}, \tag{10.2.44}$$

we obtain to the limit

$$n(x_0, y_0) = -\mathrm{e}^{-\mathrm{i}\omega \mathrm{x}_0} \left[\frac{2\omega}{\mathrm{r}} \mathrm{K}_1(\omega \mathrm{r}) - \int_{\mathrm{x}_0}^{\infty} \left(\mathrm{i}\omega \mathrm{M} + \frac{\beta^2}{\mathrm{R}_{0\tau}} \right) \frac{\mathrm{e}^{\mathrm{i}\overline{\omega}(\tau - \mathrm{M}\mathrm{R}_{0\tau})}}{\mathrm{R}_{0\tau}^2} \mathrm{d}\,\tau \right], \tag{10.2.45}$$

where

$$R_{0\tau} = \sqrt{\tau^2 + \beta^2 y_0^2}\,. \tag{10.2.46}$$

This is another expression of the kernel.

10.2.4 The Structure of the Kernel

The identity (10.2.41) makes possible to deduce from (10.2.32) the structure of the kernel in the vicinity of the singular line $y_0 = 0$. Indeed for a small r we have [1.40]

$$\begin{aligned} &K_1(\omega r) = \frac{1}{\omega r} + \frac{\omega r}{2} \ln r + \ldots\,, \\ &I_1(\omega r) = \omega r + \ldots\,,\ L_1(\omega r) = \frac{2}{\pi} + \ldots\,, \end{aligned} \tag{10.2.47}$$

such that

$$\int_{-\infty}^{X} \frac{\mathrm{e}^{\mathrm{i}\omega\lambda}}{(\lambda^2 + y_0^2)^{3/2}}\,\mathrm{d}\,\lambda = \int_0^X \frac{\cos\omega\lambda}{(\lambda^2 + y_0^2)^{3/2}}\,\mathrm{d}\,\lambda + \mathrm{i} \int_0^X \frac{\sin\omega\lambda}{(\lambda^2 + y_0^2)^{3/2}}\,\mathrm{d}\,\lambda + $$
$$+ \frac{1}{r^2} - \frac{\mathrm{i}\omega}{r} + \frac{\omega^2}{2} \ln r + \ldots\,, \tag{10.2.48}$$

the points representing series of integer powers of r.

If $X > 0$ the last two integrals have no singularities. Indeed, using the expansion formulas for

$$\int_{-\infty}^{X} \frac{\mathrm{e}^{\mathrm{i}\omega\lambda}}{(\lambda^2 + y_0^2)^{\nu+1/2}}\,\mathrm{d}\,\lambda \ldots$$

given by Ueda [10.81], we deduce

$$\begin{aligned} &\int_0^X \frac{\cos\omega\lambda}{(\lambda^2 + r^2)^{3/2}}\,\mathrm{d}\,\lambda = \sum_{n=0}^{\infty} \frac{(-1)^n \omega^{2n}}{(2n)!} I_{2n}\,, \\ &\int_0^X \frac{\sin\omega\lambda}{(\lambda^2 + r^2)^{3/2}}\,\mathrm{d}\,\lambda = \sum_{n=0}^{\infty} \frac{(-1)^n \omega^{2n+1}}{(2n+1)!} I_{2n+1}\,, \end{aligned} \tag{10.2.49}$$

where

$$I_m = \int_0^X \frac{\lambda^m}{(\lambda^2 + r^2)^{3/2}}\,\mathrm{d}\,\lambda =$$

$$= -\frac{X^{m-1}}{\sqrt{X^2+r^2}} + (m-1)X^{m-3}\sqrt{X^2+r^2} - (m-1)(m-3)J_{m-4}\,. \tag{10.2.50}$$

Using the notations

$$J_m = \int_0^X \lambda^m \sqrt{\lambda^2+r^2}\,\mathrm{d}\,\lambda = X^{m-1}(X^2+r^2)^{3/2} - (m-1)r^2 J_{m-2}\,, \tag{10.2.51}$$

we may calculate the coefficients I_m step by step, noticing that

$$I_0 = \frac{X}{r^2\sqrt{X^2+r^2}} = \frac{1}{r^2} + \ldots\,,$$

$$I_1 = \sqrt{X^2+r^2} - r\,.$$

The formulas (10.2.48) and (10.2.49) give the structure of the kernel.

10.2.5 The Sonic Flow

If the velocity of the stream equals the sound velocity ($U_\infty = c_\infty$), then one obtains the solution of the problem from (10.2.6), (10.2.7) and (10.2.12) setting $M \to 1$. We have shown in (10.2.27), that in this case there are perturbations only in the $x > 0$ zone. When we shall pass to limit we shall consider therefore $x_0 > 0$. We have

$$\lim_{M\to 1} G_0(x_0, y_0, z) = \frac{1}{4\pi x_0}\exp\left[\mathrm{i}\omega \lim \frac{M^2 x_0 - MR_1}{\beta^2}\right] =$$

$$= \frac{1}{4\pi x_0}\exp[-\mathrm{i}\omega\frac{x_0^2+y_0^2+z^2}{2x_0}] \equiv G_0^1\,, \tag{10.2.52}$$

$$\lim_{M\to 1} G(x_0, y_0, z) = \frac{1}{4\pi x_0}\exp\left[-\mathrm{i}\omega\frac{x_0^2-y_0^2-z^2}{2x_0}\right] \equiv G^1\,,$$

$$\lim_{M\to 1} X = \lim_{M\to 1}\frac{x_0 - MR}{\beta^2} = \frac{1}{2}(x_0 - \frac{y_0^2}{x_0})\,,$$

$$\lim_{M\to 1} E_2 = \int_0^{x_0}\exp\left[\frac{\mathrm{i}\omega}{2}(\tau - \frac{y_0^2}{\tau})\right]\frac{\mathrm{d}\,\tau}{\tau}\,. \tag{10.2.53}$$

From (10.2.16) it obviously results $\lim E_1 = 0$. Hence, the representation obtained from (10.2.6), (10.2.7) and (10.2.52), gives the perturbation in the sonic flow and (10.2.12) gives the kernel of the integral equation. This is

$$n_1^1+n_2^2 = \frac{\mathrm{i}\omega}{x_0}\exp\left[\frac{\mathrm{i}\omega}{2}(x_0-\frac{y_0^2}{x_0})\right]+\frac{1}{2}(\omega^2-\frac{\partial^2}{\partial y^2})\int_0^{x_0}\exp\left[\frac{\mathrm{i}\omega}{2}(\tau-\frac{y_0^2}{\tau})\right]\frac{\mathrm{d}\,\tau}{\tau}. \tag{10.2.54}$$

10.2.6 The Plane Flow

Acting like in 5.1 we obtain the formulas for the plane flow from the formulas which characterize the three - dimensional flow. We assume that in the equation of the perturbing surface (10.2.2), $h_0(x,y)$ has the form $h_0(x)$, i.e. it has the same form for every y. With a suitable choice of the reference length L_0, the domain D will be rectangular with $-1 < x < 1, -b < y < b$. In (10.2.6) and (10.2.7) we shall have $f = f(\xi)$. Considering $b \to \infty$ we obtain

$$g_0(x_0,z) = \int_{-\infty}^{+\infty} G_0(x_0,y_0,z)\mathrm{d}\,\eta =$$

$$= \frac{1}{4\pi}\int_{-\infty}^{+\infty}\frac{\mathrm{e}^{\mathrm{ik(Mx_0-R_0)}}}{R_0}\mathrm{d}\,\eta = \frac{\mathrm{e}^{\mathrm{iax_0}}}{4\mathrm{i}\beta}H_0^{(2)}(k\sqrt{x_0^2+\beta^2z^2}) =$$

$$g(x_0,z) = \int_{-\infty}^{+\infty} G(x_0,y_0,z)\mathrm{d}\,\eta = \frac{\mathrm{e}^{\mathrm{i}\bar{\omega}\mathrm{x_0}}}{4\mathrm{i}\beta}H_0^{(2)}(k\sqrt{x_0^2+\beta^2z^2}),$$

$$e(x_0,z) = \int_{-\infty}^{+\infty}(E_1+E_2)\mathrm{d}\,\eta = \int_{-\infty}^{+\infty} g(\tau,z)\mathrm{d}\,\tau,$$

$$\int_{-\infty}^{+\infty}\frac{\partial^2}{\partial y^2}G(x_0,y_0,z)\mathrm{d}\,\eta = \int_{-\infty}^{+\infty}\frac{\partial^2}{\partial \eta^2}G(x_0,y_0,z)\mathrm{d}\,\eta = 0. \tag{10.2.55}$$

Using these results we obtain from (10.2.6) and (10.2.7) the representations (10.1.12) and (10.1.13).

Taking into account the formulas of Fresnel

$$\int_0^\infty \sin x^2 \mathrm{d}\,x = \int_0^\infty \cos x^2 \mathrm{d}\,x = \frac{1}{2}\sqrt{\frac{\pi}{2}}, \tag{10.2.56}$$

we deduce

$$\int_{-\infty}^{+\infty} \mathrm{e}^{-\mathrm{i}\Omega u^2} \mathrm{d}\, u = \sqrt{\frac{\pi}{2\Omega}}(1-\mathrm{i})\,, \tag{10.2.57}$$

and then, with (10.2.54)

$$n(x_0) = \int_{-\infty}^{+\infty} (n_1^1 + n_2^2)\mathrm{d}\,\eta =$$

$$= (1+\mathrm{i})\sqrt{\pi\omega}\left[\frac{1}{\sqrt{x_0}}\exp(\frac{\mathrm{i}\omega}{2}x_0) - \frac{\omega \mathrm{i}}{2}\int_0^{x_0} \exp(\frac{\mathrm{i}\omega}{2}\tau)\frac{\mathrm{d}\,\tau}{\tau}\right]. \tag{10.2.58}$$

This is the kernel of the two-dimensional sonic flow.

10.3 The Theory of the Oscillatory Profile in a Supersonic Stream

10.3.1 The General Solution

One considers that Garrick and Rubinow [10.24] have investigated for the first time this problem. They have utilized the method of the pulsating sources potential. The presentation which follows relies on the fundamental solutions method [10.17]. The integral equation is solved explicitly. One obtains finite formulas for the lift and moment coefficients.

One utilizes the dimensionless variables (x, y) defined like in (2.1.1) and the representations (2.1.3) for the velocity and the pressure, in order to deduce that in (2.4.8), (2.4.18) and (2.4.21) the perturbation pressure and velocity determined by the action of an oscillatory force having the form $\boldsymbol{f}\mathrm{e}^{\mathrm{i}\omega t}$, applied in the origin of the axes of coordinates, have the form (10.1.4), where

$$p(x,y) = -(f \cdot \nabla)G_0(x,y)\,, \tag{10.3.1}$$

$$\begin{aligned} v(x,y) = f_1\mathrm{e}^{-\mathrm{i}\omega x}\left[G_y(x,y) - \mathrm{i}\omega\int_{-\infty}^{x} G_y(\tau,y)\mathrm{d}\,\tau\right] + \\ +f_2\mathrm{e}^{-\mathrm{i}\omega x}\left[2\mathrm{i}\omega G + k^2G_x + \omega^2\int_{-\infty}^{x} G(\tau,y)\mathrm{d}\,\tau\right]. \end{aligned} \tag{10.3.2}$$

Utilizing the notations $\overline{\omega} = \omega/k^2$, $\nu = \overline{\omega}M$ and H for the function of

Heaviside, in (10.3.1) and (10.3.2) we have

$$G_0(x,y) = H(x-k|y|)g_0(x,y)\,,$$
$$G(x,y) = H(x-k|y|)g(x,y)\,, \tag{10.3.3}$$

where

$$a = \nu M$$

$$2kg_0(x,y) = J_0\left(\nu\sqrt{x^2-k^2y^2}\right)\mathrm{e}^{-\mathrm{i}ax}\,, \tag{10.3.4}$$

$$2kg(x,y) = J_0\left(\nu\sqrt{x^2-k^2y^2}\right)\mathrm{e}^{-\mathrm{i}\overline{\omega}x}\,.$$

As it is shown in (2.4.22) the perturbation (10.3.1) and (10.3.2) is zero outside Mach's dihedron with the edge on the Oz axis and the opening 2μ. This is the fundamental solution of the problem.

Let us consider now that the uniform flow having the Mach number M, is perturbed by the presence of a profile whose equation is

$$y = h(x)\exp(\mathrm{i}\omega t)\,. \tag{10.3.5}$$

Taking the origin of the reference frame on the leading edge and the length of the chord as reference length L_0, the function $h(x)$ will be defined on the interval $[0,1]$. Replacing the profile by a forces distribution $(0,f)(\xi)\mathrm{e}^{\mathrm{i}\omega t}$ defined on $[0,1]$, one obtains the following general representation of the perturbation

$$p(x,y) = -\int_0^1 f(\xi)\frac{\partial}{\partial y}G_0(x_0,y)\mathrm{d}\,\xi\,,$$
$$v(x,y) = \int_0^1 f(\xi)\mathrm{e}^{-\mathrm{i}\omega x_0}\left[2\mathrm{i}\omega G(x_0,y)+\right. \tag{10.3.6}$$
$$\left.+k^2\frac{\partial}{\partial x}G(x_0,y)+\omega^2\int_{-\infty}^{x}G(\tau,y)\mathrm{d}\,\tau\right]\mathrm{d}\,\xi\,.$$

Taking into account the definition of the function $H(x_0-k|y_0|)$ and the formulas (A.3.15), with the notation

$$X = x-k|y|\,, \tag{10.3.7}$$

we deduce

$$\int_0^1 f(\xi)\frac{\partial}{\partial y}G_0(x_0,y)\mathrm{d}\,\xi = \frac{\partial}{\partial y}\left[H(x)\int_0^X f(\xi)g_0(x_0,y)\mathrm{d}\,\xi\right] =$$

$$= H(X)\frac{\partial}{\partial y}\int_0^X f(\xi)g_0(x_0,y)\mathrm{d}\,\xi = H(X)\int_0^X f(\xi)\frac{\partial}{\partial y}g_0(x_0,y)\mathrm{d}\,\xi\,,$$

$$\int_0^1 f(\xi)\mathrm{e}^{-\mathrm{i}\omega x_0}G(x_0,y)\mathrm{d}\,\xi = H(X)\int_0^X f(\xi)\mathrm{e}^{-\mathrm{i}\omega x_0}g(x_0,y)\mathrm{d}\,\xi\,,$$

$$\int_0^1 f(\xi)\mathrm{e}^{-\mathrm{i}\omega x_0}G_x(x_0,y)\mathrm{d}\xi = \int_0^1 f(\xi)\mathrm{e}^{-\mathrm{i}\omega x_0}\frac{\partial}{\partial x}\left[H(x_0-k|y|)g(x_0,y)\right]\mathrm{d}\xi =$$

$$= \int_0^1 f(\xi)\mathrm{e}^{-\mathrm{i}\omega x_0}g(x_0,y)\delta(x_0-k|y|)\mathrm{d}\xi + H(X)\int_0^1 f(\xi)\mathrm{e}^{-\mathrm{i}\omega x_0}g_x(x_0,y)\mathrm{d}\xi,$$

$$\int_{-\infty}^{x_0} G(\tau,y)\mathrm{d}\,\tau = H(X)\int_{k|y|}^{x_0} g(\tau,y)\mathrm{d}\,\tau\,. \tag{10.3.8}$$

Hence the solution (10.3.6) is

$$p=0\,,\ v=0\,,\quad \text{if}\, X<0\,, \tag{10.3.9}$$

$$p(x,y) = -\int_0^X f(\xi)\frac{\partial}{\partial y}g_0(x_0,y)\mathrm{d}\,\xi\,,\quad \text{if}\, X>0\,,$$

$$v(x,y) = \int_0^X f(\xi)\mathrm{e}^{-\mathrm{i}\omega x_0}n_1(x_0,y)\mathrm{d}\,\xi + \frac{k}{2}f(x)\exp(\mathrm{i}a|y|)\,,\quad \text{if}\, X>0\,, \tag{10.3.10}$$

where

$$n_1(x_0,y) = 2\mathrm{i}\omega g(x_0,y) + k^2 g_x(x_0,y) + \omega^2\int_{k|y|}^{x_0} g(\tau,y)\mathrm{d}\,\tau\,.$$

The formulas (10.3.9) show that the perturbation produced by the profile propagates only in the interior of Mach's angle with the vertex in 0, and (10.3.10) that in a point $M(x,y)$ from the interior of this angle one receives only the perturbation produced by the segment OM_0 (fig. 10.3.1).

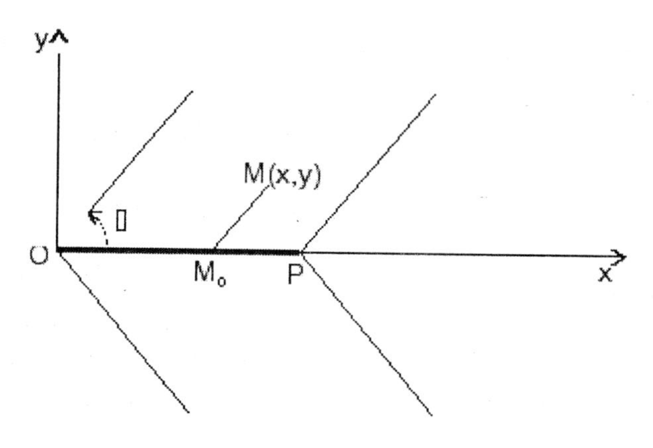

Fig. 10.3.1.

10.3.2 The Integral Equation and its Solution

For $0 < x < 1$ we deduce

$$p(x, +0) - p(x, -0) = f(x) \tag{10.3.11}$$

$$v(x, 0) = \frac{k}{2} f(x) + \frac{1}{2} \int_0^x f(\xi) N(x_0) \mathrm{d}\, \xi \,, \tag{10.3.12}$$

where

$$N(x_0) = \mathrm{e}^{-\mathrm{i}\omega \mathrm{x}_0} \lim_{y \to 0} n_1(x_0, y) =$$

$$= \frac{\mathrm{i}\omega}{k} \left[J_0(\nu x_0) + \mathrm{i} M J_1(\nu x_0) \right] \mathrm{e}^{-\mathrm{iax}_0} + \frac{\omega^2}{k} \mathrm{e}^{-\mathrm{iax}_0} \int_0^{x_0} J_0(\nu \tau) \mathrm{e}^{-\mathrm{i}\overline{\omega}\tau} \mathrm{d}\, \tau \,, \tag{10.3.13}$$

Imposing the boundary condition (10.1.11) one obtains the following integral equation

$$k f(x) + \int_0^x f(\xi) N(x_0) \mathrm{d}\, \xi = 2H(x) \,, \quad 0 < x < 1 \,. \tag{10.3.14}$$

This is a *Volterra type* integral equation of first order. We solve it using the *Laplace* transform. One knows (see for example [1.32]) that the Laplace transform of a certain function $g(x)$ is the function $\overline{g}(p)$, defined by the operator

$$\overline{g} = \mathcal{L}(g) = \int_0^\infty g(x) \mathrm{e}^{-\mathrm{px}} \mathrm{d}\, x \tag{10.3.15}$$

where p can be a complex number($p = p_1 + \mathrm{i}p_2$) whose real part is positive.

Applying the operator $\mathcal{L}$ in (10.3.14) we obtain

$$k\overline{f} + \int_0^\infty \left[\int_0^x f(\xi)N(x_0)\mathrm{d}\,\xi\right]\mathrm{e}^{-\mathrm{p}x}\mathrm{d}\,x = 2\overline{H}(p)\,.$$

Here we shall change the order of integration. In figure 10.3.2 we observe that the domain of integration is D (for a given x , ξ goes from 0 to x). But D can be also covered integrating at first with respect to x from ξ to ∞ and then with respect to ξ. We have therefore

$$k\overline{f} + \int_0^\infty f(\xi)\left[\int_\xi^\infty \mathrm{e}^{-\mathrm{p}x}\mathrm{N}(\mathrm{x}_0)\mathrm{d}\,x\right]\mathrm{d}\,\xi = 2\overline{H}(p)\,. \tag{10.3.16}$$

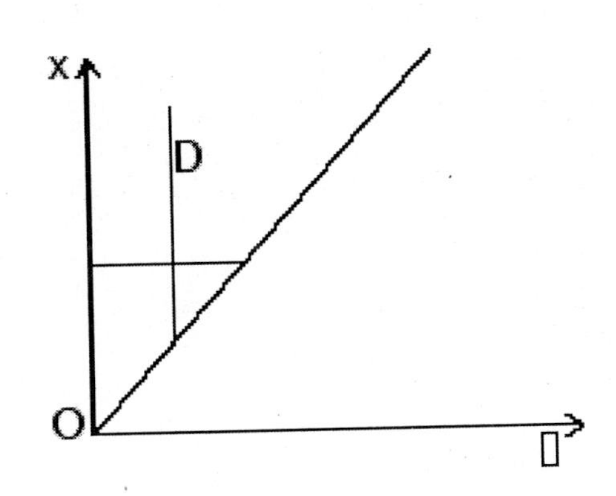

Fig. 10.3.2.

Using the change of variable $x \to u : x - \xi = u$, we deduce from (10.3.16)

$$(k + \overline{N})\,\overline{f} = 2\overline{H}\,. \tag{10.3.17}$$

In order to determine the transformation $\overline{N}$ we shall utilize the formula

$$\mathcal{L}[J_\lambda(\nu x)] = \frac{1}{\sqrt{p^2+\nu^2}}\frac{\nu^\lambda}{(p+\sqrt{p^2+\nu^2})^\lambda}\,, \tag{10.3.18}$$

which may be found in the tables with Laplace transforms [1.16], [1.32], [1.33]. So, using the notations

$$\begin{gathered} N_1(x) = J_0(\nu x)\mathrm{e}^{-\mathrm{i}ax}\,,\ N_2(x) = J_1(\nu x)\mathrm{e}^{-\mathrm{i}ax}\,, \\ N_3(x) = \mathrm{e}^{-\mathrm{i}\omega x}\int_0^x J_0(\nu\tau)\mathrm{e}^{-\mathrm{i}\overline{\omega}\tau}\mathrm{d}\,\tau\,, \end{gathered} \tag{10.3.19}$$

with $a = \nu M$, we obtain

$$\mathcal{L}(N_1) = \int_0^\infty J_0(\nu x)\mathrm{e}^{-(\mathrm{p}+\mathrm{i}a)\mathrm{x}}\mathrm{d}\,x = \frac{1}{\sqrt{(p+\mathrm{i}a)^2+\nu^2}}\,,$$

$$\mathcal{L}(N_2) = \frac{1}{\sqrt{(p+\mathrm{i}a)^2+\nu^2}}\,\frac{\nu}{p+\mathrm{i}a+\sqrt{(p+\mathrm{i}a)^2+\nu^2}}\,,$$

$$\mathcal{L}(N_3) = \int_0^\infty \mathrm{e}^{-\mathrm{px}}\mathrm{e}^{-\mathrm{i}\omega\mathrm{x}}\left[\int_0^x J_0(\nu\tau)\mathrm{e}^{-\mathrm{i}\overline{\omega}\tau}\mathrm{d}\,\tau\right]\mathrm{d}\,x =$$

$$= \int_0^\infty J_0(\nu\tau)\mathrm{e}^{-\mathrm{i}\overline{\omega}\tau}\left[\int_\tau^\infty \mathrm{e}^{-(\mathrm{p}+\mathrm{i}\omega)\mathrm{x}}\mathrm{d}\,x\right]\mathrm{d}\,\tau \overset{x=u+\tau}{=} \tag{10.3.20}$$

$$= \int_0^\infty J_0(\nu\tau)\mathrm{e}^{-\mathrm{i}\nu\mathrm{M}\tau}\mathrm{d}\,\tau\int_0^\infty \mathrm{e}^{-(\mathrm{p}+\mathrm{i}\omega)\mathrm{u}}\mathrm{d}\,u =$$

$$= \frac{1}{\sqrt{(p+\mathrm{i}\nu M)^2+\nu^2}}\,\frac{1}{p+\mathrm{i}\omega}\,.$$

With the notations

$$p_1 = \frac{\mathrm{i}\omega M}{M+1}\,,\quad p_2 = \frac{\mathrm{i}\omega M}{M-1}\,, \tag{10.3.21}$$

it results

$$\sqrt{(p+\mathrm{i}\nu M)^2+\nu^2} = \frac{1}{\sqrt{(p+p_1)(p+p_2)}}\,,$$

$$k(k+\overline{N})\left[(p+\mathrm{i}\omega)\sqrt{(p+p_1)(p+p_2)}\right] = k(p+p_1)(p+p_2)\,,$$

$$\overline{N} = \frac{1}{\sqrt{(p+p_1)(p+p_2)}}\left\{\frac{\mathrm{i}\omega}{k}+k\left[p+\mathrm{i}\nu M-\sqrt{(p+p_1)(p+p_2)}\right]+\right.$$

$$\left.+\frac{\omega^2}{k}\frac{1}{p+\mathrm{i}\omega}\right\}\,, \tag{10.3.22}$$

and then

$$\overline{g} \overset{\text{not}}{=} \frac{k}{k+\overline{N}} = \frac{p+\mathrm{i}\omega}{\sqrt{(p+p_1)(p+p_2)}} = \frac{p+p_2-p_2+\mathrm{i}\omega}{\sqrt{(p+p_1)(p+p_2)}} =$$

$$= (p+\frac{\mathrm{i}\omega M}{M+1})^{1/2}(p+\frac{\mathrm{i}\omega M}{M+1})^{-1/2} -$$

$$-\frac{\mathrm{i}\omega}{M-1}\left(p+\frac{\mathrm{i}\omega M}{M+1}\right)^{-1/2}\left(p+\frac{\mathrm{i}\omega M}{M+1}\right)^{1/2}$$

whence

$$k\overline{f} = \overline{H}_+\overline{g}\,. \tag{10.3.23}$$

From the tables with Laplace transforms [10.58] it results $g = \mathcal{L}^{-1}(\overline{g})$ and then with the convolution theorem

$$\begin{aligned} kf(x) &= 2\int_0^x H(x_0)g(\xi)\mathrm{d}\,\xi \\ &= 2H(x) - 2\mathrm{i}\overline{\omega}\int_0^x H(x_0)\left[J_0(\nu\xi) + \mathrm{i}MJ_1(\nu\xi)\right]\mathrm{e}^{-\mathrm{i}a\xi}\mathrm{d}\,\xi\,. \end{aligned} \tag{10.3.24}$$

This is the solution of the integral equation (10.3.14). It was given in [10.17].

10.3.3 Formulas for the Lift and Moment Coefficients

The lift and moment coefficients have the form

$$C_L = c_L\exp(\mathrm{i}\omega t)\,,\ C_M = c_M\exp(\mathrm{i}\omega t)\,, \tag{10.3.25}$$

where, because of the formula (10.3.11),

$$c_L = -2\int_0^1 f(x)\mathrm{d}\,x\,,\ c_M = -2\int_0^1 xf(x)\mathrm{d}\,x\,. \tag{10.3.26}$$

We considered that the length of the chord L_0 is the reference length and we defined

$$c_L = \frac{P}{(1/2)\rho_0 U_\infty^2 L_0}\,,\ c_M = \frac{M}{(1/2)\rho_0 U_\infty^2 L_0}\,, \tag{10.3.27}$$

P being the lift and M the moment on the direction Oz.

Utilizing (10.3.24) we find

$$c_L = -\frac{4}{k}\int_0^1 H(x)\mathrm{d}\,x + \frac{4\mathrm{i}\nu}{kM}\int_0^1 H(\xi)\left[J_0(\nu\xi) - \mathrm{i}MJ_1(\nu\xi)\right]\mathrm{e}^{-\mathrm{i}a\xi}\mathrm{d}\,\xi\,,$$

$$c_M = -\frac{4}{k}\int_0^1 xH(x)\mathrm{d}\,x + \frac{4\mathrm{i}\nu}{kM}\int_0^1 G(\xi)\left[J_0(\nu\xi) - \mathrm{i}MJ_1(\nu\xi)\right]\mathrm{e}^{-\mathrm{i}a\xi}\mathrm{d}\,\xi\,, \tag{10.3.28}$$

where

$$H(\xi) = \int_\xi^1 H(x-\xi)\mathrm{d}\,x\,,\; G(\xi) = \int_\xi^1 xH(x-\xi)\mathrm{d}\,x\,. \tag{10.3.29}$$

The coefficients (10.3.28) may be calculated numerically on a computer. Another method consists in approximating the function $h(x)$ by polynomials whence one deduces that c_L and c_M may be expressed by means of the terms having the form

$$f_n(M,a) = \int_0^1 \xi^n J_0(\nu\xi)\mathrm{e}^{-\mathrm{i}a\xi}\mathrm{d}\,\xi\,,$$

$$g_n(M,a) = \int_0^1 \xi^n J_1(\nu\xi)\mathrm{e}^{-\mathrm{i}a\xi}\mathrm{d}\,\xi\,, \tag{10.3.30}$$

Taking into account that $J_1(z) = -J_0'(z)$ and integrating by parts one obtains that

$$\nu g_n = -J_0(\nu)\mathrm{e}^{-\mathrm{i}a} + nf_{n-1} - \mathrm{i}af_n\,,\; n = 1,2,\ldots\,,$$

$$\nu g_0 = -J_0(\nu)\mathrm{e}^{-\mathrm{i}a} + 1 - \mathrm{i}af_0\,. \tag{10.3.31}$$

These formulas show that g_n may be expressed in terms of f_n. Integrating f_n by parts, we deduce

$$\mathrm{i}af_n = -J_0(\nu)\mathrm{e}^{-\mathrm{i}a} + nf_{n-1} - \nu\int_0^1 \xi^n J_1(\nu\xi)\mathrm{e}^{-\mathrm{i}a\xi}\mathrm{d}\,\xi\,, \tag{10.3.32}$$

$$\mathrm{i}\nu M\int_0^1 \xi^n J_1(\nu\xi)\mathrm{e}^{-\mathrm{i}a\xi}\mathrm{d}\,\xi = -J_1(\nu)\mathrm{e}^{-\mathrm{i}a} + \nu f_n +$$

$$+(n-1)\int_0^1 \xi^{n-1}J_1(\nu\xi)\mathrm{e}^{-\mathrm{i}a\xi}\mathrm{d}\,\xi\,, \tag{10.3.33}$$

Substituting (10.3.33) in (10.3.32) we find for f_n an expression which contains the last term from (10.3.33). This may be eliminated with the aid of the relation (10.3.32) where n was replaced by $n-1$. After all one obtains

$$\omega f_n = \left(\mathrm{i} - \frac{n-1}{a}\right) J_0(\nu)\mathrm{e}^{-\mathrm{i}a} - \frac{1}{M} J_1(\nu)\mathrm{e}^{-\mathrm{i}a} + \tag{10.3.34}$$

$$+\frac{(n-1)^2}{a} f_{n-2} + \mathrm{i}(1-2n) f_{n-1} .$$

This formula shows that all the terms f_n may be expressed by means of f_0. This result was given for the first time by Schwartz in [10.70]. In the same paper one gives the following expansion for f_0

$$f_0 = \mathrm{e}^{-\mathrm{i}\nu M} \sum_{n=0}^{\infty} \frac{J_n(a) + \mathrm{i}J_{n+1}(a)}{2^n n!(2n+1)} \omega^n . \tag{10.3.35}$$

In [10.70] one gives tables with the numerical values of f_0, with eight exact decimals, for $1 \leq M \leq 10$ and $0 \leq a \leq 5$. In [10.33] one gives the numerical values of the functions f_n for $n = 0, \ldots, 11$.

10.3.4 The Flat Plate

For the flat plate having the angle of attack $-\epsilon$ ($h = -\epsilon x$) we deduce $H = -2\epsilon(1 + \mathrm{i}\omega x)$ such that it results

$$c_L = -\frac{4\epsilon}{k}\left[\frac{\omega^2}{2} f_2 + \mathrm{i}\omega(2+\mathrm{i}\omega) f_1 - (1 + 2\mathrm{i}\omega - \frac{\omega^2}{2}) f_0\right],$$

$$c_M = -\frac{2\epsilon}{k}\left[\frac{\omega^2}{3} f_3 + 2\mathrm{i}\omega f_2 - (\omega^2+2) f_1 - 2(\mathrm{i}\omega - \frac{\omega^2}{3}) f_0\right]. \tag{10.3.36}$$

These formulas are sufficient if we utilize the tables for f_0, f_1, f_2, f_3. For $\omega \to 0$ one obtains the well known formulas of Ackeret

$$c_L = \frac{4\epsilon}{k}, \quad c_M = \frac{2\epsilon}{k}.$$

Obviously, c_L and c_M may be expressed only by means of f_0 if one utilizes (10.3.34).

Noticing that

$$f_n = f_n^r + \mathrm{i} f_n^{\mathrm{i}}, \tag{10.3.37}$$

where

$$f_n^r = \int_0^1 \xi^n J_0(\nu\xi)\cos(a\xi)\mathrm{d}\xi, \quad f_n^{\mathrm{i}} = -\int_0^1 \xi^n J_0(\nu\xi)\sin(a\xi)\mathrm{d}\xi,$$

we deduce from (10.3.36)

$$c_L = c_L^r + \mathrm{i}c_L^{\mathrm{i}}\,, c_M = c_M^r + \mathrm{i}c_M^{\mathrm{i}}\,, \tag{10.3.38}$$

If the equation of the plate has the form

$$y = -\epsilon x \cos \omega t = \mathrm{Re}\left[-\epsilon x \exp(\mathrm{i}\omega t)\right]\,, \tag{10.3.39}$$

then

$$\begin{aligned} C_L &= c_L^r \cos \omega t - c_L^{\mathrm{i}} \sin \omega t\,, \\ C_M &= c_M^r \cos \omega t - c_M^{\mathrm{i}} \sin \omega t\,, \end{aligned} \tag{10.3.40}$$

these formulas give the variation of the lift and moment coefficients versus the time. For example, for $\omega = \pi$ and $M = 2$, we obtain

$$\begin{aligned} C_L &= \epsilon(-9.2060 \cos \pi t + 11.8941 \sin \pi t)\,, \\ C_M &= \epsilon(-6.8779 \cos \pi t + 17.5209 \sin \pi t)\,, \end{aligned} \tag{10.3.41}$$

10.3.5 The Oscillatory Profile in the Sonic Flow

We are interested in the behaviour of the formulas of N_1 and N_2 when $M \to 1$ ($k \to 0$). It results $\nu \to \infty$ such that we shall utilize the well known asymptotic expressions

$$\begin{aligned} J_0(z) &= \sqrt{\frac{2}{\pi z}} \cos\left(z - \frac{\pi}{4}\right) + O(z^{-1})\,, \\ J_1(z) &= \sqrt{\frac{2}{\pi z}} \cos\left(z - \frac{3\pi}{4}\right) + O(z^{-1})\,, \end{aligned} \tag{10.3.42}$$

for great values of z. In this way, we deduce

$$\begin{aligned} N_1 &= \frac{1}{k}\left[J_0(\nu x_0) + \mathrm{i}MJ_1(\nu x_0)\right]\exp(-\mathrm{i}ax_0) = \\ &= \frac{1}{\sqrt{\pi\omega M x_0}}[(1-\mathrm{i}M)\cos ax_0 \cos \nu x_0 + (M-\mathrm{i})\sin ax_0 \sin \nu x_0 - \\ &\quad -\mathrm{i}(1-\mathrm{i}M)\sin ax_0 \cos \nu x_0 + \mathrm{i}(M-\mathrm{i})\cos ax_0 \sin \nu x_0]\,, \end{aligned} \tag{10.3.43}$$

and analogously

$$N_2(x) = \frac{1}{k}\int_0^{x_0} J_0(\nu u)\mathrm{e}^{-\mathrm{i}\overline{\omega}u}\mathrm{d}\,u$$

$$= \frac{1-\mathrm{i}}{2\sqrt{\pi\omega}}\int_0^{x_0}\frac{\cos\overline{\omega}(M-1)u + \mathrm{i}\sin\overline{\omega}(M-1)u}{\sqrt{Mu}}\mathrm{d}\,u + I\,, \tag{10.3.44}$$

where, with the change of variable $u \to t : u = (M-1)t$, we have

$$I = \frac{1+\mathrm{i}}{2\sqrt{\pi\omega}}\int_0^{x_0}\frac{\cos\overline{\omega}(M+1)u - \mathrm{i}\sin\overline{\omega}(M+1)u}{\sqrt{Mu}}\mathrm{d}\,u =$$

$$= \frac{1+\mathrm{i}}{2\sqrt{\pi\omega}}\sqrt{\frac{M-1}{M}}\int_0^{\frac{x_0}{M-1}}\frac{\exp(-\mathrm{i}\omega t)}{\sqrt{t}}\mathrm{d}\,t \tag{10.3.45}$$

Taking into account that we also have

$$\lim_{M\to 1}\int_0^{\frac{x_0}{M-1}}\frac{\exp(-\mathrm{i}\omega t)}{\sqrt{t}}\mathrm{d}\,t = \int_0^{\infty}\frac{\exp(-\mathrm{i}\omega t)}{\sqrt{t}}\mathrm{d}\,t = \frac{\pi}{2\omega}(1+\mathrm{i})\,,$$

it results that

$$N_0(x_0) = \lim_{M\to 1} N(x_0) =$$

$$= (\mathrm{i}+1)\sqrt{\frac{\omega}{\pi}}\,\mathrm{e}^{-\mathrm{i}\omega x_0}\left[\frac{1}{\sqrt{x_0}}\exp(\frac{\mathrm{i}\omega}{2}x_0) - \frac{\omega\mathrm{i}}{2}\int_0^{x_0}\exp(\frac{\mathrm{i}\omega}{2}u)\frac{\mathrm{d}\,u}{\sqrt{u}}\right], \tag{10.3.46}$$

i.e. exactly (10.2.54).

The integral equation (10.2.14) reduces to

$$\int_0^x f(\xi)N_0(x-\xi)\mathrm{d}\,\xi = 2H(x) \tag{10.3.47}$$

This is also a Volterra-type equation of first kind. For integrating it we shall use again the Laplace transform. Applying this transformation we deduce

$$(1+\mathrm{i})\sqrt{\omega}\,\overline{f} = 2\overline{H}\overline{g}_0\,, \tag{10.3.48}$$

where

$$\overline{g}_0 = \left(p + \frac{\mathrm{i}\omega}{2}\right)^{1/2} + \frac{\mathrm{i}\omega}{2}\left(p + \frac{\mathrm{i}\omega}{2}\right)^{-1/2} \tag{10.3.49}$$

From tables (see for example [10.58]) we have that

$$\mathcal{L}^{-1}\left[\sqrt{p+\mathrm{i}\omega/2}\right]=\exp\left(-\frac{\mathrm{i}\omega}{2}x\right)\mathcal{L}^{-1}[\sqrt{p}]=-\frac{1}{2\sqrt{\pi}}\frac{1}{x^{3/2}}\exp\left(-\frac{\mathrm{i}\omega}{2}x\right),$$

$$\mathcal{L}^{-1}\left[\frac{1}{\sqrt{p+\mathrm{i}\omega/2}}\right]=\frac{1}{\sqrt{\pi x}}\exp\left(-\frac{\mathrm{i}\omega}{2}x\right), \tag{10.3.50}$$

such that we obtain

$$g_0=\frac{1}{2\sqrt{\pi x}}(\mathrm{i}\omega-\frac{1}{x})\exp\left(-\frac{\mathrm{i}\omega}{2}x\right) \tag{10.3.51}$$

and using the convolution theorem, from (10.3.48) we deduce

$$f(x)=\frac{1-\mathrm{i}}{2\sqrt{\pi\omega}}\,{}^{*}\!\!\int_0^x\frac{H(x_0)}{\sqrt{\xi}}(\mathrm{i}\omega-\frac{1}{\xi})\exp\left(-\frac{\mathrm{i}\omega}{2}\xi\right)\mathrm{d}\,\xi\,. \tag{10.3.52}$$

After determining $f(x)$, the lift and moment coefficients result from (10.3.25) and (10.3.26). We shall give calculation formulas in 10.5.2 when we shall consider again this problem.

10.4 The Theory of the Oscillatory Wing in a Supersonic Stream

10.4.1 The General Solution

The theory of the oscillatory wing in a supersonic stream, was conceived according to the model of the theory in the subsonic stream. The papers of Küssner [10.37], [10.38] represent the starting point of this theory. We mention then, the study of Garrick and Rubinow [10.25] where the potential of the pulsating source is determined, the paper of Miles where one considers the symmetric arrow - like wing, having the leading edges outside Mach's cone [10.53], the paper of Nelson for the triangular wing [10.57], etc. But the fundamental work in this domain is the paper of Watkins and Berman [10.85]. Here one may find for the first time the integral equation of the problem and various forms of the kernel. The method is similar to the method from the subsonic case. From the potential of accelerations of a pulsating source, one obtains, deriving with respect to z the potential of accelerations of a pulsating doublet. The potential of the flow is obtained superposing the doublet potentials. The boundary condition gives the integral equation of the problem. In the following papers, due to Ashley, Windall and Landahl [10.4], Landahl [10.44], Stark [10.72], Harder and Rodden [10.29], Ueda and Dowell [10.81] the theory was developed and numerical methods for the integrations of the equation of Watkins and Berman were given.

We shall indicate in this subsection how one may also solve this problem by means of the fundamental solutions method. Assuming that the equation of the wing is (10.2.2), we shall use distributions having the shape

$$\boldsymbol{f}\mathrm{e}^{\mathrm{i}\omega t} = (0,0,f)\mathrm{e}^{\mathrm{i}\omega t}\,. \tag{10.4.1}$$

Utilizing (2.4.9)we deduce that the perturbation of the pressure determined by such a force applied in the generic point $(\xi,\eta,0)$ is given by the formula

$$p(x,y,z) = f\frac{\partial}{\partial z}G_0(x_0,y_0,z)\,. \tag{10.4.2}$$

For the component w, it results from (2.4.13)

$$\begin{aligned} \mathrm{w}(x,y,z) &= -f\mathrm{e}^{-\mathrm{i}\omega x_0}H(x_0)\delta(y_0)\delta(z)+ \\ &+f\frac{\partial^2}{\partial z^2}\mathrm{e}^{-\mathrm{i}\omega x_0}\int_{-\infty}^{x_0}G(\tau,y_0,z)\,\mathrm{d}\,\tau\,, \end{aligned} \tag{10.4.3}$$

and from (2.4.23)

$$\begin{aligned} \mathrm{w}(x,y,z) = f\mathrm{e}^{-\mathrm{i}\omega x_0}[&\left(2\mathrm{i}\omega + k^2\frac{\partial}{\partial x}\right)G(x_0,y_0,z)+ \\ &+\left(\omega^2-\frac{\partial^2}{\partial y^2}\right)\int_{-\infty}^{x_0}G(\tau,y_0,z)\,\mathrm{d}\,\tau\,]\,, \end{aligned} \tag{10.4.4}$$

where we denoted

$$\begin{aligned} G_0(x_0,y_0,z) &= \frac{1}{2\pi}\frac{H(x_0-s)}{S}\cos(\nu S)\mathrm{e}^{-\mathrm{i}a x_0}\,, \\ G(x_0,y_0,z) &= \frac{1}{2\pi}\frac{H(x_0-s)}{S}\cos(\nu S)\mathrm{e}^{-\mathrm{i}\varpi x_0}\,, \end{aligned} \tag{10.4.5}$$

$$\begin{aligned} &k=\sqrt{M^2-1}\,,\ \varpi=\omega/k^2\,,\ \nu=\varpi M\,,\ a=\nu M\,, \\ &s=k\sqrt{y_0^2+z^2}\,,\ S=\sqrt{x_0^2-s^2}\,,\ S_\tau=\sqrt{\tau^2-s^2}\,, \end{aligned} \tag{10.4.6}$$

H being the function of Heaviside.

One may prove, taking into account the formulas (2.3.35) and (2.3.36), that the perturbation given by (10.4.2)–(10.4.5) vanishes in the exterior of Mach's cone with the vertex in the point $(\xi, \eta, 0)$ and with the axis on the direction of the unperturbed stream (the Ox axis). Using a forces distribution having the form (10.4.1), applied on the domain D - the projection of the wing on the xOy plane, the perturbation will be given by the formulas

$$p(x, y, z) = \int\int_D f(\xi, \eta) \frac{\partial}{\partial z} G_0(x_0, y_0, z) \, \mathrm{d}\xi \, \mathrm{d}\eta \,, \tag{10.4.7}$$

$$\begin{aligned} \mathrm{w}(x, y, z) = & -\delta(z) \int\int_D f(\xi, \eta) \mathrm{e}^{-\mathrm{i}\omega x_0} H(x_0) \delta(y_0) \mathrm{d}\xi \, \mathrm{d}\eta + \\ & + \frac{1}{2\pi} \int\int_D f(\xi, \eta) \mathrm{e}^{-\mathrm{i}\omega x_0} n_1(x_0, y_0, z) \mathrm{d}\xi \, \mathrm{d}\eta \,, \end{aligned} \tag{10.4.8}$$

$$\mathrm{w}(x, y, z) = \frac{1}{2\pi} \int\int_D f(\xi, \eta) \mathrm{e}^{-\mathrm{i}\omega x_0} n_2(x_0, y_0, z) \mathrm{d}\xi \, \mathrm{d}\eta \,, \tag{10.4.9}$$

where

$$\begin{aligned} n_1(x_0, y_0, z) &= \frac{\partial^2}{\partial z^2} \int_{-\infty}^{x_0} H(\tau - s) \frac{\cos(\nu S_\tau)}{S_\tau} \mathrm{e}^{-\mathrm{i}\overline{\omega}\tau} \mathrm{d}\tau = \\ &= \frac{\partial^2}{\partial z^2} H(x_0 - s) \int_s^{x_0} \frac{\cos(\nu S_\tau)}{S_\tau} \mathrm{e}^{-\mathrm{i}\overline{\omega}\tau} \mathrm{d}\tau \,, \end{aligned} \tag{10.4.10}$$

$$\begin{aligned} n_2(x_0, y_0, z) = & \left(2\mathrm{i}\omega + k^2 \frac{\partial}{\partial x}\right) \frac{H(x_0 - s)}{S} \cos(\nu S) \mathrm{e}^{-\mathrm{i}\overline{\omega} x_0} + \\ & + \left(\omega^2 - \frac{\partial^2}{\partial y^2}\right) H(x_0 - s) \int_s^{x_0} \frac{\cos(\nu S_\tau)}{S_\tau} \mathrm{e}^{-\mathrm{i}\overline{\omega}\tau} \mathrm{d}\tau \,, \end{aligned} \tag{10.4.11}$$

10.4.2 The Boundary Values of the Pressure

They may be obtained writing

$$p(x, y, z) = \frac{1}{2\pi} \frac{\partial}{\partial z} \int\int_D f(\xi, \eta) H(x_0 - s) \frac{\cos(\nu S)}{S} \mathrm{e}^{-\mathrm{i}a x_0} \, \mathrm{d}\xi \, \mathrm{d}\eta \,, \tag{10.4.12}$$

and noticing that because of the presence of the factor $H(x_0 - s)$, the integrand differs from zero only in the domain D_1 defined by the inequality $x_0 > s$ for a given $\mathrm{M}(x, y, z)$. This inequality is equivalent to

$$0 < \xi < x \,, \; (\xi - x)^2 - k^2(\eta - y)^2 > k^2 z^2 \,,$$

which are solved in 8.3.3. Denoting by $M'(x,y,z)$ the projection of the point M on the xOy plane and $X=\xi-x$, $Y=\eta-y$ we deduce that D_1 is the foregoing branch of the hyperbola $X^2-k^2Y^2=k^2z^2$ (fig. 8.3.4). When M' is in D, the hyperbola degenerates into the half-lines $X=\pm kY$ (fig. 8.3.5)

Since the function f is defined only on D, we shall prolong it in the outer region taking it equal to zero. It follows that in the perturbed region from the fluid we have

$$I=\int\int_D f(\xi,\eta)G_0(x_0,y_0,z)\,\mathrm{d}\,\xi\,\mathrm{d}\,\eta=$$

$$=\frac{1}{2\pi k}\int_0^{x-k|z|}\mathrm{e}^{-\mathrm{i}ax_0}\left[\int_{Y_-}^{Y_+}f(\xi,\eta)\frac{\cos\left(\nu k\sqrt{(Y_+-\eta)(\eta-Y_-)}\right)}{\sqrt{(Y_+-\eta)(\eta-Y_-)}}\mathrm{d}\,\eta\right]\mathrm{d}\,\xi \tag{10.4.13}$$

where

$$Y_\pm=y\pm\sqrt{x_0^2k^{-2}-z^2}\,. \tag{10.4.14}$$

With the change of variable $\eta\to\theta$:

$$\eta=\frac{1}{2}(Y_++Y_-)-\frac{1}{2}(Y_++Y_-)\cos\theta=y-\sqrt{x_0^2k^{-2}-z^2}\cos\theta\,, \tag{10.4.15}$$

we deduce

$$I=\frac{1}{2\pi k}\int_0^{x-k|z|}\mathrm{e}^{-\mathrm{i}ax_0}\left[\int_0^{\pi}f(\xi,y-\sqrt{x_0^2k^{-2}-z^2}\cos\theta)\,\cdot\right.$$
$$\left.\cdot\cos\left(\nu\sqrt{x_0^2k^{-2}-z^2}\right)\sin\theta\mathrm{d}\,\theta\right]\mathrm{d}\,\xi$$

whence, if $f(x,y)$ is a continuous function,

$$p(x,y,\pm 0)=\lim_{z\to\pm 0}\frac{\partial}{\partial z}I=\mp f(x,y) \tag{10.4.16}$$

and then

$$p(x,y,+0)-p(x,y,-0)=f(x,y)\,. \tag{10.4.17}$$

Hence, like in the previous sections, f represents the jump of the pressure on the wing.

10.4.3 The Boundary Values of the Velocity. The Integral Equation

For $z \neq 0$ the first term from (10.4.8) vanishes ($\delta(z) = 0$). It has to be considered in the same way in the limit values for $z \to \pm 0$. The remaining term is the kernel given by Watkins and Berman [10.85]. Elementary calculations give

$$\frac{\partial}{\partial z} = k^2 \frac{z}{s}\frac{\partial}{\partial s}\,;\ \frac{\partial^2}{\partial z^2} = \frac{k^2}{s}(1 - \frac{k^2 z^2}{s^2})\frac{\partial}{\partial s} + \frac{k^4 z^2}{s^2}\frac{\partial^2}{\partial s^2}\,. \tag{10.4.18}$$

In the cited papers one considers that the terms which contain the factor z^2, vanish when $z \to 0$. But this is not always true (see for example (3.1.20)). This is true when the factors which multiply z^2 remain bounded when passing to the limit. In the following we shall see that for (10.4.10) the form obtained under this assumption is correct.

Hence we shall consider the kernel

$$n_1(x_0, y_0, z) \cong \frac{k^2}{s}\frac{\partial}{\partial s}\left[H(x_0 - s)\int_s^{x_0} \frac{\cos(\nu S_\tau)}{S_\tau}\mathrm{e}^{-\mathrm{i}\overline{\omega}\tau}\mathrm{d}\,\tau\right]. \tag{10.4.19}$$

The derivation is performed according to the formula (A.3.15), but we have to take care that for $s = x_0$ the integrand is unbounded. We eliminate this inconvenient writing

$$\int_s^{x_0} \frac{\cos\left(\nu\sqrt{\tau^2 - s^2}\right)}{\sqrt{\tau^2 - s^2}}\mathrm{e}^{-\mathrm{i}\overline{\omega}\tau}\mathrm{d}\,\tau = \int_s^{x_0} \frac{\cos\left(\nu\sqrt{\tau^2 - s^2}\right) - 1}{\sqrt{\tau^2 - s^2}}\mathrm{e}^{-\mathrm{i}\overline{\omega}\tau}\mathrm{d}\,\tau +$$

$$+ \int_s^{x_0} \frac{\mathrm{e}^{-\mathrm{i}\overline{\omega}\tau} - \mathrm{e}^{-\mathrm{i}\overline{\omega}s}}{\sqrt{\tau^2 - s^2}}\mathrm{d}\,\tau + \mathrm{e}^{-\mathrm{i}\overline{\omega}s}\int_s^{x_0} \frac{\mathrm{d}\,\tau}{\sqrt{\tau^2 - s^2}}\,. \tag{10.4.20}$$

After all

$$n_1(x_0, y_0, z) \simeq \frac{k^2}{s}H(x_0 - s)I\,, \tag{10.4.21}$$

where

$$I = \frac{\partial}{\partial s}\int_s^{x_0} \frac{\cos\left(\nu\sqrt{\tau^2 - s^2}\right)}{\sqrt{\tau^2 - s^2}}\mathrm{e}^{-\mathrm{i}\overline{\omega}\tau}\mathrm{d}\,\tau\,, \tag{10.4.22}$$

the derivation being possible if we utilize the equality (10.4.20), but we have no interest to do it. The integral may be calculated with the substitution $\tau \to \lambda : \tau = s\cosh\lambda$. Deriving one obtains

$$I = -\frac{x_0}{s}\frac{\cos(\nu\sqrt{x_0^2 - s^2})}{\sqrt{x_0^2 - s^2}}\mathrm{e}^{-\mathrm{i}\bar{\omega}x_0} -$$

$$-\frac{\mathrm{i}}{sM}\int_s^{x_0}\mathrm{e}^{-\mathrm{i}\bar{\omega}\tau}\frac{\mathrm{d}}{\mathrm{d}\,\tau}[\sin(\nu\sqrt{\tau^2 - \mathrm{s}^2})]\mathrm{d}\,\tau - \tag{10.4.23}$$

$$-\frac{\nu}{s}\int_s^{x_0}\mathrm{e}^{-\mathrm{i}\bar{\omega}\tau}\sin(\nu\sqrt{\tau^2 - \mathrm{s}^2})\mathrm{d}\,\tau\,.$$

We integrate by parts in the second term from the right hand side. Passing to the limit in (10.4.21) we notice that, like in the steady case, it appears the singular line $y_0 = 0$. After eliminating from D the domain D_ϵ defined by the inequalities $y - \epsilon < \eta < y + \epsilon$ we shall put in the remaining domain $z = 0$. One obtains the following singular kernel

$$n_1(x_0, y_0) = \lim_{z\to 0} n_1(x_0, y_0, z) = -\frac{H(x_0 - u)}{y_0^2}\left[x_0\frac{\cos(\nu S)}{S}\mathrm{e}^{-\mathrm{i}\bar{\omega}x_0} + \right.$$

$$\left. + \frac{\mathrm{i}}{M}\mathrm{e}^{-\mathrm{i}\bar{\omega}x_0}\sin(\nu S) + \frac{\omega}{M}\int_s^{x_0}\mathrm{e}^{-\mathrm{i}\bar{\omega}\tau}\sin(\nu S_\tau)\mathrm{d}\,\tau\right] =$$

$$= H(x_0 - u)n(x_0, y_0)\,, \tag{10.4.24}$$

where, $u = k|y_0|$ and

$$S = \sqrt{x_0^2 - u^2}\,,\ S_\tau = \sqrt{\tau^2 - u^2}\,. \tag{10.4.25}$$

For $\omega = 0$ one obtains (8.3.23). This will be the kernel of the integral equation.

In the sequel we shall give a demonstration where the terms which contain the factor z^2 are not neglected. As we have already noticed, acting in the classical manner, we have to calculate the limits for $z \to \pm 0$ of some kernels which contain derivatives with respect to this variable (see Mangler [10.52] for the subsonic steady flow, Heaslet and Loomax for the supersonic steady flow, Watkins, Runyan and Woolston [10.86] for the oscillatory subsonic flow, Watkins and Berman for the supersonic flow, etc.).

Since generally, for performing this calculation we have to evaluate at first the derivatives, the passage to the limit becomes difficult. In order to avoid this, we gave other expressions to the component w

((2.3.29), (2.3.37), (2.4.23)). In the general solutions built on the basis of these expressions it appears only the derivatives with respect to y. The passage to the limit interchanges with these derivatives. In the actual case from (10.4.11) we obtain

$$n_2(x_0, y_0) = \lim_{z\to 0} n_2(x_0, y_0, z) =$$

$$= (2\mathrm{i}\omega + k^2)\frac{H(x_0 - u)}{S}\cos(\nu S)\mathrm{e}^{-\mathrm{i}\overline{\omega}x_0} + \tag{10.4.26}$$

$$+(\omega^2 - \frac{\partial^2}{\partial y^2})H(x_0 - u)\int_u^{x_0}\frac{\cos(\nu S_\tau)}{S_\tau}\mathrm{e}^{-\mathrm{i}\overline{\omega}\tau}\,\mathrm{d}\,\tau\,,$$

where $u = k|y_0|$.

Since we have $\partial^2/\partial y^2 = k^2\partial^2/\partial u^2$, with the notation

$$J = \frac{\partial}{\partial u}\int_u^{x_0}\frac{\cos(\nu S_\tau)}{S_\tau}\mathrm{e}^{-\mathrm{i}\overline{\omega}\tau}\,\mathrm{d}\,\tau\,, \tag{10.4.27}$$

we deduce

$$\frac{\partial^2}{\partial y^2}H(x_0 - u)J = k^2\frac{\partial^2}{\partial u^2}H(x_0 - u)J = k^2\frac{\partial}{\partial u}H(x_0 - u)J \tag{10.4.28}$$

where J, calculated like I ,is

$$J = -\frac{x_0}{u}\frac{\cos(\nu S)}{S}\mathrm{e}^{-\mathrm{i}\overline{\omega}x_0} - \frac{\mathrm{i}}{Mu}\mathrm{e}^{-\mathrm{i}\overline{\omega}x_0}\sin(\nu S) - \tag{10.4.29}$$

$$-\frac{\omega}{Mu}\int_u^{x_0}\mathrm{e}^{-\mathrm{i}\overline{\omega}\tau}\sin(\nu S_\tau)\mathrm{d}\,\tau\,.$$

For determining $H(x_0 - u)J$ we take (2.3.35) into account. In this way, from (10.4.25) one obtains rigorously (10.4.24).

If the equation of the oscillatory surface is

$$z = h(x, y)\mathrm{e}^{\mathrm{i}\omega t}\,,\ (x, y) \in D$$

then one imposes the boundary condition

$$\mathrm{w}(x, y, 0) = \frac{\partial}{\partial x}h(x, y) + \mathrm{i}\omega h(x, y) \equiv G(x, y)\,,\ (x, y) \in D \tag{10.4.30}$$

One obtains the following integral equation

$$\iint_{D_1} f(\xi, \eta)\mathrm{e}^{-\mathrm{i}\omega x_0}n(x_0, y_0)\,\mathrm{d}\,\xi\,\mathrm{d}\,\eta\; = 2\pi G(x, y)\,, \tag{10.4.31}$$

D_1 being the domain marked in figure 8.3.5 (the domain where $x_0 > u$).

10.4.4 Other Expressions of the Kernel

We have

$$L = \int_u^{x_0} \mathrm{e}^{-\mathrm{i}\bar{\omega}\tau} \sin(\nu S_\tau)\mathrm{d}\,\tau = \frac{1}{2\mathrm{i}}(L_- - L_+)\,, \tag{10.4.32}$$

where we denoted

$$L_\mp = \int_u^{x_0} \mathrm{e}^{-\mathrm{i}\bar{\omega}(\tau \mp \mathrm{M}S_\tau)}\mathrm{d}\,\tau\,. \tag{10.4.33}$$

In $L_\mp$ we perform the substitutions $\tau \to \lambda$:

$$\tau \mp MS_\tau = ku\lambda\,. \tag{10.4.34}$$

Taking into account that τ is positive in both cases we deduce

$$k\tau = -u\lambda + uM\sqrt{1+\lambda^2}, \tag{10.4.35}$$

such that

$$L_\mp = |y_0| \int_{1/k}^{(x_0 \mp MS)/ku} \left(\frac{M\lambda}{\sqrt{1+\lambda^2}} - 1\right) \mathrm{e}^{-\mathrm{i}\omega|y_0|\lambda}\mathrm{d}\,\lambda \tag{10.4.36}$$

whence

$$L = -\frac{|y_0|}{2\mathrm{i}} \int_{(x_0-MS)/ku}^{(x_0+MS)/ku} \left(\frac{M\lambda}{\sqrt{1+\lambda^2}} - 1\right) \mathrm{e}^{-\mathrm{i}\omega|y_0|\lambda}\mathrm{d}\,\lambda\,. \tag{10.4.37}$$

Since, on the other side,

$$\int_{(x_0-MS)/ku}^{(x_0+MS)/ku} \mathrm{e}^{-\mathrm{i}\omega|y_0|\lambda}\mathrm{d}\,\lambda = \frac{2}{\omega|y_0|}\mathrm{e}^{-\mathrm{i}\omega x_0} \sin(\nu S)\,, \tag{10.4.38}$$

from (10.4.24) we deduce

$$n(x_0, y_0) = -\frac{x_0}{y_0^2}\frac{\cos(\nu S)}{S}\mathrm{e}^{-\mathrm{i}\bar{\omega}x_0} - \frac{\mathrm{i}\omega}{2|y_0|}\int_{(x_0-MS)/ku}^{(x_0+MS)/ku} \frac{\lambda}{\sqrt{1+\lambda^2}}\mathrm{e}^{-\mathrm{i}\omega|y_0|\lambda}\,\mathrm{d}\lambda. \tag{10.4.39}$$

This is the kernel given by Watkins [10.85]. Obviously for $\omega = 0$ one obtains the steady kernel.

Performing the change of variable $\lambda \to v : |y_0|\lambda = v$ and and integrating by parts, we deduce

$$I = \frac{\mathrm{i}\omega}{2|y_0|}\int_{(x_0-MS)/ku}^{(x_0+MS)/ku} \frac{\lambda}{\sqrt{1+\lambda^2}} \mathrm{e}^{-\mathrm{i}\omega|y_0|\lambda}\,\mathrm{d}\lambda = \frac{\mathrm{i}\omega}{2y_0^2}\int_{X_-}^{X_+} \frac{v\mathrm{e}^{-\mathrm{i}\omega v}}{\sqrt{y_0^2+v^2}}\,\mathrm{d}v =$$

$$= -\frac{1}{2y_0^2}\left(\frac{X_+}{\sqrt{y_0^2+X_+^2}}\mathrm{e}^{-\mathrm{i}\omega X_+} - \frac{X_-}{\sqrt{y_0^2+X_-^2}}\mathrm{e}^{-\mathrm{i}\omega X_-}\right) +$$

$$+\frac{1}{2}\int_{X_-}^{X_+} \frac{\mathrm{e}^{-\mathrm{i}\omega v}}{(y_0^2+v^2)^{3/2}}\mathrm{d}\,v\,, \tag{10.4.40}$$

where

$$k^2 X_\pm = x_0 \pm MS\,. \tag{10.4.41}$$

Observing now that

$$2\mathrm{e}^{-\mathrm{i}\overline{\omega}x_0}\cos(\nu S) = \mathrm{e}^{-\mathrm{i}\omega X_-} + \mathrm{e}^{-\mathrm{i}\omega X_+}\,, \tag{10.4.42}$$

utilizing (10.4.40) and the identities

$$x_0\sqrt{y_0^2+X_+^2} - SX_+ = x_0\sqrt{y_0^2+X_-^2} + SX_- = My_0^2\,, \tag{10.4.43}$$

we obtain for $n(x_0, y_0)$ the following form given by Harder and Rodden [10.29]

$$2n(x_0,y_0) = -\frac{M}{S}\left(\frac{\mathrm{e}^{-\mathrm{i}\omega X_+}}{\sqrt{y_0^2+X_+^2}} + \frac{\mathrm{e}^{-\mathrm{i}\omega X_-}}{\sqrt{y_0^2+X_-^2}}\right) - \int_{X_-}^{X_+} \frac{\mathrm{e}^{-\mathrm{i}\omega v}}{(y_0^2+v^2)^{3/2}}\mathrm{d}\,v\,. \tag{10.4.44}$$

Another form of the kernel is obtained if one utilizes the identities

$$\frac{1}{\sqrt{y_0^2+X_\pm^2}} = \frac{Mx_0 \mp S}{x_0^2+y_0^2} = \frac{M}{x_0+X_\pm}\,. \tag{10.4.45}$$

One obtains the relation

$$2n(x_0,y_0) = -\frac{M^2}{S}\left(\frac{\mathrm{e}^{-\mathrm{i}\omega X_-}}{x_0+X_-} + \frac{\mathrm{e}^{-\mathrm{i}\omega X_+}}{x_0+X_+}\right) - \int_{X_-}^{X_+} \frac{\mathrm{e}^{-\mathrm{i}\omega v}}{(y_0^2+v^2)^{3/2}}\mathrm{d}\,v\,, \tag{10.4.46}$$

utilized by Ueda and Dowell [10.81] for obtaining the numerical solution of the integral equation.

One obtains the sonic limit at once from (10.4.44), or (10.4.46) noticing that

$$\lim_{M\to 1} X_- = -\frac{1}{2}\left(x_0 - \frac{y_0^2}{x_0}\right) \equiv -X\,, \ \lim_{M\to 1} X_+ = \infty\,. \tag{10.4.47}$$

One obtains the following kernel

$$n_s(x_0, y_0) = -\frac{H(x_0)}{2}\left[\frac{2}{x_0^2 + y_0^2} + \mathrm{e}^{\mathrm{i}\omega X} + \int_{-\infty}^{X} \frac{\mathrm{e}^{\mathrm{i}\omega v}}{(y_0^2 + v^2)^{3/2}}\mathrm{d}\,v\right]\,, \tag{10.4.48}$$

which coincides with (10.4.26).

10.4.5 A New Form

We utilize the formulas (see for example [1.30], pp. 406, 422) with real parameter

$$\int_u^\infty \frac{\cos(\nu S_\tau)}{S_\tau}\cos(p\tau)\mathrm{d}\,\tau = K_0(u\sqrt{\nu^2 - p^2})\,,$$
$$\int_u^\infty \frac{\cos(\nu S_\tau)}{S_\tau}\sin(p\tau)\mathrm{d}\,\tau = 0. \tag{10.4.49}$$

For $p = \overline{\omega}$ we obtain the identity

$$\int_u^{x_0} \frac{\cos(\nu S_\tau)}{S_\tau}\mathrm{e}^{-\mathrm{i}\overline{\omega}\tau}\mathrm{d}\,\tau = K_0(\omega|y_0|) - \int_{x_0}^\infty \frac{\cos(\nu S_\tau)}{S_\tau}\mathrm{e}^{-\mathrm{i}\overline{\omega}\tau}\mathrm{d}\,\tau\,, \tag{10.4.50}$$

as follows

$$\frac{1}{2\pi}\left(\omega^2 - k^2\frac{\partial^2}{\partial u^2}\right)\int_u^{x_0} \frac{\cos(\nu S_\tau)}{S_\tau}\mathrm{e}^{-\mathrm{i}\overline{\omega}\tau}\mathrm{d}\,\tau =$$

$$= \frac{1}{2\pi}\left(\omega^2 - k^2\frac{\partial^2}{\partial u^2}\right)\left[K_0(\frac{\omega}{k}u) - \int_{x_0}^\infty \frac{\cos(\nu S_\tau)}{S_\tau}\mathrm{e}^{-\mathrm{i}\overline{\omega}\tau}\mathrm{d}\,\tau\right]\,,$$

In the last part we derive without any difficulty. Deriving, the kernel (10.4.25) becomes

$$n_2(x_0, y_0) = H(x_0 - u)n(x_0, y_0) =$$

$$n(x_0, y_0) = -\frac{k\omega}{u}K_1(\frac{\omega}{k}u) - \omega M\int_{x_0}^\infty \frac{\sin(\nu S_\tau)}{S_\tau^2}\mathrm{e}^{-\mathrm{i}\overline{\omega}\tau}\mathrm{d}\,\tau - \tag{10.4.51}$$

$$-k^2\int_{x_0}^\infty \frac{\cos(\nu S_\tau)}{S_\tau^3}\mathrm{e}^{-\mathrm{i}\overline{\omega}\tau}\mathrm{d}\,\tau\,,$$

where $u = k|y_0|$. This is the new form of the kernel. Having in view the behaviour of K_1, for small values of the argument, this is

$$\frac{1}{y_0^2} - \frac{\omega^2}{2}(\ln|y_0| + \Gamma_1)\,, \quad \Gamma_1 = \ln\frac{\omega k}{2} + \gamma - \frac{1}{2}\,, \tag{10.4.52}$$

γ being Euler's constant. An additive constant Γ_2 also appears from the two integrals (10.4.51).

The kernel of the integral equation in the case $M = 1$ is obtained from (10.4.51). We have

$$\lim_{M\to 1} \frac{k\omega}{u} K_1\left(\frac{\omega}{k}u\right) = \frac{\omega}{|y_0|} K_1(\omega|y_0|)\,.$$

Denoting

$$I_1 = \frac{1}{2\mathrm{i}} \int_{x_0}^{\infty} \exp[\mathrm{i}\overline{\omega}(MS_\tau - \tau)] \frac{\mathrm{d}\,\tau}{S_\tau^2}$$

$$I_2 = \frac{1}{2\mathrm{i}} \int_{x_0}^{\infty} \exp[-\mathrm{i}\overline{\omega}(MS_\tau + \tau)] \frac{\mathrm{d}\,\tau}{S_\tau^2}$$

we have

$$\int_{x_0}^{\infty} \frac{\sin(\nu S_\tau)}{S_\tau^2} \mathrm{e}^{-\mathrm{i}\overline{\omega}\tau} \mathrm{d}\,\tau = I_1 + I_2\,.$$

But

$$I_1 = \frac{1}{2\mathrm{i}} \int_{x_0}^{\infty} \exp\left[\mathrm{i}\overline{\omega}\frac{M^2 S_\tau^2 - \tau^2}{MS_\tau^2 + \tau}\right] \frac{\mathrm{d}\,\tau}{S_\tau^2} =$$

$$= \frac{1}{2\mathrm{i}} \int_{x_0}^{\infty} \exp\left[\mathrm{i}\omega\frac{\tau^2 - M^2 y_0^2}{M\sqrt{\tau^2 - k^2 y_0^2} + \tau}\right] \frac{\mathrm{d}\,\tau}{\tau^2 - k^2 y_0^2}\,,$$

$$\lim_{M\to 1} I_1 = \frac{1}{2\mathrm{i}} \int_{x_0}^{\infty} \exp\left[\frac{\mathrm{i}\omega}{2}\left(\tau - \frac{y_0^2}{\tau}\right)\right] \frac{\mathrm{d}\,\tau}{\tau^2}\,.$$

One obtains after all (10.6.13).

10.4.6 The Plane Problem

In this case, the density $f(\xi, \eta)$ becomes $f(\xi)$ such that in the representation (10.4.7)-(10.4.9) we can calculate the integral with respect to η. We have

$$g_0 = \int_{-\infty}^{\infty} G_0(x_0, y_0, z)\mathrm{d}\,\eta =$$

$$= \frac{1}{2\pi}\mathrm{e}^{-\mathrm{i}ax_0} \int_{-\infty}^{+\infty} H(x_0 - k\sqrt{\tau^2 + z^2}) \frac{\cos\left[\nu\sqrt{x_0^2 - k^2(\tau^2 + z^2)}\right]}{\sqrt{x_0^2 - k^2(\tau^2 + z^2)}}\mathrm{d}\,\tau\,. \tag{10.4.53}$$

Because of the presence of the function H, the integrand differs from zero only for $x_0 > k\sqrt{\tau^2 + z^2}$. This inequality implies $x_0 > 0$ and $k^2\tau^2 < x_0^2 - k^2z^2$, whence $x_0 > k|z|$ and $-c < k\tau < c$, where $c = \sqrt{x_0^2 - k^2z^2}$. After all

$$g_0 = \frac{1}{2\pi}\mathrm{e}^{-\mathrm{i}ax_0} H(x_0 - k|z|) \int_{-c}^{c} \frac{\cos\left(\nu\sqrt{c^2 - k^2\tau^2}\right)}{\sqrt{c^2 - k^2\tau^2}}\mathrm{d}\,\tau\,. \tag{10.4.54}$$

Utilizing the formula [1.16]

$$\int_0^c \frac{\cos\left(p\sqrt{c^2 - x^2}\right)}{\sqrt{c^2 - x^2}} \cos qx\mathrm{d}\,x = \frac{\pi}{2}J_0(c\sqrt{p^2 + q^2})\,, \tag{10.4.55}$$

it results

$$g_0 = \frac{1}{2k}H(x_0 - k|z|)J_0(\nu\sqrt{x_0^2 - k^2z^2})\mathrm{e}^{-\mathrm{i}ax_0} \tag{10.4.56}$$

and analogously

$$g = \int_{-\infty}^{+\infty} G\mathrm{d}\,\eta = \frac{1}{2k}H(x_0 - k|z|)J_0(\nu\sqrt{x_0^2 - k^2z^2})\mathrm{e}^{-\mathrm{i}\bar{\omega}x_0}\,, \tag{10.4.57}$$

$$e = \int_{-\infty}^{+\infty} E\mathrm{d}\,\eta = \frac{1}{2k}\int_{-\infty}^{x_0} H(\tau - k|z|)J_0(\nu\sqrt{\tau^2 - k^2z^2})\mathrm{e}^{-\mathrm{i}\bar{\omega}\tau}\mathrm{d}\,\tau =$$

$$= \frac{H(x_0 - k|z|)}{2k}\int_{k|z|}^{x_0} J_0(\nu\sqrt{\tau^2 - k^2z^2})\mathrm{e}^{-\mathrm{i}\bar{\omega}\tau}\mathrm{d}\,\tau\,. \tag{10.4.58}$$

For obtaining the results from 10.3 we have to consider the chord of the profile on the Ox axis ($0 < x < L_0$) and to take L_0 as reference length. Observing that

$$\int_{-\infty}^{+\infty} \frac{\partial^2}{\partial y^2}E(x_0, y_0, z)\mathrm{d}\,\eta = -\int_{-\infty}^{+\infty} \frac{\partial^2}{\partial y\partial \eta}E\mathrm{d}\,\eta = -\frac{\partial}{\partial y}E\,\Big|_{-\infty}^{+\infty} = 0\,,$$

we deduce

$$p(x,z) = -\int_0^1 f(\xi)\frac{\partial}{\partial z}g_0 \,\mathrm{d}\,\xi\,,$$

$$\mathrm{w}(x,z) = \int_0^1 f(\xi)\mathrm{e}^{-\mathrm{i}\omega \mathrm{x}_0}\left[(2\mathrm{i}\omega + k^2\frac{\partial}{\partial x})g + \omega^2 e\right]\mathrm{d}\,\xi\,, \tag{10.4.59}$$

which is exactly the solution (10.3.6). We obtain too

$$n(x_0) = \int_{-\infty}^{+\infty} n_2(x_0, y_0)\mathrm{d}\,\eta =$$

$$= 2\mathrm{i}\omega\int_{-\infty}^{+\infty} G(x_0, y_0, 0)\mathrm{d}\,\eta + k^2\frac{\partial}{\partial x}\int_{-\infty}^{+\infty} G(x_0, y_0, 0)\mathrm{d}\,\eta +$$

$$+\omega^2\int_{-\infty}^{+\infty} G(\tau, y_0, 0)\mathrm{d}\,\eta = \frac{k}{2}\delta(x_0)J_0(\nu x_0)\mathrm{e}^{-\mathrm{i}\overline{\omega}\mathrm{x}_0} +$$

$$+H(x_0)n_0(x_0)\,, \tag{10.4.60}$$

where

$$n_0(x_0) = \frac{\mathrm{i}\omega}{k}\left[J_0(\nu x_0) + \frac{\mathrm{i}M}{2}J_1(\nu x_0)\right]\mathrm{e}^{-\mathrm{i}\overline{\omega}\mathrm{x}_0} + \frac{\omega^2}{2k}\int_0^{x_0} J_0(\nu\tau)\mathrm{e}^{-\mathrm{i}\overline{\omega}\tau}\mathrm{d}\,\tau\,, \tag{10.4.61}$$

i.e. (10.3.13).

10.5 The Oscillatory Profile in a Sonic Stream

10.5.1 The General Solution. The Integral Equation

We proved in (10.2.54) that there exists the limit of the subsonic solution for $M \nearrow 1$, and in (10.2.45) that there exists the limit of the supersonic solution for $M \searrow 1$, and in addition, the two limits coincide. We shall prove now that there exists also the solution for $M = 1$, and this one coincides with the two limits. It will result therefore that the flow is continuous to the passage past the sonic barrier, unlike the case of the steady flow. We shall consider therefore the oscillatory profile (fig. 10.5.1) of the equation

$$y = h(x)\exp(\mathrm{i}\omega t)\,, \tag{10.5.1}$$

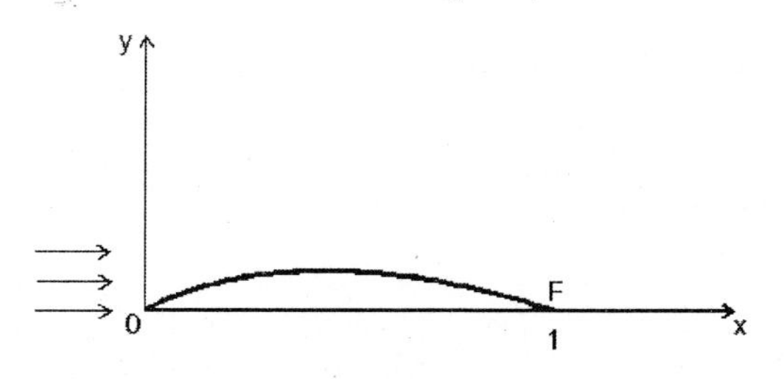

Fig. 10.5.1.

which perturbs the uniform flow which has the velocity $U_\infty = c_\infty$ ($M = 1$). With the notations (10.1.4), the boundary condition (2.1.27) gives

$$v(x, \pm 0) = h'(x) + \mathrm{i}\omega h(x)\,, 0 < x < 1\,. \tag{10.5.2}$$

The fundamental solution in the two-dimensional sonic flow is given by the formulas (2.4.32)–(2.4.34). If the profile is reduced to the skeleton like in figure 10.5.1 it is sufficient to replace it by a forces distribution having the form

$$(0, f)\exp(\mathrm{i}\omega t)\,. \tag{10.5.3}$$

It results therefore

$$\begin{aligned} p(x, y) &= -f\frac{\partial}{\partial y}G_0\,, \\ v(x, y) &= f\mathrm{e}^{-\mathrm{i}\omega x}\left[(2\mathrm{i}\omega G + \omega^2\int_0^x G(\tau, y)\mathrm{d}\,\tau\right]\,, \end{aligned} \tag{10.5.4}$$

where

$$\begin{aligned} G_0(x, y) &= H(x)\frac{\Omega}{\sqrt{x}}\exp\left[-\frac{\mathrm{i}\omega}{2}(x + \frac{y^2}{x})\right] \equiv H(x)g_0(x, y)\,, \\ G(x, y) &= H(x)\frac{\Omega}{\sqrt{x}}\exp\left[\frac{\mathrm{i}\omega}{2}(x - \frac{y^2}{x})\right] \equiv H(x)g(x, y)\,, \end{aligned} \tag{10.5.5}$$

with the notation $2\Omega\sqrt{2\pi\mathrm{i}\omega} = 1$.

A continuous superposition of forces having the shape (10.5.3) on the segment $[0, 1]$, will give the perturbation

$$\begin{aligned} p(x, y) &= -\int_0^1 f(\xi)\frac{\partial}{\partial y}G_0(x_0, y)\mathrm{d}\,\xi\,, \\ v(x, y) &= \int_0^1 f(\xi)\mathrm{e}^{-\mathrm{i}\omega x_0}\left[(2\mathrm{i}\omega G(x_0, y) + \omega^2\int_0^{x_0} G(\tau, y)\mathrm{d}\,\tau\right]\mathrm{d}\,\xi \end{aligned} \tag{10.5.6}$$

Taking the significance of the function $H(x)$ into account, it results that for $x < 0$ we have

$$p = 0\,,\ v = 0\,, \tag{10.5.7}$$

and for $x > 0$

$$p(x,y) = -\int_0^x f(\xi)\frac{\partial}{\partial y}g_0(x_0,y)\mathrm{d}\,\xi\,, \tag{10.5.8}$$

$$v(x,y) = \int_0^x f(\xi)\mathrm{e}^{-\mathrm{i}\omega x_0}\left[(2\mathrm{i}\omega g(x_0,y) + \omega^2\int_0^{x_0} g(\tau,y)\mathrm{d}\,\tau\right]\mathrm{d}\,\xi\,. \tag{10.5.9}$$

This is the general solution of the problem. It was given in [10.18].

Performing the change of variable $\xi \to u : u = y^2/x_0$ one obtains

$$p(x,y) = \mp\mathrm{i}\omega\Omega\int_{\frac{y^2}{x}}^{\infty} f(x-\frac{y^2}{u})\exp\left[-\frac{\mathrm{i}\omega}{2}(\frac{y^2}{u}+u)\right]\frac{\mathrm{d}\,u}{\sqrt{u}}$$

and then

$$p(x,\pm 0) = \mp\mathrm{i}\omega\Omega f(x)\int_0^{\infty}\exp\left(-\frac{\mathrm{i}\omega}{2}u\right)\frac{\mathrm{d}\,u}{\sqrt{u}} = \mp\frac{1}{2}f(x)\,. \tag{10.5.10}$$

It results therefore

$$p(x,-0) - p(x,+0) = f(x)\,. \tag{10.5.11}$$

For $0 < x < 1$ we deduce

$$v(x,\pm 0) = \int_0^x f(\xi)\mathrm{e}^{-\mathrm{i}\omega x_0}n(x_0)\mathrm{d}\,\xi\,, \tag{10.5.12}$$

where

$$n(x_0) = \sqrt{\frac{\mathrm{i}\omega}{2\pi}}\left[\frac{1}{\sqrt{x_0}}\exp(\frac{\mathrm{i}\omega}{2}x_0) - \frac{\mathrm{i}\omega}{2}\int_0^{x_0}\exp\left(\frac{\mathrm{i}\omega}{2}\tau\right)\frac{\mathrm{d}\,\tau}{\sqrt{\tau}}\right]\,. \tag{10.5.13}$$

Imposing the boundary condition (10.5.2) it results the following integral equation

$$\int_0^x f(\xi)n(x_0)\mathrm{d}\,\xi = A(x) \tag{10.5.14}$$

where $A(x) = h'(x) + \mathrm{i}\omega h(x)$. The kernel (10.5.13) coincides with (10.2.54) and (10.3.46).

10.5.2 Some Formulas for the Lift and Moment Coefficients

Taking into account that

$$\Omega = \frac{1}{2\sqrt{2\pi \mathrm{i}\omega}} = \frac{1-\mathrm{i}}{4\sqrt{\pi\omega}}\,, \tag{10.5.15}$$

the solution (10.3.52) may be written as follows

$$\begin{aligned} f(x) = 2\mathrm{i}\omega\Omega \int_0^x \frac{A(x_0)}{\sqrt{\xi}} \exp(-\frac{\mathrm{i}\omega}{2}\xi)\mathrm{d}\,\xi - \\ -2\Omega \,{}^*\!\!\int_0^x \frac{A(x_0)}{\xi^{3/2}} \exp(-\frac{\mathrm{i}\omega}{2}\xi)\mathrm{d}\,\xi\,, \end{aligned} \tag{10.5.16}$$

the sign $*$ indicating the Finite Part (Appendix D). Denoting

$$F = \int_0^x \frac{A(x_0)}{\sqrt{\xi}} \exp(-\frac{\mathrm{i}\omega}{2}\xi)\mathrm{d}\,\xi = \int_0^x \frac{A(\xi)}{\sqrt{x_0}} \exp(-\frac{\mathrm{i}\omega}{2}x_0)\mathrm{d}\,\xi\,, \tag{10.5.17}$$

with the definition formula (D.4.2), we deduce

$${}^*\!\!\int_0^x \frac{\partial}{\partial x}\left[\frac{A(\xi)}{\sqrt{x_0}} \exp(-\frac{\mathrm{i}\omega}{2}x_0)\right]\mathrm{d}\,\xi = \frac{\mathrm{d}\,F}{\mathrm{d}\,x}\,, \tag{10.5.18}$$

and then

$$-\,{}^*\!\!\int_0^x \frac{A(x_0)}{\xi^{3/2}} \exp(-\frac{\mathrm{i}\omega}{2}\xi)\mathrm{d}\,\xi = \mathrm{i}\omega F + 2\frac{\mathrm{d}\,F}{\mathrm{d}\,x}\,. \tag{10.5.19}$$

After all the formula (10.5.16) becomes

$$f(x) = 4\Omega\left(\mathrm{i}\omega F + \frac{\mathrm{d}\,F}{\mathrm{d}\,x}\right)\,. \tag{10.5.20}$$

The lift and moment coefficients are given by the formulas (10.3.29) with (10.3.30). Utilizing (10.5.20) we find

$$\begin{aligned} c_L &= 8\Omega \int_0^1 \frac{A(1-\xi) + \mathrm{i}\omega B(\xi)}{\sqrt{\xi}} \exp(-\frac{\mathrm{i}\omega}{2}\xi)\mathrm{d}\,\xi\,, \\ c_M &= 8\Omega \int_0^1 \frac{A(1-\xi) - B(\xi) + \mathrm{i}\omega D(\xi)}{\sqrt{\xi}} \exp(-\frac{\mathrm{i}\omega}{2}\xi)\mathrm{d}\,\xi\,, \end{aligned} \tag{10.5.21}$$

where we denoted

$$B(\xi) = \int_\xi^1 A(x_0)\mathrm{d}\,x\,,\ D(\xi) = \int_\xi^1 xA(x_0)\mathrm{d}\,x\,. \tag{10.5.22}$$

Approximating the function $h(x)$ by polynomials, we deduce that the integrals from (10.5.21) have the form

$$I_n = \int_0^1 \xi^{n-\frac{1}{2}} \exp(-\frac{i\omega}{2}\xi) d\,\xi\,,\ n = 1, 2, \ldots \tag{10.5.23}$$

Integrating by parts we obtain

$$I_n = -\frac{2}{i\omega}\exp(-\frac{i\omega}{2}) + \frac{2n-1}{i\omega} I_{n-1}\,,\ n = 1, 2, \ldots \tag{10.5.24}$$

It results that all the integrals from (10.5.21) may be expressed as functions of

$$I_0 = \int_0^1 \exp(-\frac{i\omega}{2}\xi)\frac{d\,\xi}{\sqrt{\xi}} = 2\sqrt{\frac{\pi}{\omega}}\left[C(\sqrt{\frac{\omega}{\pi}}) - iS(\sqrt{\frac{\omega}{\pi}})\right]\,, \tag{10.5.25}$$

where $C(x)$ and $S(x)$ are the integrals of Fresnel [1.30]:

$$C(x) = \frac{1}{\sqrt{2\pi}}\int_0^z \frac{\cos\xi}{\sqrt{\xi}} d\,\xi\,,\ S(x) = \frac{1}{\sqrt{2\pi}}\int_0^z \frac{\sin\xi}{\sqrt{\xi}} d\,\xi \tag{10.5.26}$$

with the notation $z = \pi x^2/2$.

In the case of the flat plate having the angle of attack ϵ ($h = -\epsilon x$) one obtains

$$\begin{aligned} c_L &= -4\Omega\epsilon(1+i\omega)\left[2\exp(-\frac{i\omega}{2}) + (1+i\omega)I_0\right]\,, \\ c_M &= -8\Omega\epsilon\left[\left(-\frac{1}{i\omega} + \frac{2}{3}i\omega + \frac{1}{3}\right)\exp(-\frac{i\omega}{2}) + \left(\frac{1}{2i\omega} + \frac{i\omega}{2} - \frac{\omega^2}{3}\right)I_0\right]\,. \end{aligned} \tag{10.5.27}$$

10.6 The Three-Dimensional Sonic Flow

10.6.1 The General Solution

In the three-dimensional sonic flow the fundamental solution is (2.4.36) and (2.4.37). A force having the shape $(0, 0, f)\exp(i\omega t)$, applied in the origin, will produce the perturbation

$$P = H(x)p\,,\ \mathrm{W} = H(x)\mathrm{w}\,, \tag{10.6.1}$$

where

$$\begin{aligned} p(x,y) &= -f\frac{\partial}{\partial z}g_0\,, \\ \mathrm{w}(x,y) &= f e^{-i\omega x}\left[2i\omega g + \left(\omega^2 - \frac{\partial^2}{\partial y^2}\right)\int_0^x g(\tau, y, z)d\,\tau\right]\,, \end{aligned} \tag{10.6.2}$$

with the notations

$$g_0(x,y,z) = \frac{1}{4\pi x}\exp\left(-\frac{\mathrm{i}\omega}{2x}r^2\right),$$
$$g(x,y,z) = \frac{1}{4\pi x}\exp\left[\frac{\mathrm{i}\omega}{2}\left(x - \frac{y^2+z^2}{x}\right)\right]. \tag{10.6.3}$$

The perturbation produced by a distribution of forces having the form

$$(0,0,f(\xi,\eta))\exp(\mathrm{i}\omega t),$$

defined on the domain D (the projection of the wing on the xOy plane), will be characterized by the formulas

$$p(x,y,z) = -\iint_D f(\xi,\eta)\frac{\partial}{\partial z}\,g_0(x_0,y_0,z)\,\mathrm{d}\xi\,\mathrm{d}\eta,$$
$$\mathrm{w}(x,y,z) = \iint_D f(\xi,\eta)\mathrm{e}^{-\mathrm{i}\omega \mathrm{x}_0}[(2\mathrm{i}\omega g(x_0,y_0,z)+$$
$$+(\omega^2 - \frac{\partial^2}{\partial y^2})\int_0^{x_0} g(\tau,y_0,z)\mathrm{d}\,\tau]\,\mathrm{d}\xi\mathrm{d}\,\eta. \tag{10.6.4}$$

10.6.2 The Integral Equation

Assuming that D is such that every parallel to the span (the Oy axis) intersects the boundary ∂D in at most two points and denoting by y_- and y_+ the ordinates of these points (fig. 10.6.1), we obtain

$$\iint_D = \int_0^x \ldots\left(\int_{\eta_-(\xi)}^{\eta_+(\xi)}\ldots\mathrm{d}\,\eta\right)\mathrm{d}\,\xi \tag{10.6.5}$$

With the change of variables $(\xi,\eta)\to u,v$:

$$\xi = x - \frac{z^2}{u},\ \eta = y + |z|v;\ \Rightarrow \mathrm{d}\,\xi\,\mathrm{d}\,\eta = \frac{z^2}{u^2}|z|\mathrm{d}\,u\,\mathrm{d}\,v, \tag{10.6.6}$$

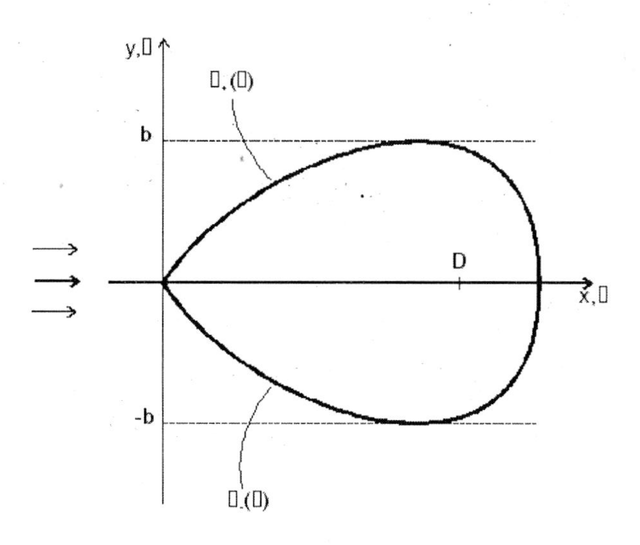

Fig. 10.6.1.

we deduce

$$p(x,y,\pm 0)=$$
$$=\frac{\mathrm{i}\omega}{4\pi}\lim_{z\to\pm 0} z\int_0^x\int_{\eta_-(\xi)}^{\eta_+(\xi)}\frac{f(\xi,\eta)}{x_0^2}\exp\left[\frac{\mathrm{i}\omega}{2}\left(x_0+\frac{y_0^2+z^2}{x_0}\right)\right]\mathrm{d}\,\xi\,\mathrm{d}\,\eta$$

$$=\frac{\mathrm{i}\omega}{4\pi}\lim_{z\to\pm 0}\frac{|z|}{z}\int_{z^2/x}^{\infty}\int_{\eta_-(\xi)-y/|z|}^{\eta_+(\xi)-y/|z|} f(x-\frac{z^2}{u},y+|z|v)\cdot$$
$$\cdot\exp\left\{-\frac{\mathrm{i}\omega}{2}\left[\frac{z^2}{u}+u\left(v^2+1\right)\right]\right\}\mathrm{d}u\,\mathrm{d}v$$

$$=\pm\frac{\mathrm{i}\omega}{4\pi}f(x,y)\int_0^{\infty}\exp\left(-\frac{\mathrm{i}\omega}{2}u\right)\left[\int_{-\infty}^{\infty}\exp\left(-\frac{\mathrm{i}\omega}{2}uv^2\right)\mathrm{d}\,v\right]\mathrm{d}\,u\,.$$

Utilizing the integrals of Fresnel

$$\int_0^{\infty}\cos x^2\mathrm{d}\,x=\frac{1}{2}\int_0^{\infty}\frac{\cos t}{\sqrt{t}}\mathrm{d}\,t=\int_0^{\infty}\sin x^2\mathrm{d}\,x=\frac{1}{2}\int_0^{\infty}\frac{\sin t}{\sqrt{t}}\mathrm{d}\,t=\frac{1}{2}\sqrt{\frac{\pi}{2}}\,, \tag{10.6.7}$$

we deduce

$$p(x,y,\pm 0)=\mp\frac{1}{2}f(x,y)\,. \tag{10.6.8}$$

Hence

$$p(x,y,-0)-p(x,y,+0)=f(x,y)\,. \tag{10.6.9}$$

Assuming that the equation of the oscillatory surface is

$$z = h(x,y)\exp(\mathrm{i}\omega t)\,, \tag{10.6.10}$$

we obtain the boundary condition

$$\mathrm{w}(x,y,\pm 0) = h'(x,y) + \mathrm{i}\omega h(x,y) \equiv A(x,y)\,, (x,y) \in D\,. \tag{10.6.11}$$

It results the following integral equation

$$\frac{1}{2\pi}\iint_D f(\xi,\eta)N(x_0,y_0)\,\mathrm{d}\,\xi\,\mathrm{d}\,\eta = A(x,y), \quad (x,y) \in D\,, \tag{10.6.12}$$

where

$$\begin{aligned} N(x_0,y_0) &= \frac{\mathrm{i}\omega}{x_0}\exp\left[\frac{\mathrm{i}\omega}{2}\left(x_0 - \frac{y_0^2}{x_0}\right)\right] + \\ &+ \frac{1}{2}\left(\omega^2 - \frac{\partial^2}{\partial y^2}\right)\int_0^{x_0}\exp\left[\frac{\mathrm{i}\omega}{2}\left(\tau - \frac{y_0^2}{\tau}\right)\right]\frac{\mathrm{d}\,\tau}{\tau}\,. \end{aligned} \tag{10.6.13}$$

We have also obtained this kernel from the kernel corresponding to the subsonic flow (10.2.58). On this formula we cannot observe yet the singular part of N We shall calculate therefore the last term. Using the formulas

$$\int_0^\infty \begin{matrix}\sin\\ \cos\end{matrix}\left(ax - \frac{b}{x}\right)\frac{\mathrm{d}\,x}{x} = \begin{cases}0\\ 2K_0(2\sqrt{ab})\,.\end{cases} \tag{10.6.14}$$

valid for $a > 0$ and $b > 0$, we may write

$$\int_0^{x_0}\exp\left[\frac{\mathrm{i}\omega}{2}\left(\tau - \frac{y_0^2}{\tau}\right)\right]\frac{\mathrm{d}\,\tau}{\tau} = 2K_0(\omega|y_0|) - \int_{x_0}^\infty \exp\left[\frac{\mathrm{i}\omega}{2}\left(\tau - \frac{y_0^2}{\tau}\right)\right]\frac{\mathrm{d}\,\tau}{\tau}.$$

The last integral may be derived with respect to y interchanging the derivation with the integration. One obtains

$$\frac{\partial^2}{\partial y^2}\left[\frac{1}{\tau}\exp\left(-\frac{\mathrm{i}\omega}{2\tau}y_0^2\right)\right] = 2\mathrm{i}\omega\frac{\partial}{\partial\tau}\left[\frac{1}{\tau}\exp\left(-\frac{\mathrm{i}\omega}{2\tau}y_0^2\right)\right] + \frac{\mathrm{i}\omega}{\tau^2}\exp\left(-\frac{\mathrm{i}\omega}{2\tau}y_0^2\right).$$

In this way we deduce the final form of the kernel

$$N(x_0,y_0) = -\frac{\omega}{|y_0|}K_1(\omega|y_0|) + \frac{\mathrm{i}\omega}{2}\int_0^{x_0}\exp\left[\frac{\mathrm{i}\omega}{2}\left(\tau - \frac{y_0^2}{\tau}\right)\right]\frac{\mathrm{d}\,\tau}{\tau^2}\,. \tag{10.6.15}$$

The principal part is in the first term. Taking into account that for small values of the argument we have

$$K_1(z) = \frac{1}{z} + z \ln z + \dots , \tag{10.6.16}$$

it results

$$N(x_0, y_0) = -\frac{1}{y_0^2} - \omega^2 \left(\ln |y_0| + \Gamma\right) + \dots . \tag{10.6.17}$$

The singularity has therefore the same order like in all the other spatial problems.

10.6.3 The Plane Problem

We remind that one obtains the solution of the plane problem if we assume that D is a rectangle having the dimensions L_0 and bL_0 and we consider $b \to \infty$. Moreover we assume that every section with a plane parallel to xOz determines the same profile, hence in (10.6.10) h depends only on x. Taking L_0 as a reference length, the domain D will be defined by $0 < x < 1$, $-b < y < b$. Considering $b \to \infty$ we have to obtain

$$n(x_0) = \int_{-\infty}^{+\infty} N(x_0, y_0)\, \mathrm{d}\,\eta , \tag{10.6.18}$$

where $n(x_0)$ must be (10.5.13), and $N(x_0, y_0)$ (10.5.15).

Indeed, utilizing the representation [1.30]

$$\frac{\omega K_1(\omega z)}{z} = \int_0^{\infty} \frac{\cos \omega t}{(t^2 + z^2)^{3/2}}\, \mathrm{d}\, t , \tag{10.6.19}$$

we deduce

$$\omega \,{}^{*}\!\!\int_{-\infty}^{+\infty} \frac{K_1(\omega |y_0|)}{|y_0|} \mathrm{d}\,\eta = 2\omega \,{}^{*}\!\!\int_0^{\infty} \frac{K_1(\omega u)}{u} \mathrm{d}\, u = 2 \,{}^{*}\!\!\int_0^{\infty} \frac{\cos \omega t}{t^2} \mathrm{d}\, t =$$

$$= 2 \,{}^{*}\!\!\int_0^{\infty} \frac{\cos \omega t - 1 + 1}{t^2} \mathrm{d}\, t = -\pi\omega + 2 \,{}^{*}\!\!\int_0^{\infty} \frac{\mathrm{d}\, t}{t^2} =$$

$$= -\pi\omega + 2 \,{}^{*}\!\!\int_0^{1} \frac{\mathrm{d}\, t}{t^2} + 2 \int_1^{\infty} \frac{\mathrm{d}\, t}{t^2} = -\pi\omega .$$

Utilizing Fresnel's formulas (10.6.7) and integrating by parts, we obtain

$$\int_{-\infty}^{+\infty}\left\{\int_{x_0}^{\infty}\exp\left[\frac{\mathrm{i}\omega}{2}\left(\tau-\frac{y_0^2}{\tau}\right)\right]\frac{\mathrm{d}\,\tau}{\tau^2}\right\}\mathrm{d}\,\eta=$$

$$=2\int_{x_0}^{+\infty}\exp\left(\frac{\mathrm{i}\omega}{2}\tau\right)\left[\int_0^{\infty}\exp\left(\frac{-\mathrm{i}\omega}{2\tau}u^2\right)\mathrm{d}\,u\right]\frac{\mathrm{d}\,\tau}{\tau^2}=$$

$$=\sqrt{\frac{\pi}{\omega}}(1-\mathrm{i})\left[\frac{2}{\sqrt{x_0}}\exp\left(\frac{\mathrm{i}\omega}{2}x_0\right)+\right.$$

$$\left.+\,\mathrm{i}\omega\int_0^{\infty}\exp\left(\frac{\mathrm{i}\omega}{2}\tau\right)\frac{\mathrm{d}\,\tau}{\sqrt{\tau}}-\mathrm{i}\omega\int_0^{x}\exp\left(\frac{\mathrm{i}\omega}{2}\tau\right)\frac{\mathrm{d}\,\tau}{\sqrt{\tau}}\right]$$

$$=2\pi\mathrm{i}+\sqrt{\frac{\pi}{\omega}}(1-\mathrm{i})\left[\frac{2}{\sqrt{x_0}}\exp\left(\frac{\mathrm{i}\omega}{2}x_0\right)-\mathrm{i}\omega\int_0^{x}\exp\left(\frac{\mathrm{i}\omega}{2}\tau\right)\frac{\mathrm{d}\,\tau}{\sqrt{\tau}}\right].$$

We deduce therefore

$$n(x_0)=(1+\mathrm{i})\sqrt{\pi\omega}\left[\frac{1}{\sqrt{x_0}}\exp\left(\frac{\mathrm{i}\omega}{2}x_0\right)-\frac{\mathrm{i}\omega}{2}\int_0^{x_0}\exp\left(\frac{\mathrm{i}\omega}{2}\tau\right)\frac{\mathrm{d}\,\tau}{\sqrt{\tau}}\right], \tag{10.6.20}$$

i.e. just (10.5.13).

10.6.4 Other Forms of the Kernel

From (2.4.13) and (2.4.35) it results the following representation of the component w of the velocity

$$\begin{aligned}\mathrm{w}(x,y,z)=&\,\delta(z)\iint_D f(\xi,\eta)H(x_0)\delta(y_0)\,\mathrm{d}\,\xi\,\mathrm{d}\,\eta+\\ &+\frac{1}{4\pi}\iint_D f(\xi,\eta)\mathrm{e}^{-\mathrm{i}\omega x_0}n(x_0,y_0,z)\,\mathrm{d}\,\xi\,\mathrm{d}\,\eta,\end{aligned} \tag{10.6.21}$$

where

$$n(x_0,y_0,z)=\frac{\partial^2}{\partial z^2}\int_{-\infty}^{x_0}\exp\left[\frac{\mathrm{i}\omega}{2}\left(\tau-\frac{y_0^2+z^2}{\tau}\right)\right]\frac{\mathrm{d}\,\tau}{\sqrt{\tau}}. \tag{10.6.22}$$

One may demonstrate that this integral is convergent. Denoting $r=|y_0|$ and performing the change of variable $\tau\to\lambda$:

$$\tau-\frac{r^2}{\tau}=2\lambda\Rightarrow\frac{\mathrm{d}\,\tau}{\tau}=\frac{\mathrm{d}\,\lambda}{\sqrt{\lambda^2+r^2}}, \tag{10.6.23}$$

one obtains

$$n(x_0, y_0, z) = \frac{\partial^2}{\partial z^2} \int_{-\infty}^{X} \frac{e^{i\omega\lambda}}{\sqrt{\lambda^2 + r^2}} d\,\lambda\,, \tag{10.6.24}$$

where

$$X = \frac{1}{2}\left(x_0 - \frac{r^2}{x_0}\right). \tag{10.6.25}$$

Observing that

$$\frac{\partial^2}{\partial z^2} = \frac{1}{r}\left(1 - \frac{z^2}{r^2}\right)\frac{\partial}{\partial r} + \frac{z^2}{r^2}\frac{\partial^2}{\partial r^2}\,,$$

we deduce that the line $r = 0$ is singular. Eliminating a vicinity of this line we obtain

$$\lim_{z \to \pm 0} \frac{\partial^2}{\partial z^2} = \lim_{z \to \pm 0} \frac{1}{r}\frac{\partial}{\partial r}$$

whence

$$\begin{aligned} n(x_0, y_0, z) &= \lim_{z \to \pm 0} n(x_0, y_0, z) = \\ &= \frac{2}{x_0^2 + r^2} e^{i\omega X} + \int_{-\infty}^{X} \frac{e^{i\omega\lambda}}{(\lambda^2 + r^2)^{3/2}} d\,\lambda\,. \end{aligned} \tag{10.6.26}$$

This kernel was obtained in (10.2.36) as a limit of the subsonic kernel and in (10.4.48) as a limit of the supersonic kernel (Ueda and Dowell [10.80]). This fact proves that the oscillatory perturbation is continuous to the passage of the sonic barrier.

Passing to the limit, when $z \to \pm 0$, the first term from (10.6.17) vanishes because $\delta(\pm 0) = 0$.

Chapter 11

The Theory of Slender Bodies

11.1 The Linear Equations and Their Fundamental Solutions

11.1.1 The Boundary Condition. The Linear Equations

In this chapter we study the aerodynamics in the presence of slender bodies (fig. 11.1.1). The axis of the body is considered the Ox axis and the Oz axis lies in the plane determined by the velocity of the unperturbed stream $\boldsymbol{V}_\infty$ and by the axis of the body. We denote by α the angle of attack of the stream and we assume that $\alpha \simeq \varepsilon$, where ε characterizes the thickness of the body. We employ the cylindrical coordinates x, r and θ which are related to the cartesian coordinates x, y, z by the formulas

$$\begin{gathered} x = x,\, y = r\cos\theta,\, z = r\sin\theta \\ x \in \mathbb{R}, r \in (0,\infty),\, \theta \in [0, 2\pi)\,. \end{gathered} \tag{11.1.1}$$

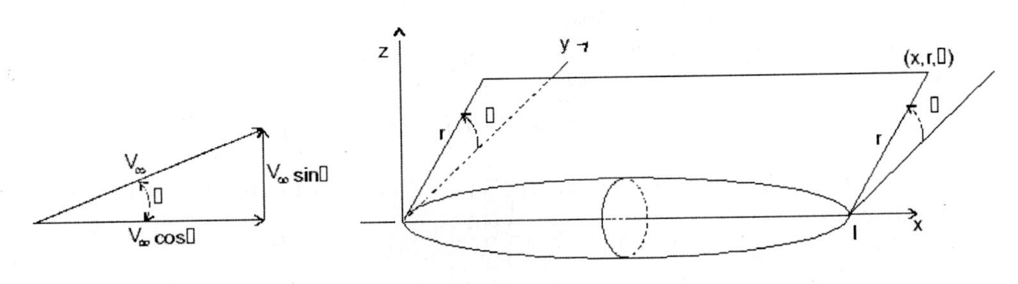

Fig. 11.1.1.

The equation of the body has the form

$$r = h(x,\theta) = \varepsilon\overline{h}(x,\theta)\,. \tag{11.1.2}$$

Denoting by $\boldsymbol{i}, \boldsymbol{j}, \boldsymbol{k}$ the versors of the Ox, Oy and Oz axes and $\boldsymbol{i}_r, \boldsymbol{i}_\theta$ (fig. 11.1.2), we shall have the following formulas

$$
\begin{aligned}
\boldsymbol{i}_r &= \boldsymbol{j}\cos\theta + \boldsymbol{k}\sin\theta & \boldsymbol{j} &= \boldsymbol{i}_r\cos\theta - \boldsymbol{i}_\theta\sin\theta \\
\boldsymbol{i}_\theta &= -\boldsymbol{j}\sin\theta + \boldsymbol{k}\cos\theta & \boldsymbol{k} &= \boldsymbol{i}_r\sin\theta + \boldsymbol{i}_\theta\cos\theta\,.
\end{aligned}
\tag{11.1.3}
$$

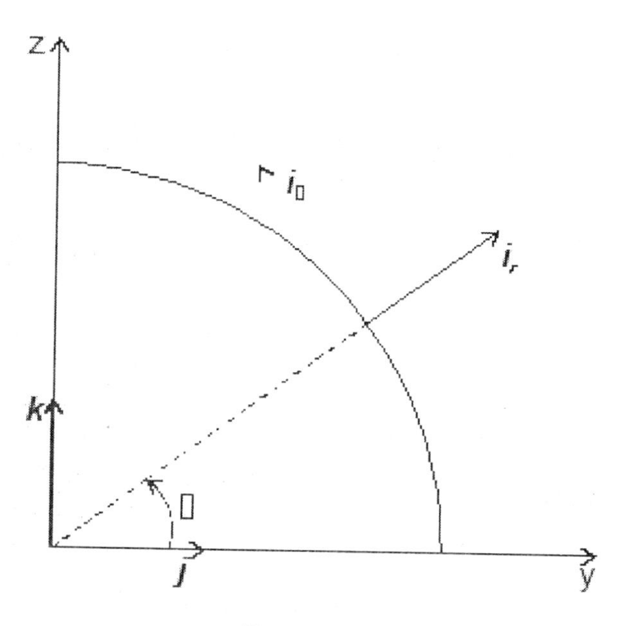

Fig. 11.1.2.

The velocity of the unperturbed stream is

$$
\boldsymbol{V}_\infty = U_\infty(\boldsymbol{i}\cos\alpha + \boldsymbol{k}\sin\alpha) = U_\infty(\boldsymbol{i} + \alpha\boldsymbol{k}) + O(\alpha^2)\,. \tag{11.1.4}
$$

Denoting by

$$
\boldsymbol{V}_1 = U_\infty\boldsymbol{V}, \quad p_1 = p_\infty + \rho_\infty U_\infty^2 p, \quad \rho_1 = \rho_\infty(1+\rho) \tag{11.1.5}
$$

the fields which characterize the perturbed flow and using the cylindrical coordinates

$$
\begin{aligned}
\boldsymbol{V} &= U\boldsymbol{i} + V\boldsymbol{i}_r + W\boldsymbol{i}_\theta \\
\boldsymbol{v} &= u\boldsymbol{i} + v\boldsymbol{i}_r + w\boldsymbol{i}_\theta,
\end{aligned}
\tag{11.1.6}
$$

we deduce

$$
U = 1 + u, \quad V = \alpha\sin\theta + v, \quad W = \alpha\cos\theta + w\,. \tag{11.1.7}
$$

Obviously, the perturbed flow will be steady because the conditions which determines it do not vary time. On the boundary we shall impose the condition

$$\boldsymbol{V} \cdot \operatorname{grad} F = 0\,, \tag{11.1.8}$$

where $F = \varepsilon\overline{h}(x,\theta) - r$. Taking into account that we have

$$\operatorname{grad} F = \frac{\partial F}{\partial x}\boldsymbol{i} + \frac{\partial F}{\partial r}\boldsymbol{i}_r + \frac{1}{r}\frac{\partial F}{\partial \theta}\boldsymbol{i}_\theta\,, \tag{11.1.9}$$

from (11.1.8) we deduce the condition

$$(1+u)\varepsilon\frac{\partial \overline{h}}{\partial x} + (\alpha\cos\theta + w)\frac{\varepsilon}{r}\frac{\partial \overline{h}}{\partial \theta} = \alpha\sin\theta + v\,, \tag{11.1.10}$$

which must be satisfied when $r = h$. Comparing the orders of magnitude we deduce:

$$v(x,h,\theta) = \varepsilon\overline{v}(x,h,\theta)\,. \tag{11.1.11}$$

We assume that this structure is valid everywhere in the fluid. We have therefore

$$v(x,r,\theta) = \varepsilon\overline{v}(x,r,\theta)\,. \tag{11.1.12}$$

From (11.1.10) and (11.1.11) we deduce the condition

$$v(x,r,\theta) + \alpha\sin\theta = h'_x(x,\theta)\,, \tag{11.1.13}$$

which will be imposed for $r = h$. In fact, this condition must be imposed for $r = 0$, but here v is not defined.

In cylindrical coordinates the equations of motion are [1.11]:

$$\begin{aligned} &M^2\left[Up_x + Vp_r + (W/r)p_\theta\right] + \\ &\quad + [1+\gamma M^2 p][U_x + V_r + (1/r)V + (1/r)W_\theta] = 0 \end{aligned} \tag{11.1.14}$$

$$(1+\rho)[UU_x + VU_r + (W/r)U_\theta] + p_x = 0 \tag{11.1.15}$$

$$(1+\rho)[UV_x + VV_r + (W/r)V_\theta - (W^2/r)] + p_r = 0 \tag{11.1.16}$$

$$(1+\rho)[UW_x + VW_r + (W/r)W_\theta + VW/r] + (1/r)p_\theta = 0\,, \tag{11.1.17}$$

where U, V, W will be replaced by (11.1.7), and $p_x = \partial p/\partial x, \ldots$

With the reasonings from 2.1, the equation (11.1.16) gives

$$p(x,r,\theta) = \varepsilon\overline{p}(x,r,\theta)\,, \quad v_x + p_r = 0\,, \tag{11.1.18}$$

the equation (11.1.17)

$$w(x,r,\theta) = \varepsilon\overline{w}(x,r,\theta), \quad rw_x + p_\theta = 0\,, \tag{11.1.19}$$

and the equation (11.1.15)

$$u(x,r,\theta)=\varepsilon\overline{u}(x,r,\theta)\,,\quad u_x+p_x=0\,. \tag{11.1.20}$$

Keeping the terms having the order of ε, from (11.1.14) we deduce

$$M^2p_x+u_x+v_r+(1/r)v+(1/r)w_\theta=0\,. \tag{11.1.21}$$

One observes that in the linearized system α does not intervene. The system coincides with the system for $\alpha=0$. It is the system (2.1.32) in cylindrical coordinates.

One may also obtain the equation of the potential. Indeed, from (11.1.18) - (11.1.20) it results:

$$p=-u,\quad v_x-u_r=0,\quad rw_x-u_\theta=0\,. \tag{11.1.22}$$

The last two equations prove the existence of the function $\varphi(x,r,\theta)$, a.î.

$$u=\varphi_x,\quad v=\varphi_r,\quad w=(1/r)\varphi_\theta\,, \tag{11.1.23}$$

and the equation (11.1.21) gives

$$(1-M^2)\frac{\partial^2\varphi}{\partial x^2}+\frac{1}{r}\frac{\partial}{\partial r}\left(r\frac{\partial\varphi}{\partial r}\right)+\frac{1}{r^2}\frac{\partial^2\varphi}{\partial\theta^2}=0\,. \tag{11.1.24}$$

11.1.2 Fundamental Solutions

We shall utilize for the solution of the system (2.3.4) the intrinsic form (2.3.8), (2.3.12) which will be written in cylindrical coordinates.

From the equality

$$f_1\boldsymbol{i}+f_2\boldsymbol{j}+f_3\boldsymbol{k}=f_x\boldsymbol{i}+f_r\boldsymbol{i}_r+f_\theta\boldsymbol{i}_\theta\,, \tag{11.1.25}$$

taking (11.1.3) into account , we deduce

$$\begin{aligned}&f_2=f_r\cos\theta-f_\theta\sin\theta,\quad f_r=f_2\cos\theta+f_3\sin\theta\\&f_3=f_r\sin\theta-f_\theta\cos\theta,\quad f_\theta=-f_2\sin\theta+f_3\cos\theta\,.\end{aligned} \tag{11.1.26}$$

Writing the inner product in cylindrical coordinates, in the *subsonic* case and Taking (11.1.26) into account, from (2.3.4) it results

$$p(x,r)=-\frac{1}{4\pi}\left(f_x\frac{\partial}{\partial x}+f_r\frac{\partial}{\partial r}\right)\left(\frac{1}{R_1}\right), \tag{11.1.27}$$

where

$$R_1 = \sqrt{x^2 + \beta^2 r^2}\,. \tag{11.1.28}$$

From (2.3.13) we deduce

$$\begin{aligned}\varphi &= \frac{1}{4\pi}\left[\frac{f_x}{R_1} - \beta^2 r f_r \int_{-\infty}^{x} \frac{\mathrm{d}\,x}{(x^2+\beta^2 r^2)^{3/2}}\right] = \\ &= \frac{1}{4\pi}\left[\frac{f_x}{R_1} - \frac{f_r}{r}\left(1 + \frac{x}{R_1}\right)\right].\end{aligned} \tag{11.1.29}$$

Taking into account the expression of the distribution $\delta(\boldsymbol{x})$ in cylindrical coordinates [A.7], [A.10], we obtain

$$v_r = f_r \frac{H(x)\delta(r)}{2\pi r} + \frac{f_x}{4\pi}\frac{\partial}{\partial r}\left(\frac{1}{R_1}\right) - \frac{f_r}{4\pi}\frac{\partial}{\partial r}\left[\frac{1}{r}\left(1+\frac{x}{R_1}\right)\right]. \tag{11.1.30}$$

In the *supersonic* flow we have

$$p(x,r) = -\frac{1}{2\pi}\left(f_x \frac{\partial}{\partial x} + f_r \frac{\partial}{\partial r}\right) E(x,r)\,, \tag{11.1.31}$$

where

$$E(x,r) = \frac{H(x-kr)}{\sqrt{x^2 - k^2 r^2}}\,. \tag{11.1.32}$$

Since

$$\frac{\partial}{\partial r}\int_{-\infty}^{x} \frac{H(\tau - kr)}{\sqrt{\tau^2 - k^2 r^2}}\mathrm{d}\,\tau = H(x-kr)\frac{\partial}{\partial r}\int_{kr}^{x}\frac{\mathrm{d}\,\tau}{\sqrt{\tau^2 - k^2 r^2}} = -\frac{x}{r}E(x,r)\,,$$

from (2.3.13) it results

$$\varphi(x,r) = \frac{1}{2\pi}\left(f_x - \frac{x}{r}f_r\right) E(x,r) \tag{11.1.33}$$

$$v_r(x,r) = f_r \frac{H(x)\delta(r)}{2\pi r} + \frac{1}{2\pi}\left(f_x \frac{\partial}{\partial r} + \frac{x}{r^2} f_r - \frac{x}{r} f_r \frac{\partial}{\partial r}\right) E. \tag{11.1.34}$$

Taking into account the formula (2.3.35), p and v_r will be:

$$\begin{aligned}p &= -\frac{H(x-kr)}{2\pi}\left(f_x\frac{\partial}{\partial x} + f_r \frac{\partial}{\partial r}\right)\frac{1}{\sqrt{x^2 - k^2 r^2}}, \\ v_r &= f_r \frac{H(x)\delta(r)}{2\pi r} + \frac{H(x-kr)}{2\pi}\left(f_x \frac{\partial}{\partial r} + \frac{x}{r^2} f_r - \right. \\ &\quad \left. - \frac{x}{r} f_r \frac{\partial}{\partial r}\right)\frac{1}{\sqrt{x^2 - k^2 r^2}}\,.\end{aligned} \tag{11.1.35}$$

These formulas show that the perturbation propagates only in the interior of the cone $x = kr$.

11.2 The Slender Body in a Subsonic Stream

11.2.1 The Solution of the Problem

In the case of slender bodies of revolution, the equation (11.1.2) has the form

$$r = h(x), \quad 0 < x < 1 . \tag{11.2.1}$$

Considering that the unperturbed flow has the angle of attack z in the xOz plane, we deduce that this plane will be the plane of symmetry of the flow. We shall replace the body with a distribution of forces defined on $[0,1]$ with $f_2 = 0$. From (11.1.26) we deduce $f_r = f \sin\theta$ (we denoted $f_3 = f$). Taking (11.1.27)–(11.1.30) into account, we deduce that perturbation produced by this distribution may be represented by means of the formulas

$$\begin{aligned} p(x,r,\theta) &= -\frac{1}{4\pi}\int_0^1 \left(f_1(\xi)\frac{\partial}{\partial x} + f(\xi)\sin\theta\frac{\partial}{\partial r}\right)\frac{1}{R}\mathrm{d}\,\xi \\ \varphi(x,r,\theta) &= \frac{1}{4\pi}\int_0^1 \frac{f_1(\xi)}{R}\mathrm{d}\,\xi - \frac{\sin\theta}{4\pi r}\int_0^1 \left(1+\frac{x_0}{R}\right) f(\xi)\mathrm{d}\,\xi \end{aligned} \tag{11.2.2}$$

$$\begin{aligned} v_r(x,r,\theta) = \delta(r)\frac{\sin\theta}{2\pi r}\int_0^x f_1(\xi)\mathrm{d}\,\xi + \frac{1}{4\pi}\int_0^1 f_1(\xi)\frac{\partial}{\partial r}\left(\frac{1}{R}\right)\mathrm{d}\,\xi - \\ -\frac{\sin\theta}{4\pi}\frac{\partial}{\partial r}\int_0^1 \frac{f(\xi)}{r}\left(1+\frac{x_0}{R}\right)\mathrm{d}\,\xi , \end{aligned} \tag{11.2.3}$$

where

$$x_0 = x - \xi , \quad R = \sqrt{x_0^2 + \beta^2 r^2} .$$

Imposing the boundary condition (11.1.13), we notice that $\delta(h) = 0$ because h does not vanish for $0 < x < 1$. Separating the variables we obtain the following integral equations:

$$\int_0^1 f_1(\xi)\frac{\partial}{\partial r}\left(\frac{1}{R}\right)\mathrm{d}\,\xi = 4\pi h'(x) , \tag{11.2.4}$$

$$\frac{\partial}{\partial r}\int_0^1 \frac{f(\xi)}{r}\left(1+\frac{x_0}{R}\right)\mathrm{d}\,\xi = 4\pi\alpha , \tag{11.2.5}$$

for $0 < x < 1$ and $r = h(x)$.

In order to solve the first equation we shall utilize the identity

$$\frac{\partial}{\partial r}\left(\frac{1}{R}\right) = -\frac{1}{r}\frac{\partial}{\partial x}\left(\frac{x_0}{R}\right) = \frac{1}{r}\frac{\partial}{\partial \xi}\left(\frac{x_0}{R}\right) . \tag{11.2.6}$$

Integrating by parts, we obtain:

$$4\pi h(x)h'(x) = f_1(1)\frac{x-1}{\sqrt{(x-1)^2+\beta^2h^2}} -$$
$$-f_1(0)\frac{x}{\sqrt{x^2+\beta^2h^2}} = -\left[f_1(1)+f_1(0)\right]\left[1+O(h^2)\right], \tag{11.2.7}$$

where

$$I = \int_0^1 \frac{x_0}{\sqrt{x_0^2+\beta^2h^2}} f_1'(\xi)\mathrm{d}\,\xi\,. \tag{11.2.8}$$

For calculating the principal part of this integral, we notice that we have

$$\frac{x_0}{\sqrt{x_0^2+\beta^2h^2}} = \frac{x_0}{|x_0|}\left[1+O(h^2)\right] \tag{11.2.9}$$

excepting the vicinity of the point $\xi = x$ where $x_0^2 = 0$. In [1.1] one utilizes this approximation on the entire interval $(0,1)$. Correctly, the integral I must be written as follows

$$I = \lim_{\eta\to 0}\left(\int_0^{x-\eta} + \int_{x-\eta}^{x+\eta} + \int_{x+\eta}^1\right). \tag{11.2.10}$$

In the first and last integrals we may utilize the approximation (11.2.9). For η small enough, in the second integral one may replace $f_1'(\xi)$ by $f_1'(x)$. We obtain therefore

$$I = \lim_{\eta\to 0}\left[\int_0^{x-\eta} f_1'(\xi)\mathrm{d}\,\xi - \int_{x+\eta}^1 f_1'(\xi)\mathrm{d}\,\xi\right]\left[1+O(h^2)\right] +$$
$$+f_1'(x)\lim_{\eta\to 0}\int_{x-\eta}^{x+\eta}\frac{x_0\mathrm{d}\,\xi}{\sqrt{x_0^2+\beta^2h^2}}\,.$$

We calculate the last integral with the substitution $x-\xi = u$ and we observe that it vanishes. Hence,

$$I = \left[f_1(x) - f_1(0) - f_1(1) + f_1(x)\right]\left[1+O(h^2)\right]. \tag{11.2.11}$$

Neglecting the terms of order h^2 with respect to 1, from (11.2.8) and (11.2.11) we obtain

$$f_1(x) = -2\pi h(x)h'(x) = -S'(x)\,, \tag{11.2.12}$$

with $S(x) = \pi h^2(x)$ the area of the cross section a of the body in the point having the abscissa x.

Deriving in (11.2.5), we find

$$\int_0^1 \left(1+\frac{x_0}{R}\right) f(\xi)\mathrm{d}\,\xi + \beta^2 h^2 \int_0^1 \frac{x_0}{R^3} f(\xi)\mathrm{d}\,\xi = -4\pi\alpha h^2 . \qquad (11.2.13)$$

Calculating the integrals by means of the formula (11.2.10), we have

$$\int_0^1 \left(1+\frac{x_0}{R}\right) f(\xi)\mathrm{d}\,\xi = 2\int_0^x f(\xi)\mathrm{d}\,\xi \left[1+O(h^2)\right] .$$

Hence, neglecting $O(h^2)$ with respect to 1, we obtain

$$\int_0^x f(\xi)\mathrm{d}\,\xi = -2\pi\alpha h^2 = -2\alpha S(x)$$

or, deriving,

$$f(x) = -2\alpha S'(x) . \qquad (11.2.14)$$

For the profile with zero angle of attack $(\alpha = 0)$ we deduce $f = 0$ whence

$$\varphi(x,r) = -\frac{1}{4\pi}\int_0^1 \frac{S'(\xi)}{\sqrt{x_0^2+\beta^2 r^2}}\mathrm{d}\,\xi . \qquad (11.2.15)$$

This representation of the potential is known in the literature [1.1], [1.38]. We have also, from (11.2.2) (for $r \neq 0$),

$$p(x,r) = -\varphi_x, v_r = \varphi_r . \qquad (11.2.16)$$

11.2.2 The Calculus of Lift and Moment Coefficients

We shall calculate at first the pressure for $r = h$. It can be obtained from (11.2.2), (11.2.12) and (11.2.14). Utilizing the identity (11.2.6) and the calculations (11.2.8) – (11.2.11), we deduce

$$\int_0^1 f(\xi)\frac{\partial}{\partial r}\left(\frac{1}{R}\right)\mathrm{d}\,\xi \frac{1}{h}\int_0^1 f(\xi)\frac{\partial}{\partial \xi}\left(\frac{x_0}{R}\right)\mathrm{d}\,\xi = -\frac{2f(x)}{h(x)}\left[1+O(\varepsilon^2)\right] . \qquad (11.2.17)$$

Hence,

$$p(x,h(x),\theta) = p_1(x,h(x)) - \alpha\frac{\sin\theta}{\pi h(x)}S'(x) , \qquad (11.2.18)$$

where

$$p_1(x, h(x)) = -\frac{1}{4\pi}\int_0^1 f_1(\xi)\frac{\partial}{\partial x}\left(\frac{1}{R}\right)\Big|_{r=h} \mathrm{d}\,\xi = \\ = -\frac{1}{4\pi}\int_0^1 \frac{x_0 S'(\xi)\mathrm{d}\,\xi}{\left(x_0^2+\beta^2 h^2\right)^{3/2}}\,. \tag{11.2.19}$$

Taking into account that $f_r = O(\varphi^2)$, we calculate the principal part of p_1 as follows:

$$4\pi p_1(x, h(x)) = \int_0^1 f_1(\xi)\frac{\partial}{\partial \xi}\left(\frac{1}{R}\right)\Big|_{r=h} \mathrm{d}\,\xi =$$

$$= \frac{f_1(1)}{\sqrt{(x-1)^2+\beta^2 h^2}} - \frac{f_1(0)}{\sqrt{x^2+\beta^2 h^2}} - {}^{*}\!\!\int_0^x \frac{f_1'(\xi)}{x-\xi}\mathrm{d}\,\xi +$$

$$+ {}^{*}\!\!\int_x^1 \frac{f_1'(\xi)}{x-\xi}\mathrm{d}\,\xi + f_1'(x)\lim_{\eta\to 0}\int_{x-\eta}^{x+\eta}\frac{\mathrm{d}\,\xi}{\sqrt{x_0^2+\beta^2 h^2}} + O(\varepsilon^4) -$$

$$-\frac{S'(1)}{1-x} + \frac{S'(0)}{x} - \int_0^x \frac{S''(x)-S''(\xi)}{x-\xi}\mathrm{d}\,\xi +$$

$$+ \int_x^1 \frac{S''(\xi)-S''(x)}{\xi - x}\mathrm{d}\,\xi + S''(x)\ln\frac{1-x}{x} + O(\varepsilon^4)\,, \tag{11.2.20}$$

the principal part which was written being $O(\varepsilon^2)$.

The lift coefficient may be calculated with the formula

$$c_L = -\iint_S p\boldsymbol{n}\cdot\boldsymbol{k}\mathrm{d}\,a\,, \tag{11.2.21}$$

where S is the surface of the body, and $\boldsymbol{n}$, the outer normal and with the notation $F = r - h(x)$, given by the formula

$$\boldsymbol{n} = \frac{\operatorname{grad} F}{|\operatorname{grad} F|} = \frac{-h'\boldsymbol{i} + \boldsymbol{i}_r}{\sqrt{1+h'^2(x)}}\,. \tag{11.2.22}$$

Taking into account the element of area on the surface S, and the relation (11.2.13), we obtain

$$c_L = -\int_0^1\int_{-\pi}^{\pi} p(x, h(x), \theta)h(x)\sin\theta\mathrm{d}\,x\mathrm{d}\,\theta = \\ = \alpha\left[S(1) - S(0)\right]\,. \tag{11.2.23}$$

For the drag coefficient one obtains

$$c_D = -\iint_S p\boldsymbol{n}\cdot\boldsymbol{i}\,\mathrm{d}\,a = \int_0^1\int_{-\pi}^{\pi} p(x,h(x),\theta)h(x)h'(x)\mathrm{d}\,x\mathrm{d}\,\theta = \tag{11.2.24}$$

$$= \int_0^1 p_1(x,h(x))S'(x)\mathrm{d}\,x = O(\varepsilon^4)\,.$$

The drag coefficient does not depend on α.

At last, the moment coefficients are the scalar components of the product

$$-2\iint_S \boldsymbol{x}\times p\boldsymbol{n}\mathrm{d}\,a\,,$$

where $\boldsymbol{x} = x\boldsymbol{i} + h(x)\boldsymbol{i}_r$. Taking into account (11.2.2) and (11.2.22), we obtain

$$c_x = 0,\quad c_z = 0,$$

$$c_y = \int_0^1\int_{-\pi}^{+\pi}(x+hh')h(x)p(x,h(x),\theta)\sin\theta\mathrm{d}\,x\,\mathrm{d}\,\theta = \tag{11.2.25}$$

$$= -2\alpha\int_0^1\left[x+\frac{S'(x)}{2\pi}\right]S'(x)\mathrm{d}\,x = O(\varepsilon^3)\,.$$

Obviously, for $\alpha = 0$ one obtains $c_y = 0$.

Neglecting the term $(S')^2$, for the moment coefficient c_y we obtain the approximate value

$$c_y = -2\alpha\int_0^1 xS'(x)\mathrm{d}\,x = 2\alpha V\,,$$

because $S(1) = 0$ and the term $\int_0^1 S(x)\mathrm{d}\,x$ represents the volume V of the body.

11.3 The Thin Body in a Supersonic Stream

11.3.1 The General Solution

In this subsection we consider the same problem like in the previous subsection, but now the unperturbed (free) flow is supersonic $(M > 1)$.

We consider again that in this case the xOz plane is a symmetry plane, such that we have $f_z = 0$ whence $f_r = f \sin\theta$. The fundamental solution is (11.1.31), (11.1.34), and the corresponding potential is (11.1.33). For a continuous superposition of forces on the segment $[0, 1]$, the perturbation will be given by

$$p(x,r,\theta) = -\frac{1}{2\pi}\int_0^1 \left[f_x(\xi)\frac{\partial}{\partial x} + f(\xi)\sin\theta\frac{\partial}{\partial r}\right]E(x_0,r)\mathrm{d}\,\xi,$$

$$\varphi(x,r,\theta) = \frac{1}{2\pi}\int_0^1 \left[f_x(\xi) - \frac{x_0}{r}f(\xi)\sin\theta\right]E(x_0,r)\mathrm{d}\,\xi,$$

$$v_r(x,r,\theta) = \delta(r)\frac{\sin\theta}{2\pi r}\int_0^x f_x(\xi)\mathrm{d}\,\xi + \frac{1}{2\pi}\int_0^1 f_x(\xi)\frac{\partial}{\partial r}\cdot$$

$$\cdot E(x_0,r)\mathrm{d}\,\xi + \frac{\sin\theta}{2\pi r}\int_0^1 x_0 f(\xi)\left(\frac{1}{r} - \frac{\partial}{\partial r}\right)E(x_0,r)\mathrm{d}\,\xi\,. \tag{11.3.1}$$

The derivatives may interchange with the integrals and, taking (A.3.15) into account, we have for example

$$\int_0^1 f(\xi)\frac{\partial}{\partial x}E(x_0,r)\mathrm{d}\,\xi = \frac{\partial}{\partial x}\int_0^1 f(\xi)E(x_0,r)\mathrm{d}\,\xi =$$

$$= \frac{\partial}{\partial x}H(x-kr)\int_0^{x-kr}\frac{f(\xi)}{\sqrt{x_0^2-k^2r^2}}\mathrm{d}\,\xi = H(x-kr)\frac{\partial}{\partial x}\int_0^{x-kr}\frac{f(\xi)}{\sqrt{x_0^2-k^2r^2}}\mathrm{d}\,\xi\,.$$

Hence, the solution (11.3.1) maybe written as follows

$$p = -\frac{1}{2\pi}\frac{\partial}{\partial x}\int_0^{x-kr}\frac{f_x(\xi)}{\sqrt{x_0^2-k^2r^2}}\mathrm{d}\,\xi - \frac{\sin\theta}{2\pi}\frac{\partial}{\partial r}\int_0^{x-kr}\frac{f(\xi)}{\sqrt{x_0^2-k^2r^2}}\mathrm{d}\,\xi,$$

$$\varphi = \frac{1}{2\pi}\int_0^{x-kr}\frac{f_x(\xi)}{\sqrt{x_0^2-k^2r^2}}\mathrm{d}\,\xi - \frac{\sin\theta}{2\pi r}\int_0^{x-kr}\frac{x_0}{\sqrt{x_0^2-k^2r^2}}f(\xi)\mathrm{d}\,\xi,$$

$$v_r = \delta(r)\frac{\sin\theta}{2\pi r}\int_0^x f(\xi)\mathrm{d}\,\xi + \frac{1}{2\pi}\frac{\partial}{\partial r}\int_0^{x-kr}\frac{f_x(\xi)}{\sqrt{x_0^2-k^2r^2}}\mathrm{d}\,\xi +$$

$$+\frac{\sin\theta}{2\pi r}\left(\frac{1}{r} - \frac{\partial}{\partial r}\right)\int_0^{x-kr}\frac{x_0}{\sqrt{x_0^2-k^2r^2}}f(\xi)\mathrm{d}\,\xi\,, \tag{11.3.2}$$

valid for $x > kr$ and

$$p = \varphi = v_r = 0 \quad \text{for } x < kr\,. \tag{11.3.3}$$

Imposing the boundary condition (11.1.13) we shall notice again that $\delta(h) = 0$, because $h \neq 0$. Separating the variables we deduce the following integral equations

$$\frac{\partial}{\partial r}\int_0^{x-kr} \frac{f_x(\xi)}{\sqrt{x_0^2 - k^2 r^2}} \mathrm{d}\,\xi \Big|_{r=h} = 2\pi h'(x) \tag{11.3.4}$$

$$\left(\frac{1}{r} - \frac{\partial}{\partial r}\right)\int_0^{x-kr} \frac{x_0}{\sqrt{x_0^2 - k^2 r^2}} f(\xi)\mathrm{d}\,\xi \Big|_{r=h} = -2\pi \alpha h(x)\,. \tag{11.3.5}$$

In order to put into evidence the principal part of the integrals for $r = h = O(\varepsilon)$, we must notice that we cannot derive directly the integrals, because the integrands become infinite for $\xi = x - kr$. For avoiding this situation, we shall perform the change of variable $\xi \to u$:

$$x - \xi = kr \cosh u\,. \tag{11.3.6}$$

One obtains

$$\begin{aligned}
\frac{\partial}{\partial r}\int_0^{x-kr} \frac{f_x(\xi)}{\sqrt{x_0^2 - k^2 r^2}} \mathrm{d}\,\xi &= \frac{\partial}{\partial r}\int_0^{\operatorname{arccosh}\frac{x}{kr}} f_x(x - kr\cosh u)\mathrm{d}u = \\
&= -\frac{x}{r}\frac{f_x(0)}{\sqrt{x^2 - k^2 r^2}} - k\int_0^{\operatorname{arccosh}\frac{x}{kr}} f_x'(x - kr\cosh u)\cosh u\,\mathrm{d}\,u = \\
&= -\frac{x}{r}\frac{f_x(0)}{\sqrt{x^2 - k^2 r^2}} - \frac{1}{r}\int_0^{x-kr} \frac{x_0}{\sqrt{x_0^2 - k^2 r^2}} f'(\xi)\mathrm{d}\,\xi = \\
&= -\frac{1}{r} f_x(0)\left[1 + O(\varepsilon^2)\right] - \frac{1}{r}\int_0^{x-kr} f_x'(\xi)\left[1 + O(\varepsilon^2)\right]\mathrm{d}\,\xi\,.
\end{aligned} \tag{11.3.7}$$

Neglecting $O(\varepsilon^2)$ with respect to 1, we deduce that the principal part of the integral from (11.3.4) is $-f_x(x)/r$. Hence, from (11.3.4) we deduce

$$f_x(x) = -2\pi h h' = -S'(x)\,, \tag{11.3.8}$$

$S(x)$ giving the area of the cross section of the body in the point x.

Similarly we obtain

$$\frac{\partial}{\partial r}\int_0^{x-kr}\frac{x_0}{\sqrt{x_0^2-k^2r^2}}f(\xi)\mathrm{d}\,\xi=-\frac{x^2}{r}\frac{f(0)}{\sqrt{x_0^2-k^2r^2}}+$$

$$+\frac{1}{r}\int_0^{x-kr}\frac{x_0}{\sqrt{x_0^2-k^2r^2}}f(\xi)\mathrm{d}\,\xi-\frac{1}{r}\int_0^{x-kr}\frac{x_0^2}{\sqrt{x_0^2-k^2r^2}}f'(\xi)\mathrm{d}\,\xi. \tag{11.3.9}$$

Neglecting $O(r^2)$ with respect to 1 we find at first

$$\left(\frac{1}{r}-\frac{\partial}{\partial r}\right)\int_0^{x-kr}\frac{x_0}{\sqrt{x_0^2-k^2r^2}}f(\xi)\mathrm{d}\,\xi\simeq\frac{1}{r}\int_0^{x-kr}f(\xi)\mathrm{d}\,\xi+\frac{x}{r}f(0)-$$

$$-\frac{1}{r}\int_0^{x-kr}f(\xi)\mathrm{d}\,\xi+\frac{x}{r}\int_0^{x-kr}f'(\xi)\mathrm{d}\,\xi-$$

$$-\frac{1}{r}\int_0^{x-kr}\xi f'(\xi)\mathrm{d}\,\xi=kf(x-kr)+\frac{1}{r}\int_0^{x-kr}f(\xi)\mathrm{d}\,\xi.$$

The equation (11.3.5) becomes

$$krf(x-kr)+\int_0^{x-kr}f(\xi)\mathrm{d}\,\xi=-2\pi\alpha rh, \tag{11.3.10}$$

where we put $r=h$. In the left hand part of the equality, we expand into series for obtaining the principal part. We have

$$krf(x)+O(r^2)+\int_0^x f(\xi)\mathrm{d}\,\xi-krf(x)+O(r^2)=-2\pi\alpha rh.$$

Hence,

$$\int_0^x f(\xi)\mathrm{d}\,\xi=-2\alpha S(x),$$

whence

$$f(x)=-2\alpha S'(x), \tag{11.3.11}$$

the analogy with the subsonic case being obvious.

11.3.2 The Pressure on the Body. The Lift and Moment Coefficients

In order to put into evidence the principal part and the expression of the pressure, we notice that utilizing the change of variable (11.3.6), we deduce

$$\frac{\partial}{\partial x}\int_0^{x-kr}\frac{f_x(\xi)}{\sqrt{x_0^2-k^2r^2}}\mathrm{d}\,\xi=\frac{f_x(0)}{\sqrt{x^2-k^2r^2}}+\int_0^{x-kr}\frac{f_x'(\xi)\mathrm{d}\,\xi}{\sqrt{x_0^2-k^2r^2}}=$$

$$=\frac{f_x(0)}{\sqrt{x^2-k^2r^2}}+\int_0^{x-kr}\frac{f_x'(\xi)-f_x'(x)}{\sqrt{x_0^2-k^2r^2}}\mathrm{d}\,\xi+$$

$$+f_x'(x)\int_0^{x-kr}\frac{\mathrm{d}\,\xi}{\sqrt{x_0^2-k^2r^2}}\simeq\frac{f_x(0)}{x}+\int_0^{x}\frac{f_x'(\xi)-f_x'(x)}{x-\xi}\mathrm{d}\,\xi+$$

$$+f_x'(x)\operatorname{arccosh}\frac{x}{kr}=-\int_0^{x}\frac{f_x'(\xi)-f_x'(\xi)}{x-\xi}\mathrm{d}\,\xi+f_x'(x)\ln\frac{x+\sqrt{x^2-k^2r^2}}{kr}=$$

$$=f_x'(x)\ln\frac{2x}{kr}-\int_0^{x}\frac{f_x'(x)-f_x'(\xi)}{x-\xi}\mathrm{d}\,\xi\,.$$

Taking into account the calculation performed at (11.3.7), we deduce

$$p(x,h,\theta)=p_1(x,h)-\alpha\frac{\sin\theta}{\pi h(x)}S'(x)\,, \tag{11.3.12}$$

where

$$p_1(x,h)=-\frac{S''(x)}{2\pi}\ln\frac{kh}{2x}-\frac{1}{2\pi}\int_0^{x}\frac{S''(x)-S''(\xi)}{x-\xi}\mathrm{d}\,\xi\,. \tag{11.3.13}$$

The lift, drag and moment coefficients may be calculated with the formulas (11.2.23) - (11.2.25). We have therefore, like in the subsonic case,

$$\begin{aligned}
c_L&=\alpha\,[S(1)-S(0)]=O(\varepsilon^3)\\
c_D&=\int_0^1 p_1(x,h)S'(x)\mathrm{d}\,x=O(\varepsilon^4)\\
c_x&=c_z=0\,,\\
c_y&=-\alpha\int_0^1\left[x+\frac{S'(x)}{2\pi}\right]S'(x)\mathrm{d}\,x=O(\varepsilon^3)\,.
\end{aligned} \tag{11.3.14}$$

11.3.3 The wing at zero angle of attack

For $\alpha = 0$, we deduce $f = 0$ whence

$$p(x,r) = \frac{1}{2\pi}\frac{\partial}{\partial x}\int_0^{x-kr}\frac{S'(\xi)}{\sqrt{x_0^2 - k^2r^2}}\mathrm{d}\,\xi = \frac{1}{2\pi}\int_0^{x-kr}\frac{S''(\xi)\mathrm{d}\,\xi}{\sqrt{x_0^2 - k^2r^2}},$$

$$\varphi(x,r) = -\frac{1}{2\pi}\int_0^{x-kr}\frac{S'(\xi)}{\sqrt{x_0^2 - k^2r^2}}\mathrm{d}\,\xi\,,$$

$$v_r(x,r) = -\frac{1}{2\pi}\frac{\partial}{\partial r}\int_0^{x-kr}\frac{S'(\xi)\mathrm{d}\xi}{\sqrt{x_0^2 - k^2r^2}} - \frac{1}{2\pi r}\int_0^{x-kr}\frac{S''(\xi)\mathrm{d}\xi}{\sqrt{x_0^2 - k^2r^2}}. \tag{11.3.15}$$

This is the solution between the wave from the leading edge and the wave from the trailing edge i.e. in the region denoted by II, where $x > kr$ (fig. 11.3.1). In I $(x < kr)$ $p = v_r = 0$, and in III the solution is

$$p(x,r) = \frac{1}{2\pi}\int_0^{x-kr}\frac{S''(\xi)\mathrm{d}\,\xi}{\sqrt{x_0^2 - k^2r^2}},$$
$$v_r(x,r) = -\frac{1}{2\pi r}\int_0^{x-kr}\frac{x_0 S''(\xi)}{\sqrt{x_0^2 - k^2r^2}}\mathrm{d}\,\xi\,. \tag{11.3.16}$$

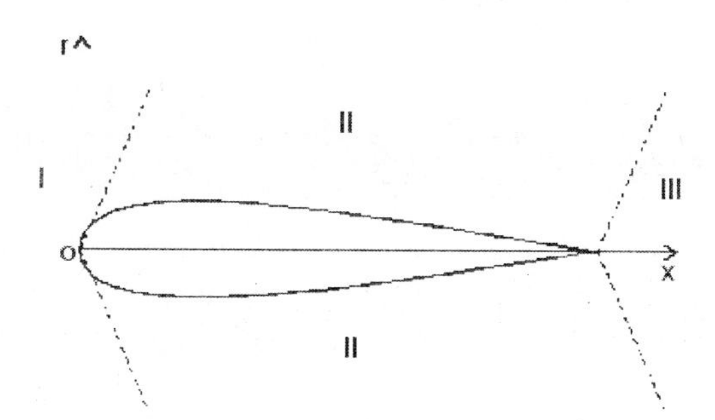

Fig. 11.3.1.

This explicit manner of representing the solution is also valid in the general case.

11.3.4 Applications

At first we shall consider the thin body having the shape of a cone (fig. 11.3.2) of equation $r = \varepsilon x$, $0 < x < 1$. We deduce $S = \pi\varepsilon^2 x^2$

whence

$$c_L = \pi\varepsilon^2\alpha, \quad c_y = -\frac{2\pi\alpha\varepsilon^2}{3}, \quad c_D = -\pi\varepsilon^4 \ln(k/2)\,. \tag{11.3.17}$$

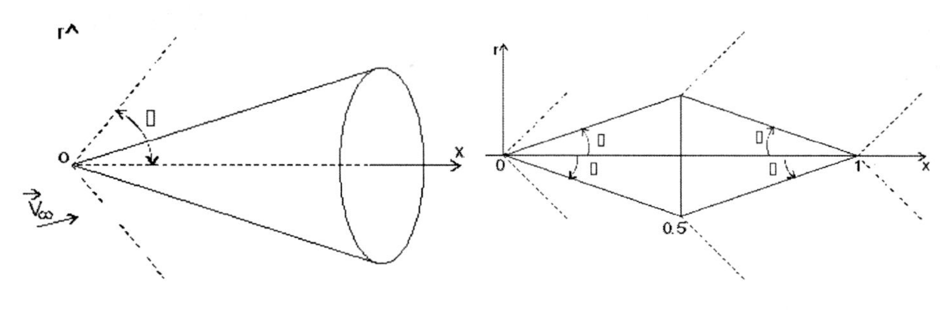

Fig. 11.3.2. Fig. 11.3.3.

For c_y and c_D we have retained only the principal part. If the angle of attack is zero, then in II we have

$$p(x,r) = \varepsilon^2 \int_0^{x-kr} \frac{\mathrm{d}\,\xi}{\sqrt{x_0^2 - k^2r^2}}\mathrm{d}\,\xi = \varepsilon^2 \mathrm{arccosh}\frac{x}{kr} =$$

$$= \varepsilon^2 \ln \frac{x + \sqrt{x^2 - k^2r^2}}{kr}\,, \tag{11.3.18}$$

$$v_r(x,r) = -\frac{\varepsilon^2}{r}\int_0^{x-kr} \frac{x_0}{\sqrt{x_0^2 - k^2r^2}}\mathrm{d}\,\xi = -\varepsilon^2 \frac{\sqrt{x^2 - k^2r^2}}{r}\,.$$

We notice that on the radius vectors $r = cx$, $c < 1/k$, the pressure and the velocity are constants. The flow is *conical.*

The second example is the double cone from figure 11.3.3 Since

$$h(x) = \begin{cases} \varepsilon x, & 0 < x < 1/2 \\ \varepsilon(1-x), & 1/2 < x < 1\,, \end{cases} \tag{11.3.19}$$

it results that S' has a discontinuity in the point $x = 1/2$. We suggest to the reader to establish the formulas of derivation for

$$\frac{\partial}{\partial x}\int_0^{x-kr} \frac{f(\xi)}{\sqrt{x_0^2 - k^2r^2}}\mathrm{d}\,\xi\,, \quad \frac{\partial}{\partial r}\int_0^{x-kr} \frac{f(\xi)}{\sqrt{x_0^2 - k^2r^2}}\mathrm{d}\,\xi \tag{11.3.20}$$

and to write the solution.

Appendix A

Fourier Transform and Notions of the Theory of Distributions

A.1 The Fourier Transform of Functions

The following definitions will be given in $\mathbb{R}^3$. Their expressions in $\mathbb{R}^1$, $\mathbb{R}^2$ or $\mathbb{R}^n$ will be easily deduced.

The Fourier transform of a function $f : \mathbb{R}^3 \to \mathbb{R}$ is denoted $\mathcal{F}[f]$, or $\widehat{f}$ and it is defined by the formula

$$\mathcal{F}[f](\boldsymbol{\alpha}) = \widehat{f}(\boldsymbol{\alpha}) = \int_{\mathbb{R}^3} f(\boldsymbol{x}) e^{\mathrm{i}\boldsymbol{\alpha}\cdot\boldsymbol{x}} \mathrm{d}\boldsymbol{x}\,, \tag{A.1.1}$$

where

$$\begin{gathered} \boldsymbol{x} = (x, y, z), \quad \boldsymbol{\alpha} = (\alpha_1, \alpha_2, \alpha_3) \\ \boldsymbol{\alpha} \cdot \boldsymbol{x} = \alpha_1 x + \alpha_2 y + \alpha_3 z, \quad d\boldsymbol{x} = \mathrm{d}x\, \mathrm{d}y\, \mathrm{d}z\,. \end{gathered} \tag{A.1.2}$$

The operation $\mathcal{F}$ is called *the Fourier transform.* We notice that if $f(\boldsymbol{x})$ is absolutely integrable on $\mathbb{R}^3$, i.e. if

$$\int_{\mathbb{R}^3} (f) \mathrm{d}\boldsymbol{x} < \infty\,, \tag{A.1.3}$$

then the Fourier transform exists. Moreover, if f satisfies certain conditions of regularity, for example, if $f \in C^0(\mathbb{R}^3)$, or if f is piecewise smooth with respect to every variable [A.10], then f may be obtained from $\widehat{f}$ using the following inversion formula

$$\begin{aligned} (2\pi)^3 f(\boldsymbol{x}) = (2\pi)^3 \mathcal{F}^{-1}[\widehat{f}](\boldsymbol{x}) &= \int_{\mathbb{R}^3} \widehat{f}(\boldsymbol{\alpha}) e^{-\mathrm{i}\boldsymbol{x}\cdot\boldsymbol{\alpha}} \mathrm{d}\boldsymbol{\alpha} = \\ &= \mathcal{F}[\widehat{f}](-\boldsymbol{x}) = \mathcal{F}[\widehat{f}(-\boldsymbol{\alpha})](\boldsymbol{x})\,, \end{aligned} \tag{A.1.4}$$

where $\mathrm{d}\boldsymbol{\alpha} = \mathrm{d}\alpha_1\, \mathrm{d}\alpha_2\, \mathrm{d}\alpha_3$. The last two expressions show that the inverse of a Fourier transform may be obtained by a direct transformation.

One proves (*Lebesque's theorem* [A.9], [A.5]) that $\widehat{f}$ tends to zero when $|\boldsymbol{\alpha}| \to \infty$.

The condition (A.1.3) is restrictive enough. For example, the function $f = 1$ does not verify this condition. The theory of distributions allows to enlarge the class of the functions which admit a Fourier transform.

Notions of the Theory of Distributions

A.2 The Spaces $\mathcal{D}$ and $\mathcal{S}$

The theory of distributions relies on the notion of space of the test functions. The space $\mathcal{D} = \mathcal{D}(\mathbb{R}^3)$ consists of the set of functions $\varphi(\boldsymbol{x})$ infinitely derivable, with compact support in $\mathbb{R}^3$. We say that a sequence of functions $\{\varphi_n\}$ from $\mathcal{D}$, converges to a function $\varphi \in \mathcal{D}$, if for every multi-index $k = (k_1, k_2, k_3)$, k_1, k_2, k_3 being non-negative integers, we have:

$$D^k \varphi_n \overset{x \in \mathbb{R}^3}{\rightrightarrows} D^k \varphi \,, \tag{A.2.1}$$

where, with the notation $|k| = k_1 + k_2 + k_3$, we have put

$$D^k = \frac{\partial^{|k|}}{\partial x^{k_1} \partial y^{k_2} \partial z^{k_3}} \,. \tag{A.2.2}$$

The set $\mathcal{D}$ endowed with this convergence law is a *linear space.*

We may see that the Fourier transform of a function from $\mathcal{D}$ is not a function from $\mathcal{D}$. Indeed, the Fourier transform is an analytical function [A.12], but the support of an analytical function which cannot be compact. There exists another space which is invariant to the Fourier transform and plays a basic role in the definition of the Fourier transform of a distribution. This is the space $\mathcal{S} = \mathcal{S}(\mathbb{R}^3)$, i.e. the space of the *rapidly decreasing* test functions. We call rapidly decreasing function an infinitely derivable function $\varphi(\boldsymbol{x})$ in $\mathbb{R}^3$ which, together with its derivatives decreases when $|\boldsymbol{x}| \to \infty$ faster then every power of $|\boldsymbol{x}|^{-1}$. This means that

$$|\boldsymbol{x}^k D^l \varphi(\boldsymbol{x})| < C_{kl} \,, \tag{A.2.3}$$

for every multi-indices k and l. We denoted $\boldsymbol{x}^k = x^{k_1} y^{k_2} z^{k_3}$. The convergence in $\mathcal{S}$ is defined in the following manner: the sequence of functions $\{\varphi_n\}$ from $\mathcal{S}$ converges to the function $\varphi \in \mathcal{S}$ if for every multi-indices k and l, we have:

$$x^k D^l \varphi_n(\boldsymbol{x}) \overset{\boldsymbol{x}\in\mathbb{R}^3}{\rightrightarrows} x^k D^l \varphi(\boldsymbol{x}) . \tag{A.2.4}$$

The set of the rapidly decreasing functions, endowed with this convergence law, is a *linear space.* The convergence in $\mathcal{D}$ implies the convergence in $\mathcal{S}$. We have $\mathcal{D} \subset \mathcal{S}$, but $\mathcal{S}$ does not coincide with $\mathcal{D}$; For example, the function $\exp.(-x^2) \in \mathcal{S}$, but $\notin \mathcal{D}$.

A.3 Distributions

The distributions are *linear and continuous functionals* on $\mathcal{D}$ or on $\mathcal{S}$. This means that a distribution f determines the correspondence between a test function φ and a number denoted by $\langle f, \varphi \rangle$ and we have:

$$\langle f, \lambda_1\varphi_1 + \lambda_2\varphi_2 \rangle = \lambda_1 \langle f, \varphi_1 \rangle + \lambda_2 \langle f, \varphi_2 \rangle , \tag{A.3.1}$$

for every two real or complex numbers λ_1, λ_2 and that if $\varphi_n \to \varphi$ in $\mathcal{D}$ (or $\mathcal{S}$), then

$$\langle f, \varphi_n \rangle \to \langle f, \varphi \rangle . \tag{A.3.2}$$

One denotes by $\mathcal{D}'$ the set of the distributions defined on $\mathcal{D}$ and by $\mathcal{S}'$ the set of the distributions defined on $\mathcal{S}$. $\mathcal{S}'$ is the space *of the temperate distributions.* Obviously, $\mathcal{S}' \subset \mathcal{D}'$.

We say that the distribution f *is equal to zero* in the domain Ω (it is denoted by $f = 0$) if $\langle f, \varphi \rangle = 0$ for every φ from $\mathcal{D}$ (or $\mathcal{S}$) with the support in Ω. The distributions f_1 and f_2 are equal in Ω if $f_1 - f_2 = 0$ in Ω, i.e. if

$$\langle f_1, \varphi \rangle = \langle f_2, \varphi \rangle , \tag{A.3.3}$$

for every $\varphi \in \mathcal{D}$.

We call *the support* of the distribution f and we denote it by $\operatorname{supp} f$, the set of the points which have a vicinity where f is not equal to zero.

Every locally integrable function $f(\boldsymbol{x})$ defines a distribution by means of the functional

$$\langle f, \varphi \rangle = \int_{\mathbb{R}^3} f(\boldsymbol{x})\varphi(\boldsymbol{x})\mathrm{d}\,\boldsymbol{x} . \tag{A.3.4}$$

A distribution f is a regular *function-type* distribution if it may have the form (A.3.4). Every other distribution is *singular.* The best known example of singular distribution is the distribution of Dirac. The distribution of Dirac with the support in $\boldsymbol{\xi}$ is denoted by $\delta_{\boldsymbol{\xi}}(\boldsymbol{x})$ or $\delta(\boldsymbol{x}-\boldsymbol{\xi})$ and is defined by the functional

$$\langle \delta(\boldsymbol{x}-\boldsymbol{\xi}), \varphi(\boldsymbol{x}) \rangle = \varphi(\boldsymbol{\xi}) . \tag{A.3.5}$$

Formally, it may be written as follows

$$\int_{\mathbb{R}^3} \varphi(\boldsymbol{x}) \delta(\boldsymbol{x}-\boldsymbol{\xi}) \mathrm{d}\, \boldsymbol{x} = \varphi(\boldsymbol{\xi}) . \tag{A.3.6}$$

If $m(\boldsymbol{x})$ is an infinitely derivable function, then $m\varphi$ is infinitely derivable and we may define *the product between the distribution* f *and the function* m, denoted by mf, by means of the formula

$$\langle mf, \varphi \rangle = \langle f, m\varphi \rangle . \tag{A.3.7}$$

It results

$$m(\boldsymbol{x}) \delta(\boldsymbol{x}-\boldsymbol{\xi}) = m(\boldsymbol{\xi}) \delta(\boldsymbol{x}) \tag{A.3.8}$$

whence,

$$m(\boldsymbol{x}) \delta(\boldsymbol{x}) = 0, \quad \text{if } m(\mathbf{0}) = 0 . \tag{A.3.9}$$

For the existence (A.3.8) the continuity of the function is sufficient m in $\boldsymbol{\xi}$.

We shall consider the three-factor c and the homothety $c\boldsymbol{x} = (c_1 x, c_2 y, c_3 z)$. If $h(\boldsymbol{x})$ is a locally integrable function, then $h(c\boldsymbol{x})$ is also locally integrable, such that the distribution corresponding to $h(c\boldsymbol{x})$ is defined by the functional

$$\langle h(c\boldsymbol{x}), \varphi(\boldsymbol{x}) \rangle = \int_{\mathbb{R}^3} h(c\boldsymbol{x}) \varphi(\boldsymbol{x}) \mathrm{d}\, \boldsymbol{x} .$$

Setting $c\boldsymbol{x} = \boldsymbol{\xi}$ and observing that the integration limits interchange when c_i is negative, we deduce, with notation $|c| = |c_1 c_2 c_3|$

$$|c| \langle h(c\boldsymbol{x}), \varphi(\boldsymbol{x}) \rangle = \int_{\mathbb{R}^3} h(\boldsymbol{\xi}) \varphi(\boldsymbol{\xi}/c) \mathrm{d}\, \boldsymbol{\xi} = \langle h(\boldsymbol{x}), \varphi(\boldsymbol{x}/c) \rangle .$$

For o distribution $f(\boldsymbol{x})$, one defines the distribution $f(c\boldsymbol{x})$ by means of the formula

$$|c| \langle f(c\boldsymbol{x}), \varphi(\boldsymbol{x}) \rangle = \langle f(\boldsymbol{x}), \varphi(\boldsymbol{x}/c) \rangle . \tag{A.3.10}$$

The definition may be extended to a non-singular linear transformation. As a particular case, it results:

$$|c| \delta(c_1 x, c_2 y, c_3 z) = \delta(x, y, z) . \tag{A.3.11}$$

The derivative of the distribution f is denoted by f' and is determined by the formula

$$\langle f', \varphi \rangle = -\langle f, \varphi' \rangle . \tag{A.3.12}$$

Defining the function of Heaviside $H(x)$ by the formula

$$H(x) = \begin{cases} 0, & x < 0 \\ 1, & x \geq 0, \end{cases} \tag{A.3.13}$$

it results

$$\langle H', \varphi \rangle = -\langle H, \varphi' \rangle = -\int_0^\infty \varphi'(x) \mathrm{d}\, x = \varphi(0) .$$

Taking into account (A.3.5) we deduce

$$H'(x) = \delta(x) . \tag{A.3.14}$$

Let us establish now the formula

$$(m(x)H(x)' = m(0)\delta(x) + m'(x)H(x) \tag{A.3.15}$$

for a function $\in C^1(\mathbb{R})$. We have

$$\langle (mH)', \varphi \rangle = -\langle mH, \varphi' \rangle = -\int_0^\infty m(x)\varphi'(x) \mathrm{d}\, x =$$

$$= m(0)\varphi(0) + \int_0^\infty m'(x)\varphi(x) \mathrm{d}\, x = \langle m(0)\delta(x) + m'(x)H(x), \varphi \rangle .$$

On the basis of (A.3.8) and (A.3.15) we deduce that the solution of the differential equation

$$Lu = \delta(x) , \tag{A.3.16}$$

where $Lu = u^{(m)}(x) + a_1(x)u^{(m-1)}(x) + \ldots + a_m(x)u(x)$, is

$$u(x) = H(x)v(x) , \tag{A.3.17}$$

where v verifies the equation $Lv = 0$ and the conditions

$$v(0) = v'(0) \ldots = v^{(m-2)}(0) = 0, v^{(m-1)}(0) = 1 .$$

We call *the direct product* of the distributions $f(\boldsymbol{x}) \in \mathcal{D}'(\mathbb{R}^n)$ and $g(\boldsymbol{y}) \in \mathcal{D}'(\mathbb{R}^m)$ the distribution $f(\boldsymbol{x}) \cdot g(\boldsymbol{y}) \in \mathcal{D}'(\mathbb{R}^{n+m})$ defined by the functional

$$\langle f(\boldsymbol{x}) \cdot g(\boldsymbol{y}), \varphi(\boldsymbol{x}, \boldsymbol{y}) \rangle = \langle f(\boldsymbol{x}), \langle g(\boldsymbol{y}), \varphi(\boldsymbol{x}, \boldsymbol{y}) \rangle \rangle , \tag{A.3.18}$$

where $\varphi \in \mathcal{D}(\mathbb{R}^{n+m})$. The definition is valid also and when $\mathcal{D}'$ is replaced by $\mathcal{S}'$. The direct product is commutative, i.e. $f(\boldsymbol{x}) \cdot g(\boldsymbol{y}) = g(\boldsymbol{y}) \cdot f(\boldsymbol{x})$. As an application, we shall prove that in $\mathbb{R}^2$, we have

$$\delta(\boldsymbol{x}) = \delta(x) \cdot \delta(y) . \tag{A.3.19}$$

Indeed, applying the definition, we have:

$$\langle \delta(\boldsymbol{x}), \varphi(x,y) \rangle = \varphi(0,0) .$$

But we have also,

$$\langle \delta(x) \cdot \delta(y), \varphi(x,y) \rangle = \langle \delta(x), \langle \delta(y), \varphi(x,y) \rangle \rangle =$$
$$= \langle \delta(x), \varphi(x,0) \rangle = \varphi(0,0) .$$

A.4 The Convolution. Fundamental Solutions

For two functions $m_1(\boldsymbol{x})$ and $m_2(\boldsymbol{x})$ absolutely integrable in $\mathbb{R}^3$, we define the convolution $m(\boldsymbol{x}) = (m_1 * m_2)(\boldsymbol{x})$ by the formula

$$(m_1 * m_2)(\boldsymbol{x}) = \int_{\mathbb{R}^3} m_1(\boldsymbol{\xi}) m_2(\boldsymbol{x} - \boldsymbol{\xi}) \mathrm{d}\,\boldsymbol{\xi} =$$
$$= \int_{\mathbb{R}^3} m_2(\boldsymbol{\xi}) m_2(\boldsymbol{x} - \boldsymbol{\xi}) \mathrm{d}\,\boldsymbol{\xi} = (m_2 * m_1)(\boldsymbol{x}) .$$

The function $m(\boldsymbol{x})$ is absolutely integrable, such that it generates a function-type distribution (A.3.4). Writing explicitly that functional, we are determined [A.5] to define the convolution $f = f_1 * f_2$, of two distributions f_1 and f_2 by means of the formula

$$\langle f_1 * f_2, \varphi \rangle = \langle f_1(\boldsymbol{x}) \cdot f_2(\boldsymbol{\xi}), \varphi(\boldsymbol{x} + \boldsymbol{\xi}) \rangle , \tag{A.4.1}$$

$f_1(\boldsymbol{x}) \cdot f_2(\boldsymbol{\xi})$ representing the direct product of the distributions f_1 and f_2. The convolution (A.4.1) exists if, for example, one of the distributions have a compact support [A.5]. The equality $f_1 * f_2 = f_2 * f_1$ results from the commutativity of the direct product.

As an application, we shall calculate $\delta * f$. We have

$$\langle \delta * f, \varphi \rangle = \langle \delta(\boldsymbol{x}) \cdot f(\boldsymbol{\xi}), \varphi(\boldsymbol{x} + \boldsymbol{\xi}) \rangle = \langle f(\boldsymbol{\xi}), \langle \delta(\boldsymbol{x}), \varphi(\boldsymbol{x} + \boldsymbol{\xi}) \rangle \rangle =$$
$$= \langle f(\boldsymbol{\xi}), \varphi(\boldsymbol{\xi}) \rangle .$$

Hence,

$$\delta * f = f * \delta = f\,. \tag{A.4.2}$$

We shall establish now the formula

$$D(f_1 * f_2) = Df_1 * f_2 = f_1 * Df_2\,, \tag{A.4.3}$$

the operator D being defined in (A.2.2). We have

$$\begin{aligned}\langle D(f_1 * f_2), \varphi\rangle &= (-1)^{|k|}\langle f_1 * f_2, D\varphi\rangle = \\ &= (-1)^{|k|}\langle f_1(\boldsymbol{x})\cdot f_2(\boldsymbol{\xi}), D\varphi(\boldsymbol{x}+\boldsymbol{\xi})\rangle = \\ &= (-1)^{|k|}\langle f_2(\boldsymbol{\xi}), \langle f_1(\boldsymbol{x}), D\varphi(\boldsymbol{x}+\boldsymbol{\xi})\rangle\rangle = \\ &= \langle f_2(\boldsymbol{\xi}), \langle Df_1(\boldsymbol{x}), D\varphi(\boldsymbol{x}+\boldsymbol{\xi})\rangle\rangle = \langle Df_1 * f_2, \varphi\rangle\,,\end{aligned}$$

and the result proves the first equality from (A.4.3). The second equality results from the commutativity of the convolution.

We shall consider now the linear differential operator of order m,

$$L = \sum_{|k|=0}^{m} a_k D^k\,,$$

where a_k are constant. We call *fundamental solution* of this operator, the distribution $\mathcal{E} \in \mathcal{D}'$ which verify the equation

$$L\mathcal{E} = \delta(\boldsymbol{x})\,. \tag{A.4.4}$$

Obviously, the fundamental solution is defined with the approximation of an arbitrary solution of the homogeneous equation. We have the following *theorem: If $f \in \mathcal{D}'$ and there exists in $\mathcal{D}'$ the convolution $f * \mathcal{E}$, then the solution of the equation*

$$Lu = f \tag{A.4.5}$$

exists in $\mathcal{D}'$, is given by the formula

$$u = f * \mathcal{E} \tag{A.4.6}$$

and it is unique in the set of distributions from $\mathcal{D}'$, for which there exists the convolution (A.4.6).

Indeed, utilizing (A.4.2) and (A.4.3), we deduce

$$Lu = L(f * \mathcal{E}) = \sum_{|k|=0}^{m} a_k D^k (f * \mathcal{E}) = f * L\mathcal{E} = f * \delta = f\,.$$

In order to demonstrate the uniqueness, we denote by $\overline{u}$ another solution. Obviously, $u_0 = u - \overline{u}$ satisfies the homogeneous equation $Lu_0 = 0$, such that we have:

$$u_0 = u_0 * \delta = u_0 * L\mathcal{E} = L(u_0 * \mathcal{E}) = (Lu_0) * \mathcal{E} = 0 .$$

A.5 The Fourier Transform of the Functions from $\mathcal{S}$

Since the functions from $\mathcal{S}$ are absolutely integrable, they have Fourier transforms $\mathcal{F}[\varphi]$ defined by (A.1.1) and these are bounded continuous functions on $\mathbb{R}^3$. Integrating by parts and taking into account (A.2.3), we deduce:

$$\mathcal{F}[\varphi_x](\boldsymbol{\alpha}) = (-i\alpha_1)\mathcal{F}[\varphi](\boldsymbol{\alpha}) . \tag{A.5.1}$$

On the basis of the same condition, we may derive inside the integral, such that it results

$$(\partial/\partial\alpha_1)\mathcal{F}[\varphi](\boldsymbol{\alpha}) = \mathcal{F}[ix\varphi](\boldsymbol{\alpha}) . \tag{A.5.2}$$

By recurrence, it results:

$$\mathcal{F}[D^k\varphi](\boldsymbol{\alpha}) = (-(\boldsymbol{\alpha})^k\mathcal{F}[\varphi](\boldsymbol{\alpha}) , \tag{A.5.3}$$

$$D^l\mathcal{F}[\varphi](\boldsymbol{\alpha}) = \mathcal{F}[(i\boldsymbol{x})^l\varphi](\boldsymbol{\alpha}) , \tag{A.5.4}$$

for every multi - indices k and l. The last relation show that $\mathcal{F}[\varphi]$ is infinitely derivable.

We shall prove in the following $\mathcal{F}[\varphi] \in \mathcal{S}$. To this aim we replace φ with $(i\boldsymbol{x})^l\varphi$ in (A.5.3) and taking into account (A.5.4), we deduce

$$\mathcal{F}[D^k((i\boldsymbol{x})^l\varphi)](\boldsymbol{\alpha}) = (-i\boldsymbol{\alpha})^k\mathcal{F}[(i\boldsymbol{x})^l\varphi](\boldsymbol{\alpha}) = \\ = (-i)^{|k|}\boldsymbol{\alpha}^k D^l\mathcal{F}[\varphi](\boldsymbol{\alpha}) ,$$

whence

$$|\boldsymbol{\alpha}^k D^l\mathcal{F}[\varphi](\boldsymbol{\alpha})| \leq \int_{\mathbb{R}^3} |D^k(\boldsymbol{x}^l\varphi)| \mathrm{d}\boldsymbol{x} < \infty ,$$

i.e. $\mathcal{F}[\varphi]$ satisfies (A.2.3).

Taking into account that the Fourier transform $\mathcal{F}[\varphi]$ is integrable and continuously differentiable, we deduce that φ may be obtained from $\widehat{\varphi}$ sing a formula like (A.1.4).

A.6 The Fourier Transform of the Temperate Distributions

The Fourier transform of the distribution $f \in \mathcal{S}'$ is denoted by $\widehat{f} = \mathcal{F}[f]$ and it is defined by formula

$$\left\langle \widehat{f}, \varphi \right\rangle = \langle f, \widehat{\varphi} \rangle \,, \tag{A.6.1}$$

for every $\varphi \in \mathcal{S}$. Since $\widehat{\varphi} \in \mathcal{S}$, it results [A.12] that the Fourier transform of a temperate distribution is also a temperate distribution.

For $f \in \mathcal{S}'$ we define the inverse Fourier transform $\mathcal{F}^{-1}$, by the formula:

$$(2\pi)^3 \mathcal{F}^{-1}[f(\boldsymbol{x})] = \mathcal{F}[f(-\boldsymbol{x})] \,. \tag{A.6.2}$$

We shall prove that

$$\mathcal{F}^{-1}[\mathcal{F}[f]] = \mathcal{F}[\mathcal{F}^{-1}[f]] = f \,. \tag{A.6.3}$$

Indeed, taking (A.3.10) into account, we deduce for $c = -1$

$$\begin{aligned}(2\pi)^3 \left\langle \mathcal{F}^{-1}[\mathcal{F}[f]], \varphi \right\rangle &= \langle \mathcal{F}[\mathcal{F}[f](-\boldsymbol{\alpha})], \varphi \rangle = \langle \mathcal{F}[f](-\boldsymbol{\alpha}), \mathcal{F}[\varphi] \rangle = \\ &= \langle \mathcal{F}[f], \mathcal{F}[\varphi](-\boldsymbol{\alpha}) \rangle = (2\pi)^3 \left\langle \mathcal{F}[f], \mathcal{F}^{-1}[\varphi] \right\rangle = \\ &= (2\pi)^3 \left\langle f, \mathcal{F}[\mathcal{F}^{-1}[\varphi]] \right\rangle = \langle f, \varphi \rangle \,.\end{aligned}$$

Similarly we prove the second equality.

We shall establish now the operations (A.5.3) and (A.5.4) for temperate distributions. We have:

$$\begin{aligned}\langle \mathcal{F}[f_x], \varphi \rangle &= \langle f_x, \widehat{\varphi} \rangle = - \langle f, \partial_x \mathcal{F}[\varphi] \rangle = \\ &= - \langle f, \mathcal{F}[\mathrm{i}\, \alpha_1 \varphi] \rangle = \langle \mathcal{F}[f], -\mathrm{i}\, \alpha_1 \varphi \rangle = \left\langle -\mathrm{i}\, \alpha_1 \widehat{f}, \varphi \right\rangle\end{aligned}$$

and we deduce,

$$\mathcal{F}[f_x] = -\mathrm{i}\, \alpha_1 \widehat{f} \,, \tag{A.6.4}$$

whence

$$\begin{aligned}&\mathcal{F}[\operatorname{grad} f] = -\mathrm{i}\, \boldsymbol{\alpha} \widehat{f}, \mathcal{F}[\operatorname{div} \boldsymbol{f}] = -\mathrm{i}\, \boldsymbol{\alpha} \cdot \widehat{\boldsymbol{f}} \\ &\mathcal{F}[\operatorname{rot} \boldsymbol{f}] = -\mathrm{i}\, \boldsymbol{\alpha} \times \widehat{\boldsymbol{f}} \,.\end{aligned} \tag{A.6.5}$$

From (A.6.4) we obtain by recurrence a similar formula to (A.5.3).

Analogously, we have:

$$\left\langle \frac{\partial}{\partial \alpha_1} \mathcal{F}[f], \varphi(\boldsymbol{\alpha}) \right\rangle = - \left\langle \mathcal{F}[f], \frac{\partial \varphi}{\partial \alpha_1} \right\rangle = - \left\langle f, \mathcal{F}\left[\frac{\partial \varphi}{\partial \alpha_1}\right] \right\rangle =$$

$$\langle f, \mathrm{i}\, x\mathcal{F}[\varphi]\rangle = \langle \mathrm{i}\, xf, \mathcal{F}[\varphi]\rangle = \langle \mathcal{F}[\mathrm{i}\, xf], \varphi\rangle \ ,$$

hence,

$$(\partial/\partial\alpha_1)\mathcal{F}[f] = \mathcal{F}[\mathrm{i}\, xf] \tag{A.6.6}$$

and, by recurrence, a similar formula to (A.5.4).

From the definition (A.6.2), it results that relations similar to (A.5.3) and (A.5.4) are available for the inverse transformation too. Hence,

$$\mathcal{F}^{-1}[D^k\widehat{f}](\boldsymbol{x}) = (\mathrm{i}\,\boldsymbol{x})^k f(\boldsymbol{x}) \tag{A.6.7}$$

$$\mathcal{F}^{-1}[(-\mathrm{i}\,\boldsymbol{\alpha})^l\widehat{f}](\boldsymbol{x}) = D^l f(\boldsymbol{x})\,. \tag{A.6.8}$$

We shall frequently utilize relations of the form

$$\mathcal{F}^{-1}[-\mathrm{i}\,\alpha_1\widehat{f}] = f_x, \quad \mathcal{F}^{-1}[\alpha_1\alpha_2\widehat{f}] = -f_{xy}\,. \tag{A.6.9}$$

We shall prove in the following the formula

$$\mathcal{F}^{-1}[\widehat{f}(c\boldsymbol{\alpha})](\boldsymbol{x}) = \frac{1}{|c|} f\left(\frac{\boldsymbol{x}}{c}\right)\,, \tag{A.6.10}$$

c representing a three-factor. Performing the change of variable $\boldsymbol{\alpha} = c\boldsymbol{\beta}$, we have for a function φ,

$$\mathcal{F}\left[\mathcal{F}^{-1}\varphi\left(\frac{\boldsymbol{\alpha}}{c}\right)\right](\boldsymbol{x}) = \int_{\mathbb{R}^3} \mathcal{F}^{-1}\varphi\left(\frac{\boldsymbol{\alpha}}{c}\right) \mathrm{e}^{\mathrm{i}\,\boldsymbol{\alpha}\cdot\boldsymbol{x}}\mathrm{d}\,\boldsymbol{\alpha} =$$

$$= |c| \int_{\mathbb{R}^3} (\mathcal{F}^{-1}\varphi)(\boldsymbol{\beta})\mathrm{e}^{\mathrm{i}\,\boldsymbol{\beta}\cdot c\boldsymbol{x}}\mathrm{d}\,\boldsymbol{\beta} = |c|\varphi(c\boldsymbol{x})\,.$$

For a distribution f we shall utilize the formula (A.3.10). We have

$$|c|\left\langle \mathcal{F}^{-1}[\widehat{f}(c\boldsymbol{\alpha})], \varphi\right\rangle = |c|\left\langle \widehat{f}(c\boldsymbol{\alpha}), \mathcal{F}^{-1}\varphi\right\rangle =$$

$$\left\langle \widehat{f}(\boldsymbol{\alpha}), \mathcal{F}^{-1}\varphi(\boldsymbol{\alpha}/c)\right\rangle = \langle f(\boldsymbol{\alpha}), \mathcal{F}^{-1}\varphi(\boldsymbol{\alpha}/c)\rangle =$$

$$= |c|\,\langle f(\boldsymbol{x}), \varphi(c\boldsymbol{x})\rangle = \langle f(\boldsymbol{x}/c), \varphi\rangle$$

whence (A.6.10).

For the direct product $f(\boldsymbol{x}\cdot g(\boldsymbol{y}))$ we have:

$$\mathcal{F}[f(\boldsymbol{x})\cdot g(\boldsymbol{y})] = \mathcal{F}[f](\boldsymbol{\alpha})\cdot\mathcal{F}[g](\boldsymbol{\beta})\,, \tag{A.6.11}$$

and for the convolution, if one of the distributions has a compact support, we deduce

$$\mathcal{F}[f * g] = \mathcal{F}[f]\mathcal{F}[g]\,. \tag{A.6.12}$$

From (A.6.11) it results

$$\mathcal{F}^{-1}[\widehat{f}(\boldsymbol{\alpha})\cdot\widehat{g}(\boldsymbol{\beta})] = f(\boldsymbol{x})\cdot g(\boldsymbol{y})\,. \tag{A.6.13}$$

A.7 The Calculus of Some Inverse Fourier Transforms

We shall calculate at first the Fourier transform of the distribution of Dirac. For $\varphi \in \mathcal{S}$ we have

$$\begin{gathered}\langle \mathcal{F}[\delta(\boldsymbol{x}-\boldsymbol{\xi})], \varphi\rangle = \langle \delta(\boldsymbol{x}-\boldsymbol{\xi}), \widehat{\varphi}\rangle = \widehat{\varphi}(\boldsymbol{\xi}) = \\ = \left\langle \mathrm{e}^{\mathrm{i}\,\boldsymbol{\alpha}\cdot\boldsymbol{\xi}}, \varphi\right\rangle ,\end{gathered} \tag{A.7.1}$$

Hence

$$\mathcal{F}[\delta(\boldsymbol{x}-\boldsymbol{\xi})] = \exp(\mathrm{i}\,\boldsymbol{\alpha}\cdot\boldsymbol{\xi}), \quad \delta(\boldsymbol{x}-\boldsymbol{\xi}) = \mathcal{F}^{-1}[\exp(\mathrm{i}\,\boldsymbol{\alpha}\cdot\boldsymbol{\xi})], \tag{A.7.2}$$

whence, for $\boldsymbol{\xi} = 0$,

$$\mathcal{F}[\delta] = 1, \quad \delta = \mathcal{F}^{-1}[1] = (2\pi)^3 \mathcal{F}[1]. \tag{A.7.3}$$

It results therefore

$$\mathcal{F}[1](\boldsymbol{\alpha}) = (2\pi)^3 \delta(\boldsymbol{\alpha}). \tag{A.7.4}$$

Hence the temperate distribution 1 has a Fourier transform.

Denoting by $\mathcal{F}_2$ the Fourier transform in $\mathbb{R}^2$ and by $\mathcal{F}_1$ the transformation in $\mathbb{R}^1$, we have on the basis of the formula (A.6.13):

$$\mathcal{F}_2^{-1}[1/\mathrm{i}\,\boldsymbol{\alpha}] = \mathcal{F}_1^{-1}[1/\mathrm{i}\,\alpha_1](x)\cdot \mathcal{F}_1^{-1}[1](y). \tag{A.7.5}$$

Applying the Fourier transform to the equation

$$\mathrm{d}u/\mathrm{d}x = \delta(x), \tag{A.7.6}$$

we deduce $-\mathrm{i}\,\alpha_1 \widehat{u} = 1$ whence:

$$u = -\mathcal{F}_1^{-1}[1/\mathrm{i}\,\alpha_1](x). \tag{A.7.7}$$

The equation (A.7.6) has the form (A.3.16) and its solution has the form (A.3.17). It results

$$u = H(x). \tag{A.7.8}$$

From (A.7.2) - (A.7.8) we deduce

$$\mathcal{F}_2^{-1}[1/\mathrm{i}\,\alpha_1] = -H(x)\delta(y). \tag{A.7.9}$$

For the determination of the fundamental solution of the equation of Laplace we need the following results (demonstrated, for example, in [A.12], §97, [A.13], §6.6, or [A.7], §5.3).

$$\mathcal{F}^{-1}\left[\frac{1}{\boldsymbol{\alpha}^2}\right] = \frac{1}{4\pi}\frac{1}{|\boldsymbol{x}|}, \quad n = 3, \tag{A.7.10}$$

$$\mathcal{F}_2^{-1}\left[FP\frac{1}{\boldsymbol{\alpha}^2}\right] = -\frac{1}{2\pi}(\ln|\boldsymbol{x}| + C), \quad n = 2, \tag{A.7.11}$$

C being a constant determined in [A.13], but without any importance here. Taking into account (A.6.10) with $c = (\beta, 1, 1.)$ we deduce, on the basis of the formulas (A.7.10) and (A.7.11):

$$\mathcal{F}^{-1}\left[\frac{1}{\beta^2\alpha_1^2 + \alpha_2^2 + \alpha_3^2}\right] = \frac{1}{4\pi}\,\frac{1}{\sqrt{x^2 + \beta^2(y^2 + z^2)}}, \tag{A.7.12}$$

$$\mathcal{F}_2^{-1}\left[FP\frac{1}{\beta^2\alpha_1^2 + \alpha_2^2}\right] = -\frac{1}{2\pi\beta}(\ln\sqrt{x^2 + \beta^2 y^2} + C - \ln\beta), \tag{A.7.13}$$

FP representing the symbol for the Finite Part (Appendix E), and β, a positive constant.

Also, for determining the fundamental solution of the wave equation we need the formulas

$$\mathcal{F}^{-1}\left[\frac{\sin a|\boldsymbol{\alpha}|t}{|\boldsymbol{\alpha}|}\right] = \frac{\delta(at - |\boldsymbol{x}|)}{4\pi at}, \quad n = 3, \tag{A.7.14}$$

$$\mathcal{F}^{-1}\left[\frac{\sin a|\boldsymbol{\alpha}|t}{|\boldsymbol{\alpha}|}\right] = \frac{1}{2\pi}\,\frac{H(at - |\boldsymbol{x}|)}{\sqrt{a^2t^2 - |\boldsymbol{x}|^2}}, \quad n = 2, \tag{A.7.15}$$

demonstrated, for example (A.7.14) in [A.5] Chapter 2, §3.4, and (A.7.15) in [A.12] §9.7. We shall prove in the sequel the following formulas

$$\mathcal{F}^{-1}\left[\frac{1}{-k^2\alpha_1^2 + \alpha_2^2 + \alpha_3^2}\right] = \frac{1}{2\pi}\,\frac{H(x - k\sqrt{y^2 + z^2})}{\sqrt{x^2 - k^2(y^2 + z^2)}}, \tag{A.7.16}$$

$$\mathcal{F}_2^{-1}\left[\frac{1}{-k^2\alpha_1^2 + \alpha_2^2}\right] = \frac{1}{2k}H(x - k|y|). \tag{A.7.17}$$

In order to prove (A.7.16), one considers the following partial differential equation:

$$k^2 u_{xx} - u_{yy} - u_{zz} = \delta(\boldsymbol{x}). \tag{A.7.18}$$

Applying the Fourier transform, we deduce $(-k^2\alpha_1^2 + \alpha_2^2 + \alpha_3^2)\widehat{u} = 1$, whence,

$$u = \mathcal{F}^{-1}\left[\frac{1}{-k^2\alpha_1^2 + \alpha_2^2 + \alpha_3^2}\right]. \tag{A.7.19}$$

It results therefore that the first member from (A.7.16) is just the solution of the equation (A.7.18). We shall determine in a different manner this solution, namely applying the Fourier transform only with respect to the variables y and z. From (A.7.18) one obtains the equation

$$(k^2 \mathrm{d}^2/\mathrm{d}x^2 + \omega^2)\widehat{u} = \delta(x), \tag{A.7.20}$$

where $\omega = \sqrt{\alpha_2^2 + \alpha_3^2}$, and $\widehat{u}$ represents the two-dimensional transformation. The equation (A.7.20) is like (A.3.16), and its solution has the form (A.3.17). We deduce

$$\widehat{u} = \frac{H(x)}{k} \frac{\sin(\omega x/k)}{\omega}.$$

Utilizing (A.7.15) one obtains (A.7.16).

Acting similarly, in the two-dimensional case, we find:

$$\widetilde{u} = \frac{H(x)}{k} \frac{\sin(\alpha_2 x/k)}{\alpha_2},$$

$\widetilde{u}$ being the notation for the Fourier transform with respect to the variable y. Utilizing the formula

$$\mathcal{F}_1\left[\frac{\sin ax}{x}\right]\begin{cases}\pi & |\alpha| < a\\ 0 & , |\alpha| > a,\end{cases} \tag{A.7.21}$$

given for example in [A.1], p.202, we obtain (A.7.17).

In the unsteady aerodynamics we shall meet the following type of formulas

$$\mathcal{F}^{-1}\left[\frac{1-\cos a|\boldsymbol{\alpha}|t}{\alpha^2}\right] = \frac{H(at-|\boldsymbol{x}|)}{4\pi|\boldsymbol{x}|}, n = 3, \tag{A.7.22}$$

$$\mathcal{F}^{-1}\left[\frac{1-\cos a|\alpha|t}{\alpha^2}\right] = \frac{H(at-|\boldsymbol{x}|)}{2\pi}\ln\left(\frac{at+\sqrt{a^2t^2-|\boldsymbol{x}|^2}}{|\boldsymbol{x}|}\right), \quad n = 2. \tag{A.7.23}$$

We shall prove these formulas (following an idea suggested by V.Iftimie) using the results concerning the Cauchy problem for the non-homogeneous wave equation. To this aim we shall denote:

$$\widehat{u}(t,\boldsymbol{\alpha}) = \frac{1-\cos a|\boldsymbol{\alpha}|t}{\alpha^2}.$$

We deduce

$$\widehat{u}(0,\boldsymbol{\alpha}) = 0, \quad \widehat{u}_t(0,\boldsymbol{\alpha}) = 0,$$

$$\widehat{u}_{tt}(t,\boldsymbol{\alpha}) = a^2\cos a|\boldsymbol{\alpha}|t = -a^2\alpha^2\widehat{u}(t,\boldsymbol{\alpha}) + a^2.$$

Applying the operator $\mathcal{F}^{-1}$, we deduce:

$$u(0,\boldsymbol{x}) = 0, \quad u_t(0,\boldsymbol{x}) = 0$$

$$u_{tt} = a^2\Delta u + a^2\delta(\boldsymbol{x})\,.$$

For determining u we have therefore to solve a Cauchy problem for the non-homogeneous wave equation. The solution of this problem in the three- and bi-dimensional cases is given by Poisson's formula [A.12]. Utilizing this formula, we find (A.7.22) and (A.7.23).

At last, replacing x by $x-t$ in the formulas (A.7.9), (A.7.14), (A.7.15), (A.7.22) and (A.7.23), we obtain in the three-dimensional case, with the notation $R=\sqrt{(x-t)^2+y^2+z^2}$:

$$\begin{aligned}
&\mathcal{F}^{-1}\left[\mathrm{e}^{\mathrm{i}\,\alpha_1 \mathrm{t}}\right] = \delta(x-t)\cdot\delta(y,z)\,,\\
&\mathcal{F}^{-1}\left[\frac{\sin a|\boldsymbol{\alpha}|t}{|\boldsymbol{\alpha}|}\mathrm{e}^{\mathrm{i}\,\alpha_1 \mathrm{t}}\right] = \frac{1}{4\pi a t}\delta(at-R)\,,\\
&\mathcal{F}^{-1}\left[\frac{1-\cos a|\boldsymbol{\alpha}|t}{\alpha^2}\mathrm{e}^{\mathrm{i}\,\alpha_1 \mathrm{t}}\right] = \frac{H(at-R)}{4\pi R}\,,
\end{aligned} \tag{A.7.24}$$

and in the two-dimensional case, with the notation $\overline{R}=\sqrt{(x-t)^2+y^2}$,

$$\begin{aligned}
&\mathcal{F}^{-1}\left[\mathrm{e}^{\mathrm{i}\,\alpha_1 \mathrm{t}}\right] = \delta(x-t)\cdot\delta(y)\,,\\
&\mathcal{F}^{-1}\left[\frac{\sin a|\boldsymbol{\alpha}|t}{|\boldsymbol{\alpha}|}\mathrm{e}^{\mathrm{i}\,\alpha_1 \mathrm{t}}\right] = \frac{1}{2\pi}\frac{H(at-\overline{R})}{\sqrt{a^2t^2-\overline{R}^2}}\,,\\
&\mathcal{F}^{-1}\left[\frac{1-\cos a|\boldsymbol{\alpha}|t}{\alpha^2}\mathrm{e}^{\mathrm{i}\,\alpha_1 \mathrm{t}}\right] = \frac{H(at-\overline{R})}{2\pi}\ln\left(\frac{at+\sqrt{a^2t^2-\overline{R}^2}}{\overline{R}}\right).
\end{aligned} \tag{A.7.25}$$

In [1.10] we may find direct demonstrations of the formulas (A.7.25).

A.8 The Fourier Transform in Bounded Domains

In this last part, we return to the Fourier transform of the functions and we give, following Homentcovschi's idea [A.6], the transformation formulas in case that the function $f(x)$ is defined on a bounded domain D. We assume that D is bounded by a surface S which closes a domain D' and by a surface of discontinuity Σ. We prolong f in D', giving it the value zero. We make the same thing in the domain D'' which closes Σ. Applying the flux-divergence formula we obtain:

$$\widehat{f_x}\big|_D = \widehat{f_x}\big|_{\mathbb{R}^3} - \int_{D'+D''} f_x e^{i\boldsymbol{\alpha}\cdot\boldsymbol{x}} d\,\boldsymbol{x} = \widehat{\mathrm{f_x}}\big|_{\mathbb{R}^3} - \int_{\mathrm{D'+D''}} \partial_{\mathrm{x}}(\mathrm{f}e^{i\boldsymbol{\alpha}\cdot\boldsymbol{x}}) d\,\boldsymbol{x} =$$

$$= \widehat{f_x}\big|_{\mathbb{R}^3} - \int_S f n_1 e^{i\boldsymbol{\alpha}\cdot\boldsymbol{x}} d\,\mathrm{a} - \int_\Sigma [\![\mathrm{f}]\!] \mathrm{n}_1 e^{i\boldsymbol{\alpha}\cdot\boldsymbol{x}} d\,\mathrm{a}. \tag{A.8.1}$$

After all, taking into account (A.6.5), we have:

$$F[\operatorname{grad} f]_D = -i\boldsymbol{\alpha}\widehat{f} - \int_S f\boldsymbol{n} e^{i\boldsymbol{\alpha}\cdot\boldsymbol{x}} d\,\mathrm{a} - \int_\Sigma [\![\mathrm{f}]\!] \boldsymbol{n} e^{i\boldsymbol{\alpha}\cdot\boldsymbol{x}} d\,\mathrm{a}$$

$$F[\operatorname{div} \boldsymbol{f}]_D = -i\boldsymbol{\alpha}\cdot\widehat{\boldsymbol{f}} - \int_S \boldsymbol{f}\cdot\boldsymbol{n} e^{i\boldsymbol{\alpha}\cdot\boldsymbol{x}} d\,\mathrm{a} - \int_\Sigma [\![\mathrm{f}]\!] \boldsymbol{n} e^{i\boldsymbol{\alpha}\cdot\boldsymbol{x}} d\,\mathrm{a}$$

$$F[\operatorname{rot} \boldsymbol{f}]_D = -i\boldsymbol{\alpha}\times\widehat{\boldsymbol{f}} - \int_S \boldsymbol{f}\times\boldsymbol{n} e^{i\boldsymbol{\alpha}\cdot\boldsymbol{x}} d\,\mathrm{a} - \int_\Sigma [\![\mathrm{f}]\!] \boldsymbol{n} e^{i\boldsymbol{\alpha}\cdot\boldsymbol{x}} d\,\mathrm{a}, \tag{A.8.2}$$

where $[\![f]\!] = f_+ - f_-$.

Appendix B

Cauchy-type Integrals. Dirichlet's Problem for the Half-Plane. The Calculus of Some Integrals

B.1 Cauchy-type Integrals

We consider in the $z = x + \mathrm{i}\,y$ complex plane a smooth curve Γ, i.e. a curve which has the parametric equations

$$x = x(s), \quad y = y(s), \quad s_1 \leq s \leq s_2\,, \tag{B.1.1}$$

where $x(s)$ and $y(s)$ are continuously differentiable functions, whose derivatives do not vanish simultaneously in the same point. The curve Γ may be closed or open; if it is closed, then $z(s_1) = z(s_2)$; if it is open, then we assume $z'(s_1) = z'(s_1 + 0)$ and $z'(s_2) = z'(s_2 - 0)$. By *definition* the *positive* sense on Γ is the sense corresponding to the increase of the parameter s.

The smooth curve is obviously rectifiable, such that we may consider as parameter s the length of the arc measured from $s_1(= 0)$ to $s_2(= l)$. In this case, we obviously have $x'^2 + y'^2 = 1$.

Let $f(t)$ be a complex function depending on the complex variable t, defined on Γ and Riemann integrable. The integral

$$F(z) = \frac{1}{2\pi\mathrm{i}} \int_\Gamma \frac{f(t)}{t - z}\mathrm{d}\,t \tag{B.1.2}$$

is called *Cauchy-type integral.* As we know from the books of complex analysis, the function $F(z)$ is holomorphic in the interior of the contour Γ. If Γ is at a finite distance, then $F(z)$ behaves at infinity like $|z|^{-1}$.

We shall investigate, in the following, what happens with the integral (B.1.2) if $z = t_0 \in \Gamma$. In this case, the integrand has obviously a non-integrable singularity in t_0 and generally the integral has no sense. There exists however a large class of functions (we are not interested here in the largest class) for which we may give a definition to the integral, namely the class of the functions which satisfy the so called *Hölder's condition.*

We say that the function $f(t)$ satisfies Hölder's condition on Γ if there exist two positive constants (different from zero) A and $\mu (\mu \leq 1)$, such that, for every two points t_1 and $t_2 \in \Gamma$ să we have

$$|f(t_1) - f(t_2)| \leq \wedge |t_1 - t_2|^{\mu} . \tag{B.1.3}$$

Obviously, the functions f which satisfy Hölder's condition are continuous on Γ. If $\mu = 1$, the functions satisfy Lipschitz's condition.

B.2 The Principal Value in Cauchy's Sense

We shall give now the definition that we have mentioned before. We assume at first that t_0 does not coincide with any extremity of the arc Γ (if it is open). We consider the arc of circle with the center in t_0 and the radius ε which cuts the curve Γ in two points t_1 and t_2 and we denote by γ the arc $t_1 t_2$. If for $\varepsilon \to 0$ the integral $\int_{\Gamma-\gamma} \frac{f(t)}{t - t_0} \mathrm{d}t$ has a finite limit, then this limit will be called *the principal value in Cauchy's sense.* We denote

$$\lim_{\varepsilon \to 0} \int_{\Gamma-\gamma} \frac{f(t)}{t - t_0} \mathrm{d}t = \int_{\Gamma}' \frac{f(t)}{t - t_0} \mathrm{d}t . \tag{B.2.1}$$

The principal value is a distribution [A.12], [A.14].

We shall prove in the sequel the following *theorem:* "*If* $f(t)$ *satisfies Hölder's condition in the vicinity of the point* t_0, *then the limit* (B.2.1) *exists and it is unique.* For the proof we shall write:

$$\int_{\Gamma-\gamma} \frac{f(t)}{t - t_0} \mathrm{d}t = \int_{\Gamma-\gamma} \frac{f(t) - f(t_0)}{t - t_0} \mathrm{d}t + f(t_0) \int_{\Gamma-\gamma} \frac{\mathrm{d}t}{t - t_0} . \tag{B.2.2}$$

Having (B.1.3) in view, the limit of the first integral from the right hand member exists and it equals the usual improper integral on Γ. The last integral is calculated as follows:

$$\int_{\Gamma-\gamma} \frac{\mathrm{d}t}{t - t_0} = \ln(t - t)\Big|_a^{t_1} + \ln(t - t_0)\Big|_{t_2}^{b} = \ln \frac{b - t_0}{a - t_0} +$$

$$+ \ln(t_1 - t_0) - \ln(t_2 - t_0) .$$

Dar $t_1 - t_0 = |t_1 - t_0| \mathrm{e}^{\mathrm{i}\alpha}$, $\mathrm{t}_2 - \mathrm{t}_0 = |\mathrm{t}_2 - \mathrm{t}_0| \mathrm{e}^{\mathrm{i}\beta}$ and $|t_1 - t_0| = |t_2 - t_0|$.

It results $\ln(t_1 - t_0) - \ln(t_2 - t_0) = \mathrm{i}\,(\alpha - \beta)$. Passing to limit, when $\varepsilon \to 0$, $\alpha - \beta = \pi$, we get

$$\lim_{\varepsilon \to 0} \int_{\Gamma - \gamma} \frac{\mathrm{d}t}{t - t_0} = \int_{\Gamma}' \frac{\mathrm{d}t}{t - t_0} = \ln \frac{b - t_0}{a - t_0} + \mathrm{i}\,\pi = \ln \frac{b - t_0}{t_0 - a} . \tag{B.2.3}$$

Since the last integral from (B.2.2) has a well determined limit, the theorem is demonstrated.

The case when t_0 coincides with one of the extremities of the arc Γ, depends on the behaviour of the function f in that point (see for example [A.27], §29–32). If t_0 coincides with an extremity and $f(t_0) = 0$, we are in the previously considered case, because we may extend arbitrarily the contour Γ beyond t_0, setting $f = 0$ on the extension.

B.3 Plemelj's Formulas

We shall investigate the behaviour of the Cauchy-type integral in the vicinity of the curve Γ. To this aim we shall give at first the following *definition* [A.27]: we say that $F(z)$ is *continuously prolongable* on Γ in t_0 (different from the extremities) at left (right), if $F(z)$ tends to a well determined limit $F_+(t_0)(F_-(t_0))$ when $z \to t_0$ on *every* path situated at left (right).

With this definition we may give the following *fundamental theorem* : *If $f(t)$ verifies on Γ Hölder's condition, then $F(z)$ is continuously prolongable on Γ at left and at right, excepting the extremities where $f(t_0) \neq 0$ and*

$$F_{\pm}(t_0) = \pm \frac{1}{2} f(t_0) + \frac{1}{2\pi \mathrm{i}} \int_{\Gamma}' \frac{f(t)}{t - t_0} \mathrm{d}t . \tag{B.3.1}$$

The formulas (B.3.1) are called *Plemelj's formulas.* They have been given in 1908 [A.29]. Their demonstration may be found for example in [A.18], [A.27].

B.4 The Dirichlet's Problem for the Half-Plane

We shall solve in the sequel the following problem: *We seek for the function*

$$F(z) = u(x, y) + \mathrm{i}\,v(x, y), \tag{B.4.1}$$

holomorphic in the half-plane $y > 0$ *and continuously prolongable on the* Ox, *axis, which reduces at infinity, to an imaginary constant* $\mathrm{i}C$ (C *may have the value zero) and whose real part is imposed on the above mentioned axis, i.e.*

$$u(x, 0) = f(x), \tag{B.4.2}$$

where f *is a function with a compact support compact which satisfies Hölder's condition.*

At first we have to mention that there exists a single function with the above mentioned properties. Indeed, assuming that there exists two functions F_1 and F_2 with these properties, their difference $F = F_1 - F_2$ is holomorphic in the superior half -plane and it vanishes at infinity. The real part of the function F is therefore harmonic in the half-plane $y > 0$, zero on the boundary $y = 0$ and zero at infinity. According to the maximum principle for the harmonic functions, the real part of the function is identical zero. F reduces therefore to an imaginary constant which is zero because F is zero at *infinity*. We shall prove that the function

$$F(z) = \frac{1}{\pi \mathrm{i}} \int_{-\infty}^{+\infty} \frac{f(t)}{t - z} \mathrm{d}t + \mathrm{i}C \tag{B.4.3}$$

satisfies the conditions of the problem and it is therefore the solution we are looking for. Indeed, the function $F(z)$ defined by (B.4.3) is holomorphic in the superior half-plane because it is a Cauchy-type integral and it is continuously prolongable on the real axis. At infinity it reduces to the constant $\mathrm{i}C$ because, if we denote by (a, b) the support of $f(x)$, we have

$$\int_{-\infty}^{+\infty} \frac{f(t)}{t - z} \mathrm{d}t = -\frac{1}{z} \int_a^b \frac{f(t)}{1 - t/z} \mathrm{d}t = -\frac{1}{z} \int_a^b f(t) \left(1 + \frac{t}{z} + \frac{t^2}{z^2} + \ldots\right) \mathrm{d}t.$$

The integral is therefore zero at infinity.

Passing to the limit with $z \to x$ a point from the real axis, and using Plemelj's formulas we obtain:

$$u(x, 0) + \mathrm{i}\, v(x, 0) = f(x) + \frac{1}{\pi \mathrm{i}} {\int_{-\infty}^{+\infty}}' \frac{f(t)}{t - x} \mathrm{d}t + \mathrm{i}C.$$

Taking the real part of this relation we obtain (B.4.2). Hence the problem is solved.

The real part of the solution (B.4.3), i.e.

$$u(x, y) = \frac{y}{\pi} \int_{-\infty}^{+\infty} \frac{f(t)}{(t - x)^2 + y^2} \mathrm{d}t \tag{B.4.4}$$

determines the harmonic function in the half-plane $y > 0$, vanishing at infinity, continuously prolongable on the Ox axis and satisfying the condition (B.4.2). This is the solution of Dirichlet's problem for the half-plane concerning the harmonic function u.

B.5 The Calculus of Certain Integrals in the Complex Plane

At first we shall prove that

$$\frac{1}{\pi}\int_a^b \sqrt{\frac{t-a}{b-t}}\frac{\mathrm{d}t}{t-z} = 1 - \sqrt{\frac{z-a}{z-b}}, \tag{B.5.1}$$

where the determination of the radical is the positive one for $z = x > b$, and (a,b) is an interval on the real axis.

Indeed, with the mentioned determination, we have

$$\sqrt{\frac{z-a}{z-b}} = \begin{cases} \sqrt{\dfrac{x-a}{x-b}}, & x < a, \quad b < x \\ \dfrac{1}{\mathrm{i}}\sqrt{\dfrac{x-a}{b-x}}, & a < x < b\,. \end{cases} \tag{B.5.2}$$

After all,

$$\mathrm{Re}\left(\mathrm{i}\sqrt{\frac{z-a}{z-b}}\right) = \begin{cases} 0, & x < a, \quad b < x \\ \sqrt{\dfrac{x-a}{b-x}}, & a < x < b\,. \end{cases}$$

Since the real part of the holomorphic function $\mathrm{i}\sqrt{\dfrac{z-a}{z-b}}$ is known on the real axis, the function will be determined by the formula (B.4.3). We have therefore

$$\sqrt{\frac{z-a}{z-b}} = -\frac{1}{\pi}\int_a^b \sqrt{\frac{t-a}{b-t}}\,\frac{\mathrm{d}t}{t-z} + C\,.$$

Considering $z \to \infty$ it results $C = 1$ and then the formula (B.5.1).

Analogously we deduce that

$$\frac{1}{\pi}\int_a^b \sqrt{\frac{b-t}{t-a}}\,\frac{\mathrm{d}t}{t-z} = -1 + \sqrt{\frac{z-b}{z-a}}\,. \tag{B.5.3}$$

Passing to the limit and taking into account Plemelj's formulas, (B.5.1) and (B.5.3) we obtain that:

$$\frac{1}{\pi}{}' \int_a^b \sqrt{\frac{t-a}{b-t}} \frac{\mathrm{d}t}{t-x} = 1, \quad \frac{1}{\pi}{}' \int_a^b \sqrt{\frac{b-t}{t-a}} \frac{\mathrm{d}t}{t-x} = -1. \qquad \text{(B.5.4)}$$

We must notice now that the above considered function $\mathrm{i}\sqrt{\frac{z-a}{z-b}}$ reduces to an imaginary constant at infinity (this is i). If we want to apply the same procedure for the function $\sqrt{(z-a)(z-b)}$, we must consider the expression

$$\sqrt{(z-a)(z-b)} - z.$$

To this aim we have

$$\operatorname{Re}\left[\mathrm{i}\left(\sqrt{(z-a)(z-b)} - z\right)\right] = \begin{cases} 0, & x > b,\ x < a \\ -(x-a)(b-x), & a < x < b \end{cases}$$

whence

$$\sqrt{(z-a)(z-b)} - z = +\frac{1}{\pi}\int_a^b \frac{\sqrt{(t-a)(b-t)}}{t-z}\mathrm{d}t + C.$$

The integral becomes zero to the infinite. This imply

$$C = -\frac{a+b}{2}.$$

We have therefore

$$\frac{1}{\pi}\int_a^b \frac{\sqrt{(t-a)(b-t)}}{t-z}\mathrm{d}t = \sqrt{(z-a)(z-b)} - z + \frac{a+b}{z} \qquad \text{(B.5.5)}$$

and, passing to the limit and applying Plemelj's formulas,

$$\frac{1}{\pi}{}'\int_a^b \frac{\sqrt{(t-a)(b-t)}}{t-x}\mathrm{d}t = -x + \frac{a+b}{z}. \qquad \text{(B.5.6)}$$

The following example is studied starting from the expression

$$\frac{\mathrm{i}}{\sqrt{(z-a)(z-b)}},$$

which vanishes at infinity. Hence, the constant C will be zero. One obtains the integrals

$$\frac{1}{\pi}\int_a^b \frac{1}{\sqrt{(b-t)(t-a)}} \frac{\mathrm{d}t}{t-z} = -\frac{1}{\sqrt{(z-a)(z-b)}}, \qquad \text{(B.5.7)}$$

$$\frac{1}{\pi}\int_a^b{}' \frac{1}{\sqrt{(b-t)(t-a)}}\,\frac{\mathrm{d}\,t}{t-x} = 0\,. \tag{B.5.8}$$

We consider now two intervals $(a,b),(c,d)$ on the real axis. The method can be extended to an arbitrary number of intervals, or even disjoint arcs from the complex plane, in the last case the integral being solved in another way in [A.18], p.88. The result is useful in the study of the grids of profiles. Denoting

$$Q(z) = (z-a)(z-b)(z-c)(z-d)\,, \tag{B.5.9}$$

for $P(z)$ and $R(z)$ arbitrary polynomials, we have

$$\operatorname{Re}\left[\mathrm{i}\,\frac{P(z)}{\sqrt{Q(z)}} - \mathrm{i}\,R(z)\right] = \begin{cases} 0\,, & -\infty < x < a, b < x < c, d < x \\ -\dfrac{P(x)}{\sqrt{|Q|}}\,, & a < x < b \\ \dfrac{P(x)}{\sqrt{|Q|}}\,, & c < x < d\,. \end{cases}$$

The function from square brackets is holomorphic in the half-plane $y > 0$. Utilizing the formula (B.4.3), we deduce

$$\frac{P(z)}{\sqrt{Q(z)}} - R(z) = \frac{1}{\pi}\int_a^b \frac{P(t)}{\sqrt{|Q(t)|}}\,\frac{\mathrm{d}\,t}{t-z} - \frac{1}{\pi}\int_c^d \frac{P(t)}{\sqrt{|Q(t)|}}\,\frac{\mathrm{d}\,t}{t-z} + C\,. \tag{B.5.10}$$

Taking into account the behaviour of the integrals for $z \to \infty$, we deduce that at infinity we must have the expansion

$$\frac{P(z)}{\sqrt{Q(z)}} - R(z) - C = \frac{\alpha_1}{z} + \frac{\alpha_2}{z^2} + \ldots\,. \tag{B.5.11}$$

This formula allows to determine $R(Z)$ when $P(z)$ is known. For example, for

$$P(z) = (z-b)(z-d)$$

it results $R = 0$ and $C = 1$, and for

$$P(z) = z(z-b)(z-d)$$

it results

$$R = z, C = \frac{a+c-b-d}{2}\,.$$

After all, we have the results

$$\frac{1}{\pi}\int_a^b \sqrt{\frac{(b-t)(d-t)}{(t-a)(c-t)}}\,\frac{\mathrm{d}\,t}{t-z} + \frac{1}{\pi}\int_c^d \sqrt{\frac{(t-b)(d-t)}{(t-a)(t-c)}}\,\frac{\mathrm{d}\,t}{t-z} = \sqrt{\frac{(z-b)(z-d)}{(z-a)(z-c)}} - 1 \tag{B.5.12}$$

$$\frac{1}{\pi}\int_a^b \sqrt{\frac{(b-t)(d-t)}{(t-a)(c-t)}}\,\frac{t\mathrm{d}\,t}{t-z} + \frac{1}{\pi}\int_c^d \sqrt{\frac{(t-b)(d-t)}{(t-a)(t-c)}}\,\frac{t\mathrm{d}\,t}{t-z} = z\left[\sqrt{\frac{(z-b)(z-d)}{(z-a)(z-c)}} - 1\right] - \frac{a+c-b-d}{2}. \tag{B.5.13}$$

Passing to the limit for $z \to x \in (a,b)$, we apply Plemelj's formulas to the first integrals. One obtains:

$$\frac{1}{\pi}{\int_a^b}' \sqrt{\frac{(b-t)(d-t)}{(t-a)(c-t)}}\,\frac{\mathrm{d}\,t}{t-x} + \frac{1}{\pi}\int_c^d \sqrt{\frac{(t-b)(d-t)}{(t-a)(t-c)}}\,\frac{\mathrm{d}\,t}{t-x} = -1$$

$$\frac{1}{\pi}{\int_a^b}' \sqrt{\frac{(b-t)(d-t)}{(t-a)(c-t)}}\,\frac{t\mathrm{d}\,t}{t-x} + \frac{1}{\pi}\int_c^d \sqrt{\frac{(t-b)(d-t)}{(t-a)(t-c)}}\,\frac{t\mathrm{d}\,t}{t-x} = -x - \frac{a+c-b-d}{2}, \tag{B.5.14}$$

and for $z \to x \in (c,d)$, employing Plemelj's formulas for the integrals on this interval, we obtain from (B.5.12) and (B.5.13):

$$\frac{1}{\pi}\int_a^b \sqrt{\frac{(b-t)(d-t)}{(t-a)(c-t)}}\,\frac{\mathrm{d}\,t}{t-x} + \frac{1}{\pi}{\int_c^d}' \sqrt{\frac{(t-b)(d-t)}{(t-a)(t-c)}}\,\frac{\mathrm{d}\,t}{t-x} = -1$$

$$\frac{1}{\pi}\int_a^b \sqrt{\frac{(b-t)(d-t)}{(t-a)(c-t)}}\,\frac{t\mathrm{d}\,t}{t-x} + \frac{1}{\pi}{\int_c^d}' \sqrt{\frac{(t-b)(d-t)}{(t-a)(t-c)}}\,\frac{t}{t-x}\mathrm{d}\,t = -x - \frac{a+c-b-d}{2}.$$

B.6 Glauert's Integral. Its Generalization and Some Applications

We shall calculate the integral

$$I_n = \frac{1}{\pi}\int_0^{\pi} \frac{\cos n\theta}{\cos\theta - s}\,\mathrm{d}\,\theta = \frac{1}{2\pi}\int_{-\pi}^{+\pi} \frac{\cos n\theta}{\cos\theta - s}\,\mathrm{d}\,\theta\,, \qquad \text{(B.6.1)}$$

where s is a real number. For $-1 < s < +1$ the integral was calculated by Glauert in [3.19]. The general case is considered in [5.38].

The method consists in passing to the complex variable. Denoting $z = \exp(\mathrm{i}\,\theta)$ and noticing that we have

$$\int_{-\pi}^{+\pi} \frac{\sin n\theta}{\cos\theta - s}\,\mathrm{d}\,\theta = 0\,,$$

because the integrand is an even function, it results

$$I_n = \frac{1}{2\pi}\int_{-\pi}^{+\pi} \frac{e^{\mathrm{i}\,n\theta}}{\cos\theta - s}\,\mathrm{d}\,\theta = \frac{1}{\pi\mathrm{i}}\int_{|z|=1} \frac{z^n\,\mathrm{d}\,z}{z^2 - 2sz + 1}$$

$$= \frac{1}{\pi\mathrm{i}}\int_{|z|=1} \frac{z^n\,\mathrm{d}\,z}{(z-\alpha)(z-\beta)}\,, \qquad \text{(B.6.2)}$$

where

$$\alpha = s + \sqrt{s^2-1}, \quad \beta = s - \sqrt{s^2-1}\,.$$

The last integral is calculated with the residue theorem.

Since $\alpha\beta = 1$, It results that the poles of the integrand are situated either one in the interior and the other in the exterior of the circle $z = 1$, or both of them on the circle.

In the case $s < -1$, the pole $z = \alpha$ is interior and according to the residue theorem, we have:

$$I_n = 2\pi\mathrm{i}\,\frac{1}{\pi\mathrm{i}}\,\frac{\alpha^n}{\alpha - \beta} = \frac{(s+\sqrt{s^2-1})^n}{\sqrt{s^2-1}}\,, \qquad \text{(B.6.3)}$$

and for $s > 1$, the pole $z = \beta$ is interior and it results

$$I_n = -\frac{(s-\sqrt{s^2-1})^n}{\sqrt{s^2-1}}\,. \qquad \text{(B.6.4)}$$

At last, in the case $-1 < s < 1$, using the substitution $s = \cos\sigma$, we have $\alpha = \exp(\mathrm{i}\,\sigma)$, $\beta = \exp(-\mathrm{i}\,\sigma)$ and from the semi-residue theorem [A.22], p.320 (the poles are situated on the integration path)

$$\oint \frac{f(z)}{z - z_0}\,\mathrm{d}\,z = \pi\mathrm{i}\,f(z_0)\,, \qquad \text{(B.6.5)}$$

we deduce

$$I_n = \frac{\alpha^n}{\alpha - \beta} + \frac{\beta^n}{\beta - \alpha} = \frac{\sin n\sigma}{\sin \sigma}.$$

Hence, Glauert's formula is

$$\frac{1}{\pi} {}' \int_0^{\pi} \frac{\cos n\theta}{\cos\theta - \cos\sigma} \mathrm{d}\,\theta = \frac{\sin n\sigma}{\sin\sigma}, \quad n = 0, 1, 2 \tag{B.6.6}$$

Glauert's formula has many applications in aerodynamics. For example, using the substitutions

$$t = c + e\cos\theta, \quad x = c + e\cos\sigma, \tag{B.6.7}$$

where

$$2c = a + b, \quad 2e = b - a, \tag{B.6.8}$$

one obtains the formulas (B.5.4), (B.5.6), (B.5.8) and other similar ones, like for example

$${}' \int_a^b \sqrt{\frac{t-a}{b-t}} \frac{t}{t-x} \mathrm{d}\,t = \pi(x + e). \tag{B.6.9}$$

Using the substitutions

$$t = c + e\cos\theta, \quad x = c + es \tag{B.6.10}$$

and the formulas (B.6.3) and (B.6.4) we also obtain

$$\frac{1}{\pi} \int_a^b \sqrt{\frac{t-a}{b-t}} \frac{\mathrm{d}\,t}{t-x} = \begin{cases} 1 - \sqrt{\dfrac{a-x}{b-x}}, & x < a \\ 1, & a < x < b \\ 1 - \sqrt{\dfrac{x-a}{x-b}}, & x > b \end{cases}$$

$$\frac{1}{\pi} \int_a^b \sqrt{\frac{b-t}{t-a}} \frac{\mathrm{d}\,t}{t-x} = \begin{cases} \sqrt{\dfrac{b-x}{a-x}} - 1, & x < a \\ -1, & a < x < b \\ \sqrt{\dfrac{x-b}{x-a}} - 1, & x > b \end{cases}$$

$$\frac{1}{\pi} \int_a^b \frac{1}{\sqrt{(b-t)(t-a)}} \frac{\mathrm{d}\,t}{t-x} = \begin{cases} \dfrac{1}{\sqrt{(a-x)(b-x)}}, & x < a \\ 0, & a < x < b \\ -\dfrac{1}{\sqrt{(x-a)(x-b)}}, & x > b \end{cases} \tag{B.6.11}$$

etc.

B.7 Other Integrals

In the boundary elements methods one may encounter the following integrals

$$I_k = \int_0^{2\pi} \frac{\cos n\theta + \mathrm{i}\,\sin n\theta}{(a + b\cos\theta + c\sin\theta)^k}\,\mathrm{d}\,\theta \tag{B.7.1}$$

for $k = 1, 2$. They are also calculated with the residue theorem putting $z = \mathrm{e}^{\mathrm{i}\,\theta}$. It results

$$\sin\theta = \frac{1}{2\mathrm{i}}\left(z - \frac{1}{z}\right), \cos\theta = \frac{1}{2}\left(z + \frac{1}{z}\right),$$

$$I_1 = \frac{2}{\mathrm{i}\,b + c}\int_{|z|=1} \frac{z^n \mathrm{d}\,z}{(z - z_1)(z - z_2)},$$

$$I_2 = \frac{4\mathrm{i}}{(\mathrm{i} + c)^2}\int_{|z|=1} \frac{z^{n+1} \mathrm{d}\,z}{(z - z_1)^2(z - z_2)^2}, \tag{B.7.2}$$

where, with the notation $r = \sqrt{a^2 - b^2 - c^2}$, we have:

$$z_1 = \frac{-a + r}{b - \mathrm{i}\,c}, \quad z_2 = \frac{-a - r}{b - \mathrm{i}\,c}. \tag{B.7.3}$$

Obviously, $|z_1 z_2| = 1$, such that either a root is in the interior of the unit circle and the other in the exterior or both of them are on the circle.

We are interested in the case when $a^2 > b^2 + c^2$ and $a > 0$. In these conditions, the root z_1 is in the interior of the unit circle, such that, utilizing the residue theorem, one obtains:

$$I_1 = F(r), I_2 = G(r), \tag{B.7.4}$$

where

$$F(r) = \frac{2\pi(-1)^n}{r}\left(\frac{b + \mathrm{i}\,c}{a + r}\right)^n, \quad G(r) = 2\pi(-1)^n \frac{a + nr}{r^3}\left(\frac{b + \mathrm{i}\,c}{a + r}\right)^n. \tag{B.7.5}$$

Separating the real part from the imaginary one, we obtain a series of integrals, like, for example

$$
\begin{aligned}
&\int_0^{2\pi} \frac{\mathrm{d}\,\theta}{a + b\cos\theta + c\sin\theta} = \frac{2\pi}{r}\,,\\
&\int_0^{2\pi} \frac{\cos\theta}{a + b\cos\theta + c\sin\theta}\mathrm{d}\,\theta = -\frac{2\pi}{r}\,\frac{b}{a+r}\,,\\
&\int_0^{2\pi} \frac{\sin\theta}{a + b\cos\theta + c\sin\theta}\mathrm{d}\,\theta = -\frac{2\pi}{r}\,\frac{c}{a+r}\,,\\
&\int_0^{2\pi} \frac{\cos 2\theta}{a + b\cos\theta + c\sin\theta}\mathrm{d}\,\theta = \frac{2\pi}{r}\,\frac{b^2 - c^2}{(a+r)^2}\,,\\
&\int_0^{2\pi} \frac{\mathrm{d}\,\theta}{(a + b\cos\theta + c\sin\theta)^2} = \frac{2\pi a}{r^3}\,,\\
&\int_0^{2\pi} \frac{\cos\theta}{(a + b\cos\theta + c\sin\theta)^2}\mathrm{d}\,\theta = -\frac{2\pi b}{r^3}\,,\\
&\int_0^{2\pi} \frac{\sin\theta}{(a + b\cos\theta + c\sin\theta)^2}\mathrm{d}\,\theta = -\frac{2\pi c}{r^3}\,,\\
&\int_0^{2\pi} \frac{\cos 2\theta}{(a + b\cos\theta + c\sin\theta)^2}\mathrm{d}\,\theta = \frac{2\pi}{r^2}\left(2 + \frac{a}{r}\right)\frac{b^2 - c^2}{(a+r)^2}\,.
\end{aligned}
\tag{B.7.6}
$$

The case $a^2 > b^2 + c^2$ and $a < 0$ is studied in the same manner (in this situation, z_2 is in the interior of the circle). One obtains

$$I_1 = F(-r), \quad I_2 = G(-r)\,. \tag{B.7.7}$$

If $a^2 < b^2 + c^2$, the two roots are on the unit circle, such that the integrals are calculated by means of the semi-residue theorem. One obtains

$$2I_1 = F(r) + F(-r), \quad 2I_2 = G(r) + G(-r)\,. \tag{B.7.8}$$

At last, if $a^2 = b^2 + c^2$, the roots coincide and they are situated on the unit circle. The integral will be calculated using the Finite Part (see Appendix D).

If $a = -s, b = 1, c = 0$ and $k = 1$, one obtains the results from the previous section.

Appendix C

Singular Integral Equations

C.1 The Thin Profile Equation

The thin profile equation in a free stream has the form

$$\frac{1}{\pi}\int_a^{'b} \frac{f(t)}{t-x}\,\mathrm{d}t = h(x), \quad a < x < b, \tag{C.1.1}$$

where $f(t)$ is the unknown, and $h(x)$ is a given function. It is sufficient to assume that f satisfies Hölder's condition on the interval (a, b) for deducing the existence of the principal value from (C.1.1). The solution of the equation also depends on the behaviour imposed to the unknown at the extremities of the interval $[a, b]$. From the expression of the solution it follows that h also has to satisfy Hölder's condition on the interval (a, b). The equation was pointed out by Birnbaum in 1923 [3.5] and solved for the first time by Söhngen in 1939 [A.34]. An ample study devoted to this equation is due to Schröder [A.33]. The solution was found again, with different methods by many other researchers (Weissinger [3.49], Cheng [A.30], Homentcovschi [A.19], Carabineanu [A.15] etc.). In the sequel we start from the method of C. Iacob [A.20].

For solving the equation (C.1.1) we have to determine the function

$$F(z) = u(x, y) + \mathrm{i}\, v(x, y), z = x + \mathrm{i}\, y, \tag{C.1.2}$$

holomorphic in the superior half-plane $y > 0$, vanishing at infinity and continuously prolongable on the real axis, with the conditions

$$\begin{aligned} u(x, 0) &= 0, \quad \text{for } x \in (-\infty, a) \cup (b, \infty) \\ v(x, 0) &= -h(x), \quad \text{for } x \in (a, b)\,. \end{aligned} \tag{C.1.3}$$

In order to ensure the uniqueness of F(z) we have to impose its behaviour in a and b. The function F may be bounded or not in these points.

We assume, for example that F is bounded in b, and it behaves in the vicinity of a like

$$F(z) = \frac{F^*}{(z-a)^\alpha}, \quad 0 < \alpha < 1, \tag{C.1.4}$$

where F^* is bounded.

We must notice that if we have determined such a function and if we denote by $f(x)$ the boundary values of u on (a,b), then It results that the function f is just the solution of the equation (C.1.1). Indeed, the function $F(z)$, holomorphic in half-plane $y > 0$ and vanishing to infinity, whose real part on Ox is

$$\begin{aligned} u &= 0, \quad \text{for } x \in (-\infty, a) \cup (b, \infty) \\ u &= f(x), \quad \text{for } x \in (a,b), \end{aligned} \tag{C.1.5}$$

is given by the formula (B.4.3), i.e.

$$F(z) = \frac{1}{\pi \mathrm{i}} \int_a^b \frac{f(t)}{t-z} \mathrm{d}t. \tag{C.1.6}$$

Utilizing Plemelj's formulas we obtain for $a < x < b$

$$u(x,+0) + \mathrm{i}\, v(x,+0) = f(x) + \frac{1}{\pi \mathrm{i}} {\int_a^b}' \frac{f(t)}{t-x} \mathrm{d}t. \tag{C.1.7}$$

Separating the imaginary part and taking (C.1.3) into account, we obtain (C.1.1) for v.

We shall solve in the sequel the problem (C.1.2) - (C.1.4). Having (C.1.4) in view, we shall consider the function (B.5.2). With the determination of the radical precised there and with the conditions (C.1.3), we deduce:

$$\operatorname{Re}\left[F(z)\sqrt{\frac{z-a}{z-b}}\right] = \begin{cases} 0, & x \in (-\infty, a) \cup (b, \infty) \\ -h(x)\sqrt{\dfrac{x-a}{b-x}}, & x \in (a,b). \end{cases}$$

Hence, we may utilize the formula (B.4.3). It results

$$F(z)\sqrt{\frac{z-a}{z-b}} = -\frac{1}{\pi \mathrm{i}} \int_a^b \sqrt{\frac{t-a}{b-t}} \frac{h(t)}{t-z} \mathrm{d}t,$$

the constant C vanishing because F must vanish for $z \to \infty$. We deduce

$$F(z) = \frac{1}{\pi \mathrm{i}} \sqrt{\frac{z-b}{z-a}} \int_a^b \sqrt{\frac{t-a}{b-t}} \frac{h(t)}{t-z} \mathrm{d}t, \tag{C.1.8}$$

Applying Plemelj's formulas and separating the real part we obtain

$$f(x) = -\frac{1}{\pi}\sqrt{\frac{b-x}{x-a}}\,{\int_a^b}' \sqrt{\frac{t-a}{b-t}}\,\frac{h(t)}{t-x}\,\mathrm{d}t\,. \tag{C.1.9}$$

This is the solution of the integral equation (C.1.1) which satisfies the boundedness condition in $x = b$.

The solution which is bounded in a and unbounded in b may be obtained starting from the function $\sqrt{\dfrac{z-b}{z-a}}$. W

We obtain

$$f(x) = -\frac{1}{\pi}\sqrt{\frac{x-a}{b-x}}\,{\int_a^b}' \sqrt{\frac{b-t}{t-a}}\,\frac{h(t)}{t-x}\,\mathrm{d}t\,. \tag{C.1.10}$$

If we wish to obtain an unbounded solution the both end-points, using the function $\sqrt{(z-a)(z-b)}$, we have

$$\operatorname{Re}\left[F(z)\sqrt{(z-a)(z-b)}\right] =$$

$$= \begin{cases} 0\,, & x \in (-\infty, a) \cup (b, \infty) \\ h(x)\sqrt{(x-a)(b-x)}\,, & x \in (a,b) \end{cases}$$

whence

$$f(x) = -\frac{1}{\pi}\,\frac{1}{\sqrt{(x-a)(b-x)}}\,{\int_a^b}' \sqrt{(t-a)(b-t)}\,\frac{h(t)}{t-x}\,\mathrm{d}t + \\ + \frac{C}{\sqrt{(x-a)(b-x)}}\,. \tag{C.1.11}$$

At last, for the solution bounded in the both end-points, we deduce

$$F(z) = -\frac{1}{\pi \mathrm{i}}\sqrt{(z-a)(z-b)}\int_a^b \frac{h(t)}{\sqrt{(t-a)(b-t)}}\,\frac{\mathrm{d}t}{t-z}\,. \tag{C.1.12}$$

When z is great enough, this becomes

$$F(z) = -\frac{1}{\pi \mathrm{i}}\,\frac{\sqrt{(z-a)(z-b)}}{z}\int_a^b \frac{h(t)}{\sqrt{(t-a)(b-t)}}\,\frac{\mathrm{d}t}{1-t/z} =$$

$$= \frac{1}{\pi \mathrm{i}}\left(1-\frac{a}{z}\right)^{1/2}\left(1-\frac{b}{z}\right)^{1/2}\int_a^b \frac{h(t)}{\sqrt{(t-a)(b-t)}}\left(1+\frac{t}{z}+\ldots\right)\mathrm{d}t$$

such that imposing the zero value for F at infinity we have

$$\int_a^b \frac{h(t)}{\sqrt{(t-a)(b-t)}} \mathrm{d}t = 0 . \tag{C.1.13}$$

If this condition is satisfied, the solution of the equation (C.1.1) (bounded at both end-points), is obtained passing to the limit in (C.1.12). Taking into account the determination of the radical, we find

$$f(x) = -\frac{1}{\pi}\sqrt{(x-a)(b-x)} \int_a^{\prime b} \frac{h(t)}{\sqrt{(t-a)(b-t)}} \frac{\mathrm{d}t}{t-x} . \tag{C.1.14}$$

Because of the restriction (C.1.13), this solution cannot be utilized in aerodynamics.

C.2 The Generalized Equation of Thin Profiles

This has the form:

$$\frac{1}{\pi}\int_a^{\prime b} \frac{f(t)}{t-x} \mathrm{d}t + \frac{1}{\pi}\int_a^b f(t)K(t-x)\mathrm{d}t = h(x) \quad a < x < b , \tag{C.2.1}$$

where K is a non-singular kernel. We assume that $f(x), K(x)$ and $h(x)$ satisfy Hölder's condition on (a, b).

This type of equations are encountered in the theory of thin profiles in ground effects, in the theory of grids of thin profiles, in the theory of grids of thin wings etc. From the mathematical point of view, the equations (C.1.5) are extensively studied in [A.27]. In order to ensure the uniqueness of the solution we have to impose the behaviour of the unknown f at the extremities of the integration interval. The method of investigation consists in their regularization, i.e. they are reduced to Fredholm-type equations for which existence and uniqueness theorems are available. For the equation (C.2.1) this may be performed utilizing the solution (C.1.1). If we are interested by the solution of the equation (C.2.1) which vanishes in the trailing edge b, we shall utilize (C.1.9). Assuming that the last two terms in (C.2.1) are known and knowing that a singular integral interchanges with a non-singular integral [A.27], we obtain the following Fredholm-type integral equation

$$f(x) + \frac{1}{\pi}\int_a^b f(\xi)M(\xi, x)\mathrm{d}\xi = H(x), \; a < x < b , \tag{C.2.2}$$

where

$$M(\xi,x) = -\frac{1}{\pi}\sqrt{\frac{b-x}{x-a}}\,{}'\!\!\int_a^b \sqrt{\frac{t-a}{b-t}}\,\frac{k(t-\xi)}{t-x}\,\mathrm{d}t\,,$$

$$M(\xi,x) = -\frac{1}{\pi}\sqrt{\frac{b-x}{x-a}}\,{}'\!\!\int_a^b \sqrt{\frac{t-a}{b-t}}\,\frac{h(t)}{t-x}\,\mathrm{d}t\,. \tag{C.2.3}$$

Sometimes we may specify the kernel M. For example, in the case of the thin profile in ground effects, we have:

$$k(t-\xi) = \frac{t-\xi}{(t-\xi)^2+m^2}\,. \tag{C.2.4}$$

The integrals

$$I = {}'\!\!\int_a^b \sqrt{\frac{t-a}{b-t}}\,\frac{t-\xi}{(t-\xi)^2+m^2}\,\frac{\mathrm{d}t}{t-x} \tag{C.2.5}$$

may be calculated with the residue theorem [A.20], or considering the integral

$$J = {}'\!\!\int_a^b \sqrt{\frac{b-a}{b-t}}\,\frac{m}{(t-\xi)^2+m^2}\,\frac{\mathrm{d}t}{t-x} \tag{C.2.6}$$

and noticing that if we denote $\zeta = \xi + \mathrm{i}\,m$, we have:

$$I+\mathrm{i}\,J = {}'\!\!\int_a^b \sqrt{\frac{t-a}{b-t}}\,\frac{\mathrm{d}t}{(t-\zeta)(t-x)} =$$

$$= \frac{1}{\zeta-x}\left[\int_a^b \sqrt{\frac{t-a}{b-t}}\,\frac{\mathrm{d}t}{(t-\zeta)} - {}'\!\!\int_a^b \sqrt{\frac{t-a}{b-t}}\,\frac{\mathrm{d}t}{(t-x)}\right].$$

Utilizing (B.5.1) and (B.5.4) it results

$$I+\mathrm{i}\,J = \frac{\pi}{x-\zeta}\sqrt{\frac{\zeta-a}{\zeta-b}}\,, \tag{C.2.7}$$

whence

$$I = \frac{\pi}{2}\left(\frac{1}{x-\zeta}\sqrt{\frac{\zeta-a}{\zeta-b}} + \frac{1}{x-\overline{\zeta}}\sqrt{\frac{\overline{\zeta}-a}{\overline{\zeta}-b}}\right), \tag{C.2.8}$$

$$J = \frac{\pi}{2}\left(-\frac{1}{x-\zeta}\sqrt{\frac{\zeta-a}{\zeta-b}} + \frac{1}{x-\overline{\zeta}}\sqrt{\frac{\overline{\zeta}-a}{\overline{\zeta}-b}}\right), \tag{C.2.9}$$

and consequently,

$$M(\xi, x) = -\frac{1}{\pi}\sqrt{\frac{b-x}{x-a}}\left(\frac{1}{x-\zeta}\sqrt{\frac{\zeta-a}{\zeta-b}} + \frac{1}{x-\overline{\zeta}}\sqrt{\frac{\overline{\zeta}-a}{\overline{\zeta}-b}}\right). \qquad \text{(C.2.10)}$$

C.3 The Third Equation

We shall consider the equation [A.16]:

$$-\frac{1}{\pi}\int_a^b f(t)(\ln|t-x| + \Gamma_0)\mathrm{d}\,t = g(x)\,, \qquad \text{(C.3.1)}$$

defined on the interval (a, b), where f is the unknown and Γ_0 is a constant. We intend to determine the general solution and to investigate the conditions which are necessary for the solution to satisfy the restrictions

$$f(a) = f(b) = 0\,. \qquad \text{(C.3.2)}$$

With the change of variables $t, x \to \theta, \sigma$ defined by the relations

$$t = c + e\cos\theta, x = c + e\cos\sigma\,, \qquad \text{(C.3.3)}$$

where

$$c = \frac{a+b}{2}, \quad e = \frac{b-a}{2}\,, \qquad \text{(C.3.4)}$$

the equation (C.3.1) becomes

$$-\frac{1}{\pi}\int_0^\pi F(\theta)(\ln e|\cos\theta - \cos\sigma| + \Gamma_0)e\sin\theta\mathrm{d}\,\theta = G(\sigma)\,.$$

With the notation

$$F^*(\theta) = eF(\theta)\sin\theta = \sqrt{(t-a)(b-t)}f(t)\,, \qquad \text{(C.3.5)}$$

this may be written as follows

$$-\frac{1}{\pi}\int_0^\pi F^*(\theta)(\ln e|\cos\theta - \cos\theta| + \Gamma_0)\mathrm{d}\,\theta = G(\sigma)\,, \qquad \text{(C.3.6)}$$

where we have

$$\ln 2|\cos\theta - \cos\sigma| = -2\sum_{m\geq 1}\frac{1}{m}\cos m\theta\cos m\sigma\,. \qquad \text{(C.3.7)}$$

The function $F^*(\theta)$ may be prolonged on the interval $(-\pi, 0)$ such that the result is an even function on $(-\pi, +\pi)$ and it may be therefore expanded into a trigonometric series, with the aid of the even functions

$$F^*(\theta) = a_0 + \sum_{n\geq 1} a_n \cos n\theta\,, \tag{C.3.8}$$

where

$$a_n = \frac{2}{\pi}\int_0^\pi F^*(\theta)\cos n\theta \mathrm{d}\,\theta\,, \tag{C.3.9}$$

$$a_0 = \frac{1}{\pi}\int_0^\pi F^*(\theta)\mathrm{d}\,\theta = \frac{1}{\pi}\int_0^\pi F(\theta)e\sin\theta\mathrm{d}\,\theta = \frac{1}{\pi}\int_a^b f(t)\mathrm{d}\,t\,.$$

Analogously, the function $G(\sigma)$ may be expanded into the series

$$G(\sigma) = b_0 + \sum_{n\geq 1} b_n \cos n\sigma\,, \tag{C.3.10}$$

where

$$b_n = -\frac{2}{\pi}\int_0^\pi G(\sigma)\cos n\sigma \mathrm{d}\,\sigma = -\frac{2}{\pi}\int_0^\pi G'(\sigma)\frac{1}{n}\sin n\sigma\mathrm{d}\,\sigma\,,$$

$$b_0 = \frac{1}{\pi}\int_0^\pi G(\sigma)\mathrm{d}\,\sigma = \frac{1}{\pi}\int_a^b \frac{g(x)}{\sqrt{(x-a)(b-x)}}\mathrm{d}\,x.. \tag{C.3.11}$$

Substituting these series into the integral equation (C.3.6), it results

$$b_0 + \sum_{n\geq 1} b_n\cos n\sigma = -\frac{1}{\pi}\int_0^\pi \left(a_0 + \sum_{n\geq 1} a_n \cos n\theta\right)\times$$

$$\times\left(-2\sum_{m\geq 1}\frac{1}{m}\cos m\theta\cos m\sigma + \Gamma\right)\mathrm{d}\,\theta = \tag{C.3.12}$$

$$= -a_0\Gamma + \sum_{n\geq 1}\frac{1}{n}a_n\cos n\sigma\,,$$

where we denoted

$$\Gamma = \Gamma_0 + \ln\frac{e}{2} = \Gamma_0 + \ln\frac{(b-a)}{4}\,. \tag{C.3.13}$$

Identifying the coefficients we find:

$$b_0 = -a_0\Gamma, b_n = +\frac{1}{n}a_n\,. \tag{C.3.14}$$

In the sequel, we have

$$
\begin{aligned}
&\frac{1}{\pi}\,{}'\!\!\int_a^b \frac{\sqrt{(t-a)(b-t)}}{t-x}\,g'(t)\mathrm{d}\,t = \frac{1}{\pi}\int_a^b \frac{\sqrt{(t-a)(b-t)}}{t-x}\,\mathrm{d}\,(g(t)) = \\
&\quad = \frac{1}{\pi}\int_0^\pi \frac{\sin\theta}{\cos\sigma-\cos\theta}\,\mathrm{d}\,(G(\theta)) = \frac{1}{\pi}\int_0^\pi \frac{\sin\theta}{\cos\sigma-\cos\theta}\,G'(\theta)\mathrm{d}\,\theta = \\
&\quad = \frac{1}{\pi}\int_0^\pi \frac{\sin\theta}{\cos\sigma-\cos\theta}\left(-\sum_{n\geq 1} nb_n\sin n\theta\right)\mathrm{d}\,\theta = \\
&\quad = \sum_{n\geq 1} nb_n\int_0^\pi \frac{\cos(n+1)\theta-\cos(n-1)\theta}{\cos\sigma-\cos\theta}\,\mathrm{d}\,\theta\,,
\end{aligned}
$$

or, utilizing Glauert's integral (B.6.6),

$$
\begin{aligned}
&\frac{1}{\pi}\,{}'\!\!\int_a^b \frac{\sqrt{(t-a)(b-t)}}{t-x}\,g'(t)\mathrm{d}\,t = \\
&\quad = -\sum_{n\geq 1} nb_n\left[\frac{\sin(n+1)\sigma-\sin(n-1)\sigma}{\sin\sigma}\right] - \sum_{n\geq 1} nb_n\cos n\sigma = \\
&\quad = -\sum_{n\geq 1} a_n\cos n\sigma + -F^*(\sigma) + a_0 = \\
&\quad = -\sqrt{(x-a)(b-x)}\,f(x) - b_0/\Gamma\,,
\end{aligned}
\tag{C.3.15}
$$

where b_0 is (C.3.11). From (C.3.15) it results

$$
\begin{aligned}
f(x) = &-\frac{1}{\pi}\frac{1}{\sqrt{(x-a)(b-x)}}\,{}'\!\!\int_a^b \frac{\sqrt{(t-a)(b-t)}}{t-x}\,g'(t)\mathrm{d}\,t - \\
&-\frac{1}{\pi\Gamma}\frac{1}{\sqrt{(x-a)(b-x)}}\int_a^b \frac{g(t)}{\sqrt{(t-a)(b-t)}}\,\mathrm{d}\,t\,.
\end{aligned}
\tag{C.3.16}
$$

This is the first form of the general solution of the equation (C.3.1).

One obtains another form if one utilizes the identity

$$\frac{\sqrt{(t-a)(b-t)}}{\sqrt{(x-a)(b-x)}} - \frac{\sqrt{(x-a)(b-x)}}{\sqrt{(t-a)(b-t)}} = \\ = \frac{(x-t)(x+t-a-b)}{\sqrt{(x-a)(b-x)}\sqrt{(t-a)(b-t)}} . \tag{C.3.17}$$

Substituting the first ratio in (C.3.16), it results the final solution

$$f(x) = -\frac{1}{\pi}\sqrt{(x-a)(b-x)} \int_a^b \frac{g'(t)}{\sqrt{(t-a)(b-t)}} \frac{\mathrm{d}t}{t-x} + \\ +\frac{1}{\pi} \frac{1}{\sqrt{(x-a)(b-x)}} \int_a^b \left[(x+t-a-b)g'(t) - \\ - \frac{g(t)}{\Gamma} \right] \frac{\mathrm{d}t}{\sqrt{(t-a)(b-t)}} . \tag{C.3.18}$$

We notice that from the relations (C.3.9), (C.3.14) and (C.3.11) it results

$$\int_a^b f(t)\mathrm{d}t = \frac{1}{\Gamma} \int_a^b \frac{g(x)}{\sqrt{(x-a)(b-x)}} \mathrm{d}x , \tag{C.3.19}$$

which is an useful relation in applications.

We shall determine in the sequel the conditions which have to be satisfied by g, such that the relations (C.3.2) are satisfied by the solution (C.3.1). At first we notice that when the parameter Γ vanishes, a necessary condition for the existence of the solution (C.3.18) is

$$\int_a^b \frac{g(t)}{\sqrt{(t-a)(b-t)}} \mathrm{d}t = 0 . \tag{C.3.20}$$

The last integral from (C.3.18) defines for x real a polynomial of the first degree. f vanishes in a, is this polynomial has the root a (the first term from (C.3.18) vanishes for $x = a$). In this case, the last integral has the order of $(x-a)$, while the denominator of the fraction has the order of $(x-a)^{1/2}$. Hence, we must have

$$\int_a^b \left[(t-b)g'(t) - \frac{g(t)}{\Gamma} \right] \frac{\mathrm{d}t}{\sqrt{(t-a)(b-t)}} = 0 . \tag{C.3.21}$$

Analogously, $f(b)=0$ implies

$$\int_a^b \left[(t-a)g'(t)-\frac{g(t)}{\Gamma}\right]\frac{\mathrm{d}t}{\sqrt{(t-a)(b-t)}}=0\,. \tag{C.3.22}$$

Imposing the both conditions, subtracting and adding, we obtain

$$\int_a^b \frac{g'(t)}{\sqrt{(t-a)(b-t)}}\mathrm{d}t=0\,; \tag{C.3.23}$$

$$\int_a^b \left[(t-c)g'(t)-\frac{g(t)}{\Gamma}\right]\frac{\mathrm{d}t}{\sqrt{(t-a)(b-t)}}=0\,. \tag{C.3.24}$$

This is the answer for the proposed problem.

C.4 The Forth Equation

At least in magnetoaerodynamics [3.9] [1.9] p.208, in the theory of oscillatory wings [10.15] and in the theory of the wing in fluids with chemical reactions [3.10] [3.24], it intervenes the following singular integral equation [A.16]

$$\frac{1}{\pi}{\int_a^b}' \frac{f(t)}{t-x}\mathrm{d}t+\frac{\omega}{\pi}\int_a^b f(t)(\ln|t-x|+\Gamma_0)\mathrm{d}t=h(x)\,,\ a<x<b\,, \tag{C.4.1}$$

where f is the unknown, Γ_0 an arbitrary constant and h, a given function. The solution is unique if one imposes the behaviour of f in one of the end-points of the interval. We shall impose the boundedness of f să in b.

Denoting

$$g(x)=-\frac{1}{\pi}\int_a^b f(t)(\ln|t-x|+\Gamma_0)\mathrm{d}t\,, \tag{C.4.2}$$

we shall notice that

$$g'(x)=\frac{1}{\pi}{\int_a^b}' \frac{f(t)}{t-x}\mathrm{d}t\,. \tag{C.4.3}$$

So, the equation (C.4.1) is reduced to the differential equation

$$g'-\omega g=h\,, \tag{C.4.4}$$

whose general solution is

$$g(x)=A\mathrm{e}^{\omega(\mathrm{x}-\mathrm{c})}+\mathrm{g}_0(\mathrm{x})\,, \tag{C.4.5}$$

where A is an undetermined constant, c is (C.3.4), and

$$g_0(x) = \int_c^x h(\xi) \mathrm{e}^{\omega(\mathrm{x}-\xi)} \mathrm{d}\,\xi\,. \tag{C.4.6}$$

The unknown f may be found from the equation (C.4.2) which is exactly the equation (C.3.1). Its solution is therefore (C.3.18) where g will be replaced by (C.4.5). For specifying it, one observes that performing the substitutions (C.3.3), denoting by $I_0(\omega), I_1(\omega), \ldots$ Bessel's functions of imaginary argument and taking into account that $I_0'(\omega) = I_1(\omega)$, one obtains

$$\begin{aligned} T_1 &= \int_a^b \frac{\mathrm{e}^{\omega(\mathrm{t}-\mathrm{c})}}{\sqrt{(t-a)(b-t)}} \mathrm{d}\,t = \int_0^\pi \mathrm{e}^{\omega\cos\theta} \mathrm{d}\,\theta = \pi \mathrm{I}_0(\overline{\omega})\,, \\ T_2 &= \int_a^b \frac{t\mathrm{e}^{\omega(\mathrm{t}-\mathrm{c})}}{\sqrt{(t-a)(b-t)}} \mathrm{d}\,t = \int_0^\pi (c + e\cos\theta) \mathrm{e}^{\omega\cos\theta} \mathrm{d}\,\theta \\ &= \pi c I_0(\overline{\omega}) + \pi e I_1(\overline{\omega})\,. \end{aligned} \tag{C.4.7}$$

From the expansion

$$\exp\left(\overline{\omega}\cos\theta\right) = I_0(\overline{\omega}) + 2\sum_{n\geq 1} I_n(\overline{\omega}) \cos n\theta \tag{C.4.8}$$

and from Glauert's formula (B.6.6) we deduce

$$\begin{aligned} T_3 &= {\int_a^b}' \frac{\mathrm{e}^{\omega(\mathrm{t}-\mathrm{c})}}{\sqrt{(t-a)(b-t)}} \frac{\mathrm{d}\,t}{t-x} = \frac{1}{e} {\int_0^\pi}' \frac{\mathrm{e}^{\overline{\omega}\cos\theta}}{\cos\theta - \cos\sigma} \mathrm{d}\,\theta \\ &= \frac{2\pi}{e} \sum_{n\geq 1} I_n(\overline{\omega}) \frac{\sin n\sigma}{\sin\sigma} = \frac{2\pi}{\sqrt{(b-x)(x-a)}} \sum_{n\geq 1} I_n(\overline{\omega}) \sin n\sigma\,, \end{aligned} \tag{C.4.9}$$

where $\overline{\omega} = e\omega$. For T_3 and x we have the parametric representation (C.4.9) and (C.3.3).

We deduce therefore

$$\begin{aligned} f(x) &= -\frac{\omega A}{\pi} \sqrt{(x-a)(b-x)}\, T_3 + \frac{A}{\pi} \frac{1}{\sqrt{(x-a)(b-x)}} \times \\ &\times \left[\omega(x-2c) - \frac{1}{\Gamma}\right] T_1 + \frac{\omega A}{\pi} \frac{1}{\sqrt{(x-a)(b-x)}} T_2 + f_0(x)\,, \end{aligned} \tag{C.4.10}$$

f_0 being obtained from f by replacing g with g_0. In aerodynamics f represents the jump of the pressure and we need therefore the integrals

$$\int_a^b f(x)\mathrm{d}\,x, \quad \int_a^b xf(x)\mathrm{d}\,x \tag{C.4.11}$$

which may be easily calculated without utilizing (C.4.9). Directly, the first integral is obtained from (C.3.18).

In the sequel we shall determine the constant A. Imposing the condition $f(b) = 0$, we have from (C.3.22) and (C.4.5)

$$A\left[\left(a\omega + \frac{1}{\Gamma}\right)T_1 - \omega T_2\right] = G_0\,, \tag{C.4.12}$$

where

$$G_0 = \int_a^b \left[(t-a)g_0'(t) - \frac{1}{\Gamma}g_0(t)\right]\frac{\mathrm{d}\,t}{\sqrt{(t-a)(b-t)}}\,. \tag{C.4.13}$$

Now the problem is solved.

C.5 The Fifth Equation

In the theory of oscillatory wings [10.20] it intervenes the following integral equation [A.16]:

$$\frac{1}{\pi}\,{}^*\!\!\int_a^b \frac{f(t)}{(t-x)^2}\mathrm{d}\,t + \frac{\omega^2}{\pi}\int_a^b f(t)(\ln|t-x| + \Gamma_0)\mathrm{d}\,t = h(x)\,,\ a < x < b, \tag{C.5.1}$$

which has to be solved together with the following conditions

$$f(a) = f(b) = 0\,. \tag{C.5.2}$$

The sign '$*$' from (C.5.1) shows that one considers the Finite Part.

Introducing the notation (C.4.2) and taking into account the formula (D.3.6), from (C.4.3) we deduce

$$g''(x) = \frac{1}{\pi}\,{}^*\!\!\int_a^b \frac{f(t)}{(t-x)^2}\mathrm{d}\,t\,, \tag{C.5.3}$$

such that the integral equation (C.5.1) reduces to the following differential equation

$$g'' - \omega^2 g = h\,. \tag{C.5.4}$$

The homogeneous equation has the solution

$$g = A_0\cosh\omega(x-c) + B_0\sinh\omega(x-c)\,.$$

Applying Lagrange's method of variation of constants, one obtains for the solution of the differential equation (C.5.4),

$$g = A\cosh\omega(x-c) + B\sinh\omega(x-c) + g_0(x)\,, \tag{C.5.5}$$

where

$$g_0(x) = \frac{1}{\omega}\int_c^x h(\xi)\sinh\omega(x-\xi)\mathrm{d}\,\xi\,. \tag{C.5.6}$$

The solution of the integral equation (C.5.1) may be obtained from (C.3.18) where one replaces (C.5.5). For specifying it we notice that setting

$$t = c + eu, \quad ,u \in [-1,+1],$$

we deduce $(t-a)(b-t) = \mathrm{e}^2(1-\mathrm{u}^2)$ and then

$$S_0 \equiv \int_a^b \frac{\sinh\omega(t-c)}{\sqrt{(t-a)(b-t)}}\mathrm{d}\,t = \int_{-1}^{+1}\frac{\sinh(\varpi u)}{\sqrt{(1-u^2)}}\mathrm{d}\,u = 0\,,$$

the integrand being an odd function,

$$\begin{aligned}
C_0 &\equiv \int_a^b \frac{\cosh\omega(t-c)}{\sqrt{(t-a)(b-t)}}\mathrm{d}\,t = \int_{-1}^{+1}\frac{\cosh(\varpi u)}{\sqrt{1-u^2}}\mathrm{d}\,u = \pi I_0(\varpi)\\
C_1 &\equiv \int_a^b \frac{t\cosh\omega(t-c)}{\sqrt{(t-a)(b-t)}}\mathrm{d}\,t\\
&= \int_{-1}^{+1}(c+eu)\frac{\cosh(\varpi u)}{\sqrt{1-u^2}}\mathrm{d}\,u = c\pi I_0(\varpi)\\
S_1 &\equiv \int_a^b \frac{t\sinh\omega(t-c)}{\sqrt{(t-a)(b-t)}}\mathrm{d}\,t = e\int_{-1}^{+1}\frac{u\sinh(\varpi u)}{\sqrt{1-u^2}}\mathrm{d}\,u = e\pi I_1(\varpi)\,.
\end{aligned} \tag{C.5.7}$$

Also, using the change (C.3.3) and taking into account (C.4.8), Glauert's formula (B.6.6) and the relations

$$I_n(-\omega) = (-1)^n I_n(\omega)\,, \tag{C.5.8}$$

we deduce

$$T_s \equiv {\int_a^b}' \frac{\sinh \omega(t-c)}{\sqrt{(t-a)(b-t)}} \frac{\mathrm{d}\, t}{t-x} = \frac{1}{e} \int_0^\pi \frac{\sinh(\overline{\omega}\cos\theta)}{\cos\theta - cos\sigma} \mathrm{d}\,\theta$$

$$= \frac{\pi}{e} \sum_{n\geq 1} [I_n(\overline{\omega}) - I_n(-\overline{\omega})] \frac{\sin n\sigma}{\sin\sigma}$$

$$= \frac{\pi}{\sqrt{(b-x)(x-a)}} \sum_{n\geq 1} [1-(-1)^n] I_n(\overline{\omega}) \sin n\sigma \tag{C.5.9}$$

$$T_c \equiv {\int_a^b}' \frac{\cosh \omega(t-c)}{\sqrt{(t-a)(b-t)}} \frac{\mathrm{d}\, t}{t-x}$$

$$= \frac{\pi}{\sqrt{(b-x)(x-a)}} \sum_{n\geq 1} [1-(-1)^n] I_n(\overline{\omega}) \sin n\sigma\,.$$

These formulas, together with (C.3.3), give the parametric representation of the integrals T_s, T_c and the variable x.

With these results, the solution of the equation (C.5.1) is

$$f(x) = \frac{\omega}{\pi}\sqrt{(x-a)(b-x)}(AT_s + BT_c) +$$

$$+\frac{1}{\pi} \frac{1}{\sqrt{(x-a)(b-x)}} \left\{ A\left(\omega S_1 - \frac{1}{\Gamma} C_0\right) + \omega B\left[(x-2c)C_0 + C_1\right]\right\}, \tag{C.5.10}$$

where f_0 is obtained from (C.3.18), replacing $g(t)$ by $g_0(t)$ given by (C.5.6). As we have already mentioned in the case of the equation (C.4.1), in aerodynamics we need the coefficients (C.4.11). They may be obtained easier utilizing the form (C.3.18).

For determining the constants A and B we impose the conditions (C.3.23) and (C.3.24) where g' has the expression (C.5.5). One obtains the system

$$\begin{aligned} &\omega C_0 A = G_1\,, \\ &\omega(C_1 - cC_0)A + (\omega S_1 - \Gamma^{-1} C_0)B = G_2\,, \end{aligned} \tag{C.5.11}$$

where

$$G_1 = -\int_a^b \frac{g_0'(t)}{\sqrt{(t-a)(b-t)}} \mathrm{d}t,$$

$$G_2 = -\int_a^b \left[(t-c)g_0'(t) - \frac{1}{\Gamma}g_0(t)\right] \frac{\mathrm{d}t}{\sqrt{(t-a)(b-t)}}. \tag{C.5.12}$$

Taking into account (C.5.7), we deduce

$$A = \frac{G_1}{\omega\pi I_0(\overline{\omega})}, \quad B = \frac{G_2}{\pi[\overline{\omega}I_1(\overline{\omega}) - \Gamma^{-1}I_0(\overline{\omega})]}. \tag{C.5.13}$$

In many applications we encounter the situation when $h = -\varepsilon$ (the case of the flat plates with the angle of attack ε). In this situation, from (C.5.6) it results

$$\omega^2 g_0(x) = \varepsilon[1 - \cosh\omega(x-c)],$$

such that

$$G_1 = 0, \ \ \omega^2 G_2 = \varepsilon\pi[\overline{\omega}I_1(\overline{\omega}) - \Gamma^{-1}I_0(\overline{\omega}) + \Gamma^{-1}]$$

whence,

$$A = 0, \quad B = \frac{\varepsilon}{\omega^2}\left[1 + \frac{1}{\overline{\omega}\Gamma I_1(\overline{\omega}) - I_0(\overline{\omega})}\right]. \tag{C.5.14}$$

Appendix D

The Finite Part

D.1 Introductory Notions

The notion of "Finite Part" of a improper integral has been introduced by Hadamard in 1923 [A.39], in order to give a significance to the divergent integrals which appear in applications and to utilize them. Hadamard studied integrals having the form

$$\int_a^b \frac{f(x)}{(b-x)^{n+1/2}} \mathrm{d}\, x\,, \tag{D.1.1}$$

where $n = 1, 2, 3, \ldots$. There exists however many integrands with non-integrable singularities which appear in applications especially in aerodynamics. It exists therefore different manners for treating this problem. In the subsonic aerodynamics one utilizes especially the definition of Mangler [A.45], but we had not the possibility to read this paper. A less cited contribution, but very adequate to aerodynamics belongs to Ch. Fox [A.37]. Here, the notion of "Finite Part" appears like a natural extension of the concept of "Principal Value" in Cauchy's sense. We shall present in the sequel some results of this author. For the integrals having the shape (D.1.1) we shall utilize the paper of Heaslet and Lomax [A.44]. These ones appear in the supersonic aerodynamics. Important results concerning the notion may be found in the papers of Kutt [A.42] and Kaya and Erdogan [A.41]. At lasst, the theory of distributions give an unitary method for the study of this notion [A.5].

D.2 The First Integral

We shall consider at first the integral

$$I_1 \equiv \int_0^a \frac{f(x)}{x^{n+1}} \mathrm{d}\, x, \quad n = 0, 1, \ldots \,. \tag{D.2.1}$$

If f admits derivatives up to the order $n+1$ in the origin, then we may write

$$\int_0^a \frac{f(x)}{x^{n+1}}\mathrm{d}\,x = \int_0^a \left[f(x) - \sum_{i=0} \frac{x^i}{i!} f^{(i)}(0) + \sum_{i=0}^{n} \frac{x^i}{i!} f^{(i)}(0)\right] \frac{\mathrm{d}\,x}{x^{n+1}} =$$

$$= \int_0^a \left[f(x) - \sum_{i=1}^{n} \frac{x^i}{i!} f^{(i)}(0)\right] \frac{\mathrm{d}\,x}{x^{n+1}} +$$

$$+ \sum_{i=}^{n-1} \frac{x^{i-n}}{i-n} \frac{f^{(i)}(0)}{i!}\bigg|_0^a + \frac{f^{(n)}}{n!} \ln x\bigg|_0^a .$$

The integrated part for $x = 0$ becomes infinite. Neglecting these infinite constants, one obtains the so called "Finite Part" of the integral I. Hence, indicating by an "asterisk" the Finite Part, we have:

$$\begin{aligned} {}^*\!\!\int_0^a \frac{f(x)}{x^{n+1}}\mathrm{d}\,x = \int_0^a \left[f(x) - \sum_{i=0}^{n-1} \frac{x^i}{i!} f^{(i)}(0)\right] \frac{\mathrm{d}\,x}{x^{n+1}} + \\ + \sum_{i=0}^{n-1} \frac{f^{(i)}(0)}{i!} \frac{a^{i-n}}{i-n} + \frac{f^{(n)}(0)}{n!} \ln a\,. \end{aligned} \tag{D.2.2}$$

For $f = 1$, it results

$${}^*\!\!\int_0^a \frac{\mathrm{d}\,x}{x} = \ln a, \quad {}^*\!\!\int_0^a \frac{\mathrm{d}\,x}{x^{n+1}} = -\frac{1}{n}\frac{1}{a^n} \quad (n \geq 1)\,. \tag{D.2.3}$$

D.3 Integrals with Singularities in an Interval

We shall consider the integrals having the form

$$I_2 = \int_a^b \frac{f(x)}{(x-u)^{n+1}}\mathrm{d}\,x, \quad n = 0, 1, \ldots, \tag{D.3.1}$$

where $a < u < b$. For $n = 0$, we consider the "Principal Value" of the integral in Cauchy's sense.

$${}'\!\!\int_a^b \frac{f(x)}{x-u}\mathrm{d}\,x = \lim_{\varepsilon \to 0}\left(\int_a^{u-\varepsilon} + \int_{u+\varepsilon}^b\right) \frac{f(x)}{x-u}\mathrm{d}\,x\,. \tag{D.3.2}$$

We know from (B.1.2) that this limit exists if f satisfies Hölder's condition in the interval (a,b).

Let us derive now (D.3.2) with respect to the variable u. In the right hand part, the derivation is performed according to the derivation formula for integrals containing the variable in the limits. We have therefore

$$\frac{\mathrm{d}}{\mathrm{d}\,u}\,{}'\!\!\int_a^b \frac{f(x)}{x-u}\mathrm{d}\,x = \lim_{\varepsilon\to 0}\left[\left(\int_a^{u-\varepsilon}+\int_{u+\varepsilon}^b\right)\frac{f(x)}{(x-u)^2}\mathrm{d}\,x - \right.$$

$$\left. -\frac{f(u-\varepsilon)}{\varepsilon}-\frac{f(u+\varepsilon)}{\varepsilon}\right],$$

or, expanding into a Taylor series the functions $f(u-\varepsilon)$, $f(u+\varepsilon)$,

$$\frac{\mathrm{d}}{\mathrm{d}\,u}\,{}'\!\!\int_a^b \frac{f(x)}{x-u}\mathrm{d}\,x = \lim_{\varepsilon\to 0}\left[\left(\int_a^{u-\varepsilon}+\int_{u+\varepsilon}^b\right)\frac{f(x)}{(x-u)^2}\mathrm{d}\,x - 2\frac{f(u)}{\varepsilon}\right]. \tag{D.3.3}$$

If the limit from the right hand part exists, we denote it by

$$ {}^*\!\!\int_a^b \frac{f(x)}{(x-u)^2}\mathrm{d}\,x \tag{D.3.4}$$

and we have

$$ {}^*\!\!\int_a^b \frac{f(x)}{(x-u)^2}\mathrm{d}\,x \stackrel{\text{def}}{=} \lim_{\varepsilon\to 0}\left[\left(\int_a^{u-\varepsilon}+\int_{u+\varepsilon}^b\right)\frac{f(x)}{(x-u)^2}\mathrm{d}\,x - 2\frac{f(u)}{\varepsilon}\right] \tag{D.3.5}$$

$$\frac{\mathrm{d}}{\mathrm{d}\,u}\,{}'\!\!\int_a^b \frac{f(x)}{x-u}\mathrm{d}\,x = {}^*\!\!\int_a^b \frac{f(x)}{(x-u)^2}\mathrm{d}\,x\,. \tag{D.3.6}$$

The limit (D.3.5) defines the Finite Part of the integral from the left hand part. The Finite Part is a distribution [A.14]. Ex. 11. One proves [A.37] that *if there exists* $f'(x)$ *on* (a,b) *and this function satisfies Hölder's condition*, then the limit from (D.3.5) exists. We notice that this theorem constitutes the extension of the theorem of existence of the limit (D.3.2).

We indicate now how one may reduce the calculation of the integral (D.3.1) to the calculation of an integral with a weaker singularity. We consider the case $n = 1$. Hence, we demonstrate that in the same conditions like above (f' is defined and satisfies Hölder's condition on (a,b)), we have

$$ {}^*\!\!\int_a^b \frac{f(x)}{(x-u)^2}\mathrm{d}\,x = \frac{f(a)}{a-u}-\frac{f(b)}{b-u}+{}^*\!\!\int_a^b \frac{f'(x)}{x-u}\mathrm{d}\,x\,, \tag{D.3.7}$$

for every u from (a,b). Indeed, employing for the left hand side member the definition (D.3.5) and integrating by parts, we obtain

$$\int_a^{*b} \frac{f(x)}{(x-u)^2}\mathrm{d}\,x = \lim_{\varepsilon\to 0}\left\{ -\left(\int_a^{u-\varepsilon} + \int_{u+\varepsilon}^b\right)\frac{\partial}{\partial x}\left[\frac{f(x)}{x-u}\right]\mathrm{d}\,x + \right.$$

$$\left. + \left(\int_a^{u-\varepsilon} + \int_{u+\varepsilon}^b\right)\frac{f'(x)}{x-u}\mathrm{d}\,x - \frac{2f(u)}{\varepsilon}\right\} = \frac{f(a)}{a-u} - \frac{f(b)}{b-u} + \int_a^b \frac{f'(x)}{x-u}\mathrm{d}x\,.$$

The extension of the definition (D.3.5) and theorem (D.3.7) to an arbitrary value of n, is performed in [A.37]. In the same paper one gives the respective definitions in the complex plane and also Plemelj's formulas for integrals having the form

$$F(z) = \frac{1}{2\pi\mathrm{i}}\int_\Gamma \frac{f(t)}{(t-z)^{n+1}}\mathrm{d}\,t\,. \tag{D.3.8}$$

Utilizing (D.3.6) we may calculate (by derivation) the Finite Part when we know the Principal Value. So, from (D.4.1) it results the integral often used in Appendix C,

$$\frac{1}{\pi}\int_a^{*b} \frac{\sqrt{(t-a)(b-t)}}{(t-x)^2}\mathrm{d}\,t = -1\,, \quad a < x < b\,. \tag{D.3.9}$$

From (B.5.3) we deduce

$$\frac{1}{\pi}\int_a^{*b} \frac{1}{\sqrt{(b-t)(t-a)}}\,\frac{\mathrm{d}\,t}{(t-x)^2} = 0\,, \tag{D.3.10}$$

and from (B.6.9),

$$\frac{1}{\pi}\int_a^{*b} \sqrt{\frac{t-a}{b-t}}\,\frac{t}{(t-x)^2}\mathrm{d}\,t = 1\,, \tag{D.3.11}$$

and the sequence of examples may be enlarged (see also [A.41]).

We also observe that we have

$$\int_a^{*b} \frac{\mathrm{d}\,x}{(x-u)^2} = \frac{1}{a-u} - \frac{1}{b-u} \tag{D.3.12}$$

which may be obtained considering $f=1$ in (D.3.7).

D.4 Hadamard-Type Integrals

We shall consider the integrals having the form

$$I_3 = \int_a^s \frac{f(x)}{(s-x)^{n+1/2}} \mathrm{d}\,x, \quad n = 1, 2, \ldots \tag{D.4.1}$$

which intervene in the supersonic flow. As we have already seen in (D.2.1), the basic idea in the definition of the Finite Part consists in leaving apart the infinite values from the structure of the integrals. In order to see the significance of the integral denoted by (D.4.1) in the case $n = 1$, we shall observe that

$$\frac{\mathrm{d}}{\mathrm{d}s} \int_a^s \frac{f(x)}{\sqrt{s-x}} \mathrm{d}\,x = \lim_{\varepsilon \searrow 0} \frac{\mathrm{d}}{\mathrm{d}\,s} \int_a^{s-\varepsilon} \frac{f(x)}{\sqrt{s-x}}$$

$$= \lim_{\varepsilon \searrow 0} \left\{ \frac{f(s-\varepsilon)}{\sqrt{\varepsilon}} + \int_a^{s-\varepsilon} \frac{\partial}{\partial s} \left[\frac{f(x)}{\sqrt{s-x}} \right] \mathrm{d}\,x \right\} .$$

If f is continuous in s, the first term from the right hand part becomes infinite such that we must leave it apart. We shall consider by definition

$$\int_a^{*s} \frac{\partial}{\partial s} \left[\frac{f(x)}{\sqrt{s-x}} \right] \mathrm{d}\,x = \frac{\mathrm{d}}{\mathrm{d}\,s} \int_a^s \frac{f(x)}{\sqrt{s-x}} \mathrm{d}\,x \tag{D.4.2}$$

or

$$-\frac{1}{2} \int_a^{*s} \frac{f(x)}{(s-x)^{3/2}} \mathrm{d}\,x = \frac{\mathrm{d}}{\mathrm{d}\,s} \int_a^s \frac{f(x)}{\sqrt{s-x}} \mathrm{d}\,x \,, \tag{D.4.3}$$

like in (D.3.6). For $f = 1$ we deduce

$$\int_a^{*s} \frac{\mathrm{d}\,x}{(s-x)^{3/2}} = -\frac{2}{\sqrt{s-a}} \,. \tag{D.4.4}$$

Let us prove now that if f is *continuous* in s and admits *bounded first order derivatives* on $[a, s)$, then we may give a formula for calculating the member from the right hand part of (D.4.3). Indeed, we have

$$\frac{\mathrm{d}}{\mathrm{d}\,s} \int_a^s \frac{f(x) - f(s)}{\sqrt{s-x}} \mathrm{d}\,x = \lim_{x \nearrow s} \frac{f(x) - f(s)}{\sqrt{s-x}} + \int_a^s \frac{\partial}{\partial s} \left[\frac{f(x) - f(s)}{\sqrt{s-x}} \right] \mathrm{d}\,x \,. \tag{D.4.5}$$

But, with the above hypotheses, on the basis of Lagrange's formula

$$f(x) = f(s) + (x-s) f'[s + \theta(x-s)] \,,$$

we deduce that the limit from (D.4.5) is zero. Hence,

$$\frac{\mathrm{d}}{\mathrm{d}\,s}\int_a^s \frac{f(x)}{\sqrt{s-x}}\mathrm{d}\,x = \frac{\mathrm{d}}{\mathrm{d}\,s}\left[\int_a^s \frac{f(x)-f(s)}{\sqrt{s-x}}\mathrm{d}\,x + f(s)\int_a^s \frac{\mathrm{d}\,x}{\sqrt{s-x}}\right] =$$

$$= -\frac{1}{2}\int_a^s \frac{f(x)-f(s)}{(s-x)^{3/2}}\mathrm{d}\,x - \frac{f(s)}{2}\,{}^{*}\!\!\int_a^s \frac{\mathrm{d}\,x}{(s-x)^{3/2}}\,. \tag{D.4.6}$$

If we also utilize (D.4.4), then (D.4.3) becomes:

$$\overset{*}{\int_a^s} \frac{f(x)}{(s-x)^{3/2}}\mathrm{d}\,x = \int_a^s \frac{f(x)-f(s)}{(s-x)^{3/2}}\mathrm{d}\,x - \frac{2f(s)}{\sqrt{s-a}}\,, \tag{D.4.7}$$

this representing the calculation formula for the Finite Part. It is similar with the formula (D.3.5).

From (D.4.5) it also results a *derivation formula*, in fact, the formula analogous to (D.3.7), which reduces the calculation of the integral with a strong singularity to the calculation of an integral whose singularity is weaker with an unity. In the case we had in view $(n = 1)$, the weaker singularity will be integrable. Indeed, taking (D.4.4) into account, we have:

$$\frac{\mathrm{d}}{\mathrm{d}\,s}\int_a^s \frac{f(x)-f(s)}{\sqrt{s-x}}\mathrm{d}\,x = \frac{\mathrm{d}}{\mathrm{d}\,s}\int_a^s \frac{f(x)}{\sqrt{s-x}}\mathrm{d}\,x - $$

$$-f'(s)\int_a^s \frac{\mathrm{d}\,x}{\sqrt{s-x}} - \frac{f(s)}{\sqrt{s-a}}\,. \tag{D.4.8}$$

Noticing now that $\partial/\partial s = -\partial/\partial x$, integrating by parts and taking into account (D.4.6), we obtain:

$$\int_a^s \frac{\partial}{\partial s}\left[\frac{f(x)-f(s)}{\sqrt{s-x}}\right]\mathrm{d}\,x = -f'(s)\int_a^s \frac{\mathrm{d}\,x}{\sqrt{s-x}} -$$

$$-\int_a^s [f(x)-f(s)]\frac{\partial}{\partial x}\left(\frac{1}{\sqrt{s-x}}\right)\mathrm{d}\,x =$$

$$= -f'(s)\int_a^s \frac{\mathrm{d}\,x}{\sqrt{s-x}} + \frac{f(a)-f(s)}{\sqrt{s-a}} +$$

$$+\int_a^s \frac{f'(x)}{\sqrt{s-x}}\mathrm{d}\,x\,. \tag{D.4.9}$$

Equating the first members from (D.4.8) and (D.4.9) according to the formula (D.4.5), we obtain:

$$\frac{\mathrm{d}}{\mathrm{d}\,s}\int_a^s \frac{f(x)}{\sqrt{s-x}}\mathrm{d}\,x = \frac{f(a)}{\sqrt{s-a}} + \int_a^s \frac{f'(x)}{\sqrt{s-x}}\mathrm{d}\,x,. \tag{D.4.10}$$

This is the formula of reduction to a weaker singularity, analogous with (D.3.7). It also proves that the term from the right hand part of the equality (D.4.2) is finite. Analogously one obtains

$$\frac{\mathrm{d}}{\mathrm{d}\,s}\int_s^b \frac{f(x)}{\sqrt{x-s}}\mathrm{d}\,x = -\frac{f(b)}{\sqrt{b-s}} + \int_s^b \frac{f'(x)}{\sqrt{x-s}}\mathrm{d}\,x,. \tag{D.4.11}$$

The generalization of the definition (D.4.2) is performed as follows

$$\mathop{{\int\!\!\!\!\!\!\ast}}_a^s \frac{\partial^n}{\partial s^n}\left[\frac{f(x)}{\sqrt{s-x}}\right]\mathrm{d}\,x = \frac{\mathrm{d}^n}{\mathrm{d}\,s^n}\int_a^s \frac{f(x)}{\sqrt{s-x}}\mathrm{d}\,x\,, \tag{D.4.12}$$

if f is continuous in s, n times derivable and with the derivative of order n bounded in $[a,s)$ [A.40].

D.5 Generalization

In the theory of oscillatory wings one may encounter integrals having the shape (D.4.1) where f depends also of s. We shall establish for these ones the derivation formula

$$\frac{\mathrm{d}}{\mathrm{d}\,s}\int_s^b \frac{f(x,s)}{\sqrt{x-s}}\mathrm{d}\,x = -\frac{f(b,s)}{\sqrt{b-s}} + \int_s^b \frac{f_x(x,s)+f_s(x,s)}{\sqrt{x-s}}\,, \tag{D.5.1}$$

from which one obtains (D.4.11) in case that f does not depend on s. Indeed, we have

$$\frac{\mathrm{d}}{\mathrm{d}\,s}\int_s^b \frac{f(x,s)}{\sqrt{x-s}}\mathrm{d}\,x = \frac{\mathrm{d}}{\mathrm{d}\,s}\left[\int_s^b \frac{f(x,s)-f(s,s)}{\sqrt{x-s}}\mathrm{d}\,x + \right.$$
$$\left. + f(s,s)\int_s^b \frac{\mathrm{d}\,x}{\sqrt{x-s}}\right]. \tag{D.5.2}$$

But,

$$\frac{\mathrm{d}}{\mathrm{d}\,s}\int_s^b \frac{f(x,s)-f(s,s)}{\sqrt{x-s}}\mathrm{d}\,x = -\lim_{x\searrow s}\frac{f(x,s)-f(s,s)}{\sqrt{x-s}} +$$
$$+\int_s^b \frac{\partial}{\partial s}\left[\frac{f(x,s)-f(s,s)}{\sqrt{x-s}}\right]\mathrm{d}\,x\,. \tag{D.5.3}$$

If $f(x,s)$ is continuous in $x=s$ and admits the derivative f_x bounded in the interval $[s,b]$, then, from

$$f(x,s) = f(s,s) + (x-s)f_x[s, s+\theta(x-s)]$$

we deduce that the limit from (D.5.3) is zero. After all, changing $\partial/\partial s$ by $-\partial/\partial x$ and integrating by parts, we obtain

$$\frac{\mathrm{d}}{\mathrm{d}\,s}\int_s^b \frac{f(x,s)-f(s,s)}{\sqrt{x-s}}\mathrm{d}\,x = -\int_s^b [f(x,s)-f(s,s)]\frac{\partial}{\partial x}\frac{1}{\sqrt{x-s}}\mathrm{d}\,x +$$

$$+\int_s^b \frac{[f(x,s)-f(s,s)]_s}{\sqrt{x-s}}\mathrm{d}\,x = -\frac{f(x,s)-f(s,s)}{\sqrt{x-s}}\Bigg|_s^b + \int_s^b \frac{f_x(x,s)}{\sqrt{x-s}}\mathrm{d}\,x +$$

$$+\int_s^b \frac{f_s(x,s)}{\sqrt{x-s}}\mathrm{d}\,x - \frac{\partial}{\partial s}f(s,s)\int_s^b \frac{\mathrm{d}\,x}{\sqrt{x-s}}$$

$$= -\frac{f(b,s)-f(s,s)}{\sqrt{b-s}} + \int_s^b \frac{f_x(x,s)+f_s(x,s)}{\sqrt{x-s}}\mathrm{d}\,x - \frac{\partial}{\partial s}f(s,s)\int_s^b \frac{x}{\sqrt{x-s}}\,.$$

Replacing in (D.5.2) and taking into account that

$$\frac{\mathrm{d}}{\mathrm{d}\,s}\int_s^b \frac{\mathrm{d}\,x}{\sqrt{x-s}} = -\frac{1}{\sqrt{b-s}}\,,$$

we deduce (D.5.1).

The established formula (D.5.1) shows that the first member is finite. Hence, we may set by definition

$$\int_s^{*b} \frac{\partial}{\partial s}\frac{f(x,s)}{\sqrt{x-s}}\mathrm{d}\,x = \frac{\mathrm{d}}{\mathrm{d}\,s}\int_s^b \frac{f(x,s)}{\sqrt{x-s}}\mathrm{d}\,x\,, \qquad \text{(D.5.4)}$$

for every function $f(x,s)$ continuous in the point (s,s), derivable, with bounded partial derivatives in x and s, where $s < x \leq b$. In the general case we shall define

$$\int_s^{*b} \frac{\partial^n}{\partial s^n}\frac{f(x,s)}{\sqrt{x-s}}\mathrm{d}\,x = \frac{\mathrm{d}^n}{\mathrm{d}\,s^n}\int_s^b \frac{f(x,s)}{\sqrt{x-s}}\mathrm{d}\,x\,. \qquad \text{(D.5.5)}$$

Appendix E

Singular Multiple Integrals

In the euclidean space with n dimensions E_n, one considers a domain D bounded or not (D may coincide with E_n) and a function $F(\xi)$ defined on D. We denote $\xi = (\xi_1, \ldots, \xi_n)$, $x = (x_1, \ldots, x_n)$. If D is unbounded we shall assume that F tends to zero when $|\xi| \to \infty$ in a certain manner which will be precised in the sequel. We admit that there exists a point $Q(x)$ in D, such that in every vicinity D_ε (having the diameter ε) F is unbounded, while, in $D - D_\varepsilon$, F is bounded and integrable in the usual sens. Then we set

$$\int_D F \mathrm{d}\,\xi = \lim_{\varepsilon \to 0} \int_{D - D_\varepsilon} F \mathrm{d}\,\xi, \mathrm{d}\,\xi = \mathrm{d}\,\xi_1, \ldots, \mathrm{d}\,\xi_n \,. \tag{E.1}$$

If this limită exists, it is finite and does not depend on the shape of D_ε, then the integral will be *convergent.* Otherwise the integral will be *divergent.*

We consider now a divergent integral. If there exists a certain shape of D_ε (for example, sphere or cube) for which the limit exists and is finite (hence it is unique for every sequence of spheres or cubes contracting towards Q), then the integral will be called *semi-convergent.* The limit (which depends on the shape of D_ε) will be called *principal value* of the integral.

Utilizing spherical coordinates one demonstrates that the integrals having the form

$$\int_D \frac{f}{r^\alpha} \mathrm{d}\,\xi \,, \tag{E.2}$$

where $r = |\xi - x|$, and f is bounded in D, are convergent for $\alpha < n$ and divergent for $\alpha > n$. The case $\alpha = n$ will be investigated separately.

We shall consider the integral having the form

$$v(x) = \int_D u(\xi) \frac{f(x, m)}{r^n} \mathrm{d}\,\xi \,, \quad \text{where } m = \frac{\xi - x}{r} \,, \tag{E.3}$$

where f will be called the characteristic, u will be called the density and x the pole of the integral. The ratio $K(x, \xi) = f/r^n$ will be called kernel. All the integrals that we utilize in this book have the form (E.3). The first who studied this type of integrals was Tricomi [A.35] who gives some results in the case $n = 2$. An ample presentation, which will guide us in the sequel, of the theory of integrals having the form (E.3), may be found in the books of Mihlin [A.25] and [A.26]. In this presentation, D_ε will be spherical and the convergence of the integral will be investigated with respect to this form.

We assume that:

1^0 $u(\xi)$ satisfies Hölder condition in D; if D is unbounded, we assume that $u(\xi) = 0(|\xi|^{-k})$, $k > 0$. Hölder's condition means that there exists two positive constants A and α, $0 < \alpha \leq 1$, such that for every two points ξ_1 and ξ_2, from D să we have

$$|u(\xi_1) - u(\xi_2)| < A|\xi_1 - \xi_2|^\alpha ; \tag{E.4}$$

2^0 The characteristic $f(x, m)$ is bounded and for a fixed x it is continuous in m. Under these assumptions, we have the following theorem:

The necessary and sufficient condition for the existence of the integral (E.3) *is to have*

$$\int_S f(x, m) \mathrm{d}\, S = 0 \,, \tag{E.5}$$

where S *is the surface of the unit sphere centered in* x.

In order to make the demonstration, we isolate the pole with a sphere included in D, having the radius δ and the center in x. Obviously,

$$\int_D u(\xi) \frac{f(x, m)}{r^n} \mathrm{d}\, \xi = \int_{D - D_\delta} u(\xi) \frac{f(x, m)}{r^n} \mathrm{d}\, \xi +$$

$$+ \lim_{\varepsilon \to 0} \int_{\varepsilon < r < \delta} [u(\xi) - u(x)] \frac{f(x, m)}{r^n} \mathrm{d}\, \xi + u(x) \lim_{\varepsilon \to 0} \int_{\varepsilon < r < \delta} \frac{f(x, m)}{r^n} \mathrm{d}\, \xi \,.$$

The first two integrals from the right hand part of the equality are absolutely convergent. In the third one utilizes spherical coordinates relative to the pole x. Since $\mathrm{d}\, \xi = r^{n-1} \mathrm{d}\, r \mathrm{d}\, S$, we obtain

$$\int_{\varepsilon < r < \delta} \frac{f(x, m)}{r^n} \mathrm{d}\, \xi = \int_S f(x, m) \int_\varepsilon^\delta \frac{\mathrm{d}\, r}{r} = \ln \frac{\delta}{\varepsilon} \int_S f(x, m) \mathrm{d}\, S \,.$$

It results the necessary and sufficient condition (E.5).

If this condition is satisfied, the integral (E.3) may be represented as follows

$$\int_D u(\xi)\frac{f(x,m)}{r^n}\mathrm{d}\,\xi = \\ = \int_{D-D_\delta} u(\xi)\frac{f(x,m)}{r^n}\mathrm{d}\,\xi + \int_{D_\delta}[u(\xi)-u(x)]\frac{f(x,m)}{r^n}\mathrm{d}\,\xi\,. \tag{E.6}$$

As an application, we shall consider the integral:

$$\int\int_D l(\xi,n)\frac{\partial}{\partial x}\left(\frac{1}{R_0}\right)\mathrm{d}\,\xi\mathrm{d}\,\eta = \int\int_D l(\xi,n)\frac{\xi-x}{R_0^3}\mathrm{d}\,\xi\mathrm{d}\,\eta\,, \tag{E.7}$$

where R_0 is (5.1.11). With the change of variables $\xi - x = \beta\cos\theta$, $\eta - y = \sin\theta$, the condition (E.5) gives

$$\int_0^{2\pi}\frac{\xi-x}{R_0}\mathrm{d}\,\theta = \int_0^{2\pi}\cos\theta\mathrm{d}\,\theta = 0\,. \tag{E.8}$$

Hence, the integral (E.7) exists.

The second important theorem that we utilize (for the transonic flow) is the following:

If, in addition to the hypotheses 1^0 and 2^0, *we assume that* $\mathrm{grad}_x K(x,\xi) = O(r^{-n-1})$, *then the singular integral* (E.3) *(as a function of* x*) satisfies Hölder's condition, with the same exponent like* u, *in every domain which is bounded, closed and included in* D.

The theorem was proved for the first time by Giraud in 1934, and for $n = 1$ by Privalov in 1916. As we can see in [A.25] and [A.26], the demonstration is not simple.

The last theorem refers to the derivation of the integrals with weak singularities having the shape

$$v(x) = \int_D u(\xi)\frac{f(x,m)}{r^{n-1}}\mathrm{d}\,\xi\,, \tag{E.9}$$

which lead to integrals of the type (E.3). Like in the previous theory, D may be a domain bounded or not of the space E_n, or it may coincide with the entire space. We assume that *the function* $f(x,m)$ *is continuous and bounded together with its first derivatives (the first order derivatives with respect to the cartesian coördinates of the points* x *and* m*).* We also assume that $u(\xi)$ *satisfies Hölder's condition and at infinity* (if D is unbounded) $u(\xi) = o(|\xi^{-l}|)$, $l > 1$. Under these assumptions, *there exist the first derivatives* of the integral (E.9), and they are given by the formula

$$\frac{\partial v}{\partial x_k} = \int_D u(\xi)\frac{\partial}{\partial x_k}\left[\frac{f(x,m)}{r^{n-1}}\right] \mathrm{d}\,\xi - u(x)\int_S f(x,m)\cos(n,x_k)\mathrm{d}\,S\,, \tag{E.10}$$

where, like above, S is the surface of the unit sphere centered in x, and n, the outer normal to the sphere. Obviously, the first integral from the right hand part of the equality is singular.

Appendix F

Gauss-Type Quadrature Formulas

F.1 General Theorems

This appendix relies on the paper [A.49] and it is completed with some results due to Monegato [A.52]. Gauss-type quadrature formulas give exact evaluations for the integrals of polynomial functions, multiplied by a weight function w. In aerodynamics it also meet integrals with singularities. Approximating an arbitrary function (according to Weierstras's theorem) by a polynomial function, we may utilize these evaluations. Practically, the approximation is performed by a Lagrange-type interpolation formula.

We shall consider in the sequel that $w : [-1, +1] \to \mathbb{R}$ is a positive integrable function.

THEOREM 1. *We have the exact evaluation*

$$\int_{-1}^{+1} f(x)w(x)\mathrm{d}\,x = \sum_{\alpha=1}^{n} A_\alpha f(x_\alpha) \tag{F.1.1}$$

if:

1^0 f *is a polynomial of degree* $\leq 2n-1$;

2^0 *the points* $x = x_\alpha$, $\alpha = \overline{1,n}$ *are the* n *zeros of the polynomial* $P_n(x)$ *of degree* n *from the orthogonal system of the weight* $w(x)$ on $[-1, +1]$

$$\int_{-1}^{+1} w(x)P_i(x)P_j(x)\mathrm{d}\,x = 0, \quad i \neq j; \tag{F.1.2}$$

3^0 *Using the notation*

$$Q_n(t) = {\int_{-1}^{+1}}' P_n(x)\frac{w(x)}{x-t}\mathrm{d}\,x\,, \tag{F.1.3}$$

the coefficients A_α *are given by the formulas*

$$A_\alpha = \frac{Q_n(x_\alpha)}{P_n'(x_\alpha)}\,. \tag{F.1.4}$$

Proof. Taking into account that f is a polynomial of degree $2n-1$, and P_n is a polynomial of degree n, we may write

$$\frac{f}{P_n} = \sum_{\alpha=1}^{n} \frac{a_\alpha}{x - x_\alpha} + F_{n-1}\,, \tag{F.1.5}$$

where F_{n-1} is a polynomial of degree $\leq n-1$.

We determine the coefficients a_α multiplying (F.1.5) with $x - x_\alpha$ and putting $x = x_\alpha$. It results

$$a_\alpha = f(x_\alpha)/P_n'(x_\alpha) \tag{F.1.6}$$

whence

$$f = P_n\left[\sum_{\alpha=1}^{n} \frac{f(x_\alpha)}{P_n'(x_\alpha)(x - x_\alpha)} + f_{n-a}\right].$$

Since

$$\int_{-1}^{+1} \frac{P_n(x) f(x_\alpha) w(x)}{P_n'(x_\alpha)(x - x_\alpha)} \mathrm{d}\,x = \frac{f(x_\alpha)}{P_n'(x_\alpha)} Q_n(x_\alpha) = A_\alpha f(x_\alpha)\,,$$

and because F_{n-1} may be written as a linear combination of $P_0, \ldots, P_{n-1}$, such that

$$\int_{-1}^{+1} P_n(x) F_{n-1}(x) w(x) \mathrm{d}\,x = 0\,,$$

we deduce (F.1.1).

THEOREM 2. *We have the following exact evaluation:*

$$\int_{-1}^{+1}{}' f(x) \frac{w(x)}{x - t_j} \mathrm{d}\,x = \sum_{\alpha=1}^{n} A_\alpha \frac{f(x_\alpha)}{x_\alpha - t_j}\,, \tag{F.1.7}$$

where $j = 1, 2, \ldots,$ *if*

1^0 f *is a polynomial of degree* $\leq 2n$;

2^0 *the points* $x = x_\alpha, \alpha = \overline{1, n}$ *are those defined in Theorem 1;*

3^0 *the points* $t = t_j,\;\; j = 1, 2, \ldots,$ *are the zeros of the function* Q_n *defined by* (F.1.3);

4^0 *the coefficients* A_α *are defined by* (F.1.4).

Proof. Reasoning like before, we have:

$$\frac{f}{P_n} = \sum_{\alpha=1}^{n} \frac{a_\alpha}{x - x_\alpha} + F_n\,, \tag{F.1.8}$$

where the degree of $F_n \leq n$. It results that a_α has the form (F.1.6). Setting $F_n = (x - t_j)F_{n-1}(x) + A$, where the degree of $F_{n-1} \leq n - 1$, we deduce

$$f = P_n \left[\sum_{\alpha=1}^{n} \frac{f(x_\alpha)}{P_n'(x_\alpha)(x - x_\alpha)} + (x - t_j)F_{n-1} + A \right].$$

whence

$$\int_{-1}^{+1}{}' \frac{P_n(x) f(x_\alpha) w(x)}{P_n'(x_\alpha)(x - x_\alpha)(x - t_j)} \mathrm{d}\,x =$$

$$\frac{f(x_\alpha}{P_n'(x_\alpha)} \int_{-1}^{+1}{}' \frac{P_n(x) w(x)}{x_\alpha - t_j} \left[\frac{1}{x - x_\alpha} - \frac{1}{x - t_j} \right] \mathrm{d}\,x =$$

$$A_\alpha \frac{f(x_\alpha)}{x_\alpha - t_j} - \frac{f(x_\alpha) Q_n(t_j)}{P_n'(x_\alpha)(x_\alpha - t_j)} = A_\alpha \frac{f(x_\alpha)}{x_\alpha - t_j}\,,$$

because $Q_n(t_j) = 0$.

THEOREM 3. *We have the exact evaluation*

$$\int_{-1}^{+1}{}^{*} f(x) \frac{w(x)}{(x - t_j)^2} \mathrm{d}\,x = \sum_{\alpha=1}^{n} A_\alpha \frac{f(x_\alpha)}{(x_\alpha - t_j)^2} + A f(t_j)\,, \tag{F.1.9}$$

where the "asterisk" is for the Finite Part (D.3.5), *if:*

1^0 f *is a polynomial of degree* $\leq 2n + 1$*;*

2^0 *the points* $x = x_\alpha$, $\alpha = \overline{1, n}$ *are those defined in Theorem 1;*

3^0 *The points* $t = t_j$, $j = 1, 2, \ldots$ *are those defined in Theorem 1;*

4^0 *the coefficients* A_α *are given by* (F.1.4), *and*

$$A = \frac{Q_n'(t_j)}{P_n(t_j)}\,. \tag{F.1.10}$$

Proof. We shall notice at first that, on the basis of the definition (D.3.6), we have for $t \in (-1, +1)$,

$$\frac{\mathrm{d}}{\mathrm{d}\,t} \int_{-1}^{+1}{}' P_n(x) \frac{w(x)}{x - t} \mathrm{d}\,x = \int_{-1}^{+1}{}^{*} P_n(x) \frac{w(x)}{(x - t)^2} \mathrm{d}\,x\,, \tag{F.1.11}$$

whence

$$Q'_n(t_j) = \int_{-1}^{*+1} P_n(x)\frac{w(x)}{(x-t_j)^2}\mathrm{d}\,x\,. \tag{F.1.12}$$

We shall write as above

$$f = P_n\left[\sum_{\alpha=1}^{n}\frac{f(x_\alpha)}{P'_n(x_\alpha)(x-x_\alpha)} + (x-t_j)^2F_{n-1} + B(x-t_j) + C\right], \tag{F.1.13}$$

F_{n-1} being a polynomial of degree $\leq n-1$ and B , C, constants.

For $x = t_j$ it results

$$C = \frac{f(t_j)}{P_n(t_j)} - \sum_{\alpha=1}^{n}\frac{f(x_\alpha)}{P'_n(x_\alpha)(t_j-x_\alpha)}\,. \tag{F.1.14}$$

Since

$$\frac{1}{(x-x_\alpha)(x-t_j)^2} = \frac{1}{(x_\alpha-t_j)^2}\;\frac{1}{x-x_\alpha} - $$
$$-\frac{2}{(x_\alpha-t_j)^2}\;\frac{1}{x-t_j} + \frac{1}{t_j-x_\alpha}\;\frac{1}{(x-t_j)^2}\,.$$

Taking into account that $Q_n(t_j) = 0$ and that we have (F.1.12),G we deduce

$$\int_{-1}^{*+1}\frac{P_n(x)f(x_\alpha)w(x)}{P'_n(x_\alpha)(x-x_\alpha)(x-t_j)^2}\mathrm{d}\,x =$$
$$= \frac{f(x_\alpha)}{P'_n(x_\alpha)}\left\{\int_{-1}^{*+1}\frac{P_n(x)w(x)}{(x_\alpha-t_j)^2}\left(\frac{1}{x-x_\alpha} - \frac{2}{x-t_j}\right)\mathrm{d}\,x + \right.$$
$$\left. + \int_{-1}^{*+1}\frac{P_n(x)w(x)}{t_j-x_\alpha}\;\frac{\mathrm{d}\,x}{(x-t_j)^2}\right\} = \tag{F.1.15}$$
$$= \frac{f(x_\alpha)}{P'_n(x_\alpha)}\left[\frac{Q_n(x_\alpha)}{(x_\alpha-t_j)^2} + \frac{Q'_n(t_j)}{t_j-x_\alpha}\right].$$

Having in view the definitions of the Principal Value and Finite Part, it results

$$\int_{-1}^{*+1}\frac{P_n(x)w(x)}{x-t_j}\mathrm{d}\,x = \int_{-1}^{\prime+1}\frac{P_n(x)w(x)}{x-t_j}\mathrm{d}\,x = Q_n(t_j) = 0\,.$$

Utilizing (F.1.14) and (F.1.10), we deduce

$$C\int_{-1}^{*+1}P_n(x)\frac{w(x)}{(x-t_j)^2}\mathrm{d}\,x = Af(t_j) - \sum_{\alpha=1}^{n}\frac{f(x_\alpha)Q'_n(t_j)}{P'_n(x_\alpha)(t_j-x_\alpha)}\,,$$

whence it results (F.1.9).

The integrals having the form

$$\int_{-1}^{*+1} f(x)\frac{w(x)}{(x-t)^n}\mathrm{d}\,x\,, \qquad n>2 \tag{F.1.16}$$

are studied in [A.49] and [A.52].

F.2 Formulas of Interest in Aerodynamics

It is well known that on the interval $[-1,+1]$ Jacobi's polynomials $P_n^{\alpha,\beta}(x)$ constitute an orthogonal basis with the weight function $w(x) = (1-x)^\alpha(1+x)^\beta$ (see, for example, [A.56]). Obviously, the zeros x_α and t_j from the theorems from F.1 do not depend on the factor of normalization of the polynomial P_n. One can simplify this factor in A_α from (F.1.4) and A from (F.1.10). We shall utilize therefore Jacobi's polynomials without the constant factor.

1^0. For the weight $w(x) = (1-x^2)^{-1/2}$, Jacobi's polynomials reduce to Chebyshev's polynomials

$$T_n(x) = \cos n\theta\,, \quad x = \cos\theta\,. \tag{F.2.1}$$

For $0<\theta<\pi$, the polynomials $T_n (n=1,2,\ldots)$ vanish when

$$\theta_\alpha = \frac{2\alpha-1}{n}\cdot\frac{\pi}{2}\,, \quad x_\alpha = \cos\frac{2\alpha-1}{n}\frac{\pi}{2}\,, \quad \alpha=\overline{1,n}\,. \tag{F.2.2}$$

Utilizing the notation $t = \cos\sigma$ and Glauert's integral (B.6.6), we deduce:

$$Q_n(t) = \pi\frac{\sin\sigma}{\sin\sigma}\,, \tag{F.2.3}$$

with the zeros

$$\sigma_j = \frac{j\pi}{n}\,, \quad t_j = \cos\frac{j\pi}{n}\,, \quad j=\overline{1,n-1}\,. \tag{F.2.4}$$

Since

$$T_n'(x) = -\frac{1}{\sin\theta}\frac{\mathrm{d}\,T_n}{\mathrm{d}\,\theta} = n\frac{\sin n\theta}{\sin\theta}\,,$$

from (F.1.4) it results that $A_\alpha = \pi/n$. Since

$$Q_n'(t_j) = -\frac{\pi}{\sin\sigma_j}\frac{\mathrm{d}}{\mathrm{d}\,\sigma}\frac{\sin n\sigma}{\sin\sigma}\bigg|_{\sigma_j} = -\frac{n\pi\cos j\pi}{\sin^2\sigma_j}\,,$$

from (F.1.10) it results that $A = -n\pi/(1-t_j^2)$.

We deduce therefore the following formulas:

$$\int_{-1}^{+1} \frac{f(x)}{\sqrt{1-x^2}} \mathrm{d}\,x = \frac{\pi}{n} \sum_{\alpha=1}^{n} f(x_\alpha)\,, \tag{F.2.5}$$

$$\int_{-1}^{+1} \frac{f(x)}{\sqrt{1-x^2}} \frac{\mathrm{d}\,x}{x-t_j} = \frac{\pi}{n} \sum_{\alpha=1}^{n} \frac{f(x_\alpha)}{x_\alpha - t_j}\,, \tag{F.2.6}$$

$$\overset{*}{\int_{-1}^{+1}} \frac{f(x)}{\sqrt{1-x^2}} \frac{\mathrm{d}\,x}{(x-t_j)^2} = \frac{\pi}{n} \sum_{\alpha=1}^{n} \frac{f(x_\alpha)}{(x_\alpha - t_j)^2} - \frac{n\pi}{1-t_j^2} f(t_j)\,, \tag{F.2.7}$$

for $j = \overline{1, n-1}$, where x_α are given by (F.2.2) and t_j, by (F.2.4).

2^0. For the weight function $w(x) = (1-x^2)^{1/2}$, Jacobi's polynomials reduce to Chebyshev's polynomials of second order

$$U_n(x) = \frac{\sin(n+1)\theta}{\sin\theta}\,, \quad x = \cos\theta\,. \tag{F.2.8}$$

For $\theta \in (0, \pi)$, the zeros of these polynomials are

$$\theta_\alpha = \frac{\alpha\pi}{n+1}\,, x_\alpha = \cos\frac{\alpha\pi}{n+1}\,, \quad \alpha = \overline{1, n}\,. \tag{F.2.9}$$

Using the notation $t = \cos\sigma$ and Glauert's integral (B.6.6), we deduce:

$$Q_n(t) = -\pi\cos(n+1)\sigma \tag{F.2.10}$$

whence

$$t_j = \cos\frac{2j-1}{n+1}\frac{\pi}{2}\,, \quad j = \overline{1, n+1}\,. \tag{F.2.11}$$

Utilizing the formulas given in the theorems F.1 for A_α and A, on the basis of Glauert's integral, we obtain:

$$A_\alpha = \frac{\pi}{n+1}(1-x_\alpha^2)\,, \quad A = -\pi(n+1)\,.$$

Hence, the following quadrature formulas are established:

$$\int_{-1}^{+1} \sqrt{1-x^2} f(x) \mathrm{d}\,x = \frac{\pi}{n+1} \sum_{\alpha=1}^{n} (1-x_\alpha^2) f(x_\alpha)\,, \tag{F.2.12}$$

$$\int_{-1}^{+1} \sqrt{1-x^2} \frac{f(x)}{x-t_j} \mathrm{d}\,x = \frac{\pi}{n+1} \sum_{\alpha=1}^{n} \frac{1-x_\alpha^2}{x_\alpha - t_j} f(x_\alpha)\,, \tag{F.2.13}$$

$$\overset{*}{\int_{-1}^{+1}} \sqrt{1-x^2}\,\frac{f(x)}{(x-t_j)^2}\mathrm{d}\,x = \frac{\pi}{n+1}\sum_{\alpha=1}^{n}\frac{1-x_\alpha^2}{(x_\alpha-t_j)^2}f(x_\alpha) - \\ -\pi(n+1)f(t_j)\,, \tag{F.2.14}$$

where $j=\overline{1,n+1}$, x_α are given by (F.2.9) and t_j by (F.2.11).

We notice that the numbers t_j given by (F.2.4) are the zeros of the polynomials U_{n-1}, and the numbers t_j given in (F.2.11) are the zeros of the polynomials $T_{n+1}(t)$.

3^0. For the weight function $w(x)=(1-x)^{1/2}(1+x)^{-1/2}$, Jacobi's polynomials are [A.56].

$$P_n(x) = \frac{\sin[(2n+1)\theta/2]}{\sin(\theta/2)}\,, \qquad x=\cos\theta\,, \tag{F.2.15}$$

with the zeros

$$x_\alpha = \cos\frac{2\alpha\pi}{2n+1}\,, \qquad \alpha=\overline{1,n}\,. \tag{F.2.16}$$

With the aid of Glauert's integral, we deduce

$$Q_n(t) = -\pi\frac{\cos[(2n+1)\sigma/2]}{\cos(\sigma/2)}\,, \qquad t=\cos\sigma\,,$$

whence

$$t_j = \cos\frac{2j-1}{2n+1}\,, \qquad j=\overline{1,n}\,. \tag{F.2.17}$$

Utilizing (F.1.4) and (F.1.10), it results:

$$A_\alpha = \frac{2n}{2n+1}(1-x_\alpha)\,, \qquad A = -\frac{2n+1}{1+t_j}\frac{\pi}{2}\,.$$

Hence, we established the following formulas:

$$\int_{-1}^{+1}\sqrt{\frac{1-x}{1+x}}f(x)\mathrm{d}\,x = \frac{2\pi}{2n+1}\sum_{\alpha=1}^{n}(1-x_\alpha)f(x_\alpha)\,, \tag{F.2.18}$$

$$\overset{\prime}{\int_{-1}^{+1}}\sqrt{\frac{1-x}{1+x}}\,\frac{f(x)}{x-t_j}\mathrm{d}\,x = \frac{2\pi}{2n+1}\sum_{\alpha=1}^{n}\frac{1-x_\alpha}{x_\alpha-t_j}f(x_\alpha)\,, \tag{F.2.19}$$

$$\overset{*}{\int_{-1}^{+1}}\sqrt{\frac{1-x}{1+x}}\,\frac{f(x)}{(x-t_j)^2}\mathrm{d}\,x = \\ = \frac{2\pi}{2n+1}\sum_{\alpha=1}^{n}\frac{1-x_\alpha}{(x_\alpha-t_j)^2}f(x_\alpha) - \frac{2n+1}{1+t_j}\frac{\pi}{2}f(t_j)\,, \tag{F.2.20}$$

for $j = \overline{1,n}$, x_α being given by the formula (F.2.16) and t_j by (F.2.17).

4^0. For the weight function $w(x) = (1-x)^{-1/2}(1+x)^{1/2}$, Jacobi's polynomials are

$$P_n(x) = \frac{\cos[(2n+1)\theta/2]}{\cos(\theta/2)}, \quad x = \cos\theta, \tag{F.2.21}$$

with the zeros

$$x_\alpha = \cos\frac{2\alpha-1}{2n+1}\pi, \quad \alpha = \overline{1,n}. \tag{F.2.22}$$

Utilizing Glauert's integral, we deduce:

$$Q_n(t) = \pi\frac{\sin[(2n+1)\sigma/2]}{\sin(\sigma/2)}, t = \cos\sigma,$$

polynomials which have the zeros

$$t_j = \cos\frac{2j\pi}{2n+1}, \quad j = \overline{1,n}. \tag{F.2.23}$$

On the basis of the formulas (F.1.4) and (F.1.10) we obtain

$$A_\alpha = \frac{2\pi}{2n+1}(1+x_\alpha), \quad A_\alpha = -\frac{2n+1}{1-t_j}\frac{\pi}{2},$$

such that one establishes the following formulas

$$\int_{-1}^{+1}\sqrt{\frac{1+x}{1-x}}f(x)\mathrm{d}x = \frac{2\pi}{2n+1}\sum_{\alpha=1}^{n}(1+x_\alpha)f(x_\alpha), \tag{F.2.24}$$

$$'\!\!\int_{-1}^{+1}\sqrt{\frac{1+x}{1-x}}\frac{f(x)}{x-t_j}\mathrm{d}x = \frac{2\pi}{2n+1}\sum_{\alpha=1}^{n}\frac{1+x_\alpha}{x_\alpha-t_j}f(x_\alpha), \tag{F.2.25}$$

$$^*\!\!\int_{-1}^{+1}\sqrt{\frac{1+x}{1-x}}\frac{f(x)}{(x-t_j)^2}\mathrm{d}x =$$

$$= \frac{2\pi}{2n+1}\sum_{\alpha=1}^{n}\frac{1+x_\alpha}{(x_\alpha-t_j)^2}f(x_\alpha) - \frac{\pi}{2}\frac{2n+1}{1-t_j}\frac{\pi}{2}f(t_j) \tag{F.2.26}$$

for $j = \overline{1,n}$, the zeros x_α being given by the formula (F.2.22) and t_j by the formula (F.2.23).

We have to notice the relations between the formulas from 3^0 and 4^0. The points x_α from 3^0 coincide with the points t_α from 4^0, and x_α from 4^0 with t_α from 3^0.

F.3 The Modified Monegato's Formula

It is preferable sometimes to replace the formula (F.1.9) which contains the numbers $f(t_j)$ by the formula given by Monegato [A.52] p. 279.

$$\int_{-1}^{+1}{}' w(x)\frac{f(x)}{(x-t)^2}\mathrm{d}x = \sum_{\alpha=1}^{n} w'_\alpha(t) f(x_\alpha) + R_n(f), \tag{F.3.1}$$

where

$$w'_\alpha(t) = \frac{Q_n(x_\alpha) - Q_n(t) - Q'_n(t)(x_\alpha - t)}{P'_n(x_\alpha)(x_\alpha - t)^2}, \tag{F.3.2}$$

the formula containing only the numbers $f(x_\alpha)$. The formula (F.3.1) is exact, i.e. $R_n(f) = 0$, if $f(x)$ is a polynomial of degree $n-1$.

In fact, in applications one utilizes not the formula (F.3.1) [5.10], [6.5], but another one which may be obtained as follows.

In (F.3.1) we isolate the term corresponding to $\alpha = j$ and we pass to the limit $t \to x_j$. We find

$$\begin{aligned} {}^*\!\int_{-1}^{+1} w(x)\frac{f(x)}{(x-x_j)^2}\mathrm{d}x = \sum_{\alpha=1}^{n}{}' w'_\alpha(x_j) f(x_\alpha) + \\ + w'_j(x_j) f(x_j) + R_n(f), \end{aligned} \tag{F.3.3}$$

where the mark " $'$ " at $\sum$ means that one excepts the term corresponding to $\alpha = j$. The factor $w'_\alpha(x_j)$ is obtained from (F.3.2). Using the rule of l'Hospital $w'_j(x_j)$ we find that

$$w'_j(x_j) = \frac{Q''_n(x_j)}{2P'_n(x_j)}. \tag{F.3.4}$$

In applications one utilizes (F.3.3) for $w(x) = (1-x^2)^{1/2}$. The numbers x_α are given therefore by (F.2.9). From an elementary calculation it results

$$Q_n(x_\alpha) = -\pi(-1)^\alpha, \quad Q'_n(x_j) = 0,$$

$$P'_n(x_\alpha) = \frac{(n+1)(-1)^\alpha}{1-x_\alpha^2}, \quad Q''_n(x_j) = \pi(n+1)^2\frac{(-1)^j}{1-x_j^2},$$

whence

$$\frac{1}{\pi}\int_{-1}^{+1}\sqrt{1-x^2}\frac{f(x)}{(x-x_j)^2}\mathrm{d}\,x =$$

$$= \frac{1}{n+1}\sum_{\alpha=1}^{n}{}' \left[1-(-1)^{j+\alpha}\right]\frac{1-x_\alpha^2}{(x_\alpha-x_j)^2}f(x_\alpha) - \frac{n+1}{2}f(x_j) + R_n\,. \tag{F.3.5}$$

This is the modified formula.

From

$${}'\!\!\int_{-1}^{+1} w(x)\frac{f(x)}{x-t}\mathrm{d}\,x = \sum_{\alpha=1}^{n} w_\alpha(t)f(x_\alpha) + R_n(f)\,, \tag{F.3.6}$$

where [A.50] page 275,

$$w_\alpha(t) = \frac{1}{P_n'(x_\alpha)}\frac{Q_n(x_\alpha)-Q_n(t)}{x_\alpha - t}\,, \tag{F.3.7}$$

isolating in Σ the term for which $\alpha = j$ and passing to limit $t \to x_j$, one obtains the formula

$${}'\!\!\int_{-1}^{+1} w(x)\frac{f(x)}{x-x_j}\mathrm{d}\,x = \sum_{\alpha=1}^{n}{}' w_\alpha(x_j)f(x_\alpha) + \frac{Q_n'(x_j)}{P_n'(x_\alpha)}f(x_j) + R_n(f). \tag{F.3.8}$$

For the weight function $w(x) = (1-x^2)^{1/2}$ we find:

$$\frac{1}{\pi}\,{}'\!\!\int_{-1}^{+1}\sqrt{1-x^2}\frac{f(x)}{x-x_j}\mathrm{d}\,x = \sum_{\alpha=1}^{n}{}' \frac{[(-1)^{j+\alpha}-1]}{n+1}\frac{1-x_\alpha^2}{x_\alpha - x_j}f(x_\alpha)\,. \tag{F.3.9}$$

F.4 A Useful Formula

We shall establish the following series expansion:

$$\frac{1}{(\eta-y)^2} = -2\sum_{1}^{\infty}(j+1)U_j(y)U_j(\eta)\,, \tag{F.4.1}$$

where $U_n(x)$ are Chebyshev's polynomials (F.2.8). The series is divergent, but it has a first Cesaro finite sum:

$$\frac{1}{(\eta - y)^2} = -2\sum_1^n (j+1) U_j(y) U_j(\eta) \,. \tag{F.4.2}$$

Indeed, setting $\eta = \cos\theta$ and $y = \cos\sigma$ on the basis of Glauert's integral, we deduce

$$\int_{-1}^{+1} \frac{\sqrt{1-\eta^2} U_n(\eta)}{\eta - y} \mathrm{d}\,\eta = -\pi \cos(n+1)\sigma \,, \tag{F.4.3}$$

and deriving and taking (D.3.6) into account, we obtain the formula

$$\int_{-1}^{+1} \frac{\sqrt{1-\eta^2} U_n(\eta)}{(\eta - y)^2} \mathrm{d}\,\eta = -\pi (n+1) U_n(y) \,. \tag{F.4.4}$$

Expanding now the function $(\eta - y)^{-2}$ on the interval $[-1, +1]$ in a Chebyshev-type polynomial series, we have:

$$\frac{1}{(\eta - y)^2} = \sum_1^n a_j(y) U_j(\eta) \,. \tag{F.4.5}$$

Taking into account (F.4.4), and the orthogonality condition

$$2\int_{-1}^{+1} \sqrt{1-\eta^2} U_j(\eta) U_k(\eta) \mathrm{d}\,\eta = \begin{cases} 0 & , j \neq k \\ \pi & , j = k \,, \end{cases} \tag{F.4.6}$$

we obtain:

$$a_j(y) = -2(j+1) U_j(y) \,. \tag{F.4.7}$$

Replacing it in (F.4.5), we obtain the formula (F.4.1).

Bibliography

Chapter 1

[1.1] H. Ashley, M. Landahl: *Aerodynamics of wings and bodies*, Addison-Wesley Publ. Co., Readings, Mass., 1965.

[1.2] R.L. Bisplinghoff, H. Ashley, R.L. Halfman: *Aeroelasticity*, Addison-Wesley Publ. Co., Inc., Cambridge, Mass., 1955.

[1.3] R.L. Bisplinghoff, H. Ashley: *Principles of Aeroelasticity*, J. Wiley, New York, London, 1962.

[1.4] P.F. Byrd, H.D. Friedman: *Handbook of eliptic integrals for engineers and physicists*, Springer, Berlin, 1954.

[1.5] E.Carafoli, V.N. Constantinescu: *Dinamica fluidelor compresibile*, Ed. Academiei, Bucureşti, 1984.

[1.6] R. Courant: *Partial differential equations*, New York, London.

[1.7] G.G. Chernyi: *Gas Dynamics* (in Russian), Moscow, 1988.

[1.8] E.H. Dowell, H.C. Curtis Jr., R.H. Scanlan and Sisto: *A modern course in aeroelasticity*, Sijthoff and Noortdhoff, the Netherlands, 1978.

[1.9] L. Dragoş: *Magnetofluid dynamics*, Ed. Academiei, Bucharest, Abacus Press, Tunbridge Wells, Kent, 1975.

[1.10] L. Dragoş: *Principiile mecanicii mediilor continue*, Ed. Tehnică, Bucureşti, 1983.

[1.11] L. Dragoş: *Mecanica fluidelor*, I, Editura Academiei Române, Bucureşti, 1999.

[1.12] W.F. Durand (Editor): *Aerodynamic theory*, Springer, Berlin, 1935.

[1.13] Van Dyke: *Perturbation methods in fluid mechanics*, Academic Press, N.Y., 1964.

[1.14] F.L. Frankl, E.A. Karpovich: *Gas Dynamics of slender bodies*, Moskow, 1948.

[1.15] J.P. Gilly, L. Rosenthal, Y. Sèmèzis: *Aerodynamique hypersonique*, Gauthier Villars, Paris, 1970.

[1.16] I.S. Gradshteyn, I.N. Ryznik: *Tables of integrals, series and products*, Academic Press, New York, 1965; in Russian 5th edition Moskow, Nauka, 1971.

[1.17] W. Gröbner, N. Hofreiter: *Integraltafel*, Teil II, Bestimte integrale, Wien, Innsbruck, Springer-Verlag, 1958.

[1.18] L. Howarth: *Modern developments in Fluid Dynamics*, High Speed Flow I, Oxford, 1953.

[1.19] W.R. Hayes, R.F. Probstein: *Hypersonic flow theory*, Academic Press, New York, 1962, 1966.

[1.20] C. Iacob: *Introducere matematică în mecanica fluidelor*, Ed. Academiei R.P.R., Bucureşti, 1952.

[1.21] C. Iacob: *Introduction mathématique à la mécanique des fluides*, Ed. de l'Academie de R.P.R., Bucharest, Gauthier-Villars, Paris, 1959.

[1.22] C. Iacob: *Matematici clasice şi moderne*, Editura Tehnică, Bucureşti, Vol 2, 1979; vol 3, 1981.

[1.23] Janke-Emde-Lösch: *Tafeln höherer funktionen*, B.G. Teubner, Stuttgart, 1960.

[1.24] W.P. Jones (Gen. Ed.): *Manual on Aeroelasticity*, vol. II, NATO AGARD.

[1.25] K. Karamcheti: *Principles of ideal fluid aerodynamics*, John Wiley and Sons, Inc. N.Y., 1966; Krieger, Malabar, FL, 1980.

[1.26] Th. von Kármán and J.M. Burgers, General Aerodynamic theory-perfect fluids, in [1.11].

[1.27] R. von Mises: *Mathematical theory of compressible fluid flow*, Academic Press, N.Y. 1958.

[1.28] J. Katz, A. Plotkin: *Low-speed aerodynamics (from wing theory to panel methods)*, Mc Grow-Hill, N.Y., 1991.

[1.29] I.K. Lifanov: *Singular integral equations and discrete vortices*, VSP, Utrecht, The Netherlands, 1996.

[1.30] W. Magnus, F. Oberhettinger, R.P. Soni: *Formulas and theorems for the special functions of Mathematical Physics*, third Ed., Springer-Verlag, Berlin, 1966.

[1.31] L.M. Milne-Thomson: *Theoretical hydrodynamics*, Mc Millan, London, 1949.

[1.32] F. Oberhettinger, L. Badü: *Tables of Laplace Transforms*, Springer-Verlag, Berlin Heidelberg, New York, 1973.

[1.33] A.N. Pancenkov: *Ghidrodinamika podvodnovo krâla*, Kiev, Naukovo Dumka, 1965.

[1.34] A.N. Pancenkov: *Teorija potenţiala. uskorenii*, Nauka, Novosibirsk, 1975.

[1.35] A.P. Prudnikov, Yu.A. Brychkov, D.I. Maricev: *Integrals and Series* (in Russian), Nauka, Moskow, 1981; in English, Gordon and Breach Sci. Publ., New York, 1986.

[1.36] H. Schlichting, E. Truckenbrodt: *Aerodynamik der Flugzenge*, Springer, Berlin, 1960.

[1.37] W.R. Sears (ed): *General theory of high speed aerodynamics, in High Speed Aerodynamics and Jet Propulsion*, vol. 6, Princ. Univ. Press, 1959.

[1.38] L.I. Sedov: *Plane problems of hydrodynamics and aerodynamics* (in Russian), Ed. 2, Nauka, Moskva, 1966.

[1.39] J. Serrin: *Mathematical principles of classical fluid mechanics, in Handbuch der Physik*, VIII/1, Springer, Berlin, 1959.

[1.40] B. Thwaites: *Incompressible aerodynamics*, Oxford Univ. Press, Oxford, 1960.

[1.41] A.N. Tihonov, A.A. Samarski: *Ecuaţiile fizicii matematice*, Ed. Tehnică, Ştiinţifică şi Enciclopedică, Bucureşti, 1956.

[1.42] G.N. Watson: *A treatise on the theory of Bessel functions*, 2nd ed., Univ. Press Cambridge, Mac Millan Co., New York, 1945.

[1.43] J. Weissinger: *Theorie des Tragflügels* in *Handbuch der Physik*, B 8/2, Springer-Verlag, Berlin 1963.

[1.44] J. Zierep: *Vorlesung über theoretische Gasdynamik*, Verlag, G.Braun, Karlsruhe, 1962.

Chapter 2

[2.1] H.R. Aggarwal: *Derivation of the fundamental equation of sound generated by moving aerodynamic surfaces*, AIAA Journ. **21**(1983) 1048; Reply by the author to F. Farassat, idem **22**(1984) 1184.

[2.2] R. Belasubramanyam: *Aerodynamic sound due to a point source near a half plane*, IMA Journ of Appl. Math., **33**(1984) 71.

[2.3] A. Barsony-Nagy, M. Hanin: *Aerodynamics of wings in subsonic shear flow*, AIAA Journ, **20**(1972) 451.

[2.4] A. Barsony-Nagy, M. Hanin: *Aerodynamics of wings in supersonic shear flow*, AIAA Journ, **22**(1984) 453.

[2.5] P. Ciobotaru: *Asupra ecuaţiilor fluidelor în distribuţii*, Studii şi Cercetări Matematice, **35**(1983) 173.

[2.6] L. Dragoş: *Fundamental matrix for the equations of ideal fluids*, Acta Mechanica, **33**(1999) 163.

[2.7] L. Dragoş: *Fundamental solutions for ideal fluids in uniform motion*, Quart. Appl. Math., **36**(1979) 401.

[2.8] L. Dragoş, D. Homentcovschi: *Stationary fundamental solution for an ideal fluid in uniform motion*, Zeitschr. f. Angew. Math. a Mech., **60**(1980) 343.

[2.9] L. Dragoş: *The fundamental plane matrix in the uniform flow of the ideal compressible fluids*, An. St. Univ. "Al.I. Cuza", Iaşi, Supl t. **27**(1981) 143.

[2.10] L. Dragoş: *Fundamental solutions in linear aerodynamics*, Rev. Roum. Math. Pures et Appl., **27**(1982) 313.

[2.11] L. Dragoş: *Fundamental solutions of the linearized Steichen equation*, Rev. Roum. Sci. Techn., Méc. Appl. **30**(1985) 173.

[2.12] F.G. Friedlander: *Sound pulses*, Cambridge Univ. Press, 1958.

[2.13] H.F. Goldstein: *Aeroacoustics*, Mc Graw-Hill, New York, 1980.

[2.14] M. Hanin, A. Bársony-Nagy: *Slender wing theory for nonuniform stream*, AIAA Journ. **18**(1980) 381.

[2.15] D. Homentcovschi, A. Bársony-Nagy: *A linearized theory of three - dimensional airfoils in nonuniform flow*, Acta Mechanica, **24**(1976) 63.

[2.16] M.J. Lighthill: *The fundamental solutions for small steady three - dimensional disturbances to a two-dimensional parallel shear flow*, J. of Fluid Mech., **3**(1957) 113.

[2.17] D.S. Jones: *Aerodynamic sound due to a source near a half plane*, J. Inst. Math. Appl. **9**(1972) 114.

[2.18] M. Morse, K.U. Ingard: *Theoretical Acousties*, Mc Graw-Hill, New York, 1968.

[2.19] N. Rott: *Das Feld einer raschbeweg Schall (The field of a moving sound source)*, Milt Inst. f. Aerod. RTH (1945) nr. 9.

[2.20] P.N. Shankar, U.N. Sinha: *Weakly sheared two-dimensional inviscid flow past bodies*, J. Mécanique, **19**(1980) 125.

[2.21] L. Sirovich: *Initial and boundary value problem in disipative gas dynamics*, Phys. of Fluids, **10**(1967) 24.

[2.22] C.S. Ventres: *Shear flow aerodynamics. Lifting surface theory*, AIAA Journ. **13**(1975) 1183.

[2.23] J. Weissinger: *Linearisierte Profiltheorie bei ungleichförmiger Anströmung*. Teil I, Unendlich dünne Profile, Acta Mechanica, **10**(1970) 207; Teil II, Schlanke Profile, Idem **13**(1972) 133.

Chapter 3

[3.1] J.A. Bagley: *Pressure distribution on two-dimensional wings near the ground*, RAE Rep.Aero 2625(1960).

[3.2] B.C. Basu, G.J. Hancock: *The unsteady motion of two-dimensional hydrofoil in incompressible inviscid flow*, J. of Fluid Mech. **87**(1978) 159.

[3.3] A. Betz: *Singularitätenverfahren zur Bestimmung der Kräfte und Momente auf Kärper in Potential strömungen*, Ingen.-Arch, **3**(1932) 454.

[3.4] A. Betz: *Applied airfoil theory*, Section J in [1.12], vol. 4.

[3.5] W. Birnbaum: *Die tragende Wirbelfläche als Hilfsmittel zur Behandlung des ebenen Problems der Tragflügeltheorie*, Z. Angew. Math. u. Mech. **3**(1923) 290.

[3.6] W. Birnbaum: *Das abene Problem des schlagenden Flügeln*, Z. Angew. Math. u. Mech. **4**(1924) 277.

[3.7] A. Carabineanu, L. Dragoş, I. Oprea: *The linearized theory for the submerged thin hydrofoil of infinite span*, Arch. Mech., **43**(1990) 25.

[3.8] D.H. Choi, L. Landweber, *Inviscid analysis of two-dimensional airfoils in unsteady motion using conformal mapping*, AIAA Journ. **28**(1990) 2025.

[3.9] L. Dragoş: *On the motion past thin airfoils of incompressible fluids of incompressible fluids with conductivity tensor*, Quart. Appl. Math., **28**(1970) 313.

[3.10] L. Dragoş: *On the theory of thin airfoils in nonequilibrium ideal fluids*, ZAMM **64**(1984) 17.

[3.11] L. Dragoş: *Method of fundamental solutions in plane steady linear aerodynamics*, Acta Mechanica **47**(1983) 277.

[3.12] L. Dragoş: *On the integration of equation of thin airfoils by quadrature formulae of Gauss type*, Bull. math. Soc. Sci. Math. Roum., **34**(1990) 227.

[3.13] L. Dragoş: *A numerical solution of the equation of thin airfoil in ground effects*, AIAA Journ. **28**(1990) 2132.

[3.14] L. Dragoş: *Subsonic flow past thin airfoil in wind tunnel*, Mech. Res. Comm. **18**(1991) 129.

[3.15] L. Dragoş: *New methods in the theory of subsonic flow past thin airfoil configurations*, Rev. Roum. Sci. Techn. Ser. Méc. Appl., **35**(1990) 179-196.

[3.16] L. Dragoş, N. Marcov: *Exact solutions of the problem of incompressible fluid flows with circulation past a circular cylinder in ground effect*, Rev. Roum. Sci. Techn., Sér. Méc. Appl., **38**(1993) 25.

[3.17] D.A. Efros: *Hydrodynamic theory of two dimensional flow with cavitation* Dokl. Akad. Nauk SSR **5**(1946) 267.

[3.18] H. Garner, E. Rogers, W. Acum and E. Maskell: *Subsonic wind tunnel wall corrections*, AGARDograph 1966, 109.

[3.19] J.P. Giesing, A.M.O. Smith: *Potential flow about two-dimensional hydrofoils*, J. Fluid Mech. **28**(1967) Part 1, pp. 113.

[3.20] D. Gilberg and J. Serrin: *Free boundaries and jets in the theory of cavitilion*, J. Math. and Phys. **29**(1950) 1.

[3.21] H. Glauert: *The elements of aerofoil and airscrew theory*, Cambridge Univ. Press 1926 (2nd ed. 1947).

[3.22] H. Glauert: *The effect of compressibility on the lift of an aerofoil*, Proceed. Roy. Soc., A, **118**(1928) 113.

[3.23] N.D. Halsey: *Potential flow analysis of multielement airfoil using conformal mapping*, AIAA Journ **17**(1979) 1281.

[3.24] D. Homentcovschi: *Non-equilibrium flow of an inviscid gas past a thin profile*, ZAMM, **57**(1977) 461.

[3.25] W.H. Isay: *Kavitation*, Schiffahrst-Verlag "Hansa" Hamburg 1981.

[3.26] D.C. Ives: *A modern look at conformal mapping including multiple connected regions*, AIA Journal **14**(1976) 1006.

[3.27] R.M. James: *Analytical studies of two-element airfoil systems*, Douglas Aircr. Comp. Rep MDC JS 831, 1974;

A new look at two-dimensional incompressible airfoil theory, Dougl. A. Comp. Rep MDC JO 918/01.1971.

[3.28] D.S. Jones: *Note on the steady flow of a fluid past a thin aerofoil*, Quart. J. Math. **6**(1955) 4.

[3.29] M.J. Lighthill: *A Mathematical method of cascade design*, Rep. Memor. aero. Res. Coun. London. 2104(1945); *A new method of two-dimensional aerodynamic design*, Idem 2112(1945).

[3.30] M.J. Lighthill: *A new approach to thin airfoil theory*, Aeronaut. Quart. **3**(1951) 193.

[3.31] L.C. Malvard: *The use of rheo-electrical analogies in certain aerodynamic problems*, J.R. aero Soc. **51**(1947) 734; *Recent develop-*

ments in the method of the rheo-electric analogy applied to aerodynamics, J. Aero. Sci. **24**(1957) 321.

[3.32] N. Marcov: *Mişcări potenţiale ale fluidelor ideale şi incompresibile în prezenţa a două obstacole circulare*, St. Cerc. Mec. Apl. **55**(1996) 29.

[3.33] M. Mokry, D.J. Peake, A.J. Bowker: *Wall interference on two-dimensional supercritical airfoils, using wall pressure measurement to determine the porosity factors for tunnel floor and ceiling*, High Speed Aerodynamics Laboratory, L.H. Ohman.

[3.34] M.M. Munk: *General theory of thin wing sections*, Rep. Nat. Adv. Com. Aero., Washington **142**(1922).

[3.35] T. Nishiyama: *Hydrodynamical investigation in the submerged hydrofoil*, ASNE pI, August 1958, pp. II Nov. 1958, pp. III Febr. 1959.

[3.36] A.N. Pancenkov, V.M. Ivcenko: *Zadaci i metodii ghidro - dinamiki podvodnih krilov i vintov*, Naukova Dumka, Kiev, 1966.

[3.37] A. Plotkin, C. Kennel: *Thickness-induced lift on a thin airfoil in ground effect*, AIAA Journ. **19**(1981) 1984.

[3.38] A. Plotkin: *Thickness and camber effects in slender wing theory*, AIAA Journ **21**(1983) 1755.

[3.39] A. Plotkin, S.S. Doubele: *Slender wing in ground effect*, AIAA Journ, **26**(1988) 493.

[3.40] L. Prandtl: *Über Stromungen deren Geschwindegkeiten mit der Schallgeschwindigkeit vergheichbar sind*, J. Aero. Res. Inst., Tokyo University **63**(1930) 14.

[3.41] E. Reissner: *On the general theory of thin airfoils for non-uniform motion*, Tech. Notes NACA, Washington **946**(1944).

[3.42] M. Shiffman: *On free boundaries of an ideal fluid I*, Comm. Pure & Appl. Math. 1948, pp. 89.

[3.43] M. Shiffman: *On free boundaries of an ideal fluid II*, Comm. Pure & Appl. Math. 1949, pp. 1.

[3.44] S. Tomotika, T. Nagamiya, Y. Takenouti: *The lift on a flat plate placed near a plane wall with special reference to the effect on the ground upon the lift of a monoplane aerofoil*, Rep. Aero Res. Inst. Tokyo **97**(1933).

[3.45] S. Tomotika, Z. Hasimoto, K. Urano: *The forces adding on an aerofoil of approximate Joukowski type in a stream bounded by a plane wall*, Q.J.Mech. Appl. Math., **4**(1951) 289.

[3.46] E.O. Tuck: *Steady flow and static stability of airfoils in extreme ground effect*, J. Engng. Math. **15**(1981) 89.

[3.47] J.S. Uhlman, Jr.: *The surface singularity method appl. to partially cavitating hydrofoils*, 5th Lips Propeller Sypp. Drunen, The Netherlands, May 1983.

[3.48] B.R. Williams: *An exact test case for the plane potential flow about two adjacent lifting airfoils*, RAE Tech. Rept. 71,1971.

[3.49] J. Weissinger: *Ein Satz über Fourierreihen und seine Anwendung auf die Tragflügeltheorie*, Math. Zeitschr. **47**(1941) 16.

[3.50] L.C. Woods: *Generalized aerofoil theory*, Proc. Roy. Soc. A, **238**(1956) 358.

Chapter 4

[4.1] P. Bassanini, C.M. Casciola, M.R. Lancia and R. Piva: *A boundary integral formulation for the kinetic field in aerodynamics*, Part.I. Mathematical Analysis, Eur.J.Mech.B/Fluids, **10**(1990) 605;
Part.II. Application to unsteady 2 D flows, idem **11**(1991) 69.

[4.2] G. Bindolino, P. Mantegazza and L. Visintini: *A comparison of panel methods for potential flow calculation*, Proc. of 8th Nat. Sympos. AIDAA (1985) 59.

[4.3] C.V. Camp: *Direct evaluation of singular boundary integrals in 2D biharmonic analysis*, Int.J.Numer Methods Engng. **35**(1992) 2067.

[4.4] A. Carabineanu, A. Dinu: *The study of the incompressible flow past a smooth obstacle in a channel by the boundary element method*, Rev. Roum. Sci. Tech.-Méc. Appl. **38**(1993) 601.

[4.5] V. Cardoş: *A method with boundary elements to the study the subsonic flows*, ZAMM **75**(1995) 483.

[4.6] L. Dragoş: *Boundary Element Methods (BEM) in the theory of thin airfoils*, Rev. Roum. Math. Pures et Appl., **34**(1989) 523.

[4.7] L. Dragoş, A. Dinu: *Application of the Boundary Element Method to the thin airfoil theory*, AIAA Journ. **28**(1990) 1822-1924.

[4.8] L. Dragoş, A. Dinu: *Application of the Boundary Integral Equations Method to subsonic flow past thin smooth airfoil in wind tunnel*, Proceed. of the 2nd Nat. Conf. on Boundary and Finite Elements 13-15 May 1992 Sibiu, pp. 75-84.

[4.9] L. Dragoş, A. Dinu: *The application of the Boundary Integral Equations Method, to subsonic flow with circulation past thin airfoils in wind tunnel*, Acta Mechanica **103**(1994) 17-30.

[4.10] L. Dragoş, A. Dinu: *A direct Boundary Integral Equations Methods to subsonic flow with circulation past thin airfoils in ground effect*, Comput. Methods Appl. Mech. Engrg., **121**(1995) 163.

[4.11] L. Dragoş, A. Dinu: *A direct Boundary Integral Method for the two-dimensional lifting flow*, Arch. Mech., **47**(1995) 813.

[4.12] L. Dragoş, N. Marcov: *Solutions of the boundary integral equations in the theory of incompressible fluids*, Analele Univ. Bucureşti, **46**(1997) 111-119.

[4.13] L. Dragoş, A. Dinu: *A boundary regularized integral method for plane potential flow past airfoil configurations* (in press).

[4.14] J.P. Giesing: *Potential flow about two-dimensional airfoils*, Douglas Aircraft Comp. Rep. No LB 31946, May 1968.

[4.15] C. Griello, G. Iannelli and L. Tulumello: *An alternative boundary element method approach to the 2−D potential flow problem around airfoils*, Eur.J.Mech.B/Fluids **9**(1990) 527.

[4.16] J. Hess, A.M.O. Smith: *Calculation of potential flows about arbitrary bodies*, Progress in Aeronautical Sci., **8**, Ed.D. Küchemann, Pergamon Press, N.Y. 1967.

[4.17] J. Katz, A. Plotkin: *Low-speed aerodynamics, From wing theory to panel methods*, McGrow-Hill Inc., N.Y., 1991.

[4.18] S.V. Kozlov, I.K. Lifanov, A.A. Mikhailov: *A new approach to mathematical modelling of flow of ideal fluid around bodies*, Sov. J. Numer. Anal. Math. Modelling, **6**(1991) 209.

[4.19] M. Mokry: *Integral equation method for subsonic flow past airfoils in ventilated wind tunnels*, AIAA J.**13**(1975) 47.

[4.20] L. Morino, C.C. Luo: *Subsonic potential aerodynamics for complex configurations*, A general theory, AIAA J.**12**(1974) 191.

[4.21] S.W. Zhang, Y.Z. Chen: *New integral equation in the thin airfoil problem*, Acta Mech.**87** (1991) 123.

Chapter 5

[5.1] H. Ashley, S. Windnall and M.T. Landahl: *New directions in lifting surface theory*, AIAA Journ., **3**(1965) 3.

[5.2] S.M. Belotserkovskii: *The theory of thin wings in subsonic flow* (translated from Russian), Plenum Press, New York 1967.

[5.3] J. Boersma: *Note on the lifting-surface problem for a circular wing in incompressible flow*, Q.J. Mech., Appl. Math., **42**(1989) 55.

[5.4] Von M. Borja, H. Brakhage: *Uber die numerische Behandlung der Tragflächengleichung*, Z. Angew. Math. u. Mech., **47**(1967) T.102.

[5.5] M. Borja, H. Brakhage: *Zur numerischen Behandlung der Tragflächengleichung*, Zeitschr. f. Flügwissensch. **16**(1968) 349.

[5.6] P. Cocârlan: *Cercetări asupra teoriei aripii portante*, teză de doctorat, Universitatea din Bucureşti, 1971.

[5.7] L. Dragoş: *Method of fundamental solutions. A novel theory of lifting surface in subsonic flow*, Arch. of Mechanics, **35**(1983) 579.

[5.8] L. Dragoş: *Subsonic flow post thick wing in ground effect. Lifting line theory*, Acta Mechanica, **82**(1990) 49.

[5.9] L. Dragoş: *Arbitrary wings of low aspect ratio in subsonic flow*, Z. Angew, Math. u. Phys, **38**(1987) 648.

[5.10] L. Dragoş, A. Carabineanu: *New quadrature formulae for the integration of the lifting surface equation*, (in press).

[5.11] L. Dragoş, A. Carabineanu: *A numerical solution for the equation of the lifting surface in ground effects*, Commun. Numer. Meth. Engng. 2002, **18** 2002, 177.

[5.12] H. Flax, H.R. Lawrence: *The aerodynamics of low aspect ratio wings and wing body combinations*, Proc. of the 3rd Anglo-Amer. Aero. Conf. R. Aero. Soc. (1951) 363.

[5.13] J.L. Hess: *The problem of three-dimensional lifting potential flow and its solutions by means of surface singularity distribution*, Comp. Meth. in Appl. Mech. & Engng., **4**(1974) 283.

[5.14] J.L. Hess: *Review of integral equation techniques for solving potential flow problems with emphasis on the surface-source method*, Comp. Meth. Appl. Mech. & Engng., **5**(1975) 145.

[5.15] P.T. Hsu: *Some recent developments in the flutter analysis of low - aspect - ratio wings*, Proc. Nat. Spec. Meeting on Dynamics and Aeroelasticity, Forth Worth, Texas 1958, pp. 7-26.

[5.16] D. Homentcovschi: *Aerodynamique stationaire linéarisée*, Arch. of Mechanics, **27**(1975) 325.

[5.17] R.T. Jones: *Properties of low-aspect-ratio pointed wings at speed below and above the speed of sound*, NACA Rep. **835**(1446).

[5.18] R.T. Jones: *Some recent developments in the aerodynamics of wings for high speeds*, Z. Flugwiss, **4**(1956) 257.

[5.19] P.F. Jordan: *On lifting wings with parabolic tips*, Z. Angew., Math. u. Mech., **54**(1974) 463.

[5.20] M. Landahl and V.J.E. Stark: *Numerical lifting surface theory-Problems and Progress*, AIAA Journ, **6**(1968) 2049.

[5.21] H.R. Lawrence: *The lift distribution on low aspect ratio wings at subsonic speeds*, J. Aero. Sci., **18**(1951) 683;

The aerodynamic characteristics of low-aspect-ratio wing-body combinations in steady subsonic flow, idem **20**(1953) 541.

[5.22] K.W. Mangler, B.F.R. Spencer: *Some remarks on Multhopp's subsonic lifting-surface theory*, Aero. Res. Counc. R. & M. 2926 (1952).

[5.23] L. Morino, K.C. Chang: *Subsonic potential aerodynamic for complex configuration: a) general theory*, AIAA Journ., **12**(1974) 191.

[5.24] H. Multhopp: *Methods for calculating the lift distribution of wings* (subsonic lifting - surface theory), Aero. Res. Coun. R & M 2884 (1950,1955).

[5.25] J.N. Newman: *Analysis of small-aspect-ratio. Lifting surfaces in ground effect*, J. Fluid Mech., **117**(1982) 305.

[5.26] L. Prandtl: *Theorie des Flugzengtrag flügels im zusammen - drückbaren Medium*, Luftfahrtforschung **13**(1936) 313.

[5.27] L. Prandtl: *Beitrag zur theorie der tragenden Fläche*, Z. Angew., Mat. u. Mech., **16**(1939) nr.6.

[5.28] J.W. Purvis, J.E. Burkhalter: *Simplified solution of the compressible lifting surface problems*, AIAA Journ., **20**(1982) 589.

[5.29] E. Reissner: *Note on the theory of lifting surface*, Proc. Nat. Acad. Sci., USA, **35**(1949) 208.

[5.30] A. Rosen, A. Isser: *Recent studies on the lifting surface theory*, Computers Math. Appl. **17**(1989) 1115.

[5.31] G.H. Sounders: *Aerodynamic characteristics of wings in ground proximity*, Canad. Aeronaut a. Space Journ. 1965, pp. 185.

[5.32] E. Truckenbrodt: *Das Geschwindigkeits potential der tragenden Fläche bei Inkompressiblen Strömung*, Z. Angew. Math. u. Mech., **33**(1953) 165.

[5.33] A.I. Van de Vooren: *An approach to lifting surface theory*, Rep. Nat. Luchtv. Lab., Amsterdam, F 129(1953).

[5.34] J. Weber: *The calculation of the pressure distribution of thick wings of small aspect ratio at zero lift in subsonic flow*, Rep. Memor. Aero. Res. Counc., London 17522(1954).

[5.35] J. Weissinger: *The lift distribution of swept - back wings*, NACA TN 1120(1947).

[5.36] J. Weissinger: *Theorie des Tragflügels bei stationären Bewegung in reibungslesen Inkompressiblen Medien*, in Handbuck der Physik, bd 8/2, Springer-Verlag, Berlin, 1963.

[5.37] S. Widnall, T.M. Barows: *An analytic solution for two and three - dimensional wing in ground effect*, J. Fluid Mech, **41**(1970) 769.

[5.38] I.T. Wu: *Hydrofoils of finite span*, Math. Phys., **33**(1954).

Chapter 6

[6.1] E. Carafoli: *Sur les caractéristiques des ailes trapezoidales et rectangulares*, Com. Academiei de Ştiinţe din România 1943; *Théorie des ailes monoplanes d'envergure fine*, Ann. de l'Académie Roumaine 1945.

[6.2] H.K. Cheng: *Lifting-line theory of oblique wings*, AIAA Journ., **16**(1978) 1211.

[6.3] H.K. Cheng, R. Chow, R.E. Melnik: *Lifting-line theory of swept wings based on the full potential theory*, Z. Angew Math. u. Phys. **32**(1981) 481.

[6.4] L. Dragoş: *Asupra ecuaţiei integro - diferenţiale a lui Prandtl*, Com. Acad. R.P.Române, **8**(1958) 451.

[6.5] L. Dragoş: *Integration of Prandtl's equation by the aid of quadrature formulae of Gauss type*, Quart. Appl. Math. **52**(1994) 23.

[6.6] L. Dragoş: *A colocation method for the integration of Prandtl's equation*, Z. Angew. Math. u. Mech., **74**(1994), 289

[6.7] I. Filimon: *On Prandtl's integro - differential equation*, Bul. Ştiinţific Acad. R.P.R., **9**(1957) 981.

[6.8] Jean-Luc Guermond: *A generalized lifting-line theory for curved and swept wings*, J. Fluid Mech., **211**(1990) 497.

[6.9] D. Homentcovschi: *On the deduction of Prandtl's equation*, Z. Angew. Math. u. Mech. **57**(1977) 115.

[6.10] Th von Kármán, H.S. Tsien: *Lifting-line theory for a wing in non-uniform flow*, Q.Appl. Math. **3**(1945) 1.

[6.11] H.R. Lawrence, A.H. Flax: *Wing-body interference at subsonic and supersonic speeds. Survey and new developments*, J.Aero. Sci., **21**(1954) 289.

[6.12] J.H. de Leeuw, W. Echans, A.I. van de Veoren: *The solution of the generalized Prandtl equation for swept wings*, Rep. Nat. Luchtv Lab., Amsterdam F. 156(1954).

[6.13] J. Legras: *Méthodes et techniques de l'analyse numérique*, Dunod, Paris 1971.

[6.14] I. Lighthill: *Hydromechanics of aquatic animal propulsion*, An, Rev. Fluid Mech., **1**(1969) 413; *Aquatic animal propulsion of high hydromechanical efficiency*, J. Fluid Mech., **44**(1970) 265.

[6.15] I. Lotz: *Berechnung der Auftriebsverteilung beliebig geformter Flügel*, Z. Flugtech, **22**(1931) 189; *The calculation of the potential flow past airship bodies in yaw*, Tech. Memor. NACA, Wachington 675(1931).

[6.16] L.G. Magnaradze: *Asupra unei noi ecuaţii integrale în teoria aripii de avion* (în lb. rusă) Coman. AN Gruz SSSR **3**(1942) 503.

[6.17] H. Multhopp: *Die Berechnung der auftriebsverteilung von Tragflügeln*, Luftfahrtforschung, **15**(1938) 153, Translation ARC 8516.

[6.18] T. Nishiyama: *Lifting - line theory of the submerged hydrofoil at finite span*, ASNE p. I, Aug. 1959, p. II, Nov. 1959, p. III, Febr. 1960, p. IV, May 1960.

[6.19] A. Pflüger: *Ein Spannungsgleichnis zum Problem der tragenden Linie*, Z. Angew. Math. u. Mech., **25-27**(1947) 177.

[6.20] A. Plotkin, C.H. Tan: *Lifting-line solution for a symmetrical thin wing in ground effect*, AIAA Journ **24**(1986) 1193.

[6.21] L. Prandtl: *Tragflügeltheorie* I, II, Mitteilung Nachr. Ges. Wiss., Göttingen, 1918/1919.

[6.22] S. Prössdorf, D. Tordella: *On an Extension of Prandtl's Lifting Line Theory to Curved Wings*, Impact of Computing in Science and Engineering, **3**(1991) 192.

[6.23] M. Schleiff: *Untersuchung einer linearen singulären Integro-differential gleichung der Tragflügeltheorie*, Wiss. Z. Univ. Halle **17**(1968) 44; *Über Naherungsverfahren zur Lösung einer singularen linearen Integro-differentialgleichung*, Z. Angew. Math. u. Mech., **48**(1968) 477.

[6.24] H. Schmidt: *Strenge Lösungen zur Prandtlschen Theorie der tragenden Linie*, Z. Angew. Math. u. Mech., **17**(1937) 101.

[6.25] K. Schröder: *Über die Prandtlsche Integrodifferentialgleichung der Tragflügeltheorie*, Abhandl Preuss Akad. Wiss., Math.-Naturwiss Klasse 1939.

[6.26] P.D. Sclavounes: *An unsteady lifting-line theory*, J. Engng Math., **21**(1987) 201.

[6.27] E. Trefftz: *Zur Prandtlschen Tragflächen-theorie*, Math. Annalen **82**(1921) 306; Z. Angew Math. u. Mech. **18**(1938) 12.

[6.28] I.N. Vekua: *On Prandtl's integro - differential equation* (in Russian), Prikl. Mat., Mekh., **9**(1945) 143.

[6.29] A.I. van de Vooren: *The generalization of Prandtl's equation for yawed and swept wings*, Rep. Nat. Luchtv Lab., Amsterdam, F 121(1952).

[6.30] J. Weissinger: *Über eine Erweiterung der Prandtlschen Theorie der tragonden Linie*, Math. Nach., **2**(1949) 46.

[6.31] J. Weissinger: *Über Integrodifferentialgleichungen von Typ der Prandtlschen Tragflügelgleichung*, Math. Nachr., **3**(1950) 316.

[6.32] D.E. Williams: *The effect of wing - tailplane aerodynamic interaction on taie flutter*, Techn. Rep. ARC R & M nr. 3065.

[6.33] J. De Young, C.W. Harper: *Theoretical symmetrical span loading at subsonic speeds for wings having arbitrary planform*, NACA Rep. 921(1948).

Chapter 7

[7.1] J.H. Argyris and D.W. Scharph: *Two and three-dimensional potential flow by the method of singularities*, Aero.J.Roy Aero Soc. **73**(1969) 959.

[7.2] L. Dragoş, A. Dinu: *Application of the boundary integral equation method to subsonic flow past bodies and wings*, Acta Mechanica **86(1)**(1991) 83.

[7.3] L. Dragoş, A. Dinu: *A direct boundary integral method for the three-dimensional lifting flow*, Comput. Methods Appl. Mech. Engng. **127**(1995) 357.

[7.4] L. Dragoş, N. Marcov: *Solutions of the boundary integral equations in the theory of incompressible fluids*, Analele Univ. Buc. **XLVI**(1997) 111-119.

[7.5] L.J. Gray, E. Lutz: *On the treatment of corners in the boundary element method*, J.Comput.Appl.Math.**32**(1990) 369.

[7.6] E. Grodtkjaer: *A direct integral equation method for the potential flow about arbitrary bodies*, The Danish Center for Appl. Mathematics and Mechanics, Rep.21, Dec.1971.

[7.7] F. Hartmann: *Introduction to boundary elements, Theory and applications*, Springer - Verlag, Berlin, 1985.

[7.8] Heshan Liao, Zhixin Xu: *Method for direct evaluation of singular integral in direct BEM*, Int.J.Numer Methods Engng, **35**(1992) 1473.

[7.9] J.L. Hess, A.M.O. Smith: *Calculation of nonlifting potential flow about arbitrary three-dimensional bodies*, J.Ship.Res. **8**(1964) 22.

[7.10] J.L. Hess, A.M.O. Smith: *Calculation of potential flows about arbitrary bodies*, Progress in Aeronautical SCI., **8**(1967) Pergamon Press N.Y. Ed.D.Küchemann.

[7.11] J.L. Hess: *The problem of three-dimensional lifting flow and its solution by means of surface singularity distribution*, Comput. Methods Appl.Mech.Engrg., **4**(1974) 183.

[7.12] J. Katz, A. Plotkin: *Low-speed aerodynamics. From wing theory to panel methods*, Mc Graw-Hill Inc., N.Y., 1991.

[7.13] S.V. Kozlov, L.K. Lifanov and A.A. Mikhailov: *A new approach to mathematical modelling of flow of ideal fluid around bodies*, Sov. J. Numer. Anal. Math. Modelling. **6**(1991) 209.

[7.14] R.I. Lewis: *Inverse method for the design at bodies of revolution by boundary integral modelling*, Proc. Inst. Mech. Engng., **205**(1991) 91.

[7.15] R. Srivastava, D.N. Contractor: *Efficient evaluation of integrals in 3D BEM using linear shape functions over plane triangular elements*, Appl.Math.Model. **16**(1992) 282.

[7.16] James S. Uhlman, Jr.: *The surface singularity method applied to partially cavitating hydrofoils*, 5th Lips Propeller Symporium, Drunen - The Netherlands 1983.

[7.17] Wu Zhang, Huan-Ran Xu: *General and effective way for evaluating the integrals with various order of singularity in the di-*

rect boundary element method, Int. J. Numer. Methods Engng. **28**(1989) 2059.

Chapter 8

[8.1] J. Ackeret: *Über Luftkräfte auf Flügel, die mit grösseres als Schallgeschwindigkeit bewegt werden*, Zeitschr. f. Flugtechnik Motorluftschiffahrt, **16**(1925), 72.

[8.2] L. Berechet: *Lifting surface theory in supersonic flow*, Lucrare de licenţă la Facultatea de Matematică, Universitatea din Bucureşti, 1984.

[8.3] A. Busemann: *Infinitesimale hegelige Überschallstromung*, Schriften der Deutschen Akademie dr. Luftfahrtfaschung 7 B(1943) pp 105-122.

[8.4] V. Cardoş: *Sur les écoulements supersoniques autour des obstacles coniques aplatis ou de faible ouverture*, Rev. Roum. Math. Pures et Appl., **27**(1982) 285.

[8.5] V. Cardoş: *Asupra mişcărilor supersonice cu undă de şoc*, St. Cerc. Mec. Apl., **52**(1993) 355.

[8.6] V. Cardoş: *Le problème aux limites simplifié pour les mouvements coniques supersoniques*, Rev. Roum. Sci. Techn. Méc. Appl., **42**(1997) 59.

[8.7] L. Dragoş: *The method of fundamental solutions in aerodynamics II. Applications to the theory of lifting surfaces in supersonic flow*, Arch. Mechanics **37**(1985) 221-230.

[8.8] L. Dragoş, A. Carabineanu: *The supersonic flow past a thin profile in ground and tunnel effects*, ZAMM, **82**(2002) 9, 649-652.

[8.9] C. Evvard: *Use of source distributions for evaluating theoretical aerodynamics of thin finite wings at supersonic speeds*, NACA Rep. 951(1950).

[8.10] C. Ferrari: *Interaction Problems* in vol. VII of Princeton Series in High Speed Aerodynamic and Jet Propulsion (A.F. Donavan and H.R. Lawrence) Princ. Univ. Press, Princeton 1957.

[8.11] P. Germain: *La théorie générale des mouvements coniques et ses applications à l'aérodynamique supersoniques*, Office Nat. Études et Rech Aéronaut. **34**(1949).

[8.12] P. Germain: *Sur le minimum de trainée d'une aile de form en plan donnée* Compt. Rend. 244(1957) 2691.

[8.13] P. Germain: *Aile symmetrique à portance nulle et de volum donnée réalisant le minimum de trainée en ecoulement supersonique*, Compt. Rend. **244**(1957) 2691.

[8.14] H. Glauert: *Theory of thin airfoils*, British A.R.C., R & H 910(1924) sau *The elements of aerofoil and airscrew theory, sec. ed.* Cambr. Univ. Press 1947.

[8.15] M.A. Heaslet, H. Lomax: *Supersonic and transonic small perturbation theory*, in General theory of High Speed Aerodynamics ed. by W.R. Sears. Oxford University Press, London, 1955.

[8.16] D. Homentcovschi: *Aerodynamique stationnaire linéarisée II. Supersonique*, Arch. Mechanics, **29**(1977) 41-51.

[8.17] C. Iacob: *Validity conditions in plane supersonic lineal aerodynamics*, Studii şi Cercet. Mat. **32**(1980) 649.

[8.18] C. Iacob: *Teoria aripii unghiulare la viteze supersonice*, Analele Academiei R.P.R. T13, Memoriul 15.

[8.19] T. von Kármán: *Supersonic aerodynamics*, Principles and Appl. J. Aeron. Sci. **14**(1947) 373-409.

[8.20] E.A. Krasilscicova: *Wing of finite average in compressible flow* (in Russian), Fizmatghiz, Moscow, 1952.

[8.21] M.N. Kogan: *On bodies of minimum drag in supersonic gas stream*, Prikl. Mat. Mekh., **21**(1957) 207.

[8.22] P.A. Lagerstrom: *Linearized supersonic theory of conical wings*, NACA Techn. Note 1685, 1948.

[8.23] H.W. Liepmann, A. Roshko: *Elements of gas dynamics*, John Wiley, New York, (1957).

[8.24] M.J. Lighthill: *On boundary layer and upstream influence II: Supersonic flow without separation*, Proc. Roy. Soc. London, Ser A, **217**(1953) 478.

[8.25] J.M. Lighthill: *The conditions behind the trailing edge of the supersonic aerofoil*, ARC Rep and Memo. 1930, 1941.

[8.26] H. Lomax, M.A. Heaslet: *Recent developments in the theory of wing-body wave drag*, Journ Aero. Sci., **23**(1956) 1061-1074.

[8.27] W. Miles: *Potential theory of unsteady supersonic flow*, Camb. Univ. Press, London 1959.

[8.28] H. Poritsky: *Homogeneous harmonic functions*, Quart. Appl. Math., **6**(1949) 379-390.

[8.29] A.E. Puckett: *Supersonic wave drag of thin airfoils*, J. Aero. Sci. **13**(1946) 475-484.

[8.30] A. Robinson: *On source and vortex distributions in the linearized theory of steady supersonic flow*, Quart. J. Mech. Appl. Math., **1**(1948) 402.

[8.31] D.A. Spence: *Prediction of the characteristics of two-dimensional airfoils*, J. Aer. Sci., **21**(1954) 577-587.

[8.32] Trinh Dinh An: *Validity conditions in supersonic flow past a rectangular wing*, St. Cerc. Mat., **33**(1981);

The validity conditions in three-dimensional supersonic lineal aerodynamics, Rev. Roum., Math. Pures et Appl., **29**(1984) 215;

Sur les conditions de validité physique en écoulement supersonique autour d'un obstacle conique de faible ouverture, idem, 291.

[8.33] G.N. Ward: *Supersonic flow past thin wings I. General theory*, Quart. J. Mech. Appl. Math., **2**(1949) 136-152.

[8.34] G.N. Ward: *Linearized theory of steady high - speed flow*, Cambridge University Press, 1955.

[8.35] G.W. Ward: *On the minimum drag of thin lifting bodies in steady supersonic flow*, British. A.R.C. Rep. 18(1956).

Chapter 9

[9.1] F. Bauer, P. Garabedian & D. Korn: *A theory of supercritical wing section with computer programs and examples*, Springer, 1972.

[9.2] Y.T. Chan: *An asymptotic analysis of transonic wind - tunnel interference based on the full potential theory*, ZAMP **36**(1985) 89, 85.

[9.3] J.D. Cole: *Modern developments in transonic flow*, SIAM J. Appl. Math., **24**(1975) 763, **27**(1978).

[9.4] J.D. Cole: *Drag of finite wedge at high subsonic speeds*, J.Math. and Phys **30**(1951) 79-93.

[9.5] J.D. Cole, L.P. Cook: *Transonic aerodynamics*, North - Holland, Amsterdam, 1986.

[9.6] J.D. Cole and A.F. Messiter: *Expansion procedures and similarity laws for transonic flow*, Z AMP **8**(1957) 1-25.

[9.7] J.C. Crown: *Calculation of transonic flow over thick airfoils by integral methods*, AIAA Journ **6**(1968), no. 3, 413-423.

[9.8] L. Dragoş: *Theory of three - dimensional wing and lifting line theory in transonic steady flow*, Rev. Roum. Sci. Math. Pures et Appl., **32**(1987) 137-154.

[9.9] L. Dragoş: *Theory of three - dimensional wing and lifting line theory in transonic flow*, Preprint Ser in Math., INCREST nr. 62/1984.

[9.10] T. Evans: *An approximate solution for two dimensional transonic flow*, Proc. Cambr. Phil. Soc. **61**(1965) 573.

[9.11] C. Ferrari, F. Tricomi: *Aerodinamica transonica*, Rome, Edit. Cremonesse 1962, English translation: *Transonic Aerodynamics*, Acad. Press N.Y. 1968.

[9.12] F.I. Frankl: *Theoretical research on the wing of infinite span at sonic speed*, Dokl. Akad. Nauk. SSSR, **57**(1947) 661-664.

[9.13] P.R. Garabedian & D.G. Korn: *Analysis of transonic airfoils*, Comm. Pure Appl. Math. **24**(1971) 841.

[9.14] K.G. Guderley: *Theorie Schallnaher Strömungen*, Springer-Verlag 1957;

Theory of transonic flow, Internat. Series of Monographs in Aeron. a. Astron., Division II Aerodynamics, vol. 3, Pergamon, New York 1962.

[9.15] K.G. Guderley: *Theorie schallnaher Strömungen*, Springer, Berlin, 1957.

[9.16] K.G. Guderley and Yoshihara: *The flow over a wedge profile at Mach number 1*, J. Aero Sci., **17**(1950) 723 - 735.

[9.17] M.A. Heaslet, J.R. Spreiter: *Three-dimensional transonic flow theory applied to slender wings and bodies*, NACA Rep. 1318(1956).

[9.18] M.A. Heaslet: *Three-dimensional transonic flow theory applied to slender wings and bodies*, NACA Rep 1318(1956).

[9.19] D. Homentcovschi: *An introduction to BEM by integral transforms*, Preprint Series in Math. INCREST **25** 1987.

[9.20] D. Homentcovschi: *Integral equation for the lifting profile in steady transonic flows*, Preprint Series in Mathematics INCREST, 63, 1983.

[9.21] D. Homentcovschi: *Steady linearised aerodynamics III Transonic*, Arch., Mech., **42**(1990) 3.

[9.22] I. Hosokawa: *A refinement of linearised transonic flow theory*, J. Phys. Soc. Japan **15**(1960) 149.

[9.23] I. Hosokawa: *Transonic flow past a wavy wall*, J. Phys. Soc. Japan **15**(1960) 2080 - 2086.

[9.24] I. Hosokawa: *Theoretical prediction of the pressure distribution on a non - lifting, thin symmetrical aerofoil at various transonic speeds*, J. Phys Soc. Japan, **16**(1961) 546 - 558.

[9.25] I. Hosokawa: *Unified formalism of the linearized compressible flow fields*, Quart Appl. Math., **22**(1964) 133-142.

[9.26] Th. von Kárman: *The similarity low of transonic flow*, J. Math. Phys., **26**(1947), 182-190.

[9.27] M.T. Landahl: *Unsteady transonic flow*, Pergamon, 1961.

[9.28] K. Leelavathi, N.R. Subramanian: *Pressure distribution in inviscid transonic flow past axisymmetric bodies* $(M_\infty = 1)$, AIAA Journ **7**(1969) 1362 - 1363.

[9.29] H.W. Liepmann, A.E. Bryson: *Transonic flow past wedge sections* J. Aero. Sci. **17**(1950) 745.

[9.30] A.R. Manwell: *The hodograph equations*, Oliver and Boyd, Edinburgh, 1971.

[9.31] A.R. Manwell: *The Tricomi equation with applications to the theory at plane transonic flow*, Pitman Adv. Publ. Pr. San Francisco, 1979.

[9.32] C. Masson, I. Paraschivoiu: *Integral method for transonic flow*, Can. Aeronaut. Space J. **36**(1990) 18.

[9.33] C.S. Morawetz: *The mathematical approach to the sonic barrier*, Bull. Amer. Math. Soc., **6**(1982) 127 - 145.

[9.34] T.H. Moulden: *Fundamentals of transonic flow*, John Wiley & Sons, New York, 1984.

[9.35] E.M. Murmann, J.D. Cole: *Calculation of plane steady transonic flows*, AIAA Journ. **9**(1971) 114.

[9.36] P. Niyogi: *Integral equation method in transonic flow*, Springer - Verlag, Berlin 1982.

[9.37] D. Nixon: *Calculation of transonic flows using and extended integral equation method*, AIAA Journ., **15**(1977) 295.

[9.38] D. Nixon: *The transonic integral equation method with curved shock waves*, Acta Mechanica, **32**(1979) 141.

[9.39] D. Nixon(ed.): *Transonic aerodynamics*, Amer. Inst. Aeron. a. Astron., New York, 1982.

[9.40] D. Nixon: *Observations on the occurence of multiple solutions in transonic potential theory*, Acta Mechanica, **68**(1987) 43.

[9.41] H. Norstrud: *Transonic flow past lifting wings*, AIAA Journ. **11**(1973) 754.

[9.42] W. Ogana: *Numerical solutions for subcritical flow by a transonic integral equation method*, AIAA J. **15**(1977) 444.

[9.43] W. Ogana: *Derivation of an integral equation for three dimensional transonic flows*, AIAA J., **17**(1979) 305.

[9.44] W. Ogana: *Transonic integro-differential and integral equations with artificial viscosity*, Eng. Anal. Boundary Elem. **6**(1989) 3.

[9.45] W. Ogana: *Boundary element solution of the transonic integro - differential equation*, Eng. Anal. Boundary Elem.

[9.46] W. Ogana: *Analysis of transonic integral equation*, Part II Boundary element methods, AIAA J. **28**(1990) 3.

[9.47] W. Ogana: *Solution of the transonic integro - differential equation using a decay functions*, Appl. Math. Model **14**(1990) 30.

[9.48] W. Ogana, J.R Sprieter: *Derivation of an integral equation for transonic flows*, AIAA J., **15**(1977) 281.

[9.49] K. Oswatitsch: *Die Geschwindigkeitsverteilung bei localen Überschallgebeiten an flachen Profilen*, ZAMM, **30**(1950), 17-24.

[9.50] K. Oswatitsch, S.B. Berndt: *Aerodynamic similarity of axisymmetric transonic flow around slender bodies*, KTH Aero TN 15, Royal Inst. Technology, Stockholm, 1950.

[9.51] K. Oswatitsch, F. Keune: *Ein Äquivalenssatz für nichtangestellte Flügel Kleiner Spannweite in Schallnaher Strömung*, Z. Flug wiss., **3**(1955) 29 - 46.

[9.52] K. Oswatitsch, K. and F. Keune: *Ein Äquivalenssatz für nichtangestellte Flügel Kleiner Spannweite in Schallnaher Strömung*, Zeit. für Flug wissenschoften, **3**(1955) 29.

[9.53] L.V. Ovsiannikov: *Equation of the transonic flow of the gases*, Vestnik LGY, 1952, nr.6, 47 - 54.

[9.54] L. Pamela Cook: *Lifting - line theory for a swept wing at transonic speeds*, QAM, **37**(1979) 177.

[9.55] L. Pamela Cook: *Lifting - line theory for transonic flow*, SIAM J. Appl. Math. **35**(1978) 209 - 228.

[9.56] M. Schubert, M. Schleiff: *Über zwei Randwertprobleme des inhomogenen Systems der Cauchy - Rimanschen Differential - gleichungen mit einer Anwendung and ein Problem der stationären schallnahen Strömmung*, ZAMM, **49**(1969) 621.

[9.57] J.R. Spreiter: *On the application of transonic similarity rules to wings of finite span*, NACA Rep. 1153, (1953).

[9.58] J.R. Spreiter: *On alternative forms for the basic equations of transonic flow theory*, J. Aero., Sci., **21**(1954) pp 70-72, Errata: J. Aero Sci **21**(1954) 360.

[9.59] J.R. Spreiter, A.Alksne: *Theoretical pressure distributions on wings of finite span at zero incidence for Mach numbers near 1*, NASA TR R-88 (1961).

[9.60] J.R. Spreiter, A.Alksne: *Slender body theory based on approximate solution of the transonic flow equation*, NASA TR R-2 1959.

[9.61] Sympossium transonicum, Aachen 1962 ed. by K.Oswatitsch.

[9.62] L. Trilling: *Transonic flow past a wedge profile at Mach number 1*, J. Aero. Sci. **17**(1950) 723 - 735.

[9.63] W.G. Vincenti, C.B. Wagoner: *Transonic flow past a wedge profile with detached bow wave*, NACA Rep 1095(1952).

[9.64] H.J. Wirz, J.J. Smoldern: *Numerical methods in fluid dynamics*, Mc Graw-Hill, New York.

[9.65] J. Zierep: *Die Integralgleichungsmethode zur Berechnung schallnaher Strömingen*, Symposium Transonicum, Berlin, 1964.

[9.66] J. Zierep: *Der Äquivalenzsatz und die parabolische Methode für schallnahe Strömungen*, ZAMM **45**(1965) 19 - 27.

Chapter 10

[10.1] E.A. Acum: *Aerodynamic forces on the rectangular wings oscillating in a supersonic air stream*, Aero Res Council R §M **2763** (1953).

[10.2] E. Albano, W.P. Roden: *A doublet - lattice method for calculating lift distributions on oscillating surfaces in subsonic flow*, AIAA Journ. **7**(1969) 279.

[10.3] S. Ando, M. Kato: *An improved kernel function computation in subsonic unsteady lifting surface theory*, Computer Meth. in Appl. Mech. Enging, **49**(1985) 343 - 355.

[10.4] H. Ashley, S. Windall, M.T. Landahl: *New directions in lifting surface theory*, AIAA Journ. **3**(1965) 3.

[10.5] S.R. Bland: *The two dimensional oscillating airfold in a wind tunnel in subsonic flow*, SIAM Journ. Appl. Math.,**1**(1970) 830 - 848.

[10.6] S. Borbely: *Aerodynamic forces on a harmonical oscillating wing at a subsonic velocity(2-Dim. case)*, ZAMM **22**(1942) 190 - 205.

[10.7] M. Borja: *Eine neue Methode zur numerische Behandlung harmonisch schwinger Tragflügel*, doctoral thesis July (1969) Universität Karlsruhe.

[10.8] A. Carabineanu: *Incompressible flow past oscillatory wings of low aspect ratio by the integral equations method*, Int. J. Numer. Eng. **45**(1999) 1187.

[10.9] P. Cicala: *Le axioni aerodinamiche sui profili di alla oscilanti in presenza di Corrente uniforme*, Mem. R. Accad. Torino, Ser. 2, *P I* **68**(1934-1935) 73 - 98.

[10.10] H.J. Cunningham: *Improved numerical procedure for harmonically deforming lifting surfaces from the supersonic kernel function method*, AIAA Journ. **4**(1966) 1961 - 1968.

[10.11] A.M. Cunningham Jr.: *Oscillatory supersonic kernel function method for interferring surfaces*, Journ. Aircraft **11**(1974) 664.

[10.12] D.E. Davies: *Three-dimensional sonic theory*, Aero. Res. Council, Rep.§Mem.**34**(1965).

[10.13] D.E. Davies: *Calculation of unsteady generalised airforces on a thin wing oscillating harmonically in subsonic flow*, Aero. Res. Council, Rep.§Mem. No **3409** (1965).

[10.14] F. Dietze: *The air forces at the harmonically vibrating wing at subsonic velocity (plane problem)*, Luftfahrt forvach **16**(1939) 84.

[10.15] L. Dragoş: *The method of fundamental solutions. The theory of thick oscillatory airfoil in subsonic flow*, Rev. Roum. Sci. Techn.-Méc. Appl. **29**(1984) 59.

[10.16] L. Dragoş: *On the theory of oscillating thick wings in subsonic flow. Lifting line theory*, Acta Mechanica **54**(1985) 221 - 231.

[10.17] L. Dragoş: *On the theory of oscillating airfoils in supersonic and sonic flow*, J. Appl. Math. a. Phys.(ZAMP) **36**(1985) 481.

[10.18] L. Dragoş: *On the theory of oscillating airfoils at Mach number one*, Mech. Res. Comm. **13**(1986) 317 - 324.

[10.19] L. Dragoş: *Calculation formulae for lift and moment coefficients for oscillating arbitrary airfoils in supersonic flow*, Rev. Roum. Sci. Techn.-Méc. Appl. **31**(1986) 345.

[10.20] L. Dragoş: *On the theory of oscilating wings in sonic flow*, ZAMM (Z. Angew. Math. u. Mech.) **68**(1988) 373.

[10.21] L. Dragoş: *Theory of oscillating wings in supersonic flow* (unpublished).

[10.22] J.A. Fromme, M.A. Goldberg: *Reformulation of Possio's kernel with application to unsteady wind tunnel interference*, AIAA Journ. **18**(1980) 951.

[10.23] J.A. Fromme, M.A. Goldberg: *Aerodynamic interference effects on oscillating airfoils with controls in ventilated wind tunnels*, Reprint form AIAA Journ. **18**(1980) 417.

[10.24] I.E. Garrick, S.I. Rubinow: *Flutter and oscillating air-forces calculations for an airfoil in two-dimensional supersonic flow*, NACA Rep. **846**(1946).

[10.25] I.E. Garrick, S.I. Rubinow: *Theoretical study of air forces on steady thin wing in a supersonic main stream*, NACA T. N. **1383**(1947).

[10.26] I.E. Garrick: *Nonsteady wing characteristics*, in High Speed Aerodynamics and Jet Propulsion, vol. VII (ed. A. F. Donavan, H. R. Lawrence) N. Y. (1957).

[10.27] H. Glauert: *The force and moment on an oscillating airfoil*, Aero. Res. Counc. R & M **1242**(1929).

[10.28] J. Grzedzinski: *Application of the unsteady lifting-line method to arbitrary configurations of lifting surfaces*, Arch. Mech. **41**(1989) 165 - 181.

[10.29] R.L. Harder, W.P. Rodden: *Kernel function for nonplanar oscillating surfaces in supersonic flow*, J. Aircraft **8**(1971) 677.

[10.30] M.D. Haskind: *Oscillations in a subsonic gas flow*, Prikl. Mat. Mekh. **11**(1947) 129 - 146.

[10.31] D. Homentcovschi: *Chemically reacting flow of an inviscid gas past a thin body*, Acta Mechanica **33**(1979).

[10.32] V. Huckel, B.J. Durling: *Tables for wing-airleron coefficients of oscillating air forces two-dimensional supersonic flow*, NACA T. N. **2055**(1950).

[10.33] V. Huckel: *Tabulation of the f-functions which occur in aerodynamic theory of oscillating wings in supersonic flow*, NACA T. N. **3606**(1956).

[10.34] T. Ichikawa, K. Isogai: *Note on the aerodynamic theory of oscillating T-tails*, Trans of Japan Soc. for Aeron a. Spaa Sci. **16**(1973).

[10.35] K. Isogai, T. Ichikawa: *Lifting-surface theorem for a wing oscillating in Yaw and Sidelip with an angle of attack*, AIAA J. **11**(1973) 509 - 606.

[10.36] P.F. Jordan: *Numerical evaluation of three-dimensional harmonic kernel*, Z. Flugwiss. **24**(1976) 205.

[10.37] H.G. Küssner: *Allgemeine Tragflächen theorie*, Luftfahrtforschung, **17**(1940) 3.

[10.38] H.G. Küssner: *A review of the two-dimensional problem of unsteady lifting surface theory during the last thirty years*, Inst. of Fl. Dyn. and Appl. Math. Univ. Maryland, Lecture Series No. **73**(1953).

[10.39] H.G. Küssner: *A general method for solving problems of the unsteady lifting surface theorem in the subsonic range*, J A S **21**(1954) 17.

[10.40] T.P. Kalman, W.P. Rodden, J.P. Giessing: *Application of the Doublet Latice*, J. Aircraft **8**(1971) nr. 6.

[10.41] K.R. Kimble, D.D. Liu, S.Y. Ruo, J.M. Wu: *Unsteady transonic flow analysis for low aspect ratio, pointed wings*, AIAA J. **12**(1974) 516.

[10.42] M.T. Landahl: *Unsteady transonic flow*, Pergamon Press, Oxford (1961).

[10.43] M.T. Landahl: *Kernel function for nonplane oscillating surface flow*, AIAA J. **5**(1967) 1045.

[10.44] M.T. Landahl: *On the pressure loading function for oscillating wings with control surfaces*, AIAA J. **6**(1968) 345 - 348.

[10.45] B. Laschka: *Das potential und das Geschwindigkeitsfeld der harmonis schwingenden tragenden Fläche bei Unterschallströmung*, ZAMM **43**(1963) 325.

[10.46] B. Laschka: *Zur Theorie der harmonisch schwingenden tragenden Fläche bei Unterschallströmung*, Zeitschr. für Flugwissenschaften. **11**(1963) 265 - 292.

[10.47] H.R. Lawrence, E.H. Gerber: *The aerodynamic force on low aspect ratio wings oscillating in an incompressible flow*, J.A.S., **19**(1952) 769 - 781(Nov.); **20** (Apr. 1953) 296.

[10.48] D.E. Lehrian, H.G. Garner: *Theoretical calculation of generalized forces and load distribution on wings oscillating at general frequencies in a subsonic stream*, A.R.C. Rep.§Mem. **3710**(1971).

10.49] M.J. Lighthill: *Oscillating airfoils at high Mach number*, J. Aeron. Sci. **20**(1953) 402 - 406.

[10.50] D.D. Liu, W.H. Hui: *Oscillating delta wings with attached shock waves*, AIAA Journ. **15**(1977) 804.

[10.51] D.D. Liu, M.F. Platzer, S.Y. Ruo: *Unsteady linearized transonic flow analysis for slender bodies*, AIAA Journ. **15**(1977) 906 - 973.

[10.52] K.W. Mangler: *A method of calculating the short-period longitudinal stability derivatives of a wing in linearized unsteady compressible flow*, Aero. Res. Counc.Rep.§Mem., London **2924**(1952).

[10.53] J.W. Miles: *Potential theory of unsteady supersonic flow*, Cambridge Univ. Press, London (1959).

[10.54] Y. Miyazawa, K. Washizu: *A finite state aerodynamic model for a lifting surface in incompressible flow*, AIAA Journ. **21**(1983) 163.

[10.55] L. Morino, L.T. Chen, E.O. Suciu: *Steady and oscillatory subsonic and supersonic aerodynamics around complex configurations*, AIAA Journ. **13**(1975) 368.

[10.56] H.C. Nelson: *Lift and Moment on oscillating triangular and related wings with supersonic edges*, NACA T.N. **2494**(1951).

[10.57] H.C. Nelson, J.H. Berman: *Calculations on the forces and moments for an oscillating wing-aileron combination in two-dimensional potential flow at sonic speed*, NACA Rep. **1128**(1953).

[10.58] K. Oberhettinger, L. Badü: *Tables of Laplace transforms*, Springer-Verlag, Berlin (1973).

[10.59] C. Possio: *L'Azione aerodinamica sul profilo oscilante in a fluido compressible a velocita iposonora*, L'Aerotecnica, **18**(1938) 4.

[10.60] E. Reissner: *On the application of the Mathieu functions in the theory of subsonic compressible flow past oscillating airfoils*, NACA T.N. **2363**(1951).

[10.61] J.R. Richardson: *A method for calculating the lifting forces on wings (unsteady subsonic and supersonic lifting-surface theory)*, Aero. Res. Council R. §M. **3157**(1955).

[10.62] W.P. Rodden, J.P. Giesing, T.P. Kalman: *New developmnent and applications of the subsonic dublet-latice meth.*, AGARD SVMP. (1970).

[10.63] W.P. Rodden: *Comments on the new solution method for lifting surfaces in subsonic flow*, AIAA Journ. **22**(1984) 160.

[10.64] N. Rott: *Oscillating airfoils at Mach number 1*, J. Aero Sci. **16**(1949) 380.

[10.65] N. Rott : *Flügelschwingungsformen in ebener kompressible Potential strömung*, ZAMP **1**(1950) 380 - 410.

[10.66] H.L. Runyan, Ch.E. Watkins: *Considerations on the effect of the wind-tunnel walls on oscillating air forces for two - dimensional subsonic compressible flow*, NACA Rep. **1150**(1953).

[10.67] H.L. Runyan, D.S. Woolston, A.G. Rainey: *A theoretical and experimental investigation of the effect of tunnel walls on the forces on an oscillating airfoil in two - dimensional subsonic compressible flow*, NACA Rep. **1262**(1956).

[10.68] H.L. Runyan, D.S. Woolston: *Method for calculating the aerodynamic loading on an oscillating finite wing in subsonic and sonic flow*, NACA Rep. **1322**(1957).

[10.69] L. Schwarz: *Berechnung der Druckverteilung einer harmonisch sich Verformanden Tragfläche in ebener Strömung*, Luftfahrtforsch **17**(1940).

[10.70] L. Schwarz: *Undersuchung einiger mit den Zylinder Funckionen nullter Ordnung verwandten Funckionen*, Luftfahrtforsch, **20**(1944) 341.

[10.71] S.F. Shen, F.P. Crimi: *The theory for an oscillating thin airfoil as derived from the Oseen equations*, J. Fluid Mech. **23**(1965) 585-609.

[10.72] V.J.E. Stark: *General equations of motion for an elastic wing and method of solution*, AIAA J. **22**(1984) 1146.

[10.73] H.J. Stewart, Ting-Yi Li: *Source-superposition method of solution of a periodically oscillating wing at supersonic speeds*, QAM **9**(1951) 31.

[10.74] K. Stewartson: *On the linearized potential theory of unsteady supersonic motion*, Quart. J. Mech. Appl. Math., **3**(1950) 182.

[10.75] T. Theodorsen: *General theory of aerodinamic instability and the mechanism of flutter*, NACA Rep. **4**(1935) 96.

[10.76] H. Tijdeman, R. Seebass: *Transonic flow past oscillating airfoils*, Annual Review of Fluid Mech. **12**(1980) 181.

[10.77] S. Turbatu: *Das Störpotential für aperiodische instationäre schallnahe Strömungen*, in Symposium transonicum II, Göttingen (1975), Springer (1976).

[10.78] S. Turbatu, J. Zierep: *Aperiodische instationäre schallnahe Strömungen*, Acta Mechanica **21**(1975) 165.

[10.79] T. Ueda: *Lifting surface calculations in Laplace domain with application to Rool Loc*, AIAA Journ. **25**(1987) 648.

[10.80] T. Ueda, E.H. Dowell: *A new solution method for lifting surface in subsonic flow*, AIAA Journ. **20**(1982) 348.

[10.81] T. Ueda, E.H. Dowell: *Doublet-Point method for supersonic unsteady lifting surfaces*, AIAA Journ. **22**(1984) 179.

[10.82] A.I. van Vooren: *Unsteady airfoil theory*, pp 36 - 89 in Advances in Appl. Mech. vol. V,(H. L. Dvyden, Th. von Karman, eds) Academic Press N. Y. (1958).

[10.83] E.G. Yates Jr.: *A Kernel function formulation for nonplanar lifting surfaces oscillating in subsonic flow*, AIAA J. **4**(1966) 1486 - 1488.

[10.84] C.E. Watkins: *Three dimensional supersonic theory ...*, in Manual on Aeroelasticity, AGRAD.

[10.85] C.E. Watkins, J. H. Berman: *On the kernel function of the integral equation relating lift and downwash distributions of oscillating wings in supersonic flow*, NACA Rept. **1257** (1953).

[10.86] C.E. Watkins, H. L. Runyan, D. S. Woolston: *On the kernel function of the integral equation relating lift and downwash distributions of oscillating finite wings in subsonic flow*, NACA T.R. **1234** (1955).

[10.87] C.E. Watkins, H. L. Runyan, D. S. Woolston: *A systematic kernel function procedure for deteremining aerodynamic forces on oscillating of steady finite wings at subsonic speeds*, NASA Tehn. Rep. **R-48** (1959).

[10.88] M.H. Williams: *Exact solutions in oscillating airfoil theory*, AIAA Journ. **15**(1977) 875.

[10.89] M.H. Williams: *The resolvent of singular integral equations*, Quart. Applied Math. (1977).

[10.90] M.H. Williams: *Linearization of unsteady transonic flows containing shocks*, AIAA Journ. **17**(1979) 394.

[10.91] M.H. Williams: *Unsteady thin airfoil theory for transonic flow with embedded shocks*, AIAA Journ. **18**(1980) 615.

[10.92] M.H. Williams, P. Jones: *Unsteady supersonic aerodynamic theory by the method of potential gradient*, AIAA Journ. **15**(1977) 59 - 65.

Chapter 11

[11.1] Mac C.Adams, W.R. Sears: *Slender-body theory - review and extension*, J. Aero. Sci. **20**(1953) 85.

[11.2] C.E. Brown: *Aerodynamics of bodies at high speeds*, in *High Speed Aerodynamics and jet propulsion*, vol 7, Editors A.F. Donovan, H.R. Lawrence.

[11.3] V. Cardoş: *General integral method in the study of supersonic flows past slender bodies*, Rev. Roum. Math. Pures Appl. **37**(1992) 571.

[11.4] L. Dragoş, A. Carabineanu: *On the steady subsonic flow past slender bodies of revolution*, Revue Roum. Sci. Techn, Série Mécanique Appl. **34**(1989) 453.

[11.5] L. Dragoş: *A new method in slender body theory in supersonic flow*, Rev. Roum. Math. Pures et Appl. **35**(1990) 139 - 146.

[11.6] A. Ferri: *Elements of aerodynamics of supersonic flows*, MacMillan Co., New York, 1949.

[11.7] F.I. Frankl, E.A. Karpovich: *Gas dynamics of slender bodies* (in russian) Gos. Izd-Vo Tekhn. - Teoret. Lit-Ry, Moskva, 1948, *Gas dynamics of thin bodies* (in English), Intersci. Publish., Inc., New York, 1953.

[11.8] Kuo Chang Wang: *A new approach to not-so-slender wing theory*, J. Math. Phys., **47**(1968) 391.

[11.9] E.V. Laitone: *The linearized subsonic and supersonic flow about inclined slender bodies of revolution*, J. Aero. Sci. **14**(1947) 631.

[11.10] E.V. Laitone: *The subsonic flow about a body of revolution*, Quart Appl. Math. **5**(1947) 227.

[11.11] M. Lighthill: *Supersonic flow past bodies of revolution*, Rep. and Mem., **2003**(1945).

[11.12] P.R. Maeder, H.U. Thommen: *Some results of linearized transonic flow about slender airfoils and bodies at revolution*, J. Aeron. Sci., **23**(1956) 187.

[11.13] W.L. Oberkampf, L.E. Watson: *Incompressible potential flow solutions for arbitrary bodies of revolution*, AIAA Journ., **12**(1974) 409.

[11.14] K. Oswatitsch, F. Keune: *Ein Äquivalenzsatz für nicht-angestellte Flügel kleiner Spannweite in Schallnaher Strömung*, Z. Flugwiss, **3**(1955) 29.

[11.15] K. Oswatitsch, F. Keune: *The flow around bodies of revolution at Mach Number 1*, Proc. Conf. on high-speed aeron., Polyt. Inst. of Brooklyn N.Y. (1955) 113.

[11.16] H.S. Tsien: *Supersonic flow over an inclined body of revolution*, J. Aero. Sci. **5**(1938) 480.

[11.17] A. van Tuyl: *Axially symmetric flow around a new family of half - bodies*, Quart. Appl. Math., **7**(1950) 395.

[11.18] A. Weinstein: *On axially symmetric flows*, Quart. Appl. Math., **5**(1948) 429.

[11.19] G.N. Ward: *Supersonic flow past slender pointed bodies*, Journ. Mech. Appl. Math., **2**(1949) 75.

[11.20] M.F. Zedan, C. Dalton: *Potential flow around axisymmetric bodies: direct and inverse problem*, AIAA Journ., **16**(1978) 242.

Appendices

[A.1] I.A. Brychkov, A.P. Prudnicov: *Integral transforms of generalized functions* (in Russian), Nauka, Moskow, 1977.

[A.2] V.A. Ditkin, A.P. Prudnicov: *Integral transforms and oper. calculus*, 2nd ed. (Russian), Nauka, Moskow, 1974.

[A.3] L. Dragoş, Al. Nicolau: *Partial differential equations*, in [1.22], vol. 3.

[A.4] A. Erdelyi a.o.: *Tables of integral transforms*, Mc Gvaw-Hill, New York, 1954.

[A.5] I.M. Guelfand, G.E. Chilov: *Les distributions I*, Dunod, Paris, 1962.

[A.6] D. Homentcovschi: *An introduction to BEM by integral transforms*, Preprint Ser. in Mathematics 25(1987).

[A.7] D. Homentcovschi: *Elemente de teoria distribuţiilor*, in [1.22], vol. 3 and 4.

[A.8] Ed. Marie de Jager: *Applications of distributions in mathematical physics*, Math. Centrum - Amsterdam, 1964.

[A.9] L. Schwartz: *Méthodes mathématiques pour les sciences physiques*, Hermann, Paris, 1965.

[A.10] I. Stakgold: *Boundary value problems of mathematical physics*, vol II. MacMillan Co., New York, 1968.

[A.11] V. Vladimirov a.o.: *Recueil de problèmes d'équations de physique mathématique*, Mir, Moscou, 1976.

[A.12] V. Vladimirov: *Equations of mathematical physics* (translated from Russian into Romanian), Editura Ştiinţifică şi Enciclopedică, Bucureşti, 1980.

[A.13] V. Vladimirov: *Distributions en physique mathématique*, Edition Mir, Moscou, 1979.

[A.14] C. Zuily: *Distributions et équations aux dérrivèes partielles*, Hermann, Paris, 1986.

Singular Integrals. Integral Equations

[A.15] A. Carabineanu, D. Bena: *Metoda transformărilor conforme pentru domenii vecine*, Ed. Academiei Române, Bucureşti, 1993.

[A.16] L. Dragoş, N. Marcov: *About some singular integral equations of interest in aerodynamics*, Z. Angew. Math. Mech., **79**(1999) 3, 205 - 210.

[A.17] M. Eichler: *Konstruction lösender Kerne für singuläre integral gleschungen erster Art insbesondere bei Differenz. kernern*, Math. Zs. **48**(1941) 503.

[A.18] F.D. Gakhoff:*Boundary problems* (in Russian), Fizmatghiz, Moskow, 1958.

[A.19] D. Homentcovschi: *Sur la résolution explicite du problème de Hilbert. Application en calcul de la portance d'un profile mince dans un fluid électroconducteur*, Rev. Roum. Math. Pures et Appl., **14**(1969) 203;

Contributions to the study of Hilbert's problem Applications to magnetohydrodynamics, Studii Cerc. Mat., **23**(1971) 727.

[A.20] D. Homentcovschi: *Funcţii complexe cu aplicaţii în ştiinţă şi tehnică*, Ed. Tehnică, Bucureşti, 1986.

[A.21] C. Iacob: *Sur une équation intégrale singulière*, Mathematica, **23**(1948) 153.

[A.22] C. Iacob: *Funcţii complexe* in [1.22], vol. 2.

[A.23] A.I. Kalandiya: *Mathematical methods of two-dimensional elasticity*, Mir Publishers, Moscow, 1975.

[A.24] A. Levinson: *Simplified treatment of integrals of Cauchy type, the Hilbert problem and singular integral equations*, SIAM Rev. **7**(1968) 474.

[A.25] S.G. Mikhlin: *Multidimensional singular integrals and integral equations* (in Russian) Gostekhizdat, Moskow, 1962.

[A.26] S.G. Mikhlin: *Linear equations with partial derivatives* (in Russian), Moskow, 1977.

[A.27] N.I. Muskhelishvili: *Singular integral equations* (in Russian), 2nd ed. Fizmatghiz, Moskow, 1962; the first ed. is translated in English, P.Noordhoff, Groningen, 1953.

[A.28] A.S. Peters: *Abel's equations and the Cauchy singular integral equation of the second kind*, Comm. Pure Appl. Math **21**(1968) 51.

[A.29] J. Plemelj: *Ein Ergànzungssatz zur Cauchy'schen integraldarstellung analytischer Funktionen, Randwerte betreffend*, Monatsch. f. Math. u. Phys, XIX, Jahrgang, (1908) 205.

[A.30] N. Rott, H.K. Cheng: *Generalizations of the inversion formula of thin airfoil theory*, J. Rat. Mech. Anal., **3**(1954) 357.

[A.31] G. Samko, A.A. Kilbas, O.I. Marichev: *Integrals and fractionar products and some of their applications* (in Russian), Nauka i Technica, Minsk, 1987.

[A.32] W. Schmeidler: *Integral gleichungen mit Anwendungen in Physik und Technik*, Leipzig, 1955.

[A.33] K. Schröder: *Über eine Integralgleichung erster Art der Tragflügeltheorie*, Sitzungsber. d. Preuss. Akad. d. Wiss., Phys. Mat. Klasse **30**(1938) 345.

[A.34] H. Söhngen: *Die Lösungen der Integral-gleichung ... und deren Anwendung in der Tragflügeltheorie*, Math - Zeitschr., **45**(1939) 245.

[A.35] G. Tricomi: *Formula d'inversione dell'ordine di due integrazi doppie "con asterisco"*, Atti della Reale Acad. Naz. dei Lincei, **323**(1926) 53.

Finite Part

[A.36] M.P. Brandâo: *Improper integrals in theoretical aerodynamics. The problem revisited*, AIAA Journ. **25**(1987) 1258.

[A.37] Ch. Fox: *A generalization of the Cauchy principal value*, Canadian J. of Math., **9**(1957) 110.

[A.38] M. Guiggiani a.o.: *General algorithm for the numerical solution of hyper - singular boundary integral equations*, J. Appl. Mech. **59**(1992) 604.

[A.39] J. Hadamard: *Lectures on Cauchy's problem in linear partial differential equations*, Yale University Press, 1923.

[A.40] M.A. Heaslet, H. Lomax: *Supersonic and transonic small perturbation theory*, Appendix in [1.37].

[A.41] A.C. Kaya, F. Erdogan: *On the solution of integral equations with strongly singular kernel*, Quart. Appl. Math., **45**(1987) 105.

[A.42] H.R. Kutt: *The numerical evaluation of principal value integrals by Finite Part integration*, Numer Math., **24**(1975) 205.

[A.43] E.G. Ladopoulos: *Finite - Part singular integro - differential equations arising in two - dimensional aerodynamics*, Arch. of Mechanics, **41**(1989) 925-936.

[A.44] H. Lomax, M.A. Heaslet, F.B. Fuller: *Integrals and integral equations in linearized wing theory*, NACA Rep. 1054(1951).

[A.45] K.W. Mangler: *Improper integrals in Theoretical aerodynamics*, British Aero. Res. Council, R & M 2424(1951).

[A.46] D.F. Paget: *The numerical evaluation of Hadamard Finite part integrals*, Numer. Math., **36**(1981) 447.

[A.47] S. Voutsianas, G. Bergeles: *Numerical calculation of singular integrals appearing in 3-D potential flow problems*, Appl. Math. Model, **14**(1990) 618.

Quadrature Formulas

[A.48] S. Ando, A. Ichikawa: *Quadrature formulas for chardwise integrals of lifting surface theories*, AIAA Journ., **21**(1983) 466.

[A.49] L. Dragoş, M. Popescu: *Certain quadrature formulae of interest in aerodynamics*, Rev. Roum. Math. Pures et Appl, **37**(1992) 587.

[A.50] D. Elliot, D.F. Paget: *Gauss type quadrature rules for Cauchy principal value integrals*, Math. Comp. **33**(1979) 301.

[A.51] D.B. Hunter: *Some Gauss - type formulae for the evaluation of Cauchy principal value of integrals*, Numer. Math., **19**(1972) 419.

[A.52] G. Monegato: *On the weights of certain quadratures for the numerical evaluation of Cauchy principal value integrals and their derivatives*, Numer. Math. **50**(1987) 273.

[A.53] P. Natanson: *Constructive function theory* (in Russian), Gostekhizdat, Moskow - Leningrad, 1949.

[A.54] V.J.E. Stark: *A generalized quadrature formula for Cauchy integrals*, AIAA Journ., **9**(1971) 1854.

[A.55] A.H. Stroud, D. Secrest: *Gaussian quadrature formulas*, Prentice Hall, N.Y. 1966.

[A.56] G. Szegö: *Ortogonal Polynomials*, Amer. Math. Soc. Colloq. Publ., vol. 23, 4th ed. 1975.

Index